Mathematical Modelling of Fluid Dynamics and Nanofluids

Mathematical Modelling of Fluid Dynamics and Nanofluids serves as a comprehensive resource for various aspects of fluid dynamics simulations, nanofluid preparation, and numerical techniques.

The book examines the practical implications and real-world applications of various concepts, including nanofluids, magnetohydrodynamics, heat and mass transfer, and radiation. By encompassing these diverse domains, it offers readers a broad perspective on the interconnectedness of these fields.

The primary audience for this book includes researchers and graduate students who possess a keen interest in interdisciplinary studies within the realms of fluid dynamics, nanofluids, and biofluids. Its content caters to those who wish to deepen their knowledge and tackle complex problems at the intersection of these disciplines.

Katta Ramesh is currently working as a senior lecturer in the School of Mathematical Sciences, Sunway University, Malaysia. He has more than ten years of experience in the education field. He has been featured in Stanford University's list of top 2% scientists worldwide in the years 2021 and 2022. He is serving as an associate editor in the journals *Frontiers in Mechanical Engineering* and *Frontiers in Thermal Engineering.* He is a guest editor for the special issues in the journals *Sustainability (Energy Efficiency: Perspectives and Policies and the Critical Aspects for Energy Conservation and Storage Towards Sustainable Development), Mathematics (New Advances in Analytical and Numerical Techniques in Fluid Mechanics)*, and *Frontiers in Thermal Engineering (Fluid Dynamics: Recent Trends and Applications to Thermal Engineering).* He has served as a scientific advisory committee member in many international conferences. He has published more than 70 articles in reputed international journals. Besides, he published book chapters and conference proceedings. He is a reviewer for more than 40 reputed international journals. His research work is concerned on the analytical and computational modelling of fluid flow problems arising in engineering and biomedical sciences.

Fateh Mebarek-Oudina received his PhD in 2010. He has published more than 100 papers in reputed international journals. Currently, he works as a full professor at Skikda University in Algeria and regularly serves as a reviewer for more than 250 international journals. He is ranked in the Top 2% Scientists Worldwide (2020, 2021, 2022) by Stanford University. His research work is focused on heat and mass transfer, MHD, mathematical simulation and modelling, biofluids, nanofluids, hybrid nanofluids, ternary nanofluids, microfluidics, and computational fluid dynamics.

Basma Souayeh is currently working as an associate professor at King Faisal University, Saudi Arabia, and has been a researcher in the Laboratory of Fluid Mechanics and Heat Transfer (URPF Lab) at the University of Tunis El Manar, Tunisia, since 2012. She has more than ten years of experience in the education and research fields, with more than 80 peer-reviewed international papers published in ISI journals in physics specialty. She is serving as a reviewer in several international ISI scientific journals, as a speaker and chair session at several national and international conferences, and a review editor in the *Frontiers in Built Environment* and *Frontiers in Thermal Engineering* journals. She also serves as a guest editor for the journals *Sustainability* and *Energies* and is devoted to scientific research, with a distinguished record of publications in scientific and academic journals worldwide and presentations in international conferences. She is a dedicated partner to university programs, promoting learning and helping students develop their full potential in their studies. Her research skills are focused on the following areas: computational physics, mechanical engineering, thermal engineering, energy, computational fluid mechanics, numerical simulations, numerical modelling, CFD simulation, finite volume method, finite difference method, engineering thermodynamics, energy engineering, heat and mass transfer, nanofluids, CFD coding, fluid flow, and mechanics.

Mathematical Modelling of Fluid Dynamics and Nanofluids

Edited by Katta Ramesh,
Fateh Mebarek-Oudina, and Basma Souayeh

CRC Press is an imprint of the
Taylor & Francis Group, an **informa** business

Designed cover image: Shutterstock

First edition published 2024
by CRC Press
2385 Executive Center Drive, Suite 320, Boca Raton, Fl 33431

and by CRC Press
4 Park Square, Milton Park, Abingdon, Oxon, OX14 4RN

CRC Press is an imprint of Taylor & Francis Group, LLC

ISBN: 978-1-032-29016-4 (hbk)
ISBN: 978-1-032-29020-1 (pbk)
ISBN: 978-1-003-29960-8 (ebk)

DOI: 10.1201/9781003299608

Typeset in Times
by Apex CoVantage, LLC

Contents

Preface

Fluid dynamics is an area of applied science concerned with the movement of liquids and gases. This covers a wide variety of applications, such as understanding interstellar nebulae, forecasting weather patterns, and determining the mass flow rate of oil through pipelines. Fluid dynamics also offers tools to study the evolution of planets, ocean tides, weather patterns, plate tectonics, and also blood circulation. Some of the important technological applications of fluid dynamics include oil pipelines, wind turbines, rocket engines, and air-conditioning systems. Recently, nanotechnology has played a major part in multi-fields of heat transfer processes and developed a remarkable progress in energy applications. One of the most significant applications of nanotechnology is to generate nanoparticles of high thermal conductivity and mixing with the base fluids, which is called as nanofluid. Addition of nanoparticles to the base fluid produces a remarkable enhancement of the thermal properties of the base properties. Nanotechnology has greatly improved the science of heat transfer by improving the properties of the energy-transmitting fluids. A high heat transfer could be obtained through the creation of innovative fluids. This also reduces the size of heat transfer equipment and saves energy. The present book is aimed to provide a review analysis and new mathematical, numerical, and experimental works on fluid dynamics and nanofluids. The present book is organized with 32 chapters related to mathematical modelling of fluid dynamics and nanofluids. The first 14 chapters are related to the mathematical modelling and simulations of classical fluid problems, and the next 18 chapters are focused on the nanofluid models with various analytical and numerical approaches.

The first two chapters of this book deal with the problems related to fluid flows in a horizontal and square cavity with ANSYS and COMSOL platforms, respectively. The next chapter (third) provides the flow behavior and corresponding thermal transport in a discretely heated inclined porous parallelogram enclosure with finite difference method (FDM). The fourth chapter talks about the double-diffusive convection in a composite system with heat source and temperature gradients. The next chapter (fifth) explains the physiological flow with ciliated walls using semi-analytical approach. The sixth, seventh, eighth, and ninth chapters discuss the flow models of Newtonian and non-Newtonian fluids over a stretchable sheet. These chapters deal majorly with problems related to hot rolling, plastic film manufacturing, glass fiber production, wire drawing, and polymer extrusion in the melt-spinning processes. Following that, the tenth chapter deals with experimental analysis on the heat removal in a channel composed of pin-fin rows configuration with well-known numerical finite element approach. The next chapter (eleventh) provides the large-eddy simulation approach of turbulence features in transport of non-Newtonian fluids. The twelfth chapter deals with the computational fluid dynamics (CFD) approach for the investigation of the flow behavior of a vertical axis wind turbine. The next chapter (thirteenth) provides the Hiemenz flow of non-Newtonian fluids over a rotating disk with the Runge–Kutta (RK) approach. The fourteenth chapter deals with an experimental study on bioethanol production from biomass using a locally developed small-scale production facility.

Chapter 15 provides the modelling which aims to reduce refracting light and improve absorbance behavior, which leads to increased thrust force of the nano-robot. Chapters 16, 17, and 18 deal with transport through the cavities with lattice Boltzmann, FDM, and Fluent CFD techniques with water-$CuO / Cu / Al_2O_3$ nanofluids, respectively. The next four chapters deal with the different kinds of nanofluids transport over a stretching/shrinking sheet. These studies have been made with diverse techniques, such as FDM, RK, and RKF-45 techniques. Chapter 23 presents the Falkner–Skan flow and heat transport of a Newtonian hybrid liquid in a wedge with MATLAB solver. The next Chapter (24th) considers the water-copper nanofluid transport in an annular domain with a thin baffle with FDM. Chapter 25 presents the mathematical modelling of 2D annular flow of water-alumina nanofluid in an enclosure with FDM method. Next, Chapter 26 deals with the effects of alumina or titanium oxide nanoparticle shapes on the annulus nanofluid transport in a duct with SIMPLER algorithm. Chapter 27 explains the flow and heat characteristics of alumina-water nanofluid flow with Laplace transformations. Next, Chapter 28 deals with the numerical simulations of nanofluid flow over a wedge with RK approach. Chapter 29 explains the 2D MHD free convection of Cu-MgO hybrid nanofluid in a parallelogram-shaped enclosure COMSOL 5.5 Multiphysics software. Next, Chapter 30 presents the modelling of alumina-copper/water hybrid nanofluid transport in a square enclosure with a hot obstacle via LBM approach. Lastly, Chapter 31 deals with numerical simulations of Ag-MgO/water hybrid nanofluid transport in a shell using COMSOL 5.5 Multiphysics software.

This book covers the novel analysis, experimental findings, mathematical models, and numerical simulations on recent developments on fluid dynamics and nanofluids. Editors are confident enough that this book will be helpful for the readers and provide a benchmark for future developments and research scopes.

Editors

Dr. Katta Ramesh is currently working as a senior lecturer in the School of Mathematical Sciences, Sunway University, Malaysia. He has more than ten years of experience in the education field. He has been featured in Stanford University's list of top 2% scientists worldwide in the years 2021 and 2022. He is serving as an associate editor in the journals *Frontiers in Mechanical Engineering* and *Frontiers in Thermal Engineering*. He is a guest editor for the special issues in the journals *Sustainability (Energy Efficiency: Perspectives and Policies and the Critical Aspects for Energy Conservation and Storage Towards Sustainable Development), Mathematics (New Advances in Analytical and Numerical Techniques in Fluid Mechanics)*, and *Frontiers in Thermal Engineering (Fluid Dynamics: Recent Trends and Applications to Thermal Engineering)*. He has served as a scientific advisory committee member in many international conferences. He has published more than 70 articles in reputed international journals. Besides, he published book chapters and conference proceedings. He is a reviewer for more than 40 reputed international journals. His research work is concerned on the analytical and computational modelling of fluid flow problems arising in engineering and biomedical sciences.

Prof. Dr. Fateh Mebarek-Oudina received his PhD in 2010. He has published more than 100 papers in reputed international journals. Currently, he works as a full professor at Skikda University in Algeria and regularly serves as a reviewer for more than 250 international journals. He is ranked in the Top 2% Scientists Worldwide (2020, 2021, 2022) by Stanford University. His research work is focused on heat and mass transfer, MHD, mathematical simulation and modelling, biofluids, nanofluids, hybrid nanofluids, ternary nanofluids, microfluidics, and computational fluid dynamics.

Dr. Basma Souayeh is currently working as an associate professor at King Faisal University, Saudi Arabia, and a researcher in Laboratory of Fluid Mechanics and Heat Transfer (URPF Lab) at the University of Tunis El Manar, Tunisia, since 2012. She has more than ten years of experience in the education and research fields, with more than 80 peer-reviewed international papers published in ISI journals in physics specialty. She is serving as a reviewer in several international ISI scientific journals and a speaker and chair session at several national and international conferences and a review editor in the *Frontiers in Built Environment* journal and the *Frontiers in Thermal Engineering* journal. She also serves as a guest editor in *Sustainability* and *Energies* journals, devoted to scientific research, with distinguished record of publications in scientific and academic journals worldwide and presentations in international conferences. She is a dedicated partner to university programs, promoting learning and helping students develop their full potential in their studies. Her research skills are focused on the following areas: computational physics, mechanical engineering, thermal engineering, energy, computational fluid mechanics, numerical simulations, numerical modelling, CFD simulation, finite volume method, finite difference method, engineering thermodynamics, energy engineering, heat and mass transfer, nanofluids, CFD coding, fluid flow, and mechanics.

Contributors

Abha Kumari
Department of Mathematics
Nirmala College, Ranchi
Jharkhand-834002, India

Abel Olajide Olorunnisola
Department of Wood Products Engineering
University of Ibadan, Nigeria

Abuzar Ghaffari
Department of Mathematics
Division of Science and Technology
University of Education Lahore 54770, Pakistan

Adewale Allen Sokan-Adeaga
Department of Environmental Health Science
Ajayi Crowther University
Oyo, Nigeria
Department of Environmental Health Science
Lead City University
Ibadan, Nigeria
Department of Environmental Health Sciences
University of Ibadan
Ibadan, Nigeria

Alaeddine Zereg
Laboratory LPEA
Department of Physics
University of Batna 1
Batna, Algeria

Alok Kumar Pandey
Department of Mathematics
Graphic Era Deemed to Be University
Dehradun 248002
Uttarakhand, India

Arshad Riaz
Department of Mathematics
Division of science and Technology
University of Education
Lahore 54770, Pakistan

Ashwin Jacob
Department of Automobile Engineering
Sathyabama Institute of Science and Technology
Chennai-600119, India

Asogwa K. K.
Department of Mathematics
Nigeria Maritime University
Okerenkoko, Delta State, Nigeria

Bahloul Derradji
Higher National School of Renewable Energy
Environment and Sustainable Development
Batna, Algeria

Basma Souayeh
Department of Physics
College of Science
King Faisal University
PO Box 400, Al-Ahsa, 31982, Saudi Arabia
Laboratory of Fluid Mechanics
Faculty of Sciences of Tunis
University of Tunis El Manar 2092
Tunis, Tunisia

Berlin M.
Department of Civil Engineering
NIT Arunachal Pradesh
Arunachal Pradesh-791123, India

Bhardwaj S. B.
Department of Mathematics
SUS Govt College
Matak-Majri, Indri, Karnal 132001, India

Bouchmel Mliki
Research Laboratory of Technology, Energy, and Innovative Materials, TEMI
Faculty of Sciences of Gafsa
University of Gafsa
2112, Tunisia

Carlton Azeez
Department of General Requirements
University of Technology and Applied Sciences
Ibri, Oman

Chara K.
LAAAS, Department of Electronics
University of Batna 2
Batna 05000, Algeria

Chibani A.
University of Gabes
Civil Engineering Department
High Institute of Applied Sciences and Technology
Omar Ibn El Khattab Street, 6029 Gabes, Tunisia

Dambaru Bhatta
School of Mathematical and Statistical Sciences
The University of Texas Rio Grande Valley
Edinburg, Texas 78539, USA

Fateh Mebarek-Oudina
Department of Physics
Faculty of Sciences
University of 20 Août 1955-Skikda
Skikda, Algeria

Fatih Selimefendigil
Mechanical Engineering Department
Manisa Celal Bayar University
Manisa, Turkey

Fatima Zohra Bouhenni
Department of Mechanical Engineering
University Ibn Khaldoun
Tiaret 14000, Algeria

Godson R. E. E. Ana
Department of Environmental Health Sciences
University of Ibadan
Ibadan, Nigeria

Gnaneswara Reddy M.
Department of Mathematics
Acharya Nagarjuna University Campus
Ongole-523 001
Andhra Pradesh, India

Gürel Şenol
Mechanical Engineering Department
Manisa Celal Bayar University
Manisa, Turkey

Hakan F. Öztop
Mechanical Engineering Department
Technology Faculty
Fırat University, Elazığ, Turkey

Hanane Laouira
LRPCSI Laboratory
Department of Physics
Faculty of Sciences
University of 20 Août 1955-Skikda
Skikda, Algeria

Hanumesh Vaidya
Department of Mathematics
Vijayanagara Sri Krishnadevaraya University
Ballari, Karnataka, India

Himanshu Upreti
SoET
BML Munjal University
Gurugram, Haryana, 122413, India

Imene Rahmoune
Department of Physics
Faculty of Matter Sciences
Applied Energetic Physics Laboratory (LPEA)
University of Batna 1
05000 Batna, Algeria

Ines Chabani
LRPCSI Laboratory
Department of Physics
Faculty of Sciences
University of 20 Août 1955-Skikda
Skikda, Algeria

Irfan Mustafa
Department of Mathematics
Allama Iqbal Open University
H-8, Islamabad 44000, Pakistan

Jino L.
Department of Automobile Engineering
Sathyabama Institute of Science and Technology
Chennai-600119, India

Katta Ramesh
Department of Pure and Applied Mathematics
School of Mathematical Sciences
Sunway University, No. 5
Jalan Universiti
Bandar Sunway
Petaling Jaya 47500
Selangor Darul Ehsan, Malaysia

Keerthi Reddy N.
Department of Mechanical Engineering
Ulsan National Institute of Science and Technology
Ulsan 44919
Republic of Korea

Kharabela Swain
Department of Mathematics
Gandhi Institute for Technology
Bhubaneswar, Odisha-752054, India

Khurrem Shehzad
Department of Mathematics and Statistics
University of Agriculture
Faisalabad Pakistan

Kumara Swamy H. A.
Department of Mathematics
Nonlinear Dynamics and Mathematical Application center
Kyungpook National University
Daegu 41566, Republic of Korea

Kushal Sharma
Department of Mathematics
Malaviya National Institute of Technology
Jaipur 302017, India

Lalia Abir Bouhenni
Department of Mechanical Engineering
University Ibn Khaldoun
Tiaret 14000, Algeria

Magnerbi M.
University of Gabes
Civil Engineering Department
High Institute of Applied Sciences and Technology
Omar Ibn El Khattab Street, 6029 Gabes, Tunisia

Mahanthesh Basavarajappa
School of Mathematical and Statistical Sciences
The University of Texas Rio Grande Valley
Edinburg, Texas 78539, USA
Department of Mathematics, CHRIST (Deemed to be University)
Bengaluru-560029, Karnataka, India

Mahabaleshwar U. S.
Department of Studies in Mathematics
Shivagangotri
Davangere University
Davangere, India

Mahesha
Department of Mathematics
University BDT College of Engineering
Davanagere 577004
Karnataka, India

Mahesh R.
Department of Studies in Mathematics
Shivagangotri, Davangere University
Davangere, India

Maimouna Al Manthri
Department of General Requirements
University of Technology and Applied Sciences
Ibri, Oman

Manel Ait Yahia
Department of Mechanical Engineering
University Ibn Khaldoun
Tiaret 14000, Algeria

Manjunatha N.
Department of Mathematics
School of Applied Sciences
REVA University
Bengaluru-560064, India

Maranna T.
Department of Studies in Mathematics
Davangere University
Shivagangothri
Davangere, 577007, India

Mchirgui A.
University of Gabes
Chemical and Process Engineering Department
National Engineering School of Gabes
Applied Thermodynamics Laboratory
Omar Ibn El Khattab Street, 6029 Gabes, Tunisia

Meryem Ould-rouiss
Laboratoire de Modélisation et Simulation Multi Echelle, MSME

Université Gustave Eiffel
UMR 8208 CNRS, 5 bd Descartes
77454 Marne-la-Vallée, Paris, France

Mohamed Abdi
Laboratoire de génie électrique et des plasmas (LGEP) University of Tiaret, Algeria

Mohamed Ammar Abbassi
Research Laboratory of Technology, Energy and Innovative Materials, TEMI
Faculty of Sciences of Gafsa
University of Gafsa, 2112, Tunisia
National Engineering School of Gafsa
University of Gafsa, 2112, Tunisia

Mokhtari K.
Department of Industrial Engineering
University of Khenchela
Khenchela 40000, Algeria

Muhammad Jawad
Department of Mathematics
The University of Faisalabad
Faisalabad 38000 Pakistan

Mohammed Benkhedda
Physics Department
Faculty of Sciences
University of M'Hamed Bougara Boumerdes, 35000, Algeria
Coating, Materials and Environment Laboratory
University M'Hamed Bougara Boumerdes
35000, Algeria

Mounir Aksas
Higher National School of Renewable Energy
Environment and Sustainable Development
Batna, Algeria

Nadhir Lebaal
Laboratory ICB-COMM
University of Technology Belfort-Montbéliard
Belfort, France

Naveen Kumar R.
Department of Mathematics
Dayananda Sagar College of Engineering
Bengaluru-560078
Karnataka, India

Nour Elhouda Beladjine
Department of Mechanical Engineering
University Ibn Khaldoun
Tiaret 14000, Algeria

Prasannakumara B. C.
Department of Studies and Research in Mathematics
Davangere University, Davangere
Karnataka, India

Punith Gowda R. J.
Department of Mathematics
Bapuji Institute of Engineering and Technology
Davanagere 577004
Karnataka, India

Ravindra P.
Department of Mathematics
East Point College of Engineering and Technology
Avalahalli, Bengaluru-560049
Karnataka, India

Renuprava Dalai
Department of Metallurgical and Materials Engineering
Veer Surendra Sai University of Technology
Burla, Odisha-768018, India

Saadi Bougoul
Department of Physics
Faculty of Matter Sciences
Applied Energetic Physics Laboratory (LPEA)
University of Batna 1
05000 Batna, Algeria

Saghir M. Z.
Department of Mechanical and Industrial Engineering
Toronto Metropolitan University
Toronto, Canada

Saleem Iqbal M.
Department of Mathematics Islamabad College for Boys G-6/3
Islamabad 44000, Pakistan

Sajjad Hussain
Department of Mathematics
Government College University Faisalabad
Layyah Campus
Pakistan

Sanjay Kumar
Department of Mathematics
Malaviya National Institute of Technology
Jaipur 302017, India

Sankar M.
Department of General Requirements
University of Technology and Applied Sciences
Ibri, Oman

Saravanakumar G.
Department of Mechanical Engineering
B.M.S. College of Engineering
Bengaluru-560019
Karnataka, India

Shamshuddin MD.
Department of Computer Science and Artificial Intelligence
SR University
Warangal-506371
Telangana, India

Srairi F.
LAAAS, Department of Electronics
University of Batna 2
Batna 05000, Algeria

Sumithra R.
Department of UG
PG Studies and Research in Mathematics
Nrupathunga University
Bengaluru-560001, India

Swapnali Doley
Department of Basic and Applied Science
NIT Arunachal Pradesh
Arunachal Pradesh-791123, India

Tejashwini M. S.
Department of Studies in Mathematics
Shivagangotri
Davangere University
Davangere, India

Toufik Boufendi
Energy Physics Laboratory
Faculty of Exact Sciences
Univerrsité Frères Mentouri Constantine 1, Algeria
Physics Department
Faculty of Sciences
University of 20 Août 1955-Skikda
21000 Skikda, Algeria

Usman
Department of Computer Science
National University of Sciences and Technology
Balochistan Campus (NBC)
Quetta 87300, Pakistan

Vanav Kumar A.
Department of Basic and Applied Science
NIT Arunachal Pradesh
Arunachal Pradesh-791123, India

Vanishree R. K.
Department of Mathematics, Maharani's Science College for Women, Maharani's Cluster University
Bengaluru-560001, India

Vishala H. V.
Department of Studies in Mathematics
Davangere University
Shivagangothri
Davangere, 577007, India

1 Mixed Convection Assessment in an Unusual Cavity with a Heat Source Opened to a Horizontal Channel

Hanane Laouira, Fateh Mebarek-Oudina, and Hanumesh Vaidya

1.1 INTRODUCTION

The past heat transfer, which has three types—radiation, conduction, and convection [1]—is one of the most prevalent mechanisms of energy exchange [2–6] despite the fact that convection has piqued the interest of many academics in the industrial thermal field [7–10], ranging from free and forced convection [11–13] to mixed convection [14–17], which allows the investigation of a variety of thermal problems and systems involving the cooling of large amounts of electronic components, among other thermal aspects [18–21]. As a result, it is critical to combine and develop new approaches with different boundary conditions in order to comprehend and improve the thermal efficiency of different heating and cooling cases [22–27].

In this context, researchers advocate utilizing the benefits of mixed convection to guarantee that extra heat is not generated by various components, such as microchips, photovoltaic sheets, solar cavities, and printed circuit boards, hence preserving the system's performance by considering each element as a rectangular duct flow into an open cavity [28, 29]. Hossain et al. [30] investigated mixed convection in a two-dimensional trapezoidal cavity with L as a height, the left wall inclined at different angles $\phi = 45°$, $30°$, and $0°$, and the y-axis enclosed inside a non-uniform heated triangular block at various positions.

The trapezoidal cavity's bottom wall is set to a non-uniform hot temperature T_h, the left and right walls to a cold temperature T_c, and the top wall is thermally insulated, with the assumption of $T_h > T_c$, and all fluid properties and boundary conditions are constant.

Also, Öztop et al. [31] numerically investigated the stationary flow and mixed convection taking place in a two-dimensional square chamber with two vertical and heated mobile walls. The top and bottom walls are adiabatic. Depending on the direction of the wall movements, three scenarios were evaluated. The flow is governed by the Richardson number ($0.01 < Ri < 100$) and the Prandtl number ($Pr = 0.7$). The findings reveal that the Richardson number and the direction of the wall motion influence both the fluid flow and the heat transmission in the cavity; it is worth noting that heat transfer is much greater for $Ri > 1$.

Furthermore, Burgos et al. [32] examined mixed convection in a square enclosure with an open upper face to a channel; the cavity base is kept at a constant hot temperature, while the left side is uniformly exposed to a cold flow. Except for the upper wall, the channel's walls are maintained adiabatic. The findings revealed that when the Richardson number was equal to or greater than one unity, the mean values of the Nusselt number increased.

According to the prior literature assessment, no work has been done to date on mixed convection in a complicated cavity like the one this chapter analyzes. The upper wall of the chamber was exposed to a conduit throughout this study, whereas the bottom wall was exposed to a discrete heat

DOI: 10.1201/9781003299608-1

source. This unique study is the first to address this issue and provides further information on the phenomenon researched. The effects of the Richardson number, the source length, and the Reynolds number ratio are addressed.

1.2 MATHEMATICAL FORMULATION

1.2.1 Studied Geometrical Model

Figure 1.1 depicts the thermal process that is studied inside of the investigated complex cavity, where a laminar and incompressible supposed flow of air over a channel combined with the horizontal upper wall of the unusual cavity promotes heat exchange. The cavity's height is H_c, and $4H_c$ is assumed to represent the independent length of the channel beyond the cavity.

The rectangular heat source has a length L_h located at the cavity's bottom wall with a temperature T_h. A cooled airflow with a temperature T_c passes in the channel; on the other hand, the remaining sides of the supposed enclosure are thermally insulated. The open top wall of the hollow is twice as wide as the bottom wall.

With the following boundary conditions.

Channel entrance:

$$X = 0,\ H_c \le y \le H_c + D,\ \Theta = 0,\ Re = 100$$

Channel exit:

$$X = 5H_c,\ H_c \le y \le H_c + D,\ \frac{\partial \theta}{\partial X} = \frac{\partial U}{\partial X} = \frac{\partial V}{\partial Y} = 0,\ P = 0$$

Heater: $\Theta = 1$
Solid walls: $U = V = 0$

1.2.2 Governing Equations

The basic equations of continuity, Navier–Stokes, and energy are rewritten in as following: [33]

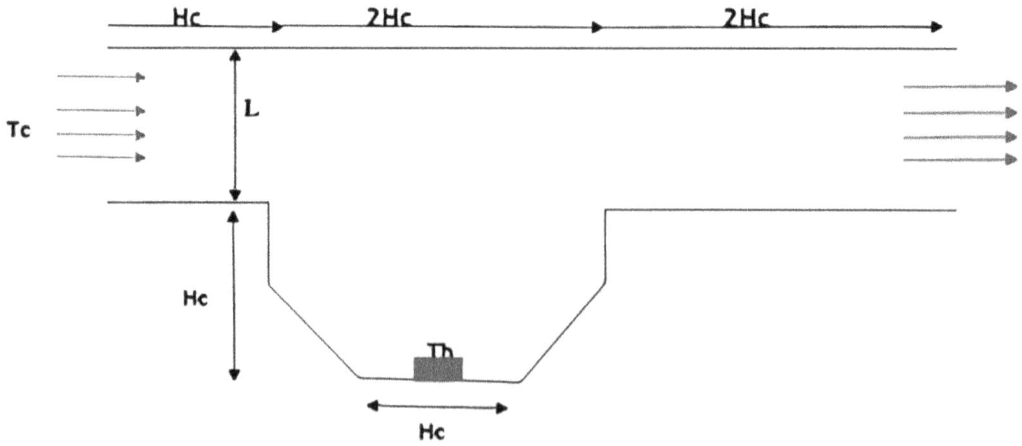

FIGURE 1.1 Physical model.

Mass conservation (continuity):

$$\frac{\partial U}{\partial X}+\frac{\partial V}{\partial Y}=0 \tag{1.1}$$

Navier–Stokes equations:

$$U\frac{\partial U}{\partial X}+V\frac{\partial U}{\partial Y}=-\frac{\partial P}{\partial X}+\frac{1}{Re_{in}}\left(\frac{\partial^2 U}{\partial X^2}+\frac{\partial^2 U}{\partial Y^2}\right) \tag{1.2}$$

$$U\frac{\partial V}{\partial X}+V\frac{\partial V}{\partial Y}=-\frac{\partial P}{\partial Y}+\frac{1}{Re_{in}}\left(\frac{\partial^2 V}{\partial X^2}+\frac{\partial^2 V}{\partial Y^2}\right)+Ri\theta \tag{1.3}$$

Energy conservation:

$$U\frac{\partial \theta}{\partial X}+V\frac{\partial \theta}{\partial Y}=-\frac{\partial P}{\partial X}+\frac{1}{Re_{in}Pr}\left(\frac{\partial^2 \theta}{\partial X^2}+\frac{\partial^2 \theta}{\partial Y^2}\right) \tag{1.4}$$

Where the dimensionless number characterizing the heat transfer problem are as follows: Re_{in}, representing the Reynolds number as $Re_{in}=\frac{\rho u_{in}H_c}{\mu}$; Ri, representing the Richardson number with $Ri=\frac{Gr}{Re_{in}^{\ 2}}=\frac{gH(T_c-T_h)}{u_{in}^{\ 2}}$; and finally, Pr is Prandtl number with $Pr=\frac{\vartheta}{\alpha}$.

Additionally, the non-dimension variables presenting this heat problem are:

$$\theta=\frac{T-Tc}{Th-T_c^{'}};\ U=\frac{u}{u_{in}};V=\frac{v}{u_{in}};X=\frac{x}{H_c^{'}};Y=\frac{y}{H_c^{'}};\ P=\frac{p}{\rho u_{in}^2};\ \varepsilon=\frac{L_h}{H}$$

1.2.3 Grid Test and Program Validation

In order to examine the dependence of the numerical results on the type of meshing used to carry on this study through the ANSYS program, a grid test has been done. **Table 1.1** demonstrates the variation of the average Nusselt number in function of the used type of mesh at Ri = 0.1, Re = 100, and Pr = 0.7. It is essential to note that the Nu_{avg} is in fact dependent on the type of mesh quality and the number of elements provided, with an enhanced and direct relationship. The Nusselt average values for the meshes M4 and M5 are nearly identical, with an error of less than 0.6%. Based on these findings, we chose the M5 mesh with 172,426 elements in order to save the computing time.

TABLE 1.1
Grid Test

Mesh	Elements Number	Nu_{avg}
M1	53,145	0,12893898
M2	67,428	0,13832584
M3	88,185	0,14483232
M4	120,004	0,15128014
M5	172,426	0,15761601
M6	269,163	0,16390136
M7	477,693	0,17024681

TABLE 1.2
Nusselt Average Comparison

D/H	Previous Work	Present Work
Nu_{avg}	1.68	1.67361609

The program's accuracy is tested and compared with the work results of Manca et al. [34]. The comparison between the previous and present work from **Table 1.2** reveals that the obtained average Nusselt number is remarkably similar.

1.3 RESULTS AND DISCUSSION

The ANSYS software is employed in the upcoming study to analyze the mixed convection due to the airflow with a fixed velocity, in a horizontal channel at the top of a complex geometry upholding a varied-length heat source generator. The Reynolds and Prandtl numbers are assumed constant, while the results of isothermal and streamlines contours, as well as the Nusselt number, are expressed as functions of the heat source length: dimensional 5 mm $\leq$ L $\leq$ 20 mm, and non-dimensional $0.25 \leq \varepsilon \leq 1$, and also the Richardson number $0.1 \leq Ri \leq 100$.

1.3.1 Impact of the Heat Source Length on Isotherms and Streamlines

The influence of different heat source lengths from $\varepsilon = 0.25$ to $\varepsilon = 1$ on the isothermal and streamlines contours is shown in Figure 1.2 and Figure 1.3. When using a shorter heat source of 5 mm, the heat exchange is limited to some sections of the lower cavity where the diffusive heat transport is prevalent, resulting in a limited convective transfer performance and only a modest change in the temperature of the upper airflow. Heat transport, on the other hand, is strengthened as the length of the heat source increases; the extension enhances the natural convective flow all along the enclosure, allowing the generated heat to ascend and reach the conduit where the cold forced convective airflow is passing. In this context, mixed convection is especially prominent, and the velocity difference in that region permits the airflow to be able to transport the heat rising along the channel.

When $L_h = 20$ mm (i.e., $\varepsilon = 1$), the isotherms are distributed towards the exit of the canal; hence, the significant difference in temperature and velocity provided great thermal transmission that has occurred between the airflow and the heat source.

1.3.2 Impact of the Heat Source Length on Nusselt

The comparison in Figure 1.4 displays what effect the length of the heat source has on the local Nusselt. Likewise, when the source length increases, so does the local Nusselt value. The findings show how the local Nusselt number increasingly fluctuates with the length of the heat source. This increments the surface exposed to the natural convection, thus boosting the buoyant forces presence, which intensifies the heat transfer rate. The results reveal that augmenting the length of the heat source from 5 mm to 20 mm strongly enhances the buoyant convection; thus, it yields to the improvement of the local Nusselt number by 120%.

1.3.3 Impact of the Richardson Number on Nusselt

Figure 1.5, which follows, illustrates the effect of the Richardson number on heat transfer along the examined cavity; the findings show that the Richardson number has a considerable influence on heat transfer—that is, as the Richardson number grows, so does the heat transfer. Considering that a rise

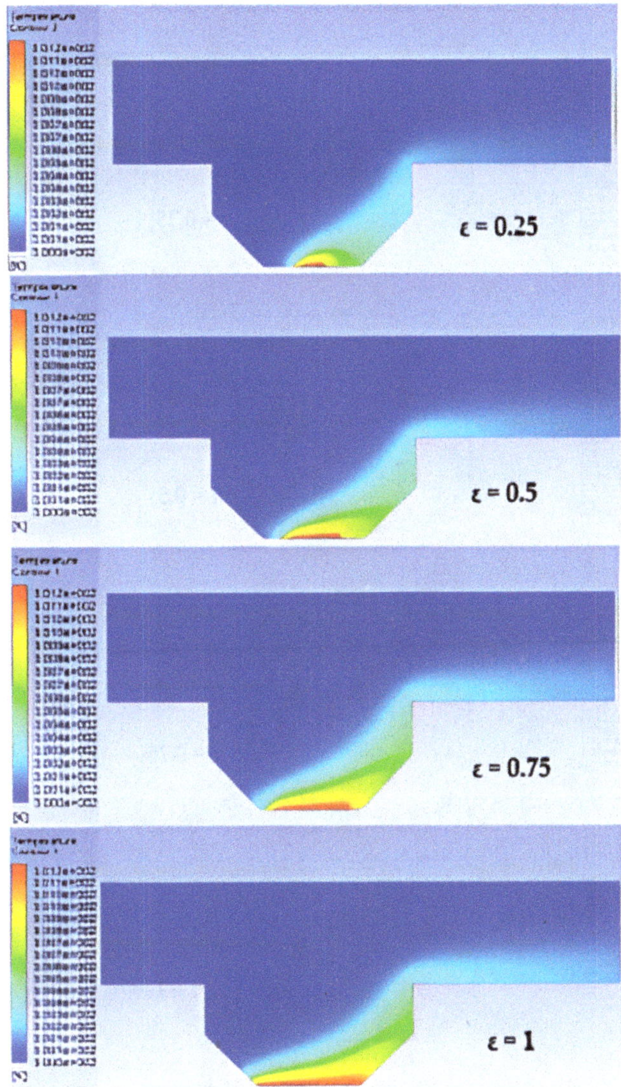

FIGURE 1.2 Isotherms for different heat source lengths.

in the Richardson number corresponds to an increase in the heat input into the enclosure, this has an overall consequence of intensifying the mixed convection and therefore the fluid movement; hence, this improvement allows the boosted levitation of the local Nusselt number.

At the end of the cavity (X = 0.004875), the local Nusselt number is enhanced from 0.5 (W/m^2K) to almost 3 (W/m^2K) when only incrementing the Richardson number from 0.1 to 1. And an enhanced proportion of 67% of the Nu_{loc} is acquired when increasing Ri from 1 to 10. Gradually augmenting Ri from 10 to 100, on the other hand, only delivers a 20% enhancement.

Figure 1.6 presents the average Nusselt number in function of the Richardson number. It is also demonstrated that these two show a proportional relativity. However, the increase slope is hugely superior when Richardson is raised up from 0.1 to 10, compared to that of from 10 to 100, which demonstrates the great impact the Richardson number has on the heat transfer characteristics, both the local and average Nusselt number.

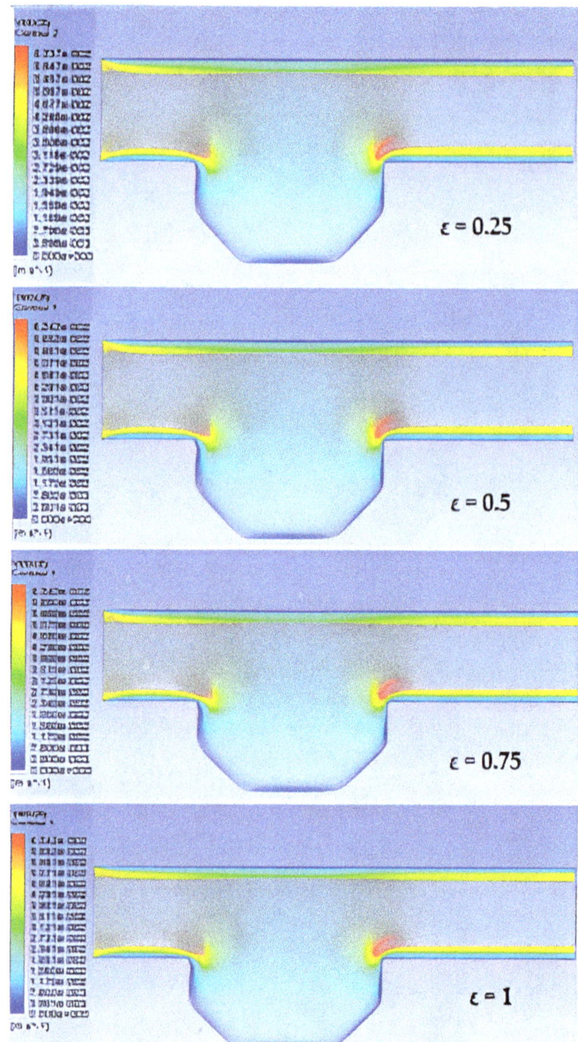

FIGURE 1.3 Streamlines contours for different heat source lengths.

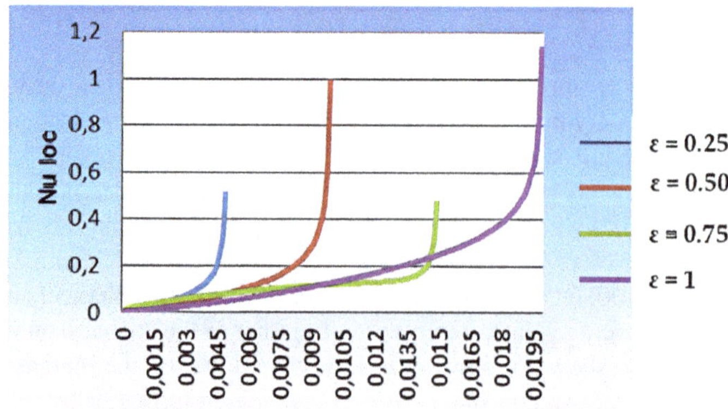

FIGURE 1.4 Comparison of the local Nusselt for different heat source lengths.

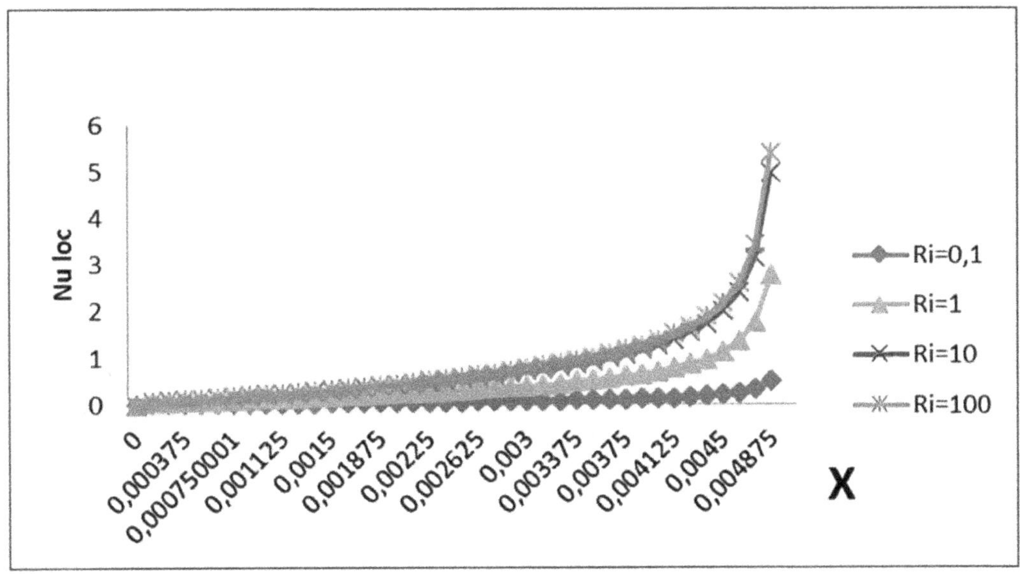

FIGURE 1.5 Comparison of the local Nusselt for different Richardson numbers.

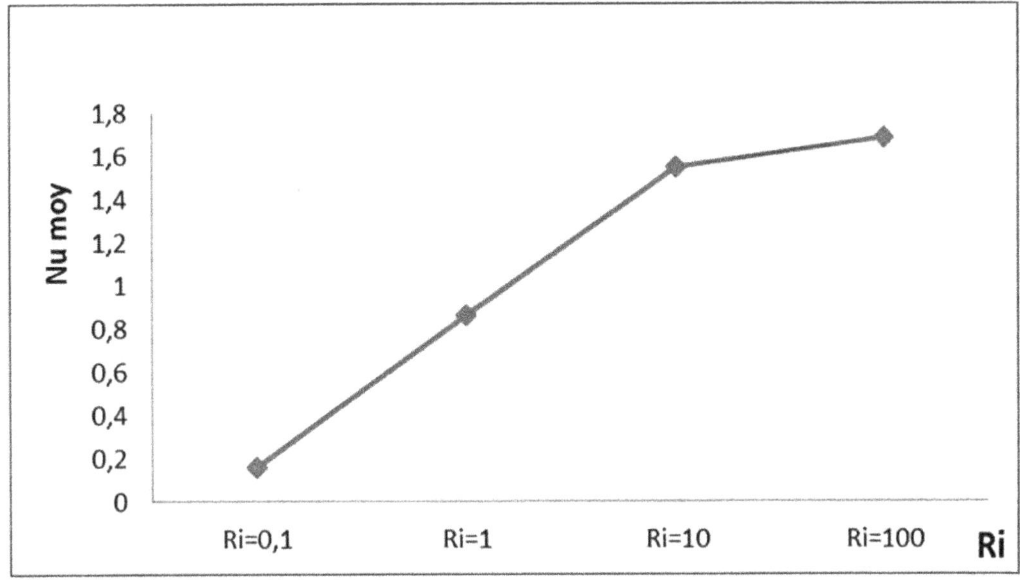

FIGURE 1.6 Average Nusselt for different Richardson numbers.

This relationship is also demonstrated in **Table 1.3**, where the average Nusselt is presented as a function of Ri and ε. The best values of Nu_{avg} are obtained for the case when Ri = 100 and ε = 1, which assures that the enhanced and extended heat transmission is highly dependent on augmenting both the Richardson number and the heat source length.

TABLE 1.3
Nusselt Average at Different Richardson Numbers and Aspect Ratio

	Ri = 0.1	Ri = 1	Ri = 10	Ri = 100
$\varepsilon = 0.25$	0,135121881	0,733070248	1,314983068	1,428366962
$\varepsilon = 0.5$	0,14989746	0,813231103	1,431737704	1,584557549
$\varepsilon = 0.75$	0,155782529	0,845159242	1,516048629	1,646768645
	0,159107292	0,863196331	1,54840385	1,681914533

1.4 CONCLUSIONS

The mixed convection is numerically examined through the ANSYS software in this chapter, featuring a horizontal channel with an open cavity holding a heat source positioned in the enclosure's bottom area by altering its length and the Richardson number. We can draw the following conclusions based on our findings:

- Heat transmission rises as source length increases.
- The local Nusselt number is strongly controlled by the length of the heating source.
- The longest length of the heat source corresponds to the greatest value of the average Nusselt.
- The pattern of isotherms and streamlines contours changes as the length of the heat source increases.

Nomenclature

D	channel height (m)
H	cavity height and length (m)
L	exit length (m)
L_H	heat source length
ε	dimensionless heat source length
P	dimensionless pressure
p	pressure (Pa)
θ	dimensionless temperature
T	temperature (°C)
U, V	dimensionless velocity components
X, Y	dimensionless Cartesian coordinates
u, v	velocity components ($m{\cdot}s^{-1}$)
x, y	Cartesian coordinates (m)
μ	dynamic viscosity (Kg $m^{-1}s^{-1}$)
υ	kinematic viscosity (m^2s^{-1})
ρ	density (Kg m^{-3})
g	gravitational acceleration (m s^{-2})

Subscripts

c	cold
h	hot
in	inlet
loc	local
avg	average
Pr	Prandtl number

Re Reynolds number
Ri Richardson number
Gr Grashof number
Nu Nusselt number

REFERENCES

[1] Bergman, T. L., Lavine, A. S., Incropera, F. P., DeWitt, D. P. *Introduction to heat transfer*. John Wiley & Sons, 2011.

[2] Kaviany, M. *Heat transfer physics*. Cambridge University Press, 2014.

[3] Welty, J. R. *Engineering heat transfer*. Wiley, 1974.

[4] Zhang, J., Zhu, X., Mondejar, M. E., Haglind, F. A review of heat transfer enhancement techniques in plate heat exchangers. *Renewable and Sustainable Energy Reviews*, 2019, *101*: 305–328.

[5] Zhang, Y., Zhang, X., Li, M., Liu, Z. Research on heat transfer enhancement and flow characteristic of heat exchange surface in cosine style runner. *Heat and Mass Transfer*, 2019, *55*(11): 3117–3131.

[6] Hassan, M., Mebarek-Oudina, F., Faisal, A., Ghafar, A., Ismail, A. I. Thermal energy and mass transport of shear thinning fluid under effects of low to high shear rate viscosity. *International Journal of Thermofluids*, 2022, *15*: 100176. https://doi.org/10.1016/j.ijft.2022.100176

[7] Ismael, M. A., Hussein, A. K., Mebarek-Oudina, F., Kolsi, L. Effect of driven sidewalls on mixed convection in an open trapezoidal cavity with a channel. *Journal of Heat Transfer*, 2020, *142*(8): 082601. https://doi.org/10.1115/1.4047049

[8] Chabani, I., Mebarek-Oudina, F., Vaidya, H., Ismail, A. I. Numerical analysis of magnetic hybrid Nanofluid natural convective flow in an adjusted porous trapezoidal enclosure. *Journal of Magnetism and Magnetic Materials*, 2022, *564*: 170142.

[9] Mebarek-Oudina, F., Laouira, H., Aissa, A., Hussein, A. K., El Ganaoui, M. Convection heat transfer analysis in a channel with an open trapezoidal cavity: Heat source locations effect. *MATEC Web of Conferences*, 2020, *330*: 01006.

[10] Reddy, Y. D., Mebarek-Oudina, F., Goud, B. S., Ismail, A. I. Radiation, velocity and thermal slips effect toward MHD boundary layer flow through heat and mass transport of Williamson nanofluid with porous medium. *Arabian Journal for Science and Engineering*, 2022, *47*(12): 16355–16369. https://doi.org/10.1007/s13369-022-06825-2

[11] Raza, J., Mebarek-Oudina, F., Ali Lund, L. The flow of magnetised convective Casson liquid via a porous channel with shrinking and stationary walls. *Pramana – Journal of Physics*, 2022, *96*: 229. https://doi.org/10.1007/s12043-022-02465-1

[12] Habib, R., Karimi, N., Yadollahi, B., Doranehgard, M. H., Li, L. K. A pore-scale assessment of the dynamic response of forced convection in porous media to inlet flow modulations. *International Journal of Heat and Mass Transfer*, 2020, *153*: 119657.

[13] Mebarek-Oudina, F. Numerical modeling of the hydrodynamic stability in vertical annulus with heat source of different lengths. *Engineering Science and Technology, an International Journal,* 2017, *20*(4): 1324–1333.

[14] Leong, J. C., Brown, N. M., Lai, F. C. Mixed convection from an open cavity in a horizontal channel. *International Communications in Heat and Mass Transfer*, 2005, *32*(5): 583–592.

[15] Manca, O., Nardini, S., Vafai, K. Experimental investigation of mixed convection in a channel with an open cavity. *Experimental Heat Transfer*, 2006, *19*(1): 53–68.

[16] Hussain, S. H., Hussein, A. K. Mixed convection heat transfer in a differentially heated square enclosure with a conductive rotating circular cylinder at different vertical locations. *International Communications in Heat and Mass Transfer*, 2011, *38*(2): 263–274.

[17] Marzougui, S., Bouabid, M., Mebarek-Oudina, F., Abu-Hamdeh, N., Magherbi, M., Ramesh, K. A computational analysis of heat transport irreversibility phenomenon in a magnetized porous channel. *International Journal of Numerical Methods for Heat & Fluid Flow*, 2021, *31*(7): 2197–2222. https://doi.org/10.1108/HFF-07-2020-0418

[18] Sahoo, S. K., Das, M. K., Rath, P. Application of TCE-PCM based heat sinks for cooling of electronic components: A review. *Renewable and Sustainable Energy Reviews*, 2016, *59*: 550–582.

[19] Gharbi, S., Harmand, S., Jabrallah, S. B. Experimental comparison between different configurations of PCM based heat sinks for cooling electronic components. *Applied Thermal Engineering*, 2015, *87*: 454–462.

[20] He, Z., Yan, Y., Zhang, Z. Thermal management and temperature uniformity enhancement of electronic devices by micro heat sinks: A review. *Energy*, 2021, *216*: 119223.

[21] Khan, U., Mebarek-Oudina, F., Zaib, A., Ishak, A., Abu Bakar, S., Sherif, E. M., Baleanu D. An exact solution of a Casson fluid flow induced by dust particles with hybrid nanofluid over a stretching sheet subject to Lorentz forces. *Waves in Random and Complex Media*, 2022. https://doi.org/10.1080/17455030.2022.2102689

[22] Farhan, M., Omar, Z., Mebarek-Oudina, F., Raza, J., Shah, Z., Choudhari, R. V., Makinde, O. D. Implementation of the one-step one-hybrid block method on the nonlinear equation of a circular sector oscillator. *Computational Mathematics and Modeling*, 2020, *31*(1): 116–132.

[23] Mebarek-Oudina, F., Chabani, I. Review on nano-fluids applications and heat transfer enhancement techniques in different enclosures. *Journal of Nanofluids*, 2022, *11*(2): 155–168.

[24] Mebarek-Oudina, F. Convective heat transfer of Titania nanofluids of different base fluids in cylindrical annulus with discrete heat source. *Heat Transfer—Asian Research*, 2019, *48*(1): 135–147.

[25] Asogwa, K., Mebarek-Oudina, F., Animasaun, I. Comparative investigation of water-based Al_2O_3 nanoparticles through Water-based CuO nanoparticles over an exponentially accelerated radiative Riga plate surface via heat transport. *Arabian Journal for Science and Engineering*, 2022, *47*(7): 8721–8738. https://doi.org/10.1007/s13369-021-06355-3

[26] Mebarek-Oudina, F., Chabani, I. Review on nano enhanced PCMs: Insight on nePCM application in thermal management/storage systems. *Energies*, 2023, *16*(3): 1066. https://doi.org/10.3390/en16031066

[27] Dharmaiah, G., Mebarek-Oudina, F., Sreenivasa Kumar, M., Chandra Kala, K. Nuclear reactor application on Jeffrey fluid flow with Falkner-Skan factor, Brownian and thermophoresis, non linear thermal radiation impacts past a wedge. *Journal of the Indian Chemical Society*, 2023, *100*(2): 100907. https://doi.org/10.1016/j.jics.2023.100907.

[28] Manca, O., Nardini, S., Khanafer, K., Vafai, K. Effect of heated wall position on mixed convection in a channel with an open cavity. *Numerical Heat Transfer: Part A: Applications*, 2003, *43*(3): 259–282.

[29] Mebarek-Oudina, F., Laouira, H., Hussein, A. K., Omri, M., Abderrahmane, A., Kolsi, L., Biswal, U. Mixed convection inside a duct with an open trapezoidal cavity equipped with two discrete heat sources and moving walls. *Mathematics*, 2022, *10*(6): 929.

[30] Hossain, M. S., Alim, M. A., Andallah, L. S. Numerical investigation of natural convection flow in a trapezoidal cavity with non-uniformly heated triangular block embedded inside. *Journal of Advances in Mathematics and Computer Science*, 2018, *28*: 1–30.

[31] Öztop, H. F., Dagtekin, I. Mixed convection in two-sided lid-driven differentially heated square cavity. *International Journal of Heat and Mass Transfer*, 2004, *47*(8–9): 1761–1769.

[32] Burgos, J., Cuesta, I., Salueña, C. Numerical study of laminar mixed convection in a square open cavity. *International Journal of Heat and Mass Transfer*, 2016, *99*: 599–612.

[33] Laouira, H., Mebarek-Oudina, F., Hussein, A. K., Kolsi, L., Merah, A., Younis, O. Heat transfer inside a horizontal channel with an open trapezoidal enclosure subjected to a heat source of different lengths. *Heat Transfer-Asian Research*, 2020, *49*(1): 406–423. https://doi.org/10.1002/htj.21618

[34] Manca, O., Nardini, S., Vafai, K. Experimental investigation of opposing mixed convection in a channel with an open cavity below. *Experimental Heat Transfer*, 2008, *21*(2): 99–114.

2 Influence of Vibrations on Thermodynamic Irreversibilities in a Flowing Fluid in a Saturated Porous Cavity

A. Mchirgui, A. Chibani, and M. Magnerbi

2.1 INTRODUCTION

In connection with its extensive applications in engineering and geophysics processes, transfer by convection has received a considerable attention in the literature. Most works in this context focus on natural, forced, or mixed convection and mainly concern either a pure fluid medium or a porous cavity saturated by a fluid. On one hand, a wide range of works about the topic of natural convection in a simple enclosure can be found in references [1–5]; on the other, various papers focus on the field of convection in porous media. Most of them are well documented in the books by Bejan et al. [5] and Vafai [6]. Recently, researchers introduced a new class of fluids, named nanofluids, as an innovative way of improving heat transfer by convection. Chabani et al. [7] performed a numerical investigation about heat transfer by Cu-TiO_2/EG hybrid nanofluid inside a Darcy–Brinkman–Forchheimer porous enclosure with zigzags. A numerical study about the buoyant convective flow and thermal transport enhancement of Cu-H_2O nanoliquid in a differentially heated upright annulus was performed by Pushpa et al. [8]. They found that the liquid flow and heat transport considerably depend on the baffle location and its length. Swain et al. [9] analyzed heat transport of nanoliquids flow over an extendable surface near a stagnation point with variable thermal conductivity under the influence of the magnetic field. A wide range of other comprehensive investigation about this topic can be found in references [10–14]. In the previous mentioned works, a great deal of attention focused on the static systems, in which convection occurs by the simple effect of gravity. However, such systems, in their reality, are subject to vibrational motions, which can be assumed to be periodic; thus, part of the research work took into consideration the existence of mechanical vibrations by studying the phenomenon of thermal convection. Fu and Shieh [15] numerically investigated the effects of vibration frequency and Rayleigh number on the thermal convection in a two-dimensional square enclosure. They divided the thermal convection into five regions: quasi-static convection, vibration convection, resonant vibration convection, intermediate convection, and high-frequency vibration convection. Bardan et al. [16] studied the effect of high-frequency vibrations on double diffusion convection in a rectangular saturated porous cell heated from the bottom. The results show that pure mechanical equilibrium exists only for vertical vibrations which stabilize the no-flow state, and vibrations shift the onset of convection to a high value of the Rayleigh number. Chung and Vafai [17] investigated vibration-induced mixed convection inside an open cavity filled with a porous medium. They noted that the vibrational effects are more pronounced at higher values of the Darcy and Reynolds numbers. Bardan and Mojtabi [18] have numerically and analytically studied double-diffusive natural convection with high-frequency vibrations in square and

DOI: 10.1201/9781003299608-2

rectangular enclosures. They noted that for vertical oscillation, increasing the vibration amplitude decreases the subcriticality of the solutions, and the reverse happens with horizontal vibrations. Guershuni et al. [19] presented a numerical study about vibrational thermal convection in a rectangular cavity. Forbes et al. [20] experimentally studied the effect of mechanical vibrations on heat transfer by natural convection in a rectangular enclosure. The results show that vibrations can have a significant effect on heat transfer characteristics, in particular, near the natural resonance frequency of the column of fluid contained within the enclosure. Zidi et al. [21] performed a numerical investigation about vertical vibration effect on thermosolutal convection in a rectangular enclosure partially filled by a Darcy–Brinkmann porous layer. The effect of ultrasonic vibrations on the melting process of a phase-change material was experimentally studied by Oh et al. [22]. They found that the process is accelerated by ultrasonic vibration. Khudhair et al. [23] performed an experimental study about the mechanical vibration effect on natural convection in a 3D cubic enclosure filled with air. The study leads to the competition between vibrational and gravitational convection.

Generally, the amount of energy is a factor of great importance for any engineering process; entropy generation as a measure of the energy losses is related to the efficiency of any system containing irreversible factors, such as heat transfer and friction. Therefore, many researches related to transport phenomena focused on entropy generation and its minimization in order to raise the performance of the system. In this context, Hooman et al. [24] analyzed the first and second laws characteristics of a fully developed forced convection in a porous medium. Hidouri et al. [25] and Magherbi et al. [26] analyzed the influence of cross effects of Soret and Dufour on entropy generation for a fluid saturating a square cavity submitted to double-diffusive convection. They concluded that thermo-diffusion and Dufour effects considerably affect the magnitude of entropy generation, especially at higher Grashof numbers. The effect of magnetic field on entropy generation encountered in mixed convection was investigated by Marzougui et al. [27, 28] for a Newtonian fluid saturating a porous horizontal channel heated from below [27] and for nanofluid-saturated lid-driven porous medium [28]. Other studies about entropy generation in different configurations were reported in the literature due to its wide range of applications [29–32].

This chapter concerns entropy generation determination in natural convection for the case of a saturated square porous cavity filled with a binary perfect gas mixture and subjected to vertical vibration. The Darcy–Brinkman model with Boussinesq approximation is employed. The set of coupled equations of mass, momentum, and energy conservation is solved using a numerical program developed under the COMSOL platform. Particular attention is paid to the study of the influence of the vibration frequency, Darcy number, and the vibrational Grashof number on steady-state entropy generation.

2.2 MATHEMATICAL FORMULATION

As shown in Figure 2.1, the considered physical model consists of a saturated porous cavity driven by a cover and filled with an incompressible Newtonian fluid. The porous matrix is assumed to be homogeneous, isotropic, and in thermodynamic equilibrium with the fluid. The left and right walls are submitted to different but uniform temperatures T_h and T_c, respectively, while the two horizontal walls are insulated and adiabatic. The cavity is subjected to a sinusoidal vertical vibration, which is characterized by its angular frequency ω and with its displacement a.

All the physical properties of the fluid are assumed to be constant, except its density, which satisfies the Boussinesq approximation, such that:

$$\rho(T) = \rho_0\left[1 - \beta_T\left(T - T_0\right)\right] \tag{2.1}$$

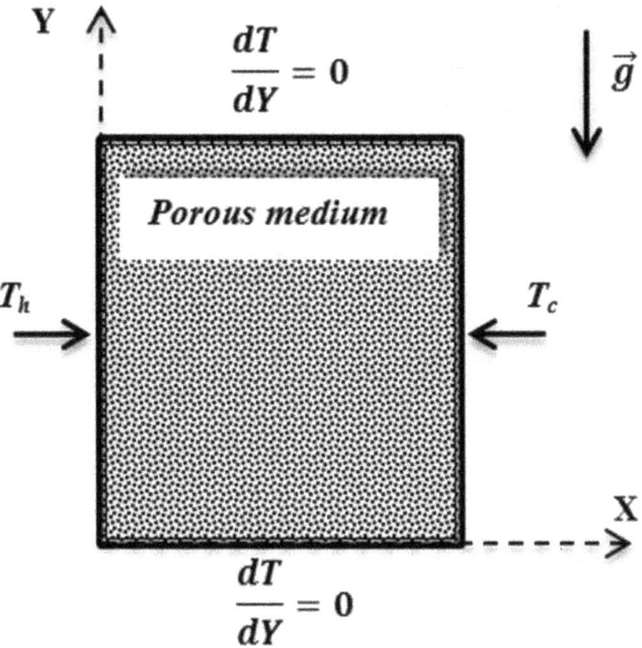

FIGURE 2.1 Physical model.

ρ_0 and T_0 are respectively the reference density and the reference temperature. The term β_T is the coefficient of thermal expansion, given by:

$$\beta_T = \frac{1}{\rho_0}\left(\frac{\partial p}{\partial T}\right)_p \tag{2.2}$$

Using the Darcy–Brinkman formulation and taking into consideration that the presence of vibration changes the gravitational acceleration in the momentum balance, the conservation equations of mass, momentum, and energy describing the phenomenon inside the cavity are given as follows:

$$\frac{\partial u}{\partial x} + \frac{\partial v}{\partial y} = 0 \tag{2.3}$$

$$\rho_0\left[\frac{1}{\epsilon}\frac{\partial u}{\partial t} + \frac{1}{\epsilon^2}\left(u\frac{\partial u}{\partial x} + v\frac{\partial u}{\partial y}\right)\right] = -\frac{\partial p}{\partial x} + \mu_{eff}\left(\frac{\partial^2 u}{\partial x^2} + \frac{\partial^2 u}{\partial y^2}\right) - \frac{\mu}{k}u \tag{2.4}$$

$$\rho_0\left[\frac{1}{\epsilon}\frac{\partial v}{\partial t} + \frac{1}{\epsilon^2}\left(u\frac{\partial v}{\partial x} + v\frac{\partial v}{\partial y}\right)\right] = -\frac{\partial p}{\partial x} + \mu_{eff}\left(\frac{\partial^2 v}{\partial x^2} + \frac{\partial^2 v}{\partial y^2}\right) - \frac{\mu}{k}v - \rho_0\beta_T\Delta T\left(g - b\omega^2\sin(\omega t)\right) \tag{2.5}$$

$$\sigma\frac{\partial T}{\partial t} + \left(u\frac{\partial T}{\partial x} + v\frac{\partial T}{\partial y}\right) = \alpha_{eff}\left(\frac{\partial^2 T}{\partial x^2} + \frac{\partial^2 T}{\partial y^2}\right) \tag{2.6}$$

K, μ, μ_{eff}, ε, α_{eff}, and σ are the medium permeability, the fluid dynamic viscosity, the effective viscosity (i.e., the viscosity in the Brinkman model), the medium porosity, the effective thermal diffusivity, and the specific heat capacities ratio, respectively.

The effective thermal diffusivity is the ratio of the saturated porous medium thermal conductivity (k_m) by the specific heat capacity of the fluid ($(\rho c)_f$). c is the specific heat. Subscript f refers to fluid properties, while subscript m refers to the fluid–solid mixture.

The dimensionless macroscopic conservation equations can therefore be written as follows:

$$\frac{\partial U}{\partial X}+\frac{\partial V}{\partial Y}=0 \tag{2.7}$$

$$\left[\frac{1}{\epsilon}\frac{\partial U}{\partial \tau}+\frac{1}{\epsilon^2}\left(U\frac{\partial U}{\partial X}+V\frac{\partial U}{\partial Y}\right)\right]=-\frac{\partial P}{\partial X}+Pr\left(\frac{\partial^2 U}{\partial X^2}+\frac{\partial^2 U}{\partial Y^2}\right)-\frac{Pr}{Da}U \tag{2.8}$$

$$\left[\frac{1}{\epsilon}\frac{\partial V}{\partial \tau}+\frac{1}{\epsilon^2}\left(U\frac{\partial V}{\partial X}+V\frac{\partial V}{\partial Y}\right)\right]=-\frac{\partial P}{\partial Y}+Pr\left(\frac{\partial^2 V}{\partial X^2}+\frac{\partial^2 V}{\partial Y^2}\right)-\frac{Pr}{Da}V$$
$$+\left(Gr_T+Pr.\omega.\sqrt{2.Gr_v}.sin(\omega\tau)\right) \tag{2.9}$$

$$\sigma\frac{\partial \theta}{\partial \tau}+\left(U\frac{\partial \theta}{\partial X}+V\frac{\partial \theta}{\partial Y}\right)=k\left(\frac{\partial^2 \theta}{\partial X^2}+\frac{\partial^2 \theta}{\partial Y^2}\right) \tag{2.10}$$

The governing equations are established using the following dimensionless variables:

$$X=\frac{x}{H},Y=\frac{y}{H},U=\frac{uH}{\alpha},V=\frac{vH}{\alpha},\theta=\frac{T-T_0}{\Delta T},\tau=\frac{t}{H^2\alpha},W=\frac{wH^2}{\alpha},Pr=\frac{\nu}{\alpha},Da=\frac{K}{H^2},$$

$$Ra=\frac{\beta g\Delta TH^3}{\alpha\nu},\quad Gr_v=\frac{\left(\beta bw\left(T_h-T_c\right)H\right)^2}{2\nu^2},Gr_T=Ra.Pr. \tag{2.11}$$

In dimensionless form, the initial and boundary conditions are:

$$For\ X=0,\ 0\le Y\le 1,\ \theta=1;\ for\ X=1,\ 0\le Y\le 1,\ \theta=0$$

$$For\ Y=0,\ 0\le Y\le 1,\ \frac{\partial \theta}{\partial y}=0;\ for\ Y=1,\ 0\le Y\le 1,\ \frac{\partial \theta}{\partial y}=0 \tag{2.12}$$

2.3 ENTROPY GENERATION

In the present problem, entropy is generated through thermal gradients and viscous dissipation. In such problem, the Darcy dissipation term should not be neglected compared to the clear fluid term in the contribution of fluid friction irreversibility. Following Magherbi et al. [26], Hooman et al. [24], and Mchirgui et al. [29], the expression of the volumetric entropy generation in natural convection in 2D approximation is given by:

$$Si_l=\frac{k_m}{T_0^2}\left(\left(\frac{\partial T}{\partial x}\right)^2+\left(\frac{\partial T}{\partial y}\right)^2\right)+\frac{\mu}{T_0.k}\left(u^2+v^2\right)+\frac{\mu}{T_0}\left[2\left(\frac{\partial u}{\partial x}\right)^2+2\left(\frac{\partial v}{\partial x}\right)^2+\left(\frac{\partial u}{\partial y}+\frac{\partial v}{\partial x}\right)^2\right] \tag{2.13}$$

Taking into account the system of adimensional variables (equation (11)), the dimensionless local entropy generation is obtained as:

$$Si_{l,a}=Si_{Th}+Si_D+Si_F \tag{2.14}$$

Where:

$$Si_{Th} = \left(\frac{\partial\theta}{\partial X}\right)^2 + \left(\frac{\partial\theta}{\partial Y}\right)^2 \tag{2.15}$$

$$Si_D = \frac{Br^*}{Da}\left(U^2 + V^2\right) \tag{2.16}$$

$$Si_F = Br^*\left[2\left(\frac{\partial U}{\partial X}\right)^2 + 2\left(\frac{\partial V}{\partial Y}\right)^2 + \left(\frac{\partial U}{\partial Y} + \frac{\partial V}{\partial X}\right)^2\right] \tag{2.17}$$

Br^* is the modified Brinkman number, and it is given by:

$$Br^* = \frac{Br}{\Omega} \tag{2.18}$$

Br and Ω are respectively the Brinkman number and the dimensionless temperature difference. They are given by:

$$Br = \frac{\mu U_0^2}{k_m \Delta T} \text{ and } \Omega = \frac{\Delta T}{T_0} \tag{2.19}$$

The dimensionless total entropy production for the entire cavity is obtained by numerical integration over the cavity of the dimensionless local entropy generation. It is given by:

$$S_{Tot} = \iiint_{Volume} Si_{l,a} dV. \tag{2.20}$$

2.4 NUMERICAL PROCEDURE

2.4.1 Numerical Method

The system of conservation equations is solved using the commercial software COMSOL Multiphysics based on the Galerkin finite element. The used software allows the simulation of several physical problems with coupled or uncoupled phenomena. Two modules related to fluid flow, namely, the "laminar fluid flow" module and the "heat transfer in fluids" module, were used in the simulation. The conservation equations of continuity, momentum, and energy were solved by a "time-dependent" solver using the damped Newton method. Linear iterations of Newton's method are accelerated by state-of-the-art algebraic multi-grid methods especially designed for transportation problems. The test function is designed to stabilize the hyperbolic terms and the pressure term in transport equations. Shock capture techniques further reduce spurious oscillations. Additionally, discontinuous Galerkin formulations are used to conserve momentum, mass, and energy across internal and external limits. We employed the "$P_2 + P_1$" Lagrange diagram for the velocity–pressure coupling, while the "Lagrange-linear" diagram was chosen during the discretization of the energy equation. The "free tetrahedral" elements of the uniform grid were generated using the built-in mesh function in COMSOL. To verify the degree of refinement of the necessary mesh, tests were carried out. The chosen mesh is 31,185 tetrahedral elements, 3,912 triangular elements, 228 edge elements, and 8 vertex elements.

TABLE 2.1
Average Nusselt Number for Pr = 0.71, G_{rT} × Pr × Da = 100, G_{rV} = 0

Da	10^{-1}	10^{-2}	10^{-3}	10^{-4}	10^{-5}
This study	1,06	1,72	2,42	4,84	3,04
Lauriat and Prasad [33]	—	1,7	2,41	2,84	3,02
Kramer et al. [34]	1,08	1,7	2,43	2,83	2,99

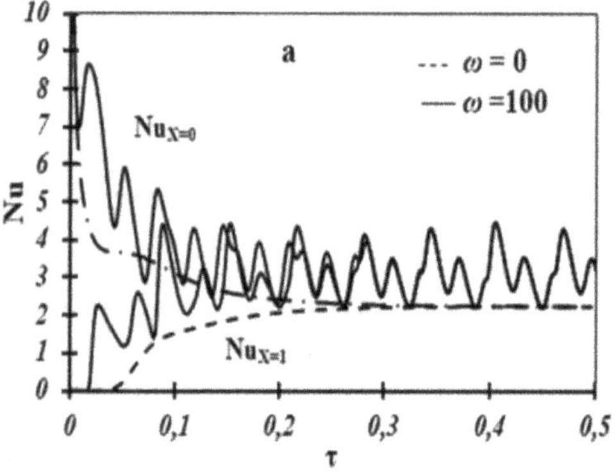

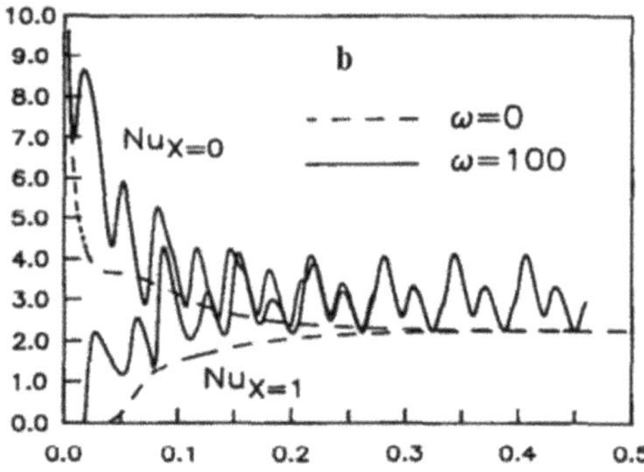

FIGURE 2.2 Average Nusselt numbers at the hot and cold walls: (a) present work; (b) Shung et al. [19].

2.4.2 Numerical Validation

First, the accuracy of the present numerical study has been performed for natural convection in a porous medium. In Table 2.1, the average values of Nusselt number are compared with those given by Lauriat and Prasad [33] and Kramer et al. [34]

Second, considering a clear continuous medium with ε = 1 and for a high Darcy number, evolution of Nusselt number in transient state is compared with results given by Shung et al. [35] for two values of vibration frequency, as shown in Figure 2.2.

As can be seen, the results are in good agreement with those given by the literature. Moreover, one can see that in the absence of vibration, the hot average Nusselt number is important at the onset of natural convection, indicating a significant heat exchange between the hot wall and the fluid and then an important thermal irreversibility. This result has been found by Magherbi et al. [36]. As time increases, the hot Nusselt number decreases and finally takes a constant value. On the other hand, the average cold Nusselt number is minimal at the beginning, then it increases and takes the same constant value. This observation indicates that as the steady state is reached, the heat taken up by the fluid from the hot wall is immediately evacuated by the cold wall. The same result is obtained for ω = 100, although their oscillating behavior and hot and cold Nusselt numbers become totally confused in steady state.

2.5 RESULTS AND DISCUSSION

In the present work, the considered medium is a square porous cavity filled with a binary perfect gas mixture characterized by Pr = 0,7 and subjected to vertical vibration with constant displacement. Due to the large number of variables related to this study, the operating parameters are in the following ranges: $10^{-5} \leq Da \leq 10^{-3}$, $10^5 \leq Gr_T \leq 10^6$, $10^4 \leq Gr_V \leq 10^6$. The porosity is fixed at 0,8, and the Brinkman and modified Brinkman numbers are assumed to be constant and are equal to 5.10^{-3} and 10^{-3}, respectively.

2.5.1 Influence of Darcy Number and Vibrational Frequency

This section concerns the effect of sinusoidal vibration frequency on the entropy generation for different values of Darcy number. The thermal and vibratory Grashof numbers are fixed at 10^5 and 10^6, respectively. The Darcy number and the vibrational frequency of the exciting force range from 10^{-3} to 10^{-5} and from 300 to 1,700, respectively. Figures 2.3 to 2.5 give the variations of entropy

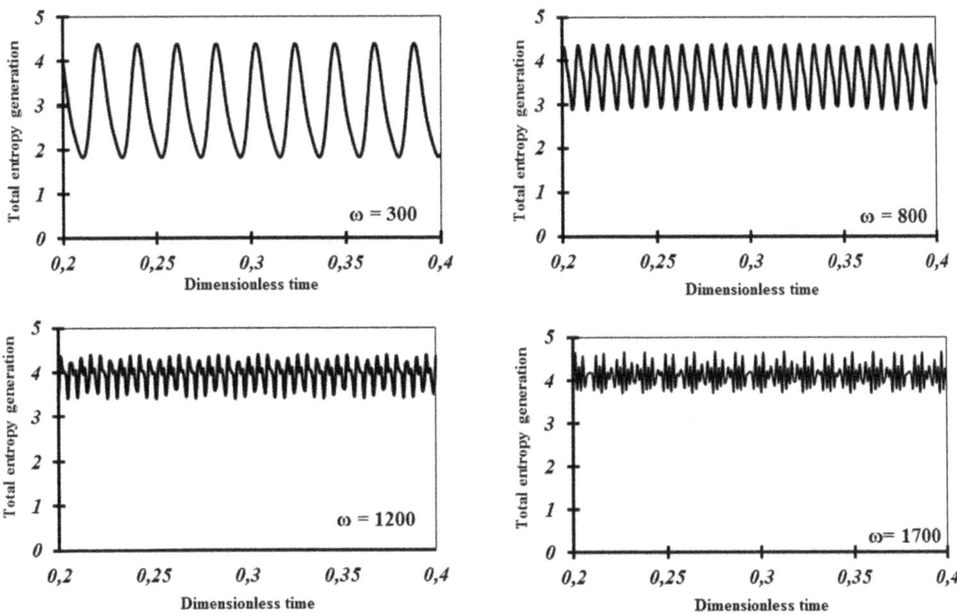

FIGURE 2.3 Total entropy generation versus time for different vibration frequencies (Da = 10^{-3}, $Gr_T = 10^5$, and $Gr_V = 10^6$).

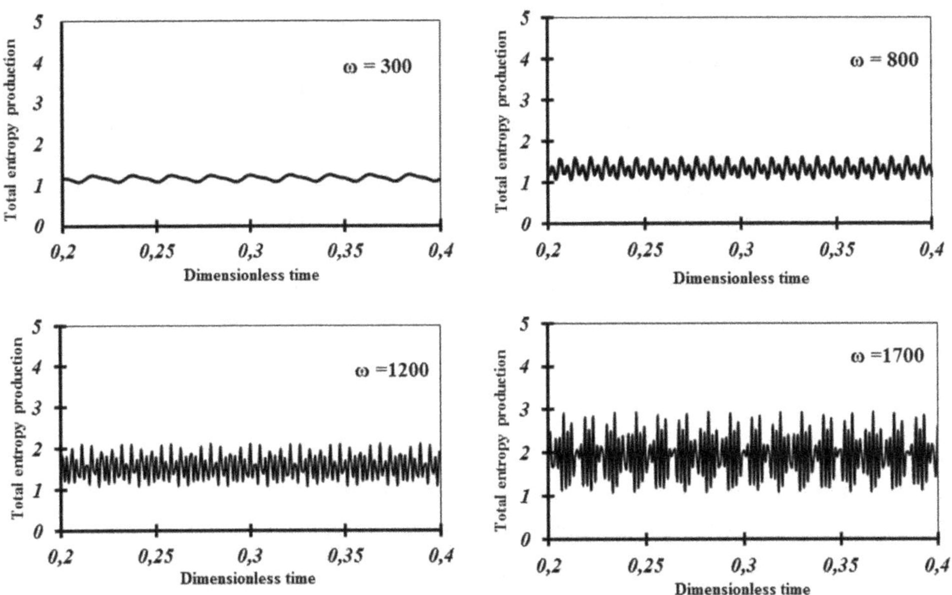

FIGURE 2.4 Total entropy generation versus time for different vibration frequencies for **Da = 10^{-4}, $Gr_T = 10^5$, and $Gr_V = 10^6$**.

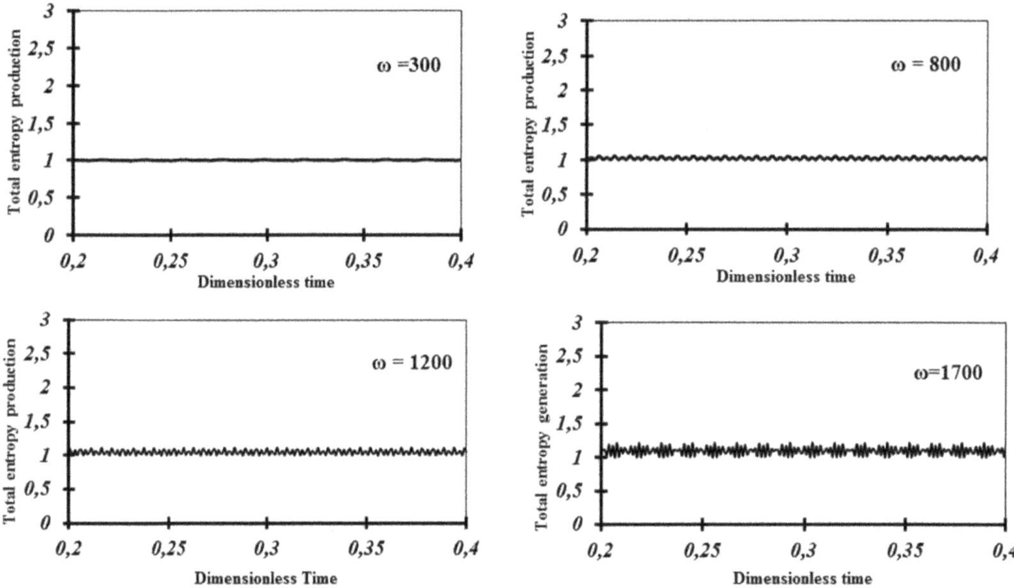

FIGURE 2.5 Total entropy generation versus time for different vibration frequencies for $\mathbf{Da = 10^{-5}}$, $\mathbf{Gr_T = 10^5}$, and $\mathbf{Gr_V = 10^6}$.

generation versus dimensionless time for different Darcy numbers. As can be seen, a sinusoidal variation of entropy generation is obtained for frequencies under 800. As frequency increases, the behavior becomes non-sinusoidal while remaining periodic.

It is important to notice that the average entropy production increases with the vibration frequency regardless of the Darcy number value. This increase is a consequence of the enhancement of convection phenomenon as a result of the cavity vibration. The improvement in convection leads to the importance of thermal gradients, which in turn induce an increase of irreversibilities within the cavity. One can note that the magnitude of entropy generation decreases as the Darcy number decreases; the average value of the entropy generation goes from 3.1 to 1.15 and to 1 for Darcy numbers equal to 10^{-3} to 10^{-4} and 10^{-5}, respectively, which means a reduction of about 70%.

For the used low values of Brinkman and modified Brinkman numbers, viscous irreversibility is neglected; consequently, entropy generation is mainly due to thermal effects. In this context, as the Darcy number decreases, the permeability of the fluid decreases and the porous matrix tends to resist the flow. Thus, the porous cavity is physically close to a solid medium, where the convection phenomenon gradually disappears. The isotherms become less and less distorted, giving rise to a reduction in the vertical and horizontal thermal gradients and, consequently, in entropy generation. It should also be noted that the decrease of entropy generation for low permeability of the porous medium is accompanied by a decrease of oscillation's amplitude. This can be explained by the fact that the effect of the vibrating force is insignificant when the porous medium is compared to a solid matrix.

2.5.2 Influence of Vibrational Grashof Number and Vibrational Frequency on Average Entropy Production

In this paragraph, our interest is focused on the effect of vibratory Grashof number on entropy generation. Thus, the Prandt, Darcy, and thermal Grashof numbers are fixed at 0.7, 10^{-3}, and 10^6, respectively. The frequency of vibration ranges from 300 to 1,700. Figures 2.6 to 2.8 illustrate the variations of total entropy generation versus time for different values of vibrational Grashof number.

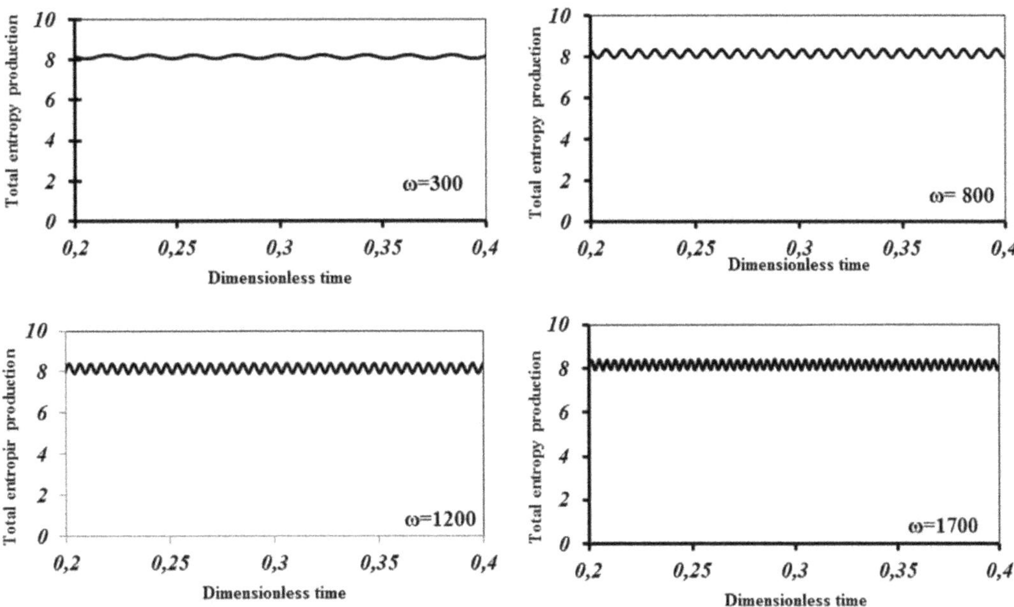

FIGURE 2.6 Total entropy generation versus time for different vibration frequencies for $\mathbf{Gr_V = 10^4}$ ($\mathbf{Da = 10^{-3}, Gr_T = 10^6}$).

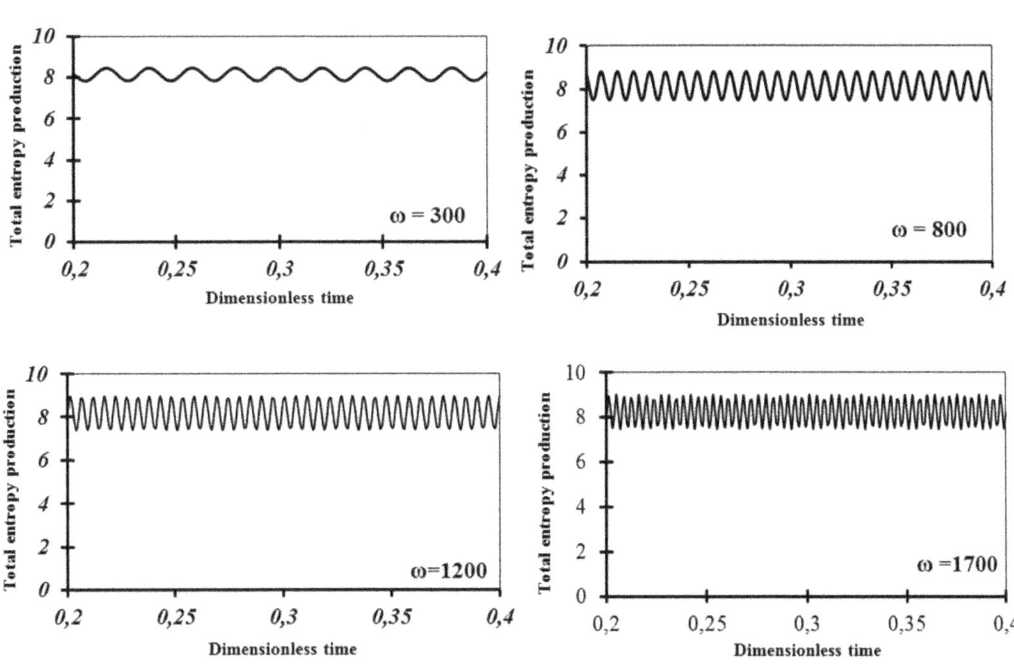

FIGURE 2.7 Total entropy generation versus time for different vibration frequencies for $\mathbf{Gr_V = 10^5}$ ($\mathbf{Da = 10^{-3}, Gr_T = 10^6}$).

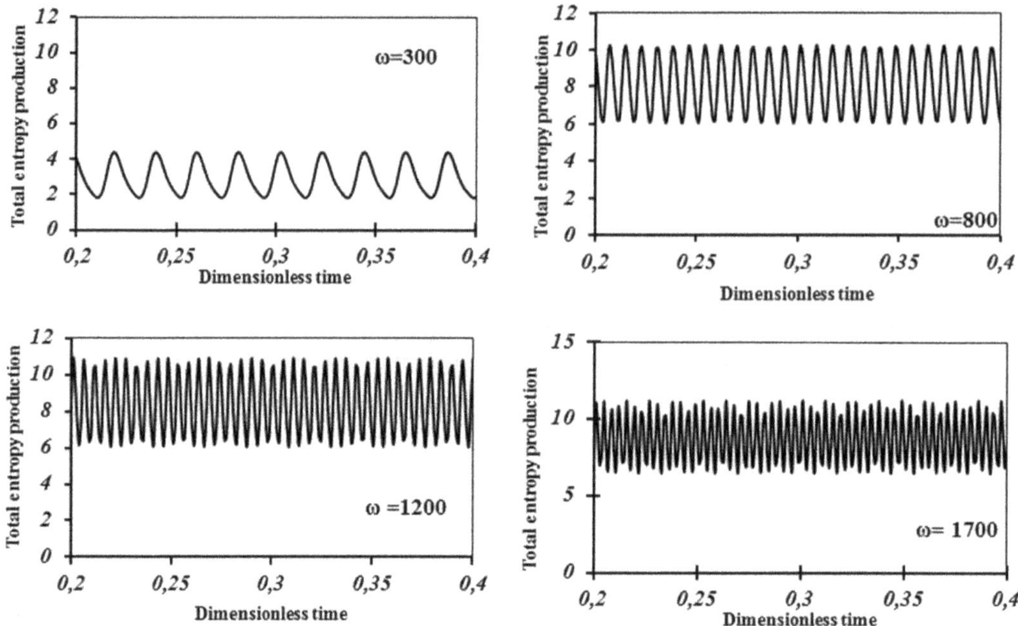

FIGURE 2.8 Total entropy generation versus time for different vibration frequencies for $\mathbf{Gr_V = 10^6}$ ($\mathbf{Da = 10^{-3}, Gr_T = 10^6}$).

For $G_{rV} = 10^4$ (Figure 2.6), we note a sinusoidal behavior of transient entropy generation. A low amplitude of oscillations of the total entropy generation is obtained, and its average value as a function of the vibration frequency is pratically constant. For this value of vibratory Grashof number, the amplitude of the oscillations of the cavity, vibration generates a weak effect on the convection and the hydrodynamics of the fluid. Thus, the variations affecting the temperature and velocity gradients and, consequently, entropy generation remain very limited. Practically the same observations are obtained for Grv = 10^5 with slight increase in terms of the oscillation amplitude of the entropy production. It is only from Grv = 10^6 (Figure 2.8) that the change becomes palpable. For this case, the solution in steady state remains periodic, but the sinusoidal regime is destructed from the frequency $\omega = 800$.

The amplitude of the oscillations (or the maximum value of the oscillating quantity) becomes significant and increases with the frequency of vibration. Similarly, the average value of the total entropy production has been found to increase with the frequency of vibration. This can be explained by the fact that in the case of a relatively large vibratory Grashof, the effect exerted by the vibrating force on the cavity becomes quite significant; hence, it generates a disturbance on the flow which leads to an improvement in convection and, consequently, an increase of irreversibilities.

2.5.3 Influence of Thermal Grashof Number and Vibration Frequency on Average Entropy Production

In this part, Prandtl and Darcy numbers are maintained at 0.7 and 10^{-3}, respectively. The vibration frequency ranges from 300 to 1,700, and the thermal Grashof number ranges from 10^4 to 10^6. Figures 9 to 11 illustrate transient total entropy generation for different thermal Grashof number values. For $Gr_T = 10^4$, evolution of entropy generation is no longer sinusoidal but remains periodic or even quasi-periodic in certain cases.

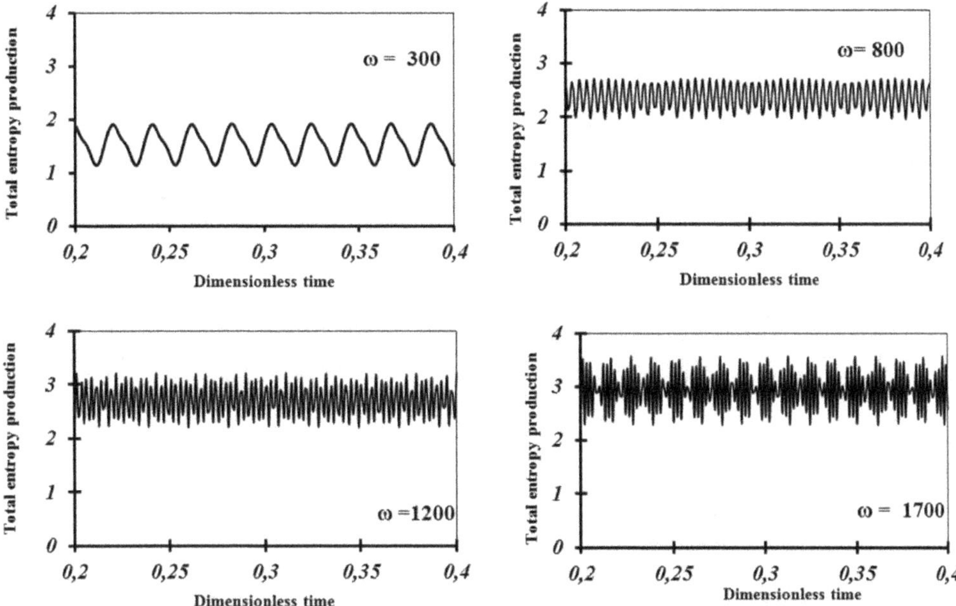

FIGURE 2.9 Total entropy generation versus time for different vibration frequencies for $\mathbf{Gr_T = 10^4}$ ($\mathbf{Da = 10^{-3}}$, $\mathbf{Gr_T = 10^6}$).

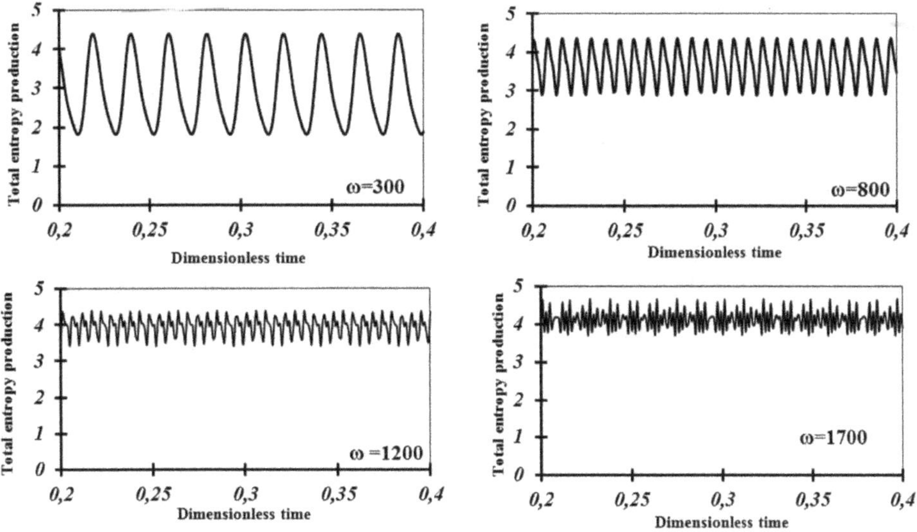

FIGURE 2.10 Total entropy generation versus time for different vibration frequencies for $\mathbf{Gr_T = 10^5}$ ($\mathbf{Da = 10^{-3}}$, $\mathbf{Gr_T = 10^6}$).

One can note that the response of the system approaches a sinusoidal regime for low frequencies (ω less than 300). In this case, the disturbance is not very significant, and the variations of the thermal gradients remain practically in phase. As the frequency of the vibrating force increases, the disturbance of the system becomes more and more significant, generating phase shifts between the different local temperature and velocity gradients, which ultimately induce a non-sinusoidal entropic response. It should be noted that the average value of entropy generation increases when the frequency of vibration increases. This is mainly due to the increasing effect of the disturbance

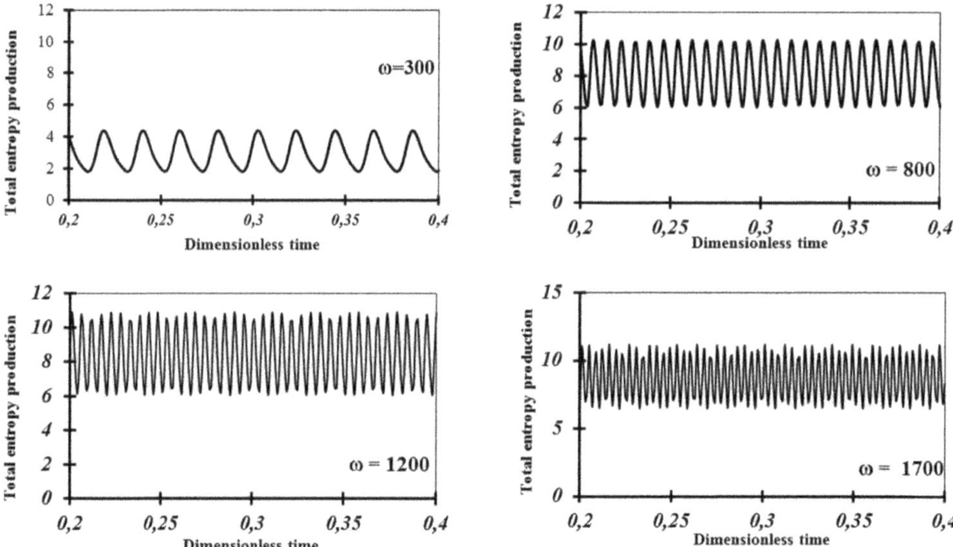

FIGURE 2.11 Total entropy generation versus time for different vibration frequencies for $\mathbf{Gr_T = 10^6}$ ($\mathbf{Da = 10^{-3}}$, $\mathbf{Gr_T = 10^6}$).

generated by the vibrating force, which is the result of the importance of the disorder in the physical system considered. Note that in the case where $Gr_T = 10^4$, the response is pseudo-periodic for low frequencies. This behavior persists for relatively large frequencies when the Grashof number increases (Figure 2.11).

2.5.4 Entropy Production Resonance

In this section, Prandtl, vibratory Grashof, and Darcy numbers are fixed at 0.7, 10^6, and 10^{-2}, respectively. The same values of Brinkman and modified Brinkman numbers are used. The vibration frequency and the Rayleigh number range from 300 to 2,100 and from 10^4 to 10^7, respectively.

Figure 2.12 illustrates the variations of the average entropy generation versus the vibration frequency for different values of Rayleigh number. It should be noted that average entropy generation was calculated for the case of periodic regime (sinusoidal or not). As can be seen, average entropy generation takes a maximum value for a critical vibration frequency ω_c, which depends on the Rayleigh number. The response of the system in terms of average entropy production to a sinusoidal excitation presents a resonant state which depends on the Rayleigh number. This observation can be confirmed by results of Fu and Shieh [15]; they found that a resonant vibration phenomenon occurs which causes the heat transfer rate to be enhanced remarkably. Consequently, maximum entropy generation is the result of a relatively large enhancement of heat transfer under the effect for a specific frequency of vibration.

Table 2.2 gives maximum entropy generation values when the resonant vibration phenomenon occurs for different Rayleigh numbers. The gap calculated at the bottom of the table represents the difference in percentage between average entropy produced by resonance and its value for the initial frequency $\omega = 300$ taken as a reference frequency.

In order to better perceive the variations of the different calculated quantities, the results given by Table 2.2 are plotted in Figures 2.13 to 2.16.

One can see that the maximum entropy generation is as important as the Rayleigh number increases. This can be explained by the increase of the convective phenomenon and therefore the importance of thermal gradients and, consequently, of entropy generation.

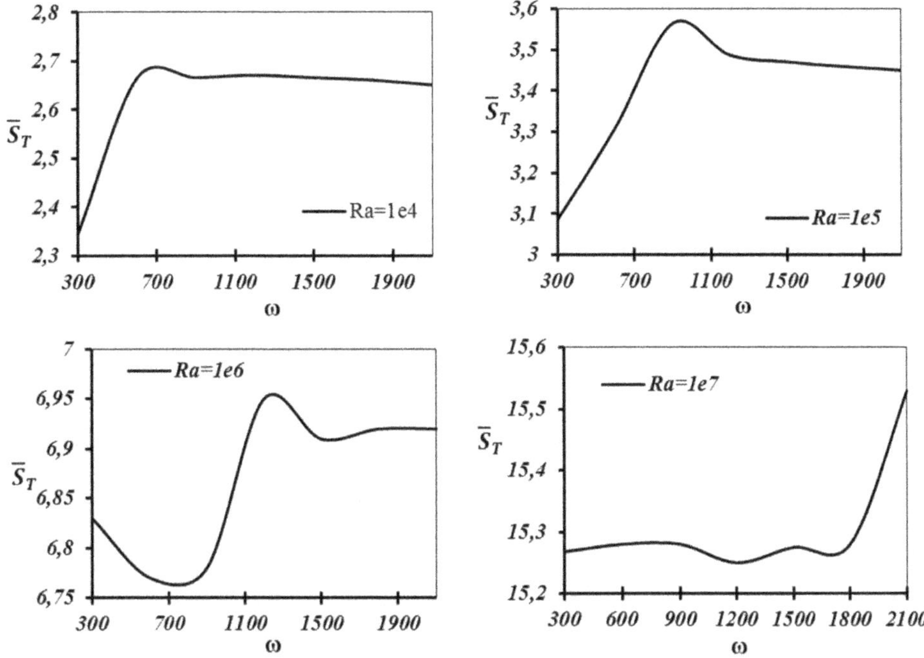

FIGURE 2.12 Average entropy production versus vibration frequency for different Rayleigh numbers.

TABLE 2.2
Resonance Entropy Production and Gap as a Function of the Rayleigh Number

Ra	10^4	10^5	10^6	10^7
ω_c	700	950	1250	2100
S_{Max}	2.7	3.6	6.95	15.55
$G = \%\ (S_{Max} - S_{\omega=300})$	17%	16%	2%	1.6%

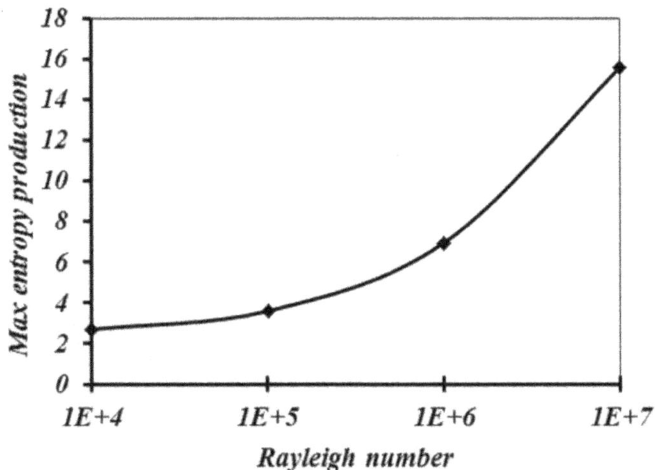

FIGURE 2.13 Evolution of maximum entropy production as a function of the Rayleigh number.

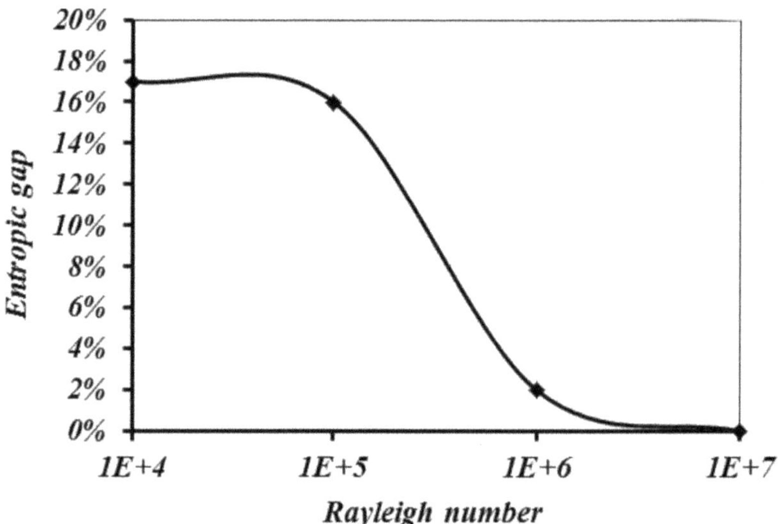

FIGURE 2.14 Evolution of the entropic gap as a function of the Rayleigh number.

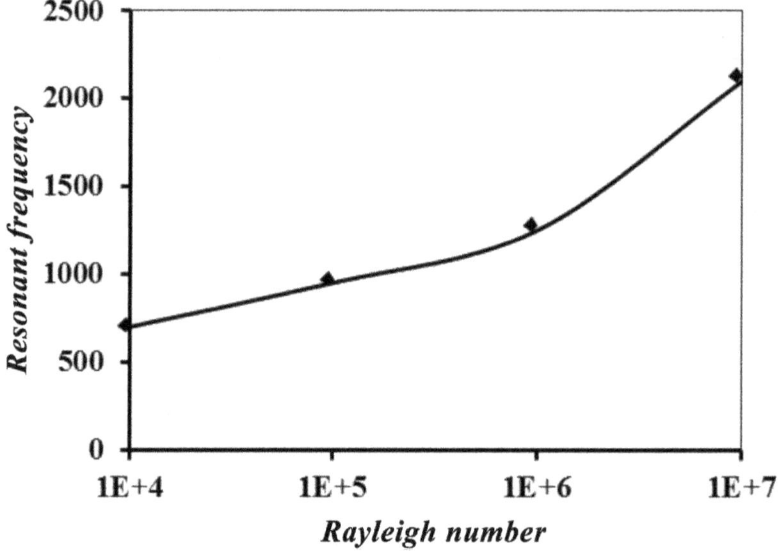

FIGURE 2.15 Evolution of the resonant frequency as a function of Rayleigh number.

It should be noted from Figure 2.14 that the gap is more important for lower Rayleigh numbers. This is due to the fact that, at resonance, the effect of vibrations on the heat transfer is more significant for low Rayleigh numbers; the vibration predominates gravity at low Rayleigh numbers. On the other hand, note that the entropic gap is reduced for high Rayleigh numbers; in this case, of gravity-driven convection, the effect of vibration becomes insignificant, the buoyancy force is relatively important, and its effect dominates the effect of the vibrating force.

As can be seen from Figures 2.15 and 2.16, the resonant frequency increases with the Rayleigh number. And consequently, the maximum value of entropy generation increases with resonant frequency. This means that entropy generation resonance requires a large vibrational frequency for

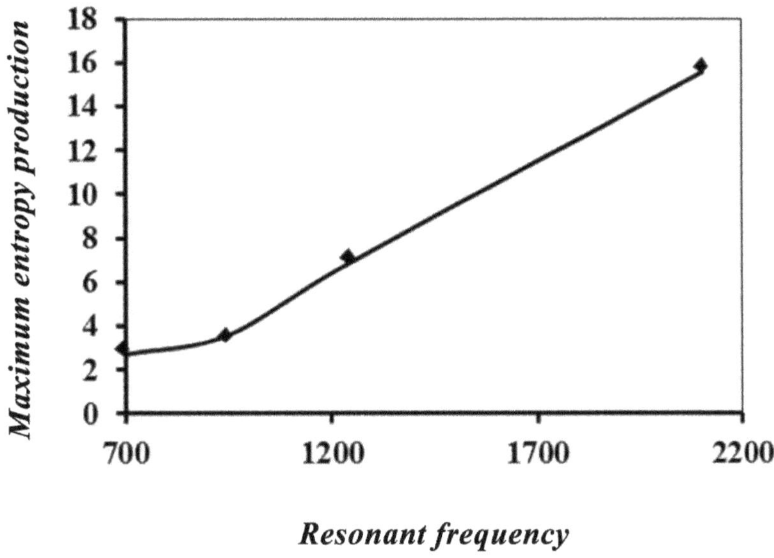

FIGURE 2.16 Evolution of maximum entropy versus resonant frequency.

important buoyancy forces. This is due to the fact that the increase in the resonant frequency is accompanied by an increase in the Rayleigh number, which in turn leads to an improvement in the convection phenomenon and then induces an increase in the irreversibility related to thermal effects.

2.6 CONCLUSION

A comprehensive study of natural convection phenomena, as well as the entropy generation rate for a fluid saturating a porous cavity submitted to mechanical vibration, is performed. By taking into account the effect of different parameters affecting the system, the following points should be noticed.

As a consequence of sinusoidal vibration, periodic oscillations of entropy generation are obtained. For low vibration frequencies, the sinusoidal regime of oscillations is maintained, especially at high-medium permeability. Average entropy generation and its oscillation's amplitude decrease when the porous medium is compared to a solid matrix.

For moderate vibrational Grashof number, sinusoidal oscillation regime of entropy generation with low amplitude is obtained. The increase of G_{rV} leads to the vibration-driven convection regime, and consequently, the sinusoidal regime is progressively destroyed and the amplitude increases.

The average value of entropy generation increases with the frequency of vibration; this can be explained by the fact that the disturbance generated by the vibrating force induces the importance of the disorder in the physical system. Since entropy is associated with the amount of the disorder in the system, the greater the disorder, the higher the entropy generation.

Average entropy generation takes a maximum value for a critical vibration frequency ω_c, which depends on the vibration frequency and on the Rayleigh number.

Nomenclature

Br	Brinkman number
Br*	modified Brinkman number
H	cavity side (m)
Da	Darcy number

G	gravitational acceleration (m/s^{-2})
G_{rT}	thermal Grashof number
G_{rV}	vibrational Grashof number
K	permeability of the porous medium (m^2)
Nu	average Nusselt number
P	pressure ($kg/m^{-1}s^{-2}$)
p	dimensionless pressure
Pr	Prandtl number
T	temperature (K)
t	time (s)
U, V	dimensionless velocity components
u, v	velocity components along x-, y-directions (m/s^{-1})
x, y	Cartesian coordinates (m)
X, Y	dimensionless coordinates

Greek symbols:

α_{eff}	thermal diffusivity (m^2s^{-1})
β_T	thermal volumetric expansion coefficients (K^{-1})
ε	porosity of the medium
μ	fluid dynamic viscosity ($kg\ m^{-1}s^{-1}$)
μ_{eff}	effective viscosity ($kg\ m^{-1}s^{-1}$)
ν	kinematic viscosity (m^2/s^{-1})
ρ	fluid density (kg/m^{-3})
σ	specific heat ratio $[(\rho c)_m/(\rho c)_f]$
θ	dimensionless temperature
τ	dimensionless time

Subscripts:

0	reference
c	cold side
f	fluid
h	hot side
m	porous medium

REFERENCES

[1] Lu Q., Qiu S., Su G., Tian W., Ye Z. Experimental research on heat transfer of natural convection in vertical rectangular channels with large aspect ratio. *Exp. Therm. Fluid Sci.*, 2010, 34(1): 73–80.

[2] Ahmed G.R., Yovanovich M.M. Numerical study of natural convection from discrete heat source in a vertical square enclosure. *Thermophys. J.*, 1992, 6: 121–127.

[3] Al-Zuhairy R.C., Alturaihi M.H., Ali F.A., Habeeb L.J. Numerical investigation of heat transfer in enclosed square cavity. *J. Mech. Eng. Res. Dev.*, 2020, 43(6): 388–403.

[4] Mahmoud A.M., Hadi J.M., Jary A.M., Habeeb L.J. Review on natural convection heat transfer in an enclosures and cavities. *J. Mech. Eng. Res. Dev.*, 2021, 44(6): 372–378.

[5] Nield D.A., Bejan A. *Convective in Porous Media*, 2nd edition, Springer, New York, 1999.

[6] Vafai K. *Handbook of Porous Media*, 2nd edition, Taylor & Francis, Boca Raton, FL, 2005.

[7] Chabani I., Mebarek-Oudina F., Ismail A.A.I. MHD flow of a hybrid nano-fluid in a triangular enclosure with zigzags and an elliptic obstacle. *Micromachines*, 2022, 13(2): 224.

[8] Pushpa B.V., Sankar M., Mebarek-Oudina F. Buoyant convective flow and heat dissipation of Cu–H_2O nanoliquids in an annulus through a thin baffle. *J. Nanofluids*, 2021, 10(2): 292–304.

[9] Swain K., Mahanthesh B., Mebarek-Oudina F. Heat transport and stagnation-point flow of magnetized nanoliquid with variable thermal conductivity, Brownian moment, and thermophoresis aspects. *Heat Trans.*, 2021, 50(1): 754–767.

[10] Chamkha A.J., Ismael M.A. Conjugate heat transfer in a porous cavity filled with nanofluids and heated by a triangular thick wall. *Int. J. Therm. Sci.*, 2013, 67: 135–151.

[11] Khan U., Mebarek-Oudina F., Zaib A., Ishak A., Abu Bakar S., Sherif S.M., Baleanu D. An exact solution of a Casson fluid flow induced by dust particles with hybrid nanofluid over a stretching sheet subject to Lorentz forces. *Waves Random Complex Media*, 2022. http://doi.org/10.1080/17455030.2022.2102689

[12] Mebarek-Oudina F. Convective heat transfer of Titania nanofluids of different base fluids in cylindrical annulus with discrete heat source. *Heat Transf.-Asian Res.*, 2019, 48(1): 135–147.

[13] Reddy Y.D., Mebarek-Oudina F., Goud B.S., Ismail A.I. Radiation, velocity and thermal slips effect toward MHD boundary layer flow through heat and mass transport of Williamson Nanofluid with porous medium. *Arab. J. Sci. Eng.*, 2022, 47: 16355–16369 https://doi.org/10.1007/s13369-022-06825-2.

[14] Asogwa A.K., Mebarek-Oudina F., Animasaun I.L. Comparative investigation of water-based Al_2O_3 nanoparticules through water-based CuO nanoparticules over an exponentially accelerated radiative riga plate surface via heat transport. *Arab. J. Sci. Eng.*, 2022, 7: 8721–8738.

[15] Fu W.S., Seieh W.J. Transient thermal convection in an enclosure induced simultaneously by gravity and vibration. *Int. J. Heat Mass Transf.*, 1993, 36: 437–452.

[16] Bardan G., Knobloch E., Mojtabi A., Khallouf H. Natural double diffusive convection with vibration. *Fluid Dyn. Res.*, 2001, 28: 159–187.

[17] Chung S., Vafai K. Vibration induced mixed convection in an open-ended obstructed cavity. *Int. J. Heat Mass Transf.*, 2010, 53: 2703–2714.

[18] Bardan G., Mojtabi A. On the Horton-Rogers-Lapwood convective instability with vertical vibration: Onset of convection. *Phys. Fluids*, 1998: 205–231.

[19] Gershuni G.Z., Zhukhovitskii E.M., Yurkov Y.S. Vibrational thermal convection in a rectangular cavity. *Fluid Dyn.*, 1982, 17: 565–569.

[20] Forbes R.E., Carley C.T., Bell C.J. Vibration effects on convective heat transfer in enclosures. *J. Heat Transf.*, 1970, 92(3): 429–437.

[21] Zidi A., Hasseine A., Moummi N. The effect of vertical vibrations on heat and mass transfers through natural convection in partially porous cavity. *Arab. J. Sci. Eng.*, 2018, 43(2): 2195–2204.

[22] Oh Y.K., Parka S.H., Cho Y.I. A study of the effect of ultrasonic vibrations on phase-change heat transfer. *Int. J. Heat Mass Trans.*, 2002, 45(23): 4631–4641.

[23] Khudhair B.K., Salh A.M., Ekaid A. Vibration effect on natural convection heat transfer in an inclosed cubic *Cavity. IOP Conf. Ser.: Mater. Sci. Eng.*, 2020, 1094: 012060.

[24] Hooman K., Gurgenci H., Merrikh A.A. Heat transfer and entropy generation optimization of forced convection in a porous-saturated duct of rectangular cross-section. *Int. J. Heat Mass Transf.*, 2007, 50: 2051–2059.

[25] Hidouri N., Abbassi H., Magherbi M., Ben Brahim A. Influence of thermodiffusion effect on entropy generation in thermosolutal convection. *Far East J. Appl. Math.*, 2006, 25(2): 179–197.

[26] Magherbi M., Hidouri N., Abbassi H., Ben Brahim A. Influence of Dufour effect on entropy generation in double diffusive convection. *Int. J. Exergy*, 2007, 4(3): 227–252.

[27] Marzougui S., Bouabid M., Mebarek-Oudina F., Abu-Hamdeh N., Magherbi M., Ramesh K. A computational analysis of heat transport irreversibility phenomenon in a magnetized porous channel. *Int. J. Numer. Methods Heat Fluid Flow*, 2021, 31(7): 2197–2222.

[28] Marzougui S., Mebarek-Oudina F., Magherbi M., Mchirgui A. Entropy generation and heat transport of Cu–water nanoliquid in porous lid-driven cavity through magnetic field. *Int. J. Numer. Meth. Heat Fluid Flow*, 2021, 32: 2047–2069.

[29] Mchirgui A., Hidouri N., MouradMagherbi M., Ben Brahim A. Entropy generation in double-diffusive convection in a square porous cavity using Darcy–Brinkman formulation. *Transp. Porous Med.*, 2012, 93: 223–240.

[30] Mchirgui A., Hidouri N., Magherbi M., Ben Brahim A. Second law analysis in double diffusive convection through an inclined porous cavity. *Comp. Fluids*, 2014, 96: 105–115.

[31] Salimath P.S., Ertesvåg I.S. Local entropy generation and entropy fluxes of a transient flame during head-on quenching towards solid and hydrogen-permeable porous walls. *Int. J. Hydrog. Energy*, 2021, 46(52): 26616–26630.

[32] Ince A.C., Serincan M.F., Colpan C.O. Local entropy generation and exergy analysis of the condenser in a direct methanol fuel cell system. *Int. J. Hydrog. Energy*, 2022, 47(45): 19850–19864.
[33] Lauriat G., Prasad V. Natural convection in a vertical porous cavity: A numerical study for Brinkman-extended Darcy formulation. *J. Heat Transfer*, 1987, 109: 688–696.
[34] Kramer J., Jecl R., Skerget L. Boundary domain integral method for the study of double diffusive natural convection in porous media. *Eng. Anal. Bound. Elem.*, 2007, 31: 897–905.
[35] Fu W.S., Shieh W.J. Transient thermal convection in an enclosure induced simultaneously by gravity and vibration. *Int. J. Heat Mass Transfer*, 1993, 36(2): 437–452.
[36] Magherbi M., Abbassi H. Ben Brahim A. Entropy generation at the onset of natural convection. *Int. J. Heat Mass Transfer*, 2003, 46: 3441–3450.

3 Effect of Discrete Heat Source–Sink Pairs on Buoyancy-Driven Convection in an Inclined Parallelogrammic Porous Enclosure

Ravindra P., Mahesha, M. Sankar, and F. Mebarek-Oudina

3.1 INTRODUCTION

Buoyant convective thermal transport in differentially heated finite geometries, in particular, non-regular-shaped geometries, has been widely investigated due to its importance in heat exchangers and geothermal, nuclear, and solar applications. Among the different non-regular-shaped domains, a parallelogram-shaped geometry has vital applications involving thermal diodes and other important thermal applications. This geometry gained importance due to its relevance to many industrial applications, as it aptly portrays the physical structure of these applications. Also, from the computational point of view, this geometry poses a challenge due to its sloped sidewalls, which make an inclination angle with vertical axis. Hyun and Choi [1] were among the pioneers to study buoyant motion and the associated thermal transfers in this geometry. They made a detailed transient investigation on convective motion for different Rayleigh numbers and sidewall tilting and proposed the best parametric combinations to achieve better thermal transfer in the geometry. Baytas and Pop [2] analyzed the impact of porosity on convective motion in parallelogram-shaped domain using Darcy's model and found enhanced thermal transport for a particular choice of sidewall tilting at all values of Ra. Baïri and co-workers [3–5] performed different analysis of buoyancy-assisted flow in this geometry by considering uniform as well as discrete thermal conditions and also proposed thermal correlations for different tilt angles of vertical boundaries and arrangement of thermal sources. Further, a detailed review of convective motions in this geometry has been discussed with the drawbacks on existing works. Costa [6] analyzed thermosolutal buoyant motion and the resulting thermal and mass transport in a parallelogrammic geometry by considering a wide span of parametric values. Using Tiwari and Das's model, Ghalambaz et al. [7] explored the buoyant motion of nanofluid in the same geometry and found that the inclusion of nano-sized particles improves thermal enhancement in the geometry. Villeneuve et al. [8] related the analysis of the geometry with the insulating capacities of thermal diode. Gupta and Nayak [9] investigated the minimization of entropy production along with buoyant motion in similar geometry and proposed a suitable parameter regime at which minimum entropy is produced with maximum thermal dissipation. The impact of magnetic force on convective motion has been investigated by Mallick et al. [10] with a solid block inside the same geometry. Marzougui et al. [11] made an attempt to explore the entropy production and thermal dissipation in a lid-driven porous domain with magnetic force.

A great amount of research has been carried out on convective heat transfer in the enclosures subjected to partial heating and cooling due to its importance in cooling applications. Ho and Chang [12] made a pioneering attempt to examine the impacts of isolated thermal heating on buoyant motion in a rectangular geometry containing single to multiple sources in the form of chips.

DOI: 10.1201/9781003299608-3

Later, Deng [13] analyzed numerically the positional influences of different source–sink locations and proposed an optimum number of pairs and location to extract higher thermal dissipation rate. Sivasankaran and co-investigators [14, 15] discussed the size positional effects of thermal source(s) in a square porous geometry and identified the optimal location to extract maximum heat transport. Recently, Devanand et al. [16] applied LBM to analyze the discrete heating effects on buoyancy-assisted flow in a square geometry. Numerical analysis on buoyant convection in a porous annular region mounted with discrete thermal sources on inner cylinder was reported by Sankar and co-investigators [17, 18] to discuss the effects of heater length and location on heat transfer rate. Mebarek-Oudina [19, 20] addressed the impacts of thermal sources present in a cylindrical annulus to control the stability of fluid motion as well as the thermal transport rates.

The fundamental problem of convection in fluid-filled square enclosures has been considered to analyze the effects of enclosure inclination on the flow behavior and, in turn, to identify optimum inclination angle to extract maximum thermal transport rate. One of the earliest investigations in this direction in a tilted porous geometry is due to Moya and Ramos [21] and Hslao et al. [22]. They performed detailed numerical simulations to explore the impacts of inclination of the geometry and proposed a suitable inclination to achieve maximum thermal dissipation benefits. The heat transport characteristics in cavities filled with heat-generating porous material have also been focused on in literature [23] to cater to the need for high cooling applications. To augment or suppress the buoyant flow of nanofluid or hybrid nanofluid in a finite-shaped geometry, different mechanisms have been utilized, and successful predictions have been brought out from their analysis [24–26]. Jagadeesha et al. [27] attempted to study the thermosolutal motion in an inclined parallelogram-shaped geometry. Alsabery et al. [28] analyzed the tilting impacts of nanofluid-filled inclined porous square geometry by considering different porous structures. Some of the recent numerical investigations on buoyant motion in tilted porous geometries with an objective to identify the impacts of cavity inclination on heat transport rate can be found in [29, 30]. In recent years, focus towards the choice of nanoparticle to boost thermal dissipation rate, boundary layer formation with nanofluid, and different non-Newtonian fluids has been increased due to their potential applications [31–39].

An in-depth review of the existing research works shows that no attempt has been made to address the impact of different source–sink configurations on the flow behavior and corresponding thermal transport in a discretely heated inclined porous parallelogram-shaped geometry. In the current analysis, different source–sink pair locations with different heater lengths are considered to achieve the best combination for extracting maximum thermal dissipation.

3.2 MATHEMATICAL FORMULATION

The geometrical representation of the model, illustrated in Figure 3.1, is the parallelogrammic enclosure with sidewalls-fitted source and sink. The parallelogram-shaped geometry is tilted at an angle of α with positive axis of abscissas and is displayed in Figure 3.1. The left wall is placed with a thermal source (T_h), and the right wall has a thermal sink (T_c) and upper and lower boundaries insulated. Further, these walls are inclined at an angle ϕ with the y-axis and is displayed in Figure 3.1. Also, the unheated regions of the left and right walls are made adiabatic. The length and height of the geometry are, respectively, L and H. In this analysis, we fix $L = H$ (A = 1). Three different positions are chosen to place the source and sink. Incompressible and Newtonian fluid with laminar flow has been assumed. The thermophysical properties are considered to be constant, except the density, which is based on Boussinesq approximation.

By assuming the preceding postulates and using Darcy model for the porous media, the dimensional governing equations are [2, 23]:

$$u = -\frac{K}{\mu}\left[\frac{\partial p}{\partial x} + \rho g \sin\alpha\right], \tag{3.1}$$

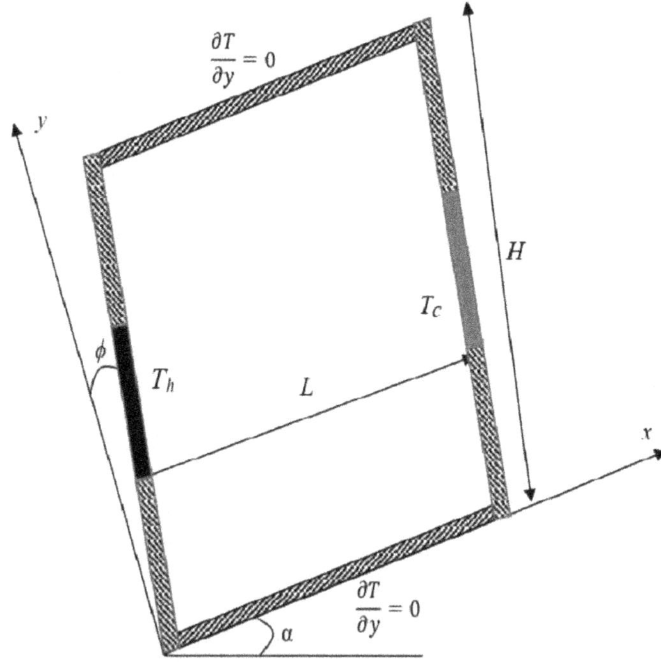

FIGURE 3.1 Physical configuration and boundary conditions with coordinate axis.

$$v = -\frac{K}{\mu}\left[\frac{\partial p}{\partial y} + \rho g \cos\alpha\right], \tag{3.2}$$

$$\sigma\frac{\partial T}{\partial t} + u\frac{\partial T}{\partial x} + v\frac{\partial T}{\partial y} = \alpha_m\left(\frac{\partial^2 T}{\partial x^2} + \frac{\partial^2 T}{\partial y^2}\right), \tag{3.3}$$

Where the heat capacity ratio σ is defined as $\sigma = \dfrac{\varepsilon(\rho c)_f + (1-\varepsilon)(\rho c)_p}{(\rho c)_f}$. Here, the subscripts f and p are respectively for fluid and porous media. In the present analysis, the stream function has been introduced to disregard the pressure gradients from the momentum equations (3.1) and (3.2), and the resulting revised Navier–Stokes equation is:

$$\frac{\partial^2 \psi^*}{\partial x^2} + \frac{\partial^2 \psi^*}{\partial y^2} = \frac{gK\beta}{\nu}\left(\frac{\partial T}{\partial y}\sin\alpha - \frac{\partial T}{\partial x}\cos\alpha\right), \tag{3.4}$$

Here, the stream function ψ^* can be taken as $u = \dfrac{\partial \psi^*}{\partial y}$ and $v = -\dfrac{\partial \psi^*}{\partial x}$.

Here, we have used the nonlinear coordinate transformation as suggested by Baytas and Pop [2] to convert the non-aligned boundaries of the geometry with the coordinate axes to boundary-fitted coordinates. For this, the following transformation is utilized:

$$X = x - y\tan\phi, \quad Y = y \tag{3.5}$$

Using this transformation, the original parallel-shaped structure has been converted to a regular-shaped geometry which is termed as computational domain. From the transformation (3.5), the model equations (3.3) and (3.4) are transformed in the computational domain as:

$$\left(\frac{\partial^2\psi^*}{\partial X^2}-2\sin\phi\cos\phi\frac{\partial^2\psi^*}{\partial X\partial Y}+\cos^2\phi\frac{\partial^2\psi^*}{\partial Y^2}\right)=$$

$$\frac{gK\beta}{v}\cos^2\phi\left[\left(\frac{\partial T}{\partial Y}-\tan\phi\frac{\partial T}{\partial X}\right)\sin\alpha-\frac{\partial T}{\partial X}\cos\alpha\right] \tag{3.6}$$

$$\sigma\frac{\partial T}{\partial t}+\frac{\partial\psi^*}{\partial Y}\frac{\partial T}{\partial X}-\frac{\partial\psi^*}{\partial X}\frac{\partial T}{\partial Y}=\frac{\alpha_m}{\cos^2\phi}\left(\frac{\partial^2 T}{\partial X^2}-2\sin\phi\cos\phi\frac{\partial^2 T}{\partial X\partial Y}+\cos^2\phi\frac{\partial^2 T}{\partial Y^2}\right) \tag{3.7}$$

Also, the preceding equations (3.6) and (3.7) are made dimensionless through the following transformation:

$$\xi=\frac{X}{L},\eta=\frac{Y}{H\cos\phi},t=\frac{t^*\alpha_m}{\sigma\ LH\cos\phi},\psi=\frac{\psi^*}{\alpha_m},\theta=\frac{(T-T_r)}{T_h-T_c},T_r=(T_h+T_c)/2. \tag{3.8}$$

$$\frac{\partial^2\psi}{\partial\xi^2}-2\frac{\sin\phi}{A}\frac{\partial^2\psi}{\partial\xi\partial\eta}+\frac{1}{A^2}\frac{\partial^2\psi}{\partial\eta^2}=Ra\cos\phi\left[\frac{\sin\alpha}{A}\frac{\partial\theta}{\partial\eta}-\cos(\phi-\alpha)\frac{\partial\theta}{\partial\xi}\right] \tag{3.9}$$

$$\frac{\partial\theta}{\partial\tau}+\frac{\partial\psi}{\partial\eta}\frac{\partial\theta}{\partial\xi}-\frac{\partial\psi}{\partial\xi}\frac{\partial\theta}{\partial\eta}=\frac{A}{\cos\phi}\left(\frac{\partial^2\theta}{\partial\xi^2}-2\frac{\sin\phi}{A}\frac{\partial^2\theta}{\partial\xi\partial\eta}+\frac{1}{A^2}\frac{\partial^2\theta}{\partial\eta^2}\right) \tag{3.10}$$

The quantified thermal transport rates are estimated from the local (Nu) and average ($\overline{Nu}$) Nusselt numbers, defined by:

$$Nu=-\frac{1}{\cos\phi}\left(\frac{\sin\phi}{A}\frac{\partial\theta}{\partial\eta}-\frac{\partial\theta}{\partial\xi}\right),\quad \overline{Nu}=\int_0^1 Nu\,d\eta$$

The thermal conditions are: the source at left-inclined boundary is kept at a higher thermal value ($\theta = 1$), and the sink at right-titled boundary is kept at a lesser thermal value ($\theta = 0$). However, the thermal condition at the unheated lower and upper surfaces is $\frac{\partial\theta}{\partial\eta}=\frac{\partial\theta}{\partial\xi}=0.$ The hydrodynamic conditions are nonslip and non-through flow at all boundaries $\psi=\frac{\partial\psi}{\partial\eta}=\frac{\partial\psi}{\partial\xi}=0.$

3.3 NUMERICAL PROCEDURE

In the current analysis, though the simplest porous media model has been used, the set of governing equations is coupled and nonlinear, and hence, an analytical solution could not be expected. Hence, they are solved by employing time-splitting and over-relaxation techniques based on an implicit FDM (finite difference method) with second-order accuracy. First, various grid sizes, from 41 × 41 to 121 × 121, are tested to check grid independency with $\overline{Nu}$ along the thermal source and sink as the sensitivity measures for grid independence. After successful tests, the grid size of 101 × 101 has been chosen to carry out the simulations. Further, the present model predictions are benchmarked with available literature predictions and found fairly excellent agreement. For shortness of the paper, validation has not been given here but can be found in our earlier works [22].

3.4 RESULTS AND DISCUSSION

Numerical experiments are performed to analyze the impacts of various pertinent parameters on flow patterns and thermal dissipation rate. In particular, three different source–sink locations (0.3, 0.5, and 1.7), three heater lengths (0.2, 0.5, and 1.0), sidewall (ϕ), and enclosure (α) inclinations are varied respectively in the ranges of $-30^0 \leq \phi \leq +30^0$ and $-45^0 \leq \alpha \leq +45^0$, and Darcy–Rayleigh number is varied $10^2 \leq \alpha \leq 10^3$.

3.4.1 Impact on Flow and Thermal Fields

The study has been conducted by fixing $A=1$. The governing key parameters are considered over a vast array, and the obtained numerical results are graphically represented in this section. The impacts of Rayleigh numbers on velocity and temperature contours are illustrated in Figure 3.2 for various magnitudes of Ra, with other parameters as fixed values. The impact of Ra is apparent on the flow pattern and thermal distribution in the geometry. For lower magnitude of Ra, flow motion appears to be very low with a regularly spaced eddies along the boundaries of the parallelogram. However, as the magnitude of Ra increases, strong thermal flow exists with a change in shape of main eddy and with higher magnitude. The impact of Ra on isotherms is also similar. For Ra = 10^2, the isothermal contours reveal the conduction mode with a minor variation among the contour lines. However, a strong thermally stratified structure can be observed for a higher magnitude of Ra along with a thin thermal boundary layer along the source and sink. The flow and thermal contours in the

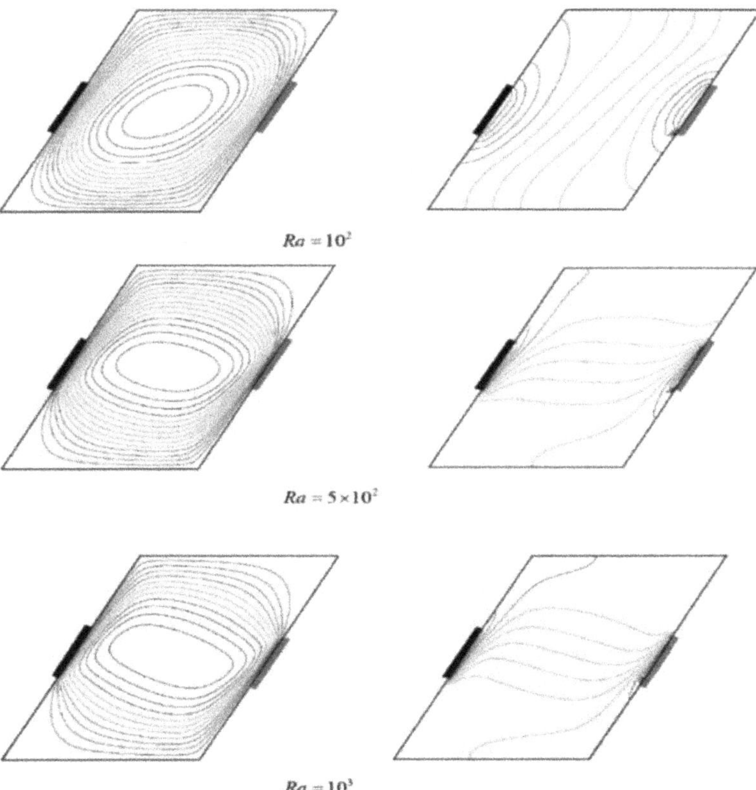

FIGURE 3.2 Flow and temperature pattern for different Rayleigh number with $\phi = 45^0$, $\alpha = 30^0$, $\varepsilon = 0.2$, $D = 0.5$.

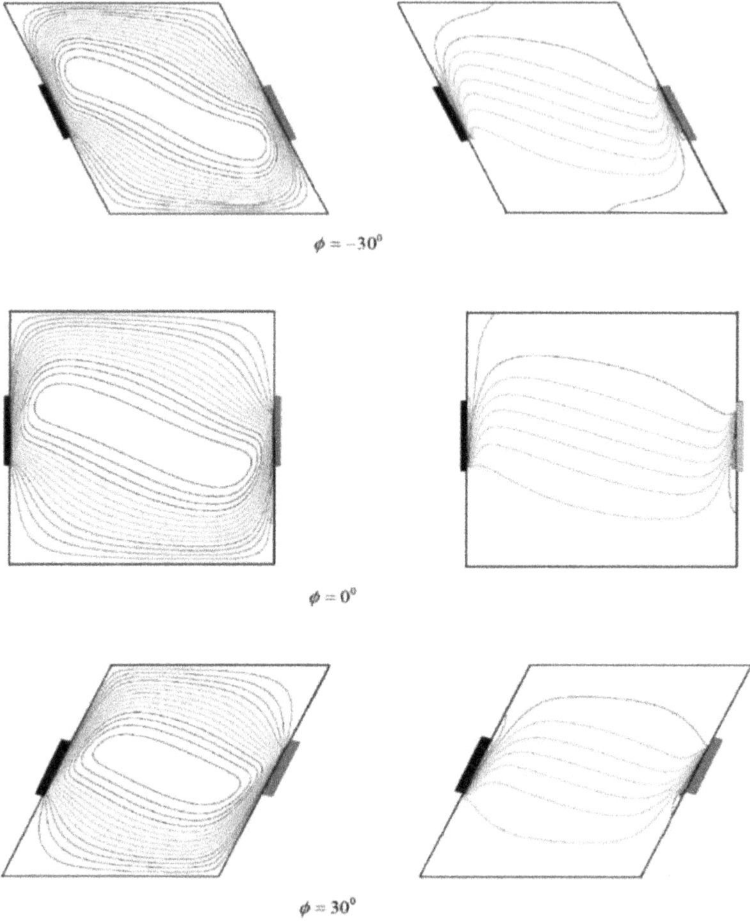

FIGURE 3.3 Flow and temperature pattern for different side angles of ϕ with $Ra = 10^3$, $\alpha = 30^0$, $\varepsilon = 0.2$, $D = 0.5$.

enclosure for different angles of sidewall inclination are shown in Figure 3.3. When the angle of sidewall is −30°, the flow pattern is more distributed in the cavity, which can be clearly observed from the streamline and isotherm contours. Also, the main flow and thermal directions greatly depend on the wall tilting. The flow with the main eddy rotating along the primary diagonal gradually changes its direction as the angle of wall tilting changes, and a similar effect could be seen from isothermal lines also.

Figure 3.4 depicts the change in velocity and temperature structure for three dissimilar geometry inclinations. It is apparent that the enclosure tilting angle has severe impacts on flow and temperature structure inside the geometry. First, the flow is bi-cellular for negative inclination (α = −45°) with two main eddies in the middle of the geometry. However, these two main eddies merged to a single vortex diagonally rotating at the center of the geometry for non-inclined geometrical condition. Further, for positive tilting of the geometry (α = +45°), the single eddy split into two main eddies in the middle and another two eddies with one eddy below the source and another eddy above the sink could be seen in the enclosure. The isotherms reveal a strong stratified structure for negative inclination and non-inclined cases. However, for the case of positive tilting, the isotherms reveal a mere conduction thermal mode with almost-parallel structure.

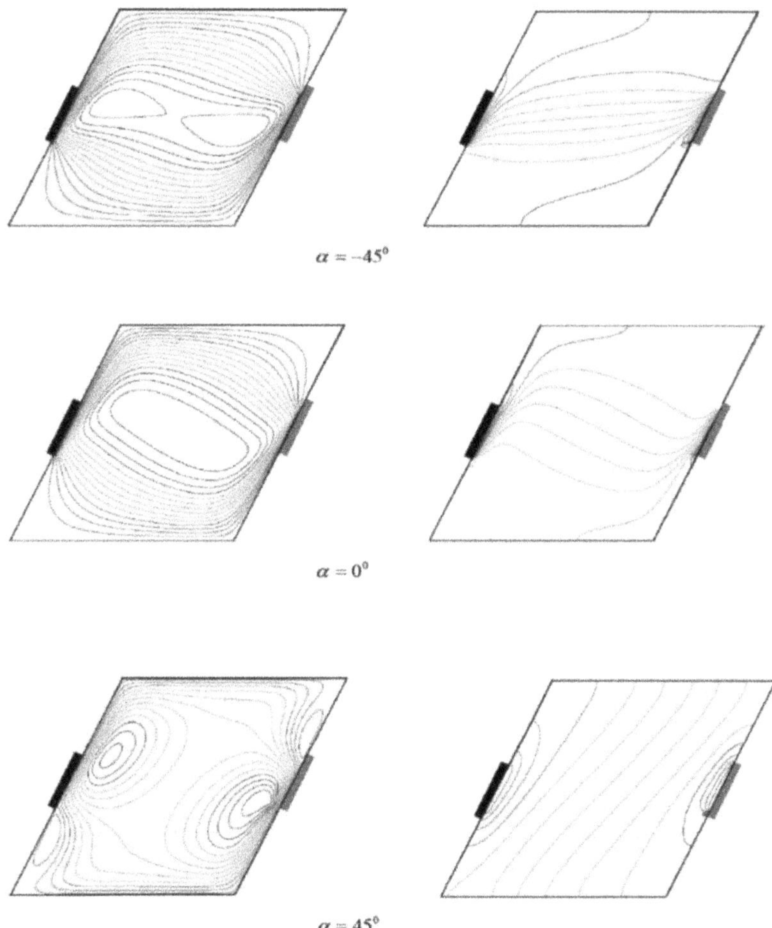

FIGURE 3.4 Flow and temperature pattern for different tilt angles of α with $Ra = 10^3, \phi = 30^0, \varepsilon = 0.2, D = 0.5$.

3.4.2 Impact on Thermal Transfer

The influence of wall inclination on the global thermal dissipation rate for three different values of Ra is illustrated in Figure 3.5 by fixing other parameters of the analysis. The impact of ϕ on thermal transport rates greatly depends on the magnitude of Ra. For lower inputs of Ra, where conduction is predominant in transporting heat, the variation of global Nu is very minimal with ϕ. For moderate magnitudes of Ra (Ra = 5 × 10^2), the variation in $\overline{Nu}$ has also been minimal. However, for larger magnitude of Ra, the impact of wall tilting on $\overline{Nu}$ could be very much visible from Figure 3.5. It could be observed from the results that the maximum thermal dissipation could be possible for the case of ϕ = 15° as compared to other inclination angles. Also, this indicates the tilt angle at which minimum thermal transport could be achieved.

Figure 3.6 elaborates the importance of parallelogram-shaped enclosure tilting angle on thermal dissipation rates. For this, the enclosure tilting is varied from −45° to +45° by fixing other parameters. In general, thermal dissipation from the source to the surrounding medium has been an increasing function of geometry inclination. That is, as the value of α is increased, the magnitude of $\overline{Nu}$ also enhances. In particular, the maximum thermal extraction from the source could

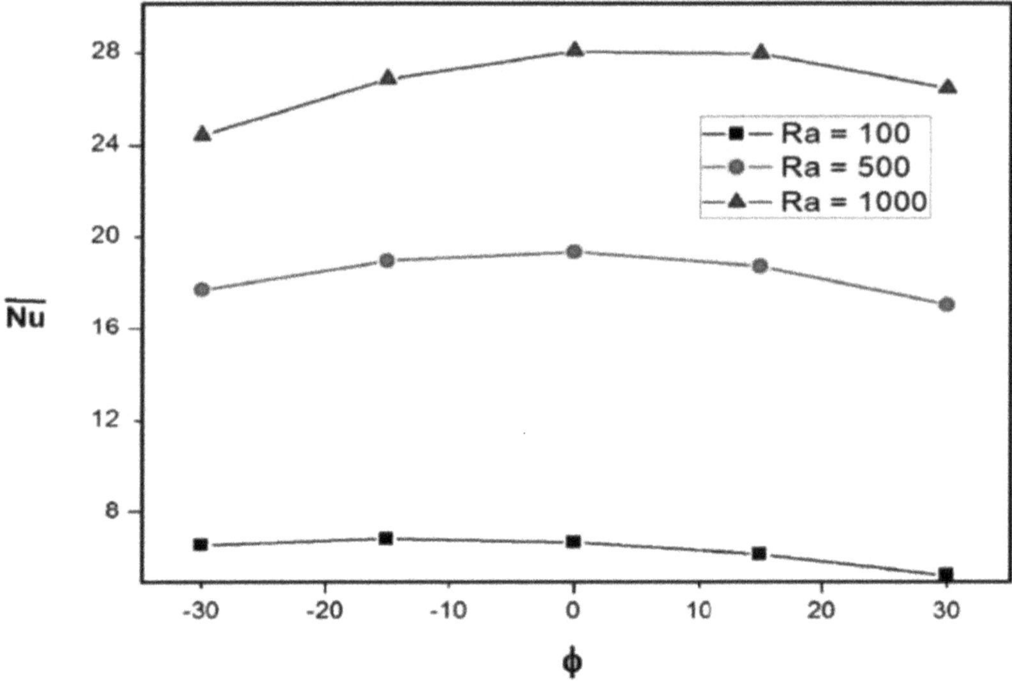

FIGURE 3.5 Effect of ϕ and Ra on average Nusselt number with $\alpha = 30^0$, $\varepsilon = 0.2$, $D = 0.5$.

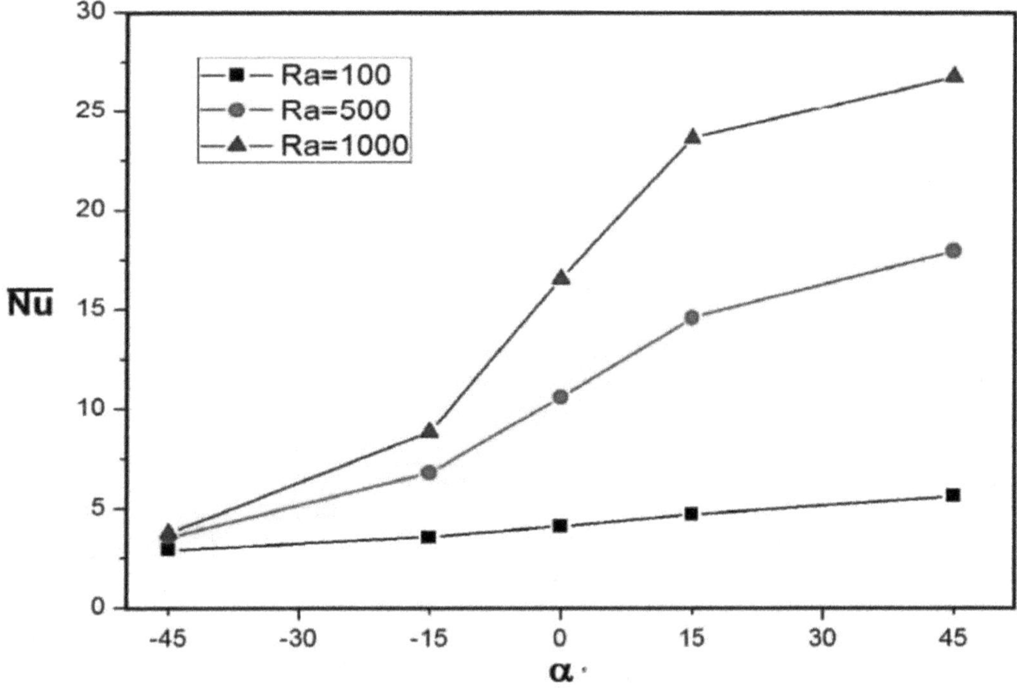

FIGURE 3.6 Impact of α and Ra on the global Nu with $\phi = 30^0$, $\varepsilon = 0.2$, $D = 0.5$.

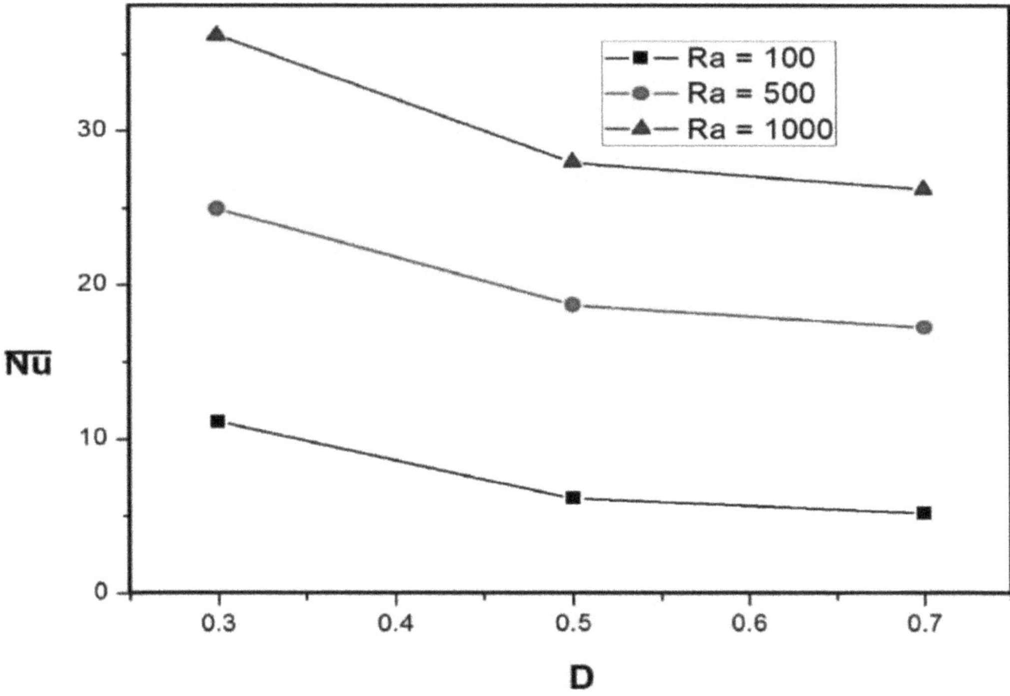

FIGURE 3.7 Impact of *D* and *Ra* on with $\overline{Nu}$ $\phi = 30^0, \alpha = 30^0, \varepsilon = 0.2$.

be possible with the geometry inclination kept at α = 45°. From the same result, we could also get an information to attain minimum thermal dissipation rate if any application requires minimum thermal transport.

Figure 3.7 discusses the positional influence of thermal source on the global thermal dissipation from the source to the surrounding medium. This would throw light into the optimal location to fix the thermal chip/source to extract maximum heat transport so that thermal mismanagement situations could be completely avoided. In this way, the design engineer could also avoid additional fan or any other mechanisms to attain maximum thermal transport. The results reveal that the source fixed between the bottom wall and mid-height of the enclosure would be more beneficial for thermal extraction from the source. In particular, this location is D = 0.3 in our analysis. It is also worth to mention that the heat transport rates from the source are lower as the location of source is elevated to the upper portion of the geometry. In addition to the locational impacts on thermal transport, our analysis also predicts the optimum dimension of the source, at which maximum heat could be extracted, and it is displayed in Figure 3.8. For this, we considered three lengths of thermal source fixed at the middle of the left wall.

Among the different lengths considered in this analysis, it has been detected that a source with smaller dimension could be utilized for maximum thermal extraction from the source. This prediction is very much consistent with the literature findings on discrete heating of boundaries of finite-sized geometries. Also, this prediction is true for all magnitudes of Ra. In inclined parallelogram-shaped geometry, the important quantity to be determined is the combination of two angles at which maximum thermal extraction could be attained, and this vital information has been obtained in Figure 3.9 by considering various tilt angles of sidewalls as well as the geometry. From the result, it could be observed that the geometry inclination greatly depends on wall tilting. For each wall

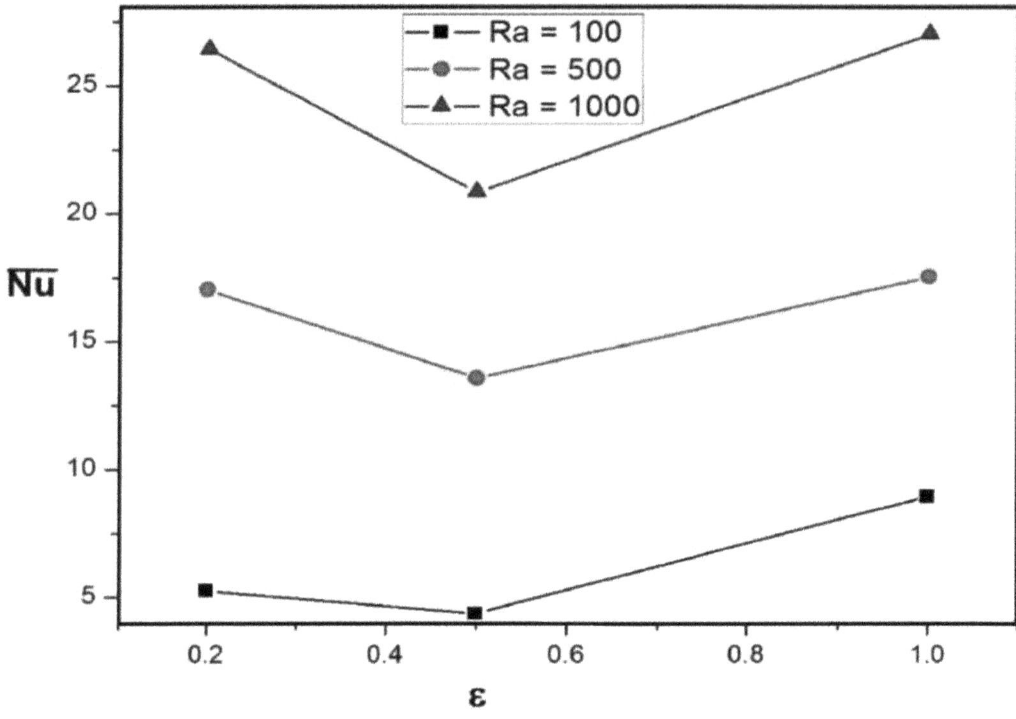

FIGURE 3.8 Impact of ε and Ra on $\overline{Nu}$ with $\phi = 30^0$, $\alpha = 30^0$, $D = 0.5$.

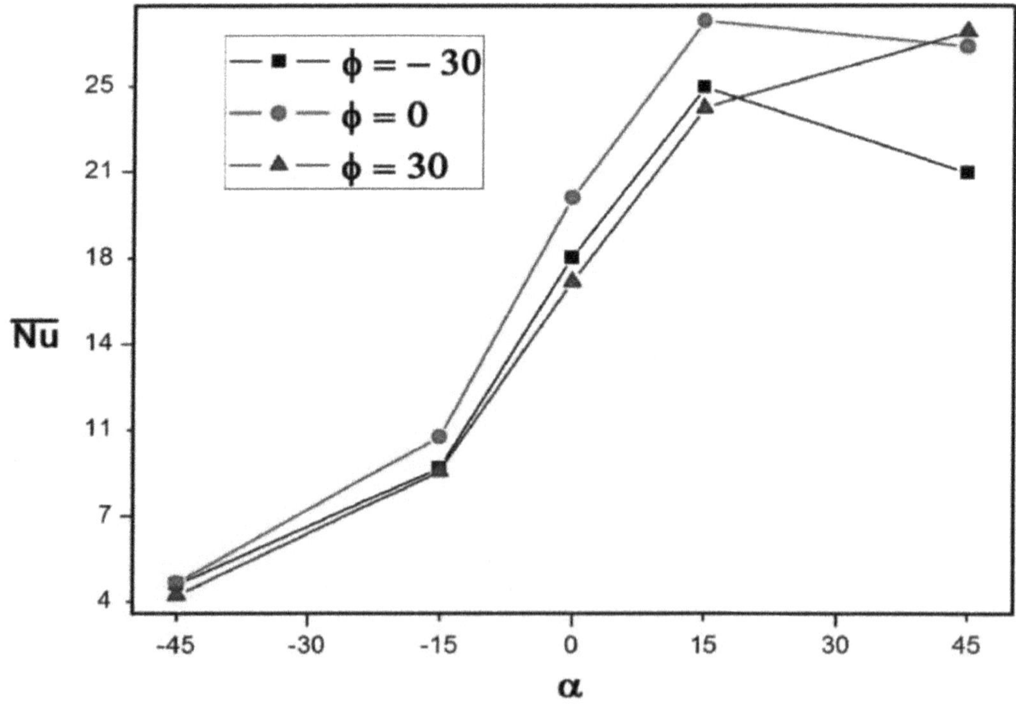

FIGURE 3.9 Effect of α and ϕ on average Nusselt number with $Ra = 10^3$, $\varepsilon = 0.2$, $D = 0.5$.

inclination, we get a cavity inclination to attain maximum thermal transport. This particular observation is also an important result for the design engineer to identify the proper combination of these two tilt angles so as to maximize the thermal transport across the parallelogram-shaped enclosure.

3.5 CONCLUSIONS

The current work reports the numerical prediction of buoyant heat transport in a porous parallelogram-shaped geometry containing source–sink pairs along the vertical walls. The influence of control parameters along with heater positions and lengths on flow and thermal patterns has been widely analyzed. The key predictions of the simulations are consolidated as follows:

- The heater position and dimension have a significant impact on enhancing thermal extraction from the source to surrounding fluid.
- Maximum thermal transport can be achieved for smaller heater length positioned near the middle portion of the wall.
- The study also identified an optimum set of angles at which thermal extraction could be maximized.

3.6 ACKNOWLEDGMENTS

All authors acknowledge their respective universities for the support and encouragement.

Nomenclature

A	aspect ratio
D	dimensionless position of source (m)
H	height of the enclosure (m)
k	thermal conductivity (W/mK)
Ra	Rayleigh number (ρgβ Ds3/να)
t	dimensionless time
T	dimensional temperature
(U,W)	velocity components
α	tilt angle of the enclosure
β	thermal expansion coefficient
ε	dimensionless source–sink length
ψ	dimensionless stream function
θ	dimensionless temperature (K−1)
ϕ	angle of sidewall inclination

REFERENCES

[1] Hyun, J.M., & Choi, B.S. Transient natural convection in a parallelogram-shaped enclosure. *International Journal of Heat and Fluid Flow*, 1990, 11: 129–134.

[2] Baytas, A.C., & Pop, I. Free convection in oblique enclosures filled with a porous medium. *International Journal of Heat and Fluid Flow*, 1999, 42: 1047–1057.

[3] Bairi, A., Garcia de Maria, J.M., Bairi, I., Laraqi, N., Zarco-Pernia, E., & Alilat, N. 2D transient natural convection in diode cavities containing an electronic equipment with discrete active bands under constant heat flux. *International Journal of Heat and Mass Transfer*, 2012, 55: 4970–4980.

[4] Baïri, A. Correlations for transient natural convection in parallelogrammic enclosures with isothermal hot wall. *Applied Thermal Engineering*, 2013, 51: 833–838.

[5] Baïri, A., Zarco-Pernia, E., & García de María, J.M. A review on natural convection in enclosures for engineering applications. The particular case of the parallelogrammic diode cavity. *Applied Thermal Engineering*, 2014, 63: 304–322.

[6] Costa, V.A.F. Double-diffusive natural convection in parallelogrammic enclosures. *Journal of Heat Mass Transfer*, 2014, 47: 2913–2926.
[7] Ghalambaz, M., Sheremet, M.A., & Pop, I. Free convection in a parallelogrammic porous cavity filled with a nanofluid using Tiwari and Das' nanofluid model. *PLoS ONE*, 2015, 10(5): e0126486.
[8] Villeneuve, T., Boudreau, M., & Dumas, G. The thermal diode and insulating potentials of a vertical stack of parallelogrammic air-filled enclosures. *International Journal of Heat and Mass Transfer*, 2017, 108: 2060–2071.
[9] Gupta, N., & Nayak, A.K. Activity of buoyancy convection and entropy generation in a parallelogrammic shaped mixed displacement ventilated system. *International Journal of Thermal Sciences*, 2019, 137: 86–100.
[10] Mallick, H., Mondal, H., Biswas, N., & Manna, N.K. Buoyancy driven flow in a parallelogrammic enclosure with an obstructive block and magnetic field. *Materials Today: Proceedings*, 2021, 44(2): 3164–3171.
[11] Marzougui, S., Mebarek-Oudina, F., Mchirgui, A., & Magherbi, M. Entropy generation and heat transport of cu-water nanoliquid in porous lid-driven cavity through magnetic field. *International Journal of Numerical Methods for Heat & Fluid Flow*, 2022, 32(6): 2047–2069.
[12] Ho, C.J., & Chang, J.Y. A study of natural convection heat transfer in a vertical rectangular enclosure with two-dimensional discrete heating: Effect of aspect ratio. *International Journal of Heat and Mass Transfer*, 1994, 37: 917–925.
[13] Deng, Q.H. Fluid flow and heat transfer characteristics of natural convection in square cavities due to discrete source-sink pairs. *International Journal of Heat and Mass Transfer*, 2008, 51: 5949–5957.
[14] Sivasankaran, S., Do, Y., & Sankar, M. Effect of discrete heating on natural convection in a rectangular porous enclosure. *Transport in Porous Media*, 2011, 86: 291–311.
[15] Sankar, M., Bhuvaneswari, M., Sivasankaran, S., & Do, Y. Buoyancy induced convection in a porous cavity with partially thermally active sidewalls. *International Journal of Heat and Mass Transfer*, 2011, 54: 5173–5182.
[16] Devanand, A., Saraei, S.H., Ghoreishi, S., & Chamka, A.J. Lattice Boltzmann simulation of natural convention in a square enclosure with discrete heating. *Mathematics and Computers in Simulation*, 2021, 179: 265–278.
[17] Sankar, M., Jang, B., & Do, Y. Numerical study of non-Darcy natural convection from two discrete heat source in a vertical annulus. *Journal of Porous Media*, 2014, 17: 373–390.
[18] Sankar, M., Park, Y., Lopez, J.M., & Do, Y. Numerical study of natural convection in a vertical porous annulus with discrete heating. *International Journal of Heat and Mass Transfer*, 2011, 54: 1493–1505.
[19] Mebarek-Oudina, F. Numerical modeling of the hydrodynamic stability in vertical annulus with heat source of different lengths. *International Journal of Engineering Science Technologies*, 2017, 20: 1324–1333.
[20] Mebarek-Oudina, F. Convective heat transfer of titania nanofluids of different base fluids in cylindrical annulus with discrete heat source. *Heat Transfer-Asian Research*, 2019, 48(1): 135–147.
[21] Moya, S.L., & Ramos, E. Numerical study of natural convection in a tilted rectangular porous material. *International Journal of Heat and Mass Transfer*, 1987, 30(4), 741–756.
[22] Hslao, S.W., Chen, C.K., & Cheng, P. A numerical solution for natural convection in an inclined porous cavity with a discrete heat source on one wall. *International Journal of Heat and Mass Transfer*, 1994, 37(15): 2193–2201.
[23] Chabani, I., Mebarek Oudina, F., & Ismail, A.I. MHD flow of a hybrid nano-fluid in a triangular enclosure with zigzags and an elliptic obstacle. *Micromachines*, 2022, 13(2): 224.
[24] Pushpa, B.V., Sankar, M., & Mebarek-Oudina, F. Buoyant convective flow and heat dissipation of Cu-H_2O nanoliquids in an annulus through a thin baffle. *Journal of Nanofluids*, 2021, 10 (2): 292–304.
[25] Dhif, K., Mebarek-Oudina, F., Chouf, S., Vaidya, H., & Chamkha, A.J. Thermal analysis of the solar collector cum storage system using a hybrid-nanofluids. *Journal of Nanofluids*, 2021, 10(4): 634–644.
[26] Tian, L., Ye, C., Xue, S.H., & Wang, G. Numerical investigation of unsteady natural convection in an inclined square enclosure with heat-generating porous medium. *Heat Transfer Engineering*, 2014, 35(6–8): 620–629.
[27] Jagadeesha, R.D., Prasanna, B.M.R., & Sankar, M. Double diffusive convection in an inclined parallelogrammic porous enclosure. *Procedia Engineering*, 2015, 127: 1346–1353.
[28] Alsabery, A.I., Chamkha, A.J., & Saleh, H. Natural convection flow of a nanofluid in an inclined square enclosure partially filled with a porous medium. *Scientific Reports*, 2017, 7: 2357.

[29] Keyhani Asl, A., Hossainpour, S., Rashidi, M.M., Sheremet, M.A., & Yang, Z. Comprehensive investigation of solid and porous fins influence on natural convection in an inclined rectangular enclosure. *International Journal of Heat and Mass Transfer*, 2019, 133: 729–744.

[30] Hadidi, N., et al. Thermosolutal natural convection across an inclined square enclosure partially filled with a porous medium. *Results in Physics*, 2021, 21: 103821.

[31] Asogwa, K., Mebarek-Oudina, F., & Animasaun, I. Comparative investigation of water-based Al_2O_3 nanoparticles through water-based CuO nanoparticles over an exponentially accelerated radiative riga plate surface via heat transport. *Arabian Journal for Science and Engineering*, 2022, 47: 8721–8738.

[32] Djebali, R., Mebarek-Oudina, F., & Choudhari, R. Similarity solution analysis of dynamic and thermal boundary layers: Further formulation along a vertical flat plate. *Physical Scripta*, 2021, 96(8): 085206.

[33] Rajashekhar, C., Mebarek-Oudina, F., Vaidya, H., Prasad, K.V., Manjunatha, G., & Balachandra, H. Mass and heat transport impact on the peristaltic flow of Ree-Eyring liquid with variable properties for hemodynamic flow. *Heat Transfer*, 2021, 50(5): 5106–5122.

[34] Shafiq, A., Mebarek-Oudina, F., Sindhu, T.N., & Rassoul, G. Sensitivity analysis for Walters' B nanoliquid flow over a radiative Riga surface by RSM. *Scientia Iranica*, 2022, 29(3): 1236–1249. https://doi.org/10.24200/SCI.2021.58293.5662

[35] Chabani, I., Mebarek-Oudina, F., Vaidya, H., & Ismail, A.I. Numerical analysis of magnetic hybrid Nano-fluid natural convective flow in an adjusted porous trapezoidal enclosure. *Journal of Magnetism and Magnetic Materials*, 2022, 564(2): 170142. https://doi.org/10.1016/j.jmmm.2022.170142

[36] Reddy, Y.D., Mebarek-Oudina, F., Goud, B.S., & Ismail, A.I. Radiation, velocity and thermal slips effect toward MHD boundary layer flow through heat and mass transport of Williamson Nanofluid with porous medium. *Arabian Journal for Science and Engineering*, 2022, 47(12): 16355–16369. https://doi.org/10.1007/s13369-022-06825-2

[37] Mebarek-Oudina, F., & Chabani, I. Review on Nano enhanced PCMs: Insight on nePCM application in thermal management/storage systems. *Energies*, 2023, 16(3): 1066. https://doi.org/10.3390/en16031066

[38] Mebarek Oudina, F., & Chabani, I. Review on Nano-fluids applications and heat transfer enhancement techniques in different enclosures. *Journal of Nanofluids*, 2022, 11(2): 155–168. https://doi.org/10.1166/jon.2022.1834

[39] Raza, J., Mebarek-Oudina, F. & Ali Lund, L. The flow of magnetised convective Casson liquid via a porous channel with shrinking and stationary walls. *Pramana – Journal of Physics*, 2022, 96: 229. https://doi.org/10.1007/s12043-022-02465-1

4 Double-Diffusive Convection in a Composite System with a Heat Source and Temperature Gradients

Manjunatha N., Sumithra R., and Vanishree R. K.

4.1 INTRODUCTION

Double-diffusive convection in two layer systems is essential in understanding the evolution of a number of systems that witness several causes for density variations. These include convections in magma chamber and in the sun (where heat and helium diffuse at differing rates). Compositional differences produced by processes such as partial crystallization, partial melting, contamination, and heat impacts cause convection in magma chambers. Understanding the stability of magma chambers under varied thermal and compositional/concentration gradients is critical to understanding magma generation. These gradients can also be found in a variety of industrial and geophysical applications, including flows in fuel cells, filtration processes, oil extraction from underground reservoirs, groundwater pollution, flow in biological materials, the manufacturing of composite materials used in the aircraft and automobile industries, water flow beneath the Earth's surface, and compound film growth in thermal chemical vapour deposition reactors.

Some authors look at the challenges of double-diffusive convection in fluid region, porous region, and fluid–porous region in the presence of temperature gradients and heat source/sink. Bennacer et al. [1] investigated thermosolutal natural convection in a fluid-filled enclosure with two saturated porous layers numerically. Using linear stability analysis, Chen and Chan [2] investigated the stability of convection in a horizontal double-diffusive fluid layer driven by the combined influences of buoyancy and surface tension. Sivasankaran et al. [3] studied the effect of discrete heating on free convection heat transfer in a rectangular porous enclosure containing a heat-generating substance. They found that the numerical results reveal that the maximum temperature decreases with the modified Rayleigh number and increases with the aspect ratio. Double-diffusive convection in a fluid-saturated vertical porous annulus subjected to discrete heat and mass fluxes from a portion of the inner wall was studied by Sankar et al. [4]. Using a regular perturbation technique, Gangadharaiah and Suma [5] investigated the Bènard–Marangoni convection in a fluid layer above an anisotropic porous layer with deformable free surface. Sheng Chen et al. [6] employ the lattice Boltzmann model to investigate double-diffusive convection in vertical annuluses with conflicting temperature and concentration gradients. Saleem et al. [7] used the successive over-relaxation technique to explore the double-diffusive Marangoni convection flow of viscous incompressible electrically conducting fluid in a square cavity. Sumithra [8] used the perturbation technique to investigate double-diffusive magneto Marangoni convection in a two-layer system. Massimo Corcione et al. [9] investigated numerically double-diffusive natural convection in vertical square enclosures caused by opposing horizontal temperature and concentration gradients. Norazam Arbin et al. [10] used the finite difference approach to quantitatively study the double-diffusive Marangoni convection in the presence of entropy generation. They discovered that as the Marangoni number increased, the heat and mass transfer patterns resembled one other. By considering a horizontal fluid layer, Akil and Fahad [11] investigated double-diffusive convection in the presence of a heat sink/source.

DOI: 10.1201/9781003299608-4

Mebarek-Oudina [12] looked into the stability of vertical annuli with various heat source lengths. Using implicit finite difference technique, Girish et al. [13] studied the annular passages for two thermal conditions. Tatyana and Ekaterina [14] investigated the commencement of double-diffusive convection in a superposed fluid and porous layer when vibrations were high-frequency and small-amplitude. By assuming rigid/free isothermal boundaries, Kanchana et al. [15] investigated Rayleigh–Bènard convection. They discovered that alumina nanoparticles in water have the same impact as alumina and copper in water. Dadheech et al. [16] investigated the flow of natural convective boundary layers in nanofluid and hybrid nanofluid across a stretching area. With viscous dissipation, a uniform inclined magnetic field was applied. For the composite layer, Sumithra et al. [17] and Vanishree et al. [18] studied single-component convection in the presence of heat source by considering adiabatic and isothermal boundaries.

Recently, Mebarek-Oudina et al. [19] used the finite volume technique to investigate the effects of the placement of a thermal source on buoyant convection of nanofluids in an annular region. Swain et al. [20] explain Brownian motion and thermophoresis diffusion. The chemical reaction's impacts, as well as the uniform internal heat source/heat sink, are taken into account. The two-dimensional magnetohydrodynamics steady boundary layer flow of a viscous magneto micropolar liquid via an extended area was examined by Warke et al. [21]. The impact of the heat sink/source as well as the chemical reaction are taken into account. Pushpa et al. [22] explore the buoyancy convection and heat transfer enhancement of nanoliquid in a differentially heated vertical annulus with a tiny baffle. Marzogui et al. [23] used the Darcy–Brinkman–Forchheimer technique to generate entropy and heat transfer of Cu-water nanoliquid in a porous lid-driven cavity. For modelling the peristaltic flow of a Ree-Eyring liquid via a uniform compliant conduit, Rajashekhar et al. [24] looked at the impact of varying thermal conductivity and viscosity. They show that the velocity of a Newtonian liquid is larger than that of a non-Newtonian liquid. Djebali et al. [25] suggest a simplified formulation of the similarity solution for the boundary layers problem that occurs along a vertical heated flat plate subjected to the buoyancy effect. Kawthar Dhif et al. [26] used a hybrid nanofluid to investigate the thermal analysis of a solar collector-cum-storage system. Manjunatha and Sumithra [27–28] and Manjunatha et al. [29] studied the impact of temperature gradients and heat source on double-diffusive convection in the presence of a magnetic field. The Walter's B nanofluid stagnation point flow produced by a Riga surface was studied by Shafiq et al. [30]. Asogwa et al. [31] compared the effects of alumina nanoparticles on cupric nanoparticles on a quick progressive Riga plate. They show that increasing the nanoparticle volume fraction in cupric nanofluid improves density over alumina nanofluid by lowering the velocity distribution. Swain et al. [32] studied the effect of a changing magnetic field on the chemical reaction of a multiwalled carbon nanotube/magnetite–water hybrid nanofluid flowing through an exponentially diminishing porous sheet with slip boundary conditions. Hybridity improves temperature and concentration profiles, according to the researchers. Sumithra et al. [33] used a perturbation technique to investigate onset convection in a combined layer with thermal diffusion.

Most of the previously cited papers either deal with single, fluid, or porous layers, but in nature and in most of the applications, the occurrence of composite layer is obvious. Also, in many of the applications, the presence of a second diffusing component is apparent; hence, in this chapter, the problem of Bènard double-diffusive Marangoni convection is investigated in a horizontally infinite composite layer system enclosed by adiabatic boundaries for the Darcy model, in the presence of heat sources, which is quite close to the realistic models. The impact of different parameters on the eigenvalue is discussed in detail.

4.2 FORMULATION OF THE PROBLEM

Consider a horizontal densely packed porous layer of thickness d_p rests on top of a two-component fluid layer of thickness d_f, each having constant heat sources Q_{hp} and Q_h. Figure 4.1 depicts the composite layer structure under research. The porous layer's lower surface is hard, while the fluid

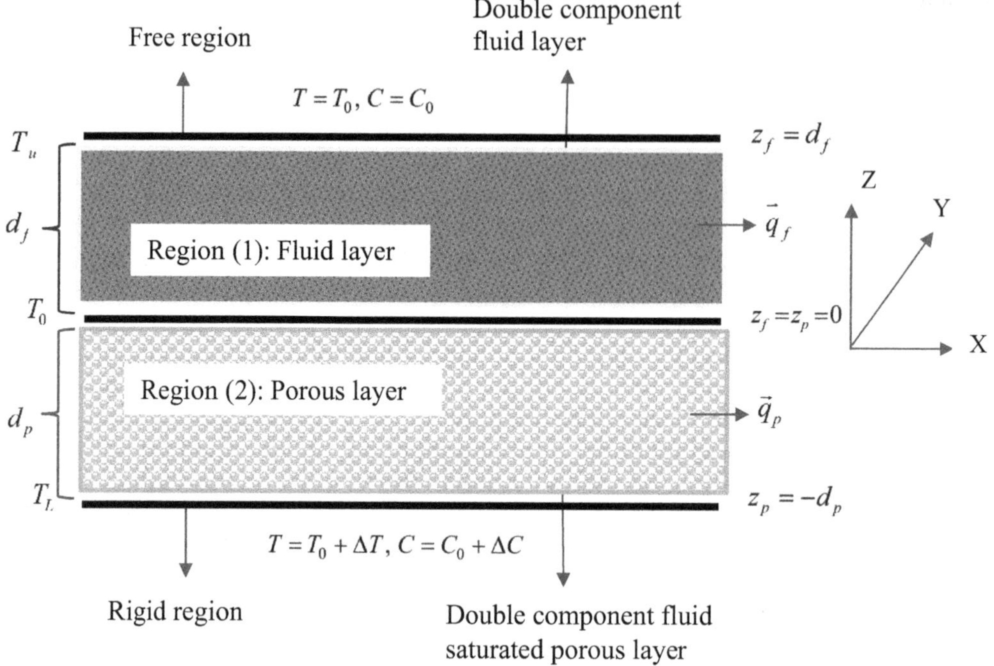

FIGURE 4.1 Physical model.

layer's top surface is free, with surface tension effects that vary with temperature and concentration. The temperature and concentration difference between the lower and higher bounds is denoted by ΔT and ΔC. Both boundaries are preserved at separate temperatures and salinities at all times. At the interface between the porous and fluid layers, we add a Cartesian coordinate system with the Z-axis pointing vertically upwards.

Under the Boussinesq approximation and microgravity conditions, the governing equations for the physical situation are (see Venkatachalappa et al. [34] and Shivakumara et al. [35]):

Fluid layer: region 1

$$\nabla_f \cdot \vec{q}_f = 0 \tag{4.1}$$

$$\frac{\partial \vec{q}_f}{\partial t} + (\vec{q}_f . \nabla_f)\vec{q}_f = \frac{1}{\rho_0}\left(-\nabla_f P_f + \mu_f \nabla_f^2 \vec{q}_f\right) \tag{4.2}$$

$$\frac{\partial T_f}{\partial t} + (\vec{q}_f . \nabla_f)T_f = \kappa_f \nabla_f^2 T_f + Q_h \tag{4.3}$$

$$\frac{\partial C_f}{\partial t} + (\vec{q}_f . \nabla_f)C_f = \kappa_{cf} \nabla_f^2 C_f \tag{4.4}$$

Porous layer: region 2

$$\nabla_p \cdot \vec{q}_p = 0 \tag{4.5}$$

$$\frac{1}{\phi}\frac{\partial \vec{q}_p}{\partial t} + \frac{1}{\phi^2}(\vec{q}_p . \nabla_p)\vec{q}_p = -\frac{1}{\rho_0}\nabla_p P_p - \frac{\mu_p}{K\rho_0}\vec{q}_p \tag{4.6}$$

$$\Lambda\frac{\partial T_p}{\partial t}+(\vec{q}_p.\nabla_p)T_p=\kappa_p\nabla_p^2T_p+Q_{hp} \tag{4.7}$$

$$\phi\frac{\partial C_p}{\partial t}+(\vec{q}_p.\nabla_p)C_p=\kappa_{cp}\nabla_p^2C_p \tag{4.8}$$

Where subscripts f and p denote the fluid and porous layer; $\vec{q}_f,\vec{q}_p$ are the velocities; ρ_0 is the density; μ_f,μ_p are the viscosities; P_f,P_p are the pressures; T_f,T_p are the temperatures; κ_f,κ_p are the thermal diffusivities; Q_h,Q_{hp} are the heat sources; κ_{cf},κ_{cp} are the salinity diffusivities; C_f,C_p are the concentrations; K is the permeability; Λ is the ratio heat capacities; and ϕ is the porosity in the region 1 and region 2, respectively.

For two regions, the basic states are (see Shivakumara et al. [36]):

Region 1:

$$\vec{q}_f=0,P_f=P_{fb}(z_f),T_f=T_{fb}(z_f),C_f=C_{fb}(z_f),\frac{-d_f}{\Delta T_f}\frac{dT_{fb}}{dz_f}=\chi(z_f) \tag{4.9}$$

Region 2:

$$\vec{q}_p=0,P_p=P_{pb}(z_p),T_p=T_{pb}(z_p),C_p=C_{pb}(z_p),\frac{-d_p}{\Delta T_p}\frac{dT_{pb}}{dz_p}=\chi(z_p) \tag{4.10}$$

Introducing (4.9) and (4.10), for two regions, the temperature distribution in the basic state is obtained as follows:

$$T_{fb}(z_f)=\frac{Q_h(z_fd_f-z_f^2)}{2\kappa_f}+\frac{(T_u-T_0)\chi(z_f)}{d_f}+T_0 \qquad 0\le z_f\le d_f \tag{4.11}$$

$$T_{pb}(z_p)=\frac{-Q_{hp}(z_p+d_p)z_p}{2\kappa_p}+\frac{(T_0-T_l)\chi(z_p)}{d_p}+T_0 \qquad -d_p\le z_p\le 0 \tag{4.12}$$

Where $\chi(z_f)$ and $\chi(z_p)$ are the temperature gradients in region 1 and region 2, respectively.

Also, the following are the distributions of salinity in the basic state for two regions:

$$C_{fb}(z_f)=C_0+\frac{(C_u-C_0)z_f}{d_f} \qquad 0\le z_f\le d_f \tag{4.13}$$

$$C_{pb}(z_p)=C_0+\frac{(C_0-C_l)z_p}{d_p} \qquad -d_p\le z_p\le 0 \tag{4.14}$$

The suffix b in the preceding phrases denotes the basic stage. At the interface, $\chi(z_f)=\chi(z_p)$, and note that $T_0=\dfrac{2(\kappa_fd_pT_u+\kappa_pd_fT_l)+d_fd_p(Q_{hp}d_p+Q_hd_f)}{2(\kappa_fd_p+\kappa_pd_f)}$ and $C_0=\dfrac{\kappa_{cf}d_pC_u+\kappa_{cp}d_fC_l}{\kappa_{cf}d_p+\kappa_{cp}d_f}$.

Perturbed state:

$$\vec{q}_f=0+\vec{q}_f{}',P_f=P_{fb}+P_f',T_f=T_{fb}(z_f)+\theta_f',C_f=C_{fb}(z_f)+S_f' \tag{4.15}$$

$$\vec{q}_p=0+\vec{q}_p{}',P_p=P_{pb}+P_p',T_p=T_{pb}(z_p)+\theta_p',C_p=C_{pb}(z_p)+S_p' \tag{4.16}$$

Where $\vec{q}_f{}', P_f{}', \theta_f{}', S_f{}'$ (velocity, pressure, temperature, and salinity, respectively) are perturbed quantities for region 1, and the corresponding quantities in region 2 are $\vec{q}_p{}', P_p{}', \theta_p{}', S_p{}'$, substituting (4.15) and (4.16) in (4.1–4.8) and omitting the primes for simplicity, linearizing the equations, and taking curl twice on the momentum equations in both the regions to eliminate pressures. In the dimensionless formulation, these are the scales for length, time, velocity, temperature, species concentration, and temperature gradient, respectively: d_f, d_p are the length, $\frac{d_f^2}{\kappa_f}, \frac{d_p^2}{\kappa_p}$ are the time, $\frac{\kappa_f}{d_f}, \frac{\kappa_p}{d_p}$ are the velocities, $T_0 - T_u, T_l - T_0$ are the temperatures, $C_0 - C_u, C_l - C_0$ are the concentrations, $\frac{\Delta T_f}{a_f\sqrt{M_t}}, \frac{\Delta T_p}{a_p\sqrt{M_t}}$ are the temperature gradients in region 1 and region 2, and M_t is the thermal Marangoni number. Choose the separate scales (see Nield [37], Chen [38], Chen and Chen [39]) so that each region is of unit depth such that $(x_f, y_f, z_f) = d_f(x_f', y_f', z_f')$ and $(x_p, y_p, z_p) = d_p(x_p{}', y_p{}', z_p{}' - 1)$.

The following perturbed equations are:

Region 1:

$$\frac{1}{P_{rf}}\frac{\partial(\nabla_f^2 q_f)}{\partial t} = \nabla_f^4 q_f \tag{4.17}$$

$$\frac{\partial \theta_f}{\partial t} - \chi(z_f)q_f + R_{Af}^*(1-2z_f)q_f = \nabla_f^2\theta_f \tag{4.18}$$

$$\frac{\partial S_f}{\partial t} = q_f + \tau_f \nabla_f^2 S_f \tag{4.19}$$

Region 2:

$$\frac{\beta^2}{P_{rp}}\frac{\partial(\nabla_p^2 q_p)}{\partial t} = -\nabla_p^2 q_p \tag{4.20}$$

$$\Lambda\frac{\partial \theta_p}{\partial t} - \chi(z_p)q_p - R_{Ap}^*(1+2z_p)q_p = \nabla_p^2\theta_p \tag{4.21}$$

$$\phi\frac{\partial S_p}{\partial t} = q_p + \tau_p \nabla_p^2 S_p \tag{4.22}$$

In region 1 and region 2, the normal mode expansion for the perturbed variables is:

$$\left[q_f{}', \theta_f{}', S_f{}'\right] = \left[V_f, \theta_f, S_f\right]\left(z_f\right) exp\left[il_f x + im_f y + n_f t\right] \tag{4.23}$$

$$\left[q_p{}', \theta_p{}', S_p{}'\right] = \left[V_p, \theta_p, S_p\right]\left(z_p\right) exp\left[il_p x + im_p y + n_p t\right] \tag{4.24}$$

With $\nabla_2^2 f = -a_f^2 f$ and $\nabla_{2p}^2 f = -a_p^2 f$ such that $a_f = \sqrt{l_f^2 + m_f^2}$, $a_p = \sqrt{l_p^2 + m_p^2}$, here a_f, a_p are horizontal wave numbers, and l_f, m_f and l_p, m_p are wave numbers in x and y direction, respectively, and n_f, n_p are frequencies in region 1 and region 2, respectively. Also, $\nabla^2 = \frac{\partial^2}{\partial x^2} + \frac{\partial^2}{\partial y^2} + \frac{\partial^2}{\partial z^2}$ is the Laplacian operator, and $\nabla_2^2 = \frac{\partial^2}{\partial x^2} + \frac{\partial^2}{\partial y^2}$ is the horizontal Laplacian operator. Introducing (4.23) and (4.24) in (4.17–4.22), the following equations are obtained.

Region 1: $0 \le z_f \le 1$

$$\left(D_f^2 - a_f^2 + \frac{n_f}{Pr}\right)(D_f^2 - a_f^2)V_f(z_f) = 0 \tag{4.25}$$

$$(D_f^2 - a_f^2 + n_f)\theta_f(z_f) + [\chi(z_f) + R_{If}^*(2z_f - 1)]V_f(z_f) = 0 \tag{4.26}$$

$$\left(\tau_f(D_f^2 - a_f^2) + n_f\right)S_f(z_f) + V_f(z_f) = 0 \tag{4.27}$$

Region 2: $-1 \le z_p \le 0$

$$\left(1 - \frac{\beta^2 n_p}{P_{rp}}\right)(D_p^2 - a_p^2)V_p(z_p) = 0 \tag{4.28}$$

$$(D_p^2 - a_p^2 + \Lambda n_p)\theta_p(z_p) + [\chi(z_p) + R_{Ip}^*(2z_p + 1)]V_p(z_p) = 0 \tag{4.29}$$

$$\left(\tau_p(D_p^2 - a_p^2) + n_p\phi\right)S_p(z_p) + V_p(z_p) = 0 \tag{4.30}$$

In the preceding expressions, $P_r = \frac{\nu_f}{\kappa_f}$, $P_{rp} = \frac{\nu_p}{\kappa_p}$ are the Prandtl numbers in region 1 and region 2, respectively. Also, $\beta = \sqrt{\frac{K}{d_p^2}}$ is the square root of Darcy number and $D_f = \frac{d}{dz_f}$ and $D_p = \frac{d}{dz_p}$.

The analysis is limited to stationary convection since the principle of exchange of stability is valid and relevant for the current instance and hence take $n_f = 0$ and $n_p = 0$ (see Shivakumara et al. [35], Nield [37], and Sumithra [40]). The eigenvalue problem (4.25–4.30) is transformed into the following stability equations:

$$(D_f^2 - a_f^2)^2 V_f(z_f) = 0 \tag{4.31}$$

$$(D_f^2 - a_f^2)\theta_f(z_f) + [\chi(z_f) + R_{Af}^*(2z_f - 1)]V_f(z_f) = 0 \tag{4.32}$$

$$\tau_f(D_f^2 - a_f^2)S_f(z_f) + V_f(z_f) = 0 \tag{4.33}$$

$$(D_p^2 - a_p^2)V_p(z_p) = 0 \tag{4.34}$$

$$(D_p^2 - a_p^2)\theta_p(z_p) + [\chi(z_p) + R_{Ap}^*(2z_p + 1)]V_p(z_p) = 0 \tag{4.35}$$

$$\tau_p(D_p^2 - a_p^2)S_p(z_p) + V_p(z_p) = 0 \tag{4.36}$$

Where $R_{Af}^* = \frac{R_{if}}{2(T_0 - T_u)}, R_{Ap}^* = \frac{R_{ip}}{2(T_l - T_0)}$ are the modified internal Rayleigh numbers (mRn's), $R_{if} = \frac{Q_h d_f^2}{\kappa_f}, R_{ip} = \frac{Q_{hp} d_p^2}{\kappa_p}$ are the internal Rayleigh numbers, $\tau_f = \frac{\kappa_{cf}}{\kappa_f}, \tau_p = \frac{\kappa_{cp}}{\kappa_p}$ are the diffusivity ratios, $V_f(z_f)$ and $V_p(z_p)$ are the velocities, $\theta_f(z_f)$ and $\theta_p(z_p)$ are the distributions of temperature, and $S_f(z_f)$ & $S_p(z_p)$ are the distributions of concentration in both the regions, respectively. Because the combined system's horizontal wave numbers must be the same, we have $a_p = \hat{d}a_f$; here, $\hat{d} = \frac{d_p}{d_f}$ is the depth ratio.

4.3 BOUNDARY CONDITIONS

Equations (4.31) to (4.36) are solved using non-dimensionalized conditions that are subjected to normal mode expansion. The velocity–temperature–salinity coupled boundary condition, which contributes the eigenvalue, the thermal Marangoni number, is:

$$D_f^2 V_f(1) + M_t a_f^2 \theta_f(1) + M_s a_f^2 S_f(1) = 0 \tag{4.37}$$

The velocity boundary conditions are:

$$\begin{aligned} &V_f(1)=0, V_p(-1)=0, D_p V_p(-1)=0, \hat{T} V_f(0)=V_p(0), \\ &\hat{T}\widehat{d_r^2}(D_f^2+a_f^2)V_f(0)=\hat{\mu}(D_p^2+a_p^2)V_p(0), \\ &\hat{T}\widehat{d_r^3}\beta^2\left(D_f^3 V_f(0)-3a_f^2 D_f V_f(0)\right)=-D_p V_p(0) \end{aligned} \tag{4.38}$$

The boundary conditions for distribution of temperature and distribution of salinity are:

$$D_f\theta_f(1)=0, \theta_f(0)=\hat{T}\theta_p(0), D_f\theta_f(0)=D_p\theta_p(0), D_p\theta_p(-1)=0 \tag{4.39}$$

$$D_f S_f(1)=0, S_f(0)=\hat{S}S_p(0), D_f S_f(0)=D_p S_p(0), D_p S_p(-1)=0 \tag{4.40}$$

Where $\hat{S}=\dfrac{C_l-C_0}{C_0-C_u}$ is the solute diffusivity ratio, $\hat{T}=\dfrac{T_l-T_0}{T_0-T_u}$ is the thermal ratio, $\hat{\mu}=\dfrac{\mu_p}{\mu_f}$ is the viscosity ratio, $M_t=\dfrac{-\partial\sigma_t}{\partial T}\dfrac{\Delta T d_f}{\mu_f \kappa_f}, M_s=\dfrac{-\partial\sigma_t}{\partial C}\dfrac{\Delta C d_f}{\mu_f \kappa_f}$ are the thermal and solute Marangoni numbers, respectively.

4.4 SOLUTION BY EXACT TECHNIQUE

4.4.1 Velocity Profile

The resultant eigenvalue problem is solved exactly with M_t as an eigenvalue. In addition, an analytical equation for the thermal Marangoni number is found.

The solutions of $V_f(z_f)$ and $V_p(z_p)$ are obtained from equations (4.31) and (4.34), in the following convenient form, as:

$$V_f(z_f)=B_1[\cosh a_f z_f + a_1 z_f \cosh a_f z_f + a_2 \sinh a_f z_f + a_3 z_f \sinh a_f z_f] \tag{4.41}$$

$$V_p(z_p)=B_1[a_4 \cosh a_p z_p + a_5 \sinh a_p z_p] \tag{4.42}$$

Where $a_1=\dfrac{a_p \coth a_p}{2a_f^3\beta^2\widehat{d_r^3}}, a_2=-1-(a_1+a_3)\tanh a_f,\ a_3=\dfrac{a_p^2\hat{\mu}-a_f^2\widehat{d_r^2}}{a_f\widehat{d_r^2}}, a_4=\hat{T}, a_5=\hat{T}\coth a_p$.

4.4.2 Salinity Profile

From (4.33) and (4.36), we get the salinity profiles $S_f(z_f)$ and $S_p(z_p)$ as:

$$S_f(z_f)=B_1[c_{13}\cosh a_f z_f + c_{14}\sinh a_f z_f + \Sigma_f(z_f)] \tag{4.43}$$

$$S_p(z_p)=B_1[c_{15}\cosh a_p z_p + c_{16}\sinh a_p z_p + \Sigma_p(z_p)] \tag{4.44}$$

Where $c_{13}-c_{16}$ are determined using (4.40). We get:

$$\Sigma_f(z_f)=\frac{-\cosh a_f z_f}{4a_f^2\tau_f}\left[2a_f z_f(a_1+\tanh a_f z_f)+a_f z_f^2(a_3+a_2\tanh a_f z_f)-z_f(a_2+a_3\tanh a_f z_f)\right]$$

$$\Sigma_p(z_p)=\frac{-z_p\cosh a_p z_p}{2a_p\tau_p}(a_5+a_4\tanh a_p z_p)$$

$$c_{13}=\hat{S}c_{15},c_{14}=\frac{1}{a_f}\left[c_{16}a_p-\frac{a_5}{2a_p\tau_p}+\frac{1}{\tau_f}\left(\frac{2a_f a_1-a_2}{4a_f^2}\right)\right]$$

$$c_{15}=\frac{\Delta_{103}\cosh a_p-\Delta_{102}\cosh a_f}{\cosh a_p\cosh a_f\left(a_p\tanh a_p+\hat{S}a_f\tanh a_f\right)},$$

$$c_{16}=\frac{\Delta_{102}\hat{S}a_f\sinh a_f+\Delta_{103}a_p\sinh a_p}{a_p\cosh a_p\cosh a_f\left(a_p\tanh a_p+\hat{S}a_f\tanh a_f\right)},$$

$$\Delta_{100}=\frac{1}{2a_f\tau_f}\left[a_f(\cosh a_f+a_1\sinh a_f)+(a_1\cosh a_f+\sinh a_f)+2a_f\Delta_{101}\right],$$

$$\Delta_{101}=\frac{\cosh a_f}{4a_f^2}\left[\left((a_f^2-1)a_2+a_3a_f\right)+\left((a_f^2-1)a_3+a_2a_f\right)\tanh a_f\right],$$

$$\Delta_{102}=\frac{\cosh a_p}{2a_p\tau_p}\left[(a_5-a_4\tanh a_p)-a_p(a_4-a_5\tanh a_p)\right],$$

$$\Delta_{103}=\Delta_{100}+\left(\frac{a_5}{2a_p\tau_p}-\frac{1}{\tau_f}\left(\frac{2a_f a_1-a_2}{4a_f^2}\right)\right).$$

4.4.3 Temperature Profiles

Introducing the preceding profiles (Table 4.1) in (4.32) and (4.35) (see Sparrow et al. [41] and Shivakumara et al. [42]), the thermal Marangoni numbers (tMn's) for linear, parabolic, and inverted parabolic profiles are respectively obtained using (4.37) as follows:

$$M_{t1}=\frac{-\Lambda_1}{a_f^2(c_1\cosh a_f+c_2\sinh a_f+\Lambda_2+\Lambda_3)} \tag{4.45}$$

$$M_{t2}=\frac{-\Lambda_1}{a_f^2(c_5\cosh a_f+c_6\sinh a_f+\Lambda_4+\Lambda_5)} \tag{4.46}$$

$$M_{t3}=\frac{-\Lambda_1}{a_f^2(c_9\cosh a_f+c_{10}\sinh a_f+\Lambda_6+\Lambda_7)} \tag{4.47}$$

TABLE 4.1
The profiles considered for present study

Profiles	Fluid Layer: Region 1	Porous Layer: Region 2
Linear	$\chi(z_f)=1$	$\chi(z_p)=1$
Parabolic	$\chi(z_f)=2z_f$	$\chi(z_p)=2z_p$
Inverted parabolic	$\chi(z_f)=2(1-z_f)$	$\chi(z_p)=2(1-z_p)$

Where $\Lambda_1 = A_{13} + A_{14}$, $\Lambda_3 = A_{15} - A_{16}$,

$$A_{13} = \cosh a_f \left[a_f^2 (1 + a_1 \tanh a_f) + a_2 (a_f^2 + 2a_f \tanh a_f) \right],$$

$$A_{14} = a_3 \cosh a_f (a_f^2 \tanh a_f + 2a_f),$$

$$\Lambda_2 = \frac{\cosh a_f}{4a_f^2} \left[a_f (b_2 + 2b_1)(a_1 + \tanh a_f) - b_2 (1 + a_1 \tanh a_f) \right],$$

$$A_{15} = \frac{\cosh a_f}{12a_f^3} \left[(2a_f^2 b_2 + 3a_f^2 b_1 + 3b_2)(a_3 + a_2 \tanh a_f) \right],$$

$$A_{16} = \frac{\cosh a_f}{4a_f^2} (b_2 + b_1)(a_3 \tanh a_f + a_2),$$

$$\Lambda_6 = \frac{\cosh a_f}{4a_f^2} \left[(b_6 + 2b_5) a_f (a_1 + \tanh a_f) - b_6 (1 + a_1 \tanh a_f) \right], \ \Lambda_7 = A_{43} - A_{44}$$

$$A_{43} = \frac{\cosh a_f (2a_f^2 b_6 + 3a_f^2 b_5 + 3b_6)}{12a_f^3} (a_3 + a_2 \tanh a_f),$$

$$A_{44} = \frac{\cosh a_f (b_6 + b_5)}{4a_f^2} (a_3 \tanh a_f + a_2),$$

$$\Lambda_6 = \frac{\cosh a_f}{4a_f^2} \left[(b_6 + 2b_5) a_f (a_1 + \tanh a_f) - b_6 (1 + a_1 \tanh a_f) \right], \ \Lambda_7 = A_{43} - A_{44},$$

$$A_{43} = \frac{\cosh a_f (2a_f^2 b_6 + 3a_f^2 b_5 + 3b_6)}{12a_f^3} (a_3 + a_2 \tanh a_f), \ A_{44} = \frac{\cosh a_f (b_6 + b_5)}{4a_f^2} (a_3 \tanh a_f + a_2)$$

For linear model:

$$\theta_f(z_f) = B_1[c_1 \cosh a_f z_f + c_2 \sinh a_f z_f + \Sigma_1(z_f)]$$

$$\theta_p(z_p) = B_1[c_3 \cosh a_p z_p + c_4 \sinh a_p z_p + \Sigma_{1p}(z_p)]$$

For parabolic model:

$$\theta_f(z_f) = A_1[c_5 \cosh a_f z_f + c_6 \sinh a_f z_f + \Sigma_2(z_f)]$$

$$\theta_p(z_p) = B_1[c_7 \cosh a_p z_p + c_8 \sinh a_p z_p + \Sigma_{2p}(z_p)]$$

For inverted parabolic model:

$$\theta_f(z_f) = B_1[c_9 \cosh a_f z_f + c_{10} \sinh a_f z_f + \Sigma_3(z_f)]$$

$$\theta_p(z_p) = B_1[c_{11} \cosh a_p z_p + c_{12} \sinh a_p z_p + \Sigma_{3p}(z_p)]$$

$$\Sigma_1(z_f) = B_1[A_1 - A_2 + A_3 - A_4], \Sigma_{1p}(z_p) = B_1[A_5 - A_6],$$

$$A_1 = \frac{\cosh a_f z_f}{4a_f} \left[(2b_1 z_f + b_2 z_f^2)(a_1 + \tanh a_f z_f) \right], \ A_2 = \frac{\cosh a_f z_f}{4a_f^2} \left(b_2 z_f (1 + a_1 \tanh a_f z_f) \right),$$

$$A_3 = \frac{\cosh a_f z_f}{12a_f^3} (3a_f^2 z_f^2 b_1 + 2a_f^2 z_f^3 b_2 + 3b_2 z_f)(a_3 + a_2 \tanh a_f z_f),$$

$$A_4 = \frac{z_f \cosh a_f z_f (b_1 + z_f b_2)}{4a_f^2}(a_2 + a_3 \tanh a_f z_f),$$

$$A_5 = \frac{z_p \cosh a_p z_p (2b_{1p} + z_p b_{2p})}{4a_p}(a_5 + a_4 \tanh a_p z_p),$$

$$A_6 = \frac{b_{2p} z_p \cosh a_p z_p}{4a_p^2}(a_4 + a_5 \tanh a_p z_p), b_1 = R_{Af}^* - 1, b_2 = -2R_{Af}^*, b_{1p} = -(R_{Ap}^* + 1), b_{2p} = -2R_{Ap}^*,$$

$$c_1 = c_3 \hat{T}, c_2 = \frac{1}{a_f}(c_4 a_f + \delta_3 - \delta_2), c_3 = \frac{\delta_8}{\delta_9}, c_4 = \frac{\delta_6}{\delta_7}, \ \delta_1 = -B_1[A_7 + A_8 + A_9 + A_{10}],$$

$$A_7 = \frac{\cosh a_f \left(2a_f^2 b_1 + b_2(a_f^2 - 1)\right)}{4a_f^2}(1 + a_1 \tanh a_f), \ A_8 = \frac{\cosh a_f (b_2 + 2b_1)}{4a_f}(a_1 + \tanh a_f),$$

$$A_9 = \frac{\cosh a_f \left((a_f^2 - 1)3b_1 + (2a_f^2 - 3)b_2\right)}{12a_f^2}(a_2 + a_3 \tanh a_f),$$

$$A_{10} = \frac{\cosh a_f}{4a_f^3}\left(a_f^2 b_1 + b_2(a_f^2 + 1)\right)(a_3 + a_2 \tanh a_f),$$

$$\delta_2 = \frac{B_1}{4a_f^3}\left((2a_f^2 a_1 - a_f a_2)b_1 + (a_3 - a_f)b_2\right), \ \delta_3 = \frac{B_1}{4a_p^2}\left(2a_p b_{1p} a_5 - a_4 b_{2p}\right), \ \delta_4 = -B_1[A_{11} + A_{12}]$$

$$A_{11} = \frac{\cosh a_p}{4a_p^2}\left(a_p^2(b_{2p} - 2b_{1p}) - b_{2p}\right)(a_4 - a_5 \tanh a_p),$$

$$A_{12} = \frac{\cosh a_p}{4a_p}\left[\left(2b_{1p} - b_{2p}\right)(a_5 - a_4 \tanh a_p)\right], \ \delta_5 = \delta_1 - (\delta_3 - \delta_2)\cosh a_f,$$

$$\delta_6 = \delta_4 a_f \hat{T} \sinh a_f + \delta_5 a_p \sinh a_p,$$

$$\delta_7 = a_p \cosh a_f \cosh a_p \left(a_f \hat{T} \tanh a_f + a_p \tanh a_p\right), \delta_8 = \delta_4 \cosh a_f - \delta_5 \cosh a_p$$

$$\delta_9 = -\cosh a_f \cosh a_p \left(a_f \hat{T} \tanh a_f + a_p \tanh a_p\right),$$

$$\Sigma_2(z_f) = B_1[A_{17} - A_{18} + A_{19} - A_{20}], \Sigma_{2p}(z_p) = B_1[A_{21} - A_{22}],$$

$$A_{17} = \frac{\cosh a_f z_f (2b_3 z_f + b_4 z_f^2)}{4a_f}(a_1 + \tanh a_f z_f), \ A_{18} = \frac{b_4 z_f \cosh a_f z_f}{4a_f^2}(1 + a_1 \tanh a_f z_f),$$

$$A_{19} = \frac{\cosh a_f z_f}{12a_f^3}\left[(3a_f^2 z_f^2 b_3 + 2a_f^2 z_f^3 b_4 + 3b_4 z_f)(a_3 + a_2 \tanh a_f z_f)\right],$$

$$A_{20} = \frac{\cosh a_f z_f (b_3 z_f + b_4 z_f^2)}{4a_f^2}(a_2 + a_3 \tanh a_f z_f),$$

$$A_{21} = \frac{\cosh a_p z_p (2b_{3p} z_p + b_{4p} z_p^2)}{4a_p}(a_5 + a_4 \tanh a_p z_p),$$

$$A_{22} = \frac{b_{4p} z_p \cosh a_p z_p}{4a_p^2}(a_4 + a_5 \tanh a_p z_p), b_3 = R_{Af}^*, b_4 = -2(R_{Af}^* + 1), b_{3p} = -R_{Ap}^*, b_{4p} = -2(R_{Ap}^* + 1),$$

$$c_5 = c_7 \hat{T}, c_6 = \frac{1}{a_f}(c_8 a_p + \delta_{12} - \delta_{11}),\ c_7 = \frac{\delta_{17}}{\delta_{18}}, c_8 = \frac{\delta_{15}}{\delta_{16}}, \delta_{10} = -B_1\left(A_{23} + A_{24} + A_{25} + A_{26}\right)$$

$$A_{23} = \frac{\cosh a_f}{4a_f^2}\left[2a_f^2 b_3 + b_4(a_f^2 - 1)(1 + a_1 \tanh a_f)\right], A_{24} = \frac{\cosh a_f\left(b_4 + 2b_3\right)}{4a_f}(a_1 + \tanh a_f)$$

$$A_{25} = \frac{\cosh a_f}{12a_f^2}\left[3(a_f^2 - 1)b_3 + (2a_f^2 - 3)b_4\right](a_2 + a_3 \tanh a_f),$$

$$A_{26} = \frac{\cosh a_f}{4a_f^3}\left[\left(a^2 b_3 + b_4(a^2 + 1)\right)(a_3 + a_2 \tanh a_f)\right],\ \delta_{11} = \frac{B_1}{4a_f^3}\left((2a_f^2 a_1 - a_f a_2)b_3 + (a_3 - a_f)b_4\right)$$

$$\delta_{12} = \frac{B_1}{4a_p^2}\left(2a_p b_{3p} a_5 - a_4 b_{4p}\right)\ \delta_{13} = -B_1\left(A_{27} + A_{28}\right),$$

$$A_{27} = \frac{\cosh a_p}{4a_p^2}\left(a_p^2(b_{4p} - 2b_{3p}) - b_{4p}\right)(a_4 - a_5 \tanh a_p), A_{28} = \cosh a_p\left(\frac{2b_{3p} - b_{4p}}{4a_p}\right)(a_5 - a_4 \tanh a_p)$$

$$\delta_{14} = \delta_{10} - (\delta_{12} - \delta_{11})\cosh a_f, \delta_{15} = \delta_{13} a_f \hat{T} \sinh a_f + \delta_{14} a_p \sinh a_p$$

$$\delta_{16} = a_p \cosh a_f \cosh a_p\left(a_f \hat{T} \tanh a_f + a_p \tanh a_p\right),\ \delta_{17} = \delta_{13} \cosh a_f - \delta_{14} \cosh a_p,$$

$$\delta_{18} = -\cosh a_f \cosh a_p\left(a_f \hat{T} \tanh a_f + a_p \tanh a_p\right)$$

$$\Sigma_3(z_f) = B_1[A_{31} - A_{32} + A_{33} - A_{34}],\ \Sigma_{3p}(z_p) = B_1[A_{35} - A_{36}]$$

$$A_{31} = \frac{\cosh a_f z_f (2b_5 z_f + b_6 z_f^2)}{4a_f}(a_1 + \tanh a_f z_f),\ A_{32} = \frac{b_6 z_f \cosh a_f z_f}{4a_f^2}(1 + a_1 \tanh a_f z_f)$$

$$A_{33} = \frac{\cosh a_f z_f}{12a_f^3}\left[(3a_f^2 z_f^2 b_5 + 2a_f^2 z_f^3 b_6 + 3b_6 z_f)(a_3 + a_2 \tanh a_f z_f)\right],$$

$$A_{34} = \frac{\cosh a_f z_f (b_5 z_f + b_6 z_f^2)}{4a_f^2}(a_2 + a_3 \tanh a_f z_f),$$

$$A_{35} = \frac{z_p \cosh a_p z_p (2b_{5p} + b_{6p} z_p)}{4a_p}(a_5 + a_4 \tanh a_p z_p)$$

$$A_{36} = \frac{z_p b_{6p} \cosh a_p z_p}{4a_p^2}(a_4 + a_5 \tanh a_p z_p),$$

$$b_5 = R_{Af}^* - 2, b_6 = 2(1 - R_{Af}^*), b_{5p} = -2 - R_{Ap}^*, b_{6p} = 2(1 - R_{Ap}^*)$$

$$c_9 = c_{11}\hat{T}, c_{10} = \frac{1}{a_f}(c_{12} a_p + \delta_{22} - \delta_{21})\ c_{11} = \frac{\delta_{26}}{\delta_{27}}, c_{12} = \frac{\delta_{24}}{\delta_{25}}, \delta_{19} = -B_1[A_{37} + A_{38} + A_{39} + A_{40}]$$

$$A_{37}=\frac{\cosh a_f\left(2a_f^2b_5+b_6(a_f^2-1)\right)}{4a_f^2}(1+a_1\tanh a_f),\ A_{38}=\frac{\cosh a_f\left(b_6+2b_5\right)}{4a_f}(a_1+\tanh a_f)$$

$$A_{39}=\frac{\cosh a_f\left(3(a_f^2-1)b_5+(2a_f^2-3)b_6\right)}{12a_f^2}(a_2+a_3\tanh a_f),$$

$$A_{40}=\frac{\cosh a_f\left(a_f^2b_5+b_6(a_f^2+1)\right)}{4a_f^3}(a_3+a_2\tanh a_f),$$

$$\delta_{20}=-B_1[A_{41}+A_{42}],A_{41}=\cosh a_p\left(\frac{b_{6p}(a_p^2-1)-2b_{5p}}{4a_p^2}\right)(a_4-a_5\tanh a_p)$$

$$A_{42}=\frac{\left(2b_{5p}-b_{6p}\right)\cosh a_p}{4a_p}(a_5-a_4\tanh a_p),$$

$$\delta_{21}=B_1\left(\frac{(2a_f^2a_1-a_fa_2)b_5+(a_3-a_f)b_6}{4a_f^3}\right),\delta_{22}=B_1\left(\frac{2a_pb_{5p}a_5-a_4b_{6p}}{4a_p^2}\right),$$

$$\delta_{23}=\delta_{19}-(\delta_{22}-\delta_{21})\cosh a_f,\delta_{24}=\delta_{20}a_f\hat{T}\sinh a_f+\delta_{23}a_p\sinh a_p,$$

$$\delta_{25}=a_p\cosh a_f\cosh a_p\left[a_f\hat{T}\tanh a_f+a_p\tanh a_p\right],\ \delta_{26}=\delta_{20}\cosh a_f-\delta_{23}\cosh a_p,$$

$$\delta_{27}=-\cosh a_f\cosh a_p\left[a_f\hat{T}\tanh a_f-a_p\tanh a_p\right].$$

4.5 RESULTS AND DISCUSSION

Double-diffusive convection in the presence of temperature profiles and a constant heat source/sink, Marangoni convection, is caused by an unstable density distribution with the surface tension effect and is investigated in a two-layer system. To execute the numerical calculations and make graphs, the Mathematica software is utilized. Because the fluids are immiscible, surface tension plays a significant effect. For various values, the influence of the thermal Marangoni number (tMn) versus the depth ratio $\hat{d}$ is depicted. The three different temperature profiles considered are linear, parabolic, and inverted parabolic profiles, with M_{t1}, M_{t2} and M_{t3} being the tMn's. The influence of porous parameter β, modified internal Rayleigh number (mRn) for porous region R^*_{Ap}, solute Marangoni number (sMn) M_s, solute diffusivity ratio $\hat{S}$, solute thermal diffusivity ratio of the fluid in the fluid region τ_f, and solute thermal diffusivity ratio of the fluid in the porous region τ_p on tMn's when $a_f=0.5$, $M_s=5$, $\hat{S}=0.5$, $\hat{T}=0.4$, $\tau_f=0.5,\tau_p=0.5$, $R^*_{Af}=R^*_{Ap}=1$, and $\beta=0.1$ are shown in Figures 4.2 to 4.7. The upward arrows in the figures denote the stability of the system, whereas the downward arrows denote the destability of the system.

The effect of β, as seen in Figure 4.2 (a, b, and c), is to stabilize the system. Physically, this means that as permeability increases, more space is available for fluid flow, delaying convection. This could be owing to the presence of a second diffusing component. For all three profiles, it is observed that that M_{t1}, M_{t2}, and M_{t3} increase with $\hat{d}$. Another note is the increase in depth ratio $\hat{d}$, indicating the porous layer's dominance, beyond the equal depths of both the layers. Furthermore, when the porous layer dominates the fluid layer, the inverted parabolic profile can be used to achieve higher stability. Figure 4.3 (a, b, c) displays the impact of mRn R^*_{Ap} for porous regions on tMn's for the three heat profiles. Very nominal values of this parameter are chosen so that the Bènard double-diffusive Marangoni convection will not supersede the effects of the heat sources. As R^*_{Ap} increases, the profiles are increased, and the onset convection is delayed as a result of

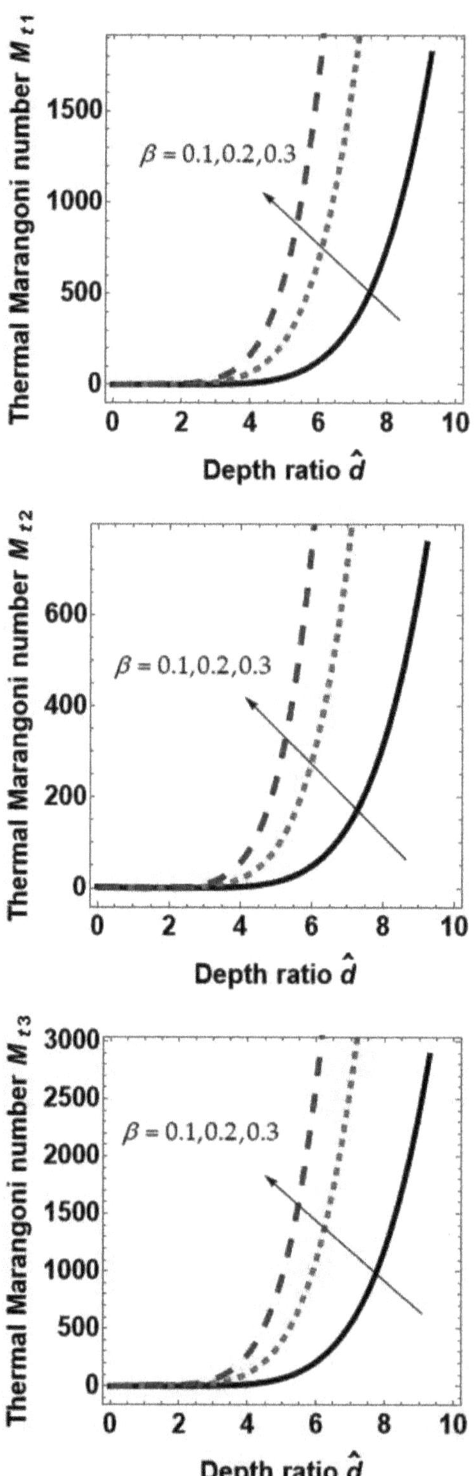

FIGURE 4.2 Impact of tMn's versus $\hat{d}$ for $\beta = 0.1, 0.2, 0.3$ when $a_f = 0.5$, $M_s = 5$, $\hat{S} = 0.5$, $\hat{T} = 0.4$, $\tau_f = 0.5, \tau_p = 0.5$, $R_{Af}^* = R_{Ap}^* = 1$.

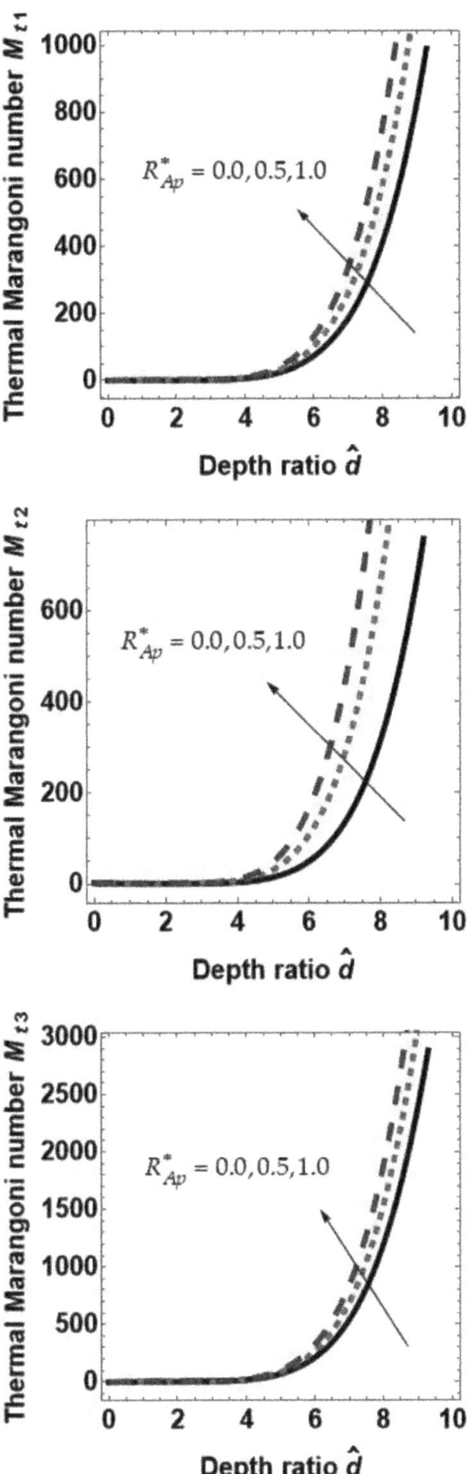

FIGURE 4.3 Impact of tMn's versus $\hat{d}$ for $R^*_{Ap} = 0.0, 0.5, 1.0$ when $a_f = 0.5$, $M_s = 5$, $\hat{S} = 0.5$, $\hat{T} = 0.4$, $\tau_f = 0.5, \tau_p = 0.5$, $R^*_{Af} = 1$, and $\beta = 0.1$.

this. Physically, this suggests that the heat source's strength has increased. This may be due to the presence of the second diffusing component, which raises the tMn. This impact can be seen regardless of temperature profiles. Figure 4.4 (a, b, c) shows the plots of tMn's M_{t1}, M_{t2}, and M_{t3} for various values of sMn M_s. For all profiles, the tMn's fall as M_s increases. This aids in the

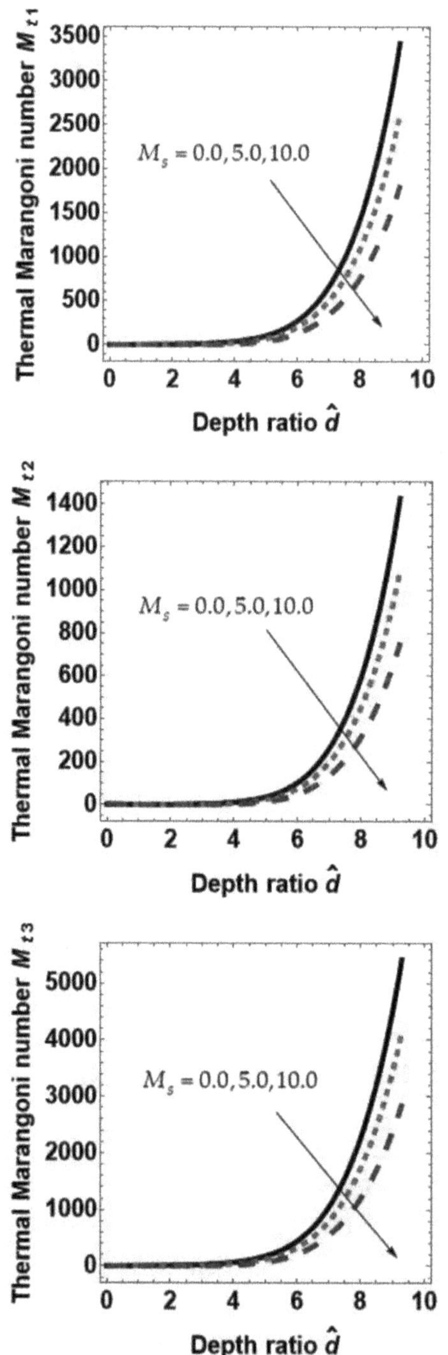

FIGURE 4.4 Impact of tMn's versus $\hat{d}$ for $M_s = 0.0, 5.0, 10.0$ when $a_f = 0.5$, $\hat{S} = 0.5$, $\hat{T} = 0.4$, $\tau_f = 0.5, \tau_p = 0.5$, $R^*_{Af} = R^*_{Ap} = 1$, and $\beta = 0.1$.

speeding up of convection. As a result, the system has become unstable. In the porous zone, we can see that liquid convective circulation is more prevalent. In Figure 4.5 (a, b, c), the effect of $\hat{S}$ is to increase the tMn's regardless of the temperature profiles. This suggests that $\hat{S}$ helps keep the system in control. The temperature profiles in the fluid layer increase when the solute thermal

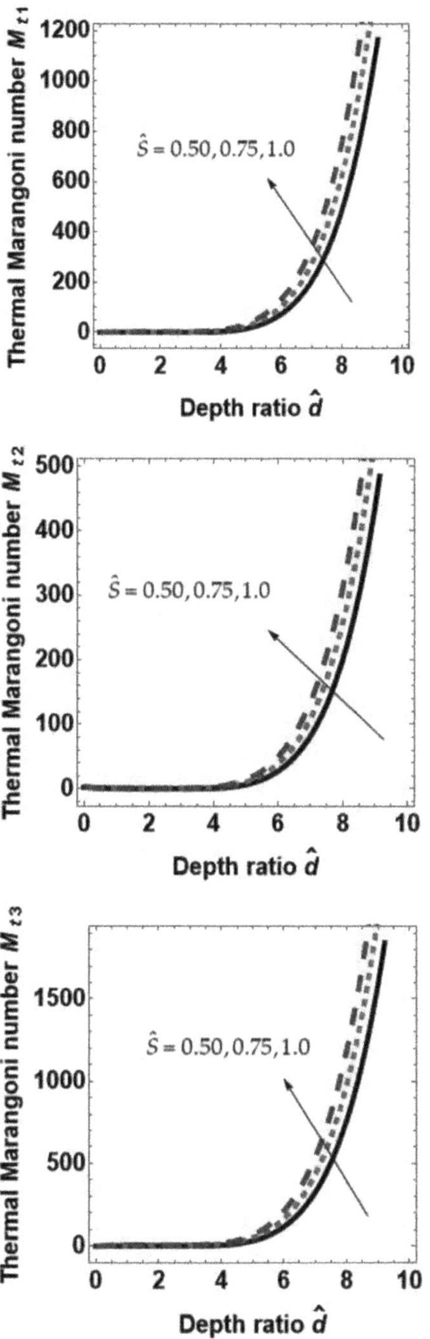

FIGURE 4.5 Impact of tMn's versus $\hat{d}$ for $\hat{S} = 0.50, 0.75, 1.0$ when $a_f = 0.5$, $M_s = 5$, $\hat{T} = 0.4$, $\tau_p = 0.5$, $\tau_p = 0.5$, $R^*_{Af} = R^*_{Ap} = 1$, and $\beta = 0.1$.

diffusivity ratio τ_f of the fluid increases, as illustrated in Figure 4.6 (a, b, c). As a result, it has the effect of stabilizing the system. In Figure 4.7 (a, b, c), it is obvious that the action of τ_p is to lower the values of tMn's, which helps speed up the commencement of convection and thereby destabilizes the system.

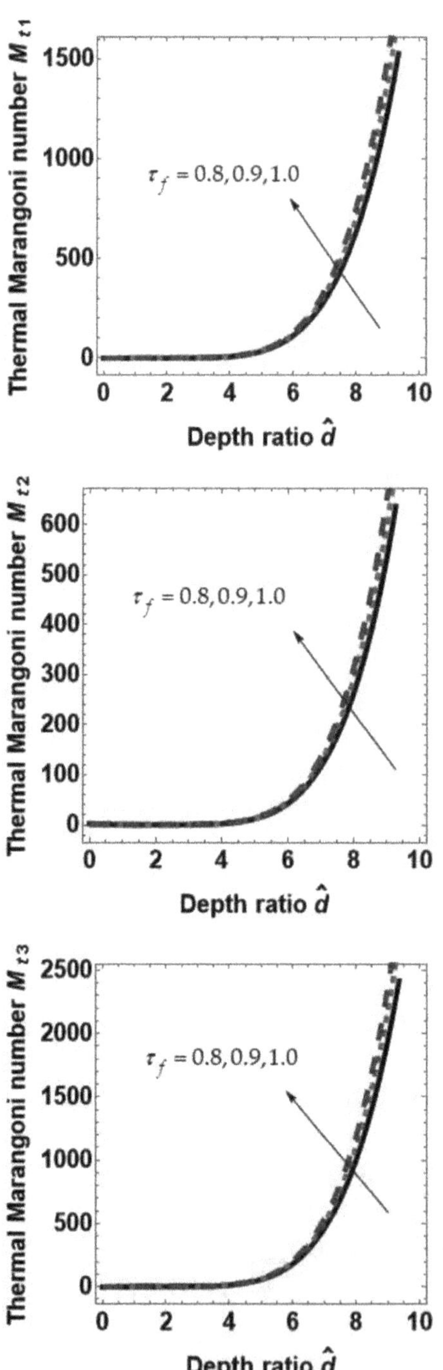

FIGURE 4.6 Impact of tMn's versus $\hat{d}$ for $\tau_f = 0.8, 0.9, 1.0$ when $a_f = 0.5$, $M_s = 5$, $\hat{S} = 0.5$, $\hat{T} = 0.4$, $\tau_p = 0.5$, $R^*_{Af} = R^*_{Ap} = 1$, and $\beta = 0.1$.

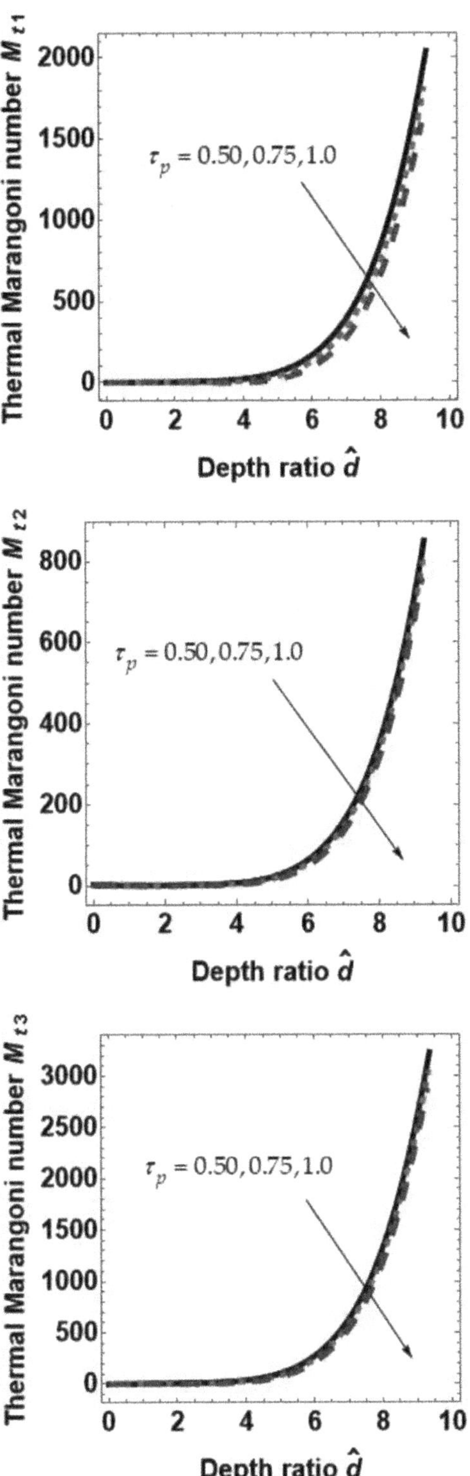

FIGURE 4.7 Impact of tMn's versus $\hat{d}$ for $\tau_p = 0.50, 0.75, 1.0$ when $a_f = 0.5$, $M_s = 5$, $\hat{S} = 0.5$, $\hat{T} = 0.4$, $\tau_f = 0.5$, $R^*_{Af} = R^*_{Ap} = 1$, and $\beta = 0.1$.

4.6 CONCLUSION

In this chapter, an attempt is made to investigate the effect of constant heat source/sink and non-uniform temperature gradients on double-diffusive natural in a composite system. With this extensive analysis of double-diffusive Marangoni (DDM) convection in a composite layer containing heat sources, subjected to uniform and non-uniform temperature profiles, when DDM convection needs to be enhanced, parabolic for moderate convection, and linear for double-diffusive convection that needs to be delayed, an inverted parabolic temperature profile might be used. In the porous dominant composite layer, the tMn rapidly increases. Controlling DDM convection using solute and thermal diffusivity parameters is possible. The beginning of DDM convection can be augmented or delayed by altering the strength of the heat sources in both layers of the composite layer. The inverted parabolic TP on DDM convection in a composite system is the most stable of all the three temperature profiles; the linear profile is the most unstable one, and the parabolic profile is the moderate one. This research will undoubtedly be useful in a variety of applications in crystal growth, engineering, geophysics, climatology, and astronomy.

Nomenclature

a_f, a_p	horizontal wave number	T_0	interface temperature
C_f, C_p	concentration	$\vec{q}_f, \vec{q}_p$	velocity vectors
C_v	specific heat	$V_f(z_f), V_p(z_p)$	velocities in fluid and porous layer
C_0	interface concentration		
$\hat{d}$	depth ratio	$\Lambda = \dfrac{(\rho_0 C_v)_f}{(\rho_0 C_v)_p}$	ratio of heat capacities
d_f, d_p	thickness of fluid and porous region		
K	permeability of the porous medium	β	porous parameter
$l_f, m_f \; l_p, m_p$	wave numbers in x- and y-direction	ϕ	porosity
M_s	solute Marangoni number	κ_f, κ_p	thermal diffusivity of the fluid and porous region
M_t	thermal Marangoni number		
M_{t1}, M_{t2}, M_{t3}	thermal Marangoni numbers for linear, parabolic, and inverted parabolic temperature profiles	κ_{cf}, κ_{cp}	solute diffusivity of the fluid and porous region
n_f, n_p	frequencies	μ_f	fluid viscosity
P_{rf}, P_{rp}	Prandtl numbers in fluid and porous layer	μ_p	effective viscosity of the fluid in the porous layer
P_f, P_p	pressure in fluid and porous region	$\hat{\mu}$	viscosity ratio
		ρ_0	fluid density
Q_h, Q_{hp}	heat source in fluid and porous region		
R^*_{Af}, R^*_{Ap}	modified internal Rayleigh numbers for fluid and porous region	τ_f, τ_p	solute thermal diffusivity ratio of the fluid and porous region
R_{if}, R_{ip}	internal Rayleigh number for fluid and porous region	$\theta_f(z_f),\ \theta_p(z_p)$	temperature distributions in fluid and porous region

$S_f(z_f), S_p(z_p)$	concentration distributions in fluid and porous region	$\chi(z_f), \chi(z_p)$	temperature profiles
		f	suffix denotes fluid region
$\hat{S}$	solute diffusivity ratio	p	suffix denotes porous region
$\hat{T}$	thermal ratio	b	suffix denotes basic state
T_f, T_p	temperature	σ_t	surface tension

REFERENCES

1. Bennacer, R., Beji, H., and Mohamad, A. A. (2003). Double diffusive convection in a vertical enclosure inserted with two saturated porous layers confining a fluid layer. *International Journal of Thermal Sciences*, 42(2), 141–151.
2. Chen, C. F., and Chan, Cho Lik. (2010). Stability of buoyancy and surface tension driven convection in a horizontal double-diffusive fluid layer. *International Journal of Heat and Mass Transfer*, 53(7–8), 1563–1569.
3. Sivasankaran, S., Do, Y., and Sankar, M. (2011). Effect of discrete heating on natural convection in a rectangular porous enclosure. *Transport in Porous Media*, 86, 261–281.
4. Sankar, M., Park, Y., Lopez, J. M., and Do, Y. (2012). Double-diffusive convection from a discrete heat and solute source in a vertical porous annulus. *Transport in Porous Media*, 91, 753–775.
5. Gangadharaiah, Y. H., and Suma, S. P. (2013). Bernard-marangoni convection in a fluid layer overlying a layer of an anisotropic porous layer with deformable free surface. *Advanced Porous Materials*, 1(2), 229–238.
6. Chen, S., Tlke, J., and Krafczyk, M. (2014). Numerical investigation of double-diffusive (natural) convection in vertical annuluses with opposing temperature and concentration gradients. *International Journal of Heat and Fluid Flow*, 31(2), 217–226.
7. Saleem, M., Hossain, M. A., and Saha, S. C. (2014). Double diffusive Marangoni convection flow of electrically conducting fluid in a square cavity with chemical reaction. *Journal of Heat Transfer*, 136(6), 1–9.
8. Sumithra, R. (2014). Double diffusive magneto Marangoni convection in a composite layer. *International Journal of Application or Innovation in Engineering & Management*, 3(2), 12–25.
9. Corcione, M., Grignaffini, S., and Quintino, A. (2015). Correlations for the double-diffusive natural convection in square enclosures induced by opposite temperature and concentration gradients. *International Journal of Heat and Mass Transfer*, 81, 811–819.
10. Arbin, N., Suhaimi, N. S., and Hashim, I. (2016). Simulation on double-diffusive Marangoni convection with the presence of entropy generation. *Indian Journal of Science and Technology*, 9(31), 1–5.
11. Harfash, A. J., and Nashmi, F. K. (2017). Triply resonant double diffusive convection in a fluid layer. *Mathematical Modelling and Analysis*, 22(6), 809–826.
12. Mebarek-Oudina, F. (2017). Numerical modeling of the hydrodynamic stability in vertical annulus with heat source of different lengths. *Engineering Science and Technology, an International Journal*, 20(4), 1324–1333.
13. Girish, N., Makinde, O. D., and Sankar, M. (2018). Numerical investigation of developing natural convection in vertical double-passage porous annuli. *Defect and Diffusion Forum*, 387, 442–460.
14. Lyubimova, T., and Kolchanova, E. (2018). The onset of double-diffusive convection in a superposed fluid and porous layer under high-frequency and small-amplitude vibrations. *Transport in Porous Media*, 122(11), 97–124.
15. Kanchana, C., Zhao, Y., and Siddheshwar, P. G. (2020). Küppers–Lortz instability in rotating Rayleigh–Bénard convection bounded by rigid/free isothermal boundaries. *Applied Mathematics and Computation*, 385, 125406.
16. Dadheech, P. K., Agrawal, P., Mebarek-Oudina, F., Abu-Hamdeh, N. H., and Sharma, A. (2020). Comparative heat transfer analysis of $MoS_2/C_2H_6O_2$ and SiO_2-$MoS_2/C_2H_6O_2$ nanofluids with natural convection and inclined magnetic field. *Journal of Nanofluids*, 9(3), 161–167.
17. Sumithra, R., Vanishree, R. K., and Manjunatha, N. (2020). Effect of constant heat source/sink on single component Marangoni convection in a composite layer bounded by adiabatic boundaries in presence of uniform & non uniform temperature gradients. *Malaya Journal of Matematik*, 8(2), 306–313.

18. Vanishree, R. K., Sumithra, R., and Manjunatha, N. (2020). Effect on uniform and non uniform temperature gradients on Benard-Marangoni convection in a superposed fluid and porous layer in the presence of heat source. *Gedrag & Organisatie Review*, 33(2), 746–758.
19. Mebarek-Oudina, F., Keerthi Reddy, N., and Sankar, M. (2021). Heat source location effects on buoyant convection of nanofluids in an annulus. In: Rushi, K. B., Sivaraj, R., and Prakash, J. (eds) *Advances in Fluid Dynamics. Lecture Notes in Mechanical Engineering*. Springer, Singapore. https://doi.org/10.1007/978-981-15-4308-1_70
20. Swain, K., Mahanthesh, B., and Mebarek-Oudina, F. (2021). Heat transport and stagnation-point flow of magnetized nanoliquid with variable thermal conductivity, Brownian moment, and thermophoresis aspects. *Heat Transfer*, 50(1), 754–767.
21. Warke, A. S., Ramesh, K., and Mebarek-Oudina, F. (2021). Numerical investigation of the stagnation point flow of radiative magnetomicropolar liquid past a heated porous stretching sheet. *Journal of Thermal Analysis and Calorimetry*, 147, 6901–6912. https://doi.org/10.1007/s10973-021-10976-z
22. Pushpa, B. V., Sankar, M., and Mebarek-Oudina, F. (2021). Buoyant convective flow and heat dissipation of Cu–H_2O nanoliquids in an annulus through a thin baffle. *Journal of Nanofluids*, 10(2), 292–304.
23. Marzougui, S., Mebarek-Oudina, F., Magherbi, M., and Mchirgui, A. (2021). Entropy generation and heat transport of Cu–water nanoliquid in porous lid-driven cavity through magnetic field. *International Journal of Numerical Methods for Heat & Fluid Flow*, 32(6), 2047–2069. https://doi.org/10.1108/HFF-04-2021-0288.
24. Rajashekhar, C., Mebarek-Oudina, F., Vaidya, H., Prasad, K. V., Manjunatha, G., and Balachandra, H. (2021). Mass and heat transport impact on the peristaltic flow of a Ree–Eyring liquid through variable properties for hemodynamic flow. *Heat Transfer*, 50(5), 5106–5122.
25. Djebali, R., Mebarek-Oudina, F., and Rajashekhar, C. (2021). Similarity solution analysis of dynamic and thermal boundary layers: Further formulation along a vertical flat plate. *Physica Scripta*, 96, 085206.
26. Dhif, K., Mebarek-Oudina, F., Chouf, S., Vaidya, H., and Chamkha, A.J. (2021). Thermal analysis of the solar collector cum storage system using a hybrid-nanofluids. *Journal of Nanofluids*, 10, 616–626.
27. Manjunatha, N., and Sumithra, R. (2021).Non-Darcian-Bènard double diffusive magneto-Marangoni convection in a two layer system with constant heat source/sink. *Iraqi Journal of Science*, 62(11), 4039–4055.
28. Manjunatha, N., and Sumithra, R. (2021). Influence of vertical magnetic field and nonuniform temperature gradients on double component Marangoni convection in a two-layer system in the presence of variable heat source/sink. *Indian Journal of Natural Science*, 12(67), 32715–32732.
29. Manjunatha, N., Sumithra, R., and Vanishree, R. K. (2021). Combined effects of nonuniform temperature gradients and heat source on double diffusive Bènard-Marangoni convection in a porous-fluid system in the presence of vertical magnetic field. *International Journal of Thermofluid Science and Technology*, 8(1), paper no: 080104.
30. Shafiq, A., Mebarek-Oudina, F., Sindhu, T. N., and Rasool, G. (2022). Sensitivity analysis for Walters' B nanoliquid flow over a radiative Riga surface by RSM. *Scientia Iranica*, 29(3), 1236–1249.
31. Asogwa, K. K., Mebarek-Oudina, F., and Animasaun, I. L. (2022). Comparative investigation of water-based al_2o_3 nanoparticles through water-based CuO nanoparticles over an exponentially accelerated radiative Riga plate surface via heat transport. *Arabian Journal for Science and Engineering*, 47, 8721–8738. https://doi.org/10.1007/s13369-021-06355-3.
32. Swain, K., Mebarek-Oudina, F., and Abo-Dahab, S. M. (2022). Influence of $MWCNT/Fe_3O_4$ hybrid nanoparticles on an exponentially porous shrinking sheet with chemical reaction and slip boundary conditions, *Journal of Thermal Analysis and Calorimetry*, 147, 1561–1570.
33. Sumithra, R., Komala, B., and Manjunatha, N. (2022).The onset of Darcy-Brinkman-Rayleigh-Benard convection in a composite system with thermal diffusion. *Heat Transfer*, 51(1), 604–620.
34. Venkatachalappa, M., Prasad, V., Shivakumara, I. S., and Sumithra, R. (1997). Hydrothermal growth due to double diffusive convection in composite materials. *Proceedings of 14th National Heat and Mass Transfer Conference and 3rd ISHMTASME Joint Heat and Mass Transfer Conference*, December 29–31, Narosa, New Delhi.
35. Shivakumara, I. S., Suma, S. P., and Krishna, C. B. (2006). Onset of surface-tension-driven convection in superposed layers of fluid and saturated porous medium. *Archives of Mechanics*, 58(1), 71–92.
36. Shivakumara, I. S., Suresh kumar, S., and Devaraju, N. (2012). Effect of non-uniform temperature gradients on the onset of convection in a couple-stress fluid-saturated porous medium. *Journal of Applied Fluid Mechanics*, 5(1), 49–55.

37. Nield, D. A. (1977). Onset of convection in a fluid layer overlying a layer of a porous medium. *Journal of Fluid Mechanics*, 81, 513. https://doi.org/10.1017/S0022112077002195
38. Chen, F. (1990). Through flow effects on convective instability in superposed fluid and porous layers. *Journal of Fluid Mechanics*, 23, 113–133.
39. Chen, F., and Chen, C. F. (1988). Onset of finger convection in a horizontal porous layer underlying a fluid layer. *Journal of Heat Transfer*, 110, 403–409. https://doi.org/10.1115/1.3250499
40. Sumithra, R. (2012). Mathematical modeling of hydrothermal growth of crystals as double diffusive convection in composite layer bounded by rigid walls. *International Journal of Engineering Science and Technology*, 4(2), 779–791.
41. Sparrow, E. M., Goldstein. R. J., and Jonson, V. K. (1964). Thermal instability in a horizontal fluid layer effect of boundary conditions and non-linear temperature profile. *Journal of Fluid Mechanics*, 18, 513.
42. Shivakumara, I. S., Rudraiah, N., and Nanjundappa, C. E. (2002). Effect of non-uniform basic temperature gradient on Rayleigh-Benard-Marangoni convection in ferrofluids. *Journal of Magnetism and Magnetic Materials*, 248, 379–395.

5 Entropy and MHD Effects on Ciliated Flow of a Williamson Fluid

Arshad Riaz

5.1 INTRODUCTION

Cilia play key role in most of the diseases. One prominent function of cilia is to push liquids over an epithelial sheet. Cilia-executed stream is always directional and spreads over large paths corresponding to a cell. Such type of flow can help in moving large rates of fluid transport and can also act as a long-sided guided signal. In human immobilized ciliary diseases such as primary ciliary dyskinesia (PCD), loss of ciliary flow prevents the transport of mucus through the airways, Causes bronchiectasis and chronic sinusitis. In female reproductive tube, ciliated cells generate flow that pushes the egg from the tube to the uterus. Detraction can cause misplacement of the egg, which may be the source of deranged pregnancies. In the brain, cilia also help in producing long-range transport of cerebrospinal fluid (CSF) inside the ventricles. One type of cilia, called motile cilia, which are phagosomes that stick out from cells and produce track, flows with small Reynolds number [1] having speeds of almost ~1–1,000 μm/s [2–5]. In the process of embryo growth, cilia-oriented fluid stream is necessary to typical left–right line forming [2]. In addition to this, evidently, cilia-type liquid flow is essential for the maturing of the inner ear [6] and central neural mechanism [7]. Several epithelial tissue types contain motile cilia, which include bronchial, mesothelium, and uterine tube epithelium [8]. The respiratory tract drives the directional flow of mucus, and defects in these cilia can lead to recurrent sinus pulmonary infections [9]. The fallopian tube cilia that line the luminal epithelium are important in the transport of the fallopian tubes [10]. Therefore, understanding the fluid flow from the ciliary physiology and the cilia is broadly relevant in medicine.

All the aforementioned studies only incorporate the viscous fluids and do not deal with the rheological properties of various types of non-Newtonian fluids. As we understand that most of the fluids in the human body and in industrial instruments are non-Newtonian in nature, showing different nonlinear viscosity characteristics, it is very important to work on the non-Newtonian models. In this regard, many researchers and scientists have shown their concern with the study of non-Newtonian fluid encountered through ciliary-generated streams. Ramesh et al. [11] have presented the flow attributes of couple stress model in a cilia-oriented channel, and they have included the effects of hydromagnetic pumping. They have handled complicated governing equations along with complex ciliary flow conditions by lubrication theory and achieved analytical solutions. They have produced the results that axial velocity varies inversely along the rising impact of cilia length, but axial pressure slope increases with the same factor. The study can be related to the magneto-hydrodynamic (MHD) biomimetic blood pumping machines. Akbar and Butt [12] have considered the heat transfer phenomenon through cilia-executed motion of fluid in human organs by considering a non-Newtonian stress model (Jeffrey). They assumed a finite-length tube as a conduit and observed the metachronal waves produced by the cilia movement. A similarity transformation pattern is employed for modelling and simplification. After successful derivation of the model, they produced exact value expressions for velocity and other key properties of the flow. Their key finding

 DOI: 10.1201/9781003299608-5

is that pressure rise is a direct function of cilia length parameter. Bhatti et al. [13] have presented multiphase fluid flow due to cilia presence in a porous planar conduit. They investigated the result that fluid travels slowly in the presence of large particle volume fraction and also due to opposing effects of magnetic field. Later on, Sadaf and Nadeem [14] have investigated theoretically the cilia-type motion of a substance with heating strategy in a curved geometry where magnetic field is also assumed vertically to the axis of the flow. The study tells us that velocity improves in left-sided cilia boundary by varying the radial magnetic field. McCash et al. [15] developed a new geometrical model for cilia flow scheme in an elliptical duct and also considered the effects of thermal variation in the transport and claimed that over a wide range of channel, the velocity profile and temperature function get distributed and both become minimum near the cilia surfaces.

After carefully noticing the preceding literature, we come to know that none of the studies discussed the channel effects when it is inclined at some angle with the basic axis. In the current chapter, the authors are keen to investigate a cilia-oriented flow of a non-Newtonian fluid (Williamson) in a channel which is inclined with some positive slope. Due to a wide range of applications in industry and medical field, magnetohydrodynamics and heat exchange effects are considered, too, along with an emerging discussion of entropy generation due to thermal and viscous attributes. On the parallel side, velocity and thermal slip conditions are imposed on the ciliary walls to make the study more practical and valid in real-life mechanism. The modelling of the problem is made regularized by injecting the coordinates into a wave frame system, and then a similarity transformation technique is utilized to produce a simple and comprehensive model relating the physical phenomenon. The finally achieved set of relations has been solved proficiently by a well-recognized perturbation approach. After getting expressions of velocity and temperature from the solution, the entropy expressions and Bejan number are also stated. In the end, the summary of the result is made on the basis of graphical findings made from the plots of various physical quantities.

5.2 MATHEMATICAL FORMULATION

Assume a magnetohydrodynamic pumping stream of a non-Newtonian Williamson fluid in a symmetric channel which is inclined at an angle ω with cilia walls under the impact of a uniform magnetic field and heat exchange. It is also considered that the transport of the liquid is driven because of rhythmic cilia beatings (with fixed speed c) that adjust the beats to produce the metachronal wave's structure along the surfaces of the conduit (see Figure 5.1).

Using a rectangular coordinate system, the model of ciliated wall in fixed frame is described as:

$$\bar{Y} = \bar{H}\left(\bar{X},t\right) = a + a\varepsilon\cos\left(\frac{2\pi}{\lambda}\left(\bar{X}-ct\right)\right) \tag{5.1}$$

The elliptical motion structure is assumed by which cilia tips move and are orthogonally located at:

$$\bar{X} = \phi\left(\bar{X},t\right) = X_0 + a\varepsilon\alpha\sin\left(\frac{2\pi}{\lambda}\left(\bar{X}-ct\right)\right), \tag{5.2}$$

Where a, α, ε, $\bar{H}$, t, λ, and X_0 stand for mean width of the geometry, eccentricity of the conical path, length representing parameter for cilia, half of the total width of the channel, time parameter, wavelength, and prescribed location of the particle.

The tangential and vertical velocity ingredients at the walls of the channel, along with the governing laws, are orderly structured as [16]:

$$U_0 = \left(\frac{\partial\bar{X}}{\partial t}\right)_{x0} = \frac{-\left(\frac{2\pi}{\lambda}\right)ac\varepsilon\alpha\cos\left(\frac{2\pi}{\lambda}\left(\bar{X}-ct\right)\right)}{1-\left(\frac{2\pi}{\lambda}\right)ac\varepsilon\alpha\cos\left(\frac{2\pi}{\lambda}\left(\bar{X}-ct\right)\right)}, \tag{5.3}$$

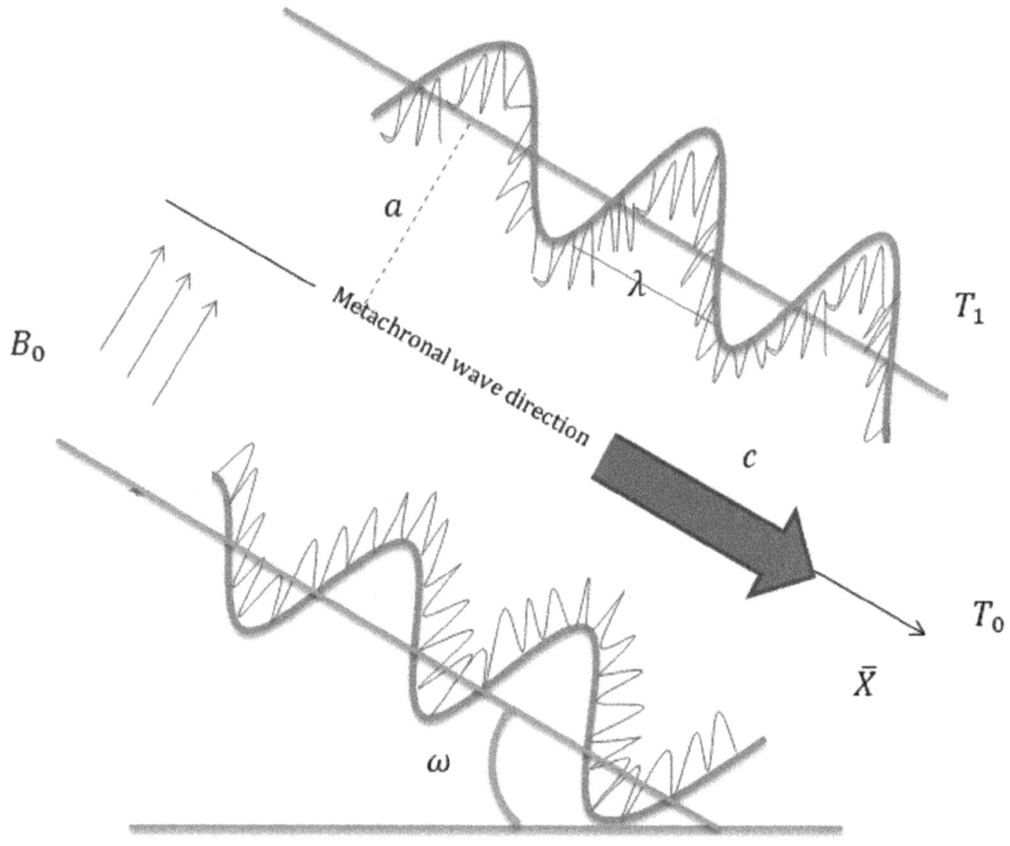

FIGURE 5.1 The schematic diagram of the ciliated inclined channel.

$$V_0 = \left(\frac{\partial \bar{Y}}{\partial t}\right)_{x0} = \frac{\left(\frac{2\pi}{\lambda}\right) ac\varepsilon \sin\left(\frac{2\pi}{\lambda}\left(\bar{X} - ct\right)\right)}{1 - \left(\frac{2\pi}{\lambda}\right) a\varepsilon\alpha \cos\left(\frac{2\pi}{\lambda}\left(\bar{X} - ct\right)\right)}, \tag{5.4}$$

$$\frac{\partial \bar{U}}{\partial \bar{X}} + \frac{\partial \bar{V}}{\partial \bar{Y}} = 0, \tag{5.5}$$

$$\rho\left(\frac{\partial}{\partial t} + \bar{U}\frac{\partial}{\partial \bar{X}} + \bar{V}\frac{\partial}{\partial \bar{Y}}\right)\bar{U} = -\frac{\partial \bar{P}}{\partial \bar{X}} + \frac{\partial \bar{\mathcal{L}}_{\overline{XX}}}{\partial \bar{X}} + \frac{\partial \bar{\mathcal{L}}_{\overline{XY}}}{\partial \bar{Y}} - \sigma B_0^2 \bar{U} + \rho g \sin(\omega), \tag{5.6}$$

$$\rho\left(\frac{\partial}{\partial t} + \bar{U}\frac{\partial}{\partial \bar{X}} + \bar{V}\frac{\partial}{\partial \bar{Y}}\right)\bar{V} = -\frac{\partial \bar{P}}{\partial \bar{Y}} + \frac{\partial \bar{\mathcal{L}}_{\overline{XY}}}{\partial \bar{X}} + \frac{\partial \bar{\mathcal{L}}_{\overline{YY}}}{\partial \bar{Y}} - \rho g \cos(\omega), \tag{5.7}$$

$$\rho\zeta\left(\frac{\partial}{\partial t} + \bar{U}\frac{\partial}{\partial \bar{X}} + \bar{V}\frac{\partial}{\partial \bar{Y}}\right)\bar{T} = \kappa\left(\frac{\partial^2 \bar{T}}{\partial \bar{X}^2} + \frac{\partial^2 \bar{T}}{\partial \bar{Y}^2}\right) + \bar{\mathcal{L}}_{\overline{XX}}\frac{\partial \bar{U}}{\partial \bar{X}} + \bar{\mathcal{L}}_{\overline{XY}}\left(\frac{\partial \bar{U}}{\partial \bar{Y}} + \frac{\partial \bar{V}}{\partial \bar{X}}\right) + \bar{\mathcal{L}}_{\overline{YY}}\frac{\partial \bar{V}}{\partial \bar{Y}}, \tag{5.8}$$

With P as the pressure, ρ the fixed density of the fluid, B_0 the applied magnetic field strength, ω the angle of channel inclination, ζ the specific heat capacity, κ the thermal conductivity, and T suggests the temperature distribution in the flow.

The stress matrix for Williamson fluid is simplified as [17]:

$$\bar{\mathcal{L}} = \mu_0\left(1+\Gamma\left|\bar{\dot{\gamma}}\right|\right)\bar{\dot{\gamma}} \tag{5.9}$$

Incorporating the following wave frame transformations for conversion of coordinates $(\bar{X},\bar{Y})$ into $(\bar{x},\bar{y})$:

$$\bar{x} = \bar{X} - ct, \bar{y} = \bar{Y}, \bar{u} = \bar{U} - c, \bar{v} = \bar{V}, \bar{p}(x,y) = \bar{P}(\bar{X},\bar{Y},t), \bar{H}(x,t) = \bar{h}. \tag{5.10}$$

The non-dimensional quantities are introduced as:

$$x = \frac{\bar{x}}{\lambda}, y = \frac{\bar{y}}{a}, u = \frac{\bar{u}}{c}, v = \frac{\lambda\bar{v}}{ac}, h = \frac{\bar{h}}{a}, p = \frac{a^2\bar{p}}{\mu c\lambda}, \mathcal{L} = \frac{a}{\mu_0 c}\bar{\mathcal{L}}, \theta = \frac{\bar{T}-T_0}{T_1-T_0}. \tag{5.11}$$

Using equations (5.10) and (5.11) into equations (5.5)–(5.9) under certain dynamical constraints [17], one faces:

$$\frac{dp}{dx} = \frac{\partial}{\partial y}\left(\frac{\partial u}{\partial y} + We\left(\frac{\partial u}{\partial y}\right)^2\right) - M^2(1+u) + \frac{Re}{Fr}\sin(\omega), \tag{5.12}$$

$$\frac{\partial^2\theta}{\partial y^2} = -Br\left(\left(\frac{\partial u}{\partial y}\right)^2 + We\left(\frac{\partial u}{\partial y}\right)^3\right). \tag{5.13}$$

The dimensionless velocity-slip and thermal-slip boundary conditions [16] are defined as:

$$\frac{\partial u}{\partial y} = 0, \quad \frac{\partial\theta}{\partial y} = 0 \quad at \quad y = 0, \tag{5.14}$$

$$u + \xi\left(\frac{\partial u}{\partial y} + \frac{1}{6}\left(\frac{\partial u}{\partial y}\right)^3\right) = -1 - \frac{2\pi\alpha\epsilon\beta\cos(2\pi x)}{1-2\pi\alpha\epsilon\beta\cos(2\pi x)}, \theta + \Omega\frac{\partial\theta}{\partial y} = 1 \; at \; y = h = 1+\varepsilon, \tag{5.15}$$

Where ξ and Ω reflect the velocity and heat slip features. The pressure per unit λ and dimensionless volume flow rate in the fixed (Q) and wave frame (F) are defined as [13]:

$$\Delta P(\lambda) = \int_0^1\left(\frac{dp}{dx}\right)dx, \quad Q = 1 + F. \tag{5.16}$$

5.3 THE PERTURBATION METHODOLOGY AND FINAL SOLUTIONS

In order to find solutions of the mixed-form nonlinear equations (5.12) and (5.13) **a**longside the nonlinear surface constraints defined in equations (5.14) and (5.15), we employ the scheme of perturbation by assuming $We << 1$ as perturbation factor. We take the series of the flow factors up to the second order as:

$$\left(u,\frac{dp}{dx},\theta\right)=\left(\sum_{m=0}^{\infty}u_m We^m,\sum_{m=0}^{\infty}\left(\frac{dp}{dx}\right)_m We^m,\sum_{m=0}^{\infty}\theta_m We^m\right). \tag{5.17}$$

Manipulating the preceding series in governing equations, we collect the subsequent systems.

5.3.1 Zeroth-Order System

$$\left(\frac{dp}{dx}\right)_0=\frac{\partial^2 u_0}{\partial y^2}-M^2\left(1+u_0\right)+\frac{Re}{Fr}\sin\left(\xi\right), \tag{5.18}$$

$$\frac{\partial^2\theta_0}{\partial y^2}=-Br\left(\frac{\partial u_0}{\partial y}\right)^2, \tag{5.19}$$

$$\frac{\partial u_0}{\partial y}=0,\frac{\partial\theta_0}{\partial y}=0\ \text{at}\ y=0, \tag{5.20}$$

$$u_0+\xi\frac{\partial u_0}{\partial y}=-1-2\pi\varepsilon\alpha\beta\cos\left(2\pi x\right),\ \theta_0+\Omega\frac{\partial\theta_0}{\partial y}=1\ at\ y=h=1+\varepsilon\cos\left(2\pi x\right). \tag{5.21}$$

5.3.2 First-Order System

$$\left(\frac{dp}{dx}\right)_1=\frac{\partial^2 u_1}{\partial y^2}+\frac{\partial}{\partial y}\left(\frac{\partial u_0}{\partial y}\right)^2-M^2u_1, \tag{5.22}$$

$$\frac{\partial^2\theta_1}{\partial y^2}=-Br\left(\left(\frac{\partial u_0}{\partial y}\right)^3+2\frac{\partial u_0}{\partial y}\frac{\partial u_1}{\partial y}\right), \tag{5.23}$$

$$\frac{\partial u_1}{\partial y}=0,\ \frac{\partial\theta_1}{\partial y}=0\ \text{at}\ y=0, \tag{5.24}$$

$$u_1+\xi\left(\frac{\partial u_1}{\partial y}+\left(\frac{\partial u_0}{\partial y}\right)^3\right)=0,\theta_1+\Omega\frac{\partial\theta_1}{\partial y}=0\qquad at\qquad y=h=1+\varepsilon\cos 2\pi x. \tag{5.25}$$

The software Mathematica is used to find the solution of the previously obtained systems. The pressure rise data is constructed by the numerical integration of equation (5.16).

The final obtained composed forms of velocity u, pressure slope dp/dx, and the heat profile θ are displayed here:

$$u=Ae^{My}+Be^{-My}-\frac{FrM^2+Frp-ReSin[\omega]}{FrM^2}+$$
$$We\left(Ce^{My}+De^{-My}-\frac{2}{3}e^{-2My}\left(-B^2+A^2e^{4My}\right)\right), \tag{5.26}$$

$$\frac{dp}{dx}=\frac{1}{3\mathrm{Fr}h}e^{-2hM}(-3\mathrm{B}e^{hM}\mathrm{Fr}M-3\mathrm{A}e^{2hM}\mathrm{Fr}M+3\mathrm{B}e^{2hM}\mathrm{Fr}M+3\mathrm{A}e^{3hM}\mathrm{Fr}M+3e^{2hM}\mathrm{Fr}M^2$$
$$-3e^{2hM}\mathrm{Fr}hM^2-3e^{2hM}\mathrm{Fr}M^2Q-3\mathrm{D}e^{hM}\mathrm{Fr}M\mathrm{We}-3\mathrm{C}e^{2hM}\mathrm{Fr}M\mathrm{We}+3\mathrm{D}e^{2hM}\mathrm{Fr}M\mathrm{We}+3\mathrm{C}e^{3hM}\mathrm{Fr}M\mathrm{We}$$
$$-\mathrm{B}^2\mathrm{Fr}M^2\mathrm{We}+\mathrm{A}^2e^{2hM}\mathrm{Fr}M^2\mathrm{We}+\mathrm{B}^2e^{2hM}\mathrm{Fr}M^2\mathrm{We}-\mathrm{A}^2e^{4hM}\mathrm{Fr}M^2\mathrm{We}+3e^{2hM}h\mathrm{ReSin}[\xi]), \tag{5.27}$$

$$\theta = E + J + Gy + Ly + \frac{1}{2}BrM\left(-\frac{B^2 e^{-2My}}{2M} - \frac{A^2 e^{2My}}{2M} + 2ABMy^2\right) + \frac{1}{9}BrM\left(-\frac{5}{3}B^3 e^{-3My}\right.$$
$$\left. -3AB^2 e^{-My} + 3A^2 Be^{My} + \frac{5}{3}A^3 e^{3My} - \frac{9BDe^{-2My}}{2M} - \frac{9ACe^{2My}}{2M} + 9BCMy^2 + ADMy^2\right) \quad (5.28)$$

Where:

$$A = -\frac{e^{hM}\left(-Frp + 2FrM^2\pi\alpha\beta\epsilon\, Cos[2\pi x] + ReSin[\omega]\right)}{FrM^2\left(1 + e^{2hM} - M\xi + e^{2hM}M\xi\right)},$$

$$B = -\frac{e^{hM}\left(-Frp + 2FrM^2\pi\alpha\beta\epsilon\, Cos[2\pi x] + ReSin[\omega]\right)}{FrM^2\left(1 + e^{2hM} - M\xi + e^{2hM}M\xi\right)},$$

$$E = 1 - \frac{1}{2}BrM\left(-\frac{B^2 e^{-2hM}}{2M} - \frac{A^2 e^{2hM}}{2M} + 2ABh^2 M\right) - \frac{1}{2}BrM,$$

$$\left(B^2 e^{-2hM} - A^2 e^{2hM} 4ABhM\right)\Omega - \frac{1}{2}Br\left(A^2 - B^2\right)M\left(h + \Omega\right),$$

$$G = \frac{1}{2}Br\left(A^2 - B^2\right)M,$$

$$C = -\frac{1}{3\left(1 + e^{2hM} - M\xi + e^{2hM}M\xi\right)} e^{-2hM} M\left(2B^2 e^{hM} - 4A^2 e^{2hM} - 4B^2 e^{2hM} - 2A^2 e^{5hM}\right.$$
$$-4B^2 e^{hM}M\xi + 4A^2 e^{2hM}M\xi + 4B^2 e^{2hM}M\xi - 4A^2 e^{5hM}M\xi - 3B^3 M^2\xi + AB^2 e^{2hM}M^2\xi$$
$$\left.-9A^2 Be^{4hM}M^2\xi + 3A^3 e^{6hM}M^2\xi\right)$$

$$D = -\frac{4}{3}\left(A^2 + B^2\right)M - \frac{1}{3\left(1 + e^{2hM} - M\xi + e^{2hM}M\xi\right)} e^{-2hM} M\left(2B^2 e^{hM} - 4A^2 e^{2hM}\right.$$
$$-4B^2 e^{2hM} - 2A^2 e^{5hM} - 4B^2 e^{hM}M\xi + 4A^2 e^{2hM}M\xi + 4B^2 e^{2hM}M\xi - 4A^2 e^{5hM}M\xi$$
$$\left.-3B^3 M^2\xi + 9AB^2 e^{2hM}M^2\xi - 9A^2 Be^{4hM}M^2\xi + 3A^3 e^{6hM}M^2\xi\right)$$

$$J = -\frac{1}{9}BrM\left(-\frac{5}{3}B^3 e^{-3hM} - 3AB^2 e^{-hM} + 3A^2 Be^{hM} + \frac{5}{3}A^3 e^{3hM} - \frac{9BDe^{-2hM}}{2M}\right.$$
$$\left.-\frac{9ACe^{2hM}}{2M} + 9BCh^2 M + 9ADh^2 M\right) - \frac{1}{9}BrM$$
$$\left(9BDe^{-2hM} - 9ACe^{2hM} + 5B^3 e^{-3hM}M + 3AB^2 e^{-hM}M + 3A^2 Be^{hM}M + 5A^3 e^{3hM}M\right.$$
$$\left.+18BChM + 18ADhM\right)\Omega + \frac{1}{9}BrM\left(-9AC + 9BD + 5A^3 M + 3A^2 BM + 3AB^2 M + 5B^3 M\right)$$
$$\left(h + \Omega\right)$$

$$L = -\frac{1}{9}BrM\left(-9AC + 9BD + 5A^3 M + 3A^2 BM + 3AB^2 M + 5B^3 M\right).$$

5.4 ENTROPY SCHEME

By following [18], the equation of total rate of entropy production is described as:

$$S'''_{gen}=\frac{\kappa}{T_0^2}\left(\frac{\partial^2\bar{T}}{\partial\bar{X}^2}+\frac{\partial^2\bar{T}}{\partial\bar{Y}^2}\right)^2+\frac{1}{T_0}\left(\bar{\mathcal{L}}_{XX}\frac{\partial\bar{U}}{\partial\bar{X}}+\bar{\mathcal{L}}_{XY}\left(\frac{\partial\bar{U}}{\partial\bar{Y}}+\frac{\partial\bar{V}}{\partial\bar{X}}\right)+\bar{\mathcal{L}}_{YY}\frac{\partial\bar{V}}{\partial\bar{Y}}\right). \tag{5.29}$$

After using equation (5.11) and the similar physical restrictions by converting $S'''_{gen}/\left(S_{G_0}\right)$ as the total entropy number $\left(N_G\right)$, we consequently receive:

$$N_G=\tau\theta'^2\left(y\right)+Br\left(u'^2\left(y\right)+Weu'^3\left(y\right)\right), \tag{5.30}$$

Where $\tau=\left(T_1-T_0\right)T_0^{-1}$ is the dimension-free temperature difference. To compare the fluid-resistive irreversibility effects with the total heat transfer irreversibility, the Bejan number is defined as [16]:

$$Be=N_{G\theta}/N_G, \tag{5.31}$$

Where $N_{G\theta}=\tau\left(\frac{\partial\theta}{\partial y}\right)^2$.

5.5 RESULTS AND DISCUSSION

In this section, we elaborate the findings of the physical quantities received from their graphical variations. Validations of the current results are justified through a comparison of the obtained date with the existing literature [18], which can be found in Table 5.1. From the table, it can be evaluated easily that the current results are in good matching with the previous work [18] with a greater limit of accuracy. From the expressions found earlier analytically, we have drawn the sketches of velocity, pressure gradient, pressure rise, temperature profile, entropy generation function, Bejan number, and the stream bolus mechanism. First of all, the effects of cilia length factor on velocity profile have been disclosed along the axial coordinate in Figure 5.2. In this graph, it is quite prominent that larger lengths of the cilia provide slower velocity in the central region of the channel, but velocity remains dominant to the cilia length near the walls containing cilia. It is also convincing that, in general, the velocity is minimum near the walls as compared to the mid idea. Figure 5.3 discusses the magnetic field impact on the fluid's flow velocity. One can observe that velocity is an increasing function of MHD towards the corners of the geometry, but in the central region, it is showing an inverse behavior. Figures 5.4, 5.5, and 5.6 tell about the pressure rise variation under some key factors of the study. Figure 5.4 suggests that the pumping rate directly varies with the Hartmann number M. It is also fetched that the co-pumping point is at $Q = 0$. If we consider the next figure (Figure 5.5), which relates the effects of cilia length parameter ε, we can have results that are quite similar to that achieved in the previous graph. On the other hand, the inclination angle of the channel ω provides totally opposite answers for pressure rise curves, as we have discussed in the cases of the last two parameters (see Figure 5.6). Figure 5.7 evaluates the pressure gradient variation along the axial direction. It is considerable over here that pressure gradient is growing downward with larger values of magnetic factor M throughout the whole domain. If we look at Figures 5.8 and 5.9, we can easily measure that there is an inverse variation of the profile in the central portion, but at the edges starting at points $x = 0.8$ and $x = 1.2$, the curves of pressure start changing their concavity and showing direct relation between dp/dx and cilia length ε and the eccentricity of the elliptical paths.

From Figures 5.10 and 5.11, we can observe that temperature difference curves are reacting inversely with the impact of magnetic field M and the Brinkman number Br. It is also noted that the change in heat becomes negligible from central phase to the walls domain. From Figure 5.12,

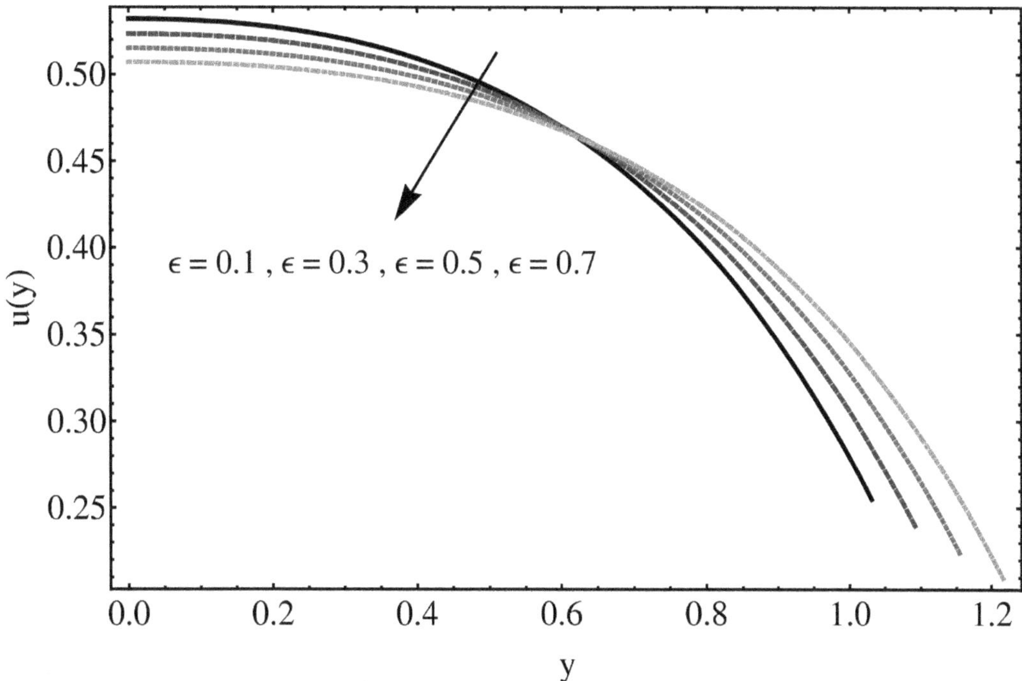

FIGURE 5.2 Axial speed for ε when $M = 1$, $\alpha = 0.5$, $We = 0.01$, $\beta = 0.1$, $\omega = \pi/3$, $Fr = 0.5$, $Re = 0.5$, $Q = 1$, $x = 0.8$.

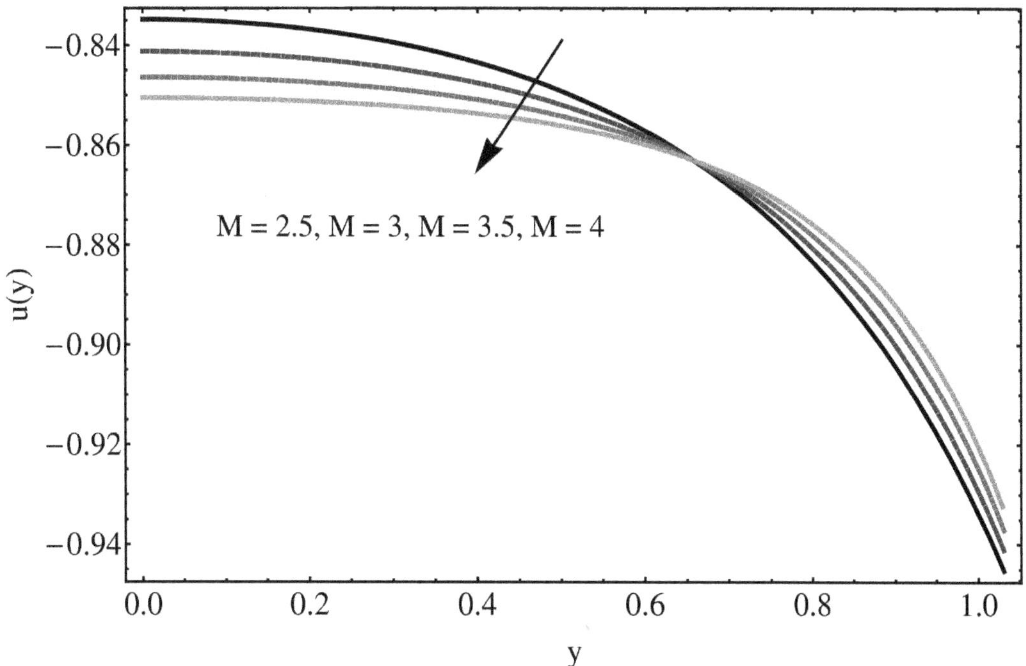

FIGURE 5.3 Axial speed M when $\alpha = 0.5$, $We = 0.05$, $\beta = 0.1$, $\omega = \pi/3$, $Fr = 0.5$, $Re = 1$, $Q = 0.1$, $x = 0.8$.

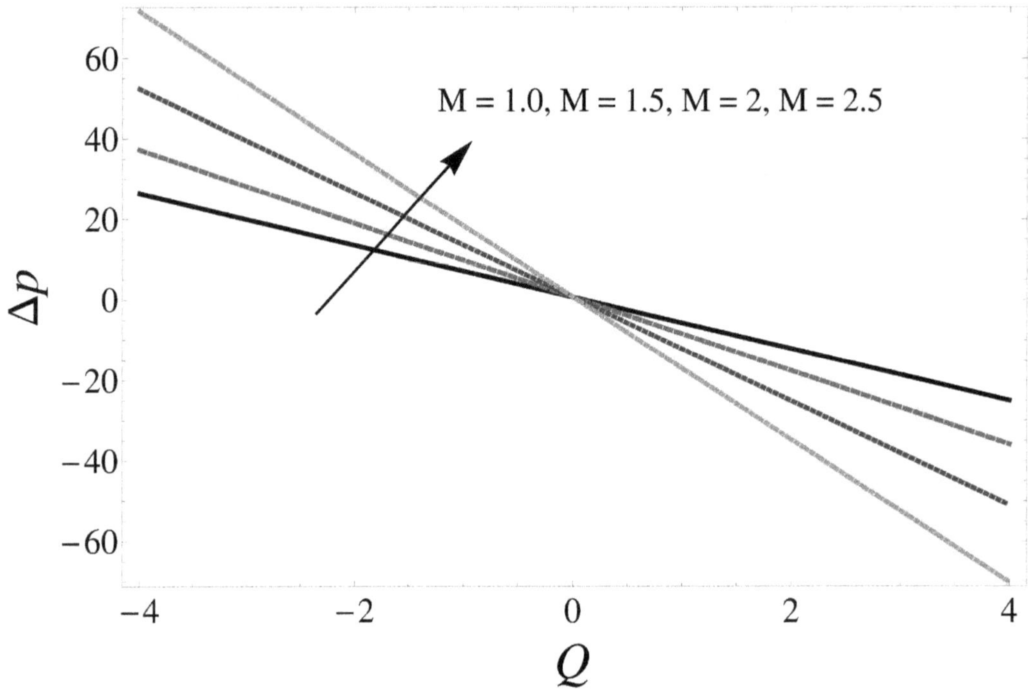

FIGURE 5.4 Pressure rise for M when $\varepsilon = \alpha = \beta = 0.1$, $Re = 1$, $Fr = 0.5$, $We = 0.05$, $\omega = \pi / 3$.

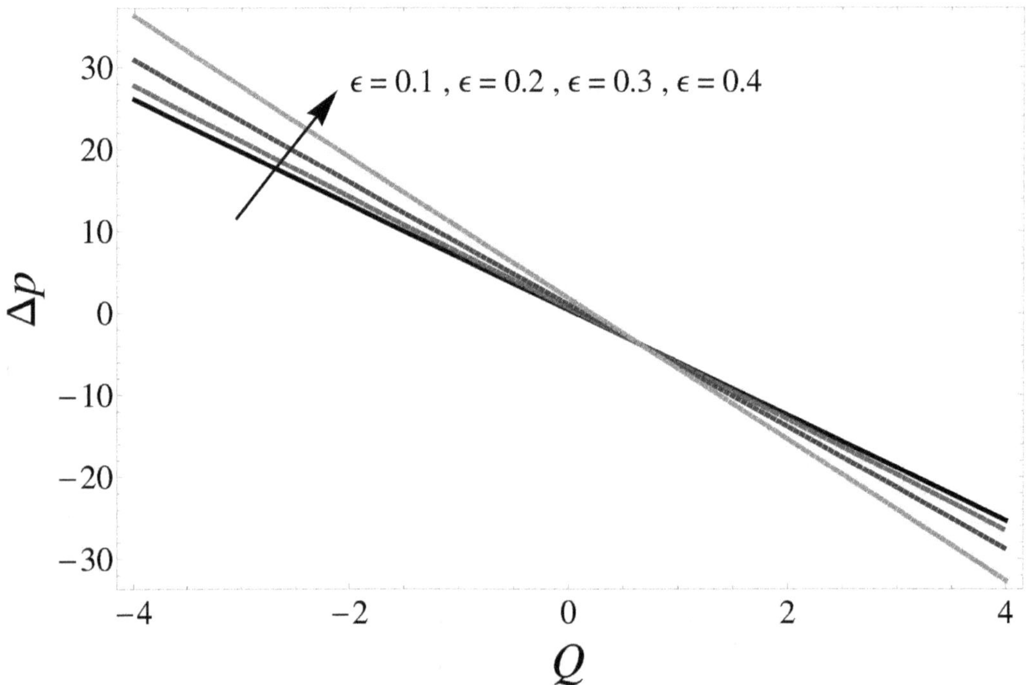

FIGURE 5.5 Pressure rise for ε when $M = 1, \alpha = \beta = 0.1$, $Re = 0.5$, $Fr = 0.5$, $We = 0.05$, $\omega = \pi / 3$.

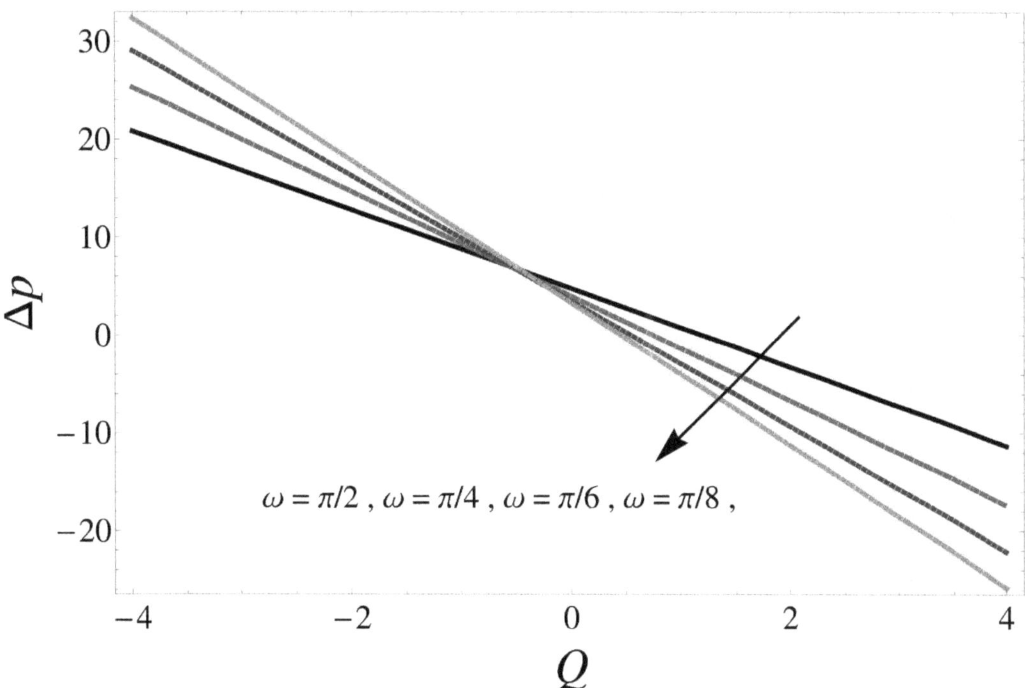

FIGURE 5.6 Pressure rise for ω when $M = 1$, $\varepsilon = 0.5$, $\alpha = \beta = 0.1$, $Re = 1$, $Fr = 0.5$, $We = 0.05$.

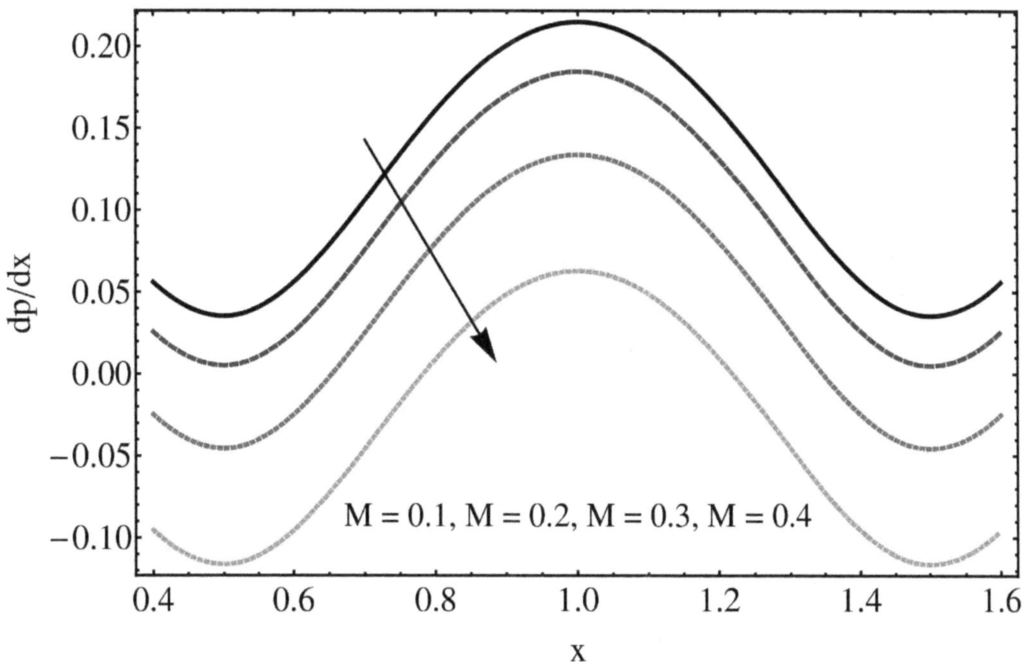

FIGURE 5.7 Pressure slope for M when $\varepsilon = 0.1$, $\alpha = 0.2$, $\beta = 0.01$, $Re = 0.5$, $Fr = 0.5$, $We = 0.03$, $Q = 1$, $\omega = \pi/3$.

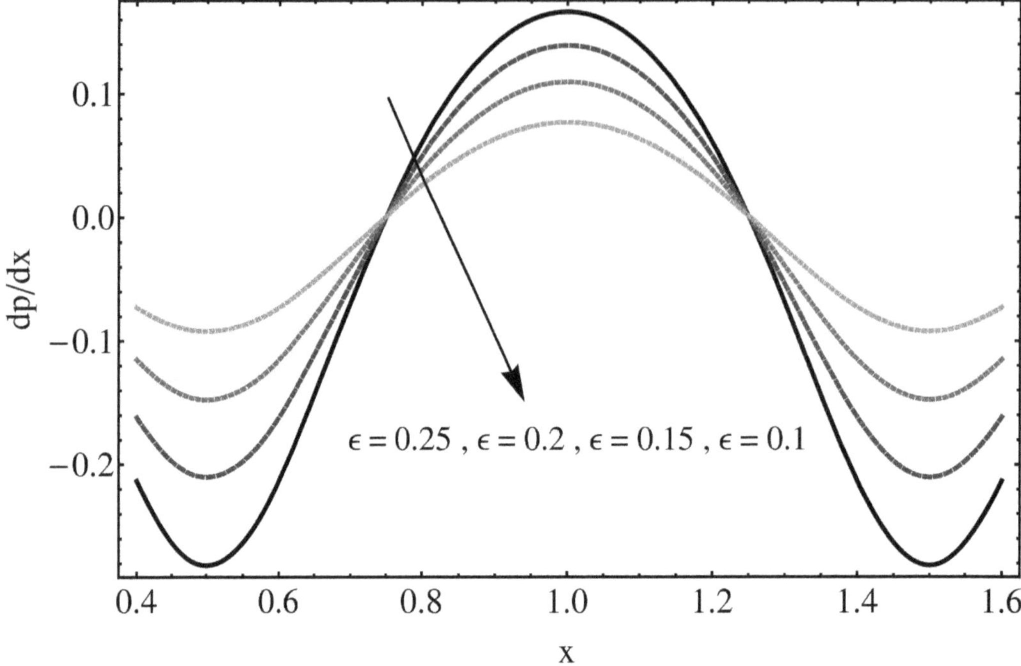

FIGURE 5.8 Pressure slope for ε when $M = 0.1, \alpha = 0.1, \beta = 0.1, Re = 1, Fr = 0.5, \omega = \pi / 3$.

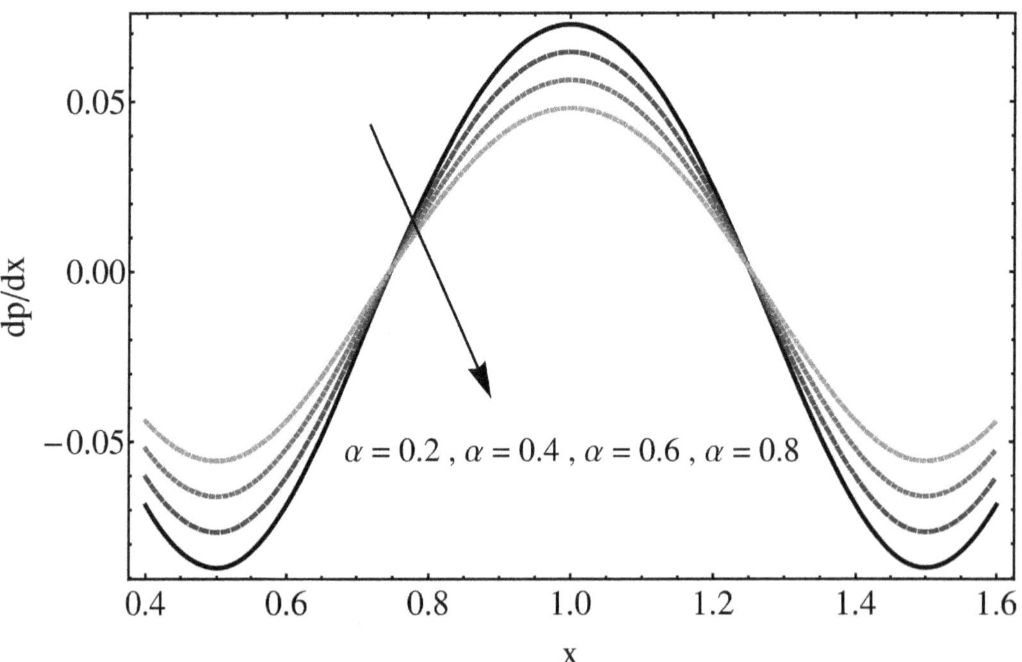

FIGURE 5.9 Pressure slope for α when $M = 1, \varepsilon = 0.1, \beta = 0.1, Q = 1, Re = 1, Fr = 0.5, \omega = \pi / 3$.

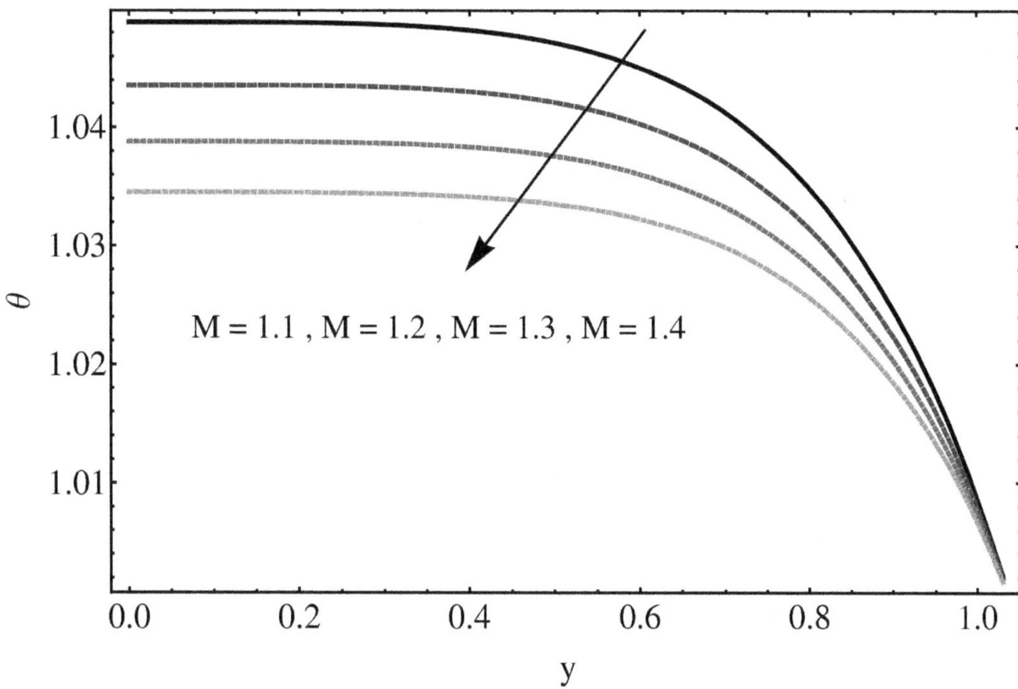

FIGURE 5.10 Temperature for different values of M when $\varepsilon = 0.1$, $\alpha = 0.2$, $\beta = 0.01$, $x = 0.8$, $Re = 1$, $Br = 8$, $Fr = 0.5$, $We = 0.03$, $Q = 0.5$, $\omega = \pi / 3$, $\Omega = 0.01$.

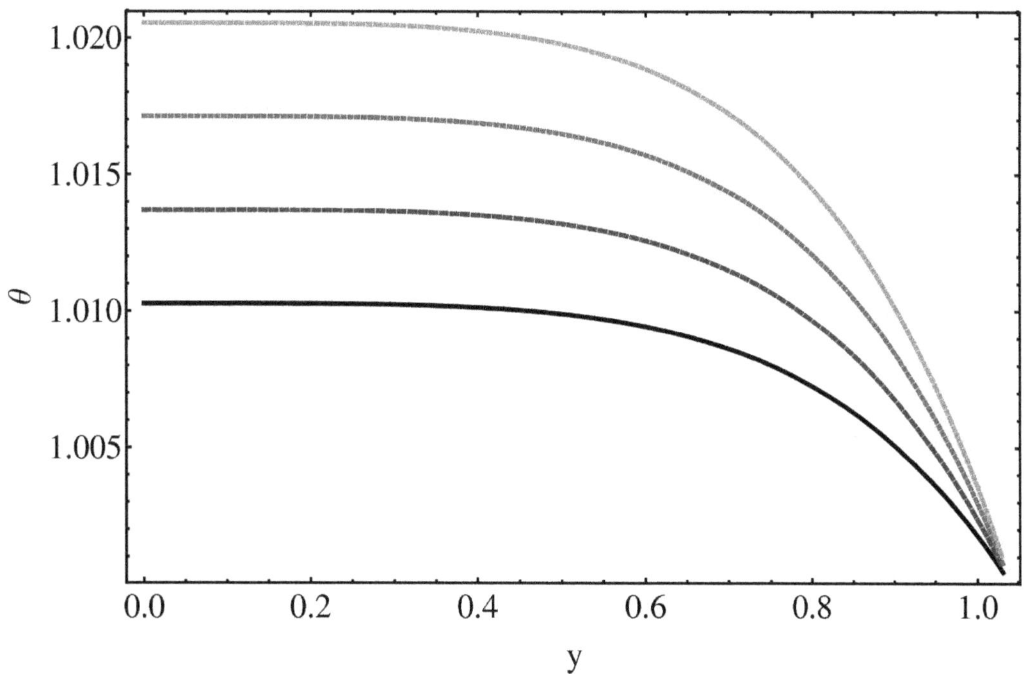

FIGURE 5.11 Temperature for different values of Br when $\varepsilon = 0.1$, $\alpha = 0.2$, $\beta = 0.01$, $x = 0.8$, $Re = 1$, $M = 1$, $Fr = 0.5$, $We = 0.05$, $Q = 0.5$, $\omega = \pi / 3$, $\Omega = 0.01$.

it is easily verified that the Bejan number profile is increasing with Brinkman number accordingly, and Figure 5.13 gives the idea that the magnetic field factor imposes opposite impact on the data of Bejan number. It can also be concluded that entropy rate due to heat profile is greater than that of the total aspects of the flow in case of Brinkman number, but quite the inverse scenario is noticed for magnetic impact. It is also pertinent to add that to reduce the entropy rate, a short range of Brinkman number values should be adopted. From entropy generation graphs in Figures 5.14 and 5.15, we can see that entropy rate slows down with the larger influence of the magnetic field and rises up with the Brinkman number. It is also valuable to mention that entropy is generated at its highest peak in the middle of the flow region as compared to the cilia walls, which also gives us an idea that if large cilia lengths occur, the entropy rate can be put down.

Figures 5.16 and 5.17 appear to investigate the trapping bolus mechanism. These graphs tell about the bolus formation and size variation under various values of the physical parameters. These plots provide us an imagination for the flow pattern in the geometry and wave formations due to the cilia-oriented pumping at the walls. The beating cilia walls produce an immense pressure on the fluid to forward and reverse strokes which suppress it and then move forward without any external source. In short, due to extension and compression of the walls due to cilia impact, the fluid experiences a forward push to travel alongside. Figure 5.16 is plotted for the streamlines of cilia length factor ε. It can be observed that boluses vary in size alternatively for increasing values of the cilia length. It is to be noted also that the number of boluses increases with the increase in cilia length, which clearly points out the physical evidence that large pressure will be exerted on the flow by long cilia. On the other hand, effects of magnetic field on flow pattern can be examined from Figure 5.17, where it is calculated that boluses are reduced in number and size when someone accelerates the magnetic field impact by inserting large currents in the flow body.

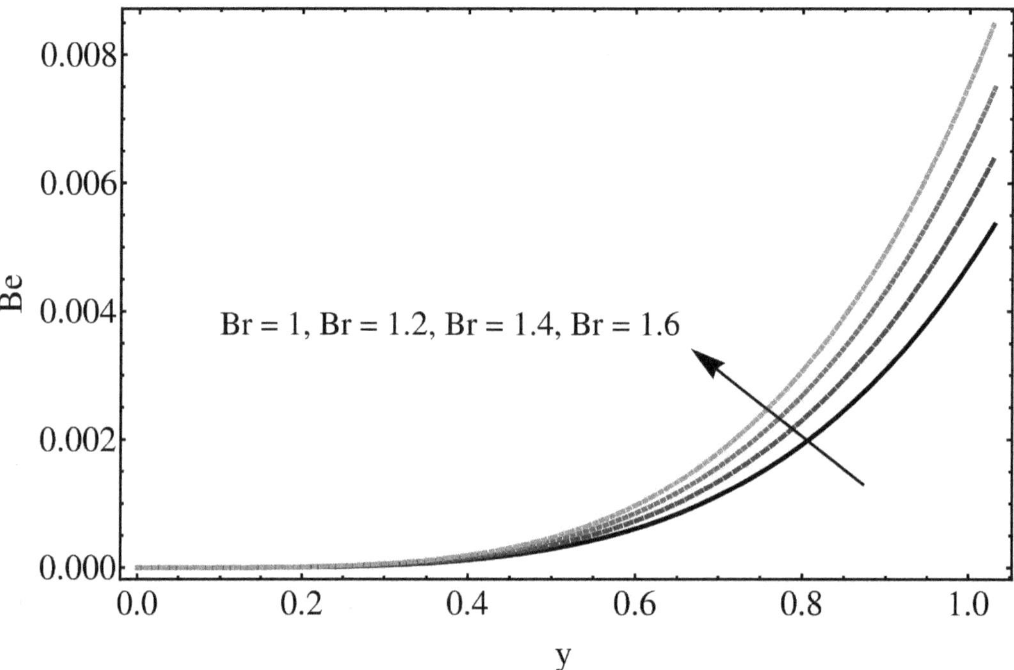

FIGURE 5.12 Effect of Br on the Bejan number when $\varepsilon = 0.1, \alpha = 0.5, \beta = 0.1, x = 0.8, Re = 1, M = 1, Fr = 0.5, We = 0.05, Q = 0.5, \omega = \frac{\pi}{3}, \Omega = 0.01, \tau = 1$.

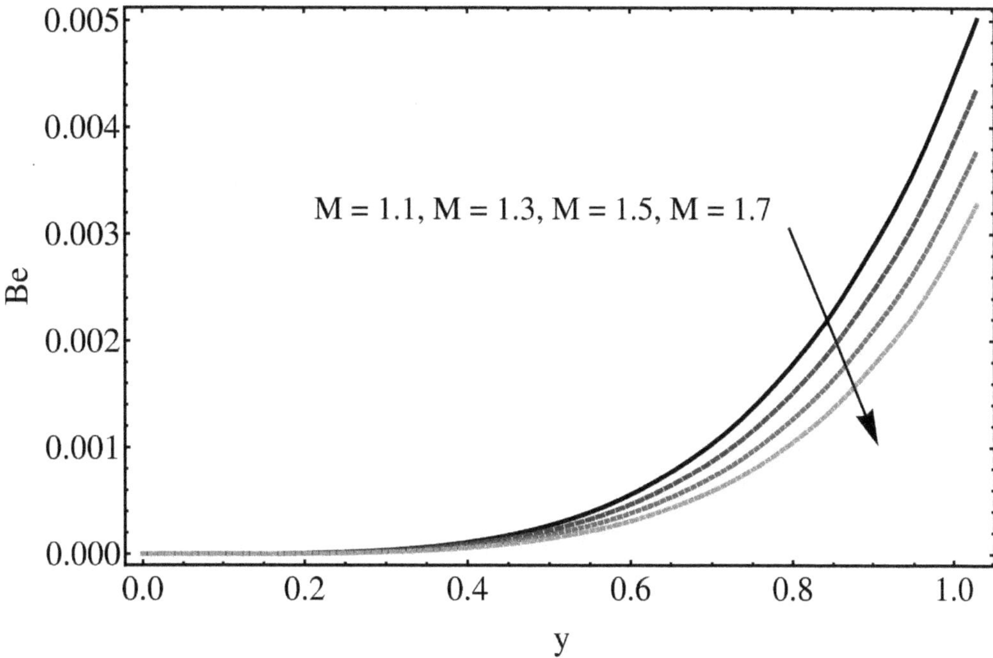

FIGURE 5.13 Effect of M on the Bejan number when $\varepsilon = 0.1$, $\alpha = 0.2$, $\beta = 0.2$, $x = 0.8$, $Re = 0.5$, $M = 1$, $Fr = 0.5$, $We = 0.01$, $Q = 0.1$, $Br = 1$, $\omega = \frac{\pi}{3}$, $\Omega = 0.01$, $\tau = 1$.

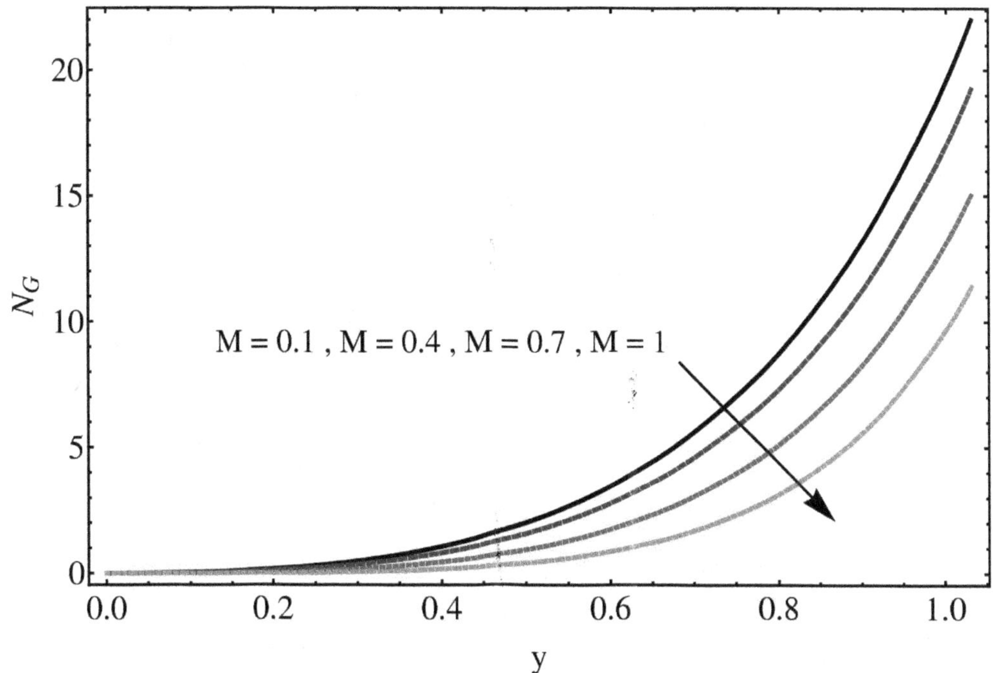

FIGURE 5.14 Effects of M on entropy generation number when $\varepsilon = 0.1$, $\alpha = 0.2$, $\beta = 0.1$, $x = 0.8$, $Re = 0.5$, $Fr = 1$, $We = 0.01$, $Q = 0.8$, $Br = 1$, $\omega = \frac{\pi}{3}$, $\Omega = 0.01$, $\tau = 1$.

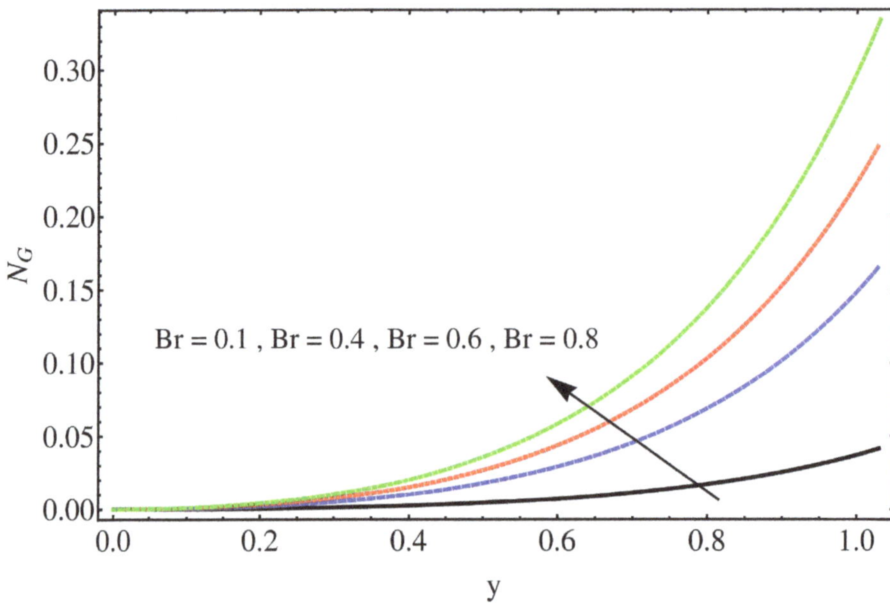

FIGURE 5.15 Effects of Br on entropy generation number when $\varepsilon = 0.1$, $\alpha = 0.2$, $\beta = 0.1$, $M = 1$, $x = 0.8$, $Re = 0.5$, $Fr = 1$, $We = 0.01$, $Q = 0.2$, $\omega = \frac{\pi}{3}$, $\Omega = 1$, $\tau = 1$.

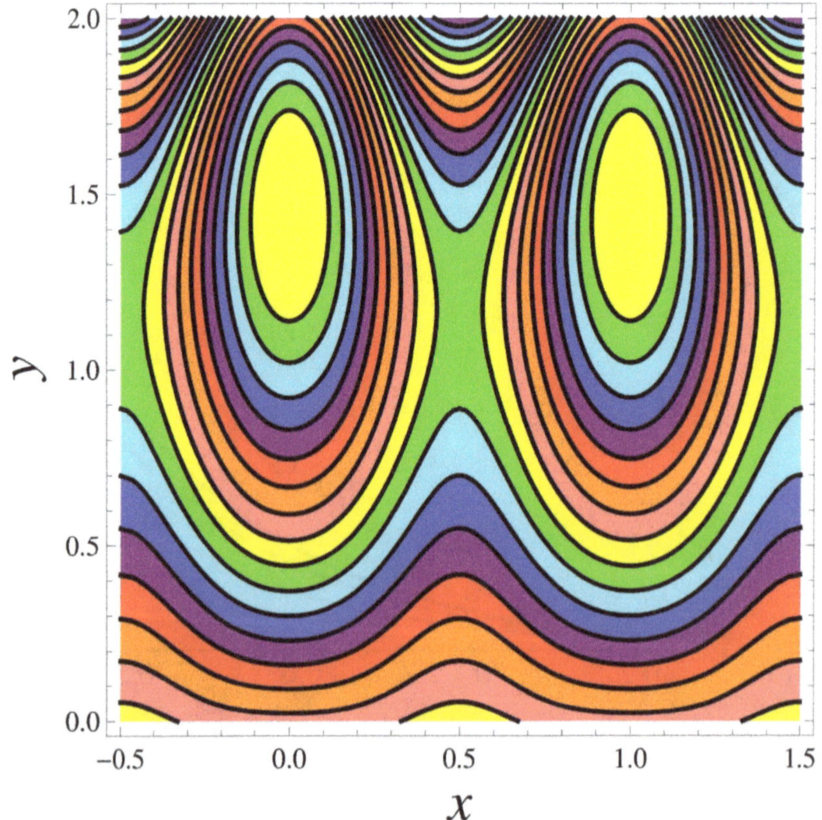

FIGURE 5.16(A, B, C) Streamlines for ε when $\alpha = 0.5$, $\omega = \pi/3$, $\beta = 0.5$, $M = 1$, $Fr = Re = 0.5$, $Q = 1$, $We = 0.05$.

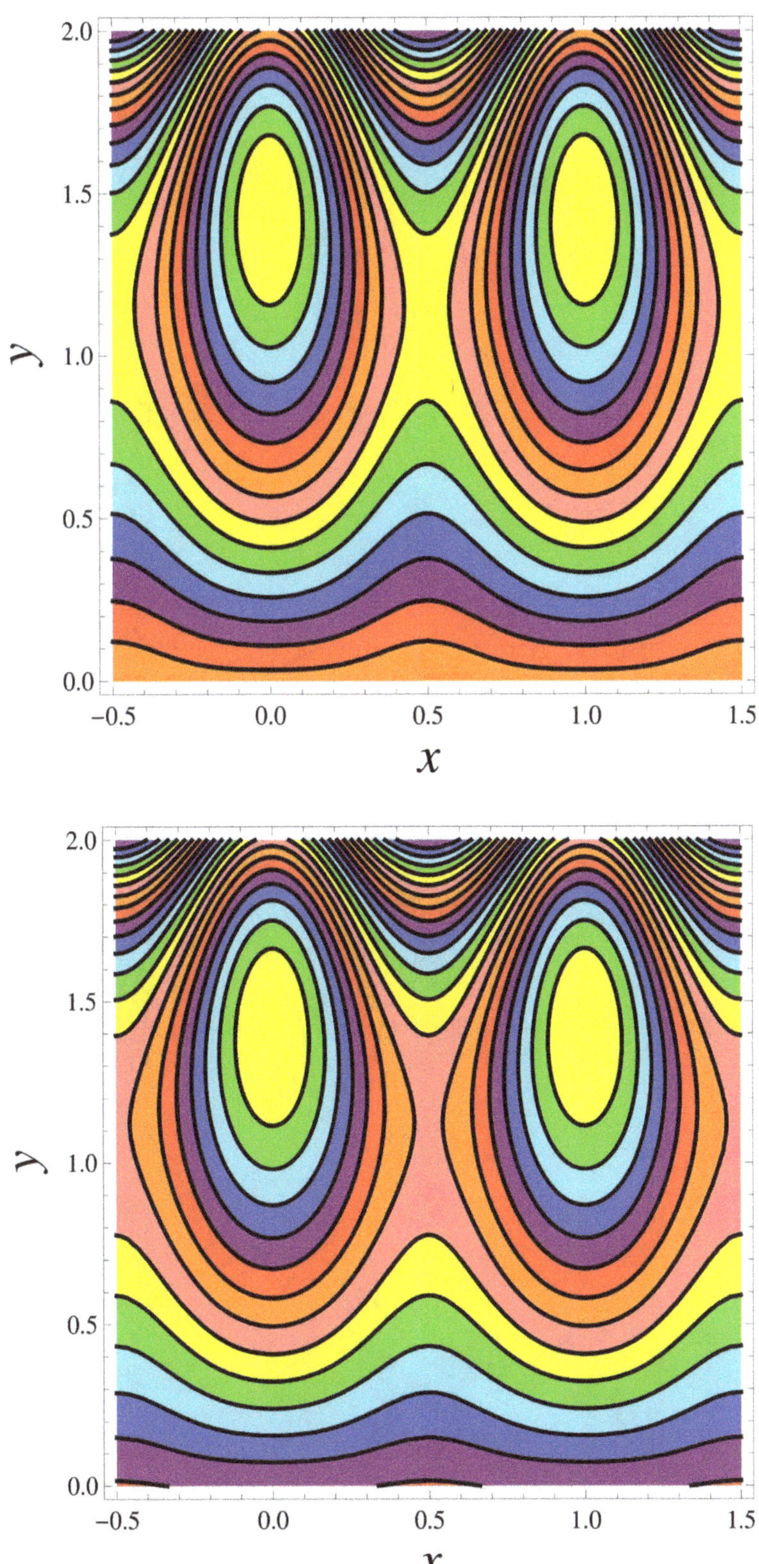

FIGURE 5.16 (Continued)

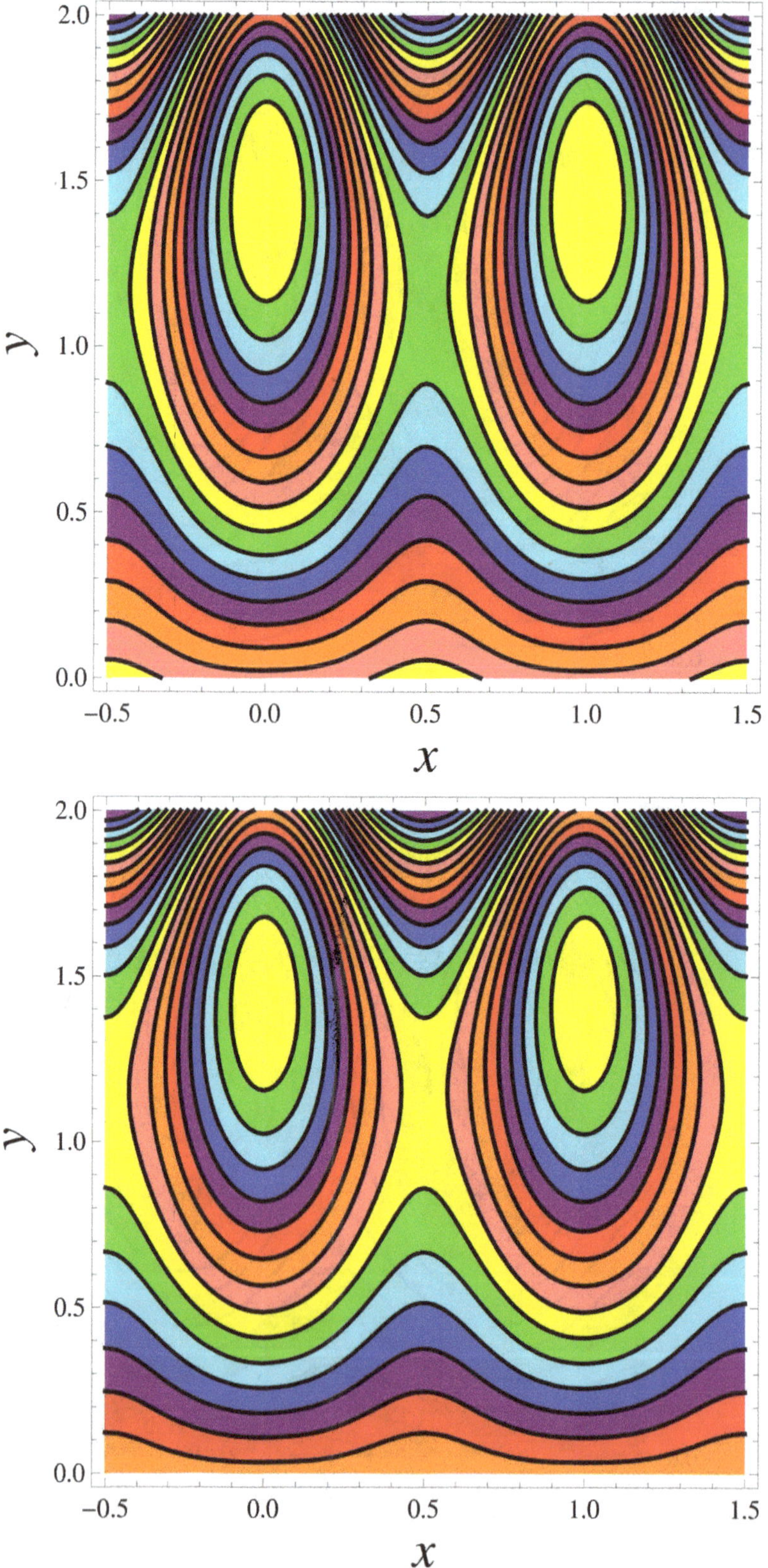

FIGURE 5.17(A, B, C) Streamlines for M when $\alpha = 0.5$, $\omega = \pi/3$, $\varepsilon = 0.1$, $\beta = 0.5$, $Re = 1$, $Fr = 0.5$, Q = 1.

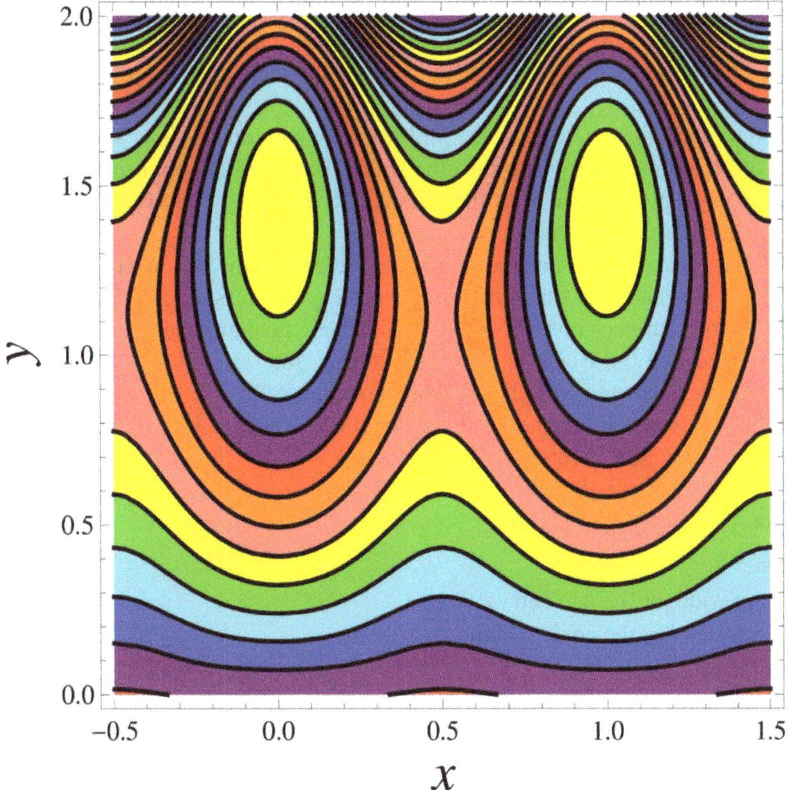

FIGURE 5.17 (Continued)

TABLE 5.1
Comparison of Current Results with [18]

y	Velocity $u(y)$ in [18]	Velocity $u(y)$ with $We = \xi = 0$	Velocity $u(y)$ with $We = 0.01, \xi = 1.05$
0.0	2.45253	2.45252	0.53198
0.1	2.44736	2.44736	0.53095
0.2	2.42287	2.42285	0.52735
03	2.36496	2.36495	0.52039
0.4	2.25811	2.25811	0.50918
0.5	2.08461	2.08460	0.49268
0.6	1.82373	1.82372	0.46970
0.7	1.45075	1.45075	0.43878
0.8	0393579	0393578	0.39818
0.9	0.24254	0.24253	0.34578
1.0	−0.67343	−0.67341	0.27900

5.6 CONCLUSIONS

In the current analysis, we have solved a mathematical model representing cilia flow of Williamson fluid in an inclined channel making an angle α with the ground. The effects of magnetic field and flow slip conditions are taken into consideration both for velocity as well as for temperature distributions. Perturbation solutions have been framed with greater efficiency and pattern for velocity, temperature, pressure curves, and also for the stream functions. From the previous graphical discussion, we can summarize our investigations in some bullets as follows:

- Cilia length and magnetic field exert an inverse impact on the flow speed in the central portion of the geometry, but near the cilia walls, the results are opposite.
- Flow pumping rate is raised with the cilia length and magnetic field, but the inclination angle produces an inverse impact on it.
- The magnetic field reduces the pressure curves throughout the region, but the eccentricity of the elliptical path and cilia length helps in increasing the pressure on the left and right sides.
- Transfer of the heat goes inversely with the magnetic field and the Brinkman number.
- It is measured that the Bejan number and entropy rate are reduced with the Hartmann number but are enhanced with the Brinkman number.
- More trapped boluses are created for large lengths of the cilia, but magnetic field impact is in inverse link with the creation of them.

Nomenclature

t	time
a	mean width of the geometry
α	eccentricity of the conical path
ε	length of cilia
λ	wavelength
c	wave propagation speed
X_0	particle location
U, U_0	tangential velocity components
V, V_0	vertical velocity components
ρ	density of the fluid
$\bar{\mathcal{L}}$	tensor of fluid model
σ	electric current density
B_0	magnetic field strength
g	gravitational acceleration
ω	angle of channel inclination
ζ	specific heat capacity
T	dimensional temperature
κ	thermal conductivity
P	pressure
T_0	lower wall temperature
T_1	upper wall temperature
μ_0	fluid viscosity
x, y	dimensionless coordinates
θ	dimensionless temperature
We	Weissenberg number
M	Hartmann number
Re	Reynolds number
Fr	Faraday number
Br	Brinkman number
ξ and Ω	velocity and heat slip factors

F and Q dimensional and non-dimensional flow rates
S'''_{gen} dimensional entropy function
N_G dimensionless entropy function
τ temperature difference
Be Bejan number
" characteristic time
$\bar{\dot{\gamma}}$ Rivlin–Ericksen tensor

REFERENCES

1. Berg, H. C. (1993). *Random Walks in Biology* (expanded Ed.). Princeton University Press, Princeton, NJ.
2. Okada, Y., Takeda, S., Tanaka, Y., Belmonte, J. C. I., & Hirokawa, N. (2005). Mechanism of nodal flow: a conserved symmetry breaking event in left-right axis determination. *Cell*, *121*(4), 633–644.
3. Mitchell, B., Jacobs, R., Li, J., Chien, S., & Kintner, C. (2007). A positive feedback mechanism governs the polarity and motion of motile cilia. *Nature*, *447*(7140), 97–101.
4. Miskevich, F. (2010). Imaging fluid flow and cilia beating pattern in Xenopus brain ventricles. *Journal of Neuroscience Methods*, *189*(1), 1–4.
5. Guirao, B., Meunier, A., Mortaud, S., Aguilar, A., Corsi, J. M., Strehl, L. & Spassky, N. (2010). Coupling between hydrodynamic forces and planar cell polarity orients mammalian motile cilia. *Nature Cell Biology*, *12*(4), 341–350.
6. Colantonio, J. R., Vermot, J., Wu, D., Langenbacher, A. D., Fraser, S., Chen, J. N., & Hill, K. L. (2009). The dynein regulatory complex is required for ciliary motility and otolith biogenesis in the inner ear. *Nature*, *457*(7226), 205–209.
7. Kramer-Zucker, A. G., Olale, F., Haycraft, C. J., Yoder, B. K., Schier, A. F., & Drummond, I. A. (2005). Cilia-driven fluid flow in the zebrafish pronephros, brain and Kupffer's vesicle is required for normal organogenesis. *Development*, *132*(8), 1907–1921.
8. Baker, K., & Beales, P. L. (2009, November). Making sense of cilia in disease: the human ciliopathies. In *American Journal of Medical Genetics Part C: Seminars in Medical Genetics* (Vol. 151, No. 4, pp. 281–295). Hoboken: Wiley Subscription Services, Inc., A Wiley Company.
9. Sawamoto, K., Wichterle, H., Gonzalez-Perez, O., Cholfin, J. A., Yamada, M., Spassky, N., & Alvarez-Buylla, A. (2006). New neurons follow the flow of cerebrospinal fluid in the adult brain. *Science*, *311*(5761), 629–632.
10. Lyons, R. A., Saridogan, E., & Djahanbakhch, O. (2006). The reproductive significance of human Fallopian tube cilia. *Human Reproduction Update*, *12*(4), 363–372.
11. Ramesh, K., Tripathi, D., & Bég, O. A. (2019). Cilia-assisted hydro magnetic pumping of biorheological couple stress fluids. *Propulsion and Power Research*, *8*(3), 221–233.
12. Akbar, N. S., & Butt, A. W. (2014). Heat transfer analysis of viscoelastic fluid flow due to metachronal wave of cilia. *International Journal of Biomathematics*, *7*(6), 1450066.
13. Bhatti, M. M., Zeeshan, A., & Rashidi, M. M. (2017). Influence of magneto hydrodynamics on metachronal wave of particle-fluid suspension due to cilia motion. *Engineering Science and Technology, an International Journal*, *20*(1), 265–271.
14. Sadaf, H., & Nadeem, S. (2020). Fluid flow analysis of cilia beating in a curved channel in the presence of magnetic field and heat transfer. *Canadian Journal of Physics*, *98*(2), 191–197.
15. McCash, L. B., Nadeem, S., Akhtar, S., Saleem, A., Saleem, S., & Issakhov, A. (2022). Novel idea about the peristaltic flow of heated Newtonian fluid in elliptic duct having ciliated walls. *Alexandria Engineering Journal*, *61*(4), 2697–2707.
16. Munawar, S., & Saleem, N. (2020). Entropy analysis of an MHD synthetic cilia assisted transport in a microchannel enclosure with velocity and thermal slippage effects. *Coatings*, *10*(4), 414.
17. Bhatti, M. M., Arain, M. B., Zeeshan, A., Ellahi, R., & Doranehgard, M. H. (2022). Swimming of gyrotactic microorganism in MHD Williamson nanofluid flow between rotating circular plates embedded in porous medium: Application of thermal energy storage. *Journal of Energy Storage*, *45*, 103511.
18. Saleem, N. (2021). Entropy analysis in ciliated inclined channel filled with hydro magnetic Williamson fluid flow induced by metachronal waves. *Thermal Science*, *25*(5 Part B), 3687–3699.

6 MHD Tangent Hyperbolic Fluid Flow Over Stretching Sheet with Cattaneo–Christov Heat Flux Model and Quadratic Convection

A Statistical Analysis

Himanshu Upreti and Alok Kumar Pandey

6.1 INTRODUCTION

The tangent hyperbolic fluid (THF) model is one of the significant models of non-Newtonian fluid which is beneficial in chemical engineering system and has additional benefits over other non-Newtonian models. Some examples of THF are whipped cream, paints, nail polish, ketchup, etc. A voluminous literature is available addressing the flow of THF; some of the notable contributions can be accessed from [1–6]. Hussain et al. [7] examined the influence of MHD on THF flow over nonlinear stretching sheet with second-order slip. Similarly, the impact of magnetic field and second-order slip on THF flow over nonlinear stretching sheet with convective boundary conditions was examined by Ibrahim [8]. Mahanthesh et al. [9] considered the impact of radiation and nonlinear convection on THF flow over convectively heated vertical surface. They declared that velocity of THF reduces with an increase in the values of nonlinear convection parameter. Kumar et al. [10] examined the effects of radiation and magnetic field on THF flow over stretching sheet. Rehman et al. [11] presented a model of regime characteristics of THF with magnetic field and heat generation. They revealed that temperature outlines of THF accelerated with heat generation parameter. Saidulu et al. [12] examined the influence of radiation on MHD THF flow over an exponential stretching surface. A numerical study illustrating the response of sundry parameters on boundary layer flow of THF over a stretching surface was reported by [13–14]. These studies employed the RKF method along with the shooting scheme, and the desired outcomes are depicted through graphs. Ibrahim and Gizewu [15] considered a non-Fourier model (Cattaneo–Christov model) to investigate the effect of mixed convection and mass diffusion on THF flow over a bidirectional stretching sheet. They obtained the solution of governing ODEs by using bvp4c. Ullah et al. [16] analyzed the impact of magnetic THF flow over a stretching surface, and this study is based upon Lie group similarity transformation. The flow analysis of hybrid nanofluid utilizing THF model over PTSC ("parabolic trough solar collector") was reported by Jamshed et al. [17]. The authors achieved the solutions of the governing flow model using a numerical scheme named the "Keller box method." Shankaralingappa et al. [18] considered the dusty Darcy–Forchheimer tangent hyperbolic fluid flow over an elongating surface with the Cattaneo–Christov model. They revealed that the temperature outlines of both dust and fluid phases depreciate with escalating values of Prandtl number. Usman et al. [19] examined the time-dependent flow of magnetic THF passed through a

DOI: 10.1201/9781003299608-6

stretching sheet with dissipation and heat generation absorption. The solutions of the existing flow model are obtained by the Legendre wavelet technique. They also proved that temperature profiles of working fluid declined with increase in Prandtl number. Moreover, the articles referenced from [20–23] are some of the recent studies which have used THF as the working fluid to analyze its flow features under different situations.

Fourier [24] explained the heat transfer phenomenon, a parabolic energy equation for temperature field with a demerit, that is, initial disturbance is felt instantly throughout the whole medium. Cattaneo [25] extended the Fourier law of heat conduction by including the concept of thermal relaxation. This concept leads to heat transportation in the form of waves with finite speed. Later on, Oldroyd's upper-convected derivative was substituted by Christov [26] in place of time derivative to attain the material–invariant formulation. This model is known as the Cattaneo–Christov heat flux model. Hayat et al. [27] examined the heat and momentum boundary layer characteristics of fluid flowing over a stretching surface with variable thickness, and the heat transfer was subject to Cattaneo–Christov model. They obtained the convergent series solutions of governing flow equations. Salahuddin et al. [28] utilized the CC model to inspect the effect of MHD in the flow of the Williamson fluid flowing over a thickness-varying stretching sheet. Dogonchi and Ganji [29] studied the effects of buoyancy forces, thermal radiation, magnetic field, and Joule heating on thermal and flow characteristics of nanofluid streaming over a sheet. Madhukesh et al. [30] used an AA7072- and AA7075-based nanofluid to examine the boundary layer flow of fluid, induced due to stretching of the curved surface with heat transfer mechanism subject to non-Fourier's law. Recently, Imtiaz et al. [31] presented the flow characteristics of CNT (carbon nanotube) based nanofluid past a stretching sheet with variable thickness. The heat transport process was explained using the CC model. Ullah et al. [32] addressed velocity and heat transfer in the unsteady flow of a Prandtl-Eyring fluid over a sheet using the non-Fourier law.

The magnetic field has various applications in the industry, medical science, and engineering field, such as in power generators, plasma studies, crystal growth, nuclear reactors, etc. Recently, the impact of the magnetic field on different fluid flow problems were stated by [33–37].

The main theme of this chapter is to present the statistical analysis for magnetic tangent hyperbolic fluid flow over a stretching surface with the Cattaneo–Christov heat flux model and quadratic convection. The statistical analysis is presented for skin friction coefficient (SFC) and local Nusselt number (LNN) with the help of correlation coefficient (*r*) and probable error (*PE*). This chapter is organized as follows: in Section 6.2, mathematical modelling as well as the discretization of PDEs to simpler ODEs are presented. In Section 6.3, the applied methodology is discussed in detail, and in Section 6.4, the obtained velocity and temperature profiles are discussed through graphs, and a statistical analysis is presented for the active parameter. In Section 6.5, the important conclusions are presented for the studied problem.

6.2 MATHEMATICAL FORMULATION OF THE PROBLEM

A two-dimensional (2D), incompressible, and steady flow of tangent hyperbolic fluid (THF) over a stretching sheet having variable thickness is considered. We assume that the surface of the sheet is non-flat, having the profile $y = \delta\left(x+b\right)^{0.5(1-n)}$. Here, b, δ is the parameter of the sheet, with the assumption that δ is sufficiently small to avoid the pressure gradient, and n is the power law index ($n = 1$, flat surface). The surface of the sheet is considered to be porous, and a non-uniform magnetic field $B(x) = B_o\left(x+b\right)^{0.5(n-1)}$ is exerted in the direction perpendicular to the x–axis. In addition, the flow of the fluid is subject to quadratic convection, and the mechanism of heat transfer is explained utilizing the Cattaneo–Christov heat flux model. Further, the surface is exposed to a quadratic thermal radiation (see fifth term on LHS of equation (6.3)). As the flow is occurring over a porous medium under the existence of a magnetic field, this physical phenomenon affects the heating mechanism (see sixth and seventh term in LHS of equation (6.3)) and is termed as resistive

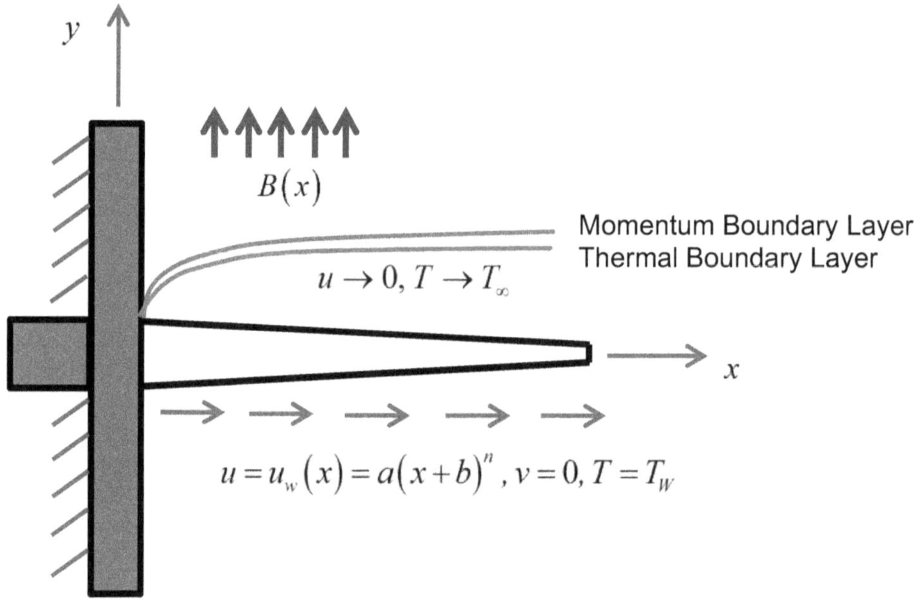

FIGURE 6.1 Physical configuration of the problem.

heating due to the porous media and Ohmic heating, respectively. The schematic representation of the problem is shown in Figure 6.1.

Based on the aforementioned assumptions, the governing equations are:

$$\frac{\partial \tilde{u}}{\partial x} + \frac{\partial \tilde{v}}{\partial y} = 0 \tag{6.1}$$

$$\begin{aligned}\tilde{u}\frac{\partial \tilde{u}}{\partial x} + \tilde{v}\frac{\partial \tilde{u}}{\partial y} &= \frac{\mu}{\rho}\left[(1-\lambda)+\lambda\Gamma\sqrt{2}\left(\frac{\partial \tilde{u}}{\partial y}\right)\right]\frac{\partial^2 \tilde{u}}{\partial y^2} - \frac{\sigma}{\rho}\left[B(x)\right]^2 \tilde{u} \\ &\quad - \frac{\vartheta}{k_{fH}}\tilde{u} + g\left[\beta_1(T-T_\infty)+\beta_2(T-T_\infty)^2\right]\end{aligned} \tag{6.2}$$

$$\begin{aligned}&(\rho C_P)\left[\tilde{u}\frac{\partial T}{\partial x} + \tilde{v}\frac{\partial T}{\partial y} + \tau_o\left(\begin{array}{c}\tilde{u}^2\dfrac{\partial^2 T}{\partial x^2} + \tilde{v}^2\dfrac{\partial^2 T}{\partial y^2} + \\ \left(\tilde{u}\dfrac{\partial \tilde{u}}{\partial x} + \tilde{v}\dfrac{\partial \tilde{u}}{\partial y}\right)\dfrac{\partial T}{\partial x} + \left(\tilde{u}\dfrac{\partial \tilde{v}}{\partial x} + \tilde{v}\dfrac{\partial \tilde{v}}{\partial y}\right)\dfrac{\partial T}{\partial y} + \\ 2\tilde{u}\tilde{v}\dfrac{\partial^2 T}{\partial x \partial y} - \dfrac{\mu}{\rho C_P k_{fH}}\tilde{u}^2 - \dfrac{\sigma}{\rho C_P}\left[B(x)\right]^2\tilde{u}^2\end{array}\right)\right] - \\ &\kappa\frac{\partial^2 T}{\partial y^2} - \frac{\partial}{\partial y}\left[\frac{4\sigma^*}{3k^*}\left(12T_\infty^2 T - 8T_\infty^3\right)\frac{\partial T}{\partial y}\right] - \frac{\mu}{k_{fH}}\tilde{u}^2 - \sigma\left[B(x)\right]^2\tilde{u}^2 = 0\end{aligned} \tag{6.3}$$

The relevant boundary conditions are:

$$\left.\begin{array}{l}\tilde{v} = 0, \tilde{u} = \tilde{u}_w(x) = a(x+b)^n, T = T_W \text{ at } y = \delta(x+b)^{0.5(1-n)} \\ \tilde{u} \to 0, T \to T_\infty \text{ as } y \to \infty\end{array}\right\} \tag{6.4}$$

Herein, the symbols $(\tilde{u}, \tilde{v})$ refer to the velocity component in $x-$ and $y-$ axes, $\tilde{u}_w(x)$ stretching velocity, λ material power law index, Γ material constant, k_{fH} permeability of the porous medium, g gravity, (β_1, β_2) linear and nonlinear thermal expansion coefficients, (T_W, T_∞) wall and ambient temperature, τ_o relaxation time of heat flux, σ^* the Stefan–Boltzmann constant, and k^* the mean absorption coefficient. Also, $\mu, \vartheta, \rho, \kappa, C_P$ and σ stand for dynamic viscosity, kinematic viscosity, density, thermal conductivity, heat capacity, and electrical conductivity of THF, respectively.

The equations describing the flow in the current scenario are coupled partial differential equations, which are solved using the numerical method; for this, we first have to reduce the governing equations to simpler form (non-dimensional form). For this, we invoke the following similarity transformations:

$$\left.\begin{aligned} &\eta = y\sqrt{\frac{n+1}{2}\frac{a}{\vartheta}(x+b)^{n-1}}, \psi(x,y) = \sqrt{\frac{2}{n+1}a\vartheta(x+b)^{n+1}}F(\eta), \theta = \frac{T-T_\infty}{T_W-T_\infty}, \\ &\tilde{u} = a(x+b)^n F'(\eta), \tilde{v} = -\sqrt{\frac{n+1}{2}a\vartheta(x+b)^{n-1}}\left(F(\eta)+\eta\frac{n-1}{n+1}F'(\eta)\right) \end{aligned}\right\} \tag{6.5}$$

On using (6.5), equation (6.1) is satisfied, and equations (6.2) and (6.3), along with boundary conditions (6.4), are reduced to the following form:

$$\begin{aligned} &\left[\left(\frac{n+1}{2}\right)(1-\lambda)+\lambda\left(\frac{n+1}{2}\right)^{3/2} WeF''(\eta)\right]F'''(\eta)-\left(Ha^2+K\right)F'(\eta)- \\ &\qquad \left[nF'(\eta)F'(\eta)-\left(\frac{n+1}{2}\right)F(\eta)F''(\eta)\right]+\gamma_1\left(\theta+\gamma_2\theta^2\right)=0 \end{aligned} \tag{6.6}$$

$$\begin{aligned} &0.5(n+1)\left[1+Rd\left(1+3\theta_r\theta(\eta)\right)-\varepsilon 0.5(n+1)\Pr\left[F(\eta)\right]^2\right]\theta''(\eta)+ \\ &(1+\varepsilon)\left(K+Ha^2\right)EcPr\left[F'(\eta)\right]^2+\begin{bmatrix} 0.5(n+1)\left[0.3Rd\theta_r\theta'(\eta)+\Pr F(\eta)\right]- \\ 0.25(n-3)\varepsilon Pr\begin{bmatrix}(n+1)F(\eta)+ \\ (n-1)\eta F'(\eta)\end{bmatrix}F'(\eta)\end{bmatrix}\theta'(\eta)=0 \end{aligned} \tag{6.7}$$

With:

$$F(\alpha)=\frac{1-n}{1+n}\alpha, F'(\alpha)=1, \theta(\alpha)=1, F'(\infty)\to 0, \theta(\infty)\to 0 \tag{6.8}$$

Now, defining $F(\eta)=\tilde{f}(\eta-\alpha)=\tilde{f}(\xi)$ and $\theta(\eta)=\tilde{\Theta}(\eta-\alpha)=\tilde{\Theta}(\xi)$ and substituting in equations (6.6) and (6.7), we get:

$$\begin{aligned} &\left[\left(\frac{n+1}{2}\right)(1-\lambda)+\lambda\left(\frac{n+1}{2}\right)^{3/2} We\tilde{f}''(\xi)\right]\tilde{f}'''(\xi)-\left(Ha^2+K\right)\tilde{f}'(\xi)- \\ &\left[n\tilde{f}'(\xi)\tilde{f}'(\xi)-\left(\frac{n+1}{2}\right)\tilde{f}(\xi)\tilde{f}''(\xi)\right]+\gamma_1\left[\tilde{\Theta}(\xi)+\gamma_2\left[\tilde{\Theta}(\xi)\right]^2\right]=0 \end{aligned} \tag{6.9}$$

$$0.5(n+1)\left[1+Rd\left(1+3\theta_r\tilde{\Theta}(\xi)\right)-\varepsilon 0.5(n+1)\Pr\left[\tilde{f}(\xi)\right]^2\right]\tilde{\Theta}''(\xi)+$$
$$(1+\varepsilon)\left(K+Ha^2\right)EcPr\left[\tilde{f}'(\xi)\right]^2+\begin{bmatrix}0.5(n+1)\left[0.3Rd\theta_r\tilde{\Theta}'(\xi)+\Pr\tilde{f}(\xi)\right]- \\ 0.25(n-3)\varepsilon Pr\begin{bmatrix}(n+1)\tilde{f}(\xi)+ \\ (n-1)\eta\tilde{f}'(\xi)\end{bmatrix}\tilde{f}'(\xi)\end{bmatrix}\tilde{\Theta}'(\xi)=0 \tag{6.10}$$

And the boundary conditions (6.8) are reduced to:

$$\tilde{f}(0)=\frac{1-n}{1+n}\alpha,\ \tilde{f}'(0)=1,\tilde{\Theta}(0)=1,\ \tilde{f}'(\infty)\to 0,\ \tilde{\Theta}(\infty)\to 0 \tag{6.11}$$

Here, the symbols $\alpha, We, Ha^2, Pr, \theta_r, K, \gamma_1, \gamma_2, Rd, \varepsilon$ and Ec stand for thickness parameter, Weissenberg number, Hartmann number, Prandtl number, temperature ratio, porosity parameter, mixed convection variable, nonlinear thermal convection parameter, radiation parameter, thermal relaxation parameter, and Eckert number, respectively, and their definitions are given as:

$$\left.\begin{aligned}&\alpha=\delta\sqrt{\frac{n+1}{2}\frac{a}{\vartheta}},\ We=\Gamma\sqrt{\frac{2\left[a^3(x+b)^{3n-1}\right]}{\vartheta}},\ Ha^2=\frac{\sigma B_o^{\ 2}}{\rho a},\ Pr=\frac{\mu C_P}{\kappa},\ \theta_r=\frac{T_W}{T_\infty}-1,\\&K=\frac{\vartheta}{k_{fH}a}(x+b)^{1-n},\gamma_1=\frac{g\left(T_W-T_\infty\right)\beta_1}{a^2(x+b)^{2n-1}},\gamma_2=\frac{\left(T_W-T_\infty\right)\beta_2}{\beta_1},\ Rd=\frac{16\sigma^* T_\infty^3}{3k^*\kappa},\\&\varepsilon=a\tau_o(x+b)^{n-1},\ Ec=\frac{a^2(x+b)^{2n}}{C_P\left(T_W-T_\infty\right)}\end{aligned}\right\} \tag{6.12}$$

The quantities of physical importance, that is, skin friction coefficient (SFC) and local Nusselt number (LNN) at the surface of the sheet in non-dimension form, is given as:

$$\left.\begin{aligned}&C_f\sqrt{Re_x}=2(1-\lambda)\left(\frac{n+1}{2}\right)^{0.5}\tilde{f}''(0)+\lambda We\left(\frac{n+1}{2}\right)\left[\tilde{f}''(0)\right]^2\\&Nu/\sqrt{Re_x}=-\left(\frac{n+1}{2}\right)^{0.5}\left[1+Rd\left(1+3\theta_r\tilde{\Theta}(0)\right)\right]\tilde{\Theta}'(0)\end{aligned}\right\} \tag{6.13}$$

Here, $Re_x=\frac{a}{\vartheta}(x+b)^{n+1}$ is the local Reynolds number.

6.3 METHODOLOGY

The expressions governing the flow in the present study are nonlinear ODEs, which consist of one third-order ODE (see equation (6.9)) and one second-order ODE (see equation (6.10)). In order to solve the system using RKF, first reduce the system of first-order ODEs. For this, the following variables are introduced $\tilde{f}=p_1, \tilde{f}'=p_2, \tilde{f}''=p_3, \tilde{\Theta}=p_4, \tilde{\Theta}'=p_5$.

Using these variables, equations (6.9)–(6.10) are reduced to a system of five first-order ODEs. The matrix representation of the same is given here:

$$\left.\begin{array}{l} p_1' = p_2;\ p_1 = \dfrac{1-n}{1+n}\alpha \\ p_2' = p_3;\ p_2 = 1 \\ p_3' = \left[\left(\dfrac{n+1}{2}\right)(1-\lambda)+\lambda\left(\dfrac{n+1}{2}\right)^{1.5} Wep_3\right]^{-1} \\ \left[\left(Ha^2+K\right)p_2+\left(n\left(p_2\right)^2-\dfrac{(n+1)}{2}p_1p_3\right)-\gamma_1p_4\left(1+\gamma_2p_4\right)\right];\ p_3 = x_1\ (\text{say}) \\ p_4' = p_5;\ p_4 = 1 \\ p_5' = \dfrac{-2\left[1+Rd\left(1+3\theta_r p_4\right)-0.5\varepsilon(n+1)\Pr\left(p_1\right)^2\right]^{-1}}{(n+1)} \\ \left[Pr\left[\begin{array}{l}(1+\varepsilon)\left(K+Ha^2\right)Ec\left(p_2\right)^2+ \\ 0.5(n+1)p_1p_5+ \\ 0.25\varepsilon\left[\begin{array}{l}\left(-n^2+2n+3\right)p_1+ \\ \left(-n^2+4n-3\right)\eta p_2\end{array}\right]p_2p_5\end{array}\right]+1.5(n+1)Rd\theta_r\left(p_4\right)^2\right];\ p_5 = x_2\ (\text{say}) \end{array}\right\} \quad (6.14)$$

To solve the preceding system, that is, equation (6.14), five initial conditions are necessary, but only three are known. The missing initial conditions, that is, p_3 and p_5, are obtained by providing some initial guess value to these unknowns and solving the preceding system of equations. This procedure is iterated until the following condition is met: $\max\left\{\left|p_2\left(\xi\to\infty\right)-0\right|,\left|p_4\left(\xi\to\infty\right)-0\right|\right\}\le 10^{-5}$ (error tolerance). After getting all the initial conditions, we solve this system of simultaneous equations by employing the RKF method.

The correctness of the technique was appraised by comparing the flouts on the skin friction coefficient among the current scheme versus the prevailing outcomes offered in Akbar et al. [2]. The consistency relation present in all the analysis is summarized in Table 6.1. However, the conclusions of the study currently underway are very correct.

6.4 RESULT AND DISCUSSION

The following discussion is based on the numerical outcomes attained from the model itemized in the last section. The effects of potential parameters λ, We, Ha^2, α, K, Ec and Rd are shown in this section, keeping the other parameters invariant. The problem is modeled for tangent hyperbolic fluid; hence, the value of $Pr = 7.38$ (Jamshed et al. [17]). The effect of the aforementioned parameters on velocity $\left(\tilde{f}'(\xi)\right)$ and temperature $\left(\tilde{\Theta}(\xi)\right)$ profile is presented by graphs, and the corresponding physical quantities, like skin friction coefficient $\left(C_f\sqrt{Re_x}\right)$ and local Nusselt number $\left(Nu/\sqrt{Re_x}\right)$, are shown through Table 6.2.

6.4.1 Analysis of Velocity and Temperature Profiles

The effect of growing λ on velocity $\tilde{f}'(\xi)$ profiles is shown in Figure 6.2. It is observed from Figure 6.2 that velocity as well as the associated boundary layer thickness decrease with growing λ. This is because with the enhancement in the values of λ, the fluid undergoes a transition from shear thinning to shear thickening; as a result, the fluid viscosity increases. This increasing viscosity of the fluid exerts a retarding force between the fluid layers, and hence, the fluid velocity decreases. Figure 6.3 shows the impact of the growing Ha^2 on velocity profile. Since increasing Ha^2 results

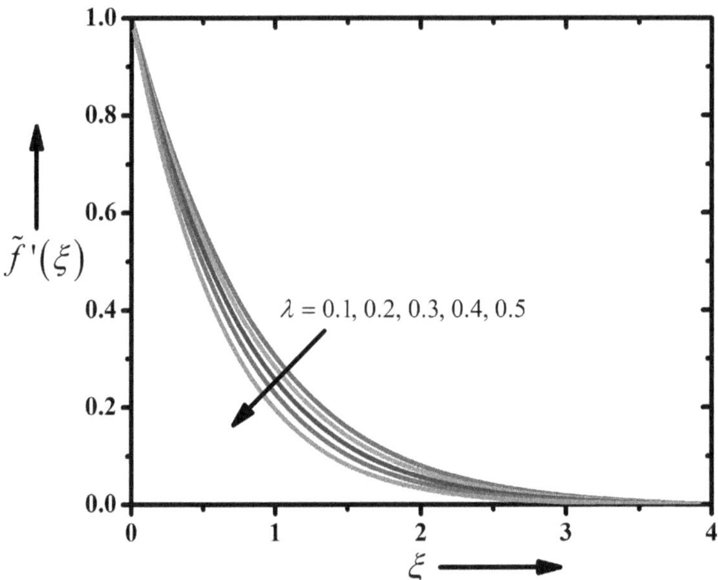

FIGURE 6.2 Velocity profiles due to λ.

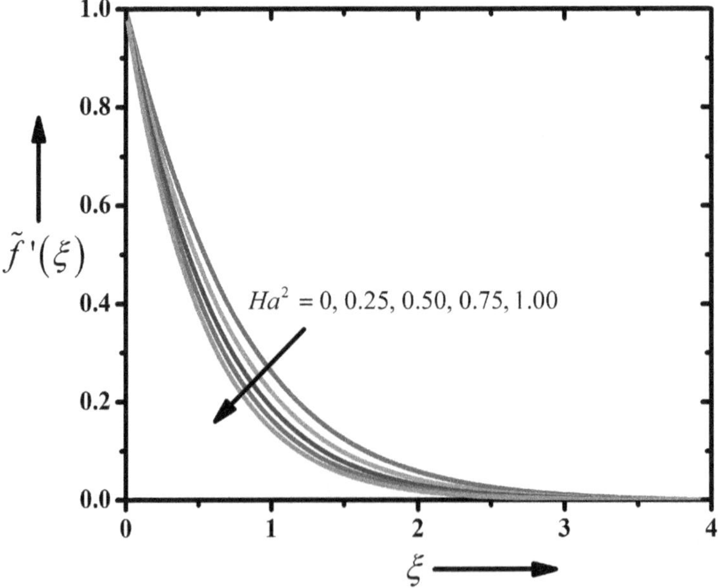

FIGURE 6.3 Velocity profiles due to Ha^2.

in the generation of Lorentz force, this force acts in the direction opposite the fluid flow and hence retards the motion of the fluid, that is, velocity decreases.

The Weissenberg number *We* reflects the deformation anisotropy level and is utilized for explaining the stretching process, like natural shear, and strengthens the non-Newtonian nature of the ephemeral fluid through frictional manipulations, which make the fluidity durable. This is depicted in Figure 6.4, that the flow slows down with the escalating values of *We*. The influence of the thickness parameter (α) on $\tilde{f}'(\xi)$ is shown in Figure 6.5. And this figure shows that velocity decreases

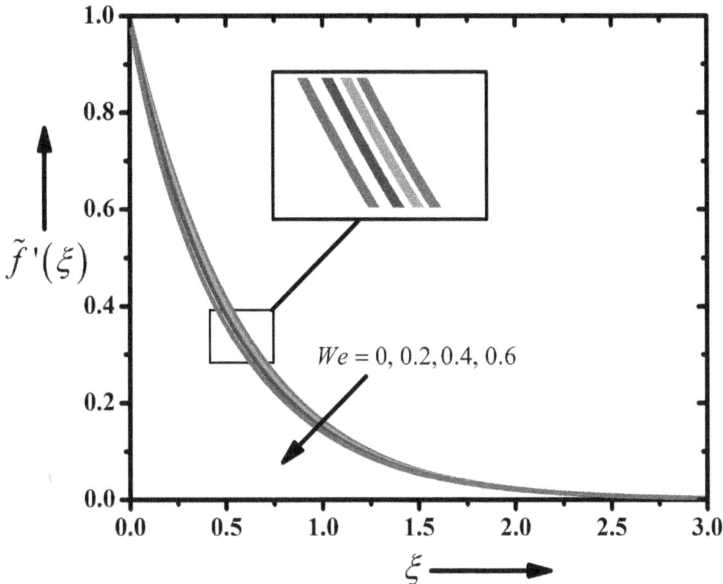

FIGURE 6.4 Velocity profiles due to *We*.

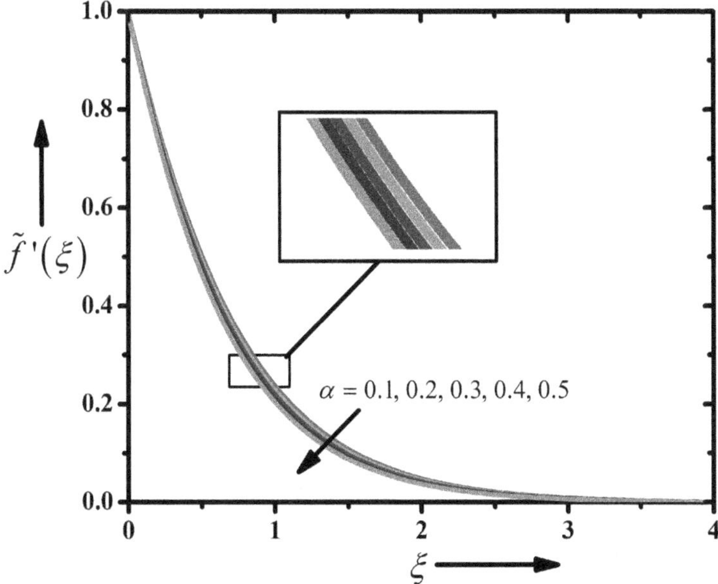

FIGURE 6.5 Velocity profiles due to α.

due to the escalation in the values of α. The impact of the porosity parameter K on $\tilde{f}'(\xi)$ is shown in Figure 6.6; it is discovered from the figure that, as the porosity of the surface increases, the fluid experiences a retarding force. Figure 6.7 reflects the effect of *Rd* on the velocity profile. Usually, the velocity profile does not undergo any variation with the fluctuation in *Rd* parameter, but the flow is subjected to quadratic convection; hence, the fluid flow increases with enhancement in the radiation parameter.

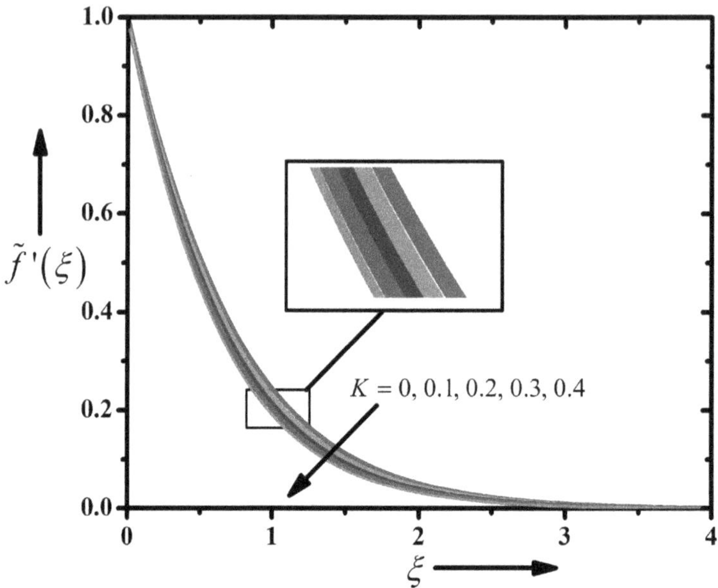

FIGURE 6.6 Velocity profiles due to *K*.

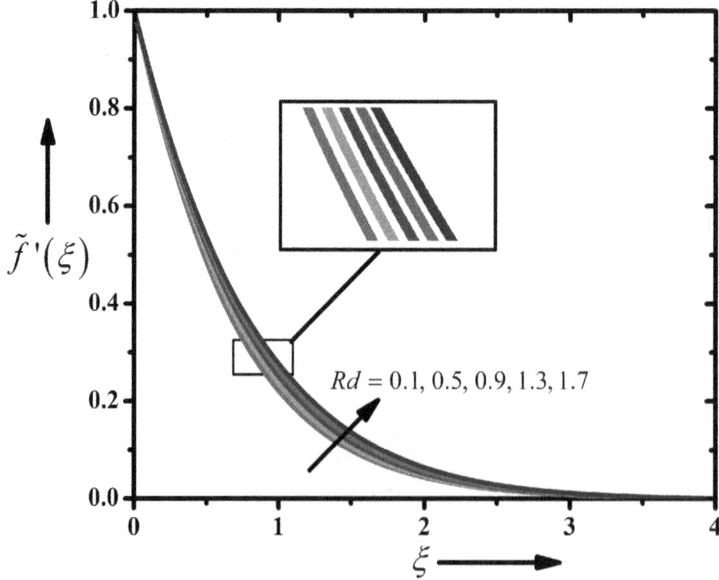

FIGURE 6.7 Velocity profiles due to *Rd*.

From Figures 6.2–6.4 we observe that fluid velocity slows down with growing values of λ, We and Ha^2, respectively. Due to this, the fluid gets enough time for thermal absorption from the surface, which leads to an increase in the temperature profile (see Figures 6.8–6.10). However, a contradictory pattern is recorded for the temperature profile corresponding to α (see Figure 6.11).

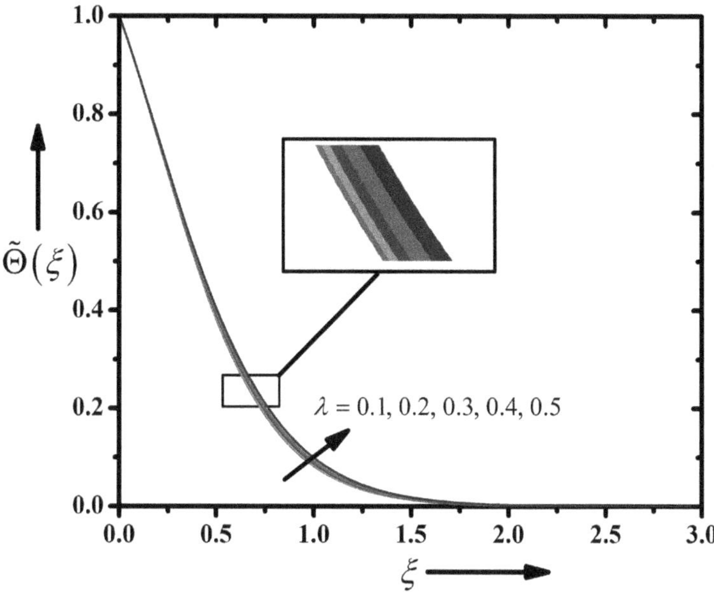

FIGURE 6.8 Temperature profiles due to λ.

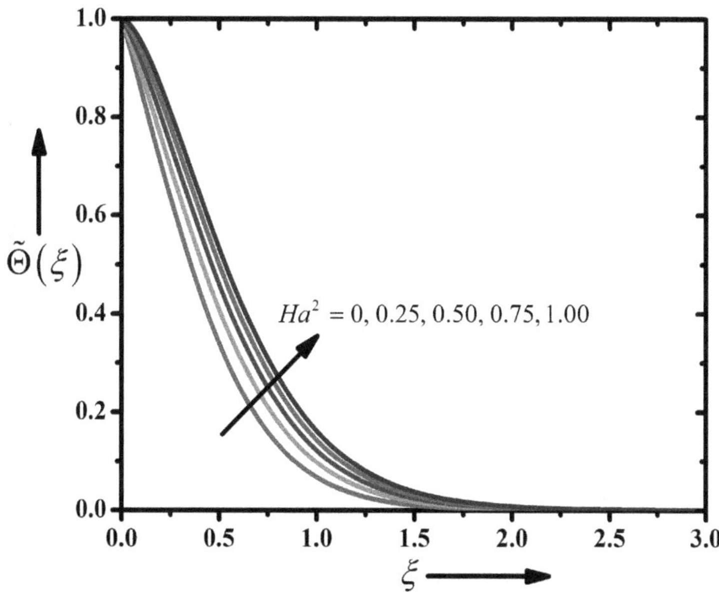

FIGURE 6.9 Temperature profiles due to Ha^2.

Figures 6.12 and 6.13 depict the influence of a growing *K* and *Rd* on temperature profiles, respectively. It is seen from these figures that $\tilde{\Theta}(\xi)$ is an escalating function of *K* and *Rd*. Figure 6.14 shows the response of the Eckert number on the temperature profile, and the figure depicts that enhancement in the E_c results in intensifying the fluid temperature. Also, the thermal boundary layer becomes wider with growing E_c.

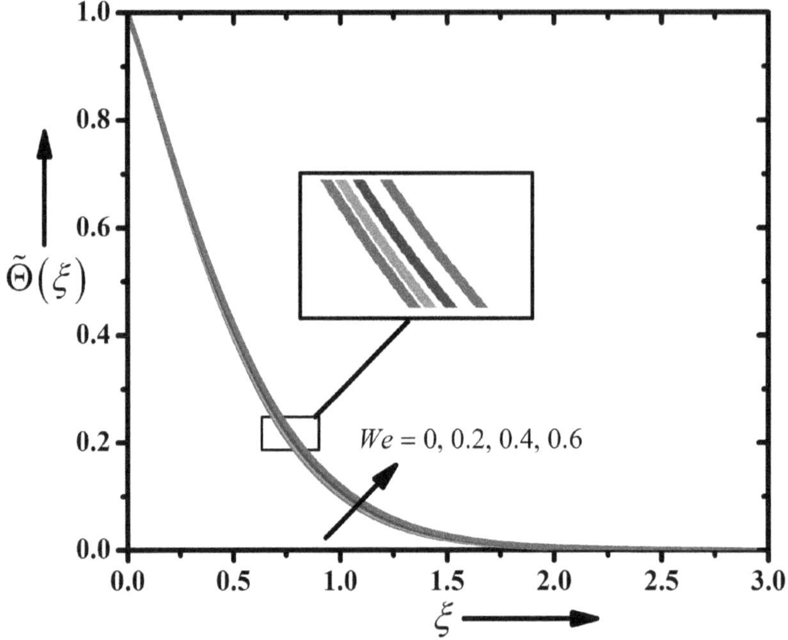

FIGURE 6.10 Temperature profiles due to *We*.

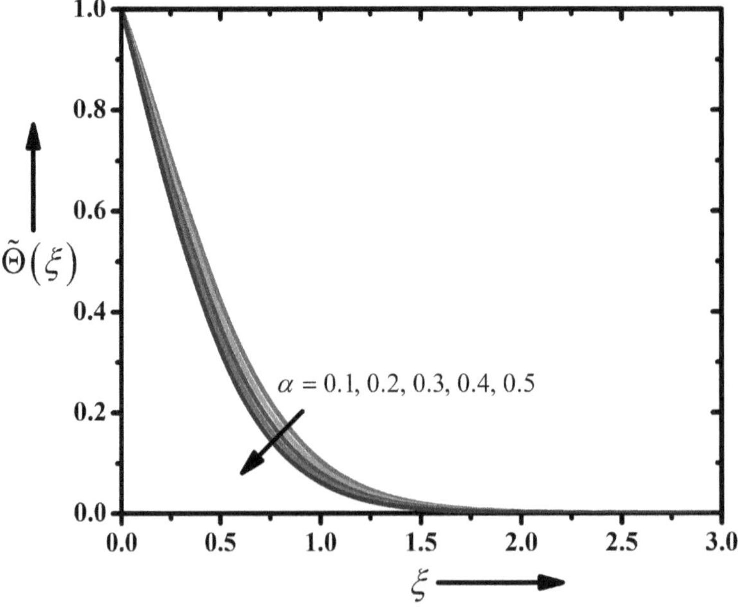

FIGURE 6.11 Temperature profiles due to α.

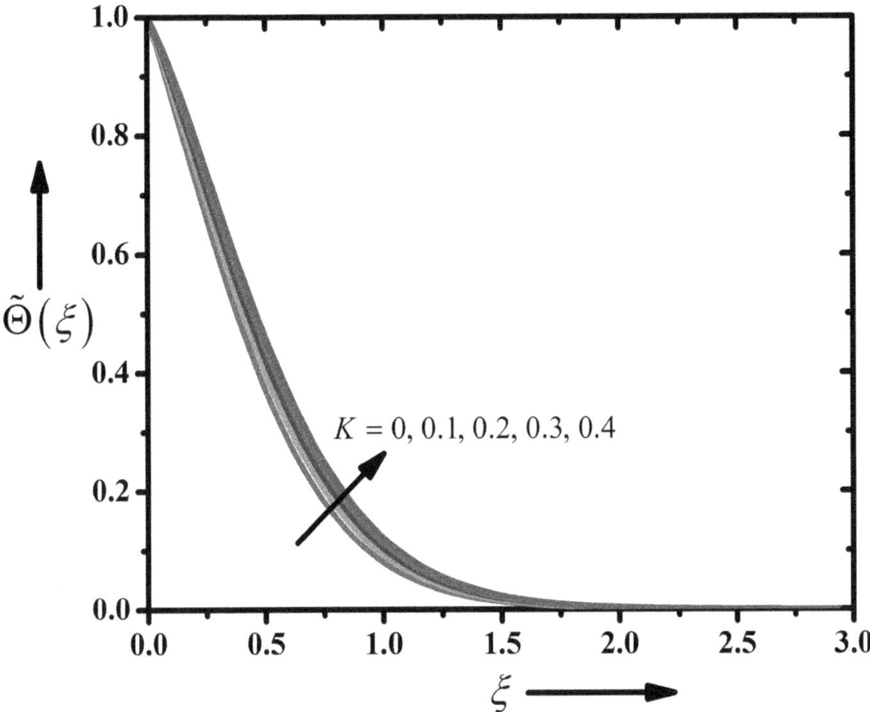

FIGURE 6.12 Temperature profiles due to K.

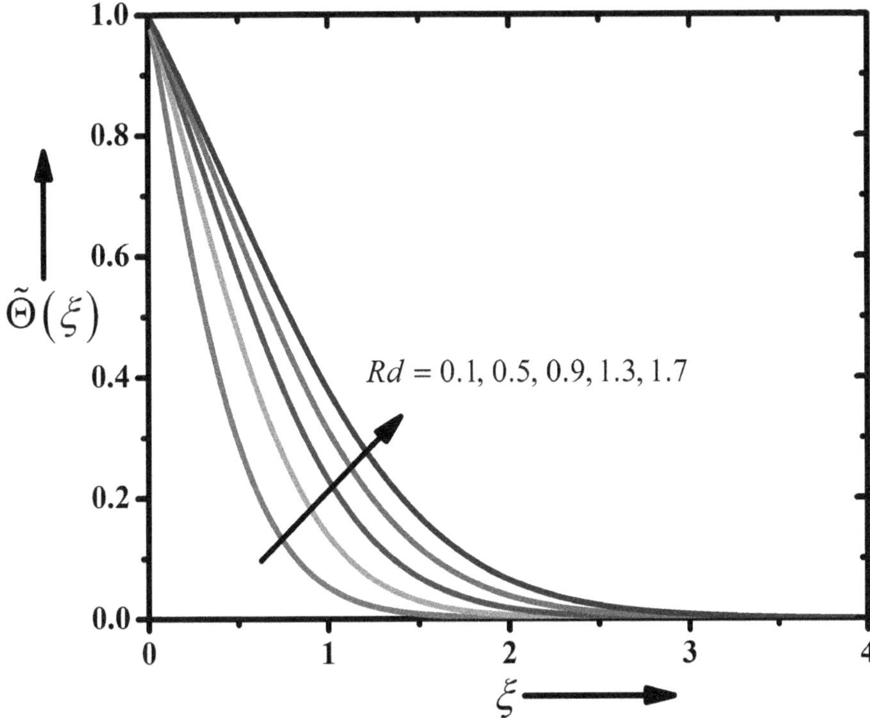

FIGURE 6.13 Temperature profiles due to Rd.

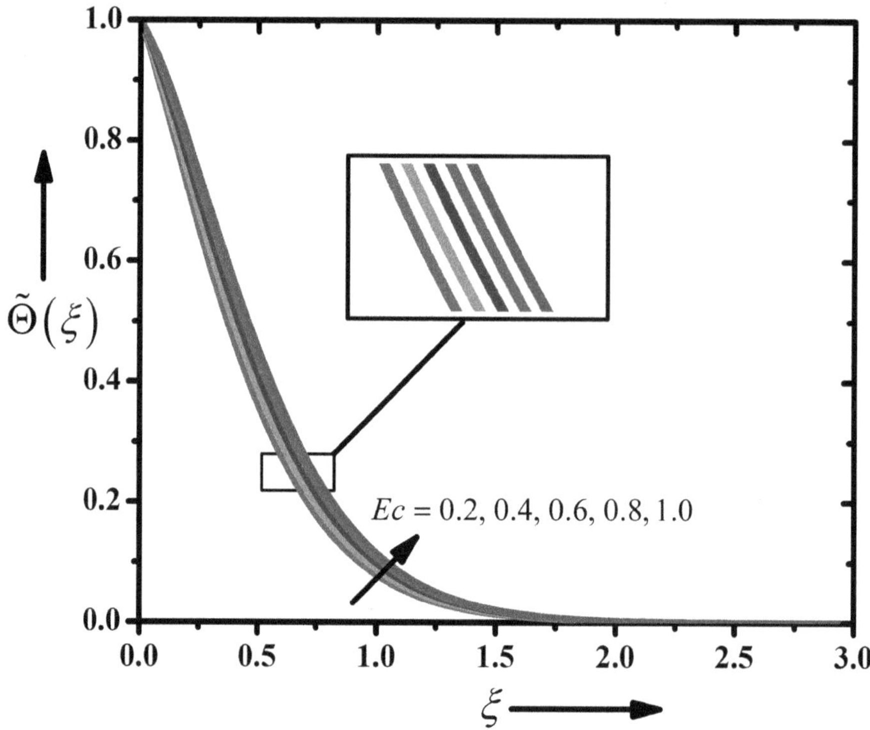

FIGURE 6.14 Temperature profiles due to *Ec*.

6.4.2 Statistical Analysis of SFC and LNN

To understand the relationship between the parameters present in the study and the physical quantities, that is, $C_f\sqrt{Re_x}$ and $Nu/\sqrt{Re_x}$, we extend our work by exploring statistical tools like correlation coefficient (r) and probable error (PE). The values of SFC and LNN for the different parameters are reported in Table 6.2, and the values of the correlation coefficient for these physical quantities and considered parameters are shown in Table 6.3. The correlation coefficient not only investigated the linear relation among the two variables but also revealed the inverse and direct correspondence among them. As the value of r varies in the interval [−1, 1], the following are the interpretations based on it:

- Perfect negative relation between two variables when $r = -1$
- Strong negative relation between two variables when $-1 < r \leq -0.7$
- No relation between the variables when $r = 0$
- Strong positive relation between two variables when $0.7 \leq r < 1$
- Perfect positive relation between two variables when $r = 1$

Based on these interpretations, it is observed that there is either a strong positive or strong negative correlation among the varying quantities and SFC as well as LNN (see Table 6.3).

The probable error (PE) is used to determine the reliability of the calculated correlation coefficient. The mathematical expression for calculating this entity is:

$$PE = \frac{0.6745\left(1-r^2\right)}{\sqrt{\text{Total number of observarions}}} \tag{6.15}$$

TABLE 6.1
Comparison of $C_f\sqrt{Re_x}$ Values with Different Values of Ha^2

	$C_f\sqrt{Re_x}$	
Ha^2	Akbar et al. 2013	Present Result
1	−1.41421	−1.4305
5	−2.44948	−2.4499
10	−3.31662	−3.3166
50	−7.14142	−7.1414

TABLE 6.2
Values of $C_f\sqrt{Re_x}$ and $Nu/\sqrt{Re_x}$ at $Pr = 7.38$, $n = 0.5$, $\varepsilon = 0.2$

λ	We	Ha^2	K	α	Rd	Ec	$C_f\sqrt{Re_x}$	$Nu/\sqrt{Re_x}$
0.1	0.2	0.2	0.1	0.2	0.3	0.5	−1.632444	1.662496
0.2							−1.523759	1.659205
0.3							−1.406407	1.653992
0.4							−1.277495	1.645962
0.5							−1.132074	1.633549
0.4	0.0	0.4	0.5			0.1	−1.760900	1.847181
	0.2						−1.691950	1.831261
	0.4						−1.609910	1.811717
	0.6						−1.496090	1.784054
	0.2	0.0	0.1			0.5	−1.125920	2.155016
		0.25					−1.313260	1.530394
		0.50					−1.481470	1.006968
		0.75					−1.634860	0.554149
		1.00					−1.776430	0.154110
		0.2	0				−1.203530	1.890441
			0.1				−1.277490	1.645962
			0.2				−1.348260	1.418827
			0.3				−1.416170	1.206521
			0.4				−1.481460	1.006862
			0.1	0.1			−1.241920	1.470027
				0.2			−1.277430	1.645644
				0.3			−1.313470	1.828178
				0.4			−1.349830	2.016308
				0.5			−1.386650	2.210599
				0.2	0.1		−1.305270	1.444542
					0.5		−1.256960	1.801773
					0.9		−1.225410	2.076749
					1.3		−1.201030	2.310559
					1.7		−1.181090	2.514669
					0.3	0.2	−1.288400	2.055082
						0.4	−1.281200	1.783527
						0.6	−1.273850	1.507906
						0.8	−1.266370	1.228219
						1.0	−1.258750	0.944372

TABLE 6.3
Numerical Values of r, PE and r / PE of SFC and LNN

	$C_f\sqrt{Re_x}$			$Nu/\sqrt{Re_x}$		
Parameter	**r**	**PE**	**r / PE**	**r**	**PE**	**r / PE**
λ	0.998324	0.001010489	987.9607	−0.969540	0.018095816	-53.57820
We	0.993289	0.004511549	220.1658	−0.991970	0.005395587	−183.8480
Ha^2	−0.998460	0.000929026	−1074.740	−0.996080	0.002359185	−422.2140
K	−0.999700	0.000182348	−5482.370	−0.999200	0.000483536	−2066.440
α	−0.999980	1.42736E-05	−70057.60	0.999808	0.000115923	8624.733
Rd	0.983792	0.009698717	101.4353	0.993712	0.00378181	262.7608
Ec	0.999939	3.68994E-05	27099.08	−0.999960	2.29364E-05	−43597.10

The values of *PE* and *r*/*PE* correspond to physical quantities (SFC and LNN), and the considered parameters are computed and shown in Table 6.3. The correlation coefficient is said to be significant if $\left|\frac{r}{PE}\right| > 6$. Thus, SFC has a significant positive correlation with λ, *We*, *Rd*, *Ec*, and a significant negative correlation is observed for Ha^2, K, α. Also, it is observed that for LNN, λ, *We*, Ha^2, K, *Ec* have a significant negative correlation, whereas there is significant positive correlation for α, *Rd*.

6.5 CONCLUSIONS

A statistical analysis is performed for the magnetic tangent hyperbolic fluid flow over a porous stretching sheet using quadratic convection. The following outcomes were obtained:

- Velocity profiles of tangent hyperbolic fluid increased with escalating *Rd*.
- Temperature profiles of working fluid declined with increase in α.
- Skin friction coefficient has significant positive correlation with λ, *We*, *Rd*, *Ec*.
- Local Nusselt number has significant positive correlation for α, *Rd*.
- Velocity outlines depreciated with increase in porosity parameter and Hartmann number.
- Temperature outlines accelerated with increase in λ, *We*, Ha^2, K, *Ec* and *Rd*.

Nomenclature

$(\tilde{u}, \tilde{v})$	velocity component in $x-$ and $y-$ axes
T	temperature
α	thickness parameter
We	Weissenberg number
Ha^2	Hartmann number
Pr	Prandtl number
θ_r	temperature ratio
K	porosity parameter
γ_1	mixed convection variable
γ_2	nonlinear thermal convection parameter
Rd	radiation parameter
λ	material power law index

k_{fH}	permeability of porous medium
τ_o	relaxation time of heat flux
(β_1, β_2)	linear and nonlinear thermal expansion coefficients
σ^*	Stefan–Boltzmann constant
k^*	mean absorption coefficient
ε	thermal relaxation parameter

REFERENCES

1. Nadeem, S., Akbar, N.S. Series solutions for the peristaltic flow of a Tangent hyperbolic fluid in a uniform inclined tube. *ZeitschriftFürNaturforschung A*, 2010, 65(11): 887–895.
2. Akbar, N.S., Nadeem, S., Haq, R.U., Khan, Z.H. Numerical solutions of magnetohydrodynamic boundary layer flow of tangent hyperbolic fluid towards a stretching sheet. *Indian Journal of Physics*, 2013, 87(11): 1121–1124.
3. Gaffar, S.A., Prasad, V.R., Bég, O.A. Numerical study of flow and heat transfer of non-Newtonian tangent hyperbolic fluid from a sphere with Biot number effects. *Alexandria Engineering Journal*, 2015, 54(4): 829–841.
4. Hayat, T., Qayyum, S., Ahmad, B., Waqas, M. Radiative flow of a tangent hyperbolic fluid with convective conditions and chemical reaction. *The European Physical Journal Plus*, 2016, 131(12): 1–13.
5. Khan, M.I., Hayat, T., Waqas, M., Alsaedi, A. Outcome for chemically reactive aspect in flow of tangent hyperbolic material. *Journal of Molecular Liquids*, 2017, 230: 143–151.
6. Khan, M., Hussain, A., Malik, M.Y., Salahuddin, T., Khan, F. Boundary layer flow of MHD tangent hyperbolic nanofluid over a stretching sheet: A numerical investigation. *Results in Physics*, 2017, 7: 2837–2844.
7. Hussain, A., Malik, M.Y., Salahuddin, T., Rubab, A., Khan, M. Effects of viscous dissipation on MHD tangent hyperbolic fluid over a nonlinear stretching sheet with convective boundary conditions. *Results in Physics*, 2017, 7: 3502–3509.
8. Ibrahim, W. Magnetohydrodynamics (MHD) flow of a tangent hyperbolic fluid with nanoparticles past a stretching sheet with second order slip and convective boundary condition. *Results in Physics*, 2017, 7: 3723–3731.
9. Mahanthesh, B., Kumar, P.S., Gireesha, B.J., Manjunatha, S., Gorla, R.S.R. Nonlinear convective and radiated flow of tangent hyperbolic liquid due to stretched surface with convective condition. *Results in Physics*, 2017, 7: 2404–2410.
10. Kumar, K.G., Gireesha, B.J., Gorla, R.S.R. Flow and heat transfer of dusty hyperbolic tangent fluid over a stretching sheet in the presence of thermal radiation and magnetic field. *International Journal of Mechanical and Materials Engineering*, 2018, 13(1): 1–11.
11. Rehman, K.U., Alshomrani, A.S., Malik, M.Y., Zehra, I., Naseer, M. Thermo-physical aspects in tangent hyperbolic fluid flow regime: A short communication. *Case Studies in Thermal Engineering*, 2018, 12: 203–212.
12. Saidulu, N., Gangaiah, T., Lakshmi, A.V. Radiation effect on MHD flow of a tangent hyperbolic nanofluid over an inclined exponentially stretching sheet. *International Journal of Fluid Mechanics Research*, 2019, 46(3): 277–293.
13. Ramzan, M., Gul, H., Sheikholeslami, M. Effect of second order slip condition on the flow of tangent hyperbolic fluid—A novel perception of Cattaneo–Christov heat flux. *PhysicaScripta*, 2019, 94(11): 115707.
14. Ganesh Kumar, K., Baslem, A., Prasannakumara, B.C., Majdoubi, J., Rahimi-Gorji, M., Nadeem, S. Significance of Arrhenius activation energy in flow and heat transfer of tangent hyperbolic fluid with zero mass flux condition. *Microsystem Technologies*, 2020, 26(8): 2517–2526.
15. Ibrahim, W., Gizewu, T. Nonlinear mixed convection flow of a tangent hyperbolic fluid with activation energy. *Heat Transfer*, 2020, 49(5): 2427–2448.
16. Ullah, Z., Zaman, G., Ishak, A. Magnetohydrodynamic tangent hyperbolic fluid flow past a stretching sheet. *Chinese Journal of Physics*, 2020, 66: 258–268.
17. Jamshed, W., Nisar, K.S., Ibrahim, R.W., Shahzad, F., Eid, M.R. Thermal expansion optimization in solar aircraft using tangent hyperbolic hybrid nanofluid: A solar thermal application. *Journal of Materials Research and Technology*, 2021, 14, 985–1006.

18. Shankaralingappa, B.M., Gireesha, B.J., Prasannakumara, B.C., Nagaraja, B. Darcy-Forchheimer flow of dusty tangent hyperbolic fluid over a stretching sheet with Cattaneo-Christov heat flux. *Waves in Random and Complex Media*, 2021, http://doi.org/10.1080/17455030.2021.1889711.
19. Usman, M., Zubair, T., Hamid, M., Haq, R.U., Khan, Z.H. Unsteady flow and heat transfer of tangent-hyperbolic fluid: Legendre wavelet-based analysis. *Heat Transfer*, 2021, 50(4), 3079–3093.
20. Das, M., Nandi, S., Kumbhakar, B., Shanker Seth, G. Soret and dufour effects on MHD nonlinear convective flow of tangent hyperbolic nanofluid over a bidirectional stretching sheet with multiple slips. *Journal of Nanofluids*, 2021, 10(2): 200–213.
21. Hussain, S.M., Jamshed, W. A comparative entropy based analysis of tangent hyperbolic hybrid nanofluid flow: Implementing finite difference method. *International Communications in Heat and Mass Transfer*, 2021, 129: 105671.
22. Ali, B., Thumma, T., Habib, D., Riaz, S. Finite element analysis on transient MHD 3D rotating flow of Maxwell and tangent hyperbolic nanofluid past a bidirectional stretching sheet with Cattaneo Christov heat flux model. *Thermal Science and Engineering Progress*, 2022, 28: 101089.
23. Zeb, S., Khan, S., Ullah, Z., Yousaf, M., Khan, I., Alshammari, N., Alam, N., Hamadneh, N.N. Lie group analysis of double diffusive MHD tangent hyperbolic fluid flow over a stretching sheet. *Mathematical Problems in Engineering*, 2022, http://doi.org/10.1155/2022/9919073
24. Fourier, J.B.J. Théorie analytique de la chaleur: Paris. *Académie des Sciences*, 1822.
25. Cattaneo, C. Sulla conduzionedelcalore. *Atti del Seminario Matematico e Fisico dell' Universita di Modena*, 1948, 3: 83–101.
26. Christov, C.I. On frame indifferent formulation of the Maxwell–Cattaneo model of finite-speed heat conduction. *Mechanics Research Communications*, 2009, 36(4): 481–486.
27. Hayat, T., Farooq, M., Alsaedi, A., Al-Solamy, F. Impact of Cattaneo-Christov heat flux in the flow over a stretching sheet with variable thickness. *AIP Advances*, 2015, 5(8): 087159.
28. Salahuddin, T., Malik, M.Y., Hussain, A., Bilal, S., Awais, M. MHD flow of Cattanneo–Christov heat flux model for Williamson fluid over a stretching sheet with variable thickness: Using numerical approach. *Journal of Magnetism and Magnetic Materials*, 2016, 401: 991–997.
29. Dogonchi, A.S., Ganji, D.D. Effect of Cattaneo–Christov heat flux on buoyancy MHD nanofluid flow and heat transfer over a stretching sheet in the presence of Joule heating and thermal radiation impacts. *Indian Journal of Physics*, 2018, 92(6): 757–766.
30. Madhukesh, J.K., Kumar, R.N., Gowda, R.P., Prasannakumara, B.C., Ramesh, G.K., Khan, M.I., Khan, S.U. and Chu, Y.M. Numerical simulation of AA7072-AA7075/water-based hybrid nanofluid flow over a curved stretching sheet with Newtonian heating: A non-Fourier heat flux model approach. *Journal of Molecular Liquids*, 2021, 335: 116103.
31. Imtiaz, M., Mabood, F., Hayat, T., Alsaedi, A. Impact of non-Fourier heat flux in bidirectional flow of carbon nanotubes over a stretching sheet with variable thickness. *Chinese Journal of Physics*, 2022, 77: 1587–1597.
32. Ullah, Z., Ullah, I., Zaman, G., Sun, T.C. A numerical approach to interpret melting and activation energy phenomenon on the magnetized transient flow of Prandtl–Eyring fluid with the application of Cattaneo–Christov theory. *Waves in Random and Complex Media*, 2022, https://doi.org/10.1080/17455030.2022.2032472.
33. Chabani, I., Mebarek Oudina, F., Ismail, A.I. MHD flow of a hybrid nano-fluid in a triangular enclosure with zigzags and an elliptic obstacle. *Micromachines*, 2022, 13(2): 224.
34. Mebarek Oudina, F., Chabani, I. Review on nano-fluids applications and heat transfer enhancement techniques in different enclosures. *Journal of Nanofluids*, 2022, 11(2): 155–168.
35. Dadheech, P.K., Agrawal, P., Mebarek-Oudina, F., Abu-Hamdeh, N.H., Sharma, A. Comparative heat transfer analysis of $MoS_2/C_2H_6O_2$ and SiO_2-$MoS_2/C_2H_6O_2$ nanofluids with natural convection and inclined magnetic field. *Journal of Nanofluids*, 2020, 9(3): 161–167.
36. Rajashekhar, C., Mebarek-Oudina, F., Vaidya, H., Prasad, K.V., Manjunatha, G., Balachandra, H. Mass and heat transport impact on the peristaltic flow of a Ree–Eyring liquid through variable properties for hemodynamic flow. *Heat Transfer*, 2021, 50(5): 5106–5122.
37. Asogwa, K.K., Mebarek-Oudina, F., Animasaun, I.L. Comparative investigation of water-based Al_2O_3nanoparticles through water-based CuO nanoparticles over an exponentially accelerated radiative Riga plate surface via heat transport. *Arabian Journal for Science and Engineering*, 2022, 47(7): 8721–8738.

7 Effect of Mass Transpiration on a Non-Newtonian Fluid Flow in a Porous Medium with the Cattaneo–Christov Heat Flux Model and Radiation

R. Mahesh, U. S. Mahabaleshwar, and Basma Souayeh

7.1 INTRODUCTION

Nowadays, fluid flow over a stretching surface is becoming increasingly important among scientists due to its manufacturing and industrial applications, such as the manufacturing of liquid films in the condensation process and extraction of rubber sheets and polymer, the production of glass fiber and paper, and so on. Crane [1] was the first to notice that a linearly stretched sheet flow could be solved analytically. Many other studies eventually confirmed Crane's solution structure, demonstrating that the exponential solution may be applied to a variety of different fluids, including non-Newtonian fluids. An exponentially closed solution was originally developed by Troy et al. [2] for elastic-viscous fluid flows induced by linear expansion plates. There are several additional explanations for the elastic-viscosity flow associated with a stretching sheet in references [3–5]. In these investigations, the authors focused on the effects of magnetic fields, porous media, and other factors on flow. There have been some studies on the uniqueness of stretched bodies [6–13].

The phenomenon of heat transfer is engaged in a wide range of industrial and technical operations, including energy generation, nuclear reactor cooling, and medicinal applications, such as heat conduction in tissues, magnetic drug targeting, and many more. Heat transfer is a natural phenomenon that arises when there is a temperature differential between two bodies or inside the same physical body. Fourier's law [14] of heat transmission has been used to examine the characteristics of heat transmission throughout the last two decades. However, this paradigm is insufficient in any initial disruption sensed quickly across the whole material. In order to address this problem, Cattaneo [15] included a thermal relaxation time into the conventional Fourier's of heat conduction, allowing heat to be transported by the circulation of thermal waves at a finite speed. Christov [16] then restructured Cattaneo rule by employing relaxation time and Oldroyd's upper-convected derivatives to get the material–invariant formulation.

Straughan [17] used gravity to test this hypothesis in an incompressible fluid layer. It was established in his study that when the Cattaneo number rises, stationary convection with synthetic cells turns into oscillatory convection with narrow cells. He went on to conduct more extensive research on acoustic waves in a Cattaneo–Christov gas [18]. The possibility of a transverse wave in heat transport was explicitly highlighted in this model. It was also demonstrated that an acoustic wave may travel alongside a thermal wave. It is worth noting that the flow under consideration is considered inviscid. Han et al. [19] show the effect of viscoelastic fluid with the Cattaneo–Christov heat flux model.

DOI: 10.1201/9781003299608-7

In Mustafa [20], the rotating flow of upper-convected Maxwell fluid exposed to the newly presented heat flux model was investigated using the homotopy analysis method. The main relevance of the study appears to be that it confirms that the relaxation time parameter is inversely linked to temperature. Using the Cattaneo–Christov model, Khan et al. [21] examined the movement of viscoelastic fluid generated by an exponentially stretching sheet. Hayat et al. [22] used homotopy to investigate the fluid flow of elastic-viscose and second-grade fluids created by a linearly extending sheet.

Liu et al. [23] investigated abnormal convection-diffusion and wave coupling transport of cells on comb frames with fractional Cattaneo–Christov heat flow. Salahuddin et al. [24] study the influence of Cattaneo–Christov heat flux on the MHD flow of a Williamson fluid across a stretched surface of varying thickness. Zehra et al. [25] investigated the double-diffusional Cattaneo–Christov heat flow model and the nonlinear Casson fluid discharged from a curved shrinking or stretching surface. It also refers to some recent research on the efficiency of the Cattaneo–Christov model [26–38].

The objective of this chapter is to extend the work of Jafarimoghaddam et al. [26] to develop a theoretical model of the Cattaneo–Christov heat flux with radiation and mass transpiration on a porous surface that expands nonlinearly under conditions of temperature and velocity distribution. The leading PDE is transformed into a highly nonlinear ODE, which is solved analytically by using the Appell's hypergeometric function of two variables to create a unique response for each temperature and velocity profile. The current findings are intended to assess their relationship. A list of correlation coefficients between the defining parameter and the physical quantity of interest in this model was calculated. The influence of the relevant parameters on flow and temperature is shown in the graphs. Finally, important findings are discussed and presented in sequence.

7.2 FORMULATION AND SOLUTION OF THE PROBLEM

The schematic exhibits the two-dimensional semi-infinite viscoelastic laminar flow of PM over a stretching sheet in Figure 7.1. The origin of the Cartesian coordinate system (x,y) is at the slot $(x,y) = (0,0)$.

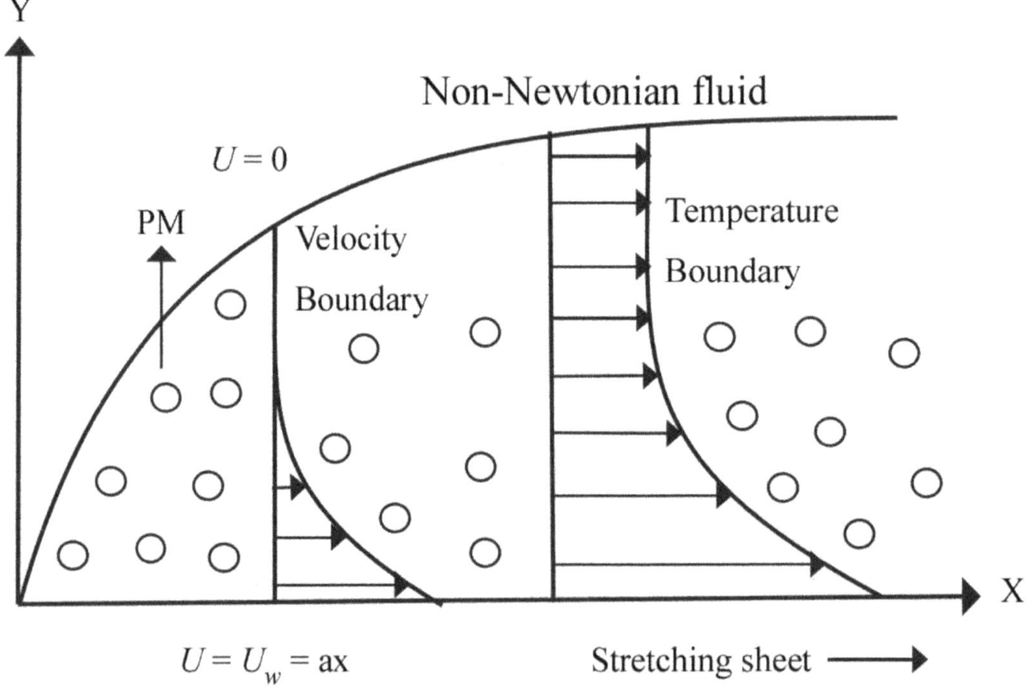

FIGURE 7.1 Physical diagram of the present flow model.

Two equal and opposing forces act along the x-axis, with the y-axis normal to the flow. The sheet is stretched at a constant pace that is proportional to the distance from the fixed origin, $x = 0$.

The current study investigates the novel properties of a steady viscoelastic fluid generated by stretching the sheet. The presence of the porous medium as well as the influence of thermal radiation in the energy equation is based on the Cattaneo–Christov model. The governing equations of two-dimensional boundary layer equations that take into account the aforementioned concerns are [26]:

$$\frac{\partial u}{\partial x}+\frac{\partial u}{\partial y}=0, \tag{7.1}$$

$$u\frac{\partial u}{\partial x}+v\frac{\partial v}{\partial y}=\upsilon\frac{\partial^2 u}{\partial y^2}-k^*\left(\frac{\partial^3 u}{\partial x\partial y^2}+v\frac{\partial^3 u}{\partial y^3}-\frac{\partial u}{\partial y}\frac{\partial^2 u}{\partial x\partial y}+\frac{\partial u}{\partial x}\frac{\partial^2 u}{\partial y^2}\right)-\frac{\mu}{\rho K}u, \tag{7.2}$$

$$\rho C_p\left(u\frac{\partial T}{\partial x}+v\frac{\partial T}{\partial y}\right)=-\nabla\cdot\mathbf{q}-q_r, \tag{7.3}$$

Here, $k^*>0$ is the elastic-viscous fluid, $k^*=0$ represents the Newtonian case, and $k^*<0$ is the second-grade fluid.

Based on the Cattaneo–Christov model [25]:

$$\mathbf{q}+\lambda\left(\frac{\partial \mathbf{q}}{\partial t}+\mathbf{V}\cdot\nabla\mathbf{q}-\mathbf{q}\cdot\nabla\mathbf{V}+(\nabla\cdot\mathbf{V})\mathbf{q}\right)=-\kappa\nabla T, \tag{7.4}$$

κ in equation (4) is the thermal conductivity, and T is the relaxation time parameter; if this is set to zero, the standard Fourier law of heat transport is obtained.

Using the Rosseland [37] approach, the radiative heat flux is given (see Mahabaleshwar et al. [6]).

$$q_r=-\frac{4\sigma^*}{3k^*}\frac{\partial T^4}{\partial y}, \tag{7.5}$$

σ^* represents the Stefan–Boltzmann constant, q_r the radiative heat flux, k^* the absorption coefficient. The term with the value T^4 is expanded using Taylor's series. It is presented by Mahabaleshwar et al. [38] as:

$$T^4={T^4}_\infty+4{T^4}_\infty(T-T_\infty)+6{T_\infty}^2(T-T_\infty)^2+\ldots \tag{7.6}$$

In the absence of higher-order terms of $(T-T_\infty)$ in the exceeding appearance absent from the primary degree, T^4 can be estimated by:

$$T^4\cong-3{T^4}_\infty+4{T^3}_\infty T, \tag{7.7}$$

By eliminating $\mathbf{q}$ from equations (7.3) and (7.4), we obtain the following:

$$u\frac{\partial T}{\partial x}+v\frac{\partial T}{\partial y}+\left\{\lambda\left(\begin{array}{l}u\frac{\partial u}{\partial x}\frac{\partial T}{\partial x}+v\frac{\partial v}{\partial y}\frac{\partial T}{\partial y}+u\frac{\partial v}{\partial x}\frac{\partial T}{\partial y}+v\frac{\partial u}{\partial y}\frac{\partial T}{\partial x}\\ +2uv\frac{\partial^2 T}{\partial x\partial y}+u\frac{\partial^2 T}{\partial x^2}+v^2\frac{\partial^2 T}{\partial y^2}\end{array}\right)\right\}=\alpha\frac{\partial^2 T}{\partial y^2}+\frac{16\sigma^* T^3}{3k^*(\rho c_p)}\frac{\partial^2 T}{\partial y^2}, \tag{7.8}$$

Here, α is the thermal diffusivity.

The boundary conditions are imposed as follows:

$$\begin{cases} u(x,0) = u_w(x,0) = ax, & v(x,0) = v_w, & u(x,y \to \infty) = 0, \\ T(x,0) = T_w, & T(x,y \to \infty) = T_\infty, & \end{cases} \tag{7.9}$$

Here, d is the stretching/shrinking constant, and a is considered a positive constant.

Now, apply the following similarity transformations [25]:

$$u = axf_\eta(\eta), \qquad v = -\sqrt{a\upsilon}\, f(\eta), \quad \theta(\eta) = \frac{T - T_w}{T_w - T_\infty}, \qquad \eta = \sqrt{\frac{a}{\upsilon}}\, y. \tag{7.10}$$

Equation (7.1) is identical prove, and equation (7.2) to equation (7.3) can be transformed into:

$$\frac{d^3 f}{d\eta^3} + f(\eta)\frac{d^2 f}{d\eta^2} - \left(\frac{df}{d\eta}\right)^2 - K_1\left[\left\{2\frac{df}{d\eta}\frac{d^3 f}{d\eta^3} - f(\eta)\frac{d^4 f}{d\eta^4} - \left(\frac{d^2 f}{d\eta^2}\right)^2\right\}\right] - Da^{-1}\frac{df}{d\eta} = 0, \tag{7.11}$$

$$(1+N_r)\frac{d^2\theta}{d\eta^2} + \Pr\left(f(\eta)\frac{d\theta}{d\eta}\right) - \gamma \Pr\left(f(\eta)\frac{df}{d\theta}\frac{d\theta}{d\eta} + (f(\eta))^2\frac{d^2\theta}{d\eta^2}\right) = 0, \tag{7.12}$$

In addition to the following boundary conditions:

$$\begin{cases} \dfrac{df}{d\eta} = 1, & f(\eta) = V_c, \quad \theta(\eta) = 1, & \text{at} & \eta = 0, \\ \dfrac{df}{d\eta} = 0, & \theta(\eta) = 0, & \text{as} & \eta \to \infty, \end{cases} \tag{7.13}$$

Where K_1 is the viscoelastic parameter, Da^{-1} is the inverse Darcy number, γ is the relaxation time parameter, N_r is the solar radiation, Pr is the Prandtl number, and V_c is mass transpiration.

7.2.1 Exact Solution of Momentum Equation

The analytical solution of equation (7.11) subjected to boundary conditions in (7.13) are assumed of the form

$$f(\eta) = V_c + \left(\frac{1 - \exp[-\beta\eta]}{\beta}\right), \tag{7.14}$$

Where V_c is the mass transpiration, β is the undetermined constant, and d is the stretching/shrinking constant.

By substituting into equation (7.11) gives the following algebraic equation for β:

$$V_c K_1 \beta^3 - (1 - K_1)\beta^2 + V_c\beta + (1 + Da^{-1}) = 0, \tag{7.15}$$

And the roots of the preceding equations:

$$\beta_1 = \frac{1}{\Gamma_2}\left[\Gamma_1 - \frac{2^{1/3}\phi_1}{\left(\phi_2 + \sqrt{4\phi_1^3 + \phi_2^2}\right)^{1/3}} + \frac{\left(\phi_2 + \sqrt{4\phi_1^3 + \phi_2^2}\right)^{1/3}}{2^{1/3}}\right],$$

$$\beta_{2,3} = \frac{1}{\Gamma_2}\left[\Gamma_1 - \frac{\left(1 \pm i\sqrt{3}\right)\phi_1}{2^{2/3}\left(\phi_2 + \sqrt{4\phi_1^{\,3} + \phi_2^{\,2}}\right)^{1/3}} - \frac{\left(1 \mp i\sqrt{3}\right)\phi_1\left(\phi_2 + \sqrt{4\phi_1^{\,3} + \phi_2^{\,2}}\right)^{1/3}}{2 * 2^{1/3}}\right],$$

where

$$\phi_1 = -\left(-1 + K_1\right) + 3V_c^{\,2}K_1,$$

$$\phi_2 = 2 - 6K_1 - 9Vc^2K_1 + 6K_1^{\,2} - 18V_c^{\,2}K_1^{\,2} - 27Da^{-1}V_c^{\,2}K_1^{\,2} - 2K_1^{\,3},$$

$$\Gamma_1 = -1 + K_1, \qquad \Gamma_2 = \frac{1}{3V_cK_1},$$

The velocity profile is determined by taking the derivative of equation (7.14):

$$\frac{df}{d\eta} = \exp\left[-\beta\eta\right],$$

7.2.2 Exact Solution for Temperature Equation

Equation (7.12) is reconstructed in the following manner:

$$\frac{d^2\theta}{d\eta^2} + g(f)\frac{d\theta}{d\eta} = 0, \tag{7.17}$$

By taking $f'(\eta) = b - \beta f(\eta)$ where $b = V_c\beta + 1$,

$$\text{where } g(f) = \frac{f(\eta)\left[\frac{Pr}{1+N_r} - \frac{Pr\gamma}{1+N_r} * b\right] + \left(f(\eta)\right)^2 \frac{\beta Pr\gamma}{1+N_r}}{1 - \frac{Pr\gamma}{1+N_r}\left(f(\eta)\right)^2}, \tag{7.18}$$

Subject to the provided boundary conditions, the solution to equation (7.17) may be expressed as follows:

$$\theta(\eta) = 1 + \left(\frac{d\theta}{d\eta}\right)_{\eta=0} \int_0^{\eta} \exp\left(-\int_0^{\eta} g(f)\,d\eta\right)d\eta, \tag{7.19}$$

$$\left(\frac{d\theta}{d\eta}\right)_{\eta=0} = -\left[\int_0^{\infty} \exp\left(-\int_0^{\eta} g(f)\,d\eta\right)d\eta\right]^{-1}, \tag{7.20}$$

$$-g(f)\,d\eta = -\int_0^{f(\eta)} \left\{\frac{g(f)}{\left(b - \beta f(\eta)\right)}\right\} d\left(f(\eta)\right) = \log\left(\frac{\left(b - \beta f(\eta)\right)^{-A}}{\left(1 - Sf(\eta)\right)^{-B}\left(1 + Sf(\eta)\right)^{-C}}\right), \tag{7.21}$$

Where:

$$\begin{cases} C_1 = \dfrac{Pr}{1+N_r}, \quad S = \sqrt{\dfrac{Pr\gamma}{1+N_r}}, \quad A = \dfrac{C_1 b}{\left((Sb)^2 - \beta^2\right)}, \\ B = \dfrac{\beta^2 S + C_1 Sb + C_1\beta - S^3 b^2}{2S\left((Sb)^2 - \beta^2\right)}, \quad C = \dfrac{-\beta^2 S - C_1 Sb + C_1\beta - S^3 b^2}{2\left((Sb)^2 - \beta^2\right)}, \end{cases} \tag{7.22}$$

Therefore, we can write this as:

$$\int_0^\eta \exp\left\{-\int_0^\eta g(f)d\eta\right\}d\eta = \int_0^{f(\eta)} \left\{\frac{\left(b-\beta f(\eta)\right)^{-A-1}}{\left(1-Sf(\eta)\right)^{-B}\left(1+Sf(\eta)\right)^{-C}}\right\}d\left(f(\eta)\right), \tag{7.23}$$

Using the AHF:

$$\begin{aligned} &\int_0^{f(\eta)} \left\{\frac{\left(b-\beta f(\eta)\right)^{-A-1}}{\left(1-Sf(\eta)\right)^{-B}\left(1+Sf(\eta)\right)^{-C}}\right\}d\left(f(\eta)\right) \\ &= \left\{\begin{array}{l} \dfrac{1}{A\beta}\left(b-\beta f(\eta)\right)^{-A}\left(1-Sf(\eta)\right)^{B}\left(1+Sf(\eta)\right)^{C}\left(\dfrac{\left(-\beta+\beta Sf(\eta)\right)}{-\beta+S}\right)^{-B}\left(\dfrac{\left(\beta+\beta Sf(\eta)\right)}{\beta+S}\right)^{-C} \\ {}_2F_1\left(-A,-B,-C,1-A;-\dfrac{S\left(b-\beta f(\eta)\right)}{\beta-bS},\dfrac{S\left(b-\beta f(\eta)\right)}{\beta+Sb}\right) \end{array}\right\} \end{aligned} \tag{7.24}$$

This is the form:

$$F_1\left(\alpha,\beta,\beta',\beta+\beta';x,y\right)=\left(1-y\right)^{-\alpha}{}_2F_1\left(\alpha,\beta,\beta+\beta',\frac{x-y}{1-y}\right), \tag{7.25}$$

Therefore, we can write:

$$\begin{aligned} &\int_0^\eta \exp\left(-\int_0^\eta g(f)d\eta\right)d\eta \\ &= \left\{\left\{\frac{\left(\dfrac{\beta-bS}{\beta}\right)^{B}\left(\dfrac{\beta+bS}{\beta}\right)^{C}}{A\beta\left(\dfrac{\beta+\beta bSf(\eta)}{\left(b-\beta f(\eta)\right)(\beta+bS)}\right)^{-A}}\right\}{}_2F_1\left(-A,-B;1-A;\frac{-2\beta^2 Sf(\eta)-2b\beta S}{(\beta-Sb)\left(\beta+\beta bsf(\eta)\right)}\right)\right\}, \end{aligned} \tag{7.26}$$

By applying the integration, we should get a finite result.

$$\lim_{\substack{\eta\to 0,\\ f(\eta)\to 0}} \int_0^{\eta} \exp\left(-\int_0^{\eta} g(f)\,d\eta\right) d\eta = \begin{cases} \dfrac{1}{A\beta}\left(\dfrac{\beta - bS}{\beta}\right)^{B}\left(\dfrac{\beta + bS}{\beta}\right)^{C-A}\left(\dfrac{b-\beta V_c}{1+bSV_c}\right)^{-A} \\ {}_2F_1\left(-A,-B,1-A;\dfrac{2S\beta^2 V_c - 2bS\beta}{(\beta - bS)(\beta + \beta bSV_c)}\right), \end{cases} \quad (7.27)$$

Consider the case where $f(\eta) \to \dfrac{1}{\beta}$. Now, it is easy to check that:

$$\lim_{\substack{\eta\to 0,\\ f(\eta)\to \frac{1}{\beta}}} \left(\int_0^{\eta} \exp\left(-\int_0^{\eta} g(f)\,d\eta\right) d\eta\right) = \begin{cases} \infty, & \text{when } A > 0, \\ 0, & \text{when } A < 0. \end{cases} \quad (7.28)$$

According to equation (7.28), it is observed that when A > 0, the solution fails from a valid physical meaning, resulting in $\left(\dfrac{d\theta}{d\eta}\right)_{\eta=0} \to V_c$. As a result, A < 0 remains the sole plausible scenario.

Using equation (7.22), we can write:

$$A < 0 \to \left(\left[\frac{d^2 f}{d\eta^2}\right]_{\eta=0}\right)^2 > \frac{\gamma\,\mathrm{Pr}}{1+N_r} \to \left(\left[\frac{d^2 f}{d\eta^2}\right]_{\eta=0}\right)^2 > \frac{a\lambda\,\mathrm{Pr}}{1+N_r}, \quad (7.29)$$

If the condition $\left(\left[\dfrac{d^2 f}{d\eta^2}\right]_{\eta=0}\right)^2 > \dfrac{\gamma \mathrm{Pr}}{1+N_r}$ is secured:

$$\left(\frac{d\theta}{d\eta}\right)_{\eta=0} = \frac{A\beta\left(\dfrac{\beta}{\beta - bS}\right)^{B}\left(\dfrac{\beta}{\beta + bS}\right)^{C-A}\left(\dfrac{1+bSV_c}{b-\beta V_c}\right)^{-A}}{{}_2F_1\left(-A,-B,1-A;\dfrac{2S\beta^2 V_c - 2bS\beta}{(\beta - bS)(\beta + \beta bSV_c)}\right)}, \quad (7.30)$$

$$\theta(\eta) = \left(\frac{(b-\beta V_c)\left(1+\dfrac{Sb}{\beta}(b-\exp(-\beta\eta))\right)}{(1-\beta b)\exp(-\beta\eta)}\right)^{A}$$

$$\frac{{}_2F_1\left(-A,-B,1-A;\dfrac{2S\beta(b-\exp(-\beta\eta)) - 2bS\beta}{(\beta - bS)(\beta + bSV_c)}\right)}{{}_2F_1\left(-A,-B,1-A;\dfrac{2S\beta^2 V_c - 2bS\beta}{(\beta - bS)(\beta + \beta bSV_c)}\right)}, \quad (7.31)$$

Where:

$$\begin{cases} C_1 = \dfrac{Pr}{1+N_r}, \quad S = \sqrt{\dfrac{Pr\gamma}{1+N_r}}, \quad A = \dfrac{C_1 b}{\left((Sb)^2 - \beta^2\right)}, \\ B = \dfrac{\beta^2 S + C_1 Sb + C_1\beta - S^3 b^2}{2S\left((Sb)^2 - \beta^2\right)}, \quad C = \dfrac{-\beta^2 S - C_1 Sb + C_1\beta - S^3 b^2}{2\left((Sb)^2 - \beta^2\right)}, \end{cases}$$

7.3 RESULTS AND DISCUSSION

In this present analysis, the study of the Cattaneo–Christov heat flux model with radiation and mass transpiration on a porous surface that expands nonlinearly under conditions of temperature and velocity distribution are presented. The leading PDE is transformed into a highly nonlinear ODE, which is solved analytically by using the Appell's hypergeometric function of two variables to create a unique response for each temperature and velocity profile. This section mainly investigates the consequences of pertinent flow factors, like the inverse Darcy number, thermal radiation, relaxation time parameters, and proportional suction parameter over a porous stretched surface.

7.3.1 Velocity Profile

Figures 7.2 and 7.3 show the effect Da^{-1} on the transverse and axial velocity profiles. It is observed that the fluid velocity decreases by increasing the value of the inverse Darcy number, which is due to the rising water in the hole of the porous structure. Figures 7.4 and 7.5 show the effect K_1 on the

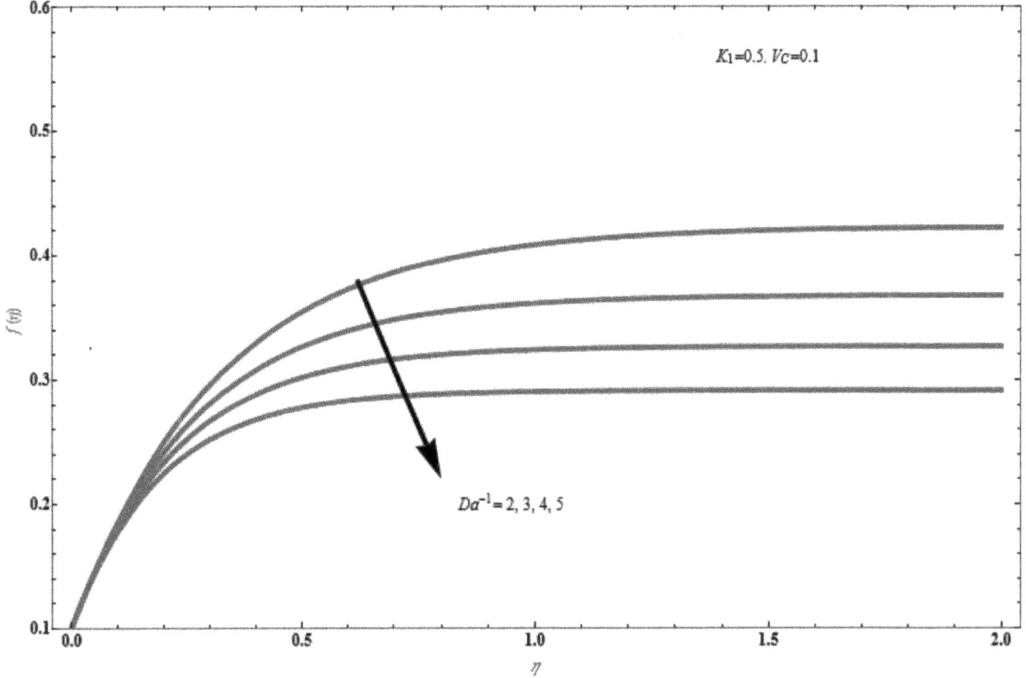

FIGURE 7.2 The effect of $f(\eta)$ versus η by varying Da^{-1} for fixed values of $K_1 = -0.5, V_c = 0.1$.

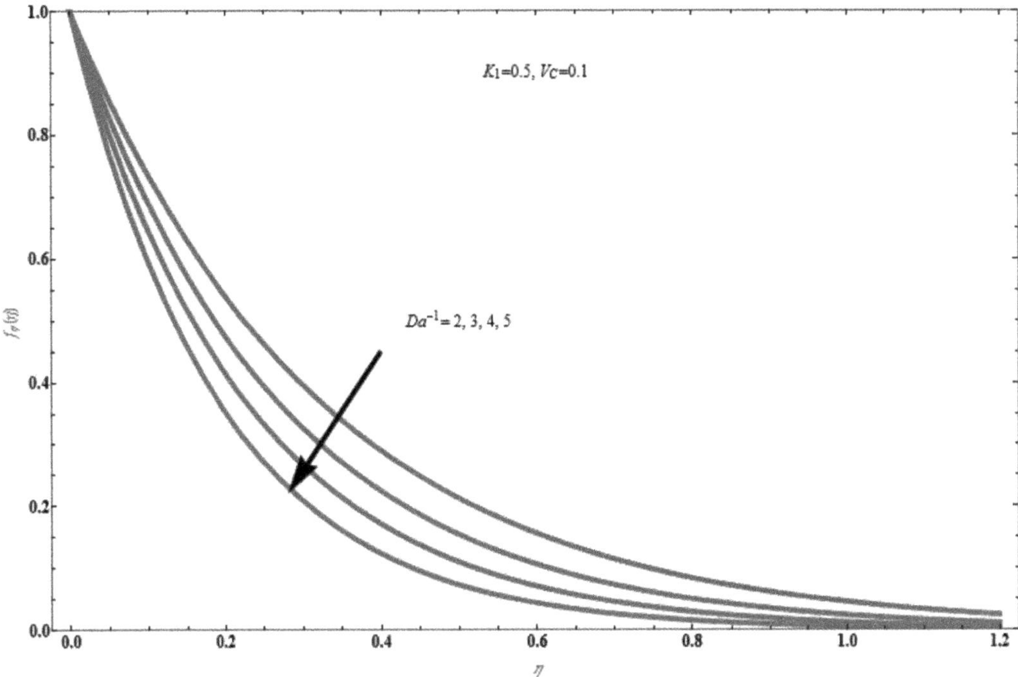

FIGURE 7.3 The effect of $f_{\eta}(\eta)$ versus η by varying Da^{-1} for fixed values of $K_1 = 0.5, V_c = 0.1$.

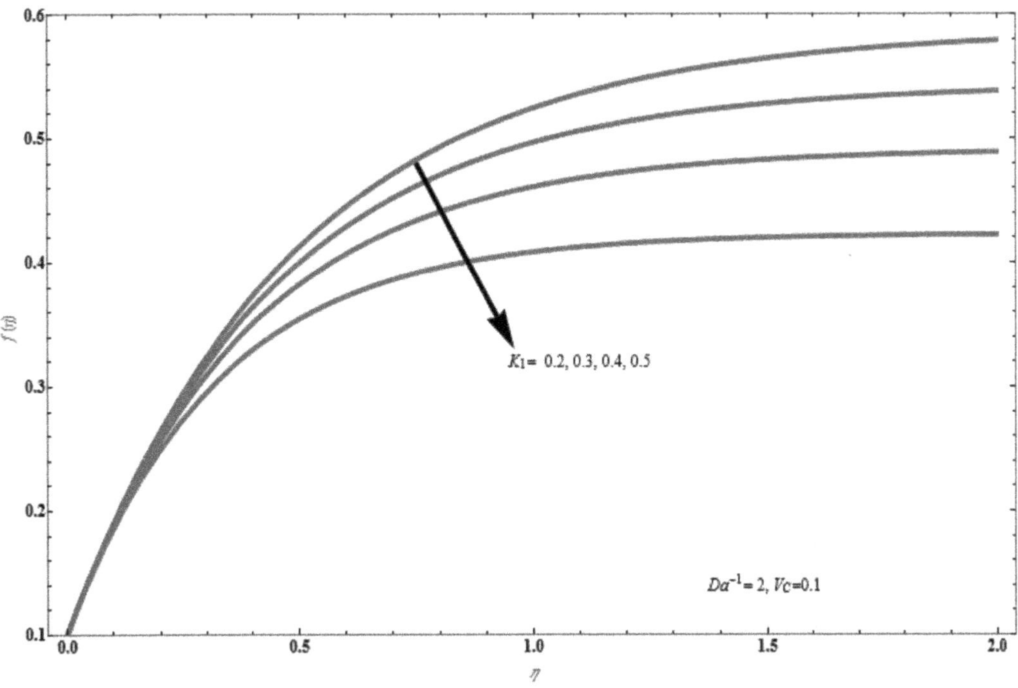

FIGURE 7.4 The effect of $f(\eta)$ versus η by varying K_1 for fixed values of $Da^{-1} = 2, V_c = 0.1$.

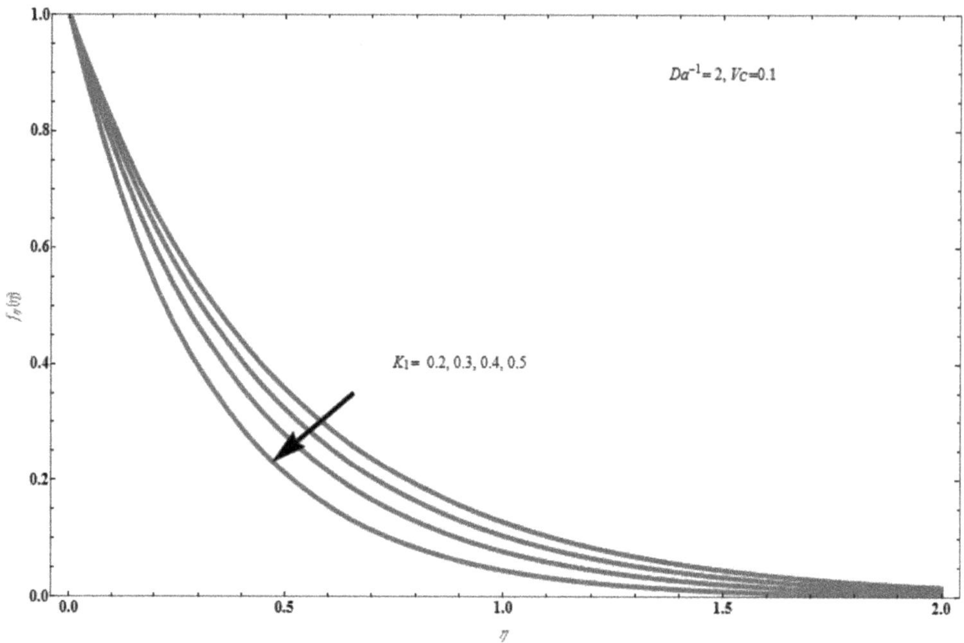

FIGURE 7.5 The effect of $f_\eta(\eta)$ versus η by varying K_1 for fixed values of $Da^{-1}=2, V_c=0.1$.

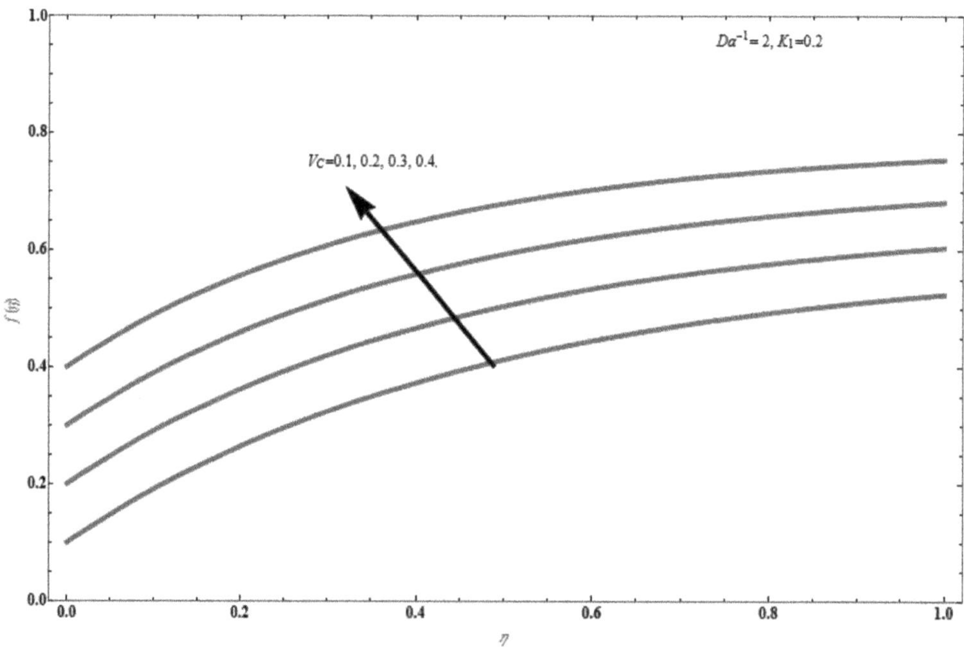

FIGURE 7.6 The effect of $f(\eta)$ versus η by varying V_c for fixed values of $Da^{-1}=0.2, K_1=0.2$.

transverse and axial velocity profiles, respectively. It is noted that the fluid velocity decreases by increasing the value of the viscoelastic parameter. Figures 7.6 and 7.7 show the effect V_c on transverse and axial velocity profiles. By increasing the value of the suction parameter, the transverse velocity increases, and the axial velocity decreases.

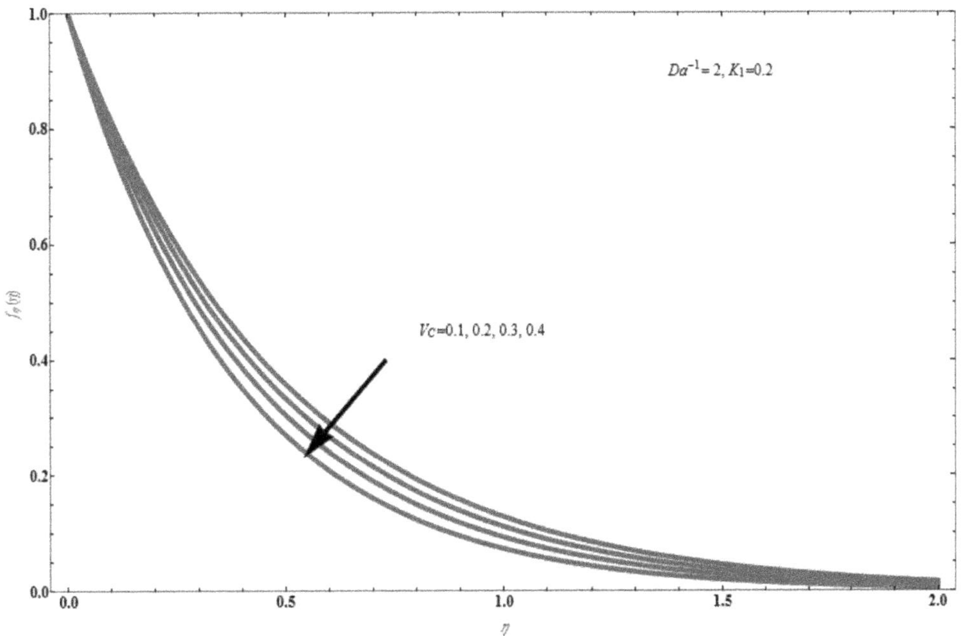

FIGURE 7.7 The effect of $f_\eta(\eta)$ versus η by varying V_c for fixed values of $Da^{-1}=0.2, K_1=0.2$.

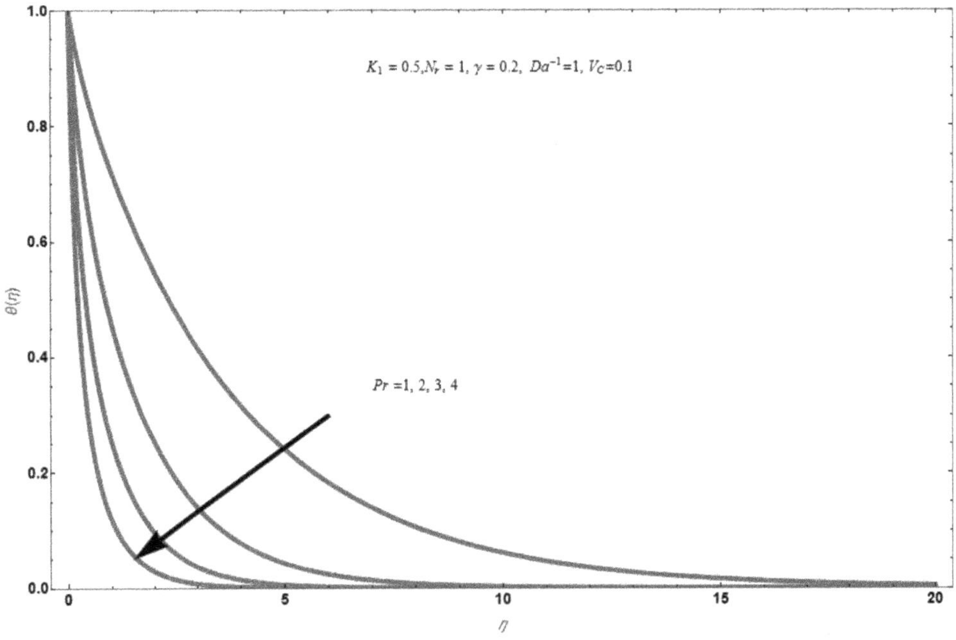

FIGURE 7.8 The effect of $\theta(\eta)$ versus η by varying Pr for fixed values of K_1 = 0.5, N_r = 1, γ = 0.2, $Da^{-1}=2, V_c=0.1$.

7.3.2 Temperature Field

Figure 7.8 shows the effect Pr on the temperature profile. It is observed that the thermal boundary layer decreases with increasing Pr. Furthermore, the temperature of the fluid decreases with increasing Pr. The effects of N_r on the temperature profile are shown in Figure 7.9. It is noted that the thermal boundary layer increases by increasing the N_r. Furthermore, fluid's temperature also increases. The effects of thermal relaxation time γ are depicted in Figure 7.10. The temperature of the field is reduced by increasing the value of the thermal relaxation time γ. It is noteworthy and important to note that the influence of thermal relaxation time γ is greater in the suction situation.

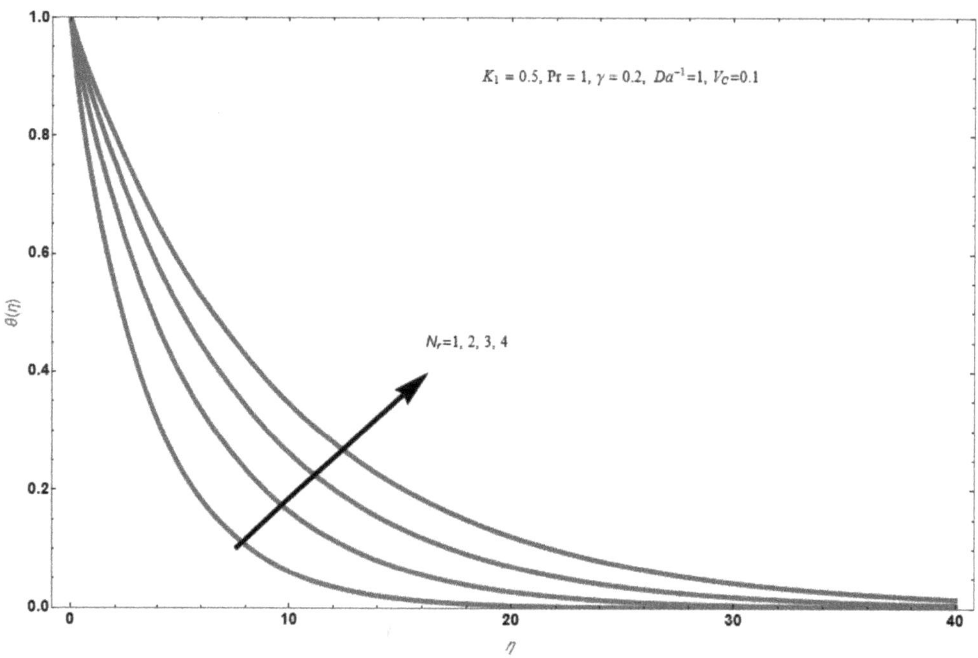

FIGURE 7.9 The effect of $\theta(\eta)$ versus η by varying N_r for fixed values of $K_1 = 0.5$, Pr = 1, $\gamma = 0.2$, $Da^{-1} = 2, V_c = 0.1$.

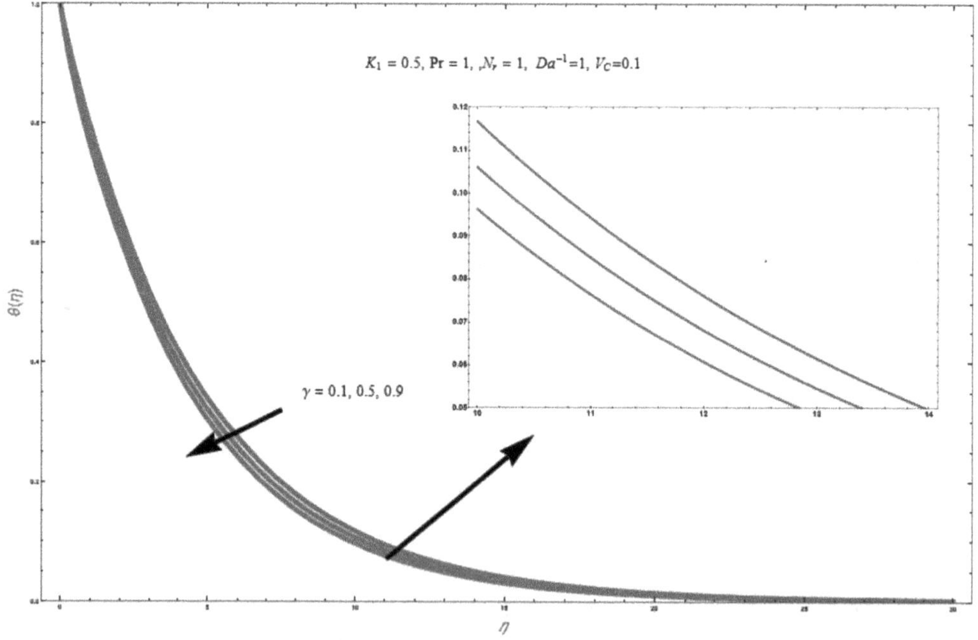

FIGURE 7.10 The effect of $\theta(\eta)$ versus η by varying γ for fixed values of $K_1 = 0.5$, $N_r = 1$, Pr = 1, $Da^{-1} = 2, V_c = 0.1$.

Thermal relaxation time is the amount of time it takes for fluid particles to disperse heat energy into their surroundings. As a result, we see a decrease in temperature distribution as the values of the thermal relaxation time γ parameter increases.

7.4 CONCLUDING REMARKS

The generalized non-Fourier Cattaneo–Christov heat flux model was used to study the flow of viscoelastic and second-grade fluids on a porous stretching sheet. The study was done in a closed-form analytical way using the Appell's hypergeometric function of two variables. The Prandtl number, thermal radiation, and mass transpiration parameter factors also influence the temperature distribution inside the porous stretching sheet. The following are the key findings of this chapter:

- The transverse and axial velocities of the fluid decrease by increasing the value of the inverse Darcy number.
- The transverse and axial velocities of the fluid are decreased by increasing the value of the viscoelastic parameter K_1.
- The transverse velocity of the fluid increases and the axial velocity decreases by increasing the suction $\left(V_c > 0\right)$ parameter.
- The temperature of the fluid decreases by increasing the value of Pr and γ.
- The fluid temperature increases by increasing the value of N_r.

Nomenclature

Latin Symbols	**Descriptions**	**S I units**
a	constant	—
C_p	constant pressure at specific heat	$(JKg^{-1} K^{-1})$
Da^{-1}	Inverse Darcy number	—
f	dimensionless stream function	—
K	permeability parameter	—
k^*	Elastic viscous	—
K_1	viscoelastic constant	—
N_r	solar radiation	—
Pr	Prandtl number	—
q_r	radiative heat flux	(wm^{-2})
ρ	density	(Kgm^{-3})
T	temperature	(K)
T_w	surface temperature	(K)
T_∞	ambient temperature	(K)
u,v	velocity component	—
V	suction/injection parameter	—
x,y	x and y coordinate axis	—

Greek Symbols

μ	dynamic viscosity	$(kgm^{-1}s^{-1})$
α	thermal diffusivity	(m^2s^{-1})
γ	relaxation time	—
σ^*	Stefan–Boltzmann constant	$(Wm^{-2}K^{-4})$
κ	thermal conductivity	$(WKg^{-1}K^{-1})$

η	similarity variable	—
υ	kinematic viscosity	(m^2s^{-1})
θ	dimensionless temperature	—

Subscript/Superscript

w	wall temperature	—
∞	ambient fluid	—
$*$	dimensionless quantities	—
$f(\eta)$	transverse velocity	—
$f_\eta(\eta)$	axial velocity	—
$\theta(\eta)$	temperature profile	—
AHF	Appell hypergeometric function of two variables	—
ODE	ordinary differential equation	—
MHD	magnetohydrodynamics	—
CCM	Cattaneo–Christov model	—

REFERENCES

[1] Crane, L. J. (1970). Flow past a stretching plate. *Journal of Applied Mathematics and Physics (ZAMP)*, 21, 645–647.

[2] Troy, W. C., Overman, E. A., Ermentrout, G. B., & Keener, J. P. (1987). Uniqueness of flow of a second-order fluid past a stretching sheet. *Quarterly of Applied Mathematics*, 44(4), 753–755.

[3] Vajravelu, K., & Rollins, D. (1991). Heat transfer in a viscoelastic fluid over a stretching sheet. *Journal of Mathematical Analysis and Applications*, 158(1), 241–255.

[4] Char, M. I. (1994). Heat and mass transfer in a hydromagnetic flow of the viscoelastic fluid over a stretching sheet. *Journal of Mathematical Analysis and Applications*, 186, 674–689.

[5] Nayak, M. K., Dash, G. C., & Singh, L. P. (2016). Heat and mass transfer effects on MHD viscoelastic fluid over a stretching sheet through porous medium in presence of chemical reaction. *Propulsion and Power Research*, 5(1), 70–80.

[6] Mahabaleswar, U. S. (2008). External regulation of convection in a weak electrically conducting non-Newtonian liquid with g-jitter. *Journal of Magnetism and Magnetic Materials*, 320(6), 999–1009.

[7] Siddheshwar, P. G., & Mahabaleswar, U. S. (2005). Effects of radiation and heat source on MHD flow of a viscoelastic liquid and heat transfer over a stretching sheet. *International Journal of Non-Linear Mechanics*, 40, 807–820.

[8] Mahabaleshwar, U. S., Nagaraju, K. R., Vinay Kumar, P. N., & Kelson, N. A. (2018). An MHD Navier's slip flow over axisymmetric linear stretching sheet using differential transform method. *International Journal of Applied and Computational Mathematics*, 4(1), 1–13.

[9] Mahabaleshwar, U. S., Vinay Kumar, P. N., & Sheremet, M. (2016). Magnetohydrodynamics flow of a nanofluid driven by a stretching/shrinking sheet with suction. *SpringerPlus*, 5, 1901.

[10] Mahabaleshwar, U. S., Nagaraju, K. R., Sheremet, M. A., Baleanu, D., & Lorenzini, E. (2020). Mass transpiration on Newtonian flow over a porous stretching/shrinking sheet with slip. *Chinese Journal of Physics*, 63, 130–137.

[11] Mahabaleshwar, U. S., Vinay Kumar, P. N., Nagaraju, K. R., Bognár, G., & Nayakar, S. N. R. (2019). A new exact solution for the flow of a fluid through porous media for a variety of boundary conditions. *Fluids*, 4(3), 125.

[12] Mahabaleshwar, U. S., Nagaraju, K. R., Vinay Kumar, P. N., Nadagouda, M. N. Bennacer, R., & Sheremet, M. A. (2020). Effects of dufour and sort mechanisms on MHD mixed convective-radiative non-Newtonian liquid flow and heat transfer over a porous sheet. *Thermal Science and Engineering Progress*, 16, 100459.

[13] Mahabaleshwar, U. S., Nagaraju, K. R., Nadagouda, M. N., Bennacer, R., & Baleanu, D. (2020). An MHD viscous liquid stagnation point flow and heat transfer with thermal radiation & transpiration. *Thermal Science and Engineering Progress*, 16, 100379.

[14] Fourier, J. B. J. (1822). *Theorie analytique de la chaleur*. Didot, Paris, 499–508.

[15] Cattaneo, C. (1948). Sulla conduzione del calore, *Seminario matematico e fisico dell'università di Modena. Modena Reggio Emilia*, 3, 83–101.

[16] Christov, C. I. (2009). On frame indifferent formulation of the Maxwell-Cattaneo model of finite speed heat conduction. *Mechanics Research Communications*, 36, 481–486.

[17] Straughan, B. (2010). Thermal convection with the Cattaneo-Christov model. *International Journal of Heat and Mass Transfer*, 53, 95–98.

[18] Straughan, B. (2010). Acoustic waves in a Cattaneo-Christov gas. *Physics Letters A*, 3742, 667–2669.

[19] Han, S., Zheng, L., Li, C., & Zhang, X. (2014). Coupled flow and heat transfer in viscoelastic fluid with Cattaneo-Christov heat flux model. *Applied Mathematics Letters*, 38, 87–93.

[20] Mustafa, M. (2015). Cattaneo-Christov heat flux model for rotating flow and heat transfer of upper-convected Maxwell fluid. *AIP Advances*, 5, 047109.

[21] Ahmad Khan, J., Mustafa, M., Hayat, T., & Alsaedi, A. (2015). Numerical study of Cattaneo-Christov heat flux model for viscoelastic flow due to an exponentially stretching surface. *PLoS ONE, 10*(9), e0137363.

[22] Hayat, T., Muhammad, T., Alsaedi, A., & Mustafa, M. (2016). A comparative study for flow of viscoelastic fluids with cattaneo-christov heat flux. *PLoS ONE, 11*(5), e0155185.

[23] Liu, L., Zheng, L., Liu, F., & Zhang, X. (2016). Anomalous convection diffusion and wave coupling transport of cells on comb frame with fractional Cattaneo-Christov flux. *Communications in Nonlinear Science and Numerical Simulations*, 38, 45–58.

[24] Salahuddin, T., Malik, M. Y., Hussain, A., Bilal, S., & Awais, M. (2016). MHD flow of Cattanneo-Christov heat flux model for Williamson fluid over a stretching sheet with variable thickness using numerical approach. *Journal of Magnetism and Magnetic Materials*, 401, 991–997.

[25] Zehra, I., Abbas, N., Amjad, M., Nadeem, S., Saleem, S., & Issakhov, A. (2021). Casson nanoliquid flow with Cattaneo-Christov flux analysis over a curved stretching/shrinking channel. *Case Studies in Thermal Engineering*, 27, 101146.

[26] Jafarimoghaddam, A., Turkyilmazoglu, M., & Pop, I. (2021). Threshold for the generalized Non-Fourier heat flux model: Universal closed form analytic solution. *International Communications in Heat and Mass Transfer*, 123, 105204.

[27] Khan, M., Amna Shahid, T., Salahuddin, M. Y., & Malik, A. H. (2020). Analysis of two dimensional Carreau fluid flow due to normal surface condition. A generalized Fourier's and Fick's laws. *Physica A*, 123024.

[28] Safwa Khashi'ie, N., Md Arifin, N., Hafidzuddin, E. H., & Wahi, N. (2019). Dual stratified nanofluid flow past a permeable shrinking/stretching sheet using a non-fourier energy model. *Applied Sciences*, 9, 2124.

[29] Sharma, B., Kumar, S., Cattani, C., & Baleanu, D. (2020). Nonlinear dynamics of Cattaneo–Christov heat flux model for third-grade power-law fluid. *Journal of Computational and Nonlinear Dynamics*, 15(1), 011009.

[30] Shankar, U., & Naduvinamani, N. B. (2019). Magnetized impacts of Cattaneo-Christov double-diffusion models on the time-dependent squeezing flow of Casson fluid a generalized perspective of Fourier and Fick's laws. *European Physical Journal—Plus*, 134, 344.

[31] Bhatti, M. M., Ellahi, R., Zeeshan, A., Marin, M., & Ijaz, N. (2019). Numerical study of heat transfer and hall current impact on peristaltic propulsion of particle-fluid suspension with compliant wall properties. *Modern Physics Letters B*, 1950439.

[32] Mahantesh, M. N., & Shakunthala, S. (2019). Impact of Cattaneo-Christov heat flux on magnetohydrodynamic flow and heat transfer of carbon nanofluid due to stretching sheet. *Journal of Nanofluids*, 8, 746–755.

[33] Sarojamma, G., Vijaya Lakshmi, R., Satya Narayana, P., & Animasaun, I. (2020). Exploration of the significance of autocatalytic chemical reaction and cattaneo-christov heat flux on the dynamics of a micropolar fluid. *Journal of Applied and Computational Mechanics*, 6(1), 77–89.

[34] Zaib, A., Khan, U., Khan, I., Seikh, A. H., & Sherif, E.-S. M. (2019). Numerical investigation of aligned magnetic flow comprising nanoliquid over a radial stretchable surface with cattaneo–christov heat flux with entropy generation. *Symmetry*, 11(12), 1520.

[35] Jain, S., & Gupta, P. (2019). Entropy generation analysis with non-linear convection on MHD williamson fluid flow through a porous stretching sheet. *Journal of Nanofluids*, 8(5), 929–937.

[36] Khan, M., & Alzahrani, F. (2020). Transportation of heat through Cattaneo-Christov heat flux model in non-Newtonian fluid subject to internal resistance of particles. *Applied Mathematics and Mechanics*, 41.

[37] Rosseland, S. (1931). *Astrophysik und atom-theoretische Grundlagen*. Springer, Berlin, 41–44.

[38] Kumar, P. N. V., Mahabaleshwar, U. S., Sakanaka, P. H., & Lorenzini, G. (2018). An MHD effect on a newtonian fluid flow due to a superlinear stretching sheet. *Journal of Engineering Thermophysics*, 27(4), 501–506.

8 Unsteady Heat and Mass Transfer in a Stagnant Flow Towards a Stretching Porous Sheet with Variable Fluid Properties

Usman, Irfan Mustafa, Abuzar Ghaffari, and M. Saleem Iqbal

8.1 INTRODUCTION

Several investigators have considered exploring the mass and heat transmission of different fluids upon a stretching sheet because of its diverse essential uses in manufacturing processes and industries. Such applications comprise hot rolling, plastic film manufacturing, glass fiber production, wire drawing, and polymer extrusion in a melt-spinning process. Other applications include cooling nuclear reactors, electronic devices, heat exchangers, and many others. The revolutionary work by Sakiadis [1, 2] was expanded by Crane [3], who studied the linear stretching sheet in a viscous fluid. Gupta and Gupta [4] considered a stretching porous sheet and investigated the mass and heat transmission with suction or blowing. Hiemenz [5] was the first one among the others who examined the 2D stagnation point flow. Eckert [6] studied the heat transfer behavior of the problem and calculated the analytical solution. Afterward, Chiam [7] merged the compositions of Hiemenz [5] and Crane [3] and investigated the stagnant flow upon a stretching plate. Ariel [8] revisited [5] the occurrence of the magnetic field. Jat and Chaudhary [9] included the upshots of MHD and examined stagnant flow upon a stretching sheet. Mustafa and co-workers [10] considered nanofluid upon a vertical plate and studied stagnation point flow. This complex model was simulated numerically, and results were captured for variations of the different parameters. In another study, Mustafa and his co-workers [11] incorporated ferroparticle within the base fluid (water) and inspected the MHD stagnant flow on a stretchable spinning disk. Mebarek-Oudina et al. [12–19] have considered various physical aspects and numerically explored heat transmission with and without nanoparticles over diverse, complex geometries.

Analyzing heat and mass transmission via a porous medium is crucial because it helps explore important features in industrial, petroleum, and geothermal engineering problems. Nield and Bejan [20] and several other studies [21–26] have included excellent work on this topic. Viscous MHD stagnant fluid flow in a porous medium has been scrutinized by Kechil and Hashim [27] using the Adomian decomposition method. Kazem et al. [28] examined the stagnant flow with heat production and calculated the problem analytically. The flow-through porous media has shown its importance to several researchers. Several studies on this topic have been published for the readers [29–32].

Numerous investigations have been made for constant thermal conductivity and viscosity. In reality, thermal conductivity and viscosity vary with temperature, and it has been verified experimentally that both thermal properties are a strong function of temperature. Many researchers have deliberated Newtonian fluid flow in different geometries with thermal conductivity and variable viscosity under various flow conditions [33–37]. The thermal boundary layer thickness and momentum

DOI: 10.1201/9781003299608-8

depend upon the thermophysical properties of the fluid. Many researchers [38–42] have recently studied boundary layer problems with variable thermophysical properties of different fluids across various geometries.

All the preceding studies are investigated for steady flows. The impulsive force investigates the unsteady flow field in a few cases. Due to this impulsive force, the inviscid flow appears instantaneously. Thus, the work in the current chapter is intended to explore the time-dependent MHD stagnant flow upon a stretching sheet saturated in a porous medium with variable properties. The variation of the Prandtl number with the temperature is also considered. The local Nusselt number, skin friction coefficient, and Sherwood numbers are also calculated for diverse parameters, and the effects are examined.

8.2 MATHEMATICAL FORMULATION

A 2D stagnant MHD fluid flow on a horizontal stretching porous sheet is considered. The flow geometry is depicted in Figure 8.1, where the y-axis is vertical to the sheet, the x-axis is chosen horizontally with the sheet, and the stagnant point O is kept fixed. The $u_W(x)$, is assumed to be the sheet's stretching velocity, T_w is the surface temperature, and C_w implies the species concentration at the surface is considered linearly in terms of x. The potential flow velocity $u_e(x)$ is taken, ambient temperature is T_∞, and concentration of species at the boundary layer edge is C_∞. A magnetic field Bo is operated perpendicular to the flow direction. Also, the magnetic Reynolds number is lesser, and the induced magnetic field is ignored. Besides thermal conductivity and dynamic viscosity, all other fluid properties are maintained.

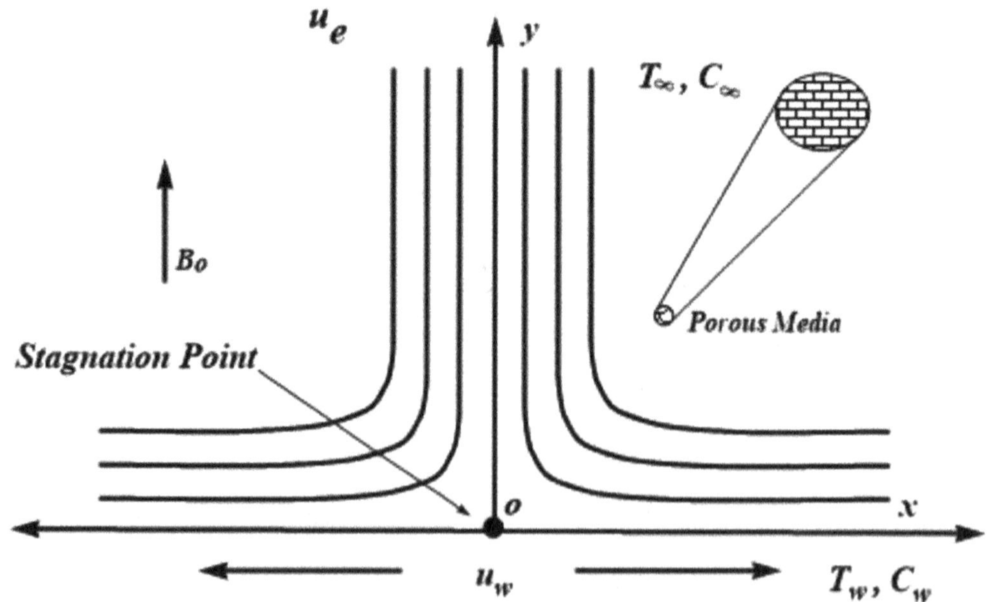

FIGURE 8.1 The flow geometry.

The governing equations within the porous medium are entertained as:

$$\frac{\partial v}{\partial y}+\frac{\partial u}{\partial x}=0, \tag{8.1}$$

$$u\frac{\partial u}{\partial x}+v\frac{\partial u}{\partial y}+\frac{\partial u}{\partial t}=u_e\frac{du_e}{dx}+\frac{1}{\rho}\frac{\partial}{\partial y}\left(\mu\frac{\partial u}{\partial y}\right)+\frac{\sigma B_o^2}{\rho}(u_e-u)+\frac{1}{\rho K}(\mu_\infty u_e-\mu u), \tag{8.2}$$

$$\frac{\partial T}{\partial t}+u\frac{\partial T}{\partial x}+v\frac{\partial T}{\partial y}=\frac{1}{\rho c_p}\frac{\partial}{\partial y}\left(k\frac{\partial T}{\partial y}\right), \tag{8.3}$$

$$u\frac{\partial C}{\partial x}+v\frac{\partial C}{\partial y}+\frac{\partial C}{\partial t}=D\frac{\partial^2 C}{\partial y^2}, \tag{8.4}$$

Where v and u are the velocities, respectively, alongside the y-axis and x-axis, ρ denotes the fluid density, μ means the dynamic viscosity, μ_∞ is the ambient dynamic viscosity, σ is the fluid electric conductivity, K is permeability parameter of the porous medium, T implies the temperature of the fluid, c_p is the effective heat capacity, k is the thermal conductivity, C is the species concentration, and D is molecular diffusivity in the fluid. The boundary conditions are:

$$\text{for } t<0;\; v(x,y)=u(x,y)=0,\; T(x,y)=T_\infty \quad \text{for all } x \text{ and } y \tag{8.5}$$

$$\text{for } t\geq 0;\; v(x,0)=0,\, u(x,0)=u_w(x)=cx,\; T(x,0)=T_w=T_\infty+Ax,\; C(x,0)=C_w=C_\infty+Bx,$$

$$T(x,y)=T_\infty,\; u(x,y)=u_e(x)=ax,\; C(x,y)=C_\infty \;\text{ as } y\to\infty,$$

Where a, c, A, and B are constants. Now, introducing the transformation [43, 44]:

$$\eta=\left(\frac{c}{v_\infty}\right)^{1/2}\xi^{-1/2}y, u(x,y,t)=cxf'(\xi,\eta), v(x,y,t)=-(cv_\infty)^{1/2}\xi^{1/2}f(\xi,\eta),$$

$$\theta(\xi,\eta)=\frac{T-T_\infty}{T_w-T_\infty}, \xi=1-\exp(-t^*), t^*=ct, g(\xi,\eta)=\frac{C-C_\infty}{C_w-C_\infty}. \tag{8.6}$$

Here, v_∞ is the ambient kinematic viscosity. There is an inverse relationship between viscosity and temperature. To analyze the heat and flow rates, the fluid viscosity is considered in the given form [45]:

$$\frac{1}{\mu}=\frac{1}{\mu_\infty}\{1+\gamma(T-T_\infty)\}, \tag{8.7}$$

Where γ denotes the fluid thermal property. Equation (8.7) implies:

$$\frac{1}{\mu}=e(T-T_r), \tag{8.8}$$

Where T_r and e depend on the reference state and the fluid thermal conductivity, in which $e > 0$ means liquid and $e < 0$ corresponds to air. The dimensionless temperature θ is:

$$\theta=\frac{T-T_r}{T_w-T_\infty}+\theta_r, \tag{8.9}$$

Where θ_r is defined as:

$$\theta_r = \frac{-1}{\gamma(T_w - T_\infty)}. \tag{8.10}$$

When $T_w - T_\infty > 0$, the physical values of θ_r are considered greater than 1 for gases and less than 0 $(\theta_r < 0)$ for liquid. It is witnessed that outside of the boundary layer, when $\gamma \to 0$, then $\mu \to \mu_\infty$ and $\theta_r \to \infty$. Following [36]:

$$k = k_\infty \{1 + \varepsilon^* (T - T_\infty)\}, \tag{8.11}$$

Where ε^* is the parameter of thermal conductivity and k_∞ is the reference thermal conductivity [46] for water $0 \le \varepsilon^* \le 0.12,$ for air $0 \le \varepsilon^* \le 6$, and for lubrication oil $-0.1 \le \varepsilon^* \le 0.$

Using equation (8.6), equations (8.2)–(8.4) are:

$$\frac{\theta_r}{\theta_r - \theta}\left(f''' + \frac{1}{\theta_r - \theta}\theta' f''\right) - K_1\xi\frac{\theta_r}{\theta_r - \theta}f' + K_1\xi\left(\frac{a}{c}\right) + M\xi\left(\frac{a}{c} - f'\right) + \xi\left(\frac{a}{c}\right)^2 - \tag{8.12}$$

$$\xi(1-\xi)\frac{\partial f'}{\partial \xi} + \frac{\eta}{2}(1-\xi)f'' - \xi f'^2 + \xi ff'' = 0,$$

$$\left(1+\varepsilon^*\theta\right)\theta'' + \varepsilon^*\theta'^2 - \mathrm{Pr}_\infty\left(\xi(1-\xi)\frac{\partial\theta}{\partial\xi} - \frac{\eta}{2}(1-\xi)\theta' + \xi\left(f'\theta - f\theta'\right)\right) = 0, \tag{8.13}$$

$$g'' - Sc\left(\xi(1-\xi)\frac{\partial g}{\partial\xi} - \frac{\eta}{2}(1-\xi)g' + \xi\left(f'g - fg'\right)\right) = 0, \tag{8.14}$$

Through the boundary conditions:

$$\left.\begin{array}{r} f(\xi,0) = 0, f'(\xi,0) = 1, \theta(\xi,0) = 1, g(\xi,0) = 1 \\ f'(\xi,¥) = \frac{a}{c}, \theta(\xi,¥) = 0, g(\xi,¥) = 0 \end{array}\right\} \tag{8.15}$$

Where $M = \sigma B_o^2 / c\rho$ is the magnetic parameter, a/c is the velocity ratio, $K_1 = v_\infty / cK$ represents the permeability parameter, ε^* is the thermal conductivity parameter, $\mathrm{Pr}_\infty = \mu_\infty c_p / k_\infty$ means the ambient Prandtl number, and $Sc = v_\infty / D$ implies the Schmidt number.

8.2.1 Variable Prandtl Number

Following [41]:

$$\mathrm{Pr} = \frac{1}{\left(1 - \frac{\theta}{\theta_r}\right)\left(1+\varepsilon^*\theta\right)}\mathrm{Pr}_\infty. \tag{8.16}$$

Using equation (8.16) in equation (8.13), we get:

$$\left(1+\varepsilon^*\theta\right)\theta'' + \varepsilon^*\theta'^2 - \mathrm{Pr}\left(1 - \frac{\theta}{\theta_r}\right)\left(1+\varepsilon^*\theta\right)\left(\xi(1-\xi)\frac{\partial\theta}{\partial\xi} - \frac{\eta}{2}(1-\xi)\theta' + \xi\left(f'\theta - f\theta'\right)\right) = 0. \tag{8.17}$$

The local Nusselt number, skin friction, and Sherwood number can be written as:

$$Nu_x = \frac{xq_w}{k(T_w - T_\infty)}, \quad C_f = \frac{\tau_w}{\rho u_w^2}, \quad Sh_x = \frac{xm_w}{D(C_w - C_\infty)}, \tag{8.18}$$

The heat flux q_w, wall shear stress τ_w, and mass flux m_w can be expressed as:

$$q_w = -k\left(\frac{\partial T}{\partial y}\right)_{y=0}, \quad \tau_w = \mu\left(\frac{\partial u}{\partial y}\right)_{y=0}, \quad m_w = -D\left(\frac{\partial C}{\partial y}\right)_{y=0}. \tag{8.19}$$

Using equation (8.19), the physical quantities are written as:

$$C_f \operatorname{Re}_x^{1/2} = \frac{1}{\sqrt{\xi}} \frac{\theta_r}{(\theta_r - 1)} f''(\xi, 0),$$

$$Nu_x \operatorname{Re}_x^{-1/2} = -\frac{1}{\sqrt{\xi}} \theta'(\xi, 0),$$

$$Sh_x \operatorname{Re}_x^{-1/2} = -\frac{1}{\sqrt{\xi}} g'(\xi, 0).$$

8.2.2 Initial Flow

For initial flow, take $t^* = 0$ ($\xi \to 0$). In this case, the governing equations reduce to:

$$\frac{\theta_r}{\theta_r - \theta}\left(f''' + \frac{1}{\theta_r - \theta}\theta' f''\right) + \frac{\eta}{2} f'' = 0, \tag{8.20}$$

$$\left(1 + \varepsilon^*\theta\right)\theta'' + \varepsilon^*\theta'^2 + \Pr\left(1 - \frac{\theta}{\theta_r}\right)\left(1 + \varepsilon^*\theta\right)\frac{\eta}{2}\theta' = 0, \tag{8.21}$$

$$g'' + Sc\frac{\eta}{2} g' = 0, \tag{8.22}$$

And BCs:

$$\begin{aligned} &f'(0) = 1, f(0) = 0, \ \theta(0) = 1, g(0) = 1, \\ &f'(\infty) = \frac{a}{c}, \ \theta(\infty) = 0, g(\infty) = 0. \end{aligned} \tag{8.23}$$

8.2.3 Steady-State Flow

In the present case, take $t^* \to \infty$ ($\xi = 1$). The governing equations and (8.17) reduce to:

$$\frac{\theta_r}{\theta_r - \theta}\left(f''' + \frac{1}{\theta_r - \theta}\theta' f''\right) - K_1 \frac{\theta_r}{\theta_r - \theta} f' + K_1\left(\frac{a}{c}\right) + M\left(\frac{a}{c} - f'\right) + \left(\frac{a}{c}\right)^2 - f'^2 + ff'' = 0, \tag{8.24}$$

$$\left(1 + \varepsilon^*\theta\right)\theta'' + \varepsilon^*\theta'^2 - \Pr\left(1 - \frac{\theta}{\theta_r}\right)\left(1 + \varepsilon^*\theta\right)\left(f'\theta - f\theta'\right) = 0, \tag{8.25}$$

$$g'' - Sc\left(f'g - fg'\right) = 0, \tag{8.26}$$

With the same boundary conditions.

8.3 NUMERICAL PROCEDURE

The scheme of governing equations is naturally nonlinear, and the analytical solution is almost impossible. We used a numerical scheme to cope with this situation [47]. First, the higher-order ODEs are lessened to a first-order mixed differential equations upon defining the parameters U, V, Q and H, such that:

$$f' = U, U' = V, \theta' = Q \text{ and } g' = H$$

Equations (8.12), (8.14), and (8.17) and equation (8.15) take the following form:

$$V' = \frac{\theta_r - \theta}{\theta_r}\left(\begin{array}{c} -\left(\frac{\theta_r}{(\theta_r - \theta)^2}\right)QV + K_1\xi\left(\frac{\theta_r}{\theta_r - \theta}\right)U - K_1\xi\left(\frac{a}{c}\right) - M\xi\left(\frac{a}{c} - U\right) - \xi\left(\frac{a}{c}\right)^2 \\ -\frac{\eta}{2}(1-\xi)V + \xi U^2 - \xi fV + \xi(1-\xi)\frac{\partial U}{\partial \xi} \end{array}\right) \tag{8.27}$$

$$Q' = \frac{1}{\left(1+\varepsilon^*\theta\right)}\left(-\varepsilon^* Q^2 + \Pr\left(1 - \frac{\theta}{\theta_r}\right)(1+\varepsilon^*\theta)\left(\xi(1-\xi)\frac{\partial \theta}{\partial \xi} - \frac{\eta}{2}(1-\xi)Q + \xi(U\theta - fQ)\right)\right), \tag{8.28}$$

$$H' = Sc\left(\xi(1-\xi)\frac{\partial g}{\partial \xi} - \frac{\eta}{2}(1-\xi)H + \xi(Ug - fH)\right), \tag{8.29}$$

$$f(\xi,0) = 0,\ U(\xi,0) = 1,\ \theta(\xi,0) = 1,\ g(\xi,0) = 1$$

$$U(\xi,\infty) = \frac{a}{c},\ \theta(\xi,\infty) = 0,\ g(\xi,\infty) = 0$$

A grid net upon the plane (η, ξ) is written as:

$$\eta_0 = 0,\ j = 1,2,\ldots N-1,\ \eta_j = \eta_{j-1} + \Delta\eta, \eta_J = \eta_\infty,$$
$$\xi^0 = 0,\ n = 1,2,\ldots,\ \xi^n = \xi^{n-1} + \Delta\xi,$$

The central difference formulae are substituted for the derivatives of the functions f, U, V, θ, Q, g and H, and their mean values replace the functions themselves on the midpoint $(\eta_{j-1/2}, \xi^{n-1/2})$, expressed as:

$$\left(\frac{\partial f}{\partial \eta}\right)_{j-1/2}^{n} = \frac{1}{\Delta\eta}(f_j^n - f_{j-1}^n),\ \left(\frac{\partial f}{\partial \xi}\right)_j^{n-1/2} = \frac{1}{\Delta\xi}(f_j^n - f_j^{n-1}),\ f_j^{n-1/2} = \frac{1}{2}(f_j^n + f_j^{n-1}), f_{j-1/2}^n = \frac{1}{2}(f_j^n + f_{j-1}^n),$$

Using equations (8.27)–(8.29) along with the Newton method, and ignoring the terms with the higher power of $\delta f_j^{n(i)}, \delta U_j^{n(i)}, \delta V_j^{n(i)}, \delta\theta_j^{n(i)}, \delta Q_j^{n(i)}, \delta g_j^{n(i)}$ and $\delta H_j^{n(i)}$, the equations are linearized as:

$$f_j^{n(i+1)} = f_j^{n(i)} + \delta f_j^{n(i)}$$

Thus, a linear algebraic equations system is achieved, which is then tackled by a tridiagonal block scheme. To find the solution at all net points (η_j, ξ^n), we first targeted the net point $(\eta_j, \xi^0), j = 0,1,2...N$ and obtained the solution at these net points, which is used as an initial guess to march further for the next value of ξ for all η.

8.4 RESULT AND DISCUSSION

The influence of pertinent parameters on flow, mass, heat transmission, and local skin friction and Sherwood and Nusselt numbers is inspected. The present numerical outcomes of $f''(0)$ in steady case ($\xi = 1$) are validated with the available data and exposed in Table 8.1 for diverse a/c and K_1, in which θ_r, ε^*, and M are neglected.

A remarkable similarity between the present outcomes originated with the available data. Figures 8.2–8.9 show air and water behavior against ξ (dimensionless time) for diverse estimations of the pertinent parameters. The selected base fluids are assumed as air and water, respectively. The behavior of skin friction and local Nusselt and Sherwood numbers is plotted against ξ (time).

The impacts of the variable parameter of viscosity θ_r upon the local skin friction, Nusselt, and Sherwood numbers are shown in Figures 8.2–8.3 for both air and water. Figure 8.2(a) illustrates that the local skin friction decreases in air and increases in water with a rise in absolute values of θ_r. It is also noted that the flow rate in air is greater than in water. In Figure 8.2(b), the local Sherwood number rises in the air and decreases in water by the increasing absolute values of θ_r, and the mass transmission rate in water is higher than in air.

Figures 8.3(a, b) illustrate the influence of variable viscosity parameter θ_r in both cases, when Pr = 0.7 (air) with $\varepsilon^* = 2$ and Pr = 7.0 (water) with $\varepsilon^* = 0.1$, as shown on the local Nusselt number. The same pattern is experienced for the local Sherwood number, but the heat transmission rate is much larger in water than in air. It is necessary to mention here that for growing estimations of θ_r, the graphs of local skin friction coefficient and Sherwood and Nusselt numbers become closer and closer, and this is only for $\theta_r \to \infty$ whenever the dynamic viscosity μ equals the viscosity of ambient fluid and it becomes a constant viscosity.

In Figure 8.4(a, b) the upshots of thermal conductivity parameter ε^* in both cases, when Pr = 0.7 (air) with $\theta_r = 1.3$ and Pr = 7.0 (water) with $\theta_r = -0.5$, are shown on the coefficient of skin friction and Nusselt number. Figure 8.4(a) witnessed that the coefficient of skin friction decays for air against ε^*. However, the variation in the local coefficient of skin friction for water is quite negligible due to ε^*. It is noticed that the flow rate in the air is greater than in water. Figure 8.4(b) depicts

TABLE 8.1
Comparison for $f''(0)$ with the Work of Salem and Fathy [34] by Considering Pr = 0.71, $\varepsilon^* = M = 0$, and $\theta_r \to \infty$.

$f''(0)$

a/c	$K_1 = 0$		$K_1 = 0.1$		$K_1 = 0.5$		$K_1 = 1.0$	
	Ref. [42]	Present	Ref. [42]	Present	Ref. [42]	Present	Ref. [42]	Present
0.1	−0.9694	−0.9695	−1.0101	−1.0099	−1.1583	−1.1584	−1.3211	−1.3212
0.2	−0.9181	−0.9182	−0.9519	−0.9519	−1.0768	−1.0769	−1.2156	−1.2158
0.5	−0.6673	−0.6673	−0.6854	−0.6855	−0.7540	−0.7541	−0.8321	−0.8322
2.0	2.0175	2.0177	2.0418	2.0421	2.1363	2.1366	2.2491	2.2494
3.0	4.7293	4.7299	4.7708	4.7716	4.9338	4.9345	5.1304	5.1312

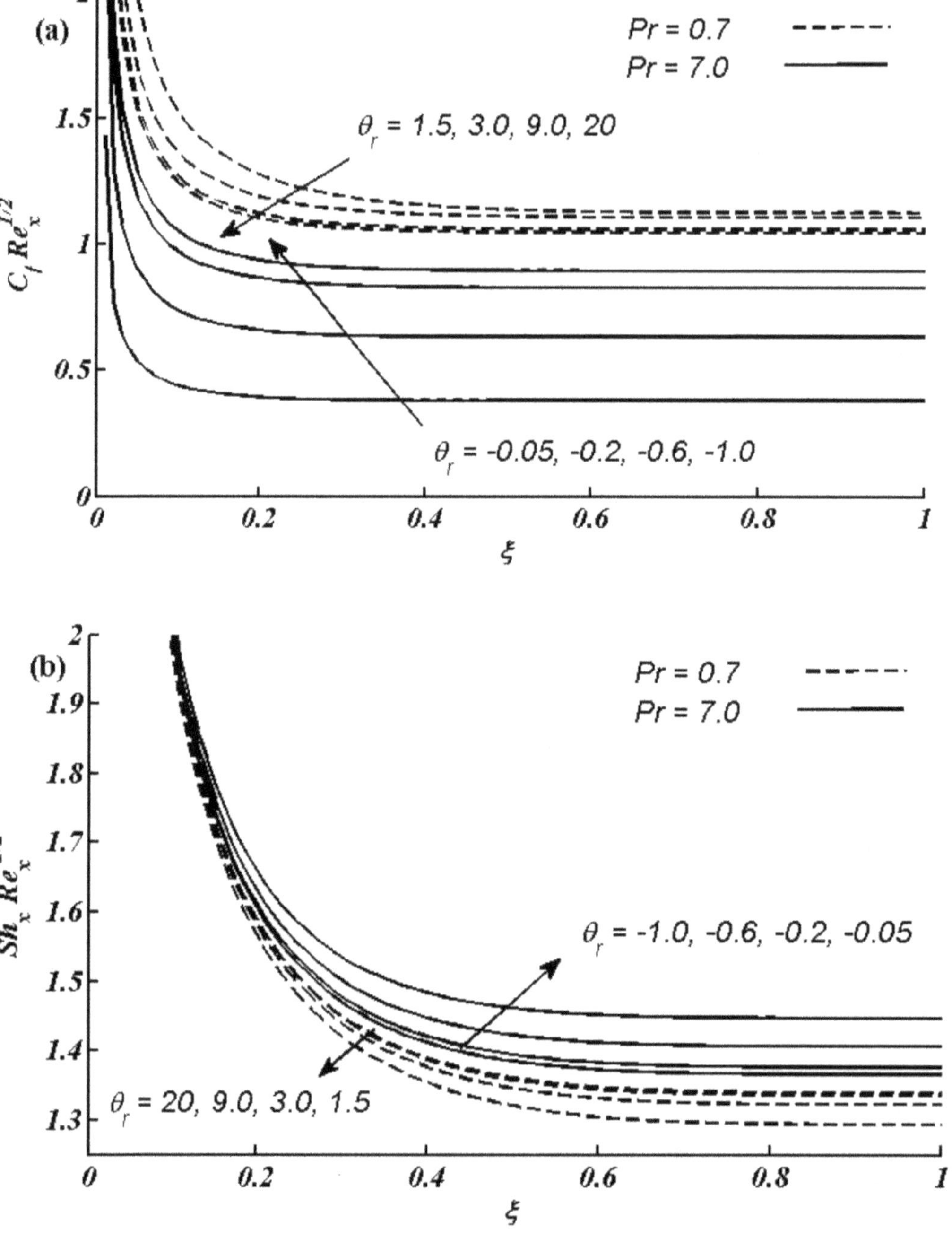

FIGURE 8.2 (a) Local skin friction; (b) local Sherwood number against ξ for various θ_r when $M = K_1 = 0.5$, $a/c = 1.5$, and $Sc = 0.94$.

the decrement in the Nusselt number by the rise in ε^* for air and water, but the heat transmission rate for water is greater than for air.

Figures 8.5(a, b) display the impact of permeability K_1 on coefficient of skin friction and Nusselt and Sherwood numbers while fixing the other parameters. Figure 8.5(a) depicts an

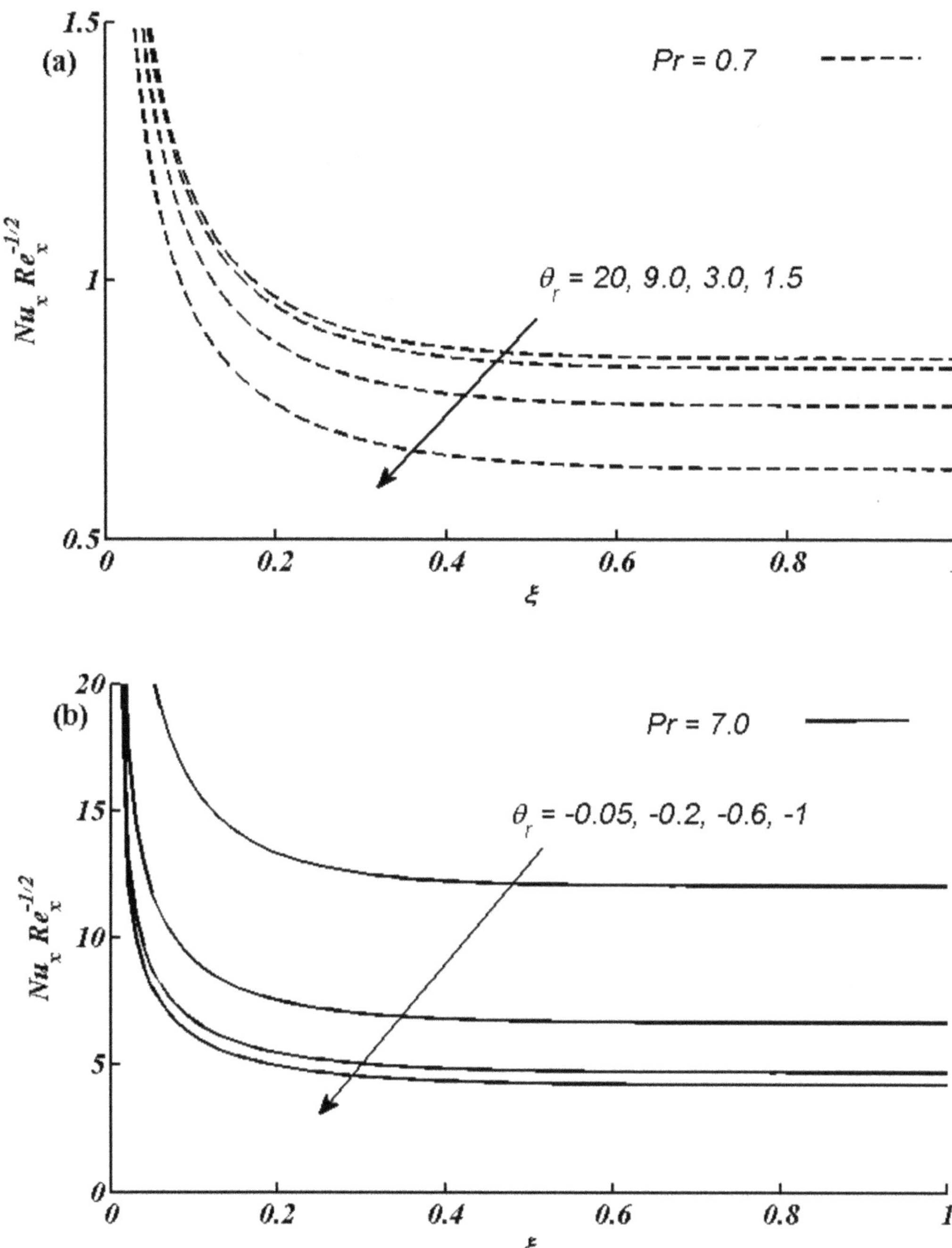

FIGURE 8.3 Local Nusselt number against ξ for various θ_r when $M = K_1 = 0.5$, $a/c = 1.5$, and $Sc = 0.94$.

escalation in the coefficient of skin friction for water and decrease for air with an increase in permeability K_1. Also, the flow rate in water becomes smaller as compared to air. The same conduct is noticed in Figure 8.5(b), with the difference being that the rate of mass transfer in water becomes greater than in air.

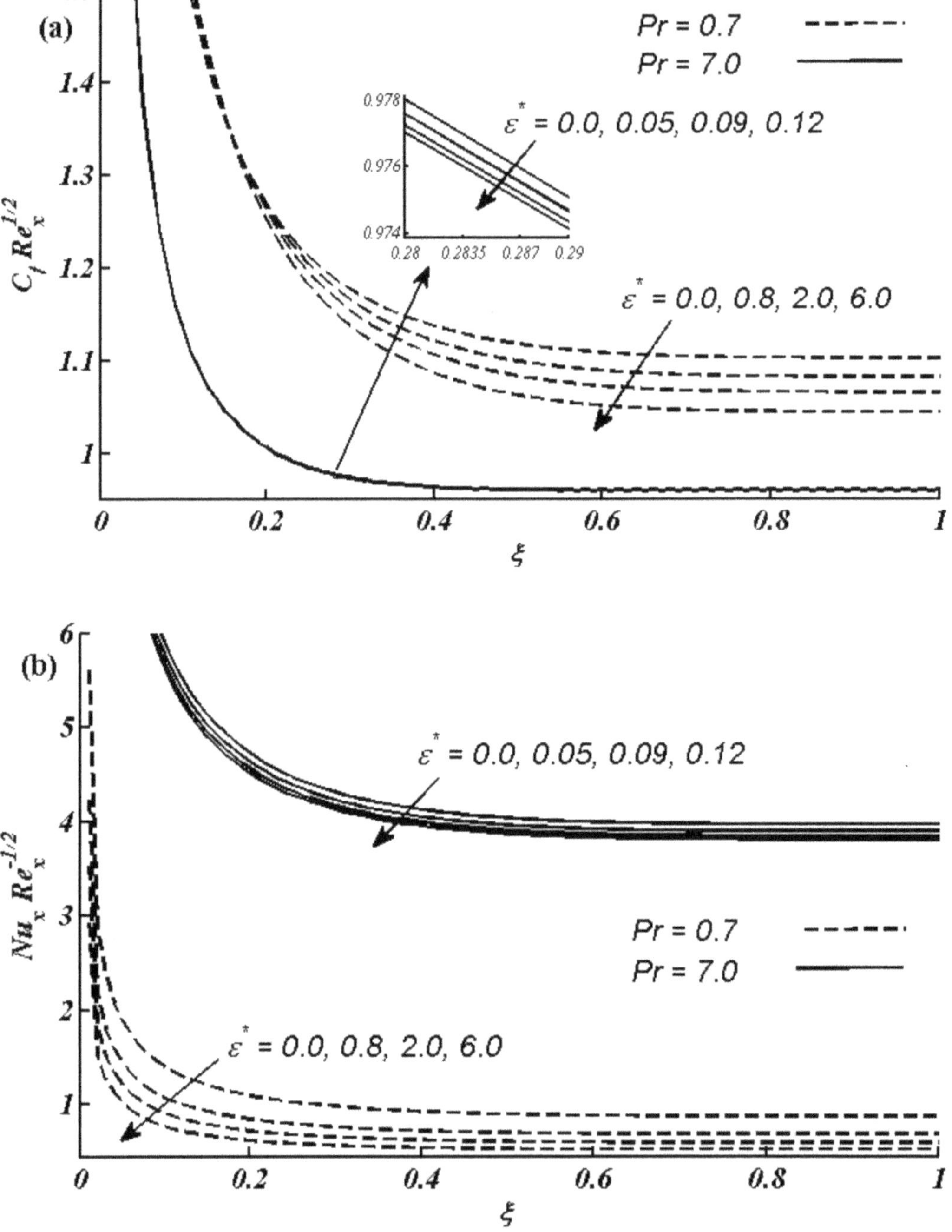

FIGURE 8.4 (a) The skin friction coefficient; (b) Nusselt number alongside ξ for various ε^* while $M = K_1 = 0.5$, $a/c = 1.5$, and $Sc = 0.94$.

Figures 8.6(a) and (b) describe the impact of K_1 on local Nusselt number for air and water, and the same behavior is experienced as that of K_1 on local Sherwood number, as portrayed in Figure 8.5(b).

The impact of several Schmidt number values on the local Sherwood number is depicted in Figure 8.7.

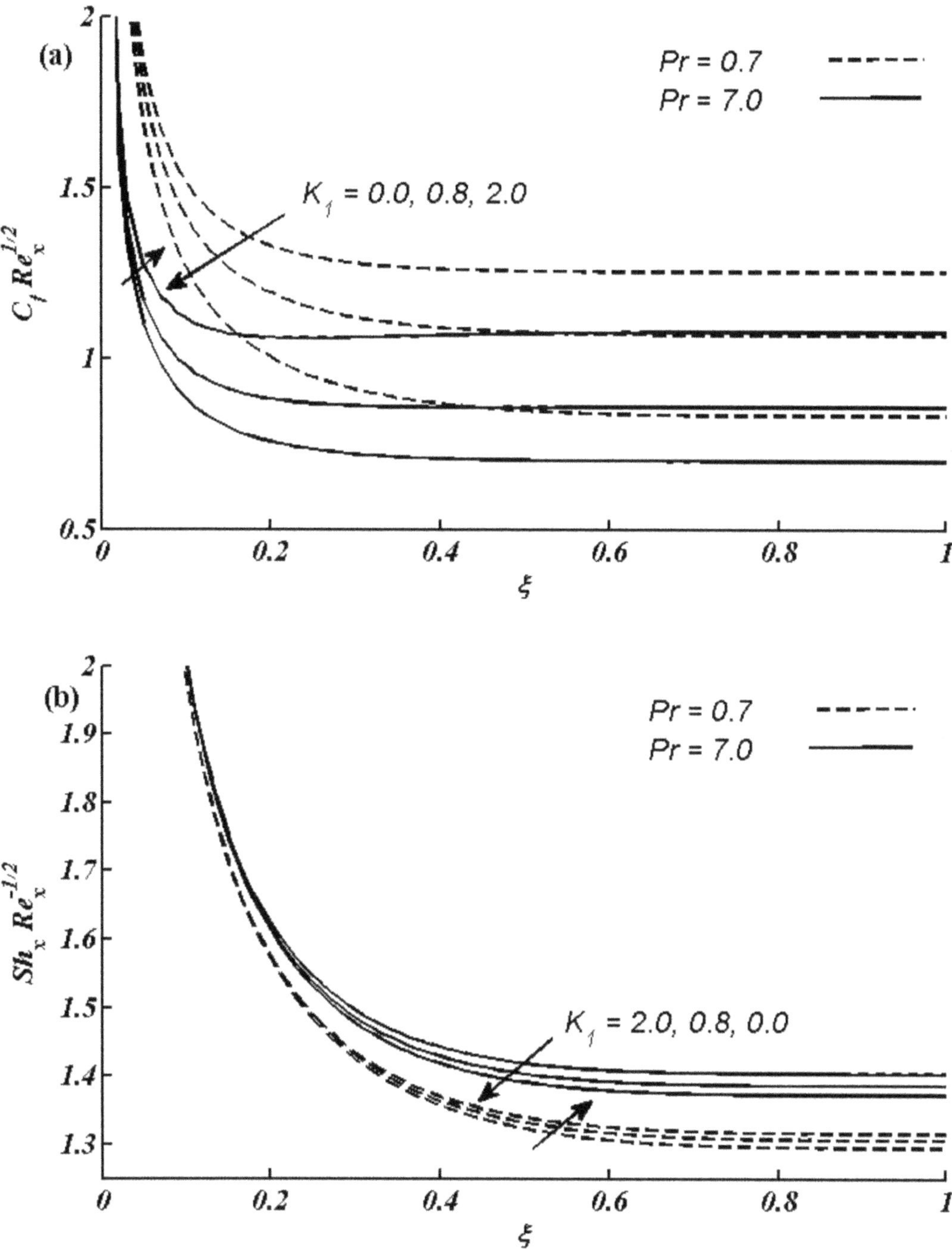

FIGURE 8.5 (a) Local coefficient of skin friction; (b) local Sherwood number alongside ξ for different K_l while $M = 0.5$, $a/c = 1.5$, and $Sc = 0.94$.

It is witnessed that the Sherwood number rises with the rise in the Schmidt number, and the mass transfer rate is greater in carbon dioxide ($Sc = 0.94$) than in other gases. For small ξ, the values of local Sherwood number in both cases are nearly equal, but for larger ξ, the rate of mass transmission is slower in the air from water.

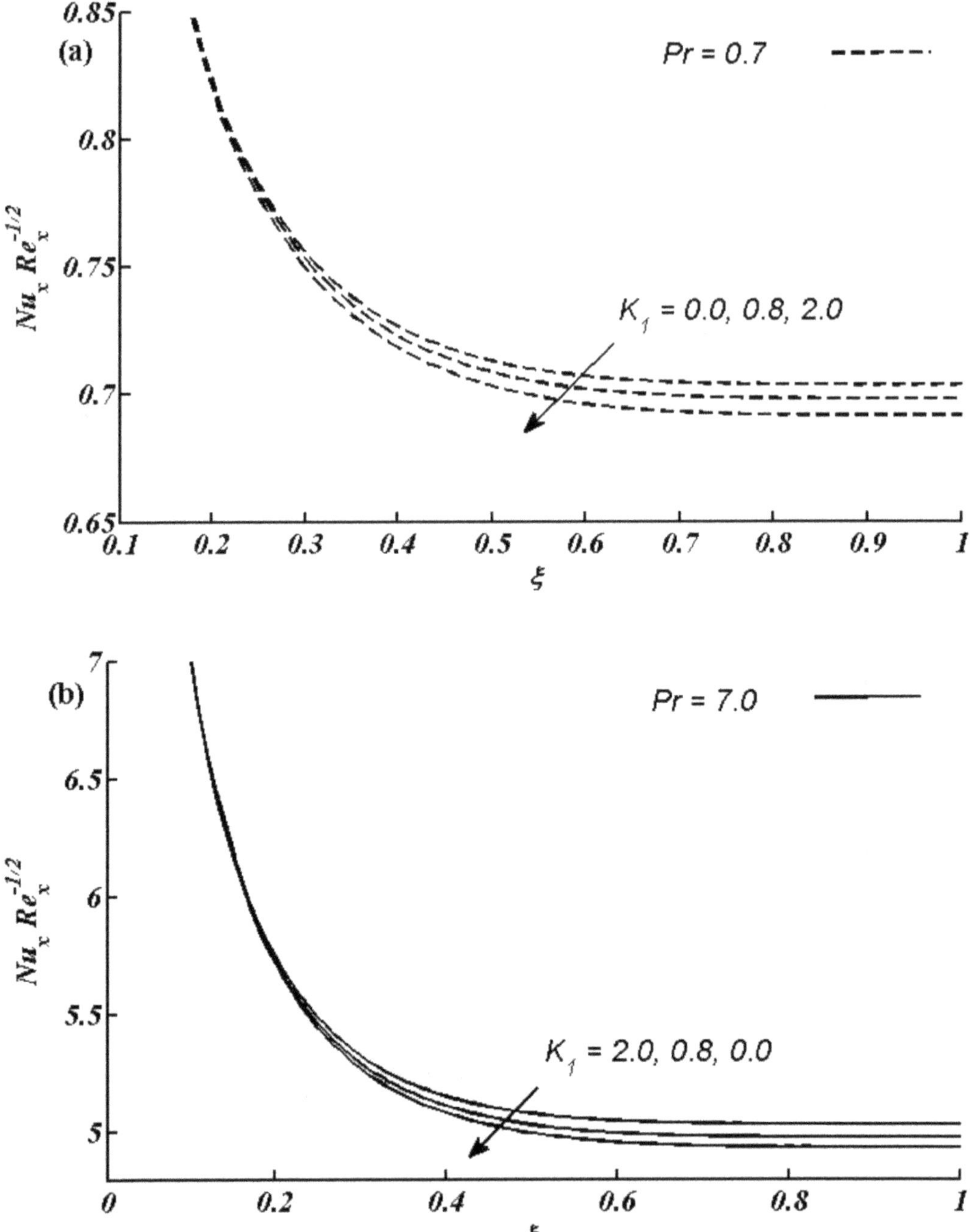

FIGURE 8.6 Local Nusselt number alongside ξ for various K_1 while $M = 0.5$, $a/c = 1.5$, and $Sc = 0.94$.

The upshots of magnetic parameter M when Pr = 0.7 (air) with $\theta_r = \varepsilon^* = 2$ and Pr = 7.0 (water) with $\theta_r = -0.5$, $\varepsilon^* = 0.1$ upon the coefficient of skin friction and Sherwood and Nusselt numbers are reflected graphically in Figures 8.8 and 8.9. Figure 8.8(a) shows that the local coefficient of skin friction rises in the air and water by boosting the magnetic parameter M. Also, the flow rate in the

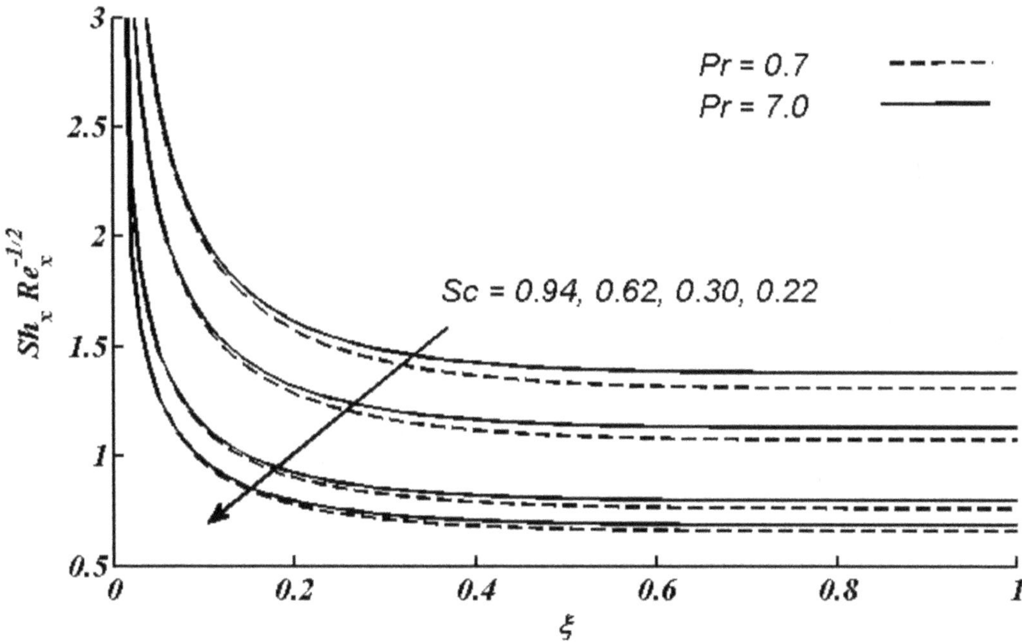

FIGURE 8.7 Local Sherwood number against ξ for various Sc while $M = K_l = 0.5$, $a/c = 1.5$.

air becomes more significant than that of water. In Figures 8.8(b) and 8.9(a, b), a similar trend is examined as in Figure 8.8(a).

Also, the mass and heat transfer rates in water become greater than in air. Figure 8.10(a, b) demonstrates the change of ambient Prandtl number inside the boundary layer against η, in which dashed and solid lines denote the solution for ξ = 0.5 and ξ = 1.0 (steady-state solution), respectively, for particular values of Pr = 0.7 (air) with $\varepsilon^* = 2$ and Pr = 7.0 (water) with $\varepsilon^* = 0.1$ for diverse values of the variable parameter of the viscosity θ_r, which takes the values 1.5, 3.0, 9.0, and 20 (air) and −0.05, −0.2, −0.6, and −1.0 (water). In the case of Pr = 0.7 (air), Figure 8.10(a) depicts that $\Pr_\infty$ increases at the surface by increasing θ_r and converges to the constant value of Pr = 0.7 at $\eta \approx 3.0$. It is noted that for ξ = 0.5 and ξ = 1.0, the values of ambient Prandtl number have the same values upon the surface, and far from the surface, the estimations of $\Pr_\infty$ for ξ = 0.5 become greater than for ξ = 1.0. Figure 8.10(b) illustrates that the ambient Prandtl number decays on the surface by the increasing absolute estimations of θ_r and converging the constant value of Pr = 7.0 (water) at $\eta \approx 0.6$.

For the thermal conductivity parameter ε^*, the change of ambient Prandtl number inside the boundary layer against η is presented in Figure 8.11(a, b). It is witnessed in Figure 8.11(a) that when Pr = 0.7 (air) with $\theta_r = 1.3$, the ambient Prandtl number increases upon the surface with the rising ε^*, and for $\varepsilon^* < \varepsilon_c^*$, the ambient Prandtl number becomes smaller than the value of particular Prandtl number Pr = 0.7 (air), and for $\varepsilon^* > \varepsilon_c^*$, the estimations of ambient Prandtl become greater than from Pr = 0.7. In Figure 8.11(b), when Pr = 7.0 with $\theta_r = -0.5$, the ambient Prandtl number escalates upon the surface with the increase of ε^* and converges to the constant Prandtl Pr = 7.0 (water) at $\eta \approx 1$.

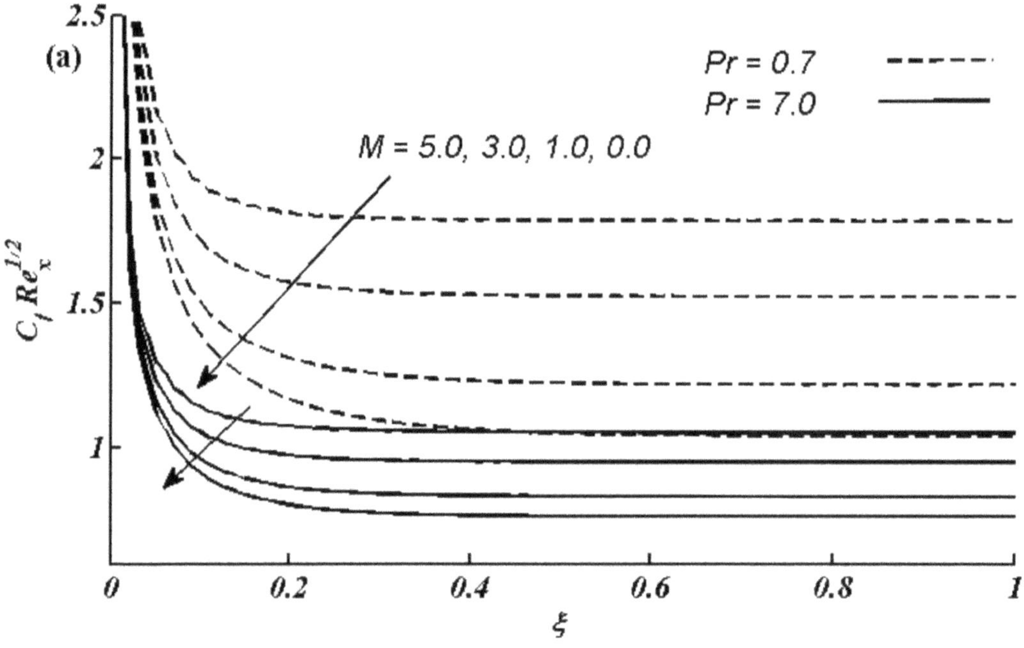

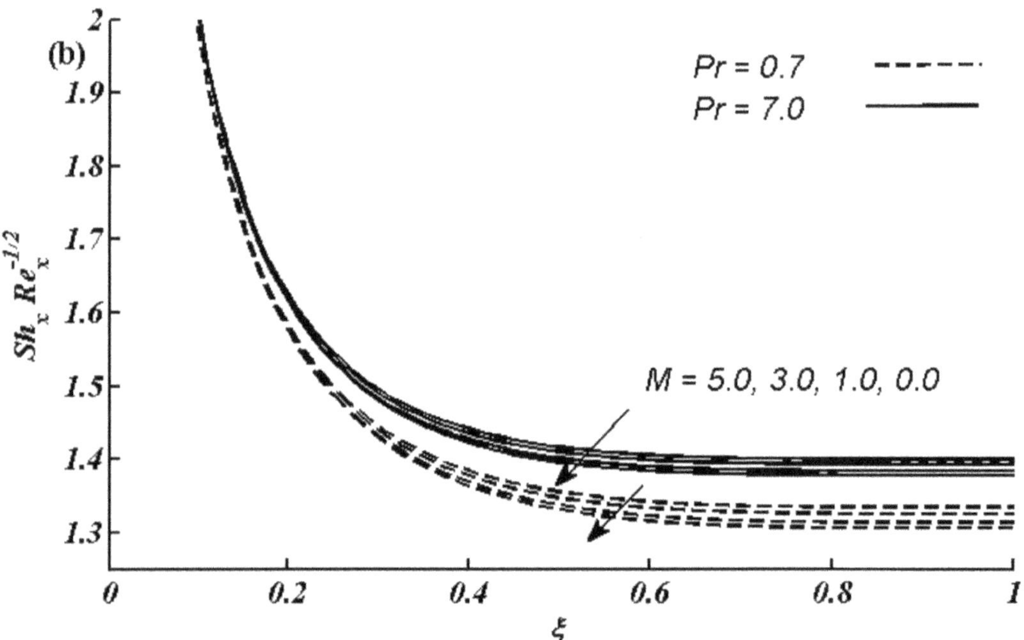

FIGURE 8.8 (a) Local coefficient of skin friction; (b) local Sherwood number against ξ for different M while $K_1 = 0.5$, $a/c = 1.5$, and $Sc = 0.94$.

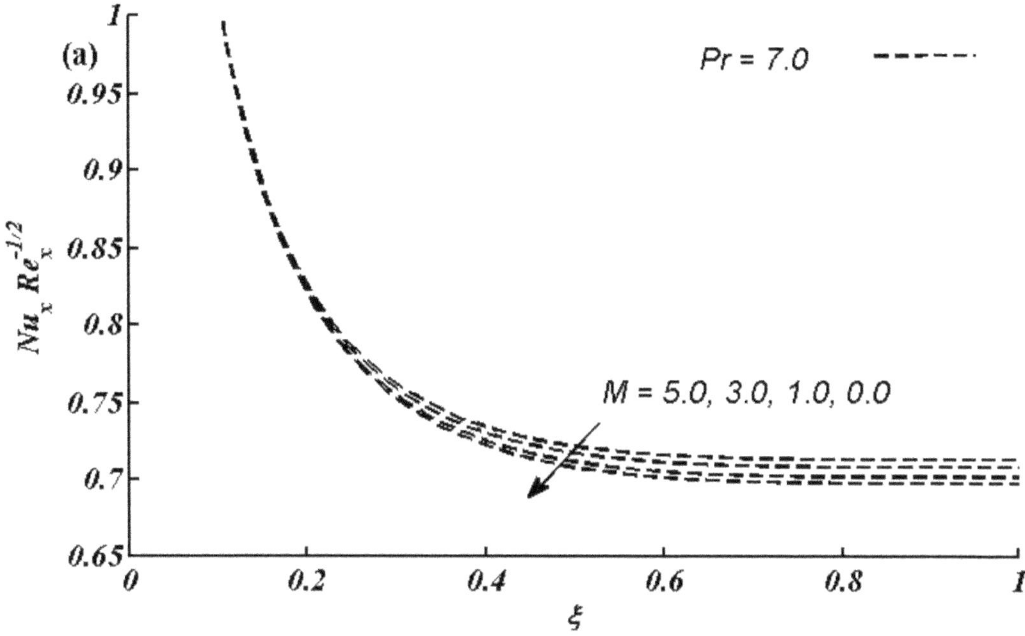

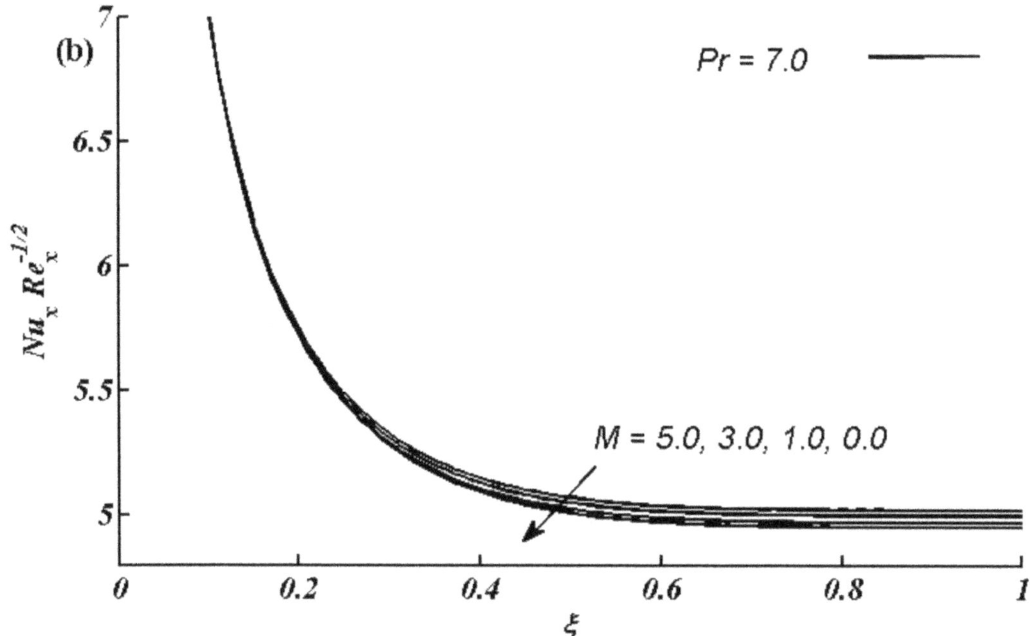

FIGURE 8.9 Local Nusselt number against ξ for diverse M while $K_1 = 0.5$, $a/c = 1.5$, and $Sc = 0.94$.

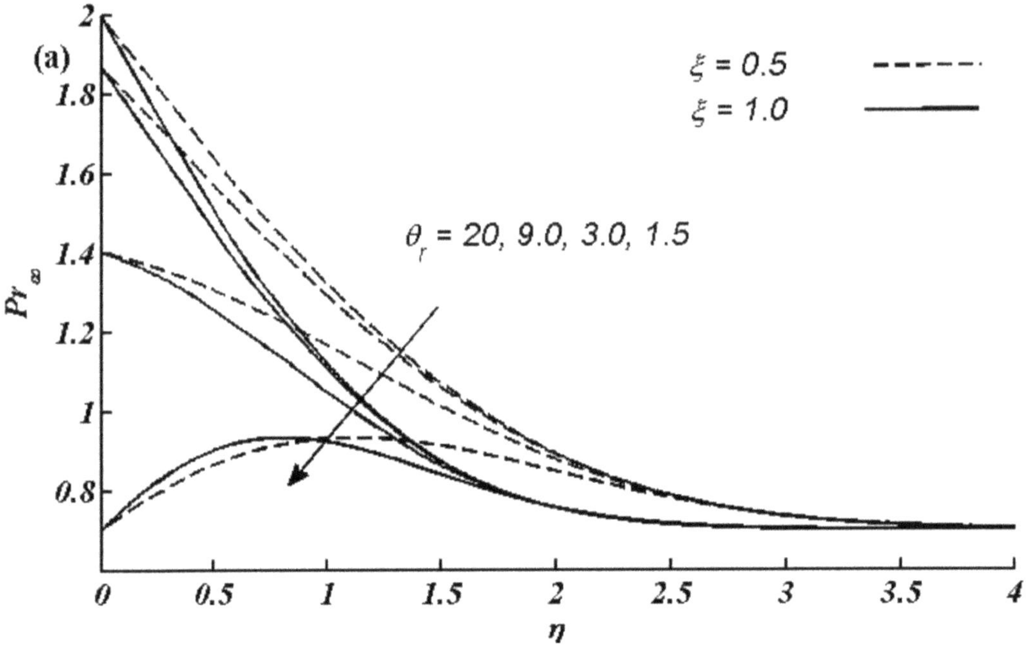

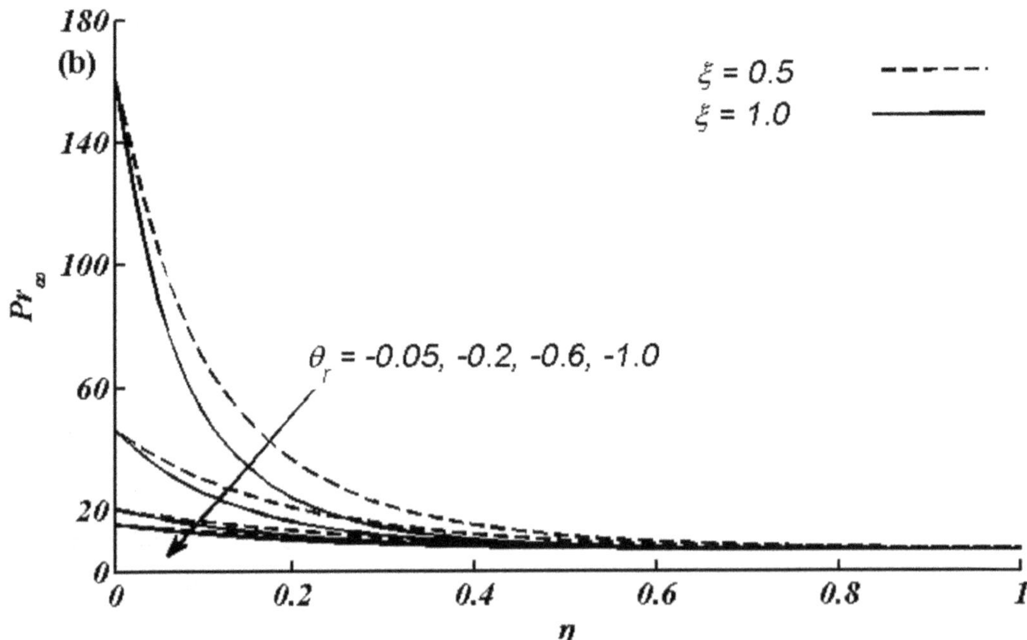

FIGURE 8.10 Alteration of ambient Prandtl number for diverse θ_r while $M = K_1 = 0.5$, $a/c = 1.5$, and $Sc = 0.94$: (a) Pr = 0.7 and $\varepsilon^* = 2$; (b) Pr = 7.0 and $\varepsilon^* = 0.1$.

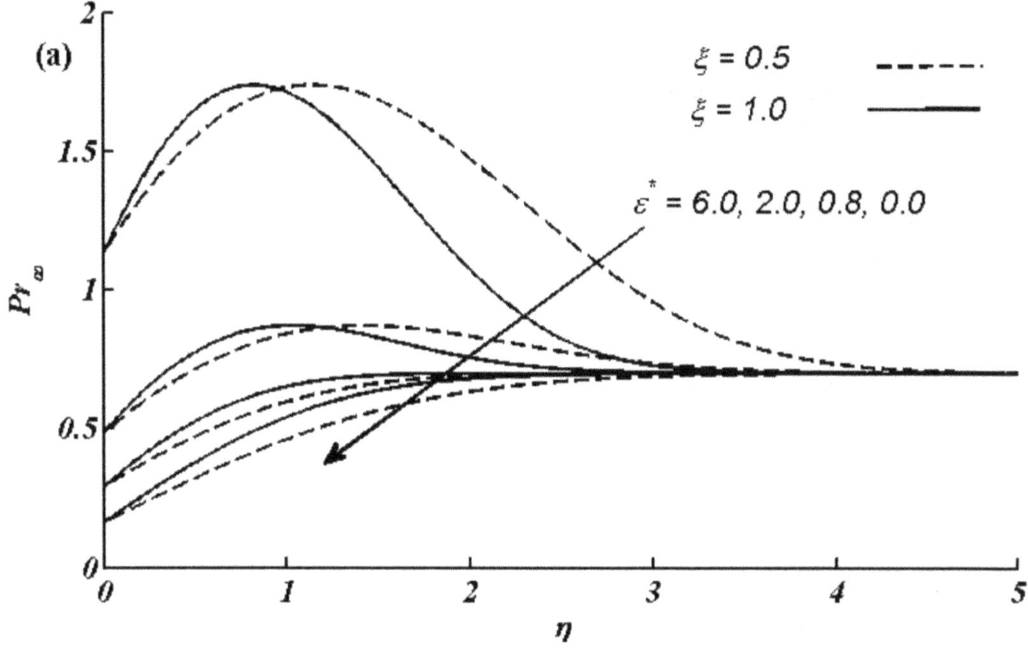

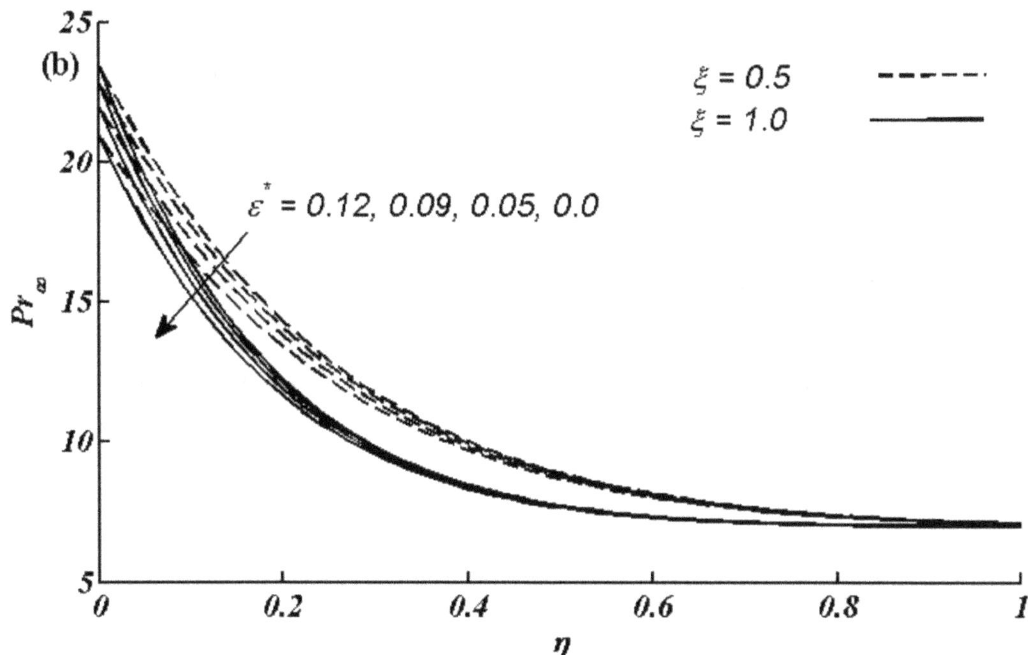

FIGURE 8.11 Variation of ambient Prandtl number for various ε^* when $M = K_1 = 0.5$, $a/c = 1.5$, and $Sc = 0.94$: (a) Pr = 0.7 and $\theta_r = 1.3$; (b) Pr = 7.0 and $\theta_r = -0.5$.

8.5 CONCLUSION

The current chapter inspects the time-dependent MHD stagnant flow with mass and heat transmission upon a stretching porous sheet. The properties of the fluid are considered in terms of temperature. Thus, the main points are:

- The influence of the parameter θ_r upon physical quantities depends upon the fluid Prandtl number.
- The parameter ε^* helps in reducing the local Nusselt number and skin friction for both fluids.
- The parameter K_1 enhances the local coefficient of skin friction and Sherwood and Nusselt numbers of water (Pr = 7.0) and decays for air (Pr = 0.7).
- A rise in Schmidt number improves the local Sherwood number for both fluids.
- The magnetic field raises the local coefficient of skin friction and decreases both local Sherwood and Nusselt numbers for both fluids.

Nomenclature

Symbols	**Description**	**Dimensions**
A,B,a,c	some constants	—
B_0	magnetic field strength	**($M^{1/2}L^{1/2}T^{-2}$)**
C	the concentration of species in the boundary layer	—
C_w	the concentration of species on the wall	—
$C_¥$	the ambient concentration species	—
C_f	coefficient of skin friction	—
c_p	heat capacity	**($L^2T^{-2}K^{-1}$)**
D	molecular diffusivity	**(L^2T^{-1})**
e	constant	—
f	non-dimensional stream function	—
g	non-dimensional concentration function	—
K	porous medium permeability	**(L^2)**
K_1	permeability parameter	—
k	thermal conductivity	**($MLT^{-3}K^{-1}$)**
$k_¥$	ambient thermal conductivity	**($MLT^{-3}K^{-1}$)**
M	magnetic parameter	—
m_w	mass flux	**($M\,L^{-2}\,T^{-1}$)**
Nu_x	Nusselt number	—
Pr	Prandtl number	—
$Pr_¥$	ambient Prandtl number	—
q_w	wall heat flux	—
Re_x	local Reynolds number	—
Sc	Schmidt number	—
Sh_x	Sherwood number	—
t	time	**(T)**
t^*	dimensionless time	—
T	the temperature of the fluid in the boundary layer	**(K)**
T_r	constant	—
T_∞	ambient fluid temperature	**(K)**

Symbols	Description	Dimensions
T_w	surface temperature	**(K)**
u, v	velocity components in x- and y-directions, respectively	**(L T^{-1})**
Greek symbols		
g	thermal property of the fluid	—
η	similarity variable	—
ξ	dimensionless variable	—
q	dimensionless temperature	—
r	fluid density	**(M L^{-3})**
m	dynamic viscosity	—
m$_¥$	ambient dynamic viscosity	—
$v_¥$	ambient kinematic viscosity	—
t_w	surface shear stress	**(M L^{-1} T^{-2})**
ε^*	thermal conductivity parameter	—
ε_c^*	the critical value of the thermal conductivity parameter	—
θ_r	variable viscosity parameter	—
ψ	dimensionless stream function	—
σ	electric conductivity	—
Subscripts		
w	condition at the surface	—
∞	condition far away from the surface	—

REFERENCES

[1] Sakiadis, B.C. (1961). Boundary-Layer Behavior on Continuous Solid Surfaces: I. Boundary-Layer Equations for Two-Dimensional and Axisymmetric Flow. *AIChE J*, **7**(1), 26–28. https://doi.org/10.1002/aic.690070108.

[2] Sakiadis, B.C. (1970) Boundary Layer Behaviour on Continuous Solid Surface: II. Boundary Layer Equations for Two Dimensional and Axisymmetric Flow. *AIChE J*, **7**, 221–225.

[3] Crane, L.J. (1970). Flow Past a Stretching Plate. *Z AngewMath Phys*, **21**, 645–647.

[4] Gupta, P.S. and Gupta, A.S. (1977). Heat and Mass Transfer on a Stretching Sheet with Suction or Blowing. *Can J ChemEng*, **55**, 744–746.

[5] Hiemenz, K. (1911). Die Grenzschicht an einem in den gleichformingen Flussigkeits-stromeingetauchtengradenKreiszylinder. *Dingler's Poly J*, **326**, 321–324.

[6] Eckert, E.R.G. (1942). Die BerechnungdesWärmeüberganges in der laminaren Grenzschicht um stromterKorper. *VDI Forchungs-heft Berlin*, 416–418.

[7] Chiam, T.C. (1994). Stagnation Point Flow Towards a Stretching Plate. *J Phys Soc Jpn*, **63**, 2443–2444.

[8] Ariel, P.D. (1994). Hiemenz Flow in Hydromagnetics. *Acta Mech*, **103**, 31–43.

[9] Jat, R.N. and Chaudhary, S. (2008). Magnetohydrodynamic Boundary Layer Flow Near the Stagnation Point of a Stretching Sheet. *I L Nuovo Cimento*, **123**, 555–566.

[10] Mustafa, I., Javed, T. and Majeed, A. (2015). Magnetohydrodynamic (MHD) Mixed Convection Stagnation Point Flow of a Nanofluid Over a Vertical Plate with Viscous Dissipation. *Can J Phys*, **93**, 1365–1374.

[11] Mustafa, I., Javed, T. and Ghaffari, A. (2016). Heat Transfer in MHD Stagnation Point Flow of a Ferrofluid Over a Stretchable Rotating Disk. *J Mol Liq*, **219**, 526–532.

[12] Pushpa, B.V., Sankar, M. and Mebarek-Oudina, F. (2021). Buoyant Convective Flow and Heat Dissipation of Cu–H_2O Nanoliquids in an Annulus Through a Thin Baffle. *J Nanofluids*, **10**, 292–304.

[13] Dhif, K., Mebarek-Oudina, F., Chouf, S., Vaidya, H. and Chamkha Ali, J. (2021). Thermal Analysis of the Solar Collector Cum Storage System Using a Hybrid-Nanofluids. *J Nanofluids*, **10**, 616–626.

[14] Swain, K., Mahanthesh, B. and Mebarek-Oudina, F. (2020). Heat Transport and Stagnation-Point Flow of Magnetized Nanoliquid with Variable Thermal Conductivity, Brownian Moment, and Thermophoresis Aspects. *Heat Transf*, **50**, 754–767.

[15] Rajashekhar, C., Mebarek-Oudina, F., Vaidya, H., Prasad, K.V., Manjunatha, G. and Balachandra, H. (2021). Mass and Heat Transport Impact on the Peristaltic Flow of a Ree–Eyring Liquid Through Variable Properties for Hemodynamic Flow. *Heat Transf*, **50**, 5106–5122.

[16] Marzougui, S., Mebarek-Oudina, F., Magherbi, M. and Mchirgui, A. (2022). Entropy Generation and Heat Transport of Cu–Water Nanoliquid in Porous Lid-driven Cavity Through Magnetic Field. *Int J Numer Method Heat Fluid Flow*, **32**, 2047–2069.

[17] Djebali, R., Mebarek-Oudina, F. and Rajashekhar, C. (2021). Similarity Solution Analysis of Dynamic and Thermal Boundary Layers: Further Formulation Along a Vertical Flat Plate. *Phys Scr*, **96**, 085206.

[18] Warke, A.S., Ramesh, K., Mebarek-Oudina, F., et al. (2022). Numerical Investigation of the Stagnation Point Flow of Radiative Magnetomicropolar Liquid Past a Heated Porous Stretching Sheet. *J Therm Anal Calorim*, **147**, 6901–6912.

[19] Chabani, I., Mebarek-Oudina, F. and Ismail, A.A.I. (2022). MHD Flow of a Hybrid Nano-Fluid in a Triangular Enclosure with Zigzags and an Elliptic Obstacle. *Micromechanics*, **13**, 224.

[20] Nield, D.A. and Bejan, A. (1992). *Convection in Porous Media*, Springer-Verlag, New York.

[21] Bejan, A., Dincer, I., Lorente, S., Miguel, A.F. and Reis, A.H. (2004). *Porous and Complex Flow Structures in Modern Technologies*, Springer, New York.

[22] Pop, I. and Ingham, D.B. (2001). *Convective Heat Transfer: Mathematical and Computational Modeling of Viscous Fluids and Porous Media*, Pergamon, Oxford.

[23] Ingham, D.B. and Pop, I. (2005). *Transport Phenomena in Porous Media III*, Elsevier, Oxford.

[24] Vafai, K. (2005). *Handbook of Porous Media*, Taylor & Francis, New York.

[25] Vadasz, P. (2008). *Emerging Topics in Heat and Mass Transfer in Porous Media*, Springer, New York.

[26] Vafai, K. (2010). *Porous Media: Applications in Biological Systems and Biotechnology*, CRC Press, Tokyo.

[27] Kechil, S.A. and Hashim, I. (2009). Approximate Analytical Solution for MHD Stagnation-Point Flow in Porous Media. *Commun Nonlinear Sci NumerSimulat*, **14**, 1346–1354.

[28] Kazem, S., Shaban, M. and Abbasbandy, S. (2011). Improved Analytical Solutions to a Stagnation-Point Flow past a Porous Stretching Sheet with Heat Generation. *J Franklin Inst*, **348**, 2044–2058.

[29] Makinde, O.D. (2012). Heat and Mass Transfer by MHD Mixed Convection Stagnation Point Flow Toward a Vertical Plate Embedded in a Highly Porous Medium with Radiation and Internal Heat Generation. *Meccanica*, **47**, 1173–1184.

[30] Hari, N., Sivasankaran, S., Bhuvaneswari, M. and Siri, Z. (2015). Effects of Chemical Reaction on MHD Mixed Convection Stagnation Point Flow Toward a Vertical Plate in a Porous Medium with Radiation and Heat Generation. *J Phys Conf Ser*, **662**, 1–9.

[31] Hari, N., Sivasankaran, S. and Bhuvaneswari, M. (2016). Analytical and Numerical Study on Magnetoconvection Stagnation-Point Flow in a Porous Medium with Chemical Reaction, Radiation, and Slip Effects. *Math Probl Eng*, 4017076, 1–12.

[32] Ghaffari, A., Javed, T. and Majeed, A. (2016). Influence of Radiation on Non-Newtonian Fluid in the Region of Oblique Stagnation Point Flow in a Porous Medium: A Numerical Study. *Transport Porous Med*, **113**, 245–266.

[33] Mukhopadhyay, S., Layek, G.C. and Samad, S.A. (2005). Study of MHD Boundary Layer Flow Over a Heated Stretching Sheet with Variable Viscosity. *Int J Heat Mass Transfer*, **48**, 4460–4466.

[34] Pop, I., Gorla, R.S.R. and Rashidi, M. (1992). The Effect of Variable Viscosity on Flow and Heat Transfer to a Continuous Moving Flat Plate. *Int J Eng Sci*, **30**, 1–6.

[35] Elbashbeshy, E.M.A. and Bazid, M.A.A. (2000). The Effect of Temperature Dependent Viscosity on Heat Transfer Over a Continuous Moving Surface. *J Phys D: Appl Phys*, **33**, 2716–2721.

[36] Chiam, T.C. (1996). Heat Transfer with Variable Conductivity in a Stagnation Point Flow Towards a Stretching Sheet. *Int Commun Heat Mass Transfer*, **23**, 239–248.

[37] Chiam, T.C. (1998). Heat Transfer in a Fluid with Variable Thermal Conductivity Over a Linearly Stretching Sheet. *Acta Mech*, **129**, 63–72.

[38] Rahman, M.M. and Salahuddin, K.M. (2010). Study of Hydromagnetic Heat and Mass Transfer Flow Over an Inclined Heated Surface with Variable Viscosity and Electric Conductivity. *Commun Nonlinear Sci NumerSimulat*, **15**, 2073–2085.

[39] Rahman, M.M., Rahman, M.A., Samad, M.A. and Alam, M.S. (2009). Heat Transfer in Micropolar Fluid Along a Nonlinear Stretching Sheet with Temperature Dependent Viscosity and Variable Surface Temperature. *Int J Thermophys*, **30**, 1649–1670.

[40] Rahman, M.M. (2010). Convective Hydromagnetic Slip Flow with Variable Properties Due to a Porous Rotating Disk. *SQU J Sci*, **15**, 55–79.

[41] Rahman, M.M. and Eltayeb, I.A. (2011). Convective Slip Flow of Rarefied Fluids Over a Wedge with Thermal Jump and Variable Transport Properties. *Int J Thermal Sci*, **50**, 468–479.

[42] Salem, A.M. and Fathy, R. (2012). Effects of Variable Properties on MHD Heat and Mass Transfer Flow Near a Stagnation Point Towards a Stretching Sheet in a Porous Medium with Thermal Radiation. *Chin Phys B*, **21**, 054701.

[43] Nazar, R., Amin, N., Filip, D. and Pop, I. (2004). Unsteady Boundary Layer Flow in the Region of the Stagnation Point on a Stretching Sheet. *Int J Eng Sci*, **42**, 1241–1253.

[44] Seshadri, R., Sreeshylan, N. and Nath, G. (2002). Unsteady Mixed Convection Flow in the Stagnation Region of a Heated Vertical Plate Due to Impulsively Motion. *Int J Heat Mass Transfer*, **45**, 1345–1352.

[45] Ling, J.X. and Dybbs, A. (1987). Forced Convection Over a Flat Plate Submersed in a Porous Medium: Variable Viscosity Case. *ASME*, Paper 87-WA/HT-23, ASME Winter Annual Meeting, Boston, Massachusetts, 13–18.

[46] Elbarbary, E.M.E. and Elgazery, N.S. (2004). Chebyshev Finite Difference Method for the Effects of Variable Viscosity and Variable Thermal Conductivity on Heat Transfer from Moving Surfaces with Radiation. *Int J Therm Sci*, **43**, 889–899.

[47] Cebeci, T. and Bradshaw, P. (1984). *Physical and Computational Aspects of Convective Heat Transfer*, Springer, New York.

9 The Impact of Radiation and Marangoni Boundary Condition on Fluid Flow through a Porous Medium with Brinkman Model

H. V. Vishala, T. Maranna, U. S. Mahabaleshwar, and B. Souayeh

9.1 INTRODUCTION

In order to fulfill a wide range of applications, such as sustaining flow behavior throughout a broad temperature as well as stress range, people have a passion for tracking the flow behavior of liquids in technological disciplines. It is classed as either a Newtonian fluid or a non-Newtonian fluid if it flows as a function of shear stress. Because of its potential applications in areas including oil exploration, building supplies, cosmetics, and blood circulation, an investigation of convection transport of heat in flowing fluid has received a lot of attention. According to Shakya et al. [1], it enhances the critical heat flux. The growing understanding of the development of nanofluids has resulted in the recognition of dimensions and forms of nanoparticles as a constant component affecting viscosity and thermal conductivity. Crane et al. [2] and Wang et al. [3] suggested the stretching sheet problem, which provides a diverse range of industrial services, as does incorporation of nanoparticles. Siddheshwar et al. [4] is one of the pioneer works on the investigation of the outcomes of transferring heat in a viscoelastic liquid due to extending plate in the appearance of dissipation.

Choi et al. [5] supported the notion of suspending nano-regime particles in water-based fluids to increase the fluids thermal conductivity in the beginning. Napolitano et al. [6–8] were the first to examine the Marangoni boundary layer in their research, and also on fields identifications in the mass and volume substances for non-Marangoni boundary layers, which contrarily depend on configuration. On thermosolutal Marangoni permeable margins, including mass transpiration as well as temperature source/sink, the impact of chemical reaction and heat penetration/formation among flow of viscous fluid was studied by Mahabaleshwar et al. [9]. Recently, research on the impact of ternary HNF on fluid flow and temperature distribution has begun when it comes to energy exchanges, ternary hybrid nanofluid exceeding normal fluids, nanofluid, hybrid nanofluid, gasoline, and acetone. HNFs have a broad range of temperature-dependent effects, including freezing in high-temperature environments. Recently, many researchers [10–13] worked on casson hybrid nanofluid, MHD flow micro-polar fluid, and an MHD nanofluid through a penetrable and also stretching/shrinking surface, a horizontal surface having a radiated effect with mass transpiration.

Animasuan et al. [14] numerically explore the influence of an induced magnetic field and various-shaped nanoparticles Ag, Al_2O_3, and Al on flow, in addition to the effect of a heat source/sink. A number of interesting numerical research on the ideal fluid Marangoni boundary layers in varied configurations are interpolated through references [15, 16]. Manjunatha et al. [17]

DOI: 10.1201/9781003299608-9

characterized the direction a $Cu\text{-}Al_2O_3\text{-}H_2O$ HNF with changing viscosity. In the existence of a suction/injection parameter with radiation, the flow of hybrid nanofluid past a stretching/shrinking sheet was studied using an accurate analytical solution by Mahabaleshwar et al. [18]. Udawattha et al. [19] awareness of numerous facts about nanofluids comprising spherical nanoparticles has resulted from the relevance of nanofluid. The researchers discovered that increasing the volume proportion of nanoparticles enhances heat transfer. A potential illustration can be used on enormous surface holes as a result of Brinkman et al. [20] innovative research. With the help of the Navier–Stokes equation, he investigated the fluid flow caused by viscous force on the surface of a thick swarm of tiny spheres.

Unsteady magnetohydrodynamic for Brinkman and fractional free convection flow was considered by Khan et al. [21, 22] and Ali et al. [23] as it passed on a continuous dish immersed via tensile force that was accelerated towards the bottom plate and perpendicular plate. With the support of an analytical technique, the movement of dusty hybrid nanoparticles owing towards permeable materials via radiating heat linearly was studied by Sneha et al. [24]. A quantitative assessment of the fluid flow consistency inside a Brinkman porous structure was developed by Shankar et al. [25]. The use of the Brinkman model to compute the free convection flow of two immiscible viscous liquids across comparable porous sheets was studied by Takhar et al. [26]. The stretching/shrinking surface embedded in a permeable material exposed to two-dimensional laminar MHD coupled stress HNF with an inclination force field was studied by Anusha et al. [27].

The novelty of the chapter is the Brinkman model used to study the impact of radiation and Marangoni convective boundary conditions on fluid flow through a penetrable media with mass transpiration and to find the exact solutions via the existence of suction/injection, consequence of radiation, together with the Marangoni convection boundary conditions upon flow and heat exchange of fluids in porous medium. Governing equations are solved by using similarity transformation. The similarity method is employed in order to gain analytical solutions for momentum as well as temperature field.

9.2 MATHEMATICAL MODELLING

As part of this study, we consider a steady two-dimensional Marangoni boundary layer flow through a porous surface. The fluid is considered to be incompressible, and the flow to be laminar. We have a look at the Cartesian coordinate system (x, y), where x and y are the coordinates measured along the sheet and normal to it, correspondingly. And the flow is positioned at $y \geq 0$. Additionally, the temperatures of the plate and the ambient fluid are, respectively, T_w and T_∞.

Formulation for fluid in the boundary layer of a steady state in the cartesian plane is given [28–30] as:

$$\frac{\partial u}{\partial x}+\frac{\partial v}{\partial y}=0, \tag{9.1}$$

$$u\frac{\partial u}{\partial x}+v\frac{\partial u}{\partial y}=\nu_{eff}\frac{\partial^2 u}{\partial y^2}-\frac{\nu}{K}u, \tag{9.2}$$

$$u\frac{\partial T}{\partial x}+v\frac{\partial T}{\partial y}=\frac{\kappa}{\left(\rho C_p\right)}\frac{\partial^2 T}{\partial y^2}-\frac{1}{\left(\rho C_p\right)}\frac{\partial q_r}{\partial y}, \tag{9.3}$$

The last term of equation (9.2) and equation (9.3) denotes the porosity (porous medium) and thermal radiation, respectively; (u, v) denotes the velocity components along the x and y-axes, and K is known to be permeability.

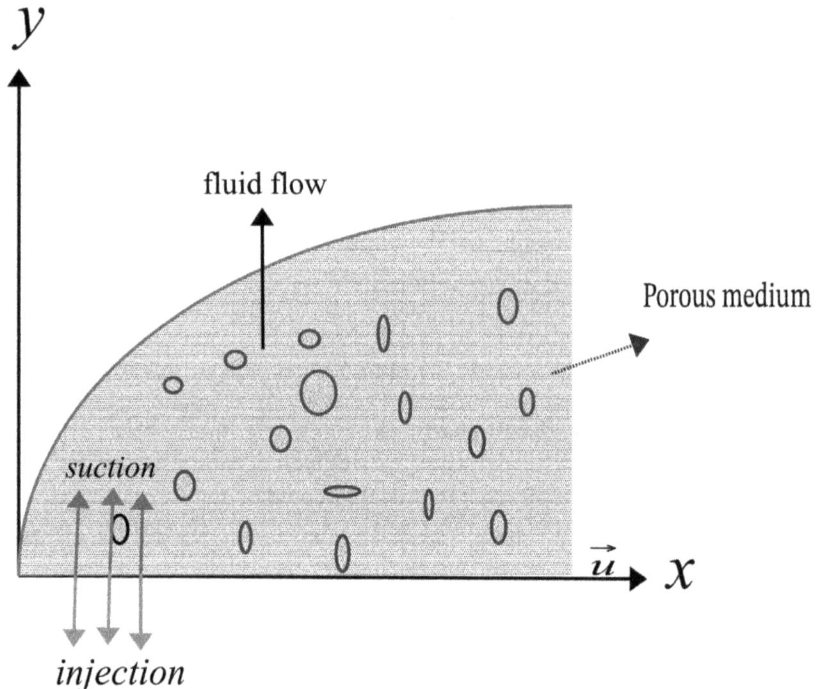

FIGURE 9.1 Diagrammatic representation of flow problem and coordinate system.

Along with imposed frontier constraints are:

$$\left.\begin{array}{llll} v = v_w, & T = T_\infty + ax^2, & \mu\dfrac{\partial u}{\partial y} = \dfrac{\partial \sigma}{\partial T}\dfrac{\partial T}{\partial y}, & \text{as} \quad y=0, \\ u=0, & T \to T_\infty, & & \text{as} \quad y \to \infty, \end{array}\right\}, \tag{9.4}$$

The stream function was then introduced with the appropriate similarity variables [31] as follows:

$$\psi(\eta) = \xi_2 x f(\eta), \quad \theta(\eta) = \frac{T - T_\infty}{T_w - T_\infty}, \quad \eta = \xi_1 y, \tag{9.5}$$

$$\text{Where } \xi_1 = \frac{\sigma_0 \gamma_a \rho_f}{\mu_f^{\,2}}, \qquad \xi_2 = \frac{\sigma_0 \gamma_a \mu_f}{\rho_f^{\,2}},$$

With velocities of the following form:

$$u = \xi_1 \xi_2 x f'(\eta), \qquad v = -\xi_2 f(\eta), \tag{9.6}$$

It is assumed that the surface tension σ is to vary linearly with temperature, as [32]:

$$\sigma = \sigma_0[1 - \gamma_T(T - T_\infty)], \tag{9.7}$$

Here, σ_0 is the surface tension at the interface, and γ_T is the rate of change of surface with temperature.

The coefficient of surface tension for temperature is given by:

$$\gamma_T = -\frac{1}{\sigma_0}\frac{\partial \sigma}{\partial T}\bigg|_T, \tag{9.8}$$

The Rosseland approximation is used to derive the formula for radiative heat flux, as stated in [33–35]:

$$q_r = -\frac{4\sigma^*}{3k^*}\frac{\partial T^4}{\partial y}, \tag{9.9}$$

Utilizing Taylor's expansion of T^4 around T_∞, which is the ambient temperature, and ignoring higher-order terms, we get:

$$T^4 \cong 4T_\infty^3 T - 3T_\infty^4, \tag{9.10}$$

Equation (9.9) and equation (9.10) give:

$$q_r = -\frac{16\sigma^* T_\infty^3}{3\kappa^*}\frac{\partial T}{\partial y}, \tag{9.11}$$

Here, σ^* is the constant of Stefan–Boltzmann, whereas k^* is the coefficient of mean absorption. Obviously, in the view of similarity transformation, continuity equation is satisfied. The basic equation (9.2) and equation (9.3) are transformed to the dimensional form of ODEs along with the boundary conditions as follows:

$$\Lambda\frac{\partial^3 f}{\partial \eta^3} + \left(f(\eta)\frac{\partial^2 f}{\partial \eta^2} - \left(\frac{\partial f}{\partial \eta}\right)^2 \right) - K_p\frac{\partial f}{\partial \eta} = 0, \tag{9.12}$$

$$\left(1+Nr\right)\frac{\partial^2 \theta}{\partial \eta^2} + \Pr\left(f(n)\frac{\partial \theta}{\partial \eta} - 2\frac{\partial f}{\partial \eta}\theta(\eta) \right) = 0, \tag{9.13}$$

Where:

$K_p = \frac{\mu_f}{\rho_f k}\frac{1}{\xi_1 \xi_2}$ is the porosity parameter, $\Pr = \frac{v_f\left(\rho C_p\right)_f}{k_f}$ is the Prandtl number, and

$Nr = \frac{16\sigma^* T_\infty^3}{3\left(\rho C_p\right)_f v_f k^*}$ is the thermal radiation parameter.

The surface velocity can be estimated by using:

$$u_w = \left[\frac{(\sigma_0 \gamma_a)^2}{\rho_f \mu_f}\right]^{\frac{1}{3}} xf'(0), \tag{9.14}$$

The imposed boundary conditions have been modified as:

$$\left(\frac{\partial^2 f}{\partial \eta^2}\right)_{\eta=0} = -2, \quad f(0) = f_w, \qquad \left(\frac{\partial f}{\partial \eta}\right)_{\eta=0} = 0,$$
$$\theta(0) = 0, \quad \theta(\infty) = 0, \tag{9.15}$$

Here, f_w is the constant mass transfer parameter, with $f_w > 0$ for suction and $f_w < 0$ for injection.

9.2.1 Analytical Solutions for Momentum

Analytical solution of $f(\eta)$:

$$f(\eta) = c_1 + c_2 \exp(-\beta\eta), \tag{9.16}$$

$$c_1 = f_w - c_2, \qquad c_2 = \frac{-2}{\beta^2}, \tag{9.17}$$

Putting equation (9.16) and equation (9.17) in equation (9.12), we get the value of β.

$$\Lambda\beta^3 - f_w\beta^2 - K_p\beta - 2 = 0, \tag{9.18}$$

Where the physical solution is achieved by essentially considering the positive resulting value.

9.2.2 Analytical Solution for Temperature by Employing Laplace Transformation

Pertaining to the benefits of Laplace transformation that were covered in the preceding section, in solving the temperature equation (9.13), we employ this strategy, and it is in good condition (9.15). Another approach to the solution can be conducted using a variable transformation technique. The new variable $t = -Exp(-\beta\eta)$, is introduced [36] and substituted in equation (9.13), yielding:

$$t\frac{\partial^2\theta}{\partial t^2} + (n - mt)\frac{\partial\theta}{\partial t} + 2mt\theta(t) = 0, \tag{9.19}$$

Where:

$$m = 2\frac{\tau}{\beta^2}, \qquad n = 1 - \frac{\tau}{\beta}(f_w + \frac{2}{\beta^2}), \tag{9.20}$$

$$\text{and } \tau = \frac{\Pr}{(1 + Nr)}, \tag{9.21}$$

Now, introduce Laplace transformation on both sides of equation (9.19). We have:

$$s(m - s)\frac{\partial\Theta}{\partial s} + [(n - 2)s + 3m]\Theta(s) = 0, \tag{9.22}$$

Here, $\Theta(s)$ is to be a Laplace transformation of $\theta(t)$. Integrating equation (9.22), we get:

$$\Theta(s) = \frac{c}{s^3(s - m)^{-(n+1)}}, \tag{9.23}$$

Where c is an integration constant as yet to determined. Now, apply inverse Laplace transform to equation (9.23), and we obtain:

$$\theta(t) = \frac{ct^2 t^{-(n+2)} e^{mt}}{2\Gamma(-(n+1))}, \tag{9.24}$$

Now, take the convolution theorem, defined by:

$$L^{-1}[\phi(s), \psi(s)] = \phi(t) * \psi(t) = \int_0^t \psi(w)\phi(t - w)dw, \tag{9.25}$$

Where * denotes the convolution property.

$$\theta(t) = \frac{c}{2\Gamma(-(n+1))}\int_0^t \frac{(t-w)^2 e^{mw}}{w^{n+2}}dw, \qquad n < -1. \tag{9.26}$$

By putting the modified boundary conditions into action, that is, $\theta(0) = 0$, it is spontaneously fulfilled. Moreover, another boundary condition, $\theta(-1) = 1$, gives c by:

$$c = -\frac{2\Gamma(-(n+1))}{\int_{-1}^{0} (1+w)^2 w^{-n-2} e^{mw} dw}, \tag{9.27}$$

Consequently, $\theta(t)$ presented in exact arrangement:

$$\theta(t) = \frac{\int_0^t (t-w)^2 w^{-n-2} e^{mw} dw}{\int_1^0 (1+w)^2 w^{-n-2} e^{mw} dw}, \tag{9.28}$$

By integrating equation (9.28) for the generalized incomplete gamma function, we obtain the precise result as follows:

$$\theta(t) = \frac{m^2 t^2 \Gamma(-n-1,0,-mt) + 2mt\Gamma(-n,0,-mt) + \Gamma(-n+1,0,-mt)}{m^2\Gamma(-n-1,0,m) - 2m\Gamma(-n,0,m) + \Gamma(-n,0,m)}, \tag{9.29}$$

In reference to this, it is stated as follows:

$$\theta(\eta) = \frac{m^2 e^{-2m\eta}\Gamma(-n-1,0,me^{-\beta\eta}) - 2me^{-\beta\eta}\Gamma(-n,0,me^{-\beta\eta}) + \Gamma(-n+1,0,me^{-\beta\eta})}{m^2\Gamma(-n-1,0,m) - 2m\Gamma(-n,0,m) + \Gamma(-n+1,0,m)}. \tag{9.30}$$

In this section, it was shown that using Laplace transform as a problem-solving tool results in simpler, unique qualities, but manipulating alternative ways results in more complex special functions.

9.3 RESULTS AND DISCUSSION

In the current chapter, we investigate the impact of radiation and Marangoni convective boundary conditions on fluid flow through a porous media with mass transpiration using the Brinkman model. A set of PDEs is used to represent the technique. These are condensed to a set of ODEs through similarity substitutions. Graphs can be used to study the exact solution in the presence of numerous physical characteristics, including the solid volume fraction parameter, mass transpiration, suction/injection, porosity Prandtl number, radiation parameter, etc.

9.3.1 Velocity Profile

Figure 9.2 depicts the upshots of porosity parameter K_p on the transverse velocity profile. It has been noted that the quantity of porosity parameter is improved with a decrease in the velocity profile; it signifies that the boundary layer's thickness decreases in both suction/injection instance with fixed other parameters. With the pore-filled structure's holes, the velocity increases. Obviously, the physical perspective concurs with this. Furthermore, it should be mentioned that $K \leq 0.1$.

The characteristics of porosity on the axial velocity are displayed in Figure 9.3. It has been noticed that the porosity parameter value increases as the velocity profile drops, which implies that the boundary layer thickness falls in both suction and injection instances with a fixed other parameter. The

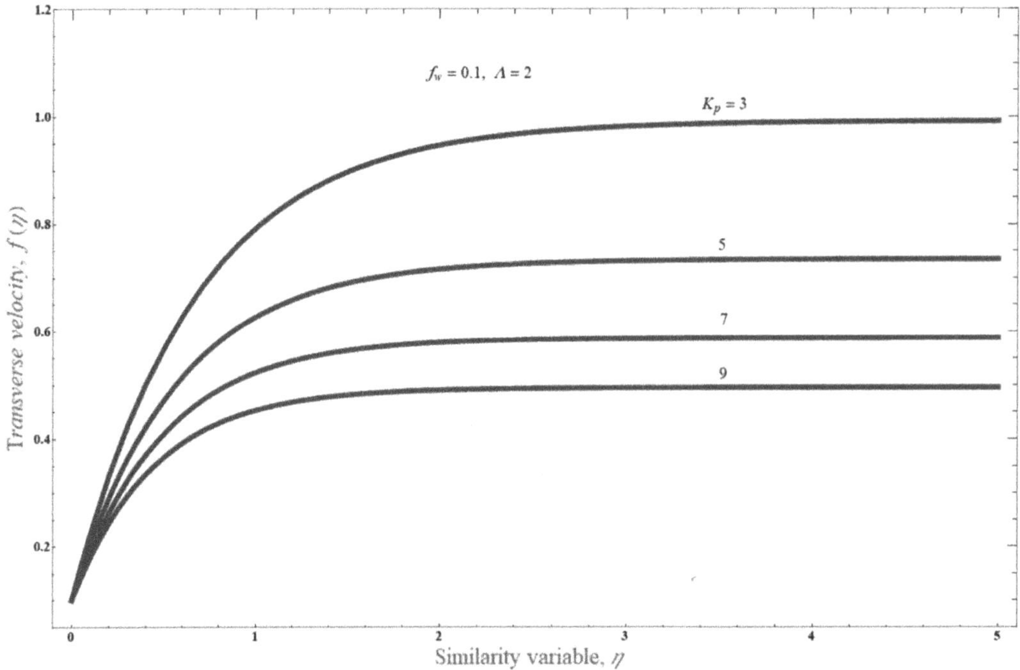

FIGURE 9.2 Transverse velocity against similarity variable with an effect of porosity parameter.

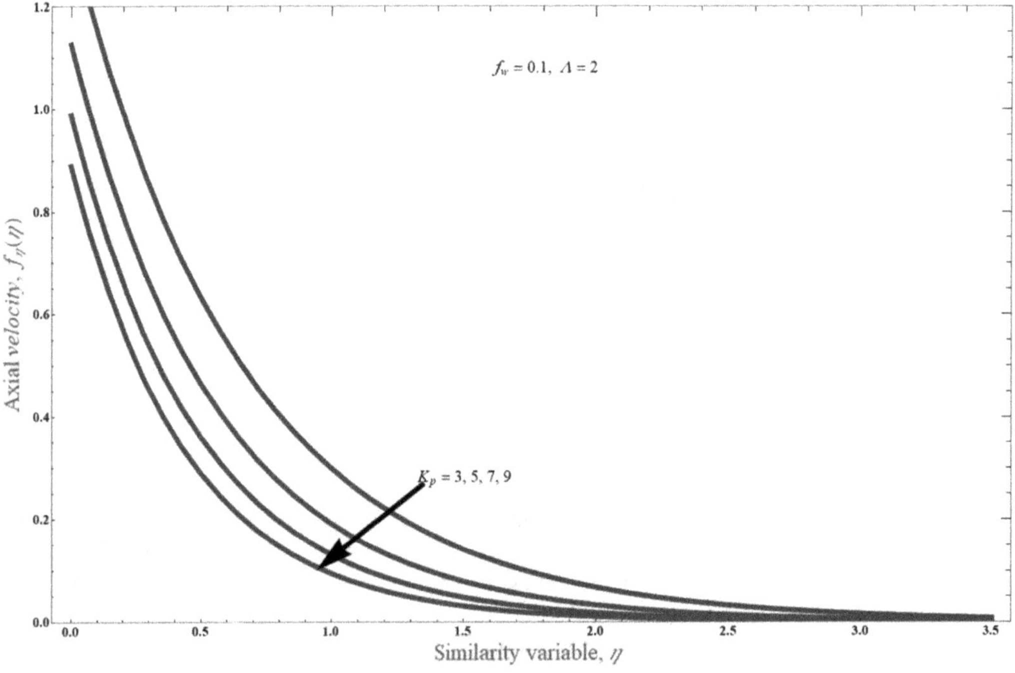

FIGURE 9.3 Axial velocity against similarity variable with an effect of porosity parameter.

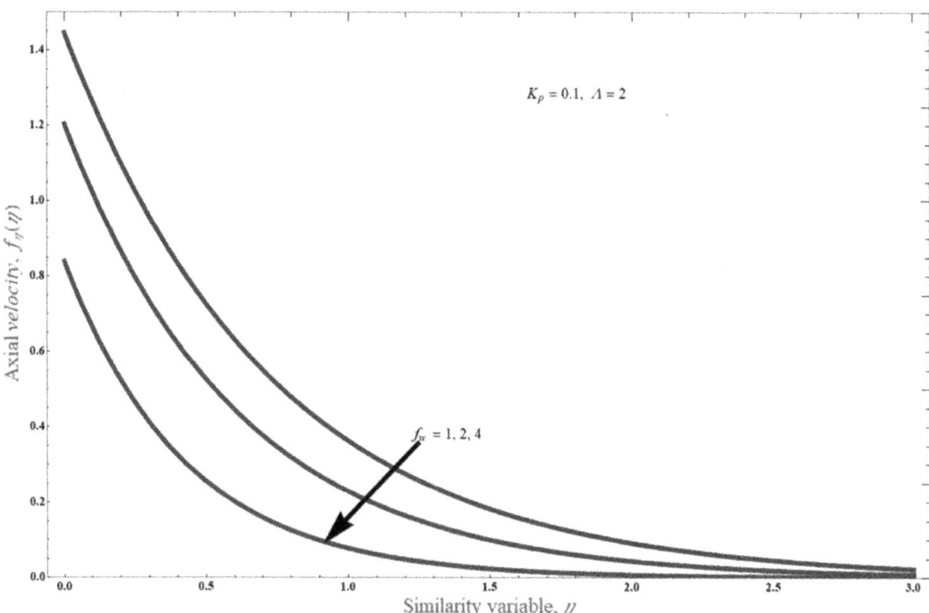

FIGURE 9.4 Axial velocity against similarity variable with an effect of mass suction/injection parameter in case of suction.

velocity increases as a result of the porous structure's holes. Of course, the physical view concurs with this. Here, it is important to highlight that. Figure 9.4 provides a different set of characteristics determining velocity values, accordingly. In Figure 9.4, the hydraulic boundary layer thickness is shown to decrease when fluid suction is raised. The opposite is true when fluid is injected; this increases fluid velocity under specific mass suction/injection circumstances. As a consequence of the notable and slightly faster velocity, as seen in Figure 9.5, the boundary layer thickness in the injection situation significantly increases, along with the quantity of the mass suction/injection parameter value.

9.3.2 Temperature Profile

When injection situations are taken into account, it is revealed that the liquid used here has remarkably similar thermal performance to that shown in Figure 9.6, in which the temperature distribution $\theta(\eta)$ is shown. The permeability parameter's value increases while fluid temperature goes up, while other variables remain constant. Figure 9.7 shows an impact of Pr on the temperature distribution. Increasing values of Pr decline the thermal boundary layer. Hence, the thermal boundary layer width is diminished by optimizing the quantities of Pr. A larger Pr has comparatively low thermal conductivity, which results in low heat absorption, and this is a primary source of the temperature drop. As a result, the thickness of the fluid's thermal boundary layer drops with increasing Pr. In order to regulate the rate of heat exchange in industrial as well as technical procedures, the optimal value of Pr is therefore very crucial. The temperature profiles and the impact of the thermal radiation parameter N_r are shown in Figure 9.8. According to the definition of the radiation parameter, rising N_r values will tend to boost the radiation intensity. Consequently, the temperature will rise everywhere as a result of the expansion of the thermal boundary layer. Because of the rise in temperature away from the wall, Figure 9.9 demonstrates the impact of porosity on temperature variation. It should be noted that movement develops in fluid flow and is identical at suction case as porosity values increase, which indicates a rise in permeability particle. The current fluid flow has nearly comparable thermal performance in suction conditions, and other parameters are fixed. Figure 9.10 describes the

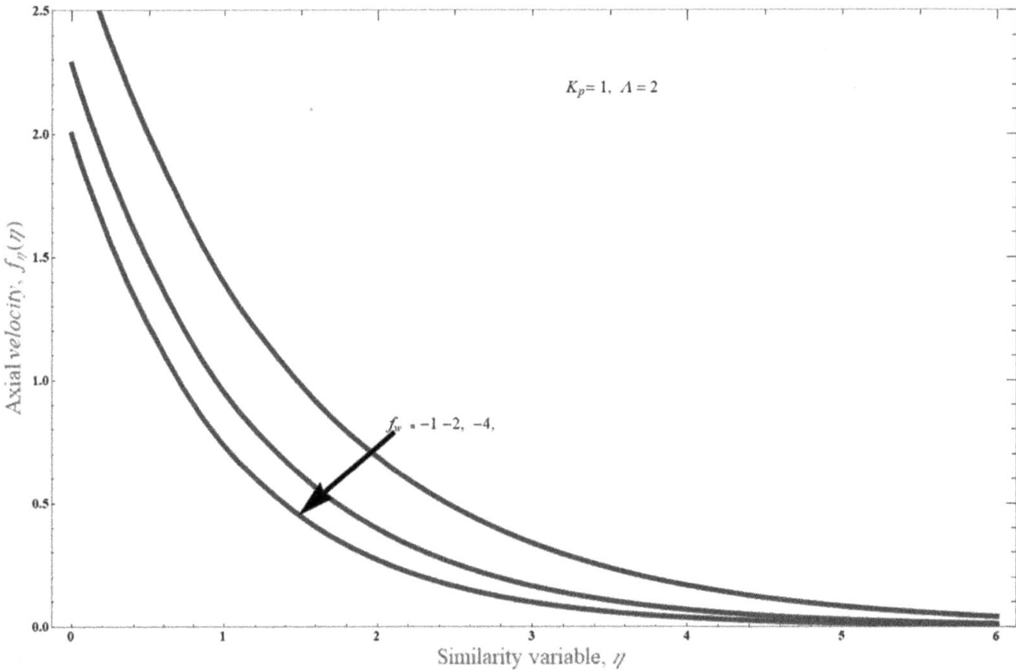

FIGURE 9.5 Axial velocity against similarity variable with an effect of mass suction/injection parameter in case of injection.

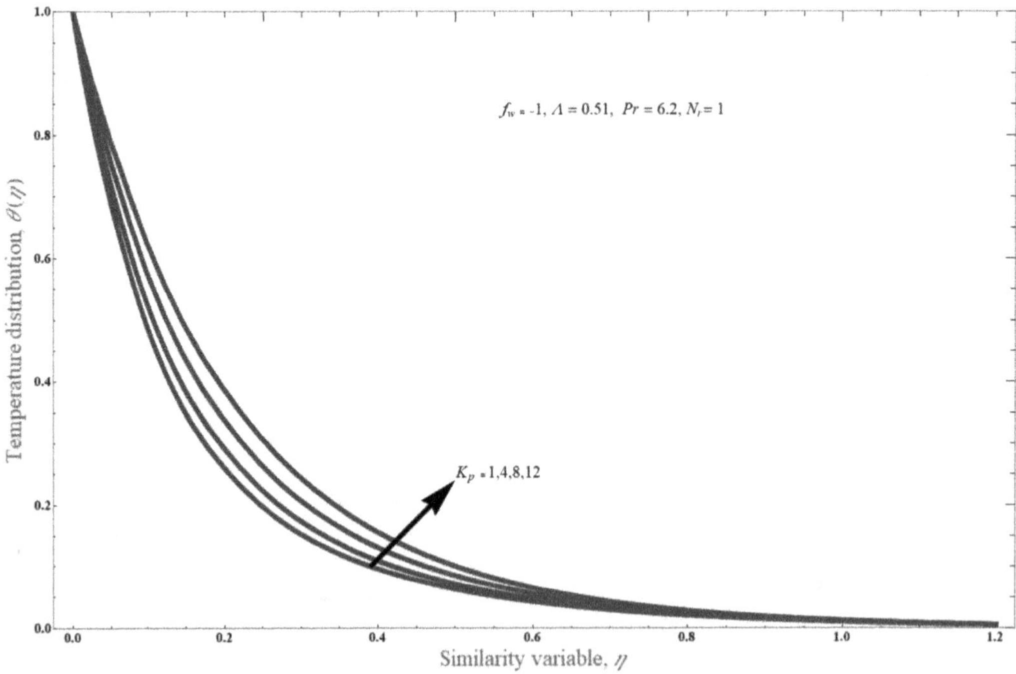

FIGURE 9.6 Temperature profile against similarity variable with an effect of porosity parameter in case of injection.

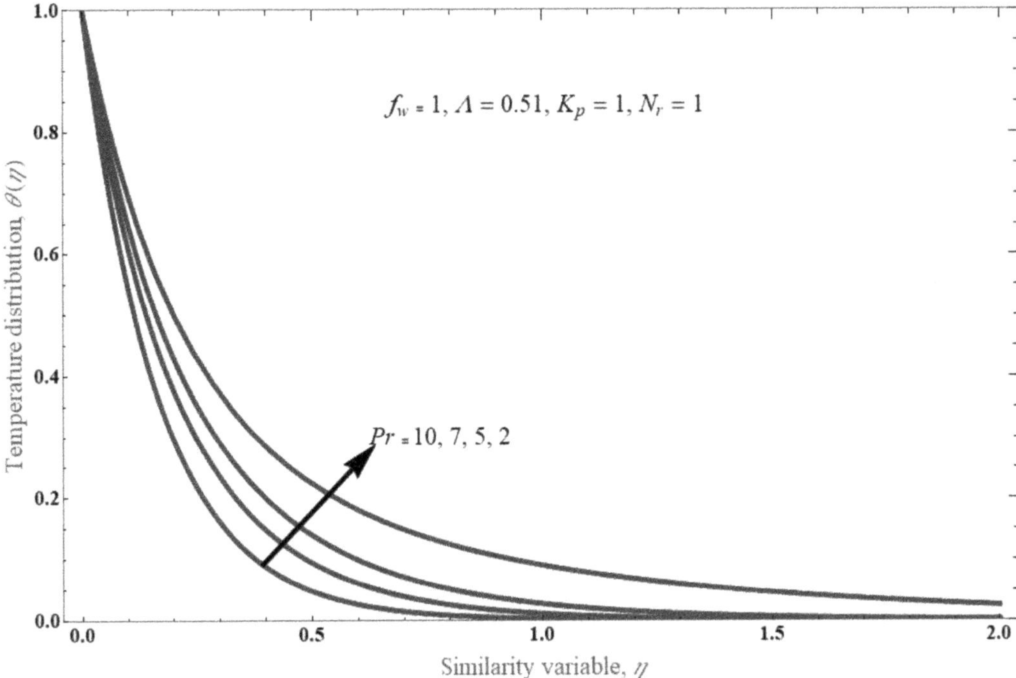

FIGURE 9.7 Temperature profile against similarity variable with an effect of Prandtl number.

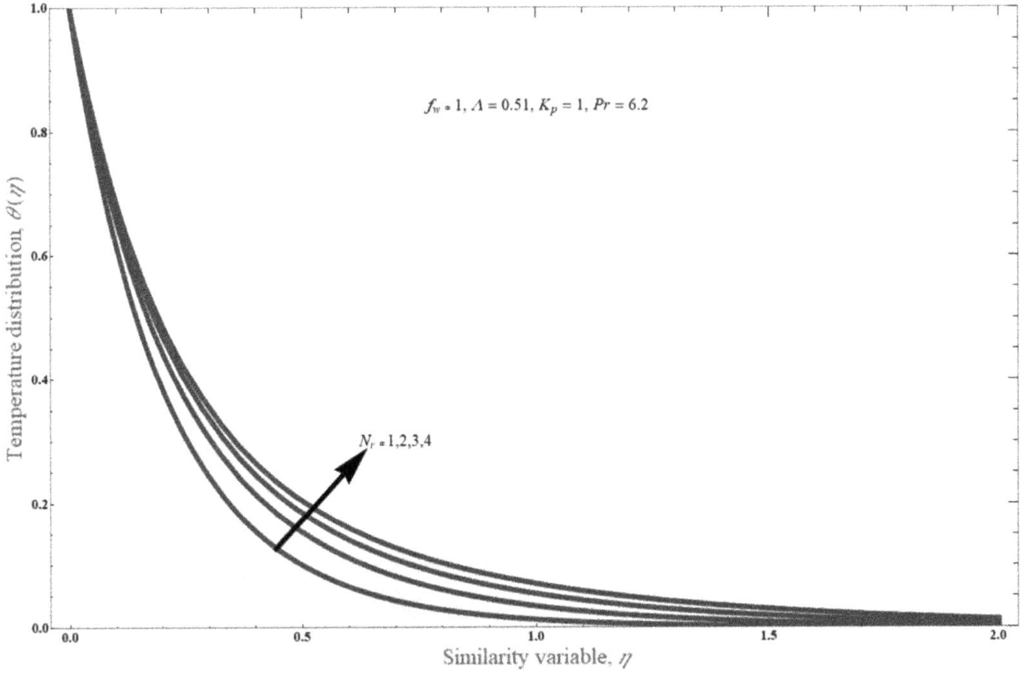

FIGURE 9.8 Temperature profile against similarity variable with an effect of radiation parameter.

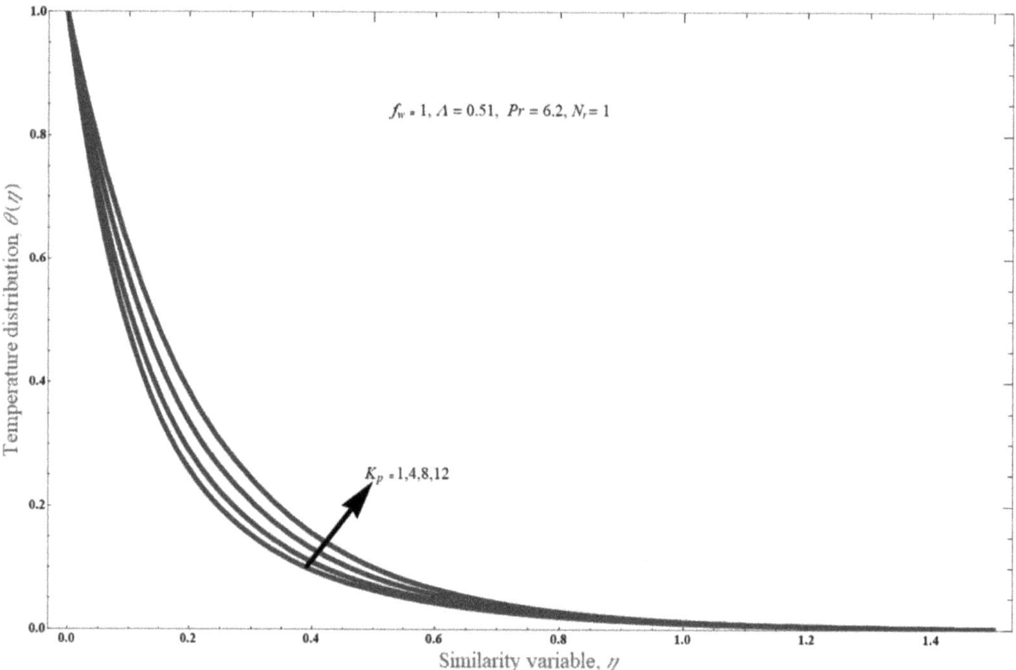

FIGURE 9.9 Temperature profile against similarity variable with an effect of porosity parameter in case of suction.

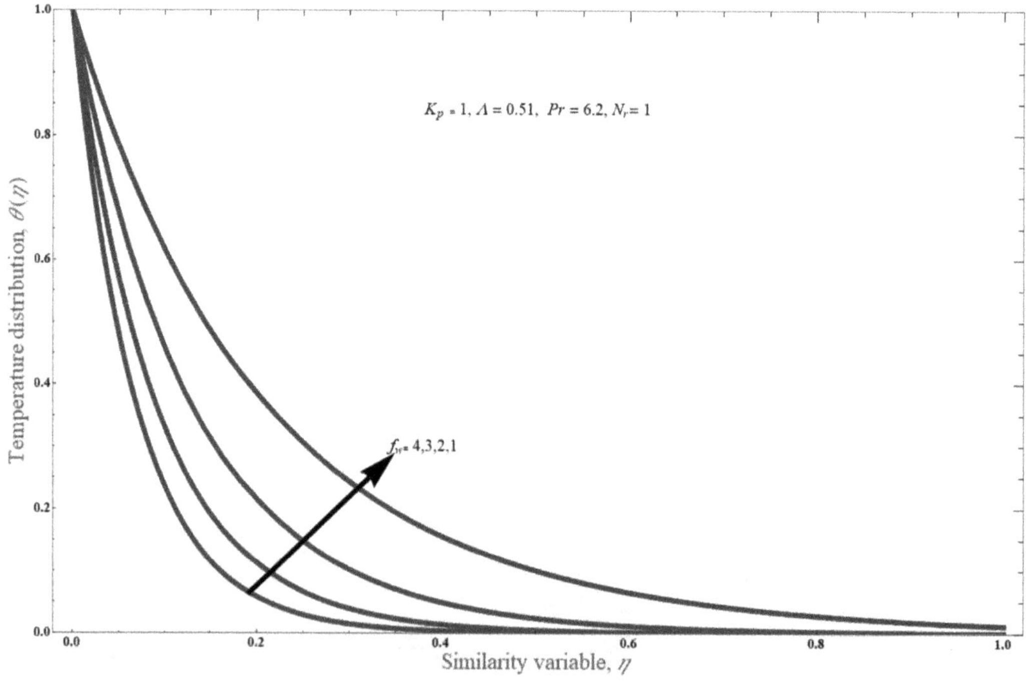

FIGURE 9.10 Temperature profile against similarity variable with an effect of mass suction/injection parameter.

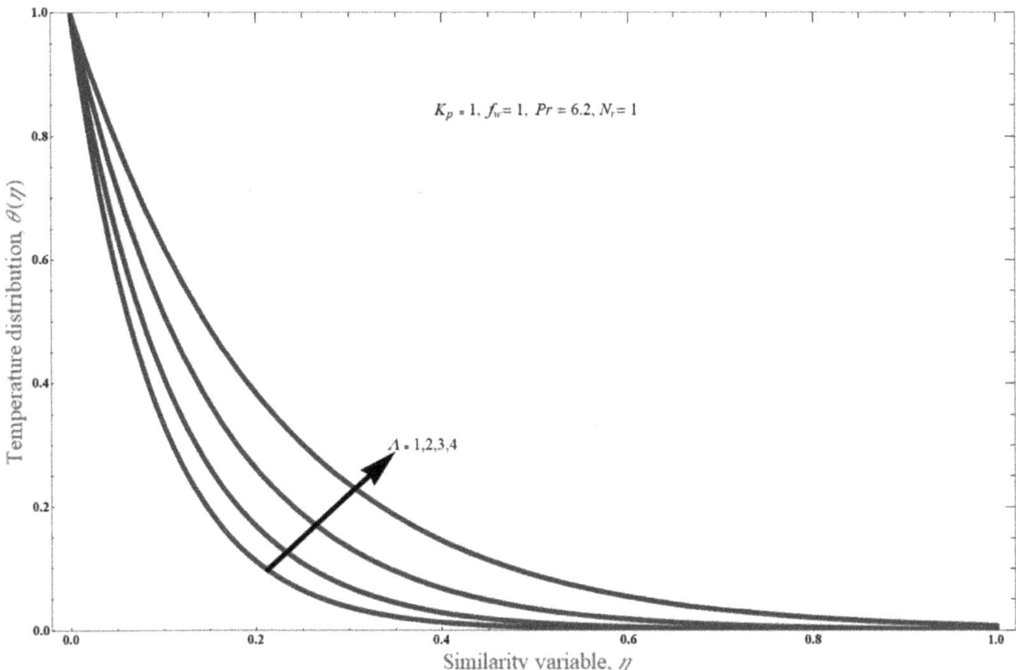

FIGURE 9.11 Temperature profile against similarity variable with an effect of Brinkman parameter.

corresponding fluctuations in temperature. It has been shown that applying fluid suction decreases with respect to fluid temperature, the heat transfer layer, flow rates, and therefore, the depth of the hydraulic boundary. An influence of the Brinkman number on temperature distribution is depicted in Figure 9.11. As Brinkman rises, the temperature distribution expands and becomes more extreme with higher values of Brinkman number.

9.4 CONCLUSIONS

The aim of the present chapter was to examine the Brinkmann equation for the impact of radiation and Marangoni convective boundary conditions on fluid flow in permeable media along mass transpiration by using similarity transformations to govern partial differential equations that have been turned into ordinary differential equations. The following are the important observations:

- Porosity enhances with decline in the boundary layer thickness, respectively.
- In temperature profiles, porosity increases as radiation is also increased.
- The mass suction/injection parameter increases with decreasing boundary layer thickness for suction case, whereas in injection case, the reverse effect will appear.
- The magnitude of the porosity parameter enhances as the thermal boundary layer also grows.
- When the temperature distribution is considerable, the Prandtl number reduces boundary layer thickness.
- Additionally, as radiation parameter R goes up, the fluid's temperature increases.
- The value of Brinkman develops with temperature distribution.

Nomenclature

Symbols	Description	SI Units
a	constant	—
ξ_1 & ξ_2	constants	—
u	velocity component along x-axis	(ms^{-1})
v	velocity component along y-axis	(ms^{-1})
C_p	heat capacitance	$(JK^{-1}Kg^{-1})$
K_p	permeability	$(N\ A^{-1})$
f_w	suction/injection parameter	—
T	temperature of the fluid	(K)
q_r	radiative heat flux	(Wm^{-2})
Pr	Prandtl number	—
Nr	radiation parameter	—
σ	surface tension	—
σ^*	Stefan–Boltzmann constant	—
u_w	surface velocity	(ms^{-1})
Greek symbols		
α	thermal diffusivity	(m^2s^{-1})
η	similarity variable	—
ψ	stream function	—
θ	temperature similarity variable	—
μ_f	dynamic viscosity	$(Ns\ m^{-2})$
v_f	kinematic viscosity	(m^2s^{-1})
ρ_f	effective density	$(Kg\ m^{-3})$
σ_0	equilibrium surface tension	—
Λ	Brinkman parameter	—
Abbreviation		
PDEs	partial differential equations	—
ODEs	ordinary differential equations	—

REFERENCES

[1] Shakya, A., Yahya, S.M., Ansari, M.A., Khan, S.A. Role of 1-butanol on critical heat flux enhancement of Tio_2, Al_2o_3, & CuO nanofluids. *Journal of Nanofluids*. 8(7) (2009), 1560–1565.

[2] Crane, L.J. Flow past a stretching sheet. *Zeitschift fur Ang-wandte mathematic and physic*. 21 (1970), 645–647.

[3] Wang, C.Y. Fluid flow due to a stretching cylinder. *Physics of Fluids*. 31 (1988), 466–468.

[4] Siddheshwar, P.G., Mahabaleshwar, U.S. Effect of radiation and heat source in MHD flow of a viscoelastic liquid and heat transfer over a stretching sheet. *International Journal of Non-linear Mechanics*. 40(6) (2005), 807–820.

[5] Choi, S.U.S., Eastman, J.A. Enhancing thermal conductivity of the fluids with nanoparticles. *American Society of Mechanical Engineers (ASME)*. 231 (1995), 99–105.

[6] Napolitano, L.G., Golia, C. Coupled Marangoni boundary layers. *Acta Astronautica*. 8 (1981), 417–434.

[7] Napolitano, L.G. Marangoni boundary layers. *Proceedings of 3rd European Symposium in Material Science in Space*. 142 (1979), 349–358.

[8] Napolitano, L.G. Surface and buoyancy driven free convection. *Acta Astronautica*. 9 (1982), 199–215.

[9] Mahabaleshwar, U.S., Nagaraju, K.R., Kumar, P.N.V., Azese, M.N. Effect of radiation and thermosolutal Marangoni convection in a porous medium with chemical reaction and heat source/sink. *Physics of Fluids*. 32(11) (2020), 113602.

[10] Anush, T., Haung, H.N., Mahabaleshwar, U.S. Two dimensional unsteady stagnation point flow of Casson hybrid nanofluid over a permeable flat surface and heat transfer. *Journal of Taiwan Institute of Chemical Engineers*. 127 (2021), 79–91.

[11] Aslani, K.E., Mahabaleshwar, U.S., Singh, J., Sarris, I.E. Combined effect of radiation and inclined MHD flow of a micro polar fluid over a porous stretching/shrinking sheet with mass transpiration. *International Journal of Applied and Computational Mathematics*. 7(3) (2021), 1–21.

[12] Mahabaleshwar, U.S., Kumar, P.N.V., Sheremet, M. MHD flow of nanofluid driven by a stretching/shrinking sheet with suction. *Springers Plus*. 5(1) (2016), 1–9.

[13] Mahabaleshwar, U.S., Sneha, K.N., Haung, H.N. An effect of MHD and radiation on CNTs-water based nanofluids due to a stretching sheet in a Newtonian fluid. *Case Studies in Thermal Engineering*. 28 (2021), 101462.

[14] Animasuan, I.L., Yook, S.J., Muhammed, T., Mathew, A. Dynamics of ternary hybrid nanofluid subjected to Magnetic flux density and heat source or sink on a convectively heated surface. *Journal of Surface and Interfaces of Materials*. 28 (2022), 101654.

[15] Magyari, E., Chamka, A.J. Exact analytical results for the thermosolutal MHD Marangoni boundary layers. *International Journal Thermal Sciences*. 47 (2008), 848–857.

[16] Nanjundappa, C.E., Shivakumar, I.S., Arunkumar, R. Benard-Marangoni ferroconvection with Magnetic field dependent viscosity. *Journal of Magnetism and Magnetic Materials*. 322 (2010), 2256–2263.

[17] Manjunatha, S., Kuttan, B.A., Jayanthi, S., Chamkha, A.J., Gireesh, B.J. Heat transfer enhancement in the boundary layer flow of hybrid nanofluids due to variable viscosity and natural convection. *Heliyon*. 5(4) (2019), e01464.

[18] Mahabaleshwar, U.S., Vishalakshi, A.B., Anderson, H.I. Hybrid nanofluid flow past a stretching/shrinking sheet with Thermal radiation and mass transpiration. *Chinese Journal of Physics*. 75 (2022), 152–168.

[19] Udawattha, D.S., Narayan, M. Development of a model for predicting the effective thermal conductivity of nanofluids: A reliable approach for nano fluids containing spherical nanoparticles. *Journal of Nanofluids*. 7(1) (2018), 129–140.

[20] Brinkmann, H.C. On the permeability of media consisting of closely packed particles. *Applied Scientific Research*. 1(1949), 81–86.

[21] Khan, Z.A., Haq, S.U., Khan, T.S., Tlili, I. Unsteady MHD flow of a Brinkmann type fluid between two side walls perpendicular to an infinite plate. *Results in Physics*. 9 (2018), 1602–1608.

[22] Khan, A., Khan, D., Khan, I., Ali, F., Ul Karim, F. MHD flow of Brinkmann type H_2O—Cu, Ag, TiO_2, and Al_2O_3 nanofluids with chemical reaction and heat generation effect in a porous medium. *Journal of Magnetics*. 24(2019), 262–270.

[23] Ali, F., Jan, S.A.A., Khan, I., Gohar, M., Sheikh, N.A. Solution with special functions for time fractional free convection flow of Brinkmann-type fluid. *The European Physical Journal Plus*. 131(2016), 1–13.

[24] Sneha, K.N., Mahabaleshwar, U.S., Bennacer, R., Ganaoui, M.E. Darcy Brinkman equations for hybrid dusty nanofluid flow with heat transfer and mass transpiration. *Computation*. 9(2021), 118.

[25] Shankar, B.M., Kumar, J., Shivakumara, I.S., Ng, C.O. Stability of fluid flow in a Brinkmann porous medium-A numerical study. *Journal of Hydrodynamics*. 26(2014), 60076.

[26] Takhar, H.S. Free convection flow of two immiscible viscous liquids through parallel permeable beds: Use of brinkman equation. *International Journal Fluid Mechanics Research*. 32 (2005), 39–56.

[27] Anusha, T., Mahabaleshwar, U.S., Sheikhnejad, Y. An MHD of nanofluid flow over a porous stretching/shrinking plate with mass transpiration and Brinkman ratio. *Transport in Porous Media*. 142 (2021), 333–352.

[28] Turkyilmazoglu, M. Analytical heat and mass transfer of the mixed hydrodynamic/thermal slip MHD viscous flow over a stretching sheet. *International Journal of Mechanical Sciences*. 53(2011), 886–896.

[29] Mahabaleshwar, U.S., Vishalakshi, A.B., Azese, M.N. The role of Brinkmann ratio on non-Newtonian fluid flow due to a porous shrinking/stretching sheet with heat transfer. *European Journal Mechanics-B/Fluids*. 92(2022), 153–165.

[30] Maranna, T., Sneha, K.N., Mahabaleshwar, U.S., Sarris, I.E., Karakasidis, T.E. An effect of radiation and MHD Newtonian fluid over a stretching/shrinking sheet with CNTs and mass transpiration. *Applied Science*. 12 (2022), 5466.

[31] Aly, E.H., Ebaid, A. Exact analysis for the effect of heat transfer on MHD and radiation Marangoni boundary layer nanofluid flow past a surface embedded in a porous medium. *Journal of Molecular Liquid*. 215 (2016), 625–659.

[32] Magyari, E., Chamkha, A.J. Analytical solution for thermosolutal Marangoni convection in the presence of heat and mass generation or consumption. *Heat and Mass Transfer*. 43 (2007), 965–974.

[33] Mahabaleshwar, U.S., Sarris, I.E., Lorenzini, G. Effect of radiation and Navier slip boundary of Walter's liquid B flow over a stretching sheet in a porous media. *International Journal of Heat and Mass Transfer*. 127 (2018), 1327–1337.

[34] Vishalakshi, A.B., Maranna, T., Mahabaleshwar, U.S., loreze, D. An effect of MHD on Non-Newtonian fluid flow over a porous stretching/shrinking sheet with heat transfer. *Applied Science*. 12(10) (2022), 4937.

[35] Kumar, M.A., Reddy, Y.D., Rao, V.S., Goud, B.S. Thermal radiation impact on MHD heat transfer natural convective nanofluid flow over an impulsively started verticle plate. *Case Studies in Thermal Engineering*. 24 (2021), 100826.

[36] Ebaid, A., Al Sharif, M. Application of Laplace transformation for the exact effect of a magnetic field on heat transfer of carbon-nanotubes suspended nanofluid. *Zeitschrift fur Naturforschung A*. 70 (2015), 471–475.

10 Forced Convection in Cylindrical and Rectangular Pin-Fins Channels Configuration

M. Z. Saghir

10.1 INTRODUCTION

Heat enhancement in channels has received considerable attention among researchers due to its high thermal efficiency. Different channel shapes have been investigated numerically and experimentally by the author [1–7]. It was found that a large channel can improve heat removal, but at the expense of a pressure drop. Different fluids were used, from water to a large set of nanofluids with water as a base fluid or ethylene glycol. The author demonstrated that heat enhancement of up to 5% is possible when using nanofluids compared to water, but at the expense of the pressure drop. Brownian motion and thermophoretic effects were also considered, and nanoparticles sedimentation is found to be the reason for some weak performance of nanofluids. All investigations focused on the laminar regime. Sertkaya et al. [8] investigated the pin-fins configuration experimentally by concentrating on the pin-fins orientation. Air is used to flow around the pin-fins. The up-facing pins enhance heat transfer more than the down-facing pins do, and the enhancement decreases with increasing orientation angle from the vertical axis.

Peng [9] is believed to be among the first researchers to study heat transfer and friction loss in pin-fins configuration. It appears that this configuration is of great interest to the turbine cooling designer because of the high heat transfer characteristics and high surface area density. The author did experimental tests to investigate the pin heights, spacing, and channel-height-to-length ratio on heat transfer enhancement. Results revealed that the pin-fins configuration provides a means to reduce friction flow and maintain an excellent heat enhancement leading to large thermal efficiency. Later, Goldstein and Chen [10] focused on pin shape by using a short pin height. Experimentally, their investigation was conducted for a range of Reynolds numbers from 3,000 to 18,000 while remaining in the turbulent regime. Different pin shapes were studied, consisting of a circular fin, then a two-step diameter circular fin. The latter is meant to be the pin's location, but all of them have a circular shape, similar to the previous author's, consisting of lower pressure drop and a high heat enhancement.

Many researchers, as stated earlier, have also investigated heat enhancement in the channel, whether it was a porous channel or a free channel. Wang et al. [11] investigated the presence of drop-shaped pins in a rectangular channel. The experiment was conducted for a range of Reynolds numbers varying between 4,800 and 8,200, thus in the turbulent regime. Three different-shaped pins were investigated, which are circular, elliptical, and drop-shaped. It appears that drop-shaped showed a promising result.

Further study by Chin et al. [12] used a perforated pin instead of a solid pin. It appears that experimentally, the authors demonstrated a lower pressure drop than in solid pin across the flow. Additional work by Ndao et al. [13], Abdoli et al. [14], Cohen and Bourell [15], Yang et al. [16], Amudhan et al. [17], Ambreen et al. [18], and Hossain et al. [19] investigated the similar performance of pin numerically and experimentally. In particular, Yang et al. [16] use air as the circulating fluid

DOI: 10.1201/9781003299608-10

between pins having a circular shape, a rectangular block shape, and mini channels. The experiment was conducted for different ranges of Reynolds numbers. It has been demonstrated that a staggering fin can produce a higher heat enhancement. From these publications so far, it appears that pin-fins can provide an excellent heat enhancement, a lower pressure drop, and their staggered location can improve the heat characteristics further. Additional researchers concentrated on using nanofluids as a means of heat extraction [20–22].

One may conclude that the pin-fin can provide heat enhancement and lower pressure drop. However, none of the researchers commented or detected any temperature fluctuation in the presence of the pin-fins. The author believes that their measurement was not taken in the correct location.

In our present study, we numerically investigated the usefulness of using pin-fins configuration instead of mini channel. We consider two different types of channels made of pins having rectangular and cylindrical shapes. The pins height varies from full height to minimum height via the aspect ratio. The novelty of this study is to investigate the importance of pins height in heat removal. To the best of the author's knowledge, no such investigation has been done yet. Section 2 presents the problem description and the boundary conditions. The formulation in the non-dimensional form is presented in Section 3, followed by mesh sensitivity. Section 4 offers a comparison with experimental data. Section 5 presents the results and discussion, followed by the conclusion in Section 6.

10.2 PROBLEM DESCRIPTION AND BOUNDARY CONDITIONS

Figure 10.1 presents the model under investigation. The rectangular and cylindrical pins are located inside an aluminum box heated below. The fluid at the inlet has a temperature T_{in} and flows at a velocity u_{in}. A more detailed information about the system could be found in reference [23]. Table 10.1 shows the physical properties of the water used in the simulation.

Figure 10.2 presents the different inserts under investigation. Figure 10.2(a) represents a different view of Figure 10.1. The two other inserts are shown in Figure 10.2(b) and Figure 10.2(c).

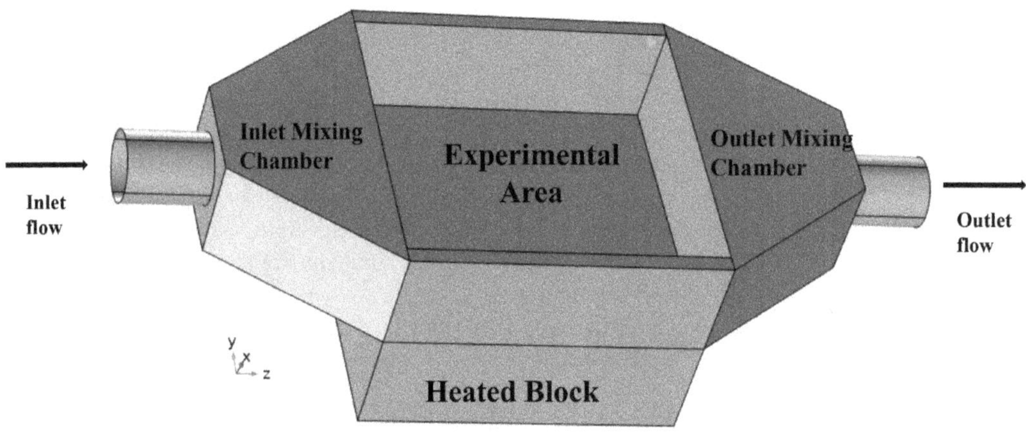

FIGURE 10.1 Experimental setup,

TABLE 10.1
Water Physical Properties

Fluids	μ(kg/m.s)	ρ(kg/m³)	cp(J/kg.K)	k(W/m.K)
Water	0.001002	998.2	4182	0.613

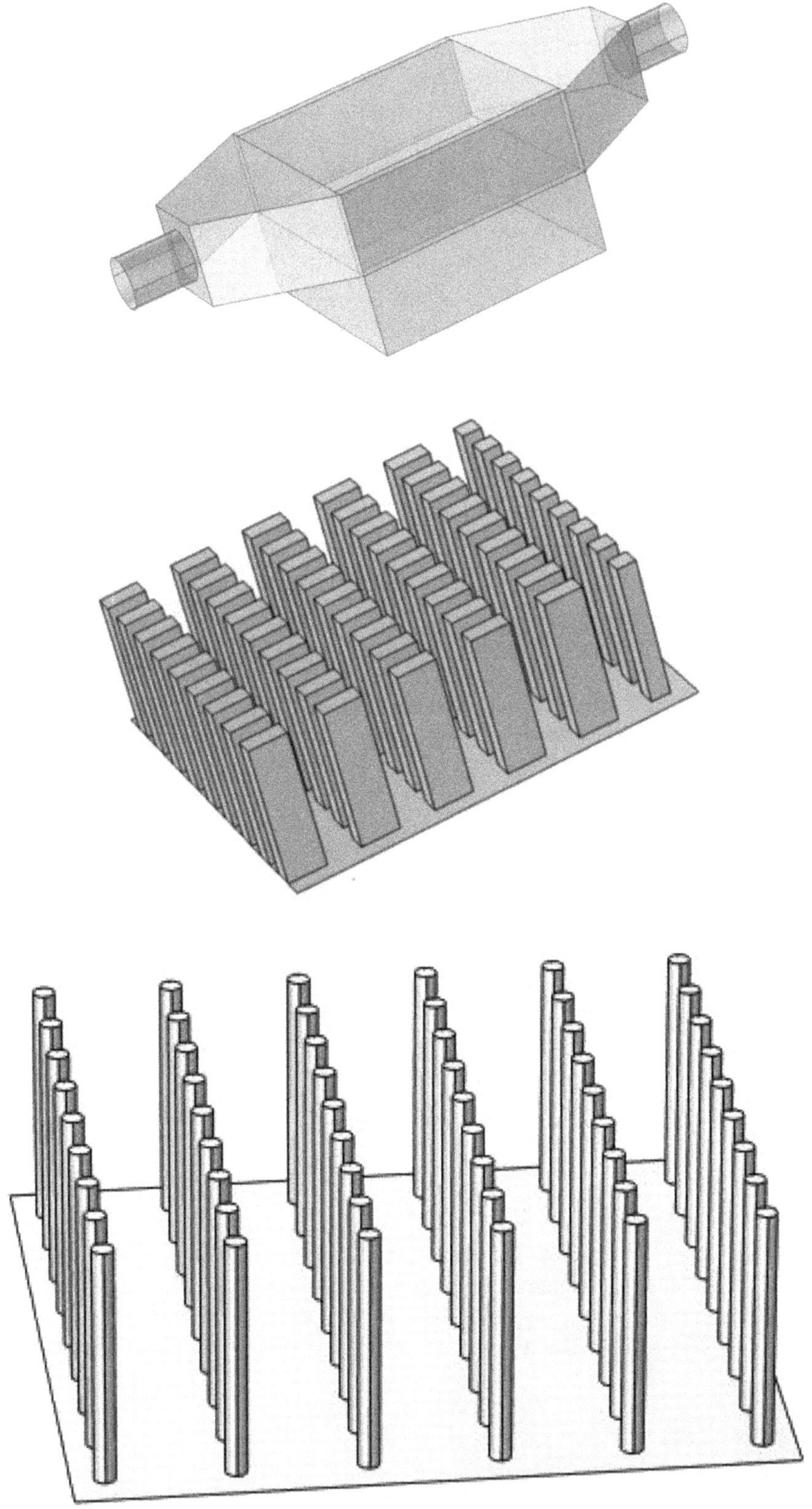

FIGURE 10.2 Geometrical model with different inserts.

Each insert base has a square shape with a length of 37.8 mm. Figure 10.2(b) represents a fin channel having a rectangular block shape. This insert is composed of nine rows, and each row contains five rectangular blocks of equal base size and one square block. For the individual rectangular block, the base dimension is 1.785 mm × 3.57 m. The square block has a base dimension of 1.785 mm. The distance between the two blocks is 1.785 mm in the x-direction and 3.75 mm in the z-direction. All of them have identical heights. Thus, four different fin heights for this insert will be investigated, which are 3.415 mm in height identified as an aspect ratio AR = 3, a height of 6.829 mm corresponding to an aspect ratio AR = 6, a height of 10.24 mm corresponding to an aspect ratio AR = 9, and finally, a height of 12.7 mm corresponding to an aspect ratio AR = 12, respectively. When the aspect ratio is zero, no insert is present, and the flow travels in an empty rectangular chamber before exiting the experimental setup, as shown in Figure 10.1.

For the second pin-fins having a cylindrical shape, the diameter is 1.382 mm, and the distance between two pin-fins is 2.52 mm in the x-direction and 5.648 mm in the z-direction. Again, here, four different heights of the pins identical to the block configuration were used. For all cases, the pressure drop is calculated between the outlet and the inlet.

It is worth noting that the number of channels for all inserts configurations is the same, thus the accurate comparison from a different model.

10.3 FINITE ELEMENT ANALYSIS

The full Navier–Stokes equations and the energy equation are solved numerically using the finite element code by COMSOL Multiphysics. Besides, the heat conduction equation was solved for the solid part of the model. The mathematical formulations in the non-dimensional form are as follows.

10.3.1 Fluid Flow Formulation

To make the governing equations of the physical model dimensionless, we define the following set of transformations.

$$X=\frac{x}{D},\ Y=\frac{y}{D},\ Z=\frac{z}{D},\ U=\frac{u}{u_{in}},\ V=\frac{v}{u_{in}},\ W=\frac{w}{u_{in}},\ P=\frac{pD}{\mu\ u_{in}},\ \tau=\frac{tu_{in}}{D},\ \theta=\frac{(T-T_{in})k_w}{q''D} \tag{10.1}$$

The dimensionless parameters, such as Reynolds number (Re) and the Prandtl number (Pr), are defined by:

$$Re=\frac{\rho u_{in}D}{\mu},\ Pr=\frac{\nu}{\alpha} \tag{10.2}$$

Here ρ, μ, D, u_{in}, ν, α, k_W represent the density, the dynamic viscosity, the characteristic diameter set equal to 18.97 mm, the inlet velocity, the kinematic viscosity, the diffusivity, and the thermal conductivity of the fluid, respectively. The heat flux *q"* is the applied heat in the heated aluminum block, as shown in Figure 10.1. The full Navier–Stokes equations in the three-dimensional form are as follows:

X-direction

$$Re\left[U\frac{\partial U}{\partial X}+V\frac{\partial U}{\partial Y}+W\frac{\partial U}{\partial Z}\right]=-\frac{\partial P}{\partial X}+\left[\frac{\partial^2 U}{\partial X^2}+\frac{\partial^2 U}{\partial Y^2}+\frac{\partial^2 U}{\partial Z^2}\right] \tag{10.3}$$

Y-direction

$$\mathrm{Re}\left[U\frac{\partial V}{\partial X}+V\frac{\partial V}{\partial Y}+W\frac{\partial V}{\partial Z}\right]=-\frac{\partial P}{\partial Y}+\left[\frac{\partial^2 V}{\partial X^2}+\frac{\partial^2 V}{\partial Y^2}+\frac{\partial^2 V}{\partial Z^2}\right] \tag{10.4}$$

Z-direction

$$\mathrm{Re}\left[U\frac{\partial W}{\partial X}+V\frac{\partial W}{\partial Y}+W\frac{\partial W}{\partial Z}\right]=-\frac{\partial P}{\partial Z}+\left[\frac{\partial^2 W}{\partial X^2}+\frac{\partial^2 W}{\partial Y^2}+\frac{\partial^2 W}{\partial Z^2}\right] \tag{10.5}$$

Here *U, V, W* are the velocities along the X-, Y-, and Z-directions. *P* is the non-dimensional pressure.

10.3.2 Heat Transfer Formulation

The energy equation for the fluid portion is as follows:

$$\mathrm{RePr}\left[U\frac{\partial \theta}{\partial X}+V\frac{\partial \theta}{\partial Y}+W\frac{\partial \theta}{\partial Z}\right]=\left[\frac{\partial^2 \theta}{\partial X^2}+\frac{\partial^2 \theta}{\partial Y^2}+\frac{\partial^2 \theta}{\partial Z^2}\right] \tag{10.6}$$

Here, θ is the non-dimensional temperature defined in equation (10.1). Finally, the temperature in the solid part of the model is studied by solving the heat conduction formulation. The local Nusselt number is the ratio of the convective heat coefficient multiplied by the characteristic length over the water conductivity (i.e., $\frac{hD}{k_w}$). Based on the non-dimensional adopted earlier, it becomes the inverse of the temperature. Thus:

$$\mathrm{Nu}=\frac{hD}{k_w}=\frac{1}{\theta} \tag{10.7}$$

The friction factor is known as the ratio of the pressure drop to the kinetic energy of the fluid. For a channel, the pipe diameter is replaced by the hydraulic diameter *D* equal to 18.97 mm. The friction factor used in our analysis is:

$$f=0.2529*\frac{\Delta P}{\mathrm{Re}} \tag{10.8}$$

To detect the most efficient fluid for heat enhancement, the performance evaluation criterion (PEC) is known to be the ratio of the average Nusselt number over the friction factor. The rationale behind this definition is that although some fluid exhibits good heat removal, sometimes it is at the expense of a significant pressure drop. This non-dimensional term is a good indicator of fluid performance. In the current chapter, the PEC is defined as follows:

$$\mathrm{PEC}=\frac{\mathrm{Nu}_{\mathrm{average}}}{f^{1/3}} \tag{10.9}$$

The amount of heat removed is represented by equation (10.10). Thus:

$$Q=\mathrm{Re}*\mathrm{Pr}*\theta_{\mathrm{out}} \tag{10.10}$$

Here, θ_{out} is the temperature at the outlet minus the temperature at the inlet. Since the inlet temperature in non-dimensional form is zero, then θ_{out} is the temperature at the outlet. All the preceding parameters will be investigated in our current study.

10.3.3 Boundary Conditions

The model is insulated all around, so it is assumed no heat leak could occur. As shown in Figure 10.1, at the inlet (i.e., where it is indicated as *inlet flow*), a non-dimensional temperature of zero is applied, and a non-dimensional normal velocity is equal to unity. At the outlet (where it is indicated as *outlet flow*), an open outflow is assumed. As shown in Figure 10.1, the heated block is subject to a heat flux perpendicular to the heated block face, having a non-dimensional value of 1. All other surfaces in the model are assumed to be insulated, that is, zero heat flux is applied.

10.3.4 Mesh Sensitivity

The mesh sensitivity is examined to determine the optimal mesh required for the analysis. The mesh levels that COMSOL supports and the element numbers for each mesh level are shown in Table 10.2. The average Nusselt numbers are evaluated 1 mm below the interface (between the insert and the heated block) from the center of the heated aluminum block.

A normal level will suit the COMSOL model since the average Nusselt number difference between the normal, fine, and finer mesh is less than 4%. Figure 10.3 presents the finite element mesh used in the current simulation.

TABLE 10.2
Different Finite Element Mesh

Extra coarse	17,900 elements
Coarser	31,000 elements
Coarse	69,000 elements
Normal	124,000 elements elements
Fine	322,000 elements
Finer	920,000 elements

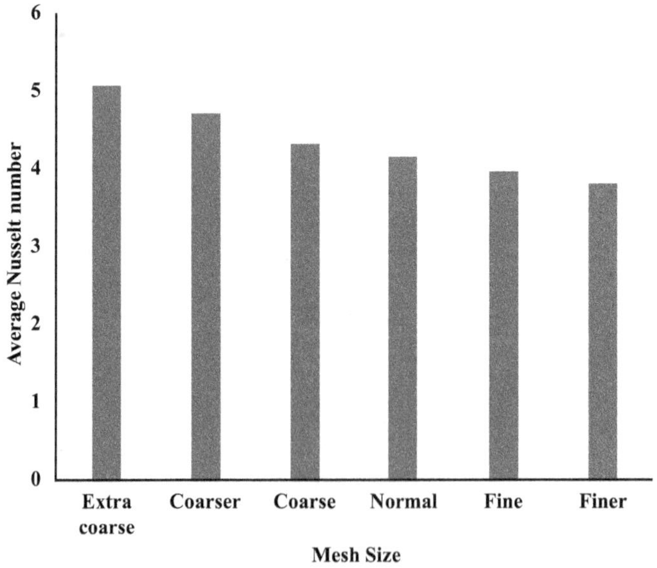

FIGURE 10.3 Finite element model.

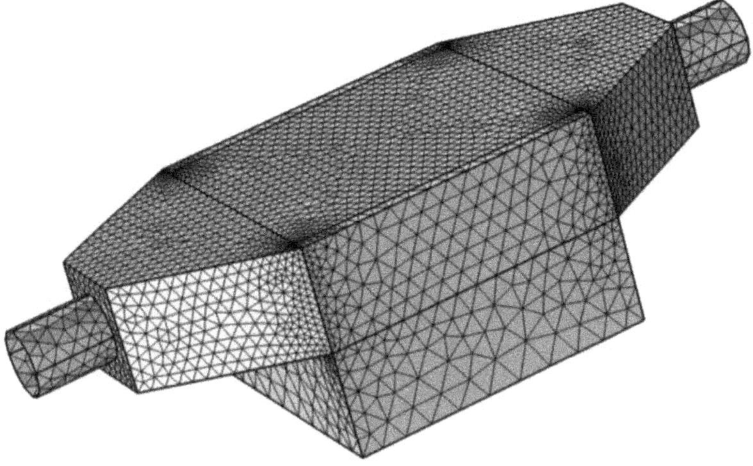

FIGURE 10.3 (Continued)

10.4 COMPARISON WITH EXPERIMENTAL DATA

To confirm the accuracy of our model, a comparison with experimental data is conducted. Plant et al. [20] conducted an experimental measurement of heat enhancement using an identical setup, as explained earlier. The only difference with the current setup is that the insert consists of three straight channels. The comparison is made for water circulating in the channels for a Reynolds number equal to 500. The temperature shown in Figure 10.4 is measured experimentally and calculated numerically at 1 mm below the interface at the exact location as our current case. A good agreement between the experimental and numerical results; thus, the proposed finite element model is accurate.

10.5 RESULTS AND DISCUSSIONS

Heat analysis will be done for two different inserts with an inlet temperature set at T_{in} = 18°C. In the first insert, water will circulate in a channel made of rectangular block pins. In the second insert, the channel is composed of a cylindrical pin. Mixing may occur behind the pin-fins and the obstacle, leading to better heat removal. The flow is assumed laminar, and a steady-state condition is applied.

10.5.1 Block Pin-Fins Insert

As shown in Figure 10.2(b), nine block rows are located equally along the test section, forming a block insert. The base size of the block in non-dimensional form is 0.0941 in width and 0.1882 in length. Thus, the length is twice the width. The height of the block varies depending on the aspect ratio, which varies between 3 and 12, as indicated earlier. The distance between two blocks, in non-dimensional form, along the Z-direction is 0.1695, and along the X-direction is 0.2823. The area occupied by a single block is 0.01771. The last column of the block has a square base of 0.0941 in size. The distance between the first block and the wall is 0.1882, allowing flow to circulate between the block and the wall. The model was tested for five different Reynolds numbers and different pin heights.

Figure 10.5 presents the temperature distribution and the Nusselt number along the flow path 1 mm below the interface. An increase in the temperature along the flow is a clear indication of the development of the boundary layer. However, what is interesting to observe in Figure 10.5(a) is a spike of

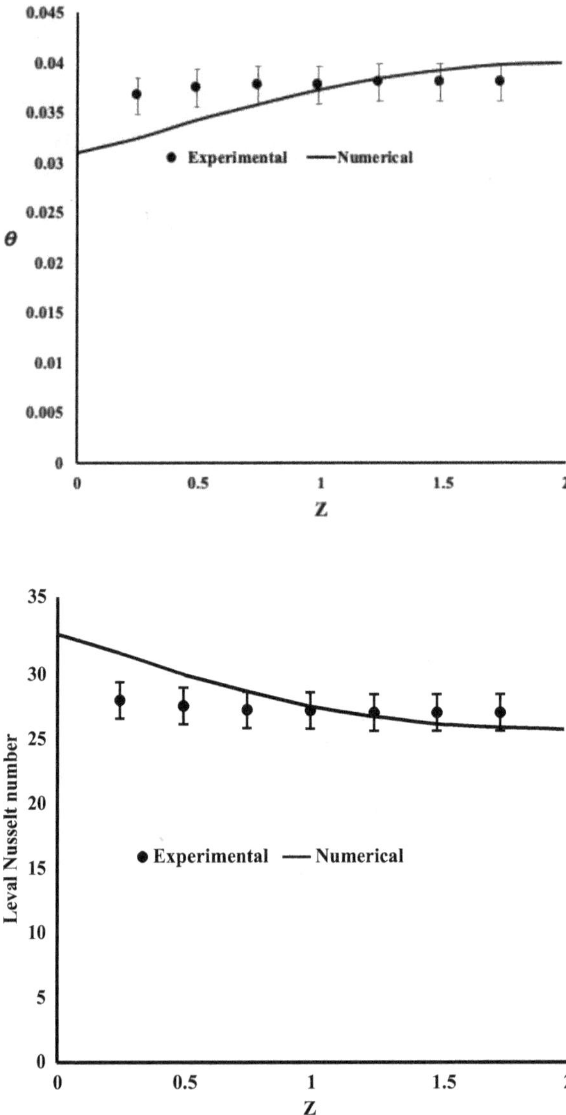

FIGURE 10.4 Comparison with experimental data for three-channel configurations.

temperature drop where the block is located. It appears that at the location of the fin, a more significant and more pronounced temperature drop is due to heat extraction. This non-smooth temperature distribution is responsible for this phenomenon. Figure 10.4(b) presents the corresponding Nusselt number. As the Reynolds number increases, more heat is extracted using this configuration.

Figure 10.6 shows the velocity distribution in two different locations, at the middle of the insert and near the wall of the experimental area, for a Reynolds number of 1,250. Figure 10.6(a) presents that when the block fin aspect ratio is 3, most of the flow overpasses the fins to the exit section. The backflow is present behind the first set of blocks, then the deflection of the flow moves the fluid to the upper region of the test chamber. As the aspect ratio increases, more backflow is, in effect, behind the first two rows of the block; thus, heat enhancement is more pronounced. The remaining flow is deflected to the upper region. As the aspect ratio increases further, as shown in Figure 10.6(c), the backflow and the reverse flow are more pronounced, and less flow is deflected to the

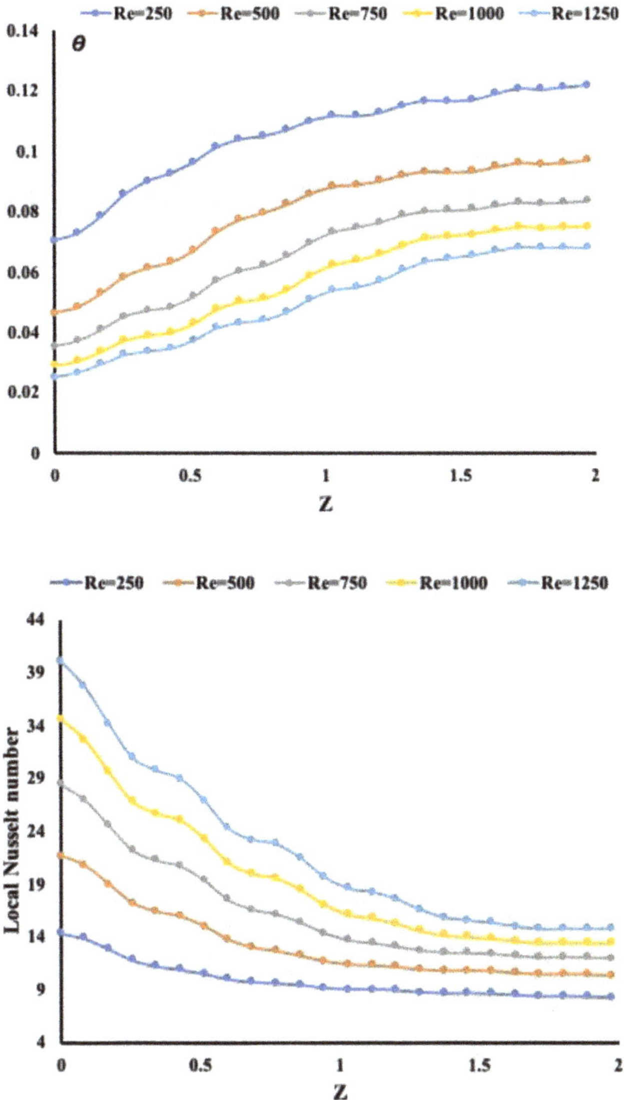

FIGURE 10.5 Temperature and local Nusselt number variation for block configuration (AR = 6).

upper region. Figure 10.6(d) presents the case when the block is full length, thus forcing the flow to interact with the pins. Very pronounced reverse flow is observed, as shown in Figure 10.6(d). Figure 10.6 also presents a similar flow, but near the wall of the experimental area. Similar behavior is observed, but higher friction near the wall is observed.

To study this heat extraction, Figure 10.7 displays the temperature distribution at the exact location as in Figure 10.6. It appears from the set of Figure 10.7(a) to Figure 10.7(d) that the large temperature gradient is noticeable for an aspect ratio of 6.

As the author [24] found, at an aspect ratio of 6, one may notice the highest heat extraction. To study further this finding, Figure 10.8 presents the average Nusselt number for all Reynolds numbers and the four cases. Based on this figure, the higher the aspect ratio, the higher the heat extraction. For all cases, as the Reynolds number increases, the average Nusselt number increases respectively.

However, one may be careful in overstating this finding, because as the block height increases, friction and pressure drop become more noticeable. Figure 10.9 presents the friction factor for all

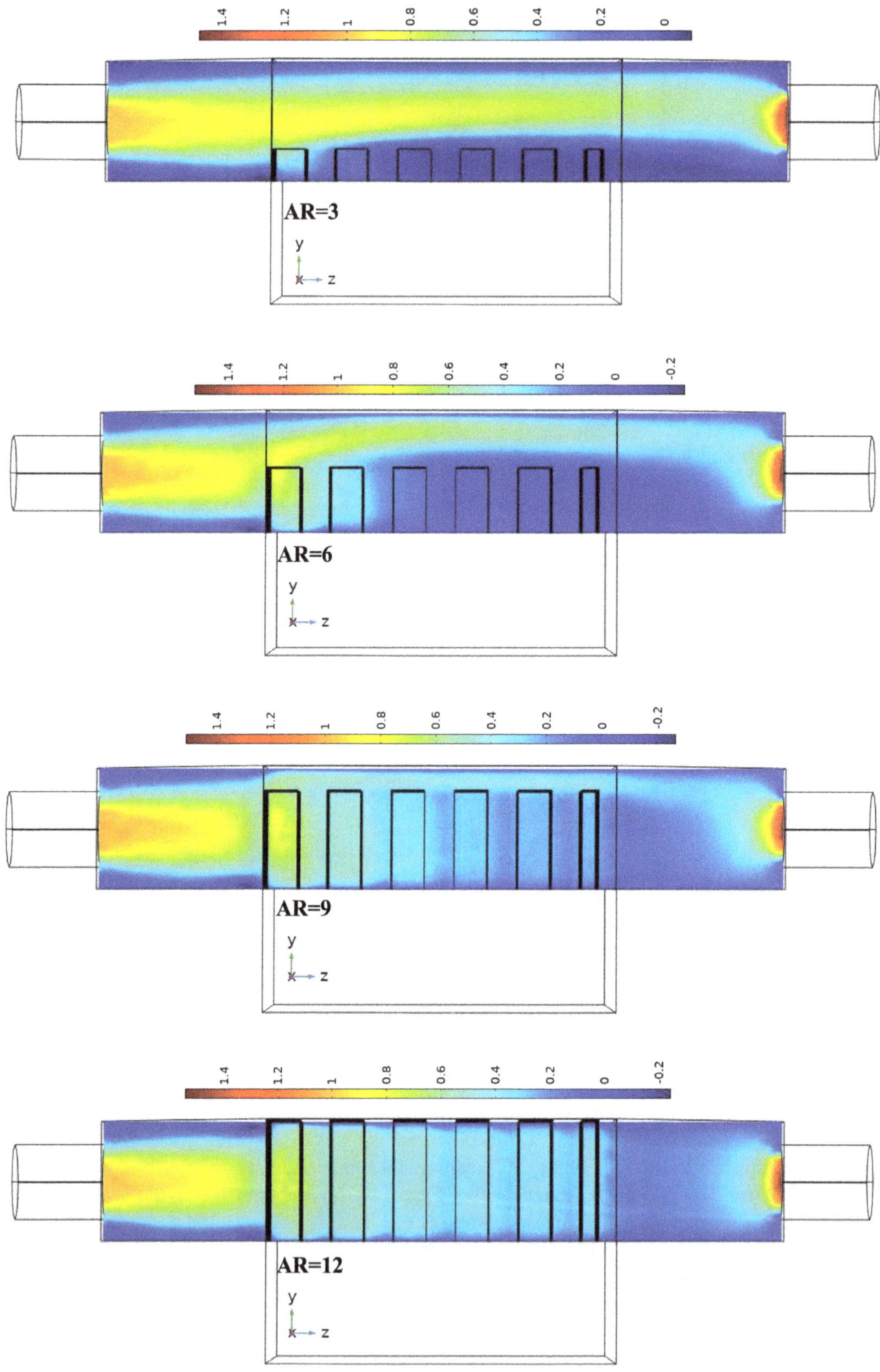

FIGURE 10.6 Velocity variation at the middle of the insert (a–d) and near the wall for blocks configuration (e–h).

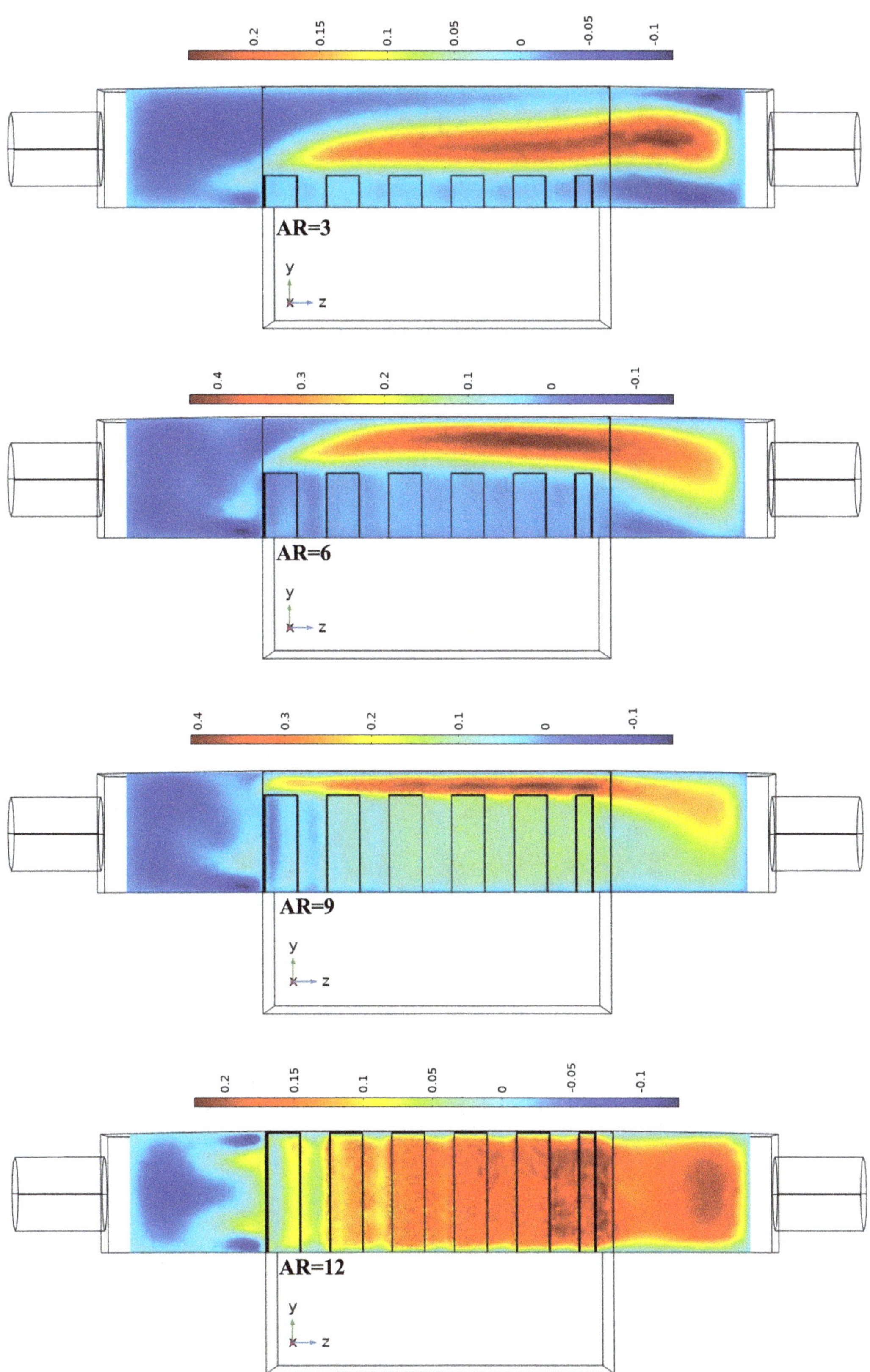

FIGURE 10.6 (Continued)

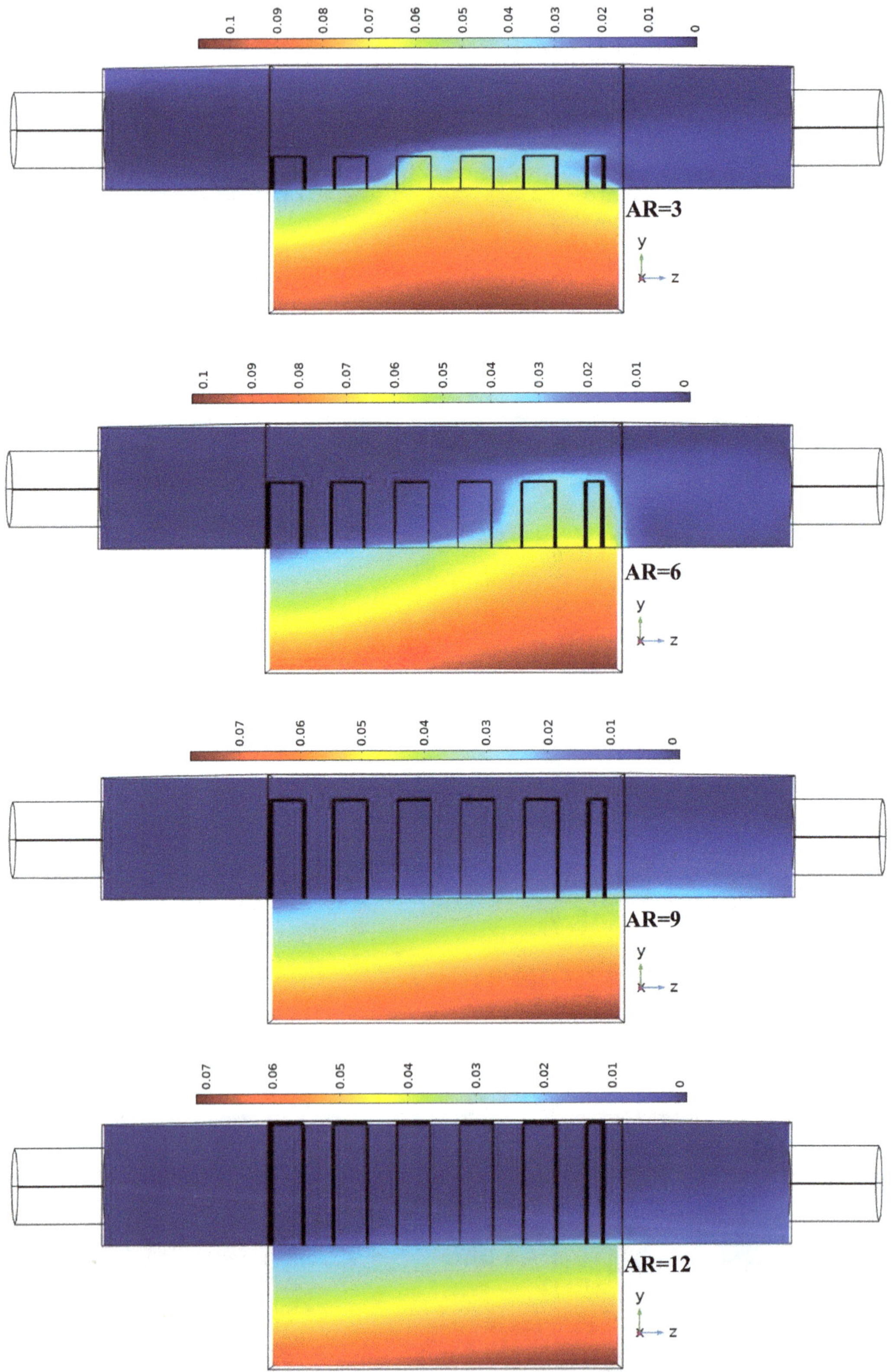

FIGURE 10.7 Temperature variation at the middle of the insert (a–d) and near the wall for blocks configuration (e–h).

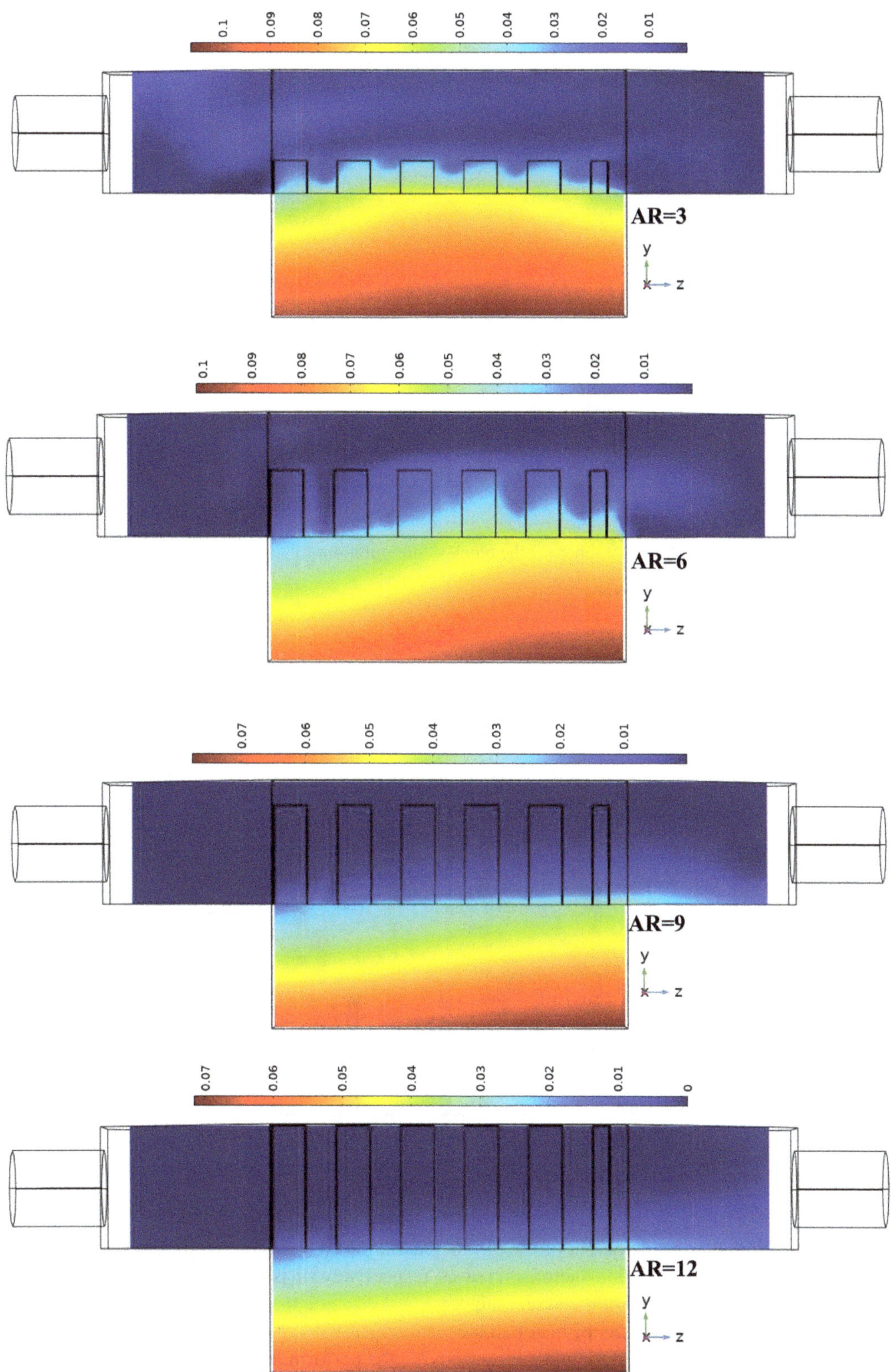

FIGURE 10.7 (Continued)

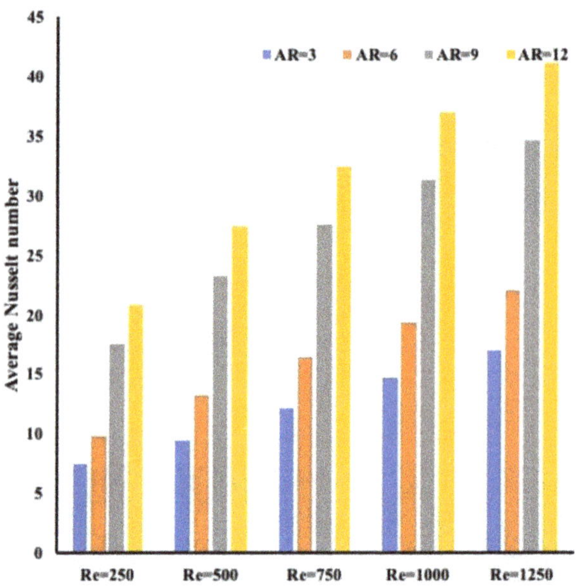

FIGURE 10.8 Average Nusselt number for the block's configuration.

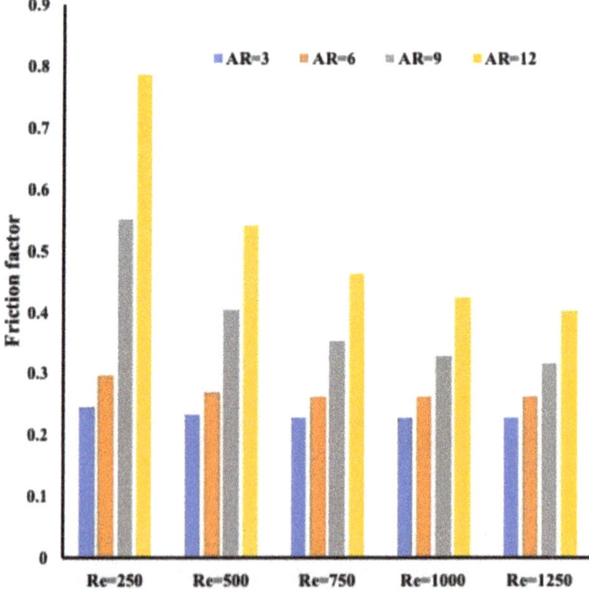

FIGURE 10.9 Friction factor for the block's configuration.

cases. It is evident that as the height of the pin increases, the change in pressure between inlet and outlet, and therefore the friction factor, increases respectively. It is true that the heat extraction increases, but at the expense of the friction factor.

One may combine the effects of heat enhancement represented by the average Nusselt number and the pressure drop via the friction factor by studying the performance evaluation criterion shown in Figure 10.10.

The optimum performance is for an aspect ratio of 6, regardless of the Reynolds number. Thus, recalling Figure 10.7, we observed that the more significant temperature gradient is obtained for an

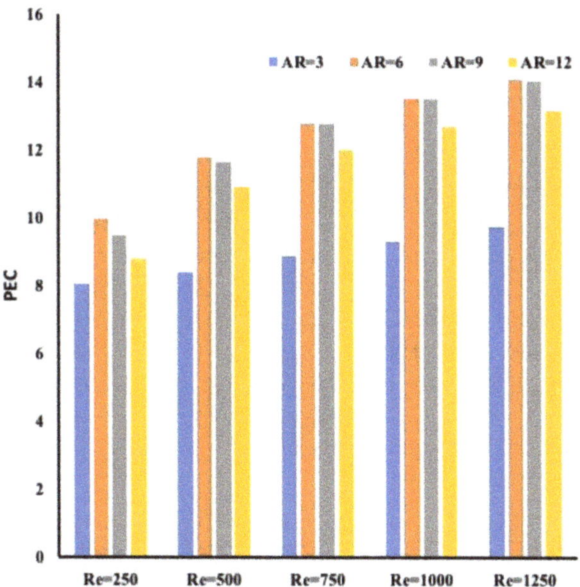

FIGURE 10.10 Performance evaluation criterion for the block's configuration

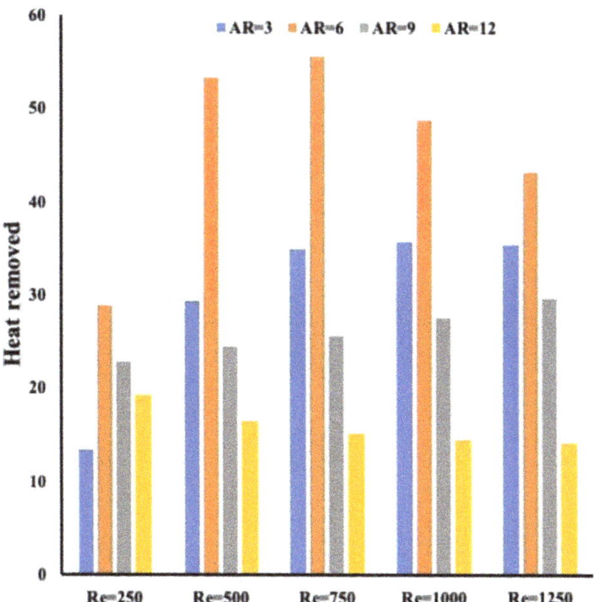

FIGURE 10.11 Heat removal for the block's configuration.

aspect ratio equal to 6. From Figure 10.7 and Figure 10.10, we confirm that the height of the optimum block pin should be for an aspect ratio of 6.

Further investigation into the case of aspect ratio equal to 6 is presented in Figure 10.11. We have calculated the amount of heat removed using equation (10.10). From Figure 10.11, the highest amount of heat removed is shown for an aspect ratio equal to 6. Thus, based on the three observations, one may conclude that an optimum aspect ratio exists for a block pin to be adopted for heat removal. The question which we can address is: Would the shape of the pin influence heat removal?

10.5.2 Cylindrical Pin-Fins Insert

The question raised in the previous section will be explored in detail by using cylindrical pin-fins. The new pin-fins shape has identical height (i.e., aspect ratio), but the spacing between the pins is more significant than in the previous case. As shown in Figure 10.2(c), nine rows of pin-fins are located equally along the test section, forming a pin insert. The base size of the pin is 0.03 in radius. The height of the pin varies depending on the aspect ratio, which varies between 3 and 12, as indicated earlier. The distance between two pins in non-dimensional form along the Z-direction is 0.2977, and along the X-direction is 0.13288. The area occupied by a single pin is 0.002827. The pin base is smaller in surface area when compared to the block pin. The pin is at a distance of 0.37982 from the wall, allowing flow to pass between wall and pin. Identical cases were repeated for this new insert, and Figure 10.12 presents the temperature variation and Nusselt number for an aspect ratio equal to 6.

Figure 10.5(a) shows similar temperature behavior as the previous case, but at a higher level of temperature. This is an indication that the cylindrical pin configuration is less effective in heat removal. Figure 10.5(b) presents the local Nusselt number variation. The fluctuation of the Nusselt

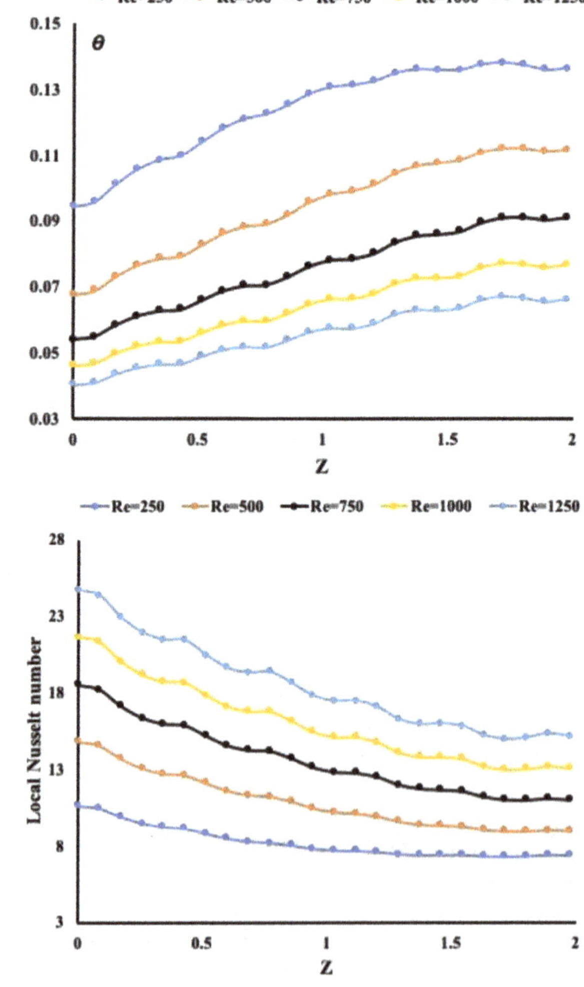

FIGURE 10.12 Temperature and local Nusselt number for aspect ratio of 6.

number is similar to the one observed in the previous insert. Thus, the pin acting as a fin can absorb more considerable heat from the hot surface. A lower Nusselt number is observed when compared to the block configuration.

Figure 10.13 presents the flow profile for two aspect ratios of 6 and 12 and the temperature distribution. Some backflow or reverse flow occurs when the aspect ratio is equal to 6, as shown in

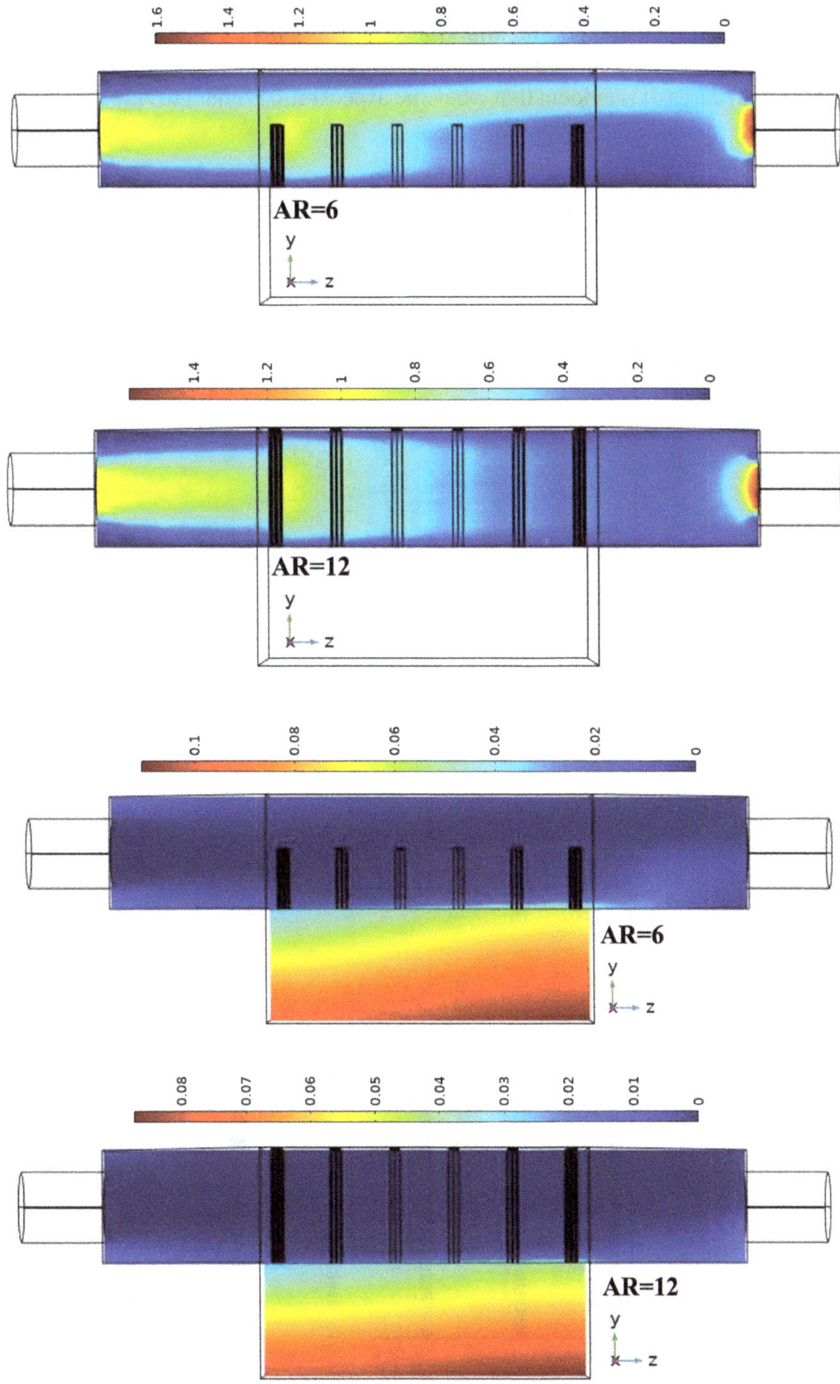

FIGURE 10.13 Temperature and velocity variation at the middle.

Figure 10.13(a), and a large temperature gradient if one compares aspect ratio 6 (i.e., Figure 10.13(a) and Figure 10.13(c)) and aspect ratio 12 (i.e., Figure 10.13(b) and Figure 10.13(d)).

Figure 10.14 represents the average Nusselt number variation. An increase in Reynolds number leads to an increase in Nusselt number, respectively. As shown in this figure, the case of an aspect ratio equal to zero indicates that no insert is present and the flow circulates in an empty cavity. The Nusselt number for this case is the lowest regardless of the Reynolds number. Also, it is interesting to notice that the average Nusselt number for aspect ratio equal to zero is identical for all Reynolds numbers. This is a clear indication of the importance of using fins for heat extraction.

Friction plays a role in the change in the pressure across the flow path. Figure 10.15 shows the friction factor for all cases. It is evident that when the aspect ratio is equal to zero, the friction factor,

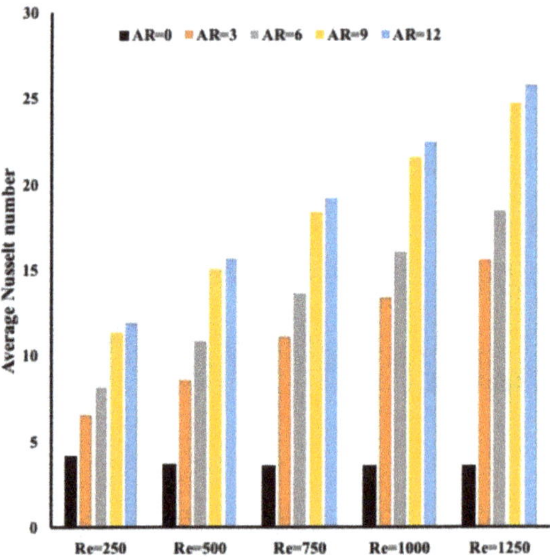

FIGURE 10.14 Average Nusselt number for all cases.

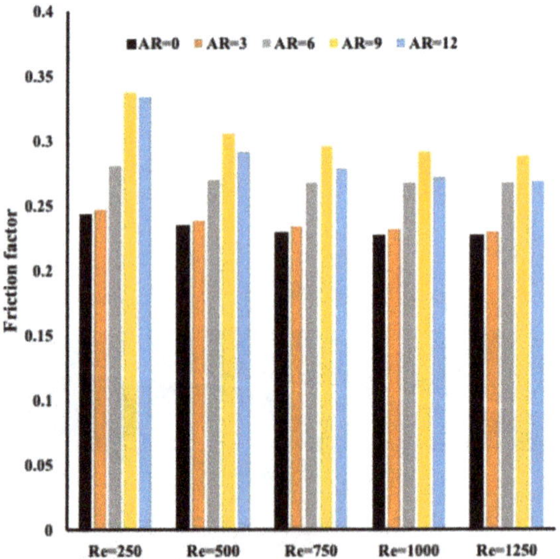

FIGURE 10.15 Friction factor for all cases.

and therefore the pressure drop, is the lowest. However, due to the shape of the pin, the highest friction factor is obtained for an aspect ratio equal to 9.

Contrary to the block configuration, it is found, as shown in Figure 10.16, that the highest performance evaluation criterion is for an aspect ratio of 12. The heat removal was evaluated for all cases to investigate this finding, as shown in Figure 10.17. It is evident in this case that the best and

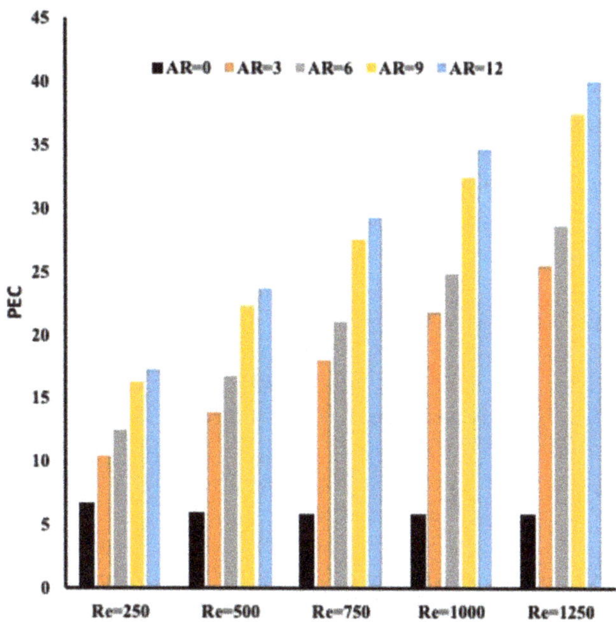

FIGURE 10.16 Performance evaluation criterion for all cases (cylindrical pins configuration).

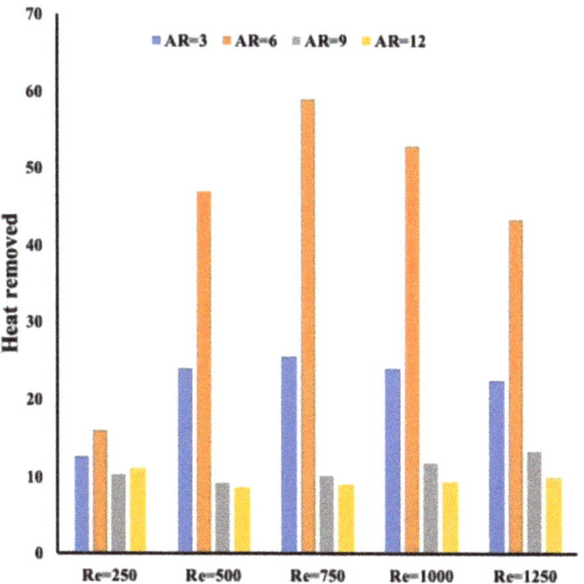

FIGURE 10.17 Heat removal for all cases.

optimum configuration for heat removal is again for an aspect ratio of 6. Thus, flow through the pin and flow above the pin are the best configurations for heat extraction.

10.6 CONCLUSIONS

In the present study, we investigated heat removal in a channel composed of a pin-fins rows configuration. The full Navier–Stokes equation, together with the heat transfer equation, was solved numerically using the finite element technique. Comparison with experimental data and an accurate mesh sensitivity provided a robust model to be used for our current investigation. Two different types of fins have been conducted. It consists of a set of nine rows of cylindrical pin-fins and rectangular pin-fins. These pin-fins helped heat extraction from the hot plate. Results revealed that:

1. In the presence of rectangular pin-fins (i.e., blocks) or cylindrical pin-fins, the temperature distribution and the Nusselt number have a wavy shape due to the enhancement of heat extraction.
2. The flow moving on top of the pin-fins created a positive temperature gradient, thus enhancing heat removal.
3. An optimum aspect ratio exists for the fins regardless of their shape, found to be equal to 6 or expressed with a fins height of 0.36 in non-dimensional form (i.e., 6.83 mm), leading to the best heat extraction.
4. Because the fins are located in a closed chamber, the interaction of the flow through the fins and above the fins helped improve heat removal.
5. Cylindrical fins provided better heat enhancement than rectangular fins.

Nomenclature

Re	Reynolds number
Pr	Prandtl number
PEC	performance evaluation criterion
Q	
Nu	Nusselt number
f	friction factor
ΔP	pressure drop in Pa

FUNDING

This research was funded by the National Science and Engineering Research Council Canada and the Faculty of Engineering and Architecture, Ryerson University.

REFERENCES

[1] Saghir, M.Z. and Bayomy, A.M., Effect of ternary nanoparticles fluid mixture in heat enhancement and heat storage. *International Journal of Aerodynamics*, **2018**, *6*(2–4):85–107.

[2] Alhajaj, Z., Bayomy, A.M., Saghir, M.Z. and Rahman, M.M., Flow of nanofluid and hybrid fluid in porous channels: experimental and numerical approach. *International Journal of Thermofluids*, **2020**, *1*:100016.

[3] Welsford, C.A., Delisle, C.S., Plant, R.D. and Saghir, M.Z., Effects of nanofluid concentration and channeling on the thermal effectiveness of highly porous open-cell foam metals: a numerical and experimental study. *Journal of Thermal Analysis and Calorimetry*, **2020**, *140*(3):1507–1517.

[4] Plant, R.D., Hodgson, G.K., Impellizzeri, S. and Saghir, M.Z., Experimental and numerical investigation of heat enhancement using a hybrid nanofluid of copper oxide/alumina nanoparticles in water. *Journal of Thermal Analysis and Calorimetry*, **2020**, *141*(5):1951–1968.

[5] Saghir, M.Z. and Rahman, M.M., Forced convection of Al_2O_3–Cu, TiO_2–SiO_2, FWCNT–Fe_3O_4, and ND–Fe_3O_4 hybrid nanofluid in porous media. *Energies*, **2020**, *13*(11):2902.

[6] Alhajaj, Z., Bayomy, A.M. and Saghir, M.Z., **2020**. A comparative study on the best configuration for heat enhancement using nanofluid. *International Journal of Thermofluids*, *7*:100041.

[7] Plant, R.D. and Saghir, M.Z., **2021**. Numerical and experimental investigation of high concentration aqueous alumina nanofluids in a two and three-channel heat exchanger. *International Journal of Thermofluids*, *9*:100055.

[8] Sertkaya, A.A., Bilir, Ş. and Kargıcı, S., Experimental investigation of the effects of orientation angle on heat transfer performance of pin-finned surfaces in natural convection. **2011**, *Energy*, *36*(3):1513–1517.

[9] Peng, Y., Heat transfer and friction loss characteristics of pin fin cooling configurations. *Transaction of the ASME*, **1984**, *106*.

[10] Goldstein, R.J. and Chen, S.B., Flow and mass transfer performance in short pin-fin channels with different fin shapes. *International Journal of Rotating Machinery*, **1998**, *4*(2):113–128.

[11] Wang, F., Zhang, J. and Wang, S., Investigation on flow and heat transfer characteristics in a rectangular channel with drop-shaped pin fins. *Propulsion and Power Research*, **2012**, *1*(1):64–70.

[12] Chin, S.B., Foo, J.J., Lai, Y.L. and Yong, T.K.K., Forced convective heat transfer enhancement with perforated pin fins. *Heat and Mass Transfer*, **2013**, *49*(10):1447–1458.

[13] Ndao, S., Peles, Y. and Jensen, M.K., Effects of pin fin shape and configuration on the single-phase heat transfer characteristics of jet impingement on micro pin fins. *International Journal of Heat and Mass Transfer*, **2014**, *70*:856–863.

[14] Abdoli, A., Jimenez, G. and Dulikravich, G.S., Thermo-fluid analysis of micro pin-fin array cooling configurations for high heat fluxes with a hot spot. *International Journal of Thermal Sciences*, **2015**, *90*:290–297.

[15] Cohen, J. and Bourell, D. Development of novel tapered pin fin geometries for additive manufacturing of compact heat exchangers. *Proceedings of the 27th Annual International SolidFreeform Fabrication Symposium*, Austin, Texas, **2016**.

[16] Yang, D., Wang, Y., Ding, G., Jin, Z., Zhao, J. and Wang, G., Numerical and experimental analysis of cooling performance of single-phase array microchannel heat sinks with different pin-fin configurations. *Applied Thermal Engineering*, **2017**, *112*:1547–1556.

[17] Amudhan, R., Kannan, G., Sakthivel, M. and Palanivel, G. Investigation of pin-fin configuration for efficient heat transfer and pressure drop in electronic components. *International Journal of Pure and Applied Mathematics*, **2018**, *119*(14):920–937.

[18] Ambreen, T., Saleem, A. and Park, C.W., Pin-fin shape-dependent heat transfer, and fluid flow characteristics of water-and nanofluid-cooled micro pin-fin heat sinks: square, circular and triangular fin cross-sections. *Applied Thermal Engineering*, **2019**, *158*:113781.

[19] Hossain, M.A., Ameri, A. and Bons, J., Conjugate heat transfer study of innovative pin-fin cooling configuration. *Journal of Propulsion and Power*, **2021**:1–11.

[20] Dadheech, P.K., Agrawal, P., Mebarek-Oudina, F., Abu-Hamdeh, N. and Sharma, A., Comparative heat transfer analysis of $MoS_2/C_2H_6O_2$ and SiO_2-$MoS_2/C_2H_6O_2$ nanofluids with natural convection and inclined magnetic field. *Journal of Nanofluids*, **2020**, *9*(3):161–167.

[21] Swain, K., Mahanthesh, B. and Mebarek-Oudina, F., Heat transport and stagnation-point flow of magnetized nanoliquid with variable thermal conductivity, Brownian moment, and thermophoresis aspects. *Heat Transfer*, **2021**, *50*(1):754–767.

[22] Asogwa, K.K., Mebarek-Oudina, F. and Animasaun, I.L., Comparative Investigation of water-based Al_2O_3 nanoparticles through water-based CuO nanoparticles over an exponentially accelerated radiative

Riga plate surface via heat transport. *Arabian Journal for Science and Engineering*, **2022**, *47*(7):8721–8738. https://doi.org/10.1007/s13369-021-06355-3

[23] Plant, R.D., Hodgson, G., Impellizzeri, S. and Saghir, M.Z., Experimental investigation of heat transfer with various aqueous mono/hybrid Nanofluids in a Multi-channel heat exchanger. *Processes*, **2021**, *9*:1932. https://doi.org/10.3390/pr9111932

[24] Saghir, M.Z. and Alhajaj, Z., Optimum multi-mini-channels height for heat enhancement under forced convection condition. *Energies*, **2021**, *14*:7020. https://doi.org/10.3390/en14217020

11 LES of Turbulence Features of Turbulent Forced Convection of Pseudoplastic and Dilatant Fluids

Mohamed Abdi, Meryem Ould-rouiss, Lalia Abir Bouhenni, Fatima Zohra Bouhenni, Nour Elhouda Beladjine, and Manel Ait Yahia

11.1 INTRODUCTION

Turbulent flow in pipes has been a famous benchmark case for testing and evaluating theories and turbulence models during the past century. Turbulence is among the most complicated issues in fluid dynamics. During the previous few decades, researchers in mechanics and engineering have shown great interest in Newtonian fluids of turbulent flow in axial pipes.

This flow is significant in the industrial domains. It involves several technical applications, including heat exchange, flow in nuclear reactors, and turbomachinery.

As part of their research, mechanics and engineering are interested in turbulent flows of non-Newtonian fluids. Various engineering and industrial applications often use this field. There are several applications where non-Newtonian fluids are used, including drilling hydraulics, wastewater transportation, slurries, pastes, suspensions of solids in liquids, emulsions, mineral oil processing, arterial blood flow, and other applications involving relatively high heat transfer rates. Fluids may also show Newtonian and non-Newtonian behavior, depending on their shear rate and their characteristic relationship to Newtonian behavior.

In these engineering applications, the turbulent flow of non-Newtonian fluids in a straight pipe is of great practical interest, where numerous studies have been performed in the last decades to better understand the behavior of this kind of fluids. A review of the literature suggests that the hydrodynamic field of fully evolved turbulent Non-Newtonian fluid flow has attracted a lot of interest in recent decades, either experimentally [1–5] or numerically by [6–17].

Abdi et al. [18] conducted LES with an extended Smagorinsky model to simulate fully developed turbulent forced convection of thermally impartial pseudoplastic fluids with a flow behavior index of 0.75 through an axially heated rotating pipe. With a rotation rate ranging from 0 to 3, the simulation Re and Pr numbers of the working fluid were assumed to be 4,000 and 1, respectively. It is observed that as the pipe wall rotates, it is seen that the temperature along its radius noticeably decreases as the rotation rate increases. This is due to a drop in apparent fluid viscosity in the pipe's core region, which causes a centrifugal force that causes the mean axial velocity profile to increase noticeably.

Convective heat transfer flows are often encountered in mechanical and chemical engineering fields, such as fluid machinery, mixing and chemical reaction enhancement in the combustion chamber, etc. Recently, several studies have been carried out on the heat transfer process of nanofluids in magnetohydrodynamic (MHD) presence. Many researchers have conducted extensive research among them to gain a deeper understanding of this phenomenon [19–24].

This literature review demonstrates that previous research has primarily concentrated on the hydrodynamic field, although the interest in forced convection heat transfer in the engineering noted

DOI: 10.1201/9781003299608-11

earlier currently lacks more details on such an issue. The overall aims of the present investigation are to explore the flow behavior index influence of the power-law fluid on the main thermal turbulent statistics, especially at the wall vicinity, to describe the thermal turbulence features, in addition to shedding further light on the transfer mechanism of the turbulent energy of this kind of fluids. To this end, a fully formed turbulent forced convection of thermally independent power-law fluids through a uniform heated pipe has been numerically investigated using LES with a standard dynamic model. The flow behavior index was chosen to be 0.75 (pseudoplastic), 1 (Newtonian), and 1.2 (dilatant) at simulation Re and Pr numbers equal to Re_s = 4,000 and Pr_s = 1, respectively, where the computations are based on a finite difference scheme, second-order accurate in space and in time, and has a numerical resolution of 65^3 gridpoints in r, θ, and z directions and with domain length of $20R$. Organizing the present chapter is as follows: governing equations and numerical procedure are presented in Section 2. The effects of flow behavior index on the main thermal turbulent statistics are presented in Section 3. This research concludes with a summary and conclusion in Section 4.

11.2 GOVERNING EQUATIONS AND NUMERICAL PROCEDURE

11.2.1 Governing Equations

A numerical analysis is presented in the present study using the LES approach with a fully developed turbulent forced convection of pseudoplastic (n = *0.75*), Newtonian (n = *1*), and dilatant (n = *1.2*) fluids through a uniform heated pipe (Figure 11.1), at Re_s equal to Re_s = 4,000 and Pr equal Pr_s = 1.

According to this definition, the dimensionless temperature is:

$$\Theta = \left(\left\langle T_w(z)\right\rangle - T(\theta, r, z, t)\right) / T_{ref} \tag{11.1}$$

Where T_w is the wall temperature and < > is the average in time and periodic directions.

Based on the filtered equations, we can write them as follows:

$$\frac{\partial \bar{u}_i}{\partial x_i} = 0 \tag{11.2}$$

$$\frac{\partial \bar{u}_j}{\partial t} + \frac{\partial \bar{u}_i \bar{u}_j}{\partial x_i} = -\frac{d\bar{P}}{\partial x_j} + \frac{1}{\mathrm{Re}_s}\frac{\partial}{\partial x_i}\left[\overline{\dot{\gamma}^{n-1}}\left(\frac{\partial \bar{u}_j}{\partial x_i} + \frac{\partial \bar{u}_i}{\partial x_j}\right)\right] + \frac{\partial \bar{\tau}_{ij}}{\partial x_i} \tag{11.3}$$

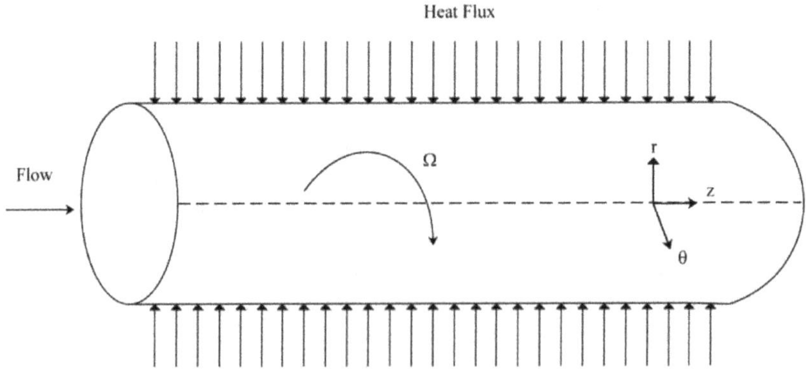

FIGURE 11.1 Physical configuration of the problem.

$$\frac{\partial \bar{\Theta}}{\partial t} + \frac{\partial}{\partial x_j}\left(\bar{u}_j \bar{\Theta} - T_{\Theta j}\right) - \bar{u}_z \frac{d}{dz}\langle T_w \rangle = \frac{1}{\mathrm{Re}_s \, \mathrm{Pr}_s} \frac{\partial^2 \bar{\Theta}}{\partial x_k \partial x_k} \tag{11.4}$$

T_{ref}, Re_s, and Pr_s denote the reference temperature, simulation Reynolds number, and simulation Prandtl number, respectively, and are defined by: $T_{ref} = q_w/\rho U_{CL} C_p$, $Re_s = \rho U_{CL}^{2-n} R^n/K$, and $Pr_s = \mu/\rho \alpha U_{CL}^{1-n} R^{n-1}$, respectively.

11.2.2 Numerical Procedure

Staggered meshes with a computational length of 20R were used to discretize the governing equations. In order to compute the numerical integration, a finite difference method with second-order spatial and temporal accuracy was employed. A fractional-step approach is used for the time progress. Convective and diffusive terms were evaluated using third-order Runge–Kutta explicit schemes and Crank–Nicolson implicit schemes, respectively. A finite difference laboratory code was used to implement the aforementioned mathematical model. It is in agreement with the available literature data that 65^3 grid points provided a reliable forecast of turbulence statics in axial, radial, and circumferential directions; however, they also provided an excellent compromise between the required CPU time and accuracy.

For the axial and circumferential directions, periodic boundary conditions are applied, while no-slip boundary conditions are usually employed for the pipe wall. An axial and circumferential uniform grid was used; otherwise, non-uniform meshes were specified in the radial direction.

11.3 RESULTS AND DISCUSSION

The findings of fully developed turbulent forced convection of thermally independent Ostwald–de Waele fluids inside a uniformly heated axially stationary pipe are analyzed and discussed in this section. The flow behavior index was chosen to be 0.75, 1, and 1.2 at simulation Reynolds and Prandtl numbers equal to Re_s = 4,000 and Pr_s = 1, respectively. This is to demonstrate the flow behavior index's impact on the main thermal turbulence statistics.

For validation purposes, the current LES predictions were compared relatively favorably with the available results of the literature: direct numerical simulation results obtained of a Newtonian fluid ($n = 1$) at Re = 5,500 reported by Redjem et al. [25]. Figure 11.2 reasonably compares the present Newtonian radial heat flux profile with Redjem et al. [25]. As seen in Figure 11.2, the present profile is in excellent agreement with the DNS data of Redjem et al. [25] throughout the pipe radius, when there are no discernible variations in the radial coordinate.

11.3.1 Root Mean Square of Temperature Fluctuations

The effects of the flow behavior index on the temperature turbulence intensities of the pseudoplastic, Newtonian, and dilatant fluids are discussed in the present section. Figure 11.3 shows the root mean square (RMS) of the temperature fluctuations distributions of pseudoplastic ($n = 0.75$), Newtonian ($n = 1$), and dilatant ($n = 1.2$) fluids against the distance from Y^+ at Re_s = 4,000 and Pr_s = 1.

As illustrated in Figure 11.3, temperature turbulence intensities profiles are almost completely independent of the flow behavior index near the pipe wall, where these profiles are consistent with each other in the viscous sublayer ($0 \leq Y^+ \leq 5$). The RMS profiles deviate from each other with the wall distance from the wall towards the core region, where the flow behavior index influence becomes significant further away from the wall. As shown in Figure 11.3, there were no appreciable differences between the RMS of the pseudoplastic and Newtonian fluids along the radial coordinate. In contrast, the RMS of the dilatant fluid lies down the pseudoplastic and Newtonian fluids along the pipe radius, where this deviation is more pronounced in the buffer ($5 \leq Y^+ \leq 30$)

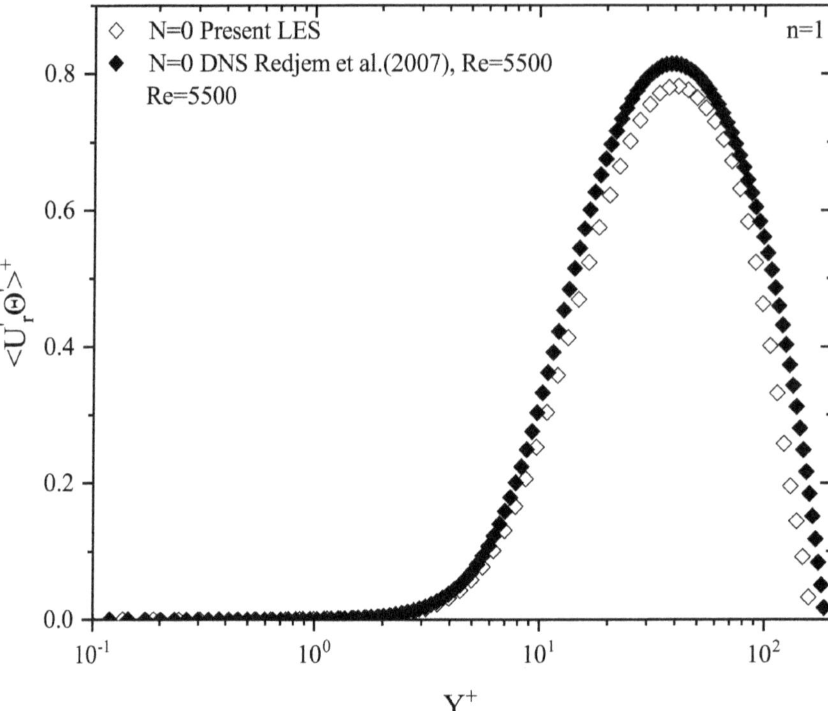

FIGURE 11.2 Validation.

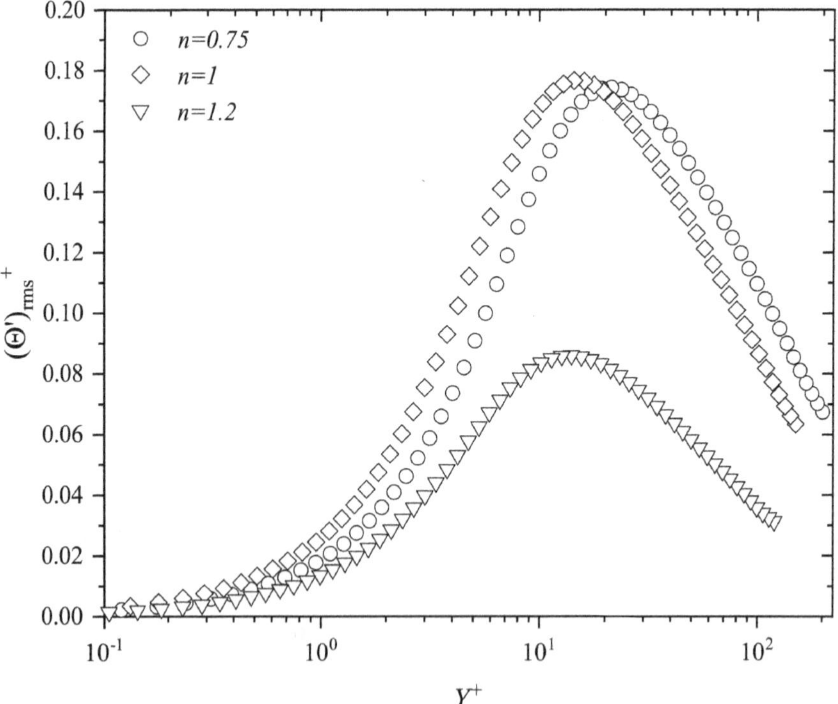

FIGURE 11.3 Temperature fluctuation RMS.

and logarithmic ($30 \leq Y^+ \leq 200$) regions. The improved flow behavior index (n) results in a noticeable attenuation in the temperature turbulence intensities of the Ostwald–de Waele fluids along the radial coordinate.

11.3.2 The Turbulent Heat Flux of Temperature Fluctuations

The next paragraphs address the effect of the flow behavior index on the turbulent heat flux of temperature changes in power-law fluids, where the radial and axial heat flux profiles of the pseudoplastic, Newtonian, and dilatant fluids are presented respectively in Figure 11.4 and Figure 11.5. Figure 11.4 and Figure 11.5 illustrate the axial and radial heat flux profiles distributions of shear-thinning ($n = 0.75$), Newtonian ($n = 1$), and shear-thickening ($n = 1.2$) fluids against the distance from the wall in wall units Y^+ at $Re_s = 4{,}000$ and $Pr_s = 1$.

As what appear in Figure 11.4 and Figure 11.5, the flow behavior index has little effect on the heat flux profiles near the wall, and the axial and radial heat flux profiles are nearly identical along the viscous sublayer, where these profiles are consistent totally with each other in the vicinity of the pipe wall. Interestingly, the axial and radial heat flux of the shear-thinning and Newtonian fluids are almost identical along the pipe radius, while the axial and radial heat flux of the shear-thickening fluids lie further bottom of the Newtonian fluid; this trend is more noticeable in the buffer region. It can be argued that the axial and radial heat flux along the pipe radius is significantly reduced due to the elevated flow behavior index.

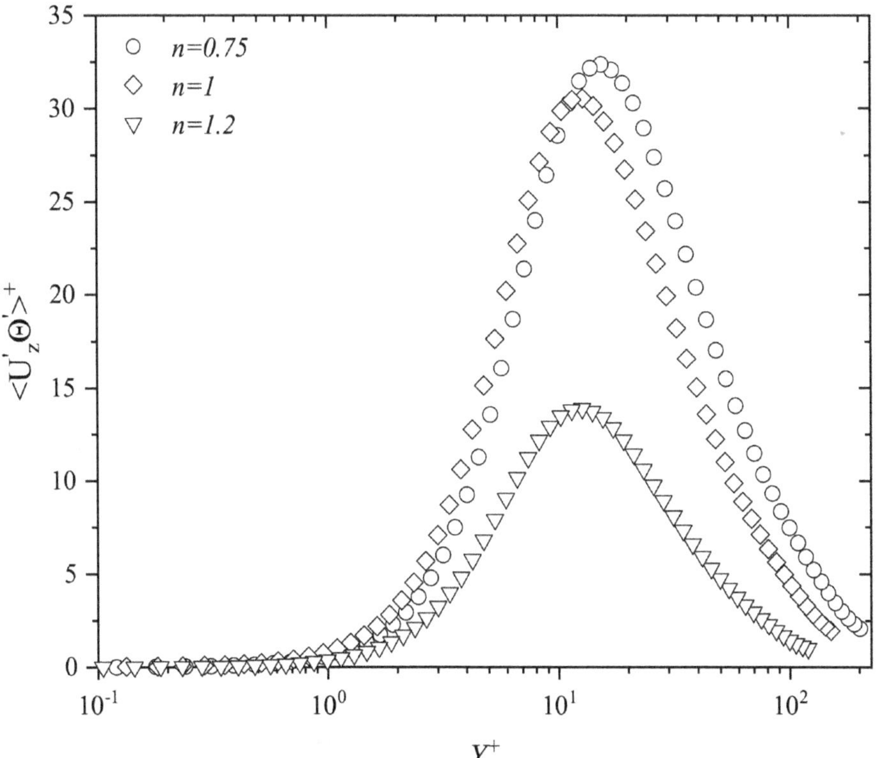

FIGURE 11.4 Heat flux axially turbulent.

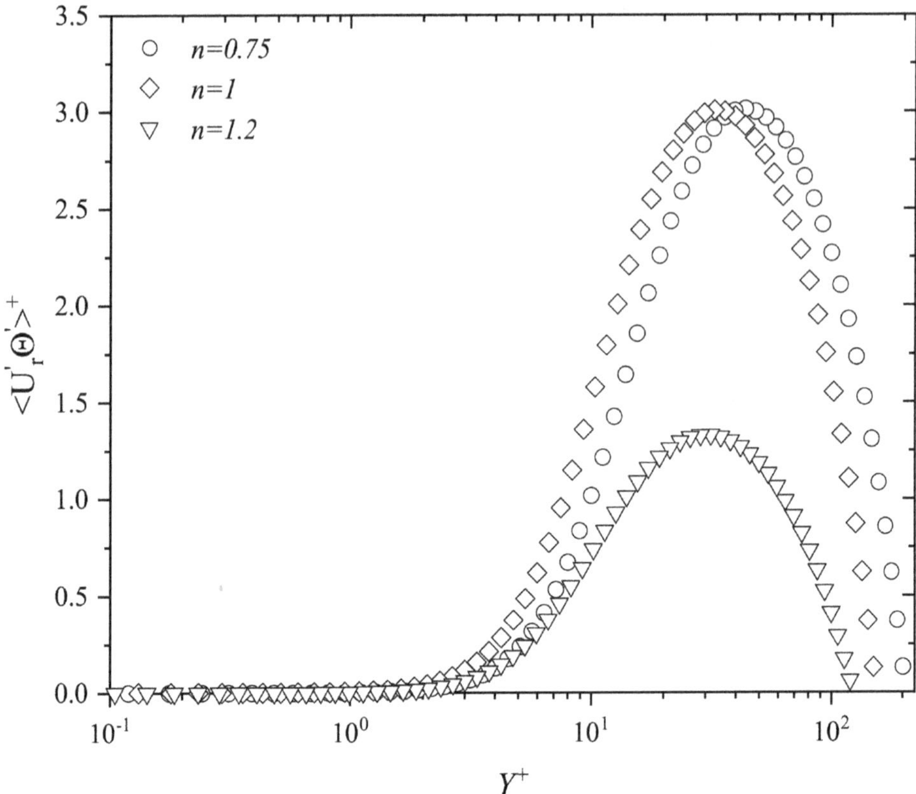

FIGURE 11.5 Radial turbulent heat flux.

11.3.3 Higher-Order Statistics

The flow behavior index and its effect on the higher-order statistics of the temperature fluctuation of the pseudoplastic, in the following paragraph, Newtonian and dilatant, fluids are analyzed and discussed. Figure 11.6 and Figure 11.7 show the skewness and flatness profiles of the temperature fluctuation of pseudoplastic (n = *0.75*), Newtonian (n = *1*), and shear-thickening (n = *1.2*) fluids against the distance from Y^+ at $Re_s = 4{,}000$ and $Pr_s = 1$.

The skewness of the temperature fluctuations is essential in the vicinity of the wall compared to that in the core region; its trend decreases rapidly near the wall towards the core region with the distance from the wall, which tends to be 0 (Gaussian value) for the flow index. The flatness of the temperature fluctuations has the same trend as the skewness one; the flatness profile drops to 4 (Gaussian value) for all flow behavior indices, as shown in Figure 11.7. It can be said that the skewness and flatness of temperature fluctuations are almost independent of flow behavior index.

11.4 CONCLUSIONS

In the present research, the large-eddy simulation (LES) approach with a standard dynamic model is numerically investigated to develop turbulent forced convection of thermally independent power-law fluids through a uniform heated pipe. The flow behavior index was chosen to be 0.75 (pseudoplastic), 1 (Newtonian), and 1.2 (dilatant) at simulation Reynolds equal to $R_{es} = 4{,}000$ and Prandtl

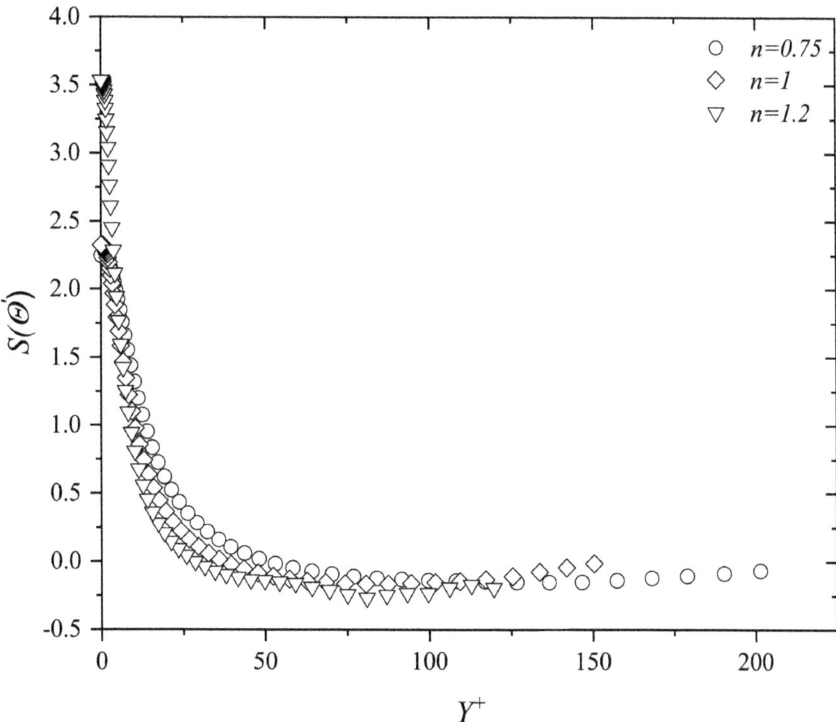

FIGURE 11.6 Skewness of temperature fluctuations.

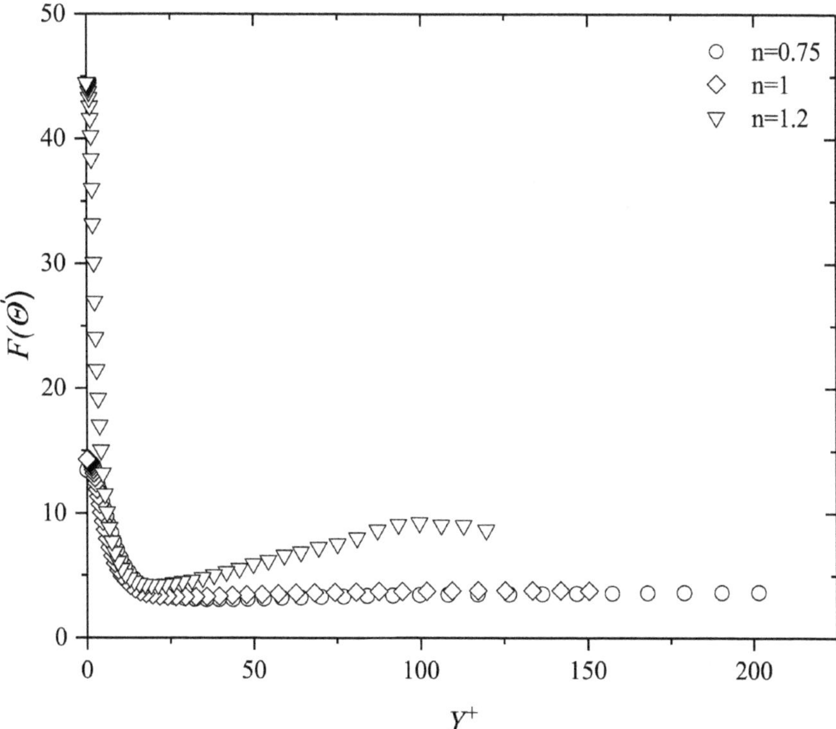

FIGURE 11.7 Flatness of temperature fluctuations.

numbers equal to $P_{rs} = 1$, where the computations are based on a finite difference scheme, second-order accurate in space and in time, and have a numerical resolution of 65^3 gridpoints in r, θ, and z directions, with computational domain length of $20R$.

The current study aimed to investigate the flow behavior index influence of the power-law fluid on the main thermal turbulent statistics to describe the thermal turbulence features of this kind of fluid. The predicted turbulent statistics agreed well with DNS data results.

The major conclusions of this research are summarized thusly:

- The increased flow behavior index produces a noticeable attenuation in the power-law fluids' temperature turbulence intensities along the radial coordinate.
- Interestingly, the axial and radial heat flux of the shear-thinning and Newtonian fluids are almost identical along the pipe radius, while the shear-thickening fluids axial and radial heat fluxes are located deeper bottom of the Newtonian fluids; this pattern is more evident in the buffer zone. It can be argued that the axial and radial heat flux along the pipe radius is significantly reduced due to the elevated flow behavior index.
- Temperature fluctuations depend almost completely on flow behavior along the radial coordinate.

Future studies will address the following subject matter:

- A fully developed turbulent flow of another non-Newtonian fluid, such as Bingham–Herschel–Bulkley and Casson fluids in pipes using LES
- The turbulent flow of Ostwald–de Waele fluid employing the direct numerical simulation (DNS)

Nomenclature

n power-law index
Re_s simulation Reynolds number
Pr_s simulation Prandtl number
η apparent viscosity
ρ density
Y^+ distance from the wall in wall units
U^+ mean axial velocity in wall units

REFERENCES

1. Metzner, A. B., & Reed, J. C. (1955). Flow of non-newtonian fluids—correlation of the laminar, transition, and turbulent-flow regions. *AIChE Journal*, 1(4), 434–440.
2. Metzner, A. B. (1957). Non-Newtonian fluid flow. Relationships between recent pressure-drop correlations. *Industrial & Engineering Chemistry*, 49(9), 1429–1432.
3. Dodge, D. W., & Metzner, A. B. (1959). Turbulent flow of non-Newtonian systems. *AIChE Journal*, 5(2), 189–204.
4. Vidyanidhi, V., & Sithapathi, A. (1970). Non-Newtonian flow in a rotating straight pipe. *Journal of the Physical Society of Japan*, 29(1), 215–219.
5. Pinho, F. T., & Whitelaw, J. H. (1990). Flow of non-Newtonian fluids in a pipe. *Journal of Non-Newtonian Fluid Mechanics*, 34(2), 129–144.
6. Malin, M. R. (1997). Turbulent pipe flow of power-law fluids. *International Communications in Heat and Mass Transfer*, 24(7), 977–988.
7. Malin, M. R. (1997). The turbulent flow of Bingham plastic fluids in smooth circular tubes. *International Communications in Heat and Mass Transfer*, 24(6), 793–804.
8. Malin, M. R. (1998). Turbulent pipe flow of Herschel-Bulkley fluids. *International Communications in Heat and Mass Transfer*, 25(3), 321–330.

9. Rudman, M., Blackburn, H. M., Graham, L. J., & Pullum, L. (2004). Turbulent pipe flow of shear-thinning fluids. *Journal of Non-Newtonian Fluid Mechanics*, 118(1), 33–48.
10. Rudman, M., & Blackburn, H. M. (2006). Direct numerical simulation of turbulent non-Newtonian flow using a spectral element method. *Applied Mathematical Modelling*, 30(11), 1229–1248.
11. Ohta, T., & Miyashita, M. (2014). DNS and LES with an extended Smagorinsky model for wall turbulence in non-Newtonian viscous fluids. *Journal of Non-Newtonian Fluid Mechanics*, 206, 29–39.
12. Gnambode, P. S., Orlandi, P., Ould-Rouiss, M., & Nicolas, X. (2015). Large-eddy simulation of turbulent pipe flow of power-law fluids. *International Journal of Heat and Fluid Flow*, 54, 196–210.
13. Gavrilov, A. A., & Rudyak, V. Y. (2016). Direct numerical simulation of the turbulent flows of power-law fluids in a circular pipe. *Thermophysics and Aeromechanics*, 23(4), 473–486.
14. Gavrilov, A. A., & Rudyak, V. Y. (2017). Direct numerical simulation of the turbulent energy balance and the shear stresses in power-law fluid flows in pipes. *Fluid Dynamics*, 52(3), 363–374.
15. Singh, J., Rudman, M., & Blackburn, H. M. (2017). The effect of yield stress on pipe flow turbulence for generalised Newtonian fluids. *Journal of Non-Newtonian Fluid Mechanics*, 249, 53–62.
16. Singh, J., Rudman, M., & Blackburn, H. M. (2017). The influence of shear-dependent rheology on turbulent pipe flow. *Journal of Fluid Mechanics*, 822, 848–879.
17. Zheng, E. Z., Rudman, M., Singh, J., & Kuang, S. B. (2019). Direct numerical simulation of turbulent non-Newtonian flow using OpenFOAM. *Applied Mathematical Modelling*, 72, 50–67.
18. Abdi, M., Noureddine, A., & Ould-Rouiss, M. (2020). Numerical simulation of turbulent forced convection of a power law fluid flow in an axially rotating pipe. *Journal of the Brazilian Society of Mechanical Sciences and Engineering*, 42(1), 1–11.
19. Hassan, M., Mebarek-Oudina, F., Faisal, A., Ghafar, A., & Ismail, A. I. (2022). Thermal energy and mass transport of shear thinning fluid under effects of low to high shear rate viscosity. *International Journal of Thermofluids*, 15, 100176.
20. Asogwa, K. K., Mebarek-Oudina, F., & Animasaun, I. L. (2022). Comparative investigation of water-based Al2O3 nanoparticles through water-based CuO nanoparticles over an exponentially accelerated radiative Riga plate surface via heat transport. *Arabian Journal for Science and Engineering*, 1–18.
21. Djebali, R., Mebarek-Oudina, F., & Rajashekhar, C. (2021). Similarity solution analysis of dynamic and thermal boundary layers: Further formulation along a vertical flat plate. *Physica Scripta*, 96(8), 085206.
22. Raza, J., Mebarek-Oudina, F., Ram, P., & Sharma, S. (2020). MHD flow of non-Newtonian molybdenum disulfide nanofluid in a converging/diverging channel with Rosseland radiation. In *Defect and Diffusion Forum* (Vol. 401, pp. 92–106). Trans Tech Publications Ltd.
23. Djebali, R., Mebarek-Oudina, F., & Rajashekhar, C. (2021). Similarity solution analysis of dynamic and thermal boundary layers: Further formulation along a vertical flat plate. *Physica Scripta*, 96(8), 085206.
24. Mebarek-Oudina, F. (2017). Numerical modeling of the hydrodynamic stability in vertical annulus with heat source of different lengths. *Engineering Science and Technology, an International Journal*, 20(4), 1324–1333.
25. Redjem-Saad, L., Ould-Rouiss, M., & Lauriat, G. (2007). Direct numerical simulation of turbulent heat transfer in pipe flows: Effect of Prandtl number. *International Journal of Heat and Fluid Flow*, 28(5), 847–861.

12 CFD Analysis of a Vertical Axis Wind Turbine

Alaeddine Zereg, Nadhir Lebaal, Mounir Aksas, Bahloul Derradji, Ines Chabani, and Fateh Mebarek-Oudina

12.1 INTRODUCTION

According to the REN evaluation [1], scientific research has developed numerous strategies to harness renewable energies, primarily wind energy, due to its accessibility, sustainability, and expansion in the upcoming years [2–4]. Thus, two main kinds of wind turbines featuring the horizontal axis wind turbine (HAWT) and the vertical axis wind turbine (VAWT) [5, 6] have been modelled and exploited. The majority of large-scale wind farms employ horizontal axis wind turbines since they are the most developed [7]. To increase power generation efficiency, the HAWT needs to constantly yaw into the direction of the local wind; hence, it needs to have yaw installations [8]. Additionally, HAWT has more stringent topographic and weather restrictions than VAWT [9, 10], which have less-stringent standards. Due to the lack of a yaw installation requirement and the straight constant section blades' high durability in comparison to the complex three-dimensional blade shape of HAWTs, the manufacturing cost for VAWTs may be lower than that for an equivalent HAWT [11–13]. Many approaches have been proposed and developed for straight-bladed vertical wind turbines because of the long-standing focus in vertical axis wind turbines [14–16].

Thus, extended studies approaches proposed and have developed several approaches for straight-bladed vertical wind turbines due to the long-standing interest in vertical axis wind turbines [17, 18]. The operational wind speeds, the range of tip speed ratio (TSR) operation, and the profile of the airfoil are only a few of the characteristics used to choose the wind turbines. All these qualities help make operations and performance more efficient, which raises the need for CFD aerodynamic study on turbines [19]. But it hasn't really taken off because of a few drawbacks, like a reduced efficiency and inability to self-start. The ideal blade shape of VAWT must be designed in order to address these innate problems [20, 21].

The fundamental issue in wind turbine engineering is the aerodynamic problem of VAWT, as the primary aerodynamics difficulties are with the test of VAWT performance. The employment of the CFD model approach (computational fluid dynamics) and experimental observations in a wind tunnel can significantly increase our understanding of VAWT [22]. The model technique has benefits in computing speed but is easy to diverge and produces huge computational errors when in the high and low tip-speed ratio [23]. The wind tunnel test needs a high cost and long cycle and is weaker in wind field simulation [24].

Numerical simulation technology based on CFD (computational fluid dynamics) is advancing quickly along with the advancement of computers, as many scholars are exploring CFD problems in several areas, such as aerodynamics [25–27] and fluid dynamics [28–32]. Additionally, the simulation's precision and visibility make it greatly suited for simulating wind turbines [33]. Numerous research has recently been conducted to solve this issue, and logical advancements have been achieved [34]. A further physical integrity study has been performed for a VAWT of NACA 0015 comprehensive knowledge of a turbine, and Cristian et al. [35] have conducted a further comparative investigation of the efficiency of the VAWT of various airfoil shapes. In addition to performing a 2D analysis on various airfoil profiles and a 3D CFD analysis on the VAWT using the selected airfoil profile, it was also decided to discuss the loads placed on the wind turbine blades as a result of pressure and wind velocity in order to gain a better understanding of the situation [36]. In their experiments, Dennis et al. [37] overlaid the VAWT performance curves for different wind speeds onto the torque curves of several generators to identify the best generator. The work advances our

 DOI: 10.1201/9781003299608-12

knowledge of the wind turbine's power-generating behavior beyond the realm of CFD analysis and offers a foundation for an effective experimental setup for proper validation.

Through the use of a two-dimensional (2D) simulation with different chord lengths, Qamar et al. [38] have examined the impact of solidity on the performance of straight camber-bladed turbines. The findings indicated poor efficiency for decreased turbine solidity throughout a wide TSR operating range. However, for short TSR ranges, increasing turbine solidity delivered higher performance. They came to the conclusion that longer chords provide more torque. However, a modification in blade count with rotors that are almost unit-solid leads in a scattered wake behind the rotor turbine, which would be detrimental to the setup of wind farms.

In order to determine how the cambered profile NACA2415 affects the behavior of an H-Darrius-type VAWT, Beri et al. [39] performed a 2D analysis, their findings showing that compared to the symmetrical profiles, cambered profiles enhanced the starting-up of the VAWT at the expense of turbine performance. The number of blades and the aspect ratio are examined in a parametric study that Gosselin et al. [40] conducted. Their investigation revealed that increasing the number of blades might smooth out variations in power performance while having no impact since the solidity of the turbine stays fixed. Furthermore, they looked at how different end plates affected the pressure distribution throughout the blades. They deduced that end plates facilitate uniform distribution and boost VAWT effectiveness. The impact of periodic intake fluctuation wind conditions on the fluid flows surrounding an H-Darrius VAWT has been investigated by Danao et al. [41] using a computational technique. The flow separation and reattachment are particularly highlighted. The findings report that the performance of the VAWT is marginally improved for a given mean TSR and fluctuation amplitude. However, with a mean TSR lower than the steady peak performance TSR, it results in a deep stall and vortex shedding under a steady inflow wind situation. Moreover, the turbine operates in a state with dominated drag when there is a significant variation in wind speed.

Maeda et al. [42] used 2D wind velocity data to investigate the properties of the flow field; the researchers have implemented an LDV system at the middle of a straight-bladed VAWT to achieve this. Using three terminal pressure devices, they have also experimentally studied the aerodynamic loading parameters. They discovered that a lower wind velocity area had developed within the turbine itself, moving downstream. When the TSR rose, the low-wind-velocity zone exhibited an expanding tendency. Yang et al. [43] used experimental and computational methods to study the impact of the tip vortex on wind turbine wake. They have observed that the support structure-generated vortex enhances the turbulence intensity. In various span-wise positions, the impact of the tip vortex on the downwind zone of the VAWT has been studied. They claimed that the center of the rotor is where the low-velocity zone is situated and that the tip vortex has a greater dissipation distance. The torque coefficient is higher in the zone upstream. The variation of energy production with variable wind speeds and the variation of power generation with varied rotational speeds are two study areas where it can be concluded that there is still a considerable need for additional research, based on the figures and experimental findings that have been published so far.

This investigation's major emphasis is a study of the coefficient of a "power anticipated" case, with a numerical method employing an enhanced computational mesh technique known as sliding mesh.

12.2 VAWT MODEL SETUP AND MESHING

12.2.1 Mathematical Model and Equations

The unstable incompressible viscous Reynolds-averaged Navier–Stokes (URANS) equations manage the flow around the VAWT, and they typically represent the conservation of mass (continuity), momentum, and velocity.

However, additional unknowns were created using the Reynolds decomposition, such as the referred-to Reynolds stresses. To mathematically solve the problem, the Reynolds stress terms should be represented. Since it is more precise and dependable for a larger class of flows, we have used the shear stress transfer (SST) k-ω model (e.g., adverse pressure gradient flows, airfoils, etc.). The common SST k-ω model has two transport equations attached, as follows [44, 45]:

Continuity equation (incompressible flow):

$$\frac{\partial \overline{u_i}}{\partial x_i} = 0 \tag{12.1}$$

Momentum equation:

$$\frac{\partial \overline{u_i}}{\partial x_i} + \overline{u_j}\frac{\partial \overline{u_i}}{\partial x_j} = -\frac{1}{\rho}\frac{\partial \overline{P}}{\partial x_i} + \nu\frac{\partial^2 \overline{u_i}}{\partial x_j^2} - u_j^{'}\frac{\partial \overline{u_i^{'}}}{\partial x_j} + g_i \tag{12.2}$$

Turbulence kinetic energy equation:

$$\frac{\partial(\rho k)}{\partial t} + \frac{\partial(\rho k u_i)}{\partial x_i} = \frac{\partial}{\partial x_j}\left(\Gamma_k \frac{\partial k}{\partial x_j}\right) + G_k - Y_k + S_k \tag{12.3}$$

Specific dissipation rate equation:

$$\frac{\partial(\rho \omega)}{\partial t} + \frac{\partial(\rho \omega u_i)}{\partial x_i} = \frac{\partial}{\partial x_j}\left(\Gamma_\omega \frac{\partial \omega}{\partial x_j}\right) + G_\omega - Y_\omega + S_\omega \tag{12.4}$$

12.2.2 Design and Meshing Methodology

Modelling the turbine to be evaluated and establishing its size and other characteristics are essential components of aerodynamic analysis. The 2D simulation, in contrast to the 3D simulation, only took into account the aerodynamic efficacy of the blade airfoil form; support arms, torque shafts, and bearing disks were not taken into account [46]. Unsteady flow across a 3D model, however, would need significantly more CPU time (several weeks per simulation). Therefore, 2D simulation may get a comparable outcome while using less CPU.

As shown in Figure 12.1, a 2D computational domain served as the foundation for the study. In order to simulate flow across the cross section of the VAWT, the three-blade model was developed.

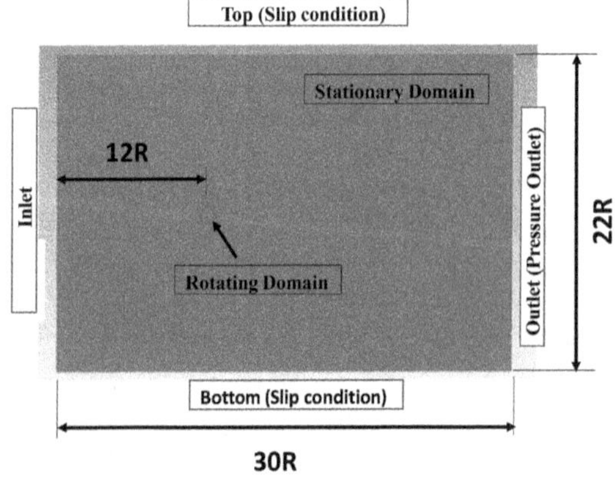

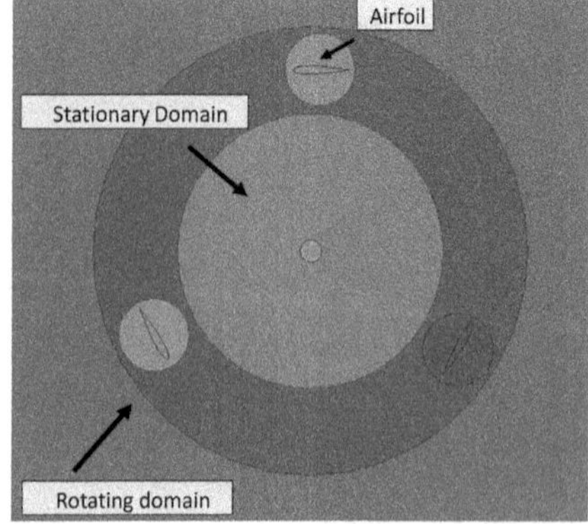

FIGURE 12.1 Schematic of computational domains.

The 2.5 m diameter of the turbine is represented by a 0.4 m chord NACA0015 airfoil. In order to obtain a complete wake development, the computational domain included a small inner rotor zone embedded with a 30R long (the inlet and outlet boundary conditions are placed respectively 12R and 18R upwind and downwind of the rotor) and 22R wide rectangular outer stationary domain. In order to verify that the boundaries are sufficiently distant from the turbine to have no impact on torque and overall turbine performance, numerous simulations were run with sequential increments in a fixed domain size [47].

The calculation was done using the finite volume method with ANSYS-FLUENT, and ANSYS-MESHING was used to create the mesh. At the boundary separating the two sub-domains, a sliding-mesh interface exists. The boundary layer surrounding the blades has a first cell height Y^+ values of less than 10 with 40 layers and a growth rate of 1.2, and both sub-domains employ an unstructured grid (Figure 12.2). By gradually reducing the airfoil surface node density until the torque coefficient from the flow solutions did not significantly change, mesh sensitivity to changes in node number was carried out.

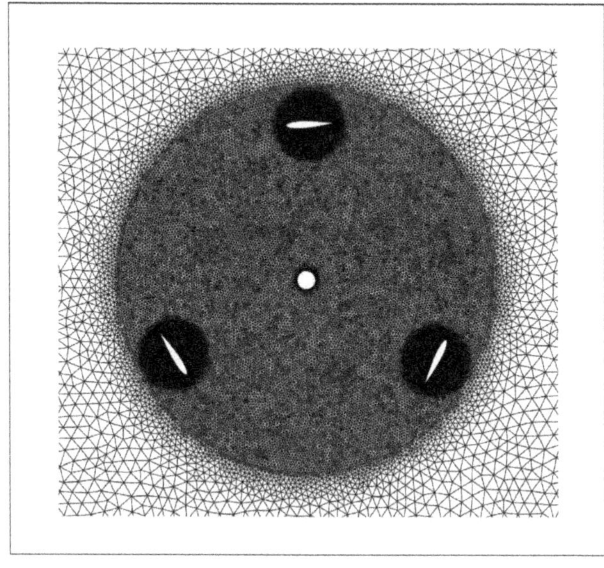

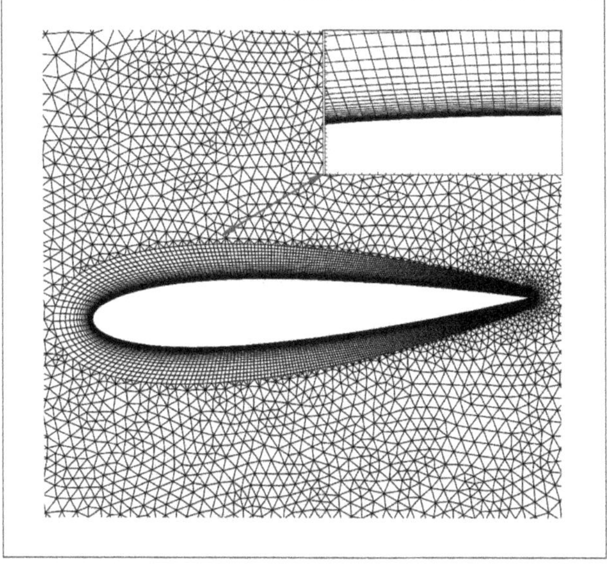

FIGURE 12.2 (a) Overview of the mesh; (b) mesh around airfoil.

12.2.3 Sliding-Mesh Implementation

The moving reference frame technique is computationally demanding, ignores flux variation, and suffers from the loss of constant time frame, which reduces the precision of the CFD solution. With the advent of high computing power, a new method has been proposed to increase the accuracy of the CFD solution and reduce the degree of error. In order to rotate mesh, the sliding-meshing technique is utilized [48].

The sliding-mesh model approach is used to represent a rotating turbine across the flow field. Two cell zones make up the channel domain. The blades of the turbine are first positioned within a rotor. For every simulation time step, this zone rotates at a certain rate. The inner rotor-encircling rectangular channel makes up the second zone, which is stationary and is shown in Figure 12.2. The mesh interface, along which these two domains move in relation to one another, is where one cell zone meets the neighboring cell zone. The flux computations across the zone interface are made simpler since the motion of the moving zone is monitored relative to the stationary frame and no moving reference frames are required. The unsteady flow field may be calculated using the sliding-mesh method. It is the technique that simulates flows in several moving reference frames with the greatest accuracy.

The alignment of the cell nodes is not necessary since the two cell zones slide in relation to one another, resulting in a non-conformal border. The intersection between the interface zones is computed using ANSYS Fluent in order to solve the flux across the non-conformal boundary at each time step. These connections are updated for the new location at each time step, and the flow is computed [49].

12.3 RESULTS AND DISCUSSION

The solution is carried out until the convergence criteria of the order of 10^{-4} for the continuity equation and convergence criteria of 10^{-4} are set for the other governing parameters, such as X-velocity, Y-velocity, and K, and the order of convergence is studied from the normalized error versus the iterations plot and report the solution is converged. This is done after initializing the aforementioned parameters into the ANSYS Fluent solver. The discretization of time and time step changes at each moment while the turbine is rotating. Therefore, it is necessary to examine a precise solution over time. Time is thus discretized in order to decrease CPU iteration time and maintain accuracy of the numerical result.

The initial conditions are based on a steady single reference frame motion (SRF). The spatial discretization of first- and second-order upwind techniques is examined in ANSYS Fluent. By comparing the two schemes, it is possible to determine how each plan affects the turbine's overall performance. It is determined that using the second-order upwind strategy increases the numerical solution's accuracy. Due to its dissipative nature, the first-order upwind scheme overlooked a few downstream flow features.

Therefore, compared to the first-order upwind method, the second-order upwind strategy is more beneficial for the present simulation. The coefficient of power as a function of the tip speed ratio is estimated, which is one of the important aerodynamic parameters. The instantaneous moment coefficient may be calculated using equation (12.5) as follows:

$$C_p = \frac{\mathrm{T}.\omega}{\frac{1}{2}\rho.U^3_\infty A} \tag{12.5}$$

The TSR is calculated as follows:

$$TSR = \frac{R.\omega}{U_\infty} \tag{12.6}$$

The estimated results are compared with the experimental and numerical findings of Castelli et al. [50] over a variety of TSR values using the NACA0021 three-bladed H-rotor.

The highest value of torque is generated at the TSR of 2.5, according to the torque versus time charts at various TSRs.

This is because the blade draws energy from the wind and creates vortices in the upwind direction. The vortices created in the upwind route have an impact on the turbine's overall performance while the blade is passing through the downwind side. The wind speed has an impact on the turbine's total power output as well. The power output of the turbine will significantly decline when the wind speed decreases. Figure 12.3 presents the CFD findings at various TSRs.

One of the most important factors in determining a turbine's performance, operating capacity, and ability to be mass-manufactured for a wind turbine is the quantity of power it produces. Angle of attack (AOA) on the turbine blades, lift and drag forces, and net torque produced by spinning turbine blades are all factors that affect how much power a wind turbine produces. As previously explained, the moment produced by the turbine blades as a result of pressure and viscous forces acting on the blades is what causes the wind turbine to produce power.

As shown in Figure 12.4, which shows the coefficient of power with TSR for wind speeds of 3 m/s, 5 m/s, and 10 m/s, respectively, the maximum power generation is observed for the TSR = 3.0 with Cp = 0.351, implying 35.1% of the available power is derived successfully. For the wind speed of 3 m/s, the coefficient of power improves progressively with increase in the TSR. The variance is shown to match the tendencies that have been anticipated in several research papers for wind turbine power output at decreasing wind velocity. When the wind speed is 5 m/s, the power generation coefficient is shown to rise steadily until the TSR, after which there is a decline in power production until Cp = 1.8. The rapid drop in wind energy production may be ascribed to flow separation brought on by the increased rotational speeds. For wind speeds of 10 m/s, the effect of dynamic

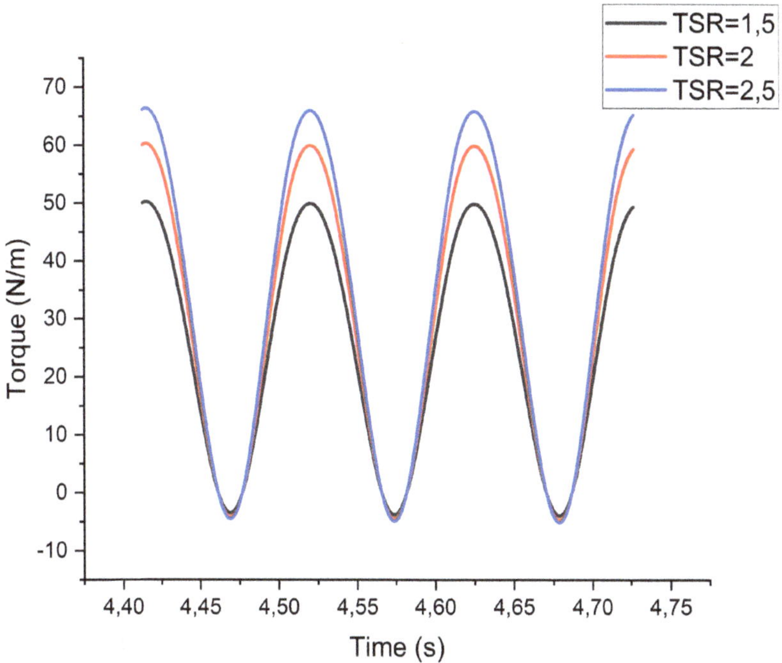

FIGURE 12.3 Periodical torque results over four cycles.

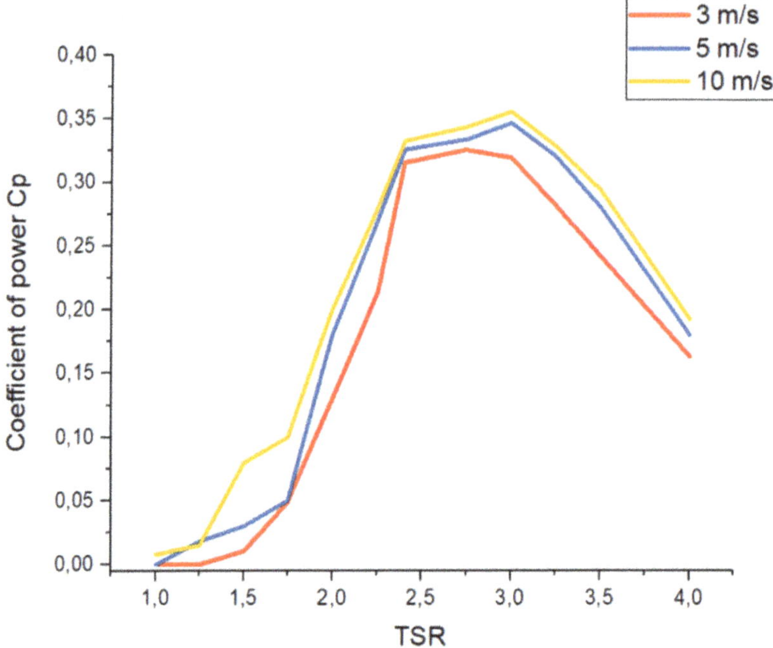

FIGURE 12.4 A comparison plots of the coefficient of power (Cp) with TSR for various wind speeds.

stall is clearly visible in the lower TSR conditions; however, for moderate TSR, the angle of attack is higher and in the upwind region, which causes the creation of turbulent bubbles in the upwind zone, which travel to the downwind zone and grow significantly in size, leading to a reduction in the power generation in the turbine.

The AOA of the vertical axis wind turbine, which changes significantly over the course of one rotation, is also another factor affecting the blade aerodynamic performance. The maximum AOA on the blades is generally lower for the upwind zone and greater TSR. As a result, the power created is larger than the other TSR condition, but a negative torque is formed in the downwind zone, which causes the power to be identified and explains why the TSR is decreasing.

The comparative analysis plots allow us to find the TSR that is generally more profitable or practical to operate for the wind speed present at the location of the wind farm, using the information from the existing study for the moderate wind speeds.

12.3.1 Velocity Distribution

The important advantages of the velocity contours are that they provide a visual representation of the development, shedding, and dissipation of produced vortex bubbles as well as a representation of the phenomenon's shedding for comparatively greater TSR. The following figures illustrate the vorticity and velocity magnitude contours for the 2D scenario. Less power is extracted in the downwind direction because of strong vortices that are shed in the upwind path. In addition, wind energy is a function of wind speed cube. As a result, even a minor decrease in wind speed might result in a significant decrease in the amount of power that can be extracted as represented in Figure 12.5 and Figure 12.6. The upwind direction allows the blades to extract the most energy. However, the wind speed has decreased and has become turbulent as they go down the downwind direction. As a

FIGURE 12.5 The velocity contours around the airfoils for wind velocity = 3m/s, TSR = 2.

FIGURE 12.6 The vorticity contours around the airfoils for wind velocity = 3m/s, TSR = 2.

consequence, compared to the upwind path, the downwind path's power extraction is minimal. The roadway of von Kármán vortex is seen in the aftermath of the turbine and blades.

The boundary layer thickness around the airfoil reduces for conditions with higher TSR because the relative velocity on the turbine blades is greater than for conditions with lower TSR, as shown in Figure 12.7 and Figure 12.8. When this happens, the viscous forces suddenly increase, which causes the pressure force that produces power to gradually lower at high TSR.

FIGURE 12.7 The velocity contours around the airfoils for wind velocity = 5m/s, TSR = 2.

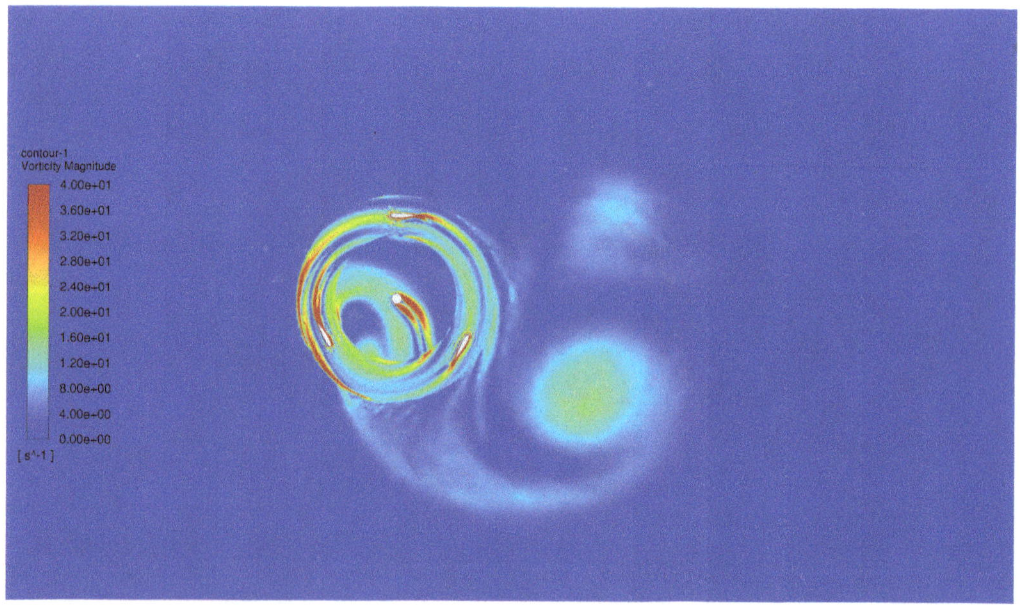

FIGURE 12.8 The vorticity contours around the airfoils for wind velocity = 5m/s, TSR = 2.

The dynamic stall of the turbine is another phenomenon that can be added to the reduced power generation abilities of the turbines. The stall phenomenon is obvious and significant in the lower-TSR, lower-wind-speed criteria, where there is a possibility for greater angle of incidences, and may result in the stalling of the turbine.

In the case of relatively higher TSR, the vortices created are dissipated along the length and are less noticeable than in the case of lower TSR; the created vortices bubbles do not disassociate from the surface, and consequently, there is not a significant drop in pressure, making sure that the effect does not interfere with the power generation.

The higher TSR requirements for the turbine show a greater power extraction because the majority of the airflow is directed more towards the upwind region of the blade and less towards the downwind region (Figure 12.9 and Figure 12.10). This provides additional evidence for the higher power generation, which is absent or not apparent at lower rotational speeds. These are some of the conversations discussing the findings and conclusions drawn from various turbine simulations that were performed for various wind speeds.

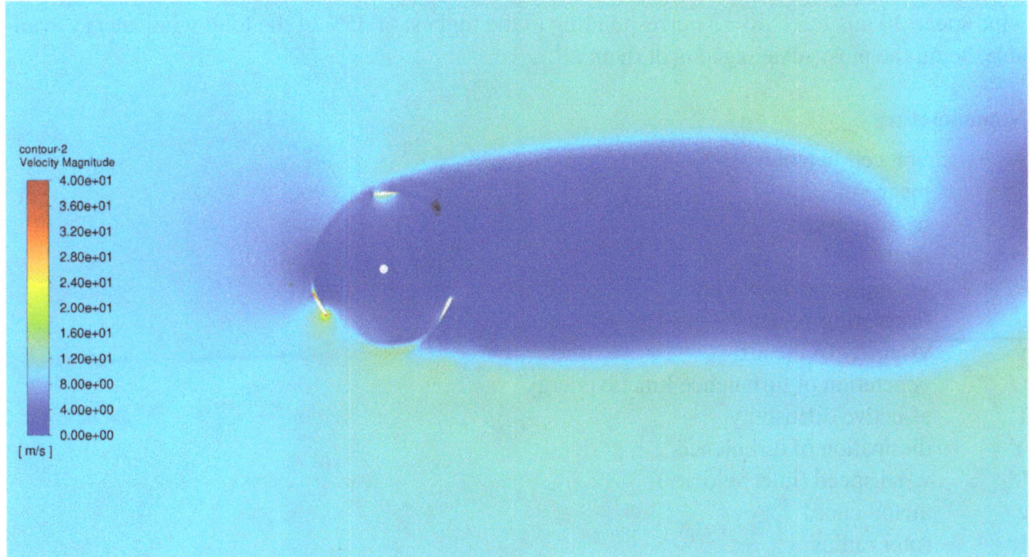

FIGURE 12.9 The velocity contours around the airfoils for wind velocity = 10m/s, TSR = 2.

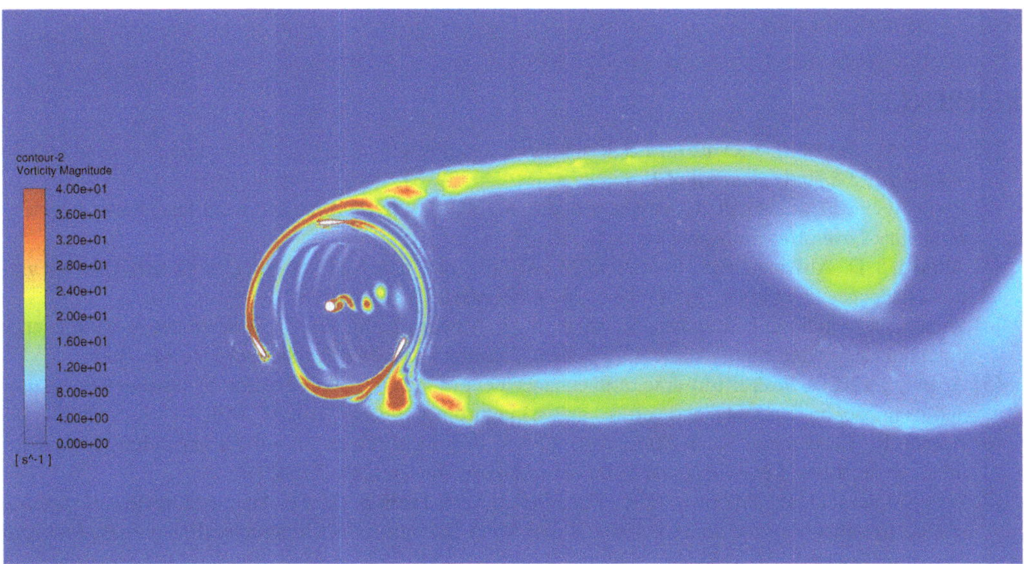

FIGURE 12.10 The vorticity contours around the airfoils for wind velocity = 10m/s, TSR = 2.

12.4 CONCLUSIONS

After looking at the data and findings for the NACA 0015 blade profile vertical axis wind turbine rotor at various wind speeds and rotational velocities, it was found that the TSR larger than 2 generated the greatest and most consistent power. Even though the high TSR parameters have strong starting capabilities, the turbine must be given a high initial power, but this is acceptable given the enhanced, reliable power output.

When compared to the other wind speed power production statistics provided, the considerably lower wind speed of 3 m/s consistently generates electricity but produces a relatively lesser amount overall. Large-scale power generation for wind speeds 4, 6, and 8 m/s in the higher TSR regions, respectively, ranges from 30 to 35% of the total wind energy available at those wind speeds, with wind speed 10 m/s for TSR = 3 corresponding to the highest of 35% of the total wind energy available, being the most advantageous of them all.

Nomenclature

y+	first cell height
$\overline{u_i}$	mean velocity
$\overline{P}$	mean pressure
ρ	density
g_i	gravitational acceleration
u_j'	fluctuating velocity
v	viscosity of the air
G	generation of turbulence kinetic energy
Γ	effective diffusivity
Y	dissipation of turbulence
U_∞	wind speed (inlet velocity)
c	airfoil chord
R	rotor radius
TSR	tip speed ratio
Cp	coefficient of power
ω	angular speed
T	average torque
A	swept rotor area

REFERENCES

[1] Abkar, M., & Dabiri, J. O. (2017). Self-similarity and flow characteristics of vertical-axis wind turbine wakes: an LES study. *Journal of Turbulence*, 18(4), 373–389.

[2] Porté-Agel, F., Bastankhah, M., & Shamsoddin, S. (2020). Wind-turbine and wind-farm flows: a review. *Boundary-Layer Meteorology*, 174(1), 1–59.

[3] Ahmed, S. D., Al-Ismail, F. S., Shafiullah, M., Al-Sulaiman, F. A., & El-Amin, I. M. (2020). Grid integration challenges of wind energy: a review. *IEEE Access*, 8, 10857–10878.

[4] Shoaib, M., Siddiqui, I., Rehman, S., Khan, S., & Alhems, L. M. (2019). Assessment of wind energy potential using wind energy conversion system. *Journal of Cleaner Production*, 216, 346–360.

[5] Acarer, S. (2020). Peak lift-to-drag ratio enhancement of the DU12W262 airfoil by passive flow control and its impact on horizontal and vertical axis wind turbines. *Energy*, 201, 117659.

[6] Kumar, R., Raahemifar, K., & Fung, A. S. (2018). A critical review of vertical axis wind turbines for urban applications. *Renewable and Sustainable Energy Reviews*, 89, 281–291.

[7] Prabowoputra, D. M., Prabowo, A. R., Bahatmaka, A., & Hadi, S. (2020). Analytical review of material criteria as supporting factors in horizontal axis wind turbines: effect to structural responses. *Procedia Structural Integrity*, 27, 155–162.

[8] Yang, J., Fang, L., Song, D., Su, M., Yang, X., Huang, L., & Joo, Y. H. (2021). Review of control strategy of large horizontal-axis wind turbines yaw system. *Wind Energy*, 24(2), 97–115.

[9] Rehman, S., Alam, M. M., Alhems, L. M., & Rafique, M. M. (2018). Horizontal axis wind turbine blade design methodologies for efficiency enhancement: a review. *Energies*, 11(3), 506.
[10] Li, S., Li, Y., Yang, C., Wang, Q., Zhao, B., Li, D., . . . & Xu, W. (2021). Experimental investigation of solidity and other characteristics on dual vertical axis wind turbines in an urban environment. *Energy Conversion and Management*, 229, 113689.
[11] Rajpar, A. H., Ali, I., Eladwi, A. E., & Bashir, M. B. A. (2021). Recent development in the design of wind deflectors for vertical axis wind turbine: a review. *Energies*, 14(16), 5140.
[12] Yosry, A. G., Fernández-Jiménez, A., Álvarez-Álvarez, E., & Marigorta, E. B. (2021). Design and characterization of a vertical-axis micro tidal turbine for low velocity scenarios. *Energy Conversion and Management*, 237, 114144.
[13] Yan, Y., Avital, E., Williams, J., & Cui, J. (2021). Aerodynamic performance improvements of a vertical axis wind turbine by leading-edge protuberance. *Journal of Wind Engineering and Industrial Aerodynamics*, 211, 104535.
[14] Dai, G., Xu, Z., HuangFu, K. L., & Zhong, Y. J. (2010). Recent research progress in the vertical axis wind turbine. *Fluid Machinery*, 38(10), 39–43.
[15] Barnes, A., Marshall-Cross, D., & Hughes, B. R. (2021). Towards a standard approach for future Vertical Axis Wind Turbine aerodynamics research and development. *Renewable and Sustainable Energy Reviews*, 148, 111221.
[16] Altmimi, A. I., Alaskari, M., Abdullah, O. I., Alhamadani, A., & Sherza, J. S. (2021). Design and optimization of vertical axis wind turbines using Qblade. *Applied System Innovation*, 4(4), 74.
[17] Lin, J., Leung, L. K., Xu, Y. L., Zhan, S., & Zhu, S. (2018). Field measurement, model updating, and response prediction of a large-scale straight-bladed vertical axis wind turbine structure. *Measurement*, 130, 57–70.
[18] Arpino, F., Scungio, M., & Cortellessa, G. (2018). Numerical performance assessment of an innovative Darrieus-style vertical axis wind turbine with auxiliary straight blades. *Energy Conversion and Management*, 171, 769–777.
[19] Elsakka, M. M., Ingham, D. B., Ma, L., & Pourkashanian, M. (2019). CFD analysis of the angle of attack for a vertical axis wind turbine blade. *Energy Conversion and Management*, 182, 154–165.
[20] Berg, D E. (1996). Vertical-axis wind turbines – The current status of an old technology. United States. https://www.osti.gov/servlets/purl/432928
[21] Hand, B., Kelly, G., & Cashman, A. (2021). Aerodynamic design and performance parameters of a lift-type vertical axis wind turbine: a comprehensive review. *Renewable and Sustainable Energy Reviews*, 139, 110699.
[22] Rezaeiha, A., Montazeri, H., & Blocken, B. (2019). On the accuracy of turbulence models for CFD simulations of vertical axis wind turbines. *Energy*, 180, 838–857.
[23] Fortunato, B., Dadone, A., & Trifoni, V. (1993, January). A theoretical model for the prediction of vertical axis wind turbineperformance. In *31st Aerospace Sciences Meeting* (p. 136).
[24] Vergaerde, A., De Troyer, T., Molina, A. C., Standaert, L., & Runacres, M. C. (2019). Design, manufacturing and validation of a vertical-axis wind turbine setup for wind tunnel tests. *Journal of Wind Engineering and Industrial Aerodynamics*, 193, 103949.
[25] Barnes, A., Marshall-Cross, D., & Hughes, B. R. (2021). Towards a standard approach for future Vertical Axis Wind Turbine aerodynamics research and development. *Renewable and Sustainable Energy Reviews*, 148, 111221.
[26] Zhang, L. X., Liang, Y. B., Liu, X. H., Jiao, Q. F., & Guo, J. (2013). Aerodynamic performance prediction of straight-bladed vertical axis wind turbine based on CFD. *Advances in Mechanical Engineering*, 5, 905379.
[27] Rezaeiha, A., Montazeri, H., & Blocken, B. (2018). Characterization of aerodynamic performance of vertical axis wind turbines: Impact of operational parameters. *Energy Conversion and Management*, 169, 45–77.
[28] Chabani, I., Mebarek-Oudina, F., & Ismail, A. A. I. (2022). MHD flow of a hybrid nano-fluid in a triangular enclosure with zigzags and an elliptic obstacle. *Micromachines*, 13(2), 224.
[29] Mebarek-Oudina, F., Redouane, F., & Rajashekhar, C. (2020). Convection heat transfer of MgO-Ag/water magneto-hybrid nanoliquid flow into a special porous enclosure. *Algerian Journal of Renewable Energy and Sustainable Development*, 2(2), 84–95.
[30] Marzougui, S., Mebarek-Oudina, F., Magherbi, M., & Mchirgui, A. (2022). Entropy generation and heat transport of Cu–water nanoliquid in porous lid-driven cavity through magnetic field. *International Journal of Numerical Methods for Heat & Fluid Flow*, 32(6), 2047–2069.

[31] Mebarek-Oudina, F., Hussein, A. K., Younis, O., Rostami, S., & Nikbakhti, R. (2021)., Natural convection enhancement in the annuli between two homocentric cylinders by using ethylene glycol/water-based Titania nanofluid. *Journal of Advanced Research in Fluid Mechanics and Thermal Sciences*, 80(2), 56–73.

[32] Hassan, M., Mebarek-Oudina, F., Faisal, A., Ghafar, A., & Ismail, A. I. (2022). Thermal energy and mass transport of shear thinning fluid under effects of low to high shear rate viscosity. *International Journal of Thermofluids*, 15, 100176.

[33] Sabaeifard, P., Razzaghi, H., & Forouzandeh, A. (2012). Determination of vertical axis wind turbines optimal configuration through CFD simulations. In *International Conference on Future Environment and Energy* (Vol. 28, pp. 109–113).

[34] Islam, M., Ting, D. S. K., & Fartaj, A. (2008). Aerodynamic models for Darrieus-type straight-bladed vertical axis wind turbines. *Renewable and Sustainable Energy Reviews*, 12(4), 1087–1109.

[35] Bottero, C. J. (2012). Aérojoules project: Vertical Axis Wind Turbine https://matheo.uliege.be/handle/2268.2/6089

[36] Chen, W. H., Chen, C. Y., Huang, C. Y., & Hwang, C. J. (2017). Power output analysis and optimization of two straight-bladed vertical-axis wind turbines. *Applied Energy*, 185, 223–232.

[37] Dennis, D., Ganesh, P. S., Joy, J., Amjith, L. R., & Bavanish, B. (2022). Computational design & analysis of vertical axis wind turbine. *Materials Today: Proceedings*, 66, 1501–1508.

[38] Qamar, S. B., & Janajreh, I. (2017). A comprehensive analysis of solidity for cambered darrieus VAWTs. *International Journal of Hydrogen Energy*, 42(30), 19420–19431.

[39] Beri, H., & Yao, Y. (2011). Effect of camber airfoil on self-starting of Vertical Axis Wind Turbine. *Journal of Environmental Science and Technology*, 4(3), 302–312.

[40] Gosselin, R., Dumas, G., & Boudreau, M. (2016). Parametric study of H-Darrieus vertical-axis turbines using CFD simulations. *Journal of Renewable and Sustainable Energy*, 8(5), 053301.

[41] Danao, L. A., Edwards, J., Eboibi, O., & Howell, R. (2014). A numerical investigation into the influence of unsteady wind on the performance and aerodynamics of a vertical axis wind turbine. *Applied Energy*, 116, 111–124.

[42] Maeda, T., Kamada, Y., Murata, J., Kawabata, T., Shimizu, K., Ogasawara, T., . . . & Kasuya, T. (2016). Wind tunnel and numerical study of a straight-bladed vertical axis wind turbine in three-dimensional analysis (Part I: For predicting aerodynamic loads and performance). *Energy*, 106, 443–452.

[43] Yang, Y., Guo, Z., Zhang, Y., Jinyama, H., & Li, Q. (2017). Numerical investigation of the tip vortex of a straight-bladed vertical axis wind turbine with double-blades. *Energies*, 10(11), 1721.

[44] Souaissa, K., Ghiss, M., Chrigui, M., Bentaher, H., & Maalej, A. (2019). A comprehensive analysis of aerodynamic flow around H-Darrieus rotor with camber-bladed profile. *Wind Engineering*, 43(5), 459–475.

[45] Belabes, B., & Paraschivoiu, M. (2021). Numerical study of the effect of turbulence intensity on VAWT performance. *Energy*, 233, 121139.

[46] Brusca, S., Lanzafame, R., & Messina, M. (2014). Design of a vertical-axis wind turbine: how the aspect ratio affects the turbine's performance. *International Journal of Energy and Environmental Engineering*, 5(4), 333–340.

[47] Yawei, C. (2016). Aerodynamic performance analysis based on FLUENT axial flow vane. *Fluid Mach*, 10, 51–54.

[48] Almohammadi, K. M., Ingham, D. B., Ma, L., & Pourkashan, M. (2013). Computational fluid dynamics (CFD) mesh independency techniques for a straight blade vertical axis wind turbine. *Energy*, 58, 483–493.

[49] Siddiqui, M. S., Durrani, N., & Akhtar, I. (2013, July). Numerical study to quantify the effects of struts and central hub on the performance of a three-dimensional vertical axis wind turbine using sliding mesh. In *ASME Power Conference* (Vol. 56062, p. V002T09A020). American Society of Mechanical Engineers.

[50] Castelli, M. R., Englaro, A., & Benini, E. (2011). The Darrieus wind turbine: Proposal for a new performance prediction model based on CFD. *Energy*, 36(8), 4919–4934.

13 Heat Transfer in MHD Hiemenz Flow of Reiner–Rivlin Fluid Over Rotating Disk with Cattaneo–Christov Heat Flux Theory

Sanjay Kumar, Kushal Sharma, and S. B. Bhardwaj

13.1 INTRODUCTION

Stagnation point flow represents the flow of a fluid in the immediate neighborhood of a stagnation point (or a stagnation line), with which the stagnation point (or the line) is identified for a potential flow or inviscid flow. Maximum heat transfer, pressure, and mass deposition rates are encountered in the stagnation region. Hiemenz [1] engineered the pioneering study of this type of flows by transforming the classical Navier–Stokes equations to nonlinear ordinary differential equations. Stagnation point flows of non-Newtonian or Newtonian fluids have vital industrial and engineering application, such as in transpiration cooling, thermal oil recovery, drag reduction, and in the design of thrust bearings. Homann [2] extended the stagnation point flows for the axisymmetric case. Engineering and environmental devices often encounter the problem of flat surface mixed convection. So it is important to study heat transfer characteristics of these type of geometries to further improve the design of heat exchangers. Some noted studies are [3–9].

The present chapter focuses on the extension of flow over rotating disk due to its various practical applications in industrial and engineering fields, such as computer storage devices, rotatory machinery, wastewater management, and many more. In practice, it is difficult to analytically solve Navier–Stokes equations due to the presence of nonlinearity in it, but under some restrictions on the flow domain, it is possible to do so. Von Kármán [10] solved the disk flow under steady and viscous nature assumption. The rotating disk flow afterwards had been extended considering various physical flow parameters. The suction effect on the disk was analyzed by Stuart [11], while the introduction of energy equation to the flow domain to study heat transfer was done by Benton [12]. The magnetic field has interesting effect on the flow domain based upon its nature of being weak or strong, as it affects drag force. El-Mistikawy [13, 14] examined the effect of weak and strong magnetic field on the flow domain. Attia [15] studied the combined effect of vertical magnetic field and suction/injection on the rotating disk flow, while Miklavvcivc and Wang [16] examined the flow when disk admits partial slip. Ram et al. [17] introduced the concept of ferrofluids flow over the rotating disk with porosity impact. Recently, many studies have been done taking the flow of these fluids over the infinite rotating disk with various physical effects, such as [18–21].

The present chapter also explores the flow over the rotating disk having stretching condition due to its importance in metal and plastic industry, such as extrusion processes during plastic transformation [22, 23]. Fang [24] initiated this type of study and solved the classical von Kármán flow problem taking stretchable condition. Turkyilmazoglu [25] extended Fang's model to study heat transfer taking magnetic field into consideration. Bhatti and Rashidi [26] investigated the Williamson fluid flow over a stretching sheet and studied the heat transfer characteristics taking thermal radiation into consideration. Hayat et al. [27] examined the radiative flow over the rotating

DOI: 10.1201/9781003299608-13

disk having non-constant thickness. Mandal and Shit [28] analyzed the entropy generation of radiative flow over a stretchable disk with time-dependent stretching rate taking biviscosity nanofluid. Magnetohydrodynamic (MHD) flow has practical significance in various engineering fields, such as MHD flow meters, pumps and power generation, etc. The intensity of the magnetic field strongly affects such types of flows. Multiple studies have been done to examine such flows, such as [29–36].

The current chapter extends the work of Turkyilmazoglu [37] to the case when the fluid considered is non-Newtonian, as this type of fluids are very common in engineering and industrial applications. The Cattaneo–Christov (C-C) model is adapted to examine the heat transfer which was missing earlier in the literature. The Reiner–Rivlin model adequately describes many food products and chemicals, biological materials, as well as geological processes, as many fluids belong to this category. The model formulation is solved by fifth-order Runge–Kutta–Fehlberg method in Maple software by appropriately transforming the equations using similarity transformations. The obtained graphical results are explained for the incorporated non-dimensional parameters that arose due to the various effects which are considered in the chapter.

13.2 MATHEMATICAL FORMULATION

We examine the MHD stagnation flow of Reiner–Rivlin fluid over a rotating disk. The disk is rotating with constant angular velocity Ω and also stretching in radial direction with rate c. The Cattaneo–Christov model of heat flux is employed to study heat transfer. The flow geometry is given in Figure 13.1.

The following relation is developed by Reiner [38] and Rivlin [39]:

$$\tau_{ij} = -p\delta_{ij} + \mu e_{ij} + \mu_c e_{ik} e_{kj}; e_{jj} = 0. \tag{13.1}$$

Where $e_{ij} = \frac{\partial u_i}{\partial x_j} + \frac{\partial u_j}{\partial x_i}$, δ_{ij}, τ_{ij}, p, and μ_c are deformation rate tensor, Kronecker symbol, stress tensor, pressure, and cross-viscosity coefficient, respectively.

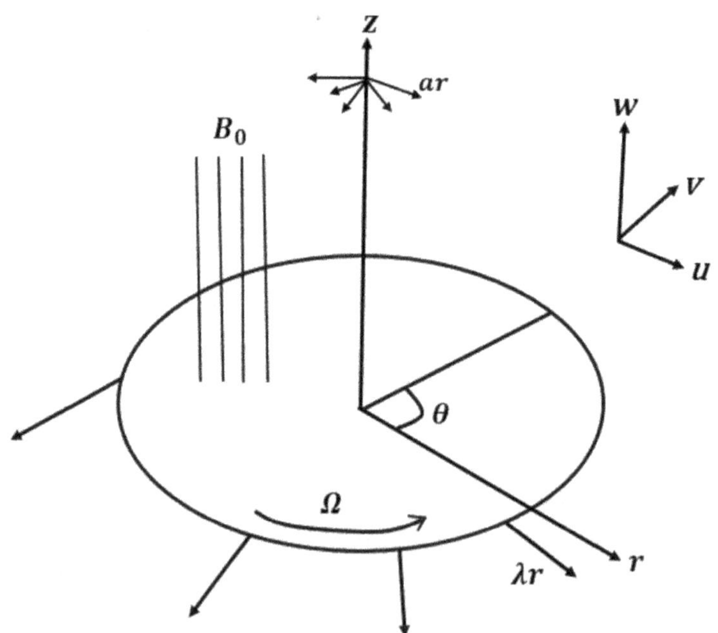

FIGURE 13.1 Geometry of the problem.

The governing Navier–Stokes equation along with energy equation for the study:

$$\frac{\partial u}{\partial r}+\frac{u}{r}+\frac{\partial w}{\partial z}=0, \tag{13.2}$$

$$\rho\left(u\frac{\partial u}{\partial r}+w\frac{\partial u}{\partial z}-\frac{v^2}{r}\right)=\frac{\partial \tau_{rr}}{\partial r}+\frac{\partial \tau_{rz}}{\partial z}+\frac{\tau_{rr}-\tau_{\phi\phi}}{r}+\sigma B_0^2\left(u_e-u\right), \tag{13.3}$$

$$\rho\left(u\frac{\partial v}{\partial r}+w\frac{\partial v}{\partial z}+\frac{uv}{r}\right)=\frac{1}{r^2}\frac{\partial}{\partial r}\left(r^2\tau_{r\phi}\right)+\frac{\partial \tau_{z\phi}}{\partial z}+\frac{\tau_{r\phi}-\tau_{\phi r}}{r}+\sigma B_0^2 v, \tag{13.4}$$

$$\rho\left(u\frac{\partial w}{\partial r}+w\frac{\partial w}{\partial z}\right)=\frac{1}{r}\frac{\partial}{\partial r}\left(r\tau_{rz}\right)+\frac{\partial \tau_{zz}}{\partial z}, \tag{13.5}$$

$$\left(\rho c_p\right)\left(\frac{\partial T}{\partial t}+u\frac{\partial T}{\partial r}+w\frac{\partial T}{\partial z}\right)=-\nabla\cdot q \tag{13.6}$$

Here, q is heat flux and satisfies the following relation [40]:

$$q+\lambda_t\left[\frac{\partial q}{\partial t}+V\cdot\nabla q-q\cdot\nabla V\right]=-k\nabla T \tag{13.7}$$

Where λ_t is thermal relaxation time. And using it into equation (13.6), we get the temperature equation as:

$$\begin{aligned}u\frac{\partial T}{\partial r}+w\frac{\partial T}{\partial z}&=\frac{k}{\rho c_p}\left(\frac{\partial^2 T}{\partial r^2}+\frac{1}{r}\frac{\partial T}{\partial r}+\frac{\partial^2 T}{\partial z^2}\right)\\ &-\lambda_t\left(u^2\frac{\partial^2 T}{\partial r^2}+w^2\frac{\partial^2 T}{\partial z^2}+2uw\frac{\partial^2 T}{\partial r\partial z}+\frac{\partial T}{\partial r}\left(u\frac{\partial u}{\partial r}+w\frac{\partial u}{\partial z}\right)+\frac{\partial T}{\partial z}\left(u\frac{\partial w}{\partial r}+w\frac{\partial w}{\partial z}\right)\right)\end{aligned} \tag{13.8}$$

The components of deformation rate tensor are [41]:

$$\begin{aligned}&e_{rr}=2\frac{\partial u}{\partial r},\ e_{\varphi\varphi}=2\frac{u}{r},\ e_{zz}=2\frac{\partial u}{\partial z},\ e_{r\phi}=e_{\phi r}=r\frac{\partial}{\partial r}\left(\frac{v}{r}\right),\\ &e_{z\phi}=e_{\phi z}=\frac{\partial v}{\partial z},\ e_{zr}=e_{rz}=\frac{\partial u}{\partial z}+\frac{\partial w}{\partial r}\end{aligned} \tag{13.9}$$

The components of stress tensor are [41]:

$$\tau_{rr}=-p+\mu\left(2\frac{\partial u}{\partial r}\right)+\mu_c\left[4\left(\frac{\partial u}{\partial r}\right)^2+\left(\frac{\partial u}{\partial z}+\frac{\partial w}{\partial r}\right)^2+\left(\frac{\partial v}{\partial r}-\frac{v}{r}\right)^2\right]$$

$$\tau_{rz}=\mu\left(\frac{\partial u}{\partial z}+\frac{\partial w}{\partial r}\right)+\mu_c\left[\left(\frac{\partial u}{\partial z}+\frac{\partial w}{\partial r}\right)\left(2\frac{\partial u}{\partial r}\right)+\left(\frac{\partial v}{\partial r}-\frac{v}{r}\right)\left(\frac{\partial u}{\partial z}\right)+\left(\frac{\partial u}{\partial z}+\frac{\partial w}{\partial r}\right)\left(2\frac{\partial w}{\partial z}\right)\right]$$

$$\tau_{\phi\phi}=-p+\mu\left(2\frac{u}{r}\right)+\mu_c\left[4\left(\frac{u}{r}\right)^2+\left(\frac{\partial v}{\partial z}\right)^2+\left(\frac{\partial v}{\partial r}-\frac{v}{r}\right)^2\right]$$

$$\tau_{r\phi}=\mu\left(\frac{\partial v}{\partial r}-\frac{v}{r}\right)+\mu_c\left[\left(\frac{\partial v}{\partial r}-\frac{v}{r}\right)\left(2\frac{\partial u}{\partial r}\right)+\left(\frac{\partial v}{\partial r}-\frac{v}{r}\right)\left(2\frac{u}{r}\right)+\left(\frac{\partial u}{\partial z}+\frac{\partial w}{\partial r}\right)\left(\frac{\partial v}{\partial z}\right)\right]$$

$$\tau_{z\phi} = \mu\frac{\partial v}{\partial z} + \mu_c\left[\left(\frac{\partial u}{\partial z} + \frac{\partial w}{\partial r}\right)\left(\frac{\partial v}{\partial r} - \frac{v}{r}\right) + 2\left(\frac{\partial v}{\partial z}\right)\left(\frac{u}{r}\right) + 2\left(\frac{\partial v}{\partial z}\right)\left(\frac{\partial w}{\partial z}\right)\right]$$

$$\tau_{zz} = -p + \mu\left(\frac{\partial w}{\partial z}\right) + \mu_c\left[4\left(\frac{\partial w}{\partial z}\right)^2 + \left(\frac{\partial v}{\partial z}\right)^2 + \left(\frac{\partial u}{\partial z} + \frac{\partial w}{\partial r}\right)^2\right]$$

The associated boundary conditions are:

$$\begin{aligned} &u = cr,\ v = r\Omega,\ w = 0,\ T = T_w \ \text{ at } z = 0 \\ &u = u_e,\ v = v_e,\ T = T_\infty \ \text{ at } z \to \infty. \end{aligned} \tag{13.10}$$

For the friction-less case, it is found:

$$u_e = ar,\ v_e = 0,\ w_e = -2az,\ p = p_0 - \frac{1}{2}\rho a^2\left(r^2 + z^2\right), \tag{13.11}$$

With a being constant and p_0 being stagnation pressure.

13.3 STABILITY ANALYSIS

Consider the following similarity transformations:

$$\begin{aligned} &u = arF(\eta),\ v = arG(\eta),\ w = 2\sqrt{av}H(\eta),\ p = -\frac{1}{2}\rho a^2\left(r^2 + P(\eta)\right), \\ &\theta(\eta) = \frac{T - T_\infty}{T_w(t) - T_\infty},\ \eta = \sqrt{\frac{a}{v}}z, \end{aligned} \tag{13.12}$$

With given transformations, system reduces to:

$$H' + F = 0 \tag{13.13}$$

$$F'' - F^2 + G^2 - HF' + K\left(F'^2 - G'^2 - 2FF''\right) - M(F - 1) + 1 = 0, \tag{13.14}$$

$$G'' - 2FG - HG' - 2K\left(FG'' - F'G'\right) - MG = 0, \tag{13.15}$$

$$\frac{1}{Pr}\theta'' - 2H\theta' - 4\lambda\left(H^2\theta'' - FH\theta'\right) = 0, \tag{13.16}$$

Along with boundary conditions:

$$\begin{aligned} &F(0) = s,\ G(0) = \omega,\ H(0) = 0,\ \theta(0) = 1 \\ &F(\infty) = 1,\ G(\infty) = \theta(\infty) = 0. \end{aligned} \tag{13.17}$$

Here, the dimensionless constants appearing in the system (13.13–13.17) are Reiner–Rivlin parameter (K), magnetic interaction parameter (M), Prandtl number (Pr), thermal relaxation parameter (λ), stretching parameter (s), rotation parameter (ω) and defined as:

$$k = \frac{\mu_c a}{\mu},\ M = \frac{\sigma B_0^2}{\rho a},\ Pr = \frac{\mu c_p}{k},\ \lambda = \lambda_t a^2,\ s = \frac{c}{a},\ \omega = \frac{\Omega}{a}. \tag{13.18}$$

Now, the quantities of engineeering interests viz radial and tangential skin friction coeffiecients are calculated using Newtonian formulas, given by:

$$C_{fr} = \frac{\tau_{rz}}{\mu a r\sqrt{\frac{a}{\nu}}}|_{z=0} = F'(0),$$
$$C_{f\phi} = \frac{\tau_{r\phi}}{\mu a r\sqrt{\frac{a}{\nu}}}|_{z=0} = G'(0), \tag{13.19}$$

Also, the measure of heat transfer, that is, local Nusselt number, is also calulated using Fourier's law of heat flux as:

$$Nu_r = -\frac{\frac{\partial T}{\partial z}}{r\sqrt{\frac{a}{\nu}}}|_{z=0} = -\theta'(0) \tag{13.20}$$

13.4 NUMERICAL PROCEDURE

The system (13.13–13.17) is tedious to solve analytically and examined numerically using R-K Fehlberg method in Maple software. The boundary value problem must be transformed into a set of initial value problems using the following set of transformations:

$$\begin{gathered} F(\eta) = y_1,\ F'(\eta) = y_2,\ G(\eta) = y_3,\ G'(\eta) = y_4, \\ H(\eta) = y_5,\ \theta(\eta) = y_6,\ \theta'(\eta) = y_7, \end{gathered} \tag{13.21}$$

Using (13.21), the system reduces to the following form as seven first-order equations:

$$y_{1'} = y_2, y_{2'} = \frac{1}{1-2Ky_1}\left[y_1^2 - y_3^2 + 2y_2y_5 - 1 - K\left(y_2^2 - y_4^2\right) + M\left(y_1 - 1\right)\right], y_{3'} = y_4,$$

$$y_{4'} = \frac{1}{1-2Ky_1}\left[2y_1y_3 + 2y_4y_5 - 2Ky_2y_4 + My_3\right],$$

$$y_{5'} = -2y_1,\ y_{6'} = y_7,\ y_{7'} = \frac{1}{1-\lambda Pry_5^2}\left[2y_5y_6 - \lambda y_1y_5y_7\right],$$

With boundary conditions:

$$\begin{gathered} y_1(0) = s,\ y_3(0) = \omega,\ y_5(0) = 0,\ y_6(0) = 1, \\ y_1(\eta_\infty) = 1,\ y_3(\eta_\infty) = 0,\ y_6(\eta_\infty) = 0. \end{gathered}$$

The initial guesses $y_2(0)$, $y_4(0)$, and $y_7(0)$ are calculated using Newton's method with tolerance limit of 10^{-7}. The end point boundary conditions are utilized to find the missing initial values sequentially until the residual error is lower than the tolerance limit, and then the system is solved by using the fifth-order Runge–Kutta–Fehlberg approach.

TABLE 13.1
Comparison of $F'(0)$, $-G'(0)$ for Variation in M and s with Fixed $\omega = 2$ and $K = 0$.

		$F'(0)$		$-G'(0)$	
M	s	Ref. [37]	Current	Ref. [37]	Current
0	0	2.29564228	2.29490511	2.39366194	2.39835711
	1	0.69507028	0.69452267	3.64222140	3.64349463
	2	−1.56932723	−1.56975032	4.61622802	4.61666718
1	0	2.45332513	2.45329149	3.04158525	3.04358585
	1	0.62200395	0.62177681	4.12536843	4.12600613
	2	−1.82583546	−1.82605503	5.01131217	5.01155748
2	0	2.62086284	2.62092811	3.60136233	3.60225047
	1	0.56673037	0.56631143	4.56593521	4.56626216
	2	−2.05876450	−2.05888118	5.38190132	5.38203964

13.5 VALIDATION OF NUMERICAL SCHEME

The applied numerical scheme is validated through the skin friction values for the reduced case when $K = 0$, $\omega = 2$ to compare them with the already-published literature [37] seen in Table 13.1.

The close correlation between the studies reveals the accurateness of the numerical scheme. Further, η_∞ is chosen appropriately to satisfy the far boundary conditions, as suggested by Pantokratoras [42]. Hence, based on the stated facts, the authors are confident that the used numerical scheme is valid and accurate.

13.6 RESULTS AND PHYSICAL INTERPRETATIONS

This section demonstrates the graphical implications of rotation, stretching, magnetic field, non-Newtonian behavior, and heat flux on velocity and temperature fields obtained using fifth-order Runge–Kutta–Fehlberg scheme in the Maple software.

Figure 13.2 shows the influences of non-Newtonian behavior of the fluid with stretching on the velocity and temperature fields. Radial velocity observes an increment with stretching for both whether the fluid is Newtonian ($K = 0$) or non-Newtonian ($K = 0.2$) with fixed rotation. But in the latter case, this increase is less prominent due to the non-Newtonian nature of the fluid. The growth is attributed to the enhancement in centrifugal force due to the stretching of the disk as more fluid particles are moving in radial direction. For no stretching ($s_p = 0$), the flow profile is parabolic type, irrespective of the nature of the fluid. The flow behavior is identical for both the cases when $0 \leq s_p \leq 1$, but with $s_p > 1$. This changes as the decrease in radial velocity with axial distance is more for Reiner–Rivlin (R-R) fluid due to an increase in the viscous nature of the fluid. An increment in the non-Newtonian parameter means more viscous fluid, implying more resistance between fluid layers. The increase in radial velocity is augmented by a decrease in azimuthal and axial velocities. The circumferential flow (G) decreases from a higher value in the case of R-R fluid, as in Figure 13.2(b), with reverse trend in axial case (Figure 13.2(c)). There is a decrease in the thermal boundary layer due to a decrease in thermal penetration depth caused by enhancement in radial flow. However, this decrease is more influential when fluid is of the R-R type. Hence, the stretching parameter cools down the disk system more efficiently when fluid is Newtonian, as in Figure 13.2(d).

The influence of magnetic field strength on radial velocity is shown in Figure 13.3(a), with disk stretching in the radial direction. Magnetic field imposition in the vertical direction gives

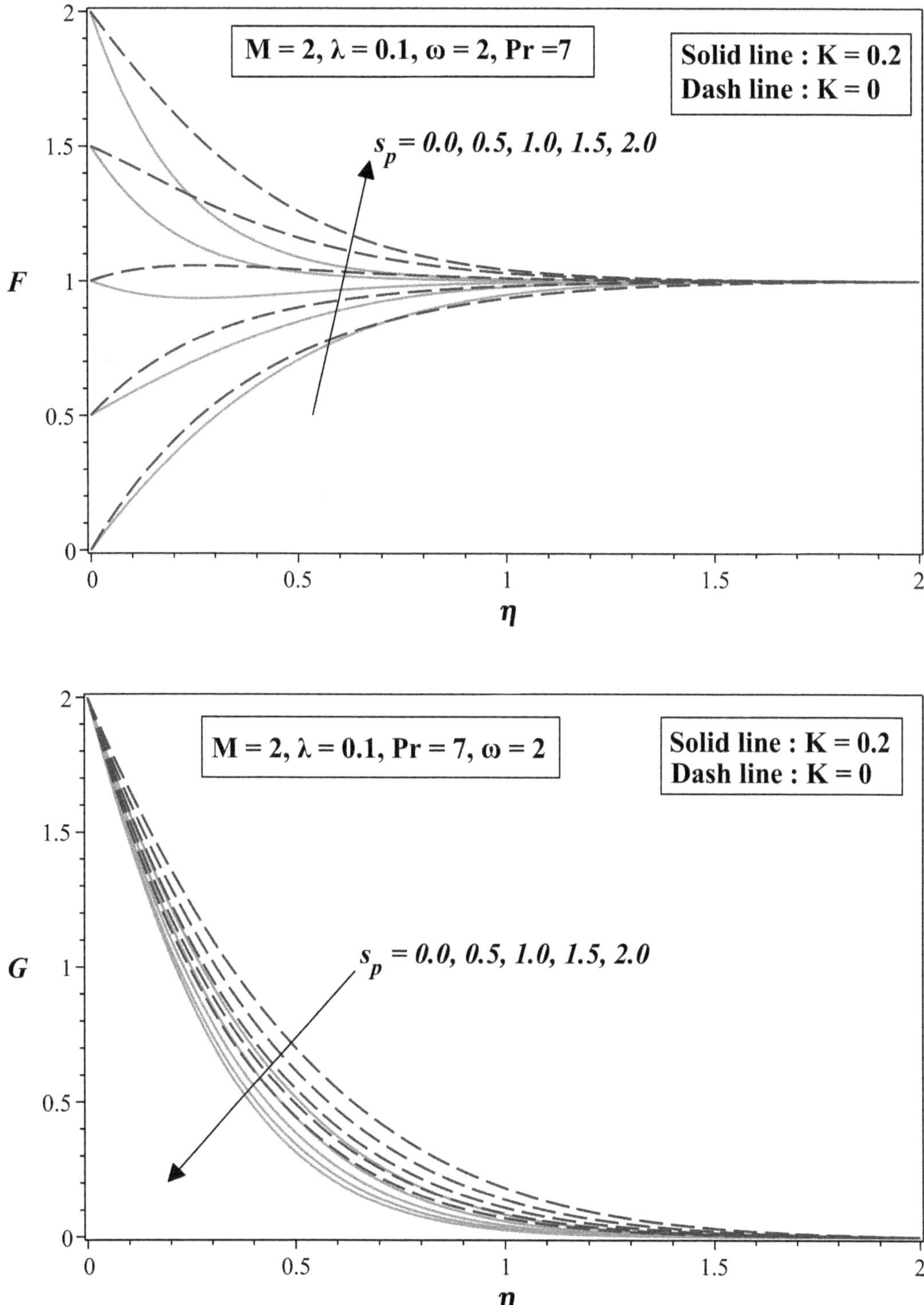

FIGURE 13.2 Impact of K on (a) radial velocity profile, (b) tangential velocity profile, (c) axial velocity profile, and (d) temperature profile.

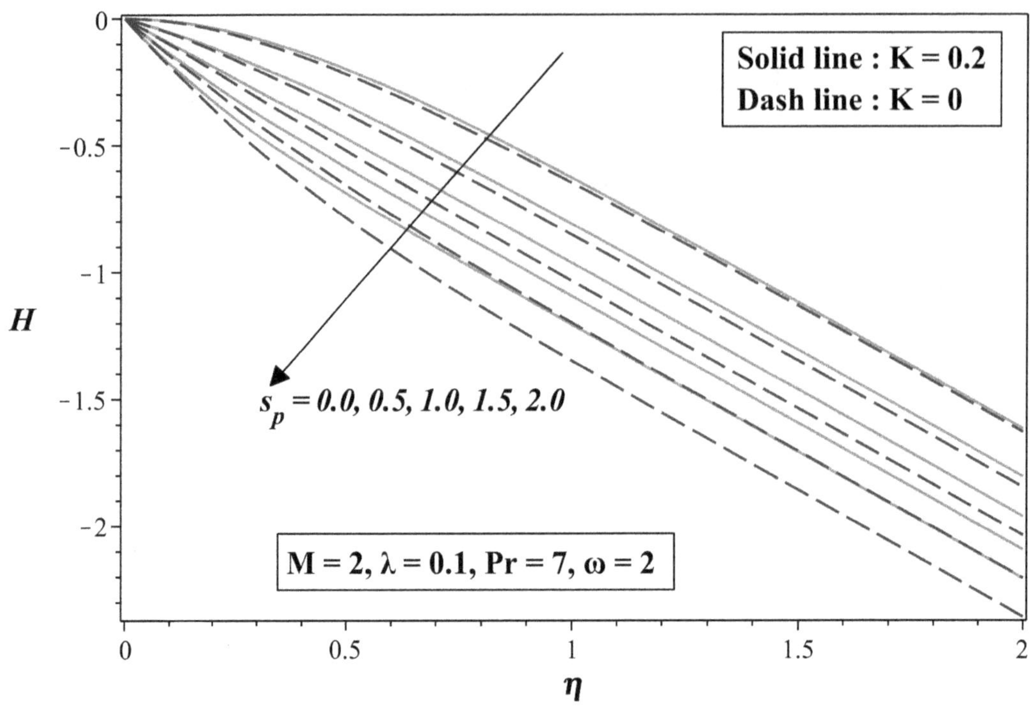

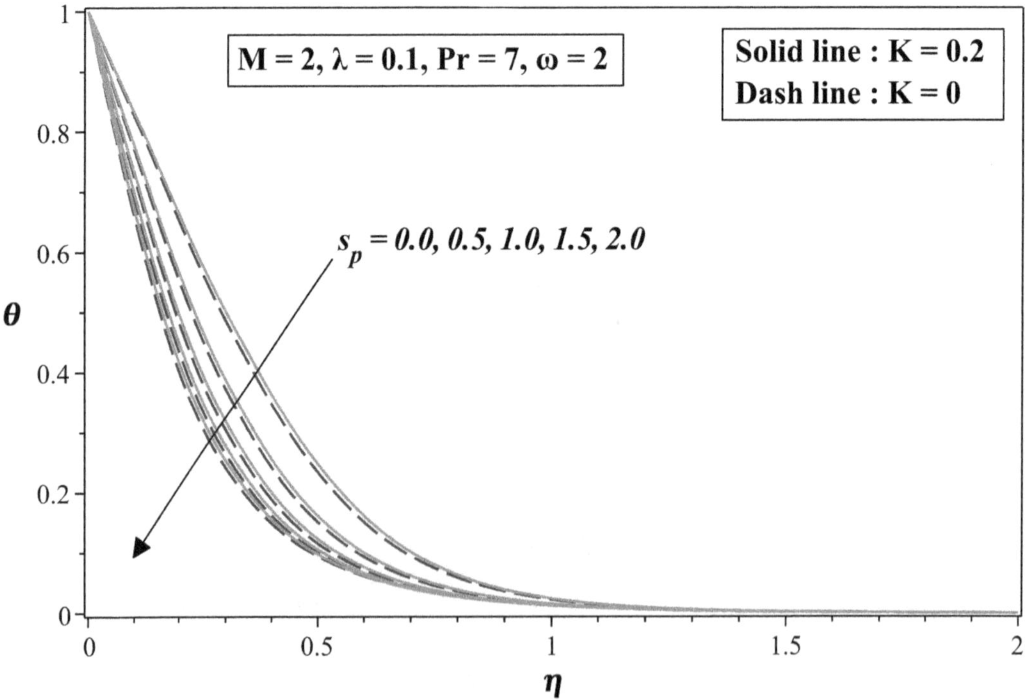

FIGURE 13.2 (Continued)

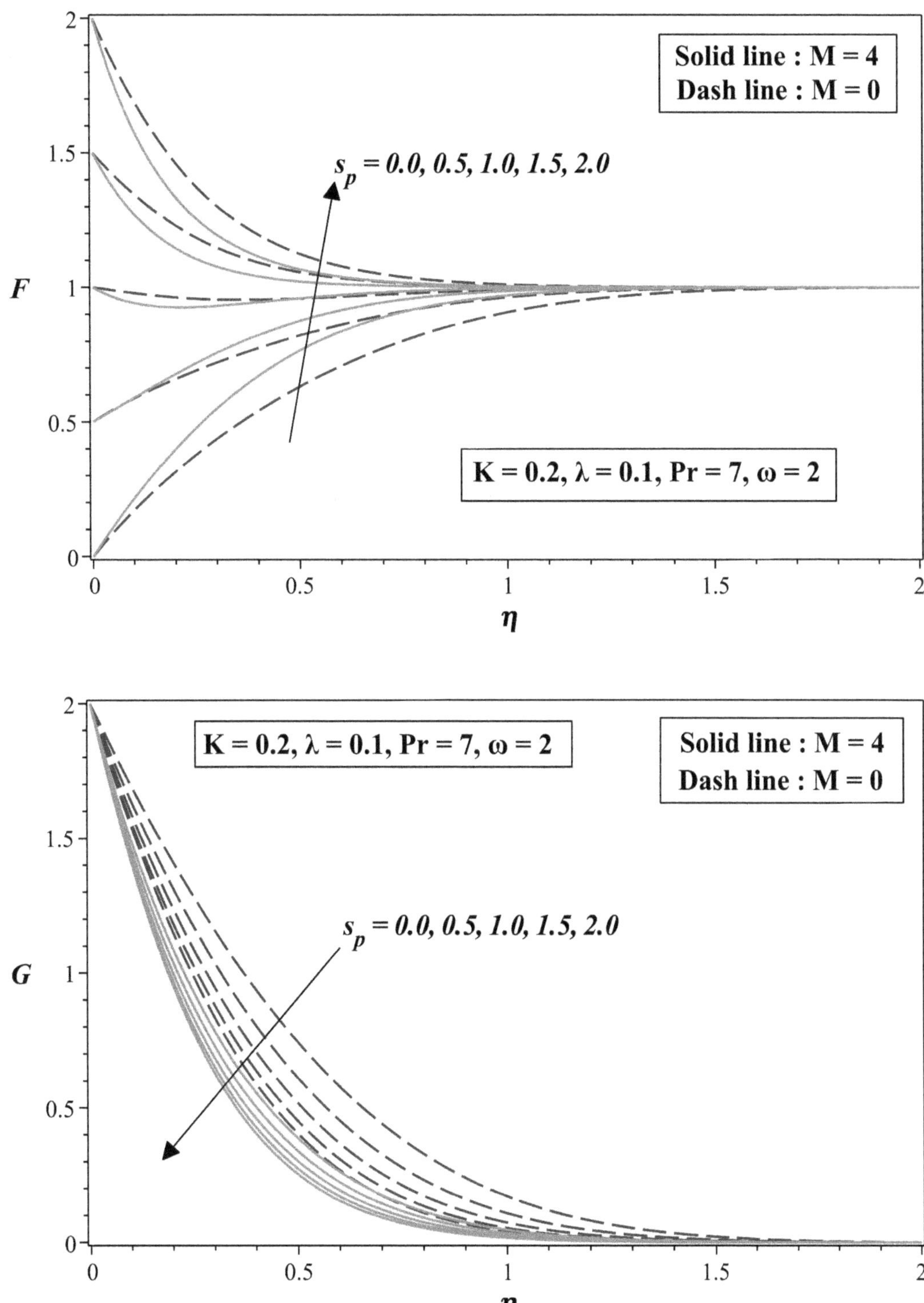

FIGURE 13.3 Impact of M on (a) radial velocity profile, (b) tangential velocity profile, (c) axial velocity profile, and (d) temperature profile.

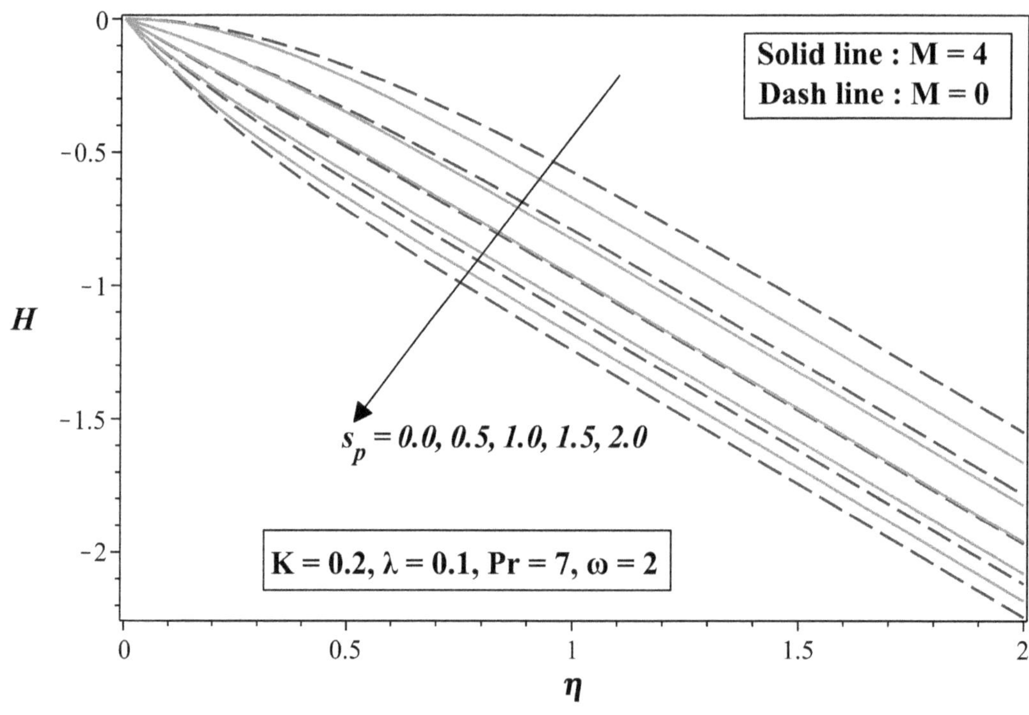

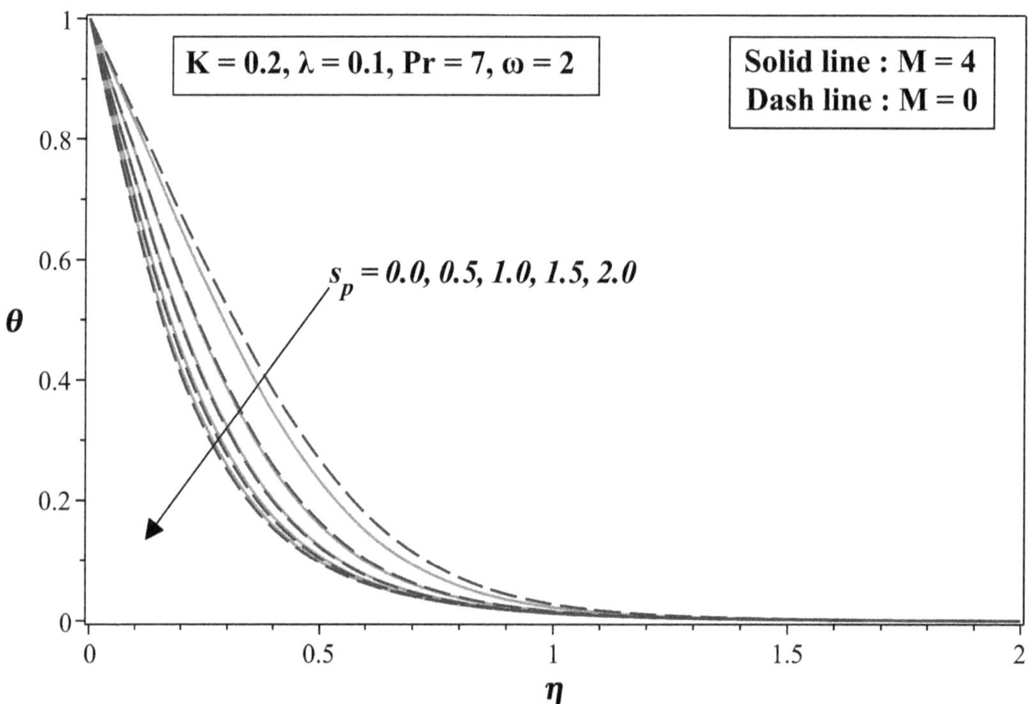

FIGURE 13.3 (Continued)

rise to a drag-like force, called Lorentz force, triggering a slowdown of fluid, as shown by the azimuthal and axial profiles at any stretching rate and for radial case when $s \geq 1$. There is a thickening of the boundary layer when $0 \leq s \leq 1$, meaning an increase in radial velocity, unlike the situation in which the flow should have behaved. This may be because the Lorentz force at low stretching fails to provide enough resistance to fluid particles to slow them down. The fluid temperature decreases with magnetic parameters with fixed stretching and rotation as drag forces slow down the fluid particles with more magnetic field application (Figure 13.3(d)). Moreover, stretching cools down the system more efficiently than no magnetic field with specified magnetic strength.

Figure 13.4 demonstrates the influence of disk rotation on velocity and temperature profiles with stretching in the radial direction. At lower rotational speeds, radial velocity increases with stretching parameters due to the influence of centrifugal force. However, this increment is less influential when disk rotation is higher as rotation and stretching counter each other, decreasing radial velocity. The azimuthal and axial flow velocities increase with disk rotation at higher speed, as both higher rotation and stretching contribute to more fluid particles in the radial direction. As observed earlier, the stretching parameter reduces the axial and azimuthal velocities with fixed rotation. Rotation upgrades the fluid temperature with fixed stretching as higher disk rotation heats the fluid particles, raising the temperature in Figure 13.4(d). Hence, for a system to cool down efficiently, the rotation should be optimal, as stretching with lesser rotation of the disk does the work elegantly.

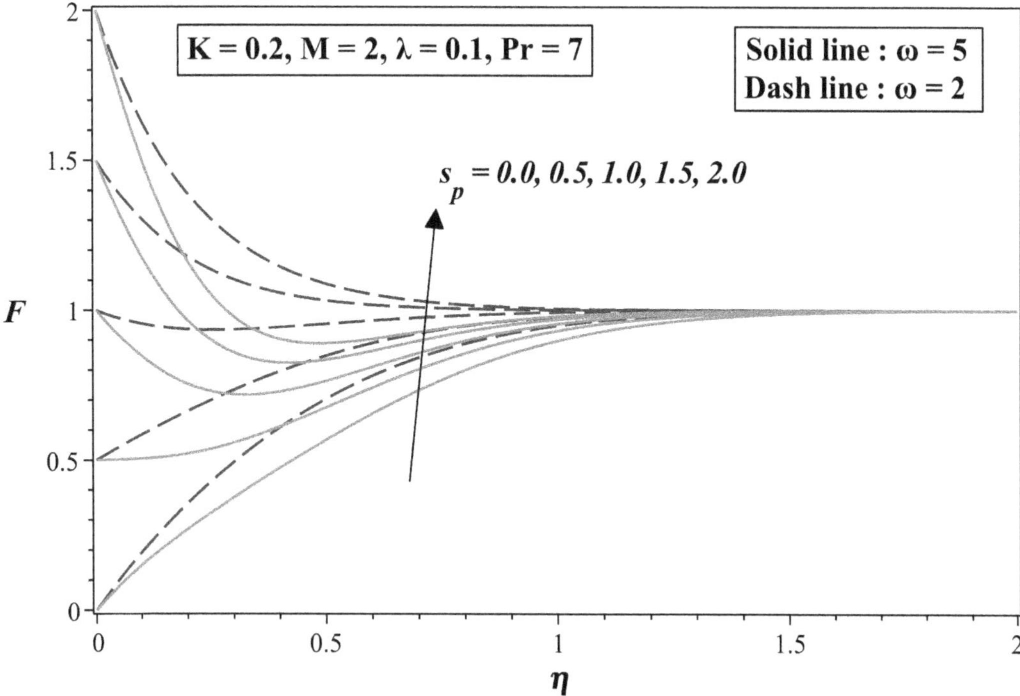

FIGURE 13.4 Impact of ω on (a) radial velocity profile, (b) tangential velocity profile, (c) axial velocity profile, and (d) temperature profile.

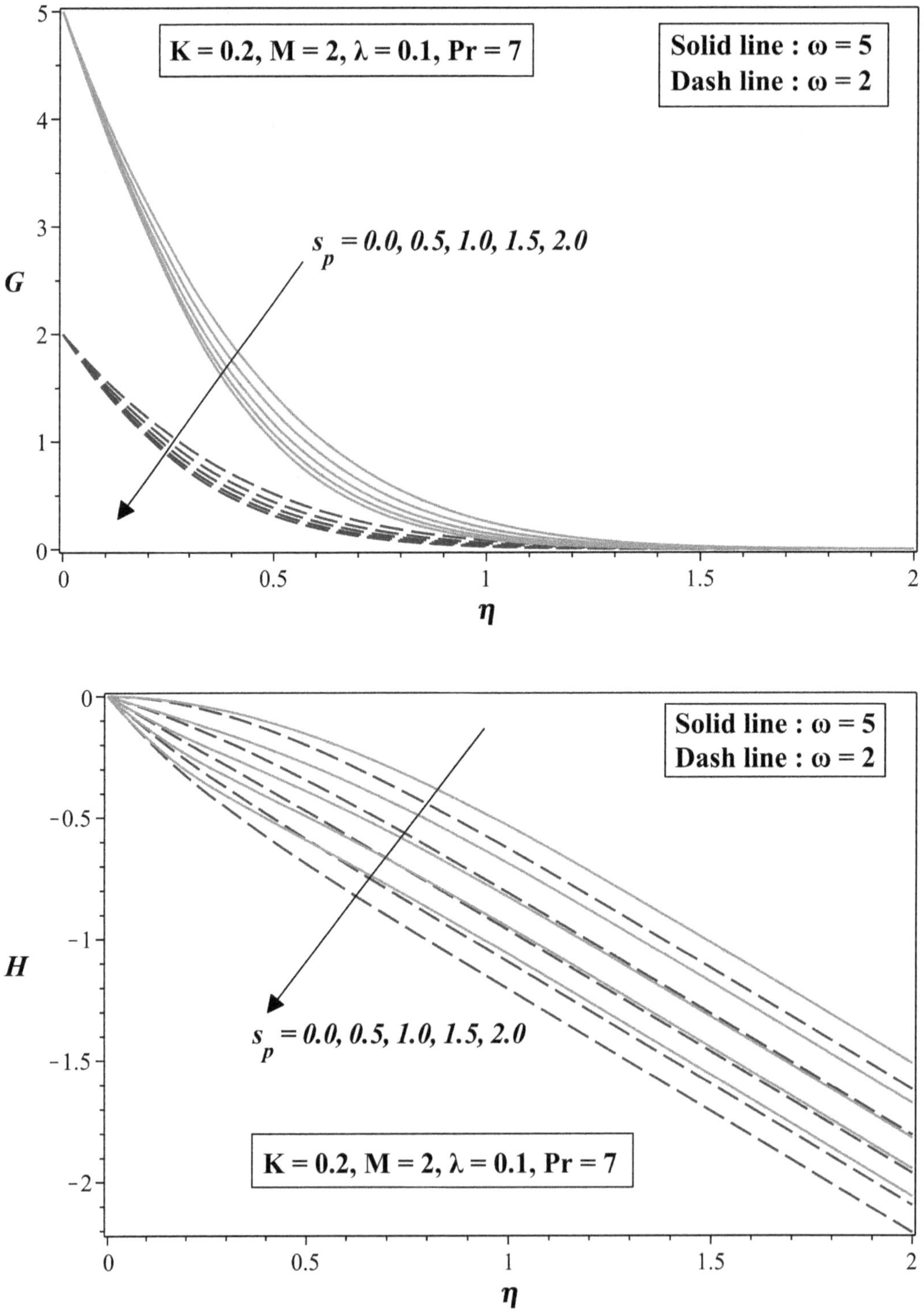

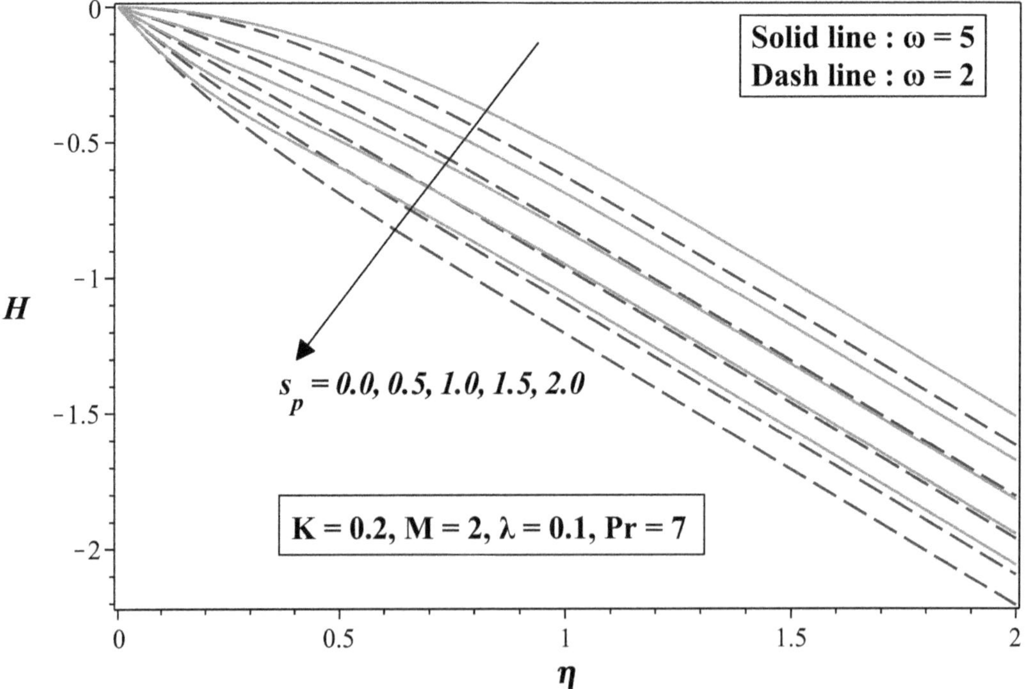

FIGURE 13.4 (Continued)

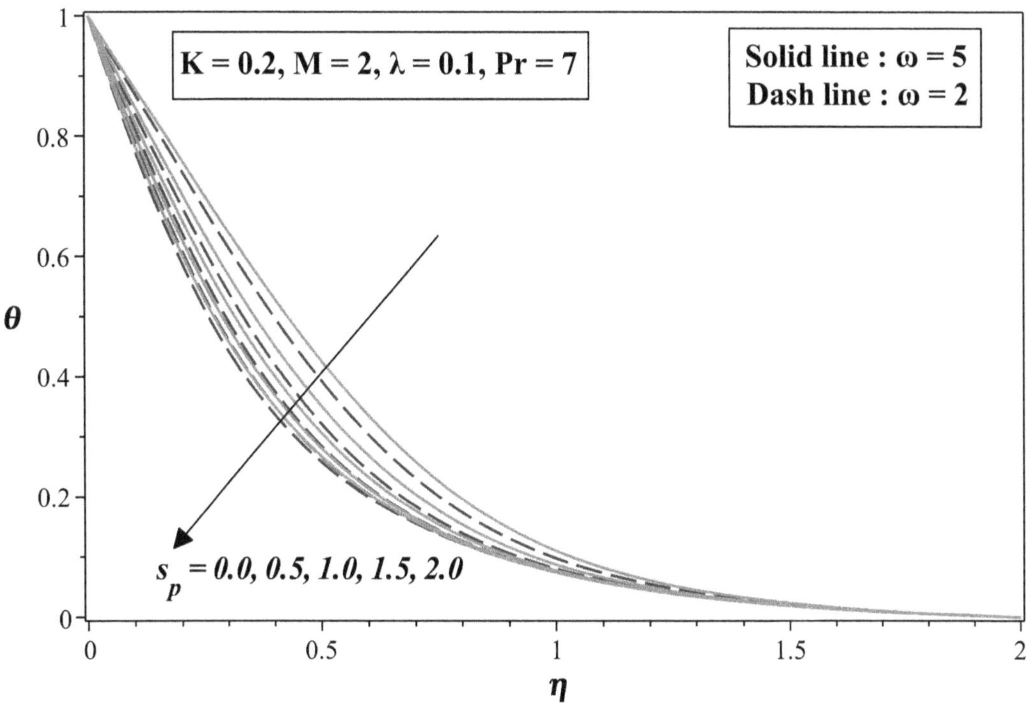

FIGURE 13.4 (Continued)

Figure 13.5(a) shows the influence of the Prandtl number (Pr) on temperature profile. An increment in Pr means a relative reduction in the thermal diffusion, causing a decrease in thermal boundary layer thickness and hence decreasing the temperature, as shown here with fixed stretching. The temperature also drops with disk stretching, but this decrease is smaller in the case of higher Pr. This behavior is more apparent when the stretching rate gets larger. The influence of thermal relaxation parameter (λ) on temperature is shown in Figure 13.5(b). An increment in λ means fluid particles take extra time to pass the heat to another, causing a decrease in the thermal field. When $\lambda = 0$, the C-C model turns into the classical Fourier model, and heat gets transferred immediately, as shown in Figure 13.5(b), where $\lambda = 0$ curves lie below that of $\lambda = 0.2$. It is also observed that the temperature reduction with stretching is less when the fluid particles take extra time to transfer heat.

Table 13.2 elaborates the impact of said parameters on local skin friction coefficients explained through the values of $F'(0)$ and $-G'(0)$. Magnetic parameter enhances the radial skin friction for $s_p < 1$, but the reverse effect is observed for $s_p \geq 1$ for fixed rotation, unlike the case of Newtonian fluid done by Turkyilmazoglu [37], where the same effect was seen at $s_p > 1$, that is, for more stretching rate. Reiner–Rivlin and rotation parameter both reduce the radial skin friction with disk stretching, unlike magnetic field. More torque is required to steady the rotating disk, as all the parameters involved increase the tangential skin friction coefficient. The numerical values for the Nusselt number for the associated parameters are given in Table 13.3. For fixed stretching of the disk, the Nusselt number decreases for all the parameters except the magnetic parameter. Magnetic field enhances the heat transfer rate for $0 \leq s_p < 1$, while the reverse is observed for $s_p \geq 1$. However, the Nusselt number observes an increment with stretching parameter for fixed values of all the other parameters involved.

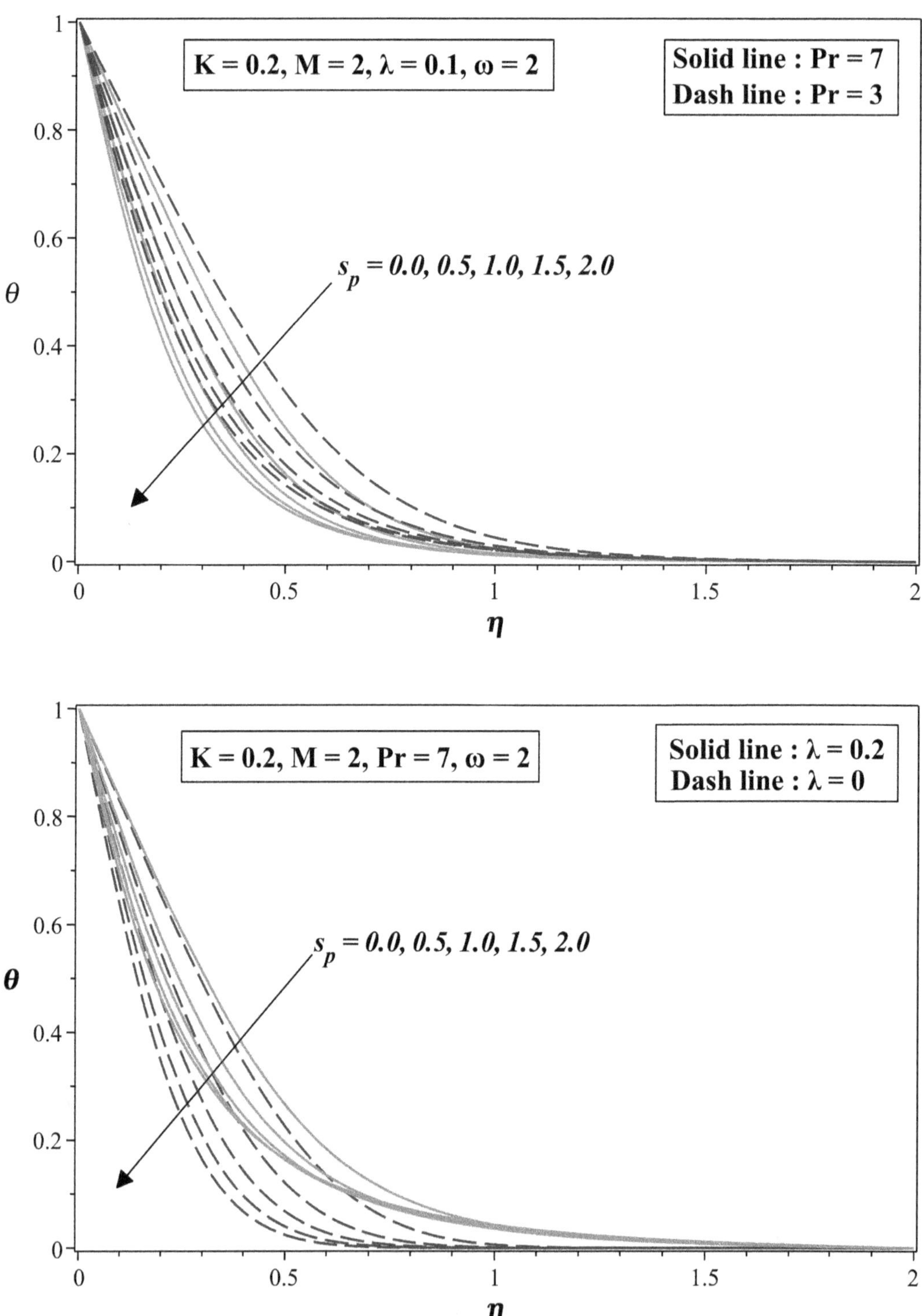

FIGURE 13.5 Impact of (a) Pr and (b) λ on temperature profiles.

TABLE 13.2
Computed Values of Skin Friction Coefficients $F'(0)$ and $-G'(0)$ for Different Parameters

			$F'(0)$			$-G'(0)$		
K	M	ω	$s_p = 2$	$s_p = 1$	$s_p = 2$	$s_p = 0$	$s_p = 1$	$s_p = 2$
0	2	2	2.620928	0.566631	−2.058881	3.602250	4.566262	5.382039
0.1			2.441327	0.167841	−2.908376	4.093140	5.071417	5.874935
0.2			2.099280	−0.595880	−4.549496	4.629649	5.523619	5.759278
0.2	0	2	1.885701	−0.215432	−3.474520	3.199442	4.390774	4.891717
	2		2.099280	−0.595880	−4.549496	4.629649	5.523619	5.759278
	4		2.331146	−0.882241	−5.435386	5.751341	6.476095	6.519151
0.2	2	2	2.099280	−0.595880	−4.549496	4.629649	5.523619	5.759278
		4	1.778361	−1.521345	−5.508173	8.744955	9.919405	9.918667
		6	1.836906	−1.812142	−5.515317	12.566344	13.652784	13.797783

TABLE 13.3
Computed Values of Nusselt Number $-\theta'(0)$ for Various Parameters Keeping Fixed $Pr = 7$

				$-\theta'(0)$		
K	M	ω	λ	$s_p = 0$	$s_p = 1$	$s_p = 2$
0	2	2	0.1	1.344363	1.985576	2.434437
0.1				1.333021	1.960726	2.398069
0.2				1.306796	1.919486	2.336680
0.2	0	2	0.1	1.259184	1.932737	2.368771
	2			1.306796	1.919486	2.336680
	4			1.347118	1.912783	2.312878
0.2	2	0	0.1	1.336592	1.954679	2.370402
		2		1.306796	1.919486	2.336680
		4		1.238844	1.810530	2.229248
0.2	2	2	0	1.353675	2.064606	2.622629
			0.1	1.306796	1.919486	2.336680
			0.2	1.298693	1.879382	2.261441

13.7 CONCLUSIONS

A study on MHD Heimenz flow of Reiner–Rivlin fluid over rotating disk stretching in the radial direction is investigated. Some of the significant observations are listed as:

- Stretching parameter increases the radial velocity irrespective of the nature of the fluid, but in the case of Reiner–Rivlin fluid, this increase is less prominent.
- The temperature degrades more in the Reiner–Rivlin fluid than in a Newtonian fluid with stretching parameter.

- Magnetic parameter at low stretching values increases the radial velocity, unlike the situation in which flow should have behaved due to ineffectiveness of the Lorentz force to slow down particles for $0 \le s_p \le 1$.
- Magnetic field with increased stretching cools down the system more efficiently, indicating the importance of magnetic field imposition in the vertical direction.
- Higher disk rotation increases the radial velocity with the disk stretching, but this enhancement is less positive than lower rotation.
- Optimal disk rotation and stretching are the critical factors for cooling the system effectively.
- Heat transfer rate enhances with the magnetic parameter for low stretching values, while a reverse trend is observed for higher values.

In the future, the current chapter can be extended in the case of other non-Newtonian fluids, like Williamson fluid, Maxwell fluid, Ree–Eyring, etc. Further, the various physical effects, like buoyancy, chemical reactions, Soret–Dufour effects, and variable thermal conductivity, can be included with the previously mentioned non-Newtonian models to strengthen the literature. Note that there is always a scope for performing experiments to validate the results.

Nomenclature

u	radial velocity
v	tangential velocity
w	axial velocity
T	temperature
q	heat flux
λ_t	thermal relaxation time
k	thermal conductivity
c	stretching rate
Ω	angular velocity
T_w	wall temperature
T_∞	ambient temperature
F	dimensionless radial velocity
G	dimensionless tangential velocity
H	dimensionless axial velocity
K	Reiner–Rivlin parameter
θ	dimensionless temperature
M	magnetic parameter
Pr	Prandtl number
λ	thermal relaxation parameter
s	stretching parameter
ω	dimensionless angular velocity

REFERENCES

[1] Hiemenz, K. (1911) Die grenzschicht an einem in den gleichformigen flussigkeitsstrom eingetauchten geraden kreiszylinder. Dinglers Polytech. J., 326, 321–324.

[2] Homann, F. (1936) Der einfluss grosser zähigkeit bei der strömung um den zylinder und um die kugel. ZAMM-Journal of Applied Mathematics and Mechanics/Zeitschrift für Angewandte Mathematik und Mechanik, 16, 153–164.

[3] Djebali, R., Oudina, F. M. and Rajashekhar, C. (2021) Similarity solution analysis of dynamic and thermal boundary layers: further formulation along a vertical flat plate. Physica Scripta, 96, 085206.

[4] Pushpa, B. V., Sankar, M. and Oudina, F. M. (2021) Buoyant convective flow and heat dissipation of cu–h2o nanoliquids in an annulus through a thin baffle. Journal of Nanofluids, 10, 292–304.
[5] Asogwa, K. K., Oudina, F. M. and Animasaun, I. L. (2022) Comparative investigation of water-based al2o3 nanoparticles through water-based cuo nanoparticles over an exponentially accelerated radiative riga plate surface via heat transport. Arabian Journal for Science and Engineering, 47, 8721–8738.
[6] Oudina, F. M. (2017) Numerical modeling of the hydrodynamic stability in vertical annulus with heat source of different lengths. Engineering science and technology, an international journal, 20, 1324–1333.
[7] Vaidya, H., Prasad, K. V., Tlili, I., Makinde, O. D., Rajashekhar, C., Khan, S. U., Kumar, R. and Mahendra, D. L. (2021) Mixed convective nanofluid flow over a non linearly stretched riga plate. Case Studies in Thermal Engineering, 24, 100828.
[8] Rajashekhar, C., Oudina, F. M., Vaidya, H., Prasad, K. V., Manjunatha, G. and Balachandra, H. (2021) Mass and heat transport impact on the peristaltic flow of a ree–eyring liquid through variable properties for hemodynamic flow. Heat Transfer, 50, 5106–5122.
[9] Swain, K., Oudina, F. M. and Abo-Dahab, S. M. (2022) Influence of mwcnt/fe3o4 hybrid nanoparticles on an exponentially porous shrinking sheet with chemical reaction and slip boundary conditions. Journal of Thermal Analysis and Calorimetry, 147, 1561–1570.
[10] Kármán, V. (1921) Uber laminare und turbulente reibung. Z. Angew. Math. Mech., 1, 233–252.
[11] Stuart, J. T. (1954) On the effects of uniform suction on the steady flow due to a rotating disk. The Quarterly Journal of Mechanics and Applied Mathematics, 7, 446–457.
[12] Benton, E. R. (1966) On the flow due to a rotating disk. Journal of Fluid Mechanics, 24, 781–800.
[13] El-Mistikawy, T. M. A. (1990) The rotating disk flow in the presence of strong magnetic field. Proc. 3rd Int. Conger. of Fluid Mechanics, 3, 1211–1222.
[14] El-Mistikawy, T. M. A. (1991) The rotating disk flow in the presence of weak magnetic field. Proc. 4th Conf. on Theoretical and Applied Mechanics, 69–82.
[15] Attia, H. A. (1998) Unsteady mhd flow near a rotating porous disk with uniform suction or injection. Fluid Dynamics Research, 23, 283.
[16] Miklavčič, M. and Wang, C. Y. (2004) The flow due to a rough rotating disk. Zeitschrift für angewandte Mathematik und Physik ZAMP, 55, 235–246.
[17] Ram, P., Sharma, K. and Bhandari, A. (2011) Effect of porosity on revolving ferrofluid flow with rotating disk. International Journal of Fluids Engineering, 3, 261–271.
[18] Sharma, K., Vijay, N., Makinde, O. D., Bhardwaj, S.B., Singh, R. M. and Mabood, F. (2021) Boundary layer flow with forced convective heat transfer and viscous dissipation past a porous rotating disk. Chaos, Solitons & Fractals, 148, 111055.
[19] Sharma, K., Vijay, N., Kumar, S. and Makinde, O. D. (2021) Hydromagnetic boundary layer flow with heat transfer past a rotating disc embedded in a porous medium. Heat Transfer Asian Research, 50, 4342–4353.
[20] Sharma, K. (2021) Rheological effects on boundary layer flow of ferrofluid with forced convective heat transfer over an infinite rotating disk. Pramana, 95, 1–9.
[21] Sharma, K. (2022) Fhd flow and heat transfer over a porous rotating disk accounting coriolis force along with viscous dissipation and thermal radiation. Heat Transfer, 51, 4377–4392.
[22] Tadmor, Z. and Klein, I. (1970) Engineering principles of plasticating extrusion. Van Nostrand Reinhold Company New York.
[23] Fisher, E. G. (1976) Extrusion of plastics. Wiley, New York.
[24] Fang, T. (2007) Flow over a stretchable disk. Physics of fluids, 19, 128105.
[25] Turkyilmazoglu, M. (2012) Mhd fluid flow and heat transfer due to a stretching rotating disk. International journal of thermal sciences, 51, 195–201.
[26] Bhatti, M. M, and Rashidi, M. M. (2016) Effects of thermo-diffusion and thermal radiation on williamson nanofluid over a porous shrinking/stretching sheet. Journal of Molecular Liquids, 221, 567–573.
[27] Hayat, T., Qayyum, S., Imtiaz, M. and Alsaedi, A. (2017) Radiative flow due to stretchable rotating disk with variable thickness. Results in Physics, 7, 156–165.
[28] Mandal, S. and Shit, G.C. (2021) Entropy analysis on unsteady mhd biviscosity nanofluid flow with convective heat transfer in a permeable radiative stretchable rotating disk. Chinese Journal of Physics, 74, 239–255.

[29] Raza, J., Oudina, F. M., Ram, P. and Sharma, S. (2020) Mhd flow of non-newtonian molybdenum disulfide nanofluid in a converging/diverging channel with rosseland radiation. Defect and Diffusion Forum, 401, 92–106.
[30] Dadheech, P. K., Agrawal, P, Oudina, F. M., Hamdeh, N. H. A. and Sharma, A. (2020) Comparative heat transfer analysis of mos2/c2h6o2 and sio2-mos2/c2h6o2 nanofluids with natural convection and inclined magnetic field. Journal of Nanofluids, 9, 161–167.
[31] Marzougui, S., Oudina, F. M., Magherbi, M. and Mchirgui, A. (2021) Entropy generation and heat transport of cu–water nanoliquid in porous lid-driven cavity through magnetic field. International Journal of Numerical Methods for Heat & Fluid Flow, 32, 2047–2069.
[32] Warke, A. S., Ramesh, K., Oudina, F. M. and Abidi, A. (2021) Numerical investigation of the stagnation point flow of radiative magnetomicropolar liquid past a heated porous stretching sheet. Journal of Thermal Analysis and Calorimetry, 147, 6901–6912.
[33] Swain, K., Mahanthesh, B. and Oudina, F. M. (2021) Heat transport and stagnation-point flow of magnetized nanoliquid with variable thermal conductivity, brownian moment, and thermophoresis aspects. Heat Transfer, 50, 754–767.
[34] Rajashekhar, C., Manjunatha, G., Oudina, F. M., Vaidya, H., Prasad, K.V., Vajravelu, K. and Wakif, A. (2021) Magnetohydrodynamic peristaltic flow of bingham fluid in a channel: An application to blood flow. Journal of Mechanical Engineering and Sciences, 15, 8082–8094.
[35] Sharma, K., Kumar, S. and Vijay, N. (2021) Numerical simulation of mhd heat and mass transfer past a moving rotating disk with viscous dissipation and ohmic heating. Multidiscipline Modeling in Materials and Structures, 18, 153–165.
[36] Chabani, I., Oudina, F. M and Ismail, A. A. I. (2022) Mhd flow of a hybrid nano-fluid in a triangular enclosure with zigzags and an elliptic obstacle. Micromachines, 13, 224
[37] Turkyilmazoglu, M. (2012) Three dimensional mhd stagnation flow due to a stretchable rotating disk. International Journal of Heat and Mass Transfer, 55, 6959–6965.
[38] Reiner, M. (1945) A mathematical theory of dilatancy. American Journal of Mathematics, 67, 350–362.
[39] Rivlin, R. S. (1947) Hydrodynamics of non-newtonian fluids. Nature, 160, 611–611.
[40] Hafeez, A., Khan, M. and Ahmed, J. (2020) Flow of oldroyd-b fluid over a rotating disk with cattaneo–christov theory for heat and mass fluxes. Computer methods and programs in biomedicine, 191, 105374.
[41] Tabassum, M. and Mustafa, M. (2018) A numerical treatment for partial slip flow and heat transfer of non-newtonian reiner-rivlin fluid due to rotating disk. International Journal of Heat and Mass Transfer, 123, 979–987.
[42] Pantokratoras, A. (2009) A common error made in investigation of boundary layer flows. Applied Mathematical Modelling, 33, 413–422.

14 Temperature Variations Effect on Bioethanol Production from Lignocellulosic Biomass Using Locally Developed Small-Scale Production Facility: A Pilot Study

Adewale Allen Sokan-Adeaga, Godson R. E. E. Ana, and Abel Olajide Olorunnisola

14.1 INTRODUCTION

The perennial dependency on non-renewable energy source has led to widespread environmental deterioration, such as climate change, rain acidification, photochemical smog formation, loss of biodiversity, air, and ecosystem pollution [1]. It is forecasted that by 2025, the global energy demand will rise to about 50%, with a lion's share of this increment emanating from the third world countries [2–4]. Currently, the globe is confronted with the twin problems of crude oil dwindlement and environmental depravity due to non-judicious extraction and usage of crude oil, which has led to a significant reduction in the underground carbon sources [5]. According to Agarwal [6], the global fossil fuels reserves are obviously limited, with the reserves estimated at 218 years for coal, 41 years for oil, and 63 years for natural gas under a business-as-usual situation. Taking into cognizance the teeming global population, rising energy demand per capital, and acute need to mitigate greenhouse gases emission, the exigent need for a sustainable, eco-friendly, and alternative renewable energy supply becomes inevitable. One of such fuel sources has been found to be bioenergy [7–9].

Bioenergy is biomass-derived energy; this includes decomposable component of agricultural, forestry, and agro-industrial wastes and also decomposable portion of municipal and industrial residue [10]. Numerous biomass sources, such as agro-wastes of organics, are a precursor for various bioenergy types [11, 12]. Examples of well-known existing alternative fuels include natural gas, methane, methanol, hydrogen, biodiesel, bioethanol, etc. However, ethanol has remarkable characteristics which distinguish it as the most promising alternative automobile fuel. Some of the distinguishable properties of ethanol include high octane number and rating, declined Reid vapor pressure, and comprises 35% oxygen, which enhances its total combustion and hence makes it emit lower amount of noxious gases [13, 5, 14]. According to Lakhfif et al. [15], the droplet size and heating property of a fuel are considered to be one of the major factors in determining its combustion performance. Ethanol has a high American Petroleum Institute (API) gravity (>31.1), which affirmed it is a light fuel with lower viscosity and carbon residue content and high heat of combustion. Thus, it performed remarkably well in an internal compression engine than gasoline [16]. Ethanol can also be utilized as automobile fuel either wholly or partly as gasohol (mixed with gasoline) [17] and thus operates as a two-phase model which enhances energy transfer and mass transfer, as reported by Gourari et al. [18].

DOI: 10.1201/9781003299608-14

Biofuel production has been a vanguard program of numerous nations across the globe. This global effort has been termed as "bioeconomy revolution" by Keeney and Muller [19]. Presently, bioethanol is produced on an industrial scale from eatable feedstocks, such as sugarcane juice and cornstarch. The United States (US) is the major producer of bioethanol worldwide, with a production target of 132.6 billion liters of ethanol in 2017 from domestically cultivated maize. Two decades ago, Brazil championed as the leading producer and consumer of biofuels, until it relinquished that position to the United States. In 2008, the production capacity of Brazil was estimated at around 21 billion liters of ethanol per year, of which over 90% is consumed locally [20]. Other emerging countries that are driving the bioethanol industry are Japan and China, with China consuming about 6.3 billion liters of ethanol in 2012, which is over 150% increment vis-à-vis 2006 [21]. Base on this breakthrough recorded by these nations, it became crucial for African and other poor-economy country to toll the same path. Nevertheless, this current production system of producing bioethanol from edible feedstocks poses a major challenge, as it negatively affects food security. To avert this dilemma, bioethanol production from non-edible feedstock, such as lignocellulosic biomass, is attracting desirous perchance [5, 22].

Lignocellulose is an organic polysaccharide complex comprising primarily of three components in the following percentages based on their dry matter: 40–60% cellulose, 20–40% hemicellulose, and 10–25% lignin [23]. Lignocellulose biomass is readily available, plenteous, and limitless, and it serves as prospective resources for the production of alternative economical fuel [24, 22]. Various sectors are engaged in lignocellulosic wastes generation yearly, thus providing viable feedstocks for biofuel. Examples of these industries are forestry, agriculture, municipal solid waste (MSW) disposal facilities, and paper manufacturing [22]. In the last few decades, the extraction of fermentable sugars from lignocellulose is considered a giant stride in bioethanol production process [25]. This conversion process involves two steps, the first stage being lignocellulose hydrolysis to produce fermentable sugars, and the second involves sugar fermentation to utilizable substances [26]. Nevertheless, the commercialization of lignocellulosic-derived bioethanol requires overcoming many economical and technical bottlenecks. In addition, for effective utilization of lignocellulosic biomass, due to its obstinate property, an appropriate pretreatment is essential. According to Wingrel et al. [27], the overall ethanol production cost is greatly influenced by the starting material price; hence, a cheaper raw material, coupled with efficient enzymes and high ethanol yield, will significantly diminish the cost of production.

Based on existing literature, there is sparsity of data on optimization of production parameters of bioethanol from lignocellulosic wastes. Previous researches either limit their studies to laboratory optimization only or used technologies with stereotypical specified conditions without optimization. In view of the aforementioned shortcomings, the authors aim to enrich the existing database by taking a step further to address these limitations, by fabricating a prototype adaptive machine that would optimize the different production process parameters (fermentation temperatures and timings) affecting bioethanol yield from typical lignocellulosic wastes, such as cassava peels. This chapter is organized as follows: in Section 14.2, we discuss the step-by-step methodology employed in the production of bioethanol from cassava peel (CP) using locally developed small-scale production facility (SSPF). We begin with the study design and other baseline information, followed by the fabrication of the SSPF, how it was used to optimize bioethanol production from the biomass (cassava peel), energy consumption of the SSPF, and cost–benefit analysis framework. In Section 14.3, we present the results and discussions of the physicochemical characterization of the cassava peels (CP) done in the lab, the hydrolysis and fermentation efficiency of the SSPF, the bioethanol concentrations and yields, the utilization and energy consumption of the optimal bioethanol produced, as well as the comparative cost–benefit analysis of the SSPF using frequency tables, charts, ANOVA tables, and correlation coefficient. In Section 14.4, the important conclusions are presented from the pilot study.

14.2 MATERIALS AND METHODS

In this section, we shall discuss the step-by-step methodology employed in the production of bioethanol from cassava peel (CP) using locally developed small-scale production facility (SSPF). We shall begin with the study design and other baseline information, followed by the fabrication of the SSPF and how it was used to optimize bioethanol production from the biomass (cassava peel).

14.2.1 Baseline Information

The study was exploratory and engineering-based. This technology involved the construction, optimization, and evaluation of bioethanol production from an adaptive small-scale bioethanol plant using separate hydrolysis and co-fermentation (SHCF) technique. We utilized cassava peels (CP) as a source of biomass for bioethanol production. Our choice of cassava peels was due to its high generation in Nigeria both from domestic and industrial activities. Cassava peel was obtained from the cassava processing center at the International Institute of Tropical Agriculture (IITA), Ibadan, Nigeria. It occupies a landmass of about 1,000 acres. The CP was obtained from the variety of *Manihot esculenta* called TMS I980581 grown on IITA cassava plantation. This breed has the following feature traits: high dry matter (25%), high CMD resistance, high yielding (>25 t/ha), ability to stay green, and drought tolerance. It has a maturity age of 6–8 months. The institute has a cassava plantation which covers about one (1) hectare of land; beside it is a cassava processing unit with a modernized processing machine for making garri. A large quantity of cassava peels is generated from this processing unit during garri production while the wastewater is subjected to anaerobic digestion in a sedimentation tank to produce biogas.

We characterize the CP for physicochemical properties, such as:

- pH of the fresh biomass
- Percentage moisture content (%) for the fresh biomass using the laboratory reference method
- Percentage moisture content (%) for the dried biomass using the laboratory reference method
- Percentage dry matter content (%) for the dried biomass laboratory reference method
- Total organic carbon (TOC) (%) by using the Walkley–Black wet oxidation method
- Total nitrogen (%) using routine semi-micro Kjeldahl technique (involves digestion, distillation, and titration)
- Total phosphorus (%) using the vanado-molybdate colorimeter method
- Carbon-to-nitrogen ratio (C:N) calculated by diving the TOC by TN

14.2.2 Engineering Section (Small-Scale Production Facility)

We ensure quality assurance in the production of the small-scale bioethanol plants by fabricating the machines according to the specifications indicated in the engineering drawing/design. Also, each production process was carried out with a high level of precision without hampering the electronic gadgets and mechanical functionality of the various components of the machines. The turnkey/system consists of the following equipment/machines: mixing machine/hydrolysis chamber, an improvised sieving apparatus, an improvise neutralization chamber, a filter press, an incubator, a distiller, and a cooling tank.

The design process involved a well sketch engine drawing and diagrammatic representation of the following:

- Mixing machine/hydrolysis chamber (0.104m^3 inner and 0.224m^3 outer chamber) (see **Appendix for Figure 14.A1**)
- Incubator (0.77m^3) (see **Appendix for Figure 14.A2**)
- Distillation machine (distiller and cooling tank) (45L/0.045m^3) (see **Appendix for Figure 14.A3**)

The engineering drawings show all the component parts and material specifications. This serves as a guide in the assembly and fabrication process of the various machines. The various component parts, the quantity of materials used, and the specifications of each of the machines are shown in the **Appendix**.

- **Power Rating of the Bioethanol Plant**

The power rating was calculated from the power output of the electric motor, electric fan, heating element, and digital temperature gauge, respectively.

Electric motor (2 hp) = 1,500 W
Electric fan = 10 W
Heating element = 2,500 W
Digital temperature guage (×2) = 10 W (×2) = 20 W
Therefore, the power rating is = (1,500 + 10 + 2500 + 20) W = **4,030 W**

- **Fabrication of the Bioethanol Plant**

Procedure. The assembly of the component parts and the fabrication process were carried out at the mechanical and electrical workshop units in the maintenance department of the University of Ibadan. This was done with utmost attention so as to attain the plant's effectiveness level and improved performance. The various mechanical components and elements were welded together to build the digester (hydrolysis machine), incubator, distiller, and cooling tank. After the mechanical construction, electrical gadget and wiring were fitted to each of the machines by an electrician to test and ascertain the functionality of the plant.

Safety measures. We employed the following safety measures during the bioethanol plant construction to protect against body injuries and damages. These include:

1. Welding shield was worn to protect the eyes from burns caused by radiant energy from oxyacetylene flame use in welding. This protective equipment also guides the eyes and face from flying sparks and metal spatter produced during welding and soldering operations.
2. A hard hat or helmet was worn to shield the head against mechanical injury from falling objects and possible contact with electrical hazards.
3. Safety shoes were put on to shield the legs and feet against heat hazards, like molten metal or welding sparks, and also from workplace electrical hazards.
4. Hand gloves were worn to protect against hazards.
5. Coveralls were worn for protection against hazardous substances.

14.2.3 Optimization and Production Process of the Small-Scale Production Facility (SSPF)

After the fabrication of the SSPF, we employed the separate hydrolysis and co-fermentation (SHCF) technique in the optimization of bioethanol production from cassava peels (CP) as developed by Farone and Cuzens [28, 29] and modified by Ana and Sokan-Adeaga [30], Sokan-Adeaga et al. [5], and Bolade et al. [31]. This acid-based technology is a generic process that consists of seven (7) basic steps and is illustrated by the following with the aid of a flowchart (Figure 14.a), as depicted below.

- Biomass pretreatment.
- Optimization layout.
- Chemical hydrolysis of lignocellulose to produce sugars.

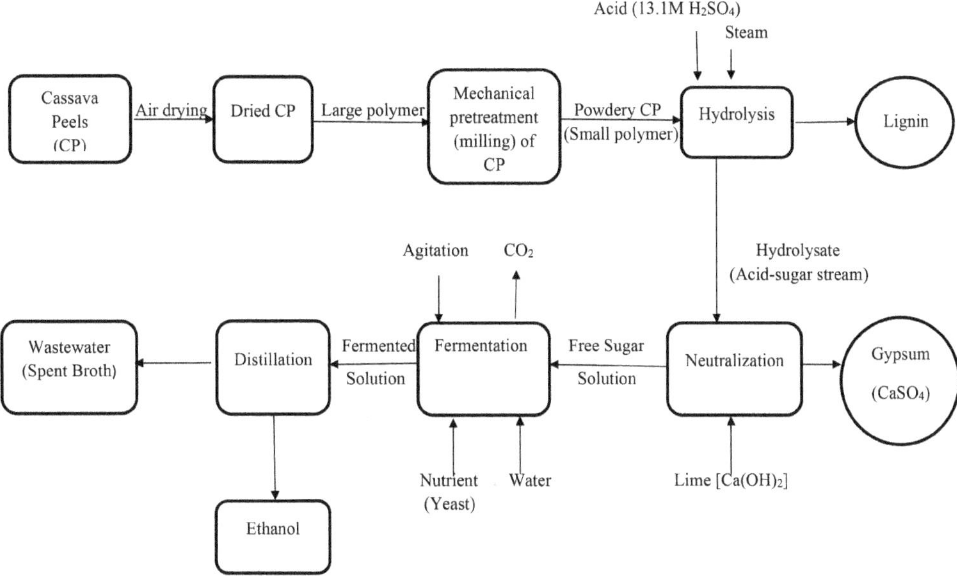

FIGURE 14.A Production flowchart

- Neutralization process (to separate the sugars from acid).
- Sugars fermentation (glucose) to ethanol.
- Distillation (to obtain pure ethanol).
- Analytical assay: the pH, total soluble solids (TSS), glucose yield, total reducing sugar (TRS), and bioethanol yield of the cassava peels were analyzed.

14.2.3.1 Biomass Pretreatment

We air-dried the cassava peels (CP) until they lost all their moisture content. The dried cassava peel was then milled to a fine powder using an electrical grinder and afterwards sieved with a +1.5 mm sieve to get a homogeneous powder. This process was to expose a large surface area of the biomass to chemical hydrolysis. The powdery CP utilized in the engineering production process was weighed using a top load weighing balance in the laboratory unit of the Environmental Health Sciences Department, University of Ibadan, Ibadan, Nigeria. Around 5 kg of the powdery biomass were utilized in the engineering phase.

14.2.3.2 Optimization Layout

The optimization phase of the bioethanol plant was divided into three (3) trials of experimental layout, as shown in Table 14.1, and a uniform mass of 5 kg of milled CP was utilized for each trial in the batch. We utilized different temperature regimes (23°C, 30°C, and 35°C) of the adaptive incubator to carry out fermentation of the substrate based on literature survey. Parameters such as total soluble solids (TSS), pH, and bioethanol concentration of the samples before and after fermentation were assessed.

14.2.3.3 Hydrolysis Stage

We mixed 5 kg each of the milled CP separately with 25 L of 13.1 M H_2SO_4 in 1:5 (w/v) in the inner stainless steel cylindrical container (104 L/0.104 m^3 capacity) of the hydrolysis chamber. The mixture was steamed to 100°C for 60 mins in the water contained in the outer cylindrical container (224 L/0.224 m^3 capacity) using a high-pressure gas burner. The reaction was aided by stirring the mixture with a mixer rotating at 1,400 rev/min. After the hydrolysis, the resulting thick gel was pressed on

TABLE 14.1
Different Treatment Groups to Assess the Effect of Fermentation Temperature on Bioethanol Production

Batch A	Mode of Treatment
Trial 1	Incubator set at 23°C of fermentation
Trial 2	Incubator set at 30°C of fermentation
Trial 3	Incubator set at 35°C of fermentation

the improvised sieve to obtain the first hydrolysate (acid–sugar stream), while the leftover solid was again subjected to second hydrolysis using 25 L of 13.1 M H_2SO_4. The second hydrolysis was performed at 100°C for 50 mins in the inner hydrolysis chamber. This also produced a thick gel which was sieved to obtain the second hydrolysate. We combined the two hydrolysates together in a stainless steel pail, and the total volume was recorded. The leftover solid was discarded in a waste bin.

Equation of the reaction:

$$\underset{\text{Lignocellulose}}{(C_6H_{10}O_5)n} \xrightarrow[\text{Conc.}H_2SO_4]{+H_2O} \underset{\text{Reducing sugar}}{C_6H_{12}O_6}$$

The glucose hydrolysis efficiency (GHE) (%), glucose productivity (g L^{-1} h^{-1}), and total reducing sugars (TRS) productivity (g L^{-1} h^{-1}) were calculated by adopting the formula described by Zhu et al. [26]:

$$\text{GHE}(\%) = \frac{Glucose\,concentration\ (gL^{-1})}{Cassava\ peels\ conc.\ \times 51.52\% \times 1.1\ (gL^{-1})} \times 100 \tag{14.1}$$

$$\text{Glucose Productivity}\,(gL^{-1}h^{-1}) = \frac{Glucose\,concentration\,(gL^{-1})}{Hydrolysis\,time\,(hr)} \tag{14.2}$$

$$\text{TRS Productivity}\,(gL^{-1}h^{-1}) = \frac{TRS\,concentration\,(gL^{-1})}{Hydrolysis\,time\,(hr)} \tag{14.3}$$

Note:

Cassava peels conc = 20%, that is, 1:5 (w/v), since 5 kg of the biomass were treated with 25 L of acid.

Total hydrolysis time = 1.833 hr

14.2.3.4 Neutralization Process

The hydolysates we obtained from the hydrolysis stage were neutralized with a well-calculated volume of $Ca(OH)_2$ solution projected from the previous laboratory works done by the authors [5, 30, 31] in a stainless steel container to give the required pH of 5.5. The content was placed under the stirrer driven by an electric motor to ensure complete neutralization during mixing. After proper mixing, the content was allowed to cool and filter using the improvised filter press to obtained a free sugar solution. The residue, which is gypsum [$CaSO_4$], was discarded into a waste bin.

We took samples from the sugar solution and analyzed for reducing sugars using Fehling's solutions and also for quantitative determination of total soluble solids (TSS), glucose, and total reducing sugars (TRS) concentrations, respectively.

14.2.3.5 Fermentation Process

We adopted the technique described by Hossain et al. [32] in the fermentation process. The sugar solution obtained was fermented with *Saccharomyces cerevisiae* under aseptic condition using the proportion of 3 g of yeast to 1 L of sugar solution (**Plate 14.2a–c**). Ethanol presence, ethanol concentration, and yield were carried out by periodically taken samples from the fermenting broth every 24 hours. The bioethanol plant optimization was carried out as follows:

- For the first trial (trial 1) in batch A, the fermentation was performed by setting the adaptive incubator at 23°C for 72 hr.
- For the second trial (trial 2) in batch A, the fermentation was performed by setting the adaptive incubator at 30°C for 72 hr.
- For the third trial (trial 3) in batch A, the fermentation was performed by setting the adaptive incubator at 35°C for 72 hr.

The fermentation efficiency and ethanol productivity were calculated using the formula described by Zhu et al. [26]:

$$\text{FE}(\%) = \frac{\textit{Ethanol concentration } (gL^{-1}) \times 1}{\textit{Cassava peels conc.} \times 51.52\% \times 1.1\ (gL^{-1}) \times 0.51} \times 100 \tag{14.4}$$

$$\text{Ethanol Productivity}\,(gL^{-1}h^{-1}) = \frac{\textit{Ethanol concentration}\,(gL^{-1})}{\textit{Fermentation time}\,(hr)} \tag{14.5}$$

Note: FE = fermentation efficiency

14.2.3.6 Distillation Process

We pour the fermented solution into the distiller (45 L capacity) of the fabricated distillation machine (**Plate 14.2(d)**), and its digital temperature gauge was set at 78°C to obtained pure ethanol solution through distillation. The content was heated at the fixed temperature and the distillate passed through the reflux line (conducting stainless pipe) into the cooling tank (water tank), where it condensed to liquid and was collected as pure ethanol solution. The volume of ethanol obtained from each distillation was measured and recorded. The following were also calculated:

- **Actual yield of ethanol** in volume or gram, that is, the volume/gram of ethanol recovered after distillation from the solution.
- The **theoretical yield of ethanol** in volume or gram, that is, the volume or mass of ethanol that would be obtained using the following balanced stoichiometric equation:

$$\underset{\text{Glucose}}{C_6H_{12}O_6} \xrightarrow{\text{yeast}} \underset{\text{Ethanol}}{2C_2H_5OH} + \underset{\text{Carbon dioxide}}{2CO_2\uparrow(g)}$$

- **Percentage yield of ethanol** was calculated viz:

$$\%\ \text{Yield} = \frac{\textit{Actual yield}}{\textit{Theoretical yield}} \times 100\% \tag{14.6}$$

- **Percentage volume/concentration of ethanol (% v/v)** was determined viz:

$$\text{Volume}\ \% = \frac{\textit{Volume of ethanol recovered}}{\textit{Total volume of solution}} \times 100\% \tag{14.7}$$

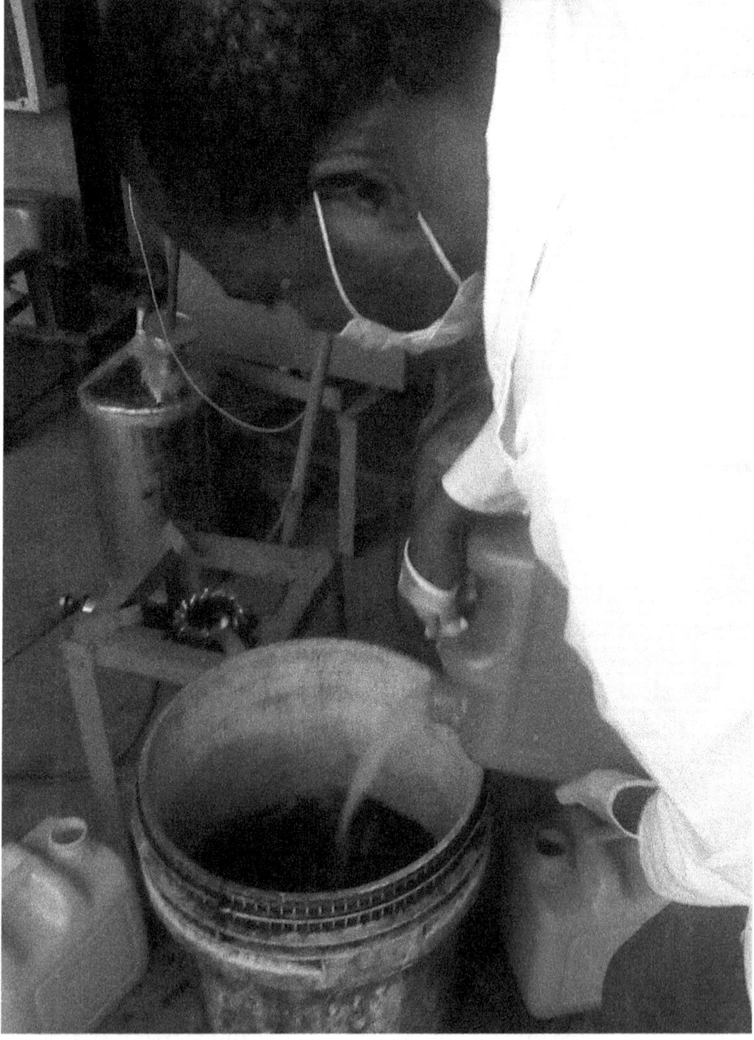

PLATE 14.2(A–B) Fermentation process: (a) hydration of the yeast solution, (b) mixing of hydrolysate with yeast solution under aseptic condtion.

PLATE 14.2(C) Samples subjected to fermentation in the constructed incubator with blower to circulate heat.

PLATE 14.2(D) Distillation setup showing the collection of the distillate (bioethanol) by the researchers.

14.2.4 Utilization and Energy Consumption of the Optimal Ethanol Produced

The optimal bioethanol produced was used to determine the efficiency of the distiller, lagging time and energy consumption by the SSPF.

1. The efficiency of the distiller in distilling the optimal bioethanol produced in this study was extrapolated from the percentage yield of the bioethanol produced.
2. The lagging time in distilling the optimal bioethanol produced was also calculated, which is the time delay before another round of distillation will take place.
    - Lagging time = time taken to reach the boiling point + time taken for distillation
3. The heat energy consumption of the distiller in distilling the optimal bioethanol produced was calculated as follow:
    - Total quantity of heat energy (J) consumed in the distillation is

$$Q = [m_e c_e + m_s c_s]\,\Delta t \tag{14.8}$$

Where:

Q = total quantity of heat energy consumed during distillation
m_e = mass of optimal bioethanol produced in kilograms
c_e = specific heat capacity of ethanol = 2,460 J/kg°C
m_s = mass of the stainless steel distiller in kilograms
c_s = specific heat capacity of the stainless steel distiller = 502.416 J/kg°C
$\Delta t = t_f - t_a$ = difference in final maximum and initial temperature of the system in °C

4. The total electrical energy consumed by the adaptive small-scale production facility (SSPF) in the production of the optimal bioethanol was calculated as follows:
    - Total electrical energy consumed by SSPF (KWhr)

    = electrical energy used by the mixer + electrical energy used by the incubator + electrical energy used by the distiller

    Note: Since energy = power × time, the electrical energy of each device is obtained by multiplying the power rating of the device by the time taken (hr) for the device to work.

14.2.5 Data Management and Statistical Analysis

We analyzed the data collected in the pilot study with the Statistical Product and Service Solution (IBM SPSS Statistics) version 27.0 software for descriptive and inferential statistics. Descriptive statistics was used to summarize data using bar charts, line graphs, means, and standard deviations. The results obtained from the physiochemical and proximate analyses of the pilot study were subjected to one-way analysis of variance (ANOVA) as described by Statistical Analysis System (SAS, 1997) and New Duncan's Multiple Range Test (Dunca, 1959) for means separation at 95% level of probability.

The study also utilized the Pearson product–moment correlation coefficient to assess the relationship between the physicochemical properties and the glucose/TRS concentrations obtained from the cassava peels in the pilot study.

14.2.6 Framework for Cost–Benefit Analysis of the Small-Scale Production Facility

A cost–benefit analysis (also known as a benefit–cost analysis) was carried out on the small-scale production facility (SSPF) used for the bioethanol production. This was carried out in order to identify the benefits of the project as well as the associated costs. Also, a comparative analysis was made between the fabricated SSPF and existing/proposed biorefineries in the country (Nigeria). This would help develop a reasonable conclusion around the feasibility of the facility, help in deciding whether to pursue the scaling up of the production facility at industrial level, weigh investment opportunities, and measure the environmental and socioeconomic benefits of adopting the proposed facility.

A framework was established to itemized the various costs involved in the SSPF. These include:

- The cost of raw materials/machines used in the fabrication of the SSPF
- The labor cost
- The cost of consumables used in bioethanol production (e.g., chemicals, yeast, cassava peels, distilled water, etc.)
- The cost of energy consumed
- The cost of maintenance or servicing of the SSPF

The overall cost of the SSPF and its short- and long-term benefits using aggregated information were outlined. This was compared with the cost of other existing/proposed facilities in the country. The results were analyzed to make an informed and final decision, as well as recommendations regarding the SSPF.

14.3 RESULTS AND DISCUSSIONS

This section deals with the results and discussions of the physicochemical characterization of the cassava peels (CP) done in the lab, the hydrolysis and fermentation efficiency of the small-scale production facility (SSPF), the bioethanol concentrations and yields, the utilization and energy consumption of the optimal bioethanol produced, as well as the comparative cost–benefit analysis of the SSPF. The results have been presented using frequency tables, bar charts, line graphs, mean, standard deviation, ANOVA table, and correlation matrix. The effect of various temperatures (23°C, 30°C, and 35°C) on fermentation efficiency and ethanol production from CP using SSPF has been discussed. Lastly, we discussed the energy consumption and calculate the cost–benefit analysis of the SSPF.

14.3.1 Physicochemical Characterization of Cassava Peels

Table 14.2 outlines the physicochemical characterization of the biomass-cassava peels (CP) in terms of pH, dry matter (%), moisture content (%) fresh biomass, moisture content (%) dried biomass, total organic carbon (TOC, %), total nitrogen (TN, %), total phosphorus (TP, %), and carbon-to-nitrogen (C:N). The results of the physicochemical parameters of the CP revealed that the pH of the CP was in the acidic range and suggest that cassava waste disposal into the environment would have a negative impact on the soil nutrient uptake by plants. This is because the acidity of the soil tends to leach most nutrients and reduce their availability to plants grown on such soil. From the experimental analysis, the percentage moisture content of the fresh cassava peels was relatively high (73.04%), as shown in Table 14.2. This was in agreement with the reports of numerous workers [33, 34] who

TABLE 14.2
Physicochemical Characterization of the Biomass-Cassava Peels

Parameters	1st	2nd	3rd	Mean ± SD
Ph	5.70	5.72	5.68	5.70 ± 0.02
Moisture content (%) fresh biomass	72.14	73.52	73.45	73.04 ± 0.78
Moisture content (%) dried biomass	10.01	10.19	10.11	10.10 ± 0.09
Dry matter (%)	87.52	87.50	87.52	87.51 ± 0.01
total organic carbon, TOC (%)	50.56	50.40	51.20	50.72 ± 0.42
Total nitrogen, TN (%)	0.78	0.75	0.81	0.78 ± 0.03
Total phosphorus, TP (%)	0.25	0.27	0.20	0.24 ± 0.04
Carbon-to-nitrogen (C:N)	64.82	67.20	63.21	65.08 ± 2.01

reported CP moisture contents of 50–80%. The relatively high moisture content of CP could contribute to its rapid decay, vectors attraction, and nauseating odor, which if inhaled over a chronic duration could be detrimental to health [35]. The percentage dry matter for CP (87.51%) obtained in this study agrees with the findings of Yoonan and Kongkiattikajorn [36], who reported dry matter content of 86.5–94.5% for cassava peels.

All the carbon atoms covalently bonded in organic molecules are collectively referred to as the total organic carbon. These organic bonded carbons in substrates/samples are usually oxidized to carbon dioxide (CO_2) and other inorganic carbon (IC), such as carbonate (CO_3^{2-}) and bicarbonate (HCO_3^-), during oxidation and other chemical reactions. In our study, analysis of the CP showed that the percentage of TOC was relatively high (50.72%), which was consistent with the findings of other researchers [37, 38]. According to McCauley et al. [39], the TOC of a plant material is directly related to its organic material, which is that part of a plant that has its tissue and structure still intact and visually recognizable and has not undergone decay. This implies that the natural decomposition of cassava peels aerobically and anaerobically contributes to greenhouse gas (CO_2, CH_4) emissions in the atmosphere. This corroborated the report of Lal and Reddy [40] that the anaerobic degradation of lignocelluloses by methanogens contributes about 25 million tons of methane gas to the atmosphere annually. Harnessing the abundant organic carbon in this compound for bioethanol production is a viable waste management option to solve the environmental challenges posed by this biomass [38]. The mean TN and TP content of the CP was relatively low. This finding is in conformity with the reports of several authors [36–38], who reported that most agricultural residues, such as cassava peels, are lignocellulosic wastes with low nitrogen and phosphorus content.

The ratio of the content of carbon and nitrogen (C/N ratio) in a biological substance is a vital parameter for the regulation of biological treatment plants. Numerous workers have investigated the influence of C/N quotient on anaerobic processes, and it has been well-established that the ideal C/N quotient is 20–30:1 [41, 42]. This is because microorganisms which anaerobically decompose organic substrates digest carbon 25 to 30 times faster than they digest nitrogen [43]. However, the C/N ratio of a substrate can be adjusted to optimal levels by mixing the organic materials with suitable contents [44]. The C/N ratio of the cassava peels (65.08) in the study was higher than the optimal value of 20–30:1 recommended for bioenergy production. A similar result was reported for CP by Sokan-Adeaga and Ana [38]. In a study conducted by Adelekan [37], he reported that CP had an elevated C/N proportion as a result of low nitrogen content but high organic carbon value. Malherbe and Cloete [45] opined that the optimal C/N ratios for different substrates are not the same but vary with the type of feedstock to be digested. Hence, to optimize the C/N ratio of a lignocellulosic biomass for bioprocessing, it must be co-digested with sewage, sludge, animal manure, or poultry litter.

14.3.2 Hydrolysis of the Small-Scale Production Facility

Figure 14.1 illustrates the various parameters assessed for in the hydrolysis stage viz glucose concentration, glucose hydrolysis efficiency (GHE) %, glucose productivity, total reducing sugars (TRS) concentration, and TRS productivity of the cassava peels, respectively.

The overall mean glucose and TRS concentrations obtained from the cassava peels using the small-scale production facility (SSPF) were 92.31 ± 0.19 and 143.51 ± 0.10 gL^{-1}. Likewise, the overall mean GHE %, glucose productivity, and TRS productivity of the cassava peels from the SSPF were 81.45 ± 0.16%, 50.36 ± 0.10, and 78.30 ± 0.05 $gL^{-1}h^{-1}$, respectively. All these results imply a promising yield of sugar production from CP using the SSPF.

The findings at the hydrolysis stage revealed that hydrolysis of 20% cassava peels with 13.1 M H_2SO_4 for 110 min at 100°C using the SSPF led to a significant increase in glucose and TRS concentrations and productions, which were approximately five (5) times greater than those results we obtained in the laboratory under the same conditions. Other authors, Zhu et al. [26], reported a glucose production as high as 66.86 gL^{-1} from 20% cassava pulp using a fed-batch hydrolysis for 120 hrs, in which enzymatic cocktails were introduced at hydrolysis timings of 24 hrs and 48 hrs gap, respectively. This is smaller to the overall mean glucose concentration of 92.31±0.19 gL^{-} obtained in this study. The high GHE (%) obtained from the study is slightly lesser than the 90% glucose hydrolysis efficiency reported by Kosugi et al. [46], but higher than the 74.32% GHE observed by Zhu et al. [26]. In industrial applications, the biomass concentration must be improved to mitigate operational cost. Conversely, highly concentrated biomass poses a serious challenge as it causes heat resistance and mass transfer and has the overall effect of decreasing the hydrolysis efficiency. Hence, it is imperative to select an optimum biomass concentration.

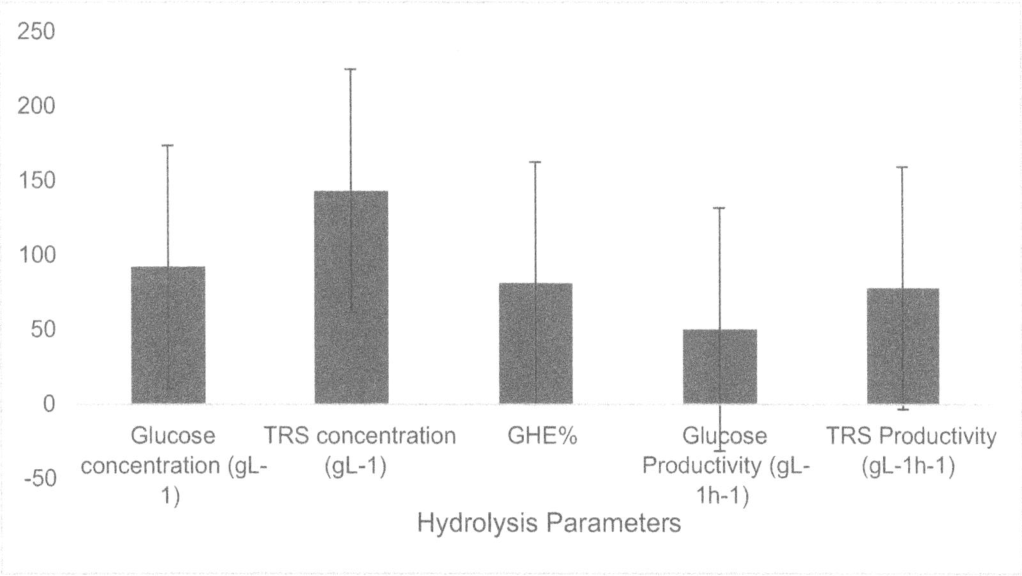

FIGURE 14.1 Hydrolysis parameters of the SSPF.

14.3.3 Optimization of Fermentation Temperature and Time on Bioethanol Concentrations and Yields

Table 14.3 has been prepared to illustrate the effect of temperature (23°C, 30°C, and 35°C) on the total soluble solids (TSS), pH, and bioethanol concentration of the cassava peel (CP) hydrolysates using the SSPF. Comparison of the TSS of the CP hydrolysate before and after fermentation at different temperatures (23°C, 30°C, and 35°C) was significant ($p < 0.05$). The peak TSS after fermentation was noticed in batch A trial 1 (23°C), followed by batch A trial 2 (30°C) and batch A trial 3 (35°C), respectively. The pH values of CP hydrolysate before fermentation were greater than those after fermentation ($p < 0.05$). The TSS and pH after fermentation across the different temperature were not statistically significant ($p > 0.05$). Figure 14.2 outlines the bioethanol concentration at different temperatures using the SSPF. Batch A trial 2 performed at 30°C gave the highest mean concentration of bioethanol of 4.36 ± 0.09% (w/v), compared with batch A trial 1 and 3 with readings of 4.06 ± 0.04 and 3.35 ± 0.13% (w/v) performed at 23°C and 35°C, respectively. The variations in bioethanol concentrations across the different fermentation timing and temperature were significantly different ($p < 0.05$), as shown in Table 14.3. Also, the peak ethanol production was observed at 72 hrs of fermentation. This ethanol concentration was significantly higher than those obtained at 24 and 48 hrs, respectively ($p < 0.05$).

We illustrate the effects of temperature (23°C, 30°C, and 35°C) on glucose, total reducing sugars (TRS), and bioethanol yields of cassava peel (CP) hydrolysates using the SSPF in Table 14.4. It can be noticed that as the fermentation duration increased from 24 hrs to 72 hrs, the glucose and TRS yield decreased proportionately. The mean glucose and TRS yields in batch A (trials 1–3) after fermentation were significantly different ($p > 0.05$). In Figure 14.3, comparison of the mean bioethanol yields across the various trials (1–3) are shown. The highest bioethanol yield for the fermented CP broths was observed at 72 hrs of fermentation. Fermentation in batch A trial 2

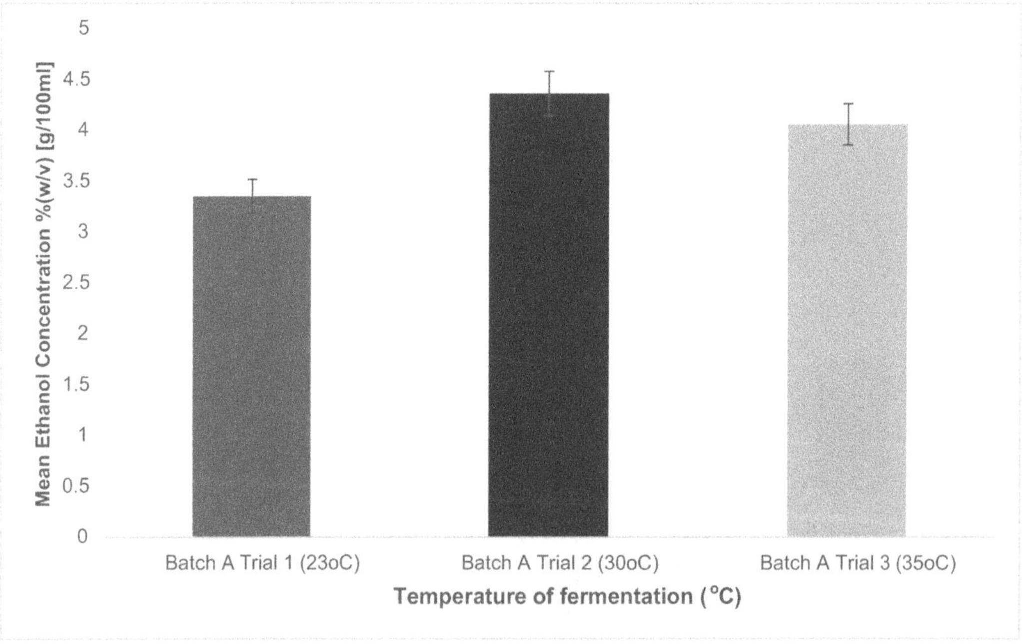

FIGURE 14.2 Comparison of mean bioethanol concentrations of cassava peels hydrolysate at various fermentation temperatures in the SSPF.

TABLE 14.3
Effect of Fermentation Temperature on Total Soluble Solids, pH, and Bioethanol Concentration of Cassava Peels in the Small-Scale Production Facility (Mean ± SD, n = 3)

Parameter	Total Soluble Solids (TSS) g/100 g (°Bx)		pH		Glucose Concentration % (w/v) (g/100 mL)				Total Reducing Sugars Concentration % (w/v) (g/100 mL)				Ethanol Concentration % (w/v) (g/100 mL)			
(Temperature)	**Initial**	**After**	**Initial**	**After**	**0hr**	**24 hrs**	**48 hrs**	**72 hrs**	**0 hr**	**24 hrs**	**48 hrs**	**72 hrs**	**0 hr**	**24 hrs**	**48 hrs**	**72 hrs**
Batch A	28.58 ±	13.95 ±	5.47 ±	4.36 ±	9.24 ±	6.21 ±	3.64 ±	1.78 ±	14.35 ±	9.70 ±	5.08 ±	3.08 ±	0.00 ±	2.09 ±	3.06 ±	3.35 ±
Trial 1 (23°C)	0.42_b	1.23_a	0.06_a	0.31_a	0.02	0.02_a	0.04_c	0.13_b	0.01_a	0.11_b	0.07_c	0.04_c	0.00	0.03_a	0.03_a	0.13_a
Batch A	28.60 ±	13.21 ±	5.47±	4.05 ±	9.23 ±	5.84 ±	3.24 ±	1.05 ±	14.35 ±	8.28 ±	4.13 ±	2.26 ±	0.00 ±	3.14 ±	4.07 ±	4.36 ±
Trial 2 (30°C)	0.35_b	0.58_a	0.06_a	0.31_a	0.01	0.04_a	0.03_a	0.02_a	0.02_a	0.01_a	0.03_a	0.02_a	0.00	0.02_c	0.03_c	0.09_c
Batch A	27.61 ±	12.36 ±	5.47 ±	4.25 ±	9.23 ±	6.09 ±	3.42 ±	1.19 ±	14.34 ±	9.86 ±	4.68 ±	2.64 ±	0.00 ±	2.63 ±	3.73 ±	4.06 ±
Trial 3 (35°C)	0.07_a	1.95_a	0.06_a	0.03_a	0.01	0.02_a	0.03_b	0.02_a	0.01_a	0.02_c	0.27_b	0.14_b	0.00	0.12_b	0.13_b	0.04_b
F value	8.21	1.01	0.00	1.14	0.47	152.71	102.53	72.37	0.08	581.85	25.50	74.17	—	173.52	132.48	95.01
p value	*0.02	0.42	1.00	0.38	0.65	*0.00	*0.00	*0.00	*0.92	*0.00	*0.00	*0.00	—	*0.00	*0.00	*0.00

Different letters (a, b, and c) indicate significant differences along the columns.

***Significant at $p = 0.05$.**

TABLE 14.4
Effect of Fermentation Temperature on Glucose, Total Reducing Sugars, and Bioethanol Yields of Cassava Peels in the Small-Scale Production Facility (Mean ± SD, n = 3)

Parameter (Temperature)	Mean Glucose Yield (g/5 kg)				Mean Total Reducing Sugars (TRS) Yield (g/5 kg)				Mean Ethanol Yield (L/5 kg)			
	0 hr	24 hrs	48 hrs	72 hrs	0 hr	24 hrs	48 hrs	72 hrs	0 hr	24 hrs	48 hrs	72 hrs
Batch A	2309.17 ±	1552.50 ±	910.83 ±	444.17 ±	3586.67	2424.17 ±	1269.17 ±	770.83 ±	0.00 ±	4.50 ±	6.59 ±	7.22 ±
Trial 1 (23°C)	3.82_a	5.00_c	11.81_c	33.29_b	± 1.44_a	26.73_b	16.27_c	10.10_c	0.00	0.23_a	0.33_a	0.31_a
Batch A	2306.67 ±	1460.00 ±	810.00 ±	263.33 ±	3586.67	2069.17 ±	1033.33 ±	564.17 ±	0.00 ±	7.55 ±	9.79 ±	10.49 ±
Trial 2 (30°C)	2.89_a	9.01_a	7.50_a	3.82_a	± 3.81_a	1.44_a	8.04_a	3.82_a	0.00	0.04_c	0.06_c	0.20_c
Batch A	2308.33 ±	1522.50 ±	854.17 ±	297.50 ±	3585.83	2464.17 ±	1170.00 ±	660.00 ±	0.00 ±	5.78 ±	7.95 ±	8.66 ±
Trial 3 (35°C)	2.87_a	5.00_b	7.64_b	5.00_a	± 2.89_a	3.82_c	67.96_b	34.37_b	0.00	0.03_b	0.09_b	0.23_b
F value	0.47	152.71	90.47	72.37	0.08	581.85	25.50	74.17	—	368.01	191.11	129.50
p value	0.65	*0.00	*0.00	*0.00	0.92	*0.00	*0.00	*0.00	—	*0.00	*0.00	*0.00

Different letters (a, b, and c) indicate significant differences along the columns.

*Significant at $p = 0.05$.

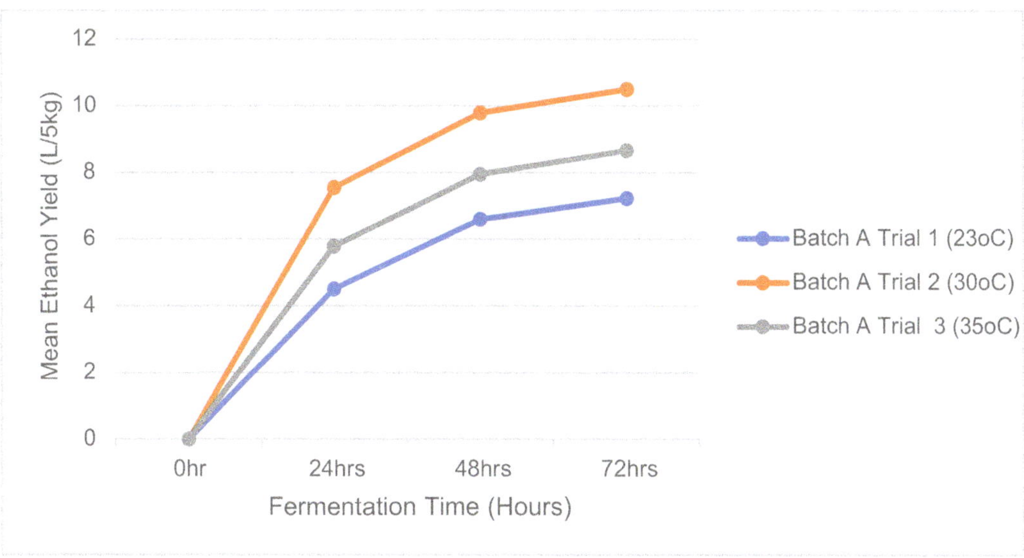

FIGURE 14.3 Comparison of mean bioethanol yield of cassava peels at different fermentation temperatures in the SSPF.

(30°C) gave the highest mean bioethanol yield of 10.49 ± 0.20 L/5 kg (2,098.00 ± 40.00 ml/kg). This was followed by batch A trial 3 (35°C) at 8.66 ± 0.23 L/5 kg (1,732.00 ± 46.00 ml/kg), and the least bioethanol yield of 7.22 ± 0.31 L/5 kg (1,444.00 ± 62.00 ml/kg) was observed in batch A trial 1 (23°C). The bioethanol yields across the different fermentation temperatures of the SSPF varied significantly ($p < 0.05$).

From this result, we inferred that the optimal temperature for peak production of ethanol from the CP is 30°C. This assertion is in concordance with the findings of Sobrinho et al. [47], who reported the highest yield within 30–34°C. It is also in consonance with that of Fakruddin et al. [48] (2013), who carried out process optimization of bioethanol production at 30°C, 35°C, and 40°C, respectively, using three (3) thermo-tolerant yeasts isolated from agro-allied by-product. From the different experiments, it was observed that a temperature of 30°C and a pH range of 5.0 to 6.0 were optimal conditions for peak ethanol yield by strains *Saccharomyces unisporous* (P), *Saccharomyces cerevisiae* (C) and (T), respectively. Fakruddin et al. [49] also reported that the highest ethanol yield by the yeast *Saccharomyces cerevisiae* IFST-072011 occurred at temperature of 30°C. In general, it may be deduced that enzymatic reaction rate rises proportionally with temperature up to a threshold value, after which the denaturation of enzymes sets in. Elevated thermal condition impedes cell growth and consequently leads to significant decline in fermentation. In this study, ethanol yield depreciated significantly as the fermentation temperature was reduced to 23°C or increased to 35°C, as illustrated in Figures 14.2 and 14.3. This observation is corroborated by reports of McMeckin et al. [50] and Phisalaphong et al. [51], respectively. Other authors, such as Vaidya et al. [52] and Balachandra et al. [53], also reported the influence of temperature on the chemical process, such as catalysis in a homogeneous and heterogeneous reaction.

From our experimental results, the highest ethanol yield from the cassava peel was obtained at 72 hrs of fermentation. It is logical that we conclude that lengthy fermentation duration has a positive impact on fermentation rates of *S. cerevisiae*, since longer time results to higher rates of formation of product. The time avails the yeast the opportunity to maximally utilize the sugars contained in the CP [54]. This outcome agrees with that of Pippo and Carlos [55], who observed a maximum ethanol yield of 8% (w/v) after 72 hrs of fermentation. Also, findings from this study agree with that of Suryawati et al. [56] and Faga et al. [57], who reported 72 hrs as the optimum time for diverse strains of thermotolerant yeasts to give maximum ethanol quantity. Conclusively, many of the available literature reported that complete bioethanol production from different feedstocks is achieved at 72 hrs of fermentation [58–60].

14.3.4 Effect of Temperature Variation on Fermentation Efficiency and Ethanol Productivity

Figure 14.4 depicts the effects of temperature variation (23°C, 30°C, and 35°C) on the mean fermentation efficiency and ethanol productivity of cassava peels by the ethanologenic organism using the SSPF in the small-scale study. The highest mean fermentation efficiency and ethanol productivity was given by batch A trial 2 (30°C) (75.37 ± 1.47%, 0.605 ± 0.012 $gL^{-1}h^{-1}$), while the least values were found in batch A trial 1 (23°C) (58.01 ± 2.23%, 0.466 ± 0.018 $gL^{-1}h^{-1}$) at $p < 0.05$. Temperature is among the crucial environmental determinants affecting yeast growth and ethanol productivity, the reason being that enzyme hydrolysis and rate of fermentation are both dependent on the ambient temperature. In most instances, an elevated temperature can enhance the rate of fermentation consequent to exponential growth of bacteria and also product formation. Conversely, a temperature increase above the optimal level tolerated by microorganisms can result in adverse effects, such as denaturation of enzymes, decrease in product formation rate, and cell mortality [61].

14.3.5 Distillation Results for the Small-Scale Production Facility

Figure 14.5 outlines the mean ethanol recovered from the various trials in batch A in the pilot study using the SSPF. We can clearly see that batch A trial 2 produced the highest mean ethanol recovery of 18.20 ± 0.10L, while batch A (trials 1 and 3) gave the least ethanol (15.16 ± 0.95 and 15.83 ± 0.66, respectively). The mean recovery from the various trials was statistically significant at $p < 0.05$. Likewise, the ethanol percentage concentrations by volume obtained from different

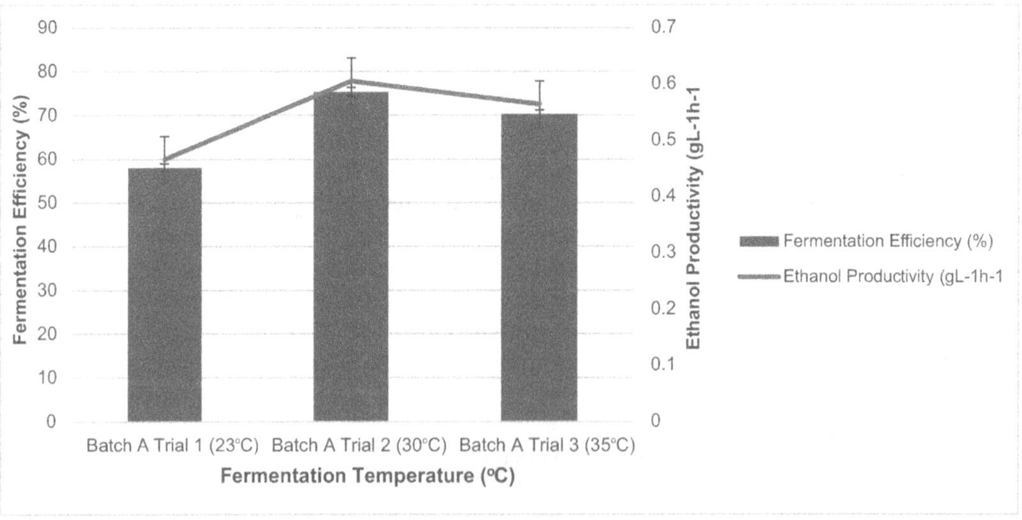

FIGURE 14.4 Effect of temperature variation on fermentation efficiency and ethanol production of cassava peels in the SSPF.

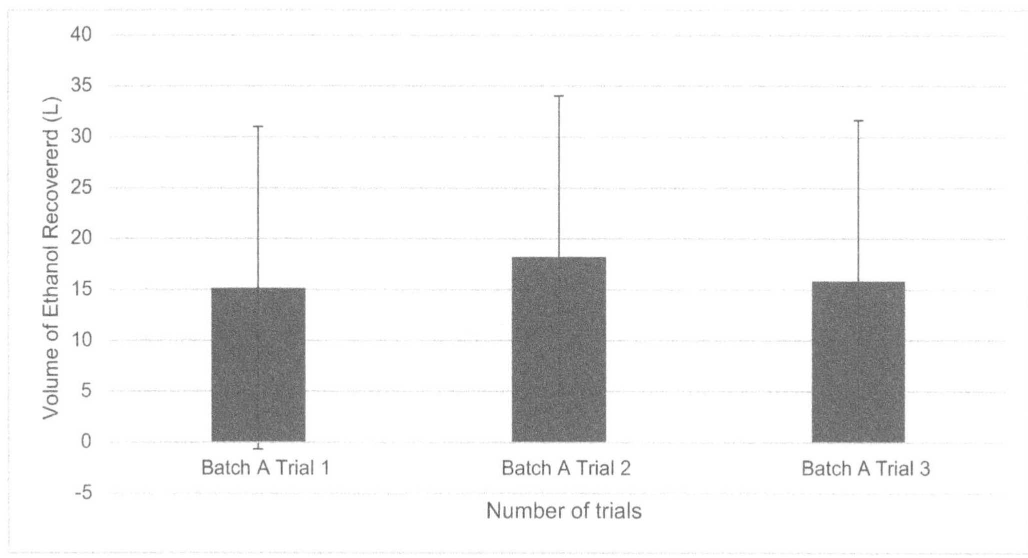

FIGURE 14.5 Mean volume of ethanol recovered (L) using the SSPF.

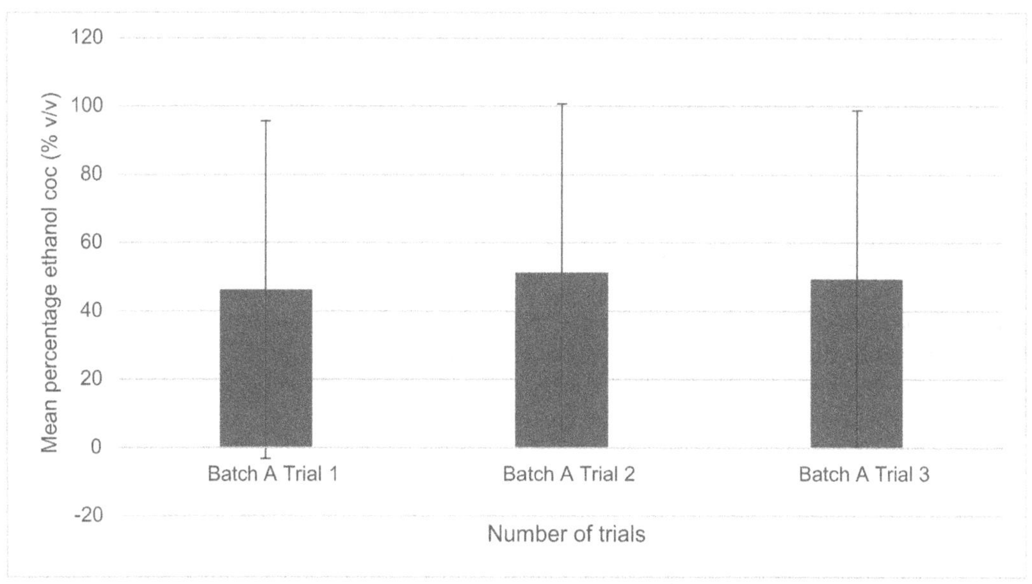

FIGURE 14.6 Mean percentage ethanol concentration (% v/v) using the SSPF.

trials are shown in Figure 14.6. We can observe that the highest ethanol concentration by volume was produced by batch A (trial 2), of 51.31 ± 0.21% v/v, while the least value (46.23 ± 0.15% v/v) was obtained in batch A (trial 1) ($p < 0.05$). Lastly, Figure 14.7 depicts the mean percentage yield ethanol obtained from the various trials. Batch A (trial 2) gave the highest yield of 59.41 ± 1.15%, followed closely by batch A (trial 3) (55.38 ± 0.52%), with the least yield of 45.73 ± 1.74% produced by batch A (trial 1).

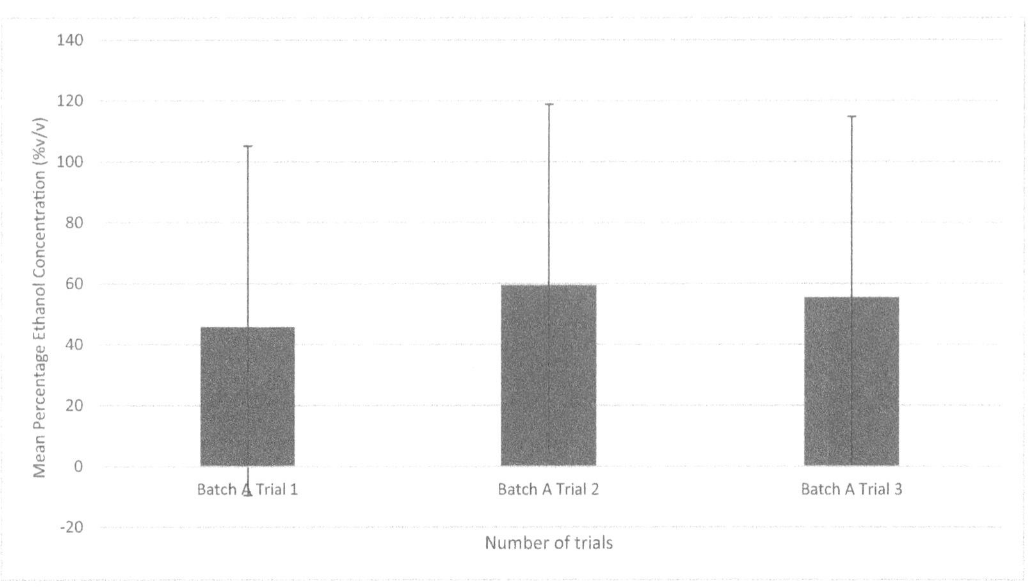

FIGURE 14.7 Mean percentage ethanol yield of various trials using the SSPF.

14.3.6 Correlation Matrix between Physicochemical Parameters of Cassava Peels and Glucose and TRS Concentrations Obtained in the Small-Scale Production Facility

In Table 14.5, we illustrate the correlation matrix between the physicochemical properties and the glucose and TRS concentrations obtained in the small-scale production facility (SSPF). Positive correlations were found between the glucose concentration and some of the physicochemical parameters of the cassava peels viz dry matter, TOC, and TN. However, we observed negative correlations between pH, TP, and C:N of the cassava peels and the glucose concentrations obtained in the SSPF. Also, there were no correlations between all the physicochemical properties of the cassava peels and TRS concentration. The correlation between the pH, TN, and TOC of the cassava peels and the glucose concentration in the SSPF was strong, that is, r tends towards unity (1).

14.3.7 Accessory Information on the Small-Scale Production Facility (SSPF)

The energy consumption of the SSPF, the efficiency and production rate of the distiller, the lagging time during distillation, and the estimated cost of the optimal bioethanol produced are indicated in Table 14.6. The total electrical energy consumed by the SSPF in the bioethanol production process was estimated at 183.49 KWhr, while the quantity of heat energy consumed in distilling the bioethanol production was 2.05 MJ. The distiller had an efficiency of ~59% and a lagging time of 1 hr, 51 mins. The production rate of the fabricated distiller was estimated at ~1,200 L per month at eight (8) working hours per day. The cost of production of the 18.20 L bioethanol was ₦ 45,141.00k, and the selling price of the produced bioethanol was estimated at ₦ 2,728.30k per liter using a 10% profit margin (see **Appendix** for workings).

14.3.8 Calculation and Comparison of Costs and Benefits of the Small-Scale Production Facility

The inventories of the various emerging biorefineries in Nigeria that proposed to utilize cassava as the main feedstock are outlined in Table 14.7. These facilities were designed to operate on the simultaneous saccharification and co-fermentation (SSCF) technique. Among all the proposed plants, the cassava ethanol plant in Taraba state was projected to produce the highest ethanol production per year (72 million liters per annum).

We then calculated the cost–benefit analysis of our fabricated small-scale production facility (SSPF) at industrial scale as depicted in Table 14.8. The SSPF was scaled up by arbitrarily factor ×12,000 to produce 172.8 million liters ethanol per annum at industrial level. The estimated overall cost of constructing the SSPF was ~₦ 6.02 billion, and the financial benefits to be accrued by adopting the SSPF in lieu of the best proposed biorefinery were also calculated as ~₦ 1.550 trillion. From the analysis, the socioeconomic, environmental, and financial benefits of the SSPF far outweighed its cost, with a huge net cost benefit in trillions of naira and a high benefit–cost ratio of approximately 257.84, as outlined in Table 14.8. The implication of this finding is that the SSPF is financially viable and the net gains from the facility can be harnessed into socioeconomic development and sustainable environmental management projects. The SSPF is economically viable because it utilized readily available, cheap, and sustainable feedstock (cassava peels), thus aiding waste to wealth. The components of the fabricated SSPF are metal scraps, which are easy to obtain. The facility consumes lesser electrical energy as it is mainly driven by natural gas, which is cheap and affordable. However, briquette may be utilized as heating source in rural communities, where natural gas is not easily accessible.

14.4 CONCLUSIONS

The global climate change and environmental deterioration arising due to the indiscriminate use of fossil fuels have triggered the drive for sustainable energy sources to either substitute or complement non-renewable energy. This development has generated significant interest in biofuels as a replenishable and green energy source. However, developing countries, such as Nigeria, are yet to explore commercial production of bioethanol. Hence, the purpose of this chapter was to optimize and characterize bioethanol production from cassava peels, a cheap and readily abundant feedstock in the country.

In this chapter we have shown that production parameters such as fermentation temperature and period greatly affect the optimum production of bioethanol from cassava peels. The bioethanol yield and quality of cassava peels using the small-scale production facility were improved at 30°C fermentation temperature for 72 hrs. The overall mean peak ethanol yield of 10.49 ± 0.20 L/5 kg (2098.00 ± 40.00 ml/kg), corresponding to a percentage yield of 59.41 ± 1.15%, was obtained in the small-scale production facility. The study clearly demonstrates that the small-scale production facility significantly scaled up the yield of the bioethanol from cassava peels. The distillation equipment is capable of distilling ~60 L of bioethanol per 8 working hours in a day, culminating in 1,200 L of bioethanol production in a month.

This chapter concludes that cassava peels are a viable and sustainable precursor for bioethanol production which can be used to complement gasoline in automobile engines to reduce greenhouse gas emission. Also, bioethanol production from cassava peels can be fully optimized and scaled up to industrial levels by adopting the fabricated small-scale production facility and employing the optimal production conditions obtained in this chapter.

Nomenclature

Q heat quantity
m mass
c specific heat capacity
Δt difference in temperature

TABLE 14.5
Pearson Product–Moment Correlation between Physicochemical Parameters of Cassava Peels and Glucose and TRS Concentrations in the Small-Scale Production Facility

	pH	Moisture Content Fresh Biomass	Moisture Content Dry Biomass	Dry Matter	TOC	TN	TP	C:N	Glucose Concentration	TRS Concentration
pH	1									
Moisture content	0.045	1								
fresh biomass	0.813									
Moisture content	0.444*	0.915**	1							
dry biomass	0.014	0.000								
Dry matter	−0.866**	−0.538**	−0.832**	1						
	0.000	0.002	0.000							
TOC	−0.945**	0.284	−0.126	0.655**	1					
	0.000	0.128	0.508	0.000						
TN	−1.000**	−0.045	−0.444*	0.866**	0.945**	1				
	0.000	0.813	0.014	0.000	0.000					
TP	0.971**	−0.916	0.215	−0.721**	−0.996**	−0.971**	1			
	0.000	0.299	0.253	0.000	0.000	0.000				
C:N	0.994**	0.155	0.540**	−0.916**	−0.903**	−0.994**	0.938**	1		
	0.000	0.412	0.002	0.000	0.000	0.000	0.000			
Glucose	−0.903**	−0.060	−0.418	0.792**	0.847**	0.903**	−0.872**	−0.900**	1	
concentration	0.000	0.796	0.059	0.000	0.000	0.000	0.000	0.000		
TRS	−0.184	−0.398	−0.431	0.354	0.046	0.184	−0.085	−0.226	0.141	1
concentration	0.424	0.074	0.051	0.115	0.842	0.424	0.000	0.324	0.542	

*Correlation is significant at the 0.05 level (two-tailed).
**Correlation is significant at the 0.01 level (two-tailed).

TABLE 14.6
Accessory Information on Small-Scale Production Facility Obtained During Optimal Bioethanol Production from Cassava Peels

Total electrical energy (Kwhr)	183.49 Kwhr (cost of electricity ₦ 57.40 per Kwh, NERC, 2019)
Quantity of heat energy (J) used in distillation	2.05MJ
Total quantity of gas consumed in distilling the optimal bioethanol	4.03 kg (estimated cost is ₦ 300 per kilogram of gas at the time of the research)
Capacity of distillation machine	45 L
Optimal volume of bioethanol produced by the distiller	18.20 L (~59% efficiency)
Production rate of distiller	~1,200 L of bioethanol per month (at the rate of ~60L of bioethanol per eight working hours in a day)
Lagging time in distillation	1 hr, 51 mins
Overall cost of production of optimal bioethanol produced (18.20 L)	₦ 45,141.00k
Estimated price of produced bioethanol per liter using a 10% profit margin	₦ 2,728.30k per liter (global market price of ethanol is \$0.98 (₦ 355.80) per liter at the time of the study) (source: www.GobalPetrolPrices.com)

Note: NERC—National Electricity Regulatory Commission.

TABLE 14.7
Emerging Biofuel Projects in Nigeria Designed to operate on Simultaneous Saccharification and Co-Fermentation (SSCF)

SN	Project	Cost	Location	Owners	Feedstocks	Feedstock Quantity (Tons/Year)	Ethanol Production/Year	Project Phase
1	Automotive biofuel project	\$125M	Ebenebe, Anambra State	NNPC/private sector	Cassava	3–4 million	40–60 million liters	Planning
2	Automotive biofuel project	\$125M	Okeluse, Ondo State	NNNPC/private sector	Cassava	3–4 million	40–60 million liters	Planning
3	Kwara Casplex Ltd.	\$90M estimated	Kwara State	Private/government	Cassava	300,000 estimated	38.86 million liters	EPIC
4	Oke-Ayedun cassava ethanol project	\$18M	Oke-Ayedun, Ekiti State	Ekiti State government/private	Cassava	238,500	38.1 million liters	EPIC
5	CrowNet green energy ethanol plant	\$122M	Iyemero, Ekiti State	Ekiti State government/private	Cassava	150,000	65 million liters	Operational (4 Sept, 2008)
6	Cassava ethanol plant	\$115M	Taraba State	Taraba State	Cassava	300,000	72 million liters	EPIC
7	Niger state government ethanol plant	\$90M estimated	Niger State	Niger State	Cassava	150,000	27 million liters	EPIC
8	Cassava bioethanol project	\$138M	Niger Delta region	NA	Cassava	0.32 million estimated	58 million liters	Conception
9	Cassava industralization project	\$16.4M	Ogun State	Private + government	Cassava	75,000	3 million liters	Conception
10	National Casskero cooking fuel programme	\$1B	36 states + Abuja	Private	Cassava	8 million	1.44 billion liters	EPIC

Source: Ohimain [62].
Note: NA, not Available; EPIC, Electronic Plant Information Centre.

TABLE 14.8
Cost–Benefit Analysis of the Small-Scale Production Facility (SSPF) for Industrial Production Projected at 172.8 Million Liters of Ethanol per Year

Costs			
Category	**Price**	**Scaling-Up Factor**	**Total**
Design and fabrication of the SSPF	₦ 456,000.00k	×12,000	₦ 5.472 billion
Consumables and raw materials used in the bioethanol production	₦ 33,400.00k	×12,000	₦ 400.8 million
Cost of electrical energy consumption (1Kwh = #57.40, NERC, 2019)	₦ 10,532.33k	×12,000	₦ 126.38796 million
Cost of gas consumption (1 kg = #300 at 2019)	₦ 1,209.00k	×12,000	₦ 14.508 million
	₦ 501,141.33k		**₦ 6,013,695,960 (~₦ 6.01 billion)**
Benefits			
Enhanced production rate in excess of 100 million liter of bioethanol			₦ 35.182 billion
Cost saved by utilising cassava peels (CP)			#₦ 1.512 trillion
Marginal profit accrued by adopting the SSPF over the cheapest SSCF process			₦ 415.6 million
Marginal profit from 10% of the Climate Trust Fund (CTF) intervention for clean energy			₦ 2.955 billion
Total Benefits			**₦ 1,550,552,600,000 (~₦ 1.550 trillion)**
Cost–Benefit Analysis of the Small-Scale Production Facility			
Total costs (C)			₦ 6,013,695,960 (~#6.01 billion)
Total benefits (B)			₦ 1,550,552,600,000 (~ #1.550 trillion)
Net cost benefit (B–C)			₦ 1,544,538,904,040 (~#1.544 trillion)
Benefit–cost ratio (rounded)			**257.84**

Note: See Appendix D for details.

APPENDIX

Energy Consumption of Machine in Optimal Bioethanol Production

1 QUANTITY OF HEAT CONSUMED BY DISTILLING THE OPTIMAL BIOETHANOL PRODUCED

- Total quantity of heat energy (J) consumed in the distillation is:

$$\mathbf{Q = [m_e c_e + m_s c_s]\ \Delta t}$$

Since the optimal bioethanol recovered was 18.20 L and the density of ethanol is 0.789 kg/L, hence, mass = density × volume = 0.789 × 18.20 = 14.36 kg.

Where:

Q = total quantity of heat energy consumed during distillation

m_e = mass of optimal bioethanol produced in kilograms = 14.36 kg
c_e = specific heat capacity of ethanol = 2,460 J/kg°C
m_s = mass of the stainless steel distiller in kilograms = 9.8 kg
c_s = specific heat capacity of the stainless steel distiller = 502.416J/kg°C
t_a = initial temperature of the distillation setup = 28°C
t_f = final temperature of the distillation setup = 79°C
$\Delta t = t_f - t_a$ = 79°C – 28°C = 51°C

Thus, $\mathbf{Q = [m_e c_e + m_s c_s]\,\Delta t}$

= [(14.36 × 2460) + (9.8 × 502.416)] × 51
= [35,325.6 + 4,923.6768] × 51
= [40,249.2768] 51 = 2,052,713.1168 J = **2.05 MJ**

2 TOTAL ELECTRICAL ENERGY CONSUMED BY THE SSPF IN THE PRODUCTION OF THE OPTIMAL BIOETHANOL

- Total electrical energy consumed by SSPF (Kwh)
  = electrical energy used by the mixer + electrical energy used by the incubator + electrical energy used by the distiller

Time of hydrolysis = 110 mins = 1.83 hrs
Time of fermentation = 72 hrs
Time of distillation = 172 mins = 2.52 hrs
Power rating of electrical devices of the SSPF
Electric motor (2 hp) of the mixer = 1,500 watts
Electric fan in the incubator = 10 watts
Heating element in the incubator = 2,500 watts
Digital temperature gauge = 10 watts (found in both incubator and distiller)

Thus, the total electrical energy consumed by SSPF (Kwh)
= electrical energy used by the mixer + electrical energy used by the incubator + electrical energy used by the distiller
(Recall, electrical energy = electrical power × time)
= (1,500 × 1.83) + (2,510 × 72) + (10 × 2.52)
= 2,745 + 18,0720 + 25.20
= 183,490.20 watts hr
= **183.49 Kwh**

3 ESTIMATED COST OF ELECTRICITY USED IN THE PRODUCTION OF THE OPTIMAL BIOETHANOL

Cost of electricity in Nigeria at the time of the study is estimated at #57.40 per Kwh (source: Nigeria Electricity Regulatory Commission, NERC).

1 Kwh = ₦ 57.40
183.49 Kwh = ₦ 10,532.33k

Hence, the total cost of electricity used by the SSPF in the production of the optimal bioethanol is **₦ 10,532.33k**.

4 LAGGING TIME, THAT IS, TIME DELAY FOR ANOTHER OF DISTILLATION TO OCCUR

= time taken to reach the BP + time taken for distillation to complete
= 43 mins + 68 mins
= 111 mins
= **1 hrs, 51 mins**

5 QUANTITY AND COST OF GAS CONSUMPTION BY THE MIXER AND DISTILLER IN THE OPTIMAL BIOETHANOL PRODUCTION AT THE TIME OF THE STUDY

Total gas consumed by the mixer and distiller

= 3.11 kg (mixer) + 0.92 kg (distiller)
= **4.03 kg**

Hence, the total quantity of gas consumed by the mixer and distiller is 4.03 kg.

Cost of gas consumption used by the mixer and distiller

1 kg of gas = ₦ 300
4.03 kg of Gas = ₦ 1,209.00k

Hence, the cost of gas consumption by the mixer and distiller is **₦ 1,209.00k**.

6 ESTIMATED PRICE OF THE PRODUCED BIOETHANOL PER LITER

Cost of consumables = ₦ 33,400.00k
Cost of electricity consumption = ₦ 10,532.33k
Cost of gas consumption = ₦ 1,209.00k

Overall cost of production of optimal bioethanol is calculated as:

= cost of consumables + cost of electricity + cost of gas consumed
= ₦ 33,400.00k + ₦ 10,532.33k + ₦ 1,209.00k
= ₦ 45,141.33
= **~₦ 45,141.00k**

Hence, the overall cost of production of **18.20 L** optimal bioethanol is **₦ 45,141.00k**.

Using a 10% profit margin (**#4514**), the estimated price of the optimal bioethanol per liter is estimated as:

selling price (SP) of bioethanol = cost production + profit

= ₦ 45,141 + #₦ 4514
= ₦ 49,655.00k

The estimated price of produced bioethanol per liter using a 10% profit margin is equal to:

$$= \frac{₦\ 49{,}655.00k}{18.20}$$
= **₦ 2,728.30k**

Cost–Benefit Analysis

The production rate of the small-scale production facility (SSPF) is 1,200 L of ethanol per month; hence, the annual rate production is 14,400 L of ethanol per annum (based on the 45 L capacity of distiller).

Scaling-up to industrial level by choosing an arbitrary 540,000 L capacity of distiller
45 L distiller can produced 14,400 L of ethanol
540,000 L distiller will produce $= \frac{540{,}000 \times 14{,}400}{45}$
= 172,800,000 liters of ethanol per year

The scaling-up factor/multiplier was calculated as $= \frac{540{,}000 \text{ L distiller capacity.}}{45 \text{ L distiller capacity}}$
= **12,000**

Hence, the multiplier (12,000) was used to multiply the costs of fabricating machine, consumables and chemicals, energy consumption, etc. to arrive at the overall cost of SSPF at industrial scale.

The maximum projected bioethanol production from the proposed biorefinery plants in Nigeria is estimated at 72 million liters per annum, while the SSPF is projected to produce 172.8 million liters of ethanol per annum. Thus, the enhanced production rate is calculated as:

Hence, excess production rate = 172.8 – 72 million liters of ethanol
= 100 million liters of ethanol

Since 1 liter = $0.98 (current global mrket price of ethanol) (source: www.GlobalPetrolPrices.com), then market price of the excess bioethanol is 100 million × $0.98 = $98 million.

= **₦ 35.182 billion** (using exchange rate at September 9, 2019, that is, $1 = ₦ 359)

Nigeria is the largest producer of cassava in the globe, producing about 54 million metric tons per annum (FAO, 2013). Hence, cost saved by utilizing cassava for the production of garri and other processed food of human consumption as against its used for ethanol production is estimated as:

Since 1 ton of cassava = ₦ 28,000 (source: www.africanprice.com),

hence, 54,000,000 tons of cassava = 54,000,000 × #28,000
= **₦ 1.512 trillion**

The marginal profit that will be accrued by adopting the SSPF in lieu of the cheapest emerging biorefinery in the country was calculated.

The cheapest of the emerging biorefineries in the country was the Cassava Industralization Project located in Ogun state, according to Ohimain (2010). This was estimated at a cost of \$16.4 million, which is equivalent to ₦ 5,887,600,000 (using \$1 = ₦ 359 as at September 9, 2019).

The cost of designing and fabricating the SSPF was estimated at #5.472 billion = ₦ 5,472,000,000.

Thus, the marginal profit is = ₦ 5,887,600,000 – ₦ 5,472,000,000
= ₦ 415,600,000
= **₦ 415.6 million**

The Climate Trust Fund (CTF) investment plan appropriated by the Nigeria government is estimated at \$250 million (source: www.climateinvestmentfunds.org).

About 10% of this fund (i.e., \$25 million) is allotted for clean energy = ₦ 8.975 billion.

The overall cost of producing 172.8 million liters of ethanol per year by scaling up the SSPF at industrial level was estimated at **₦ 6.02 billion** (see Table 4.21).

Hence, the net gain by the government in adopting the SSPF technology for clean energy production is = **(₦ 8.975 – ₦ 6.02) billion**
= **₦ 2.955 billion**

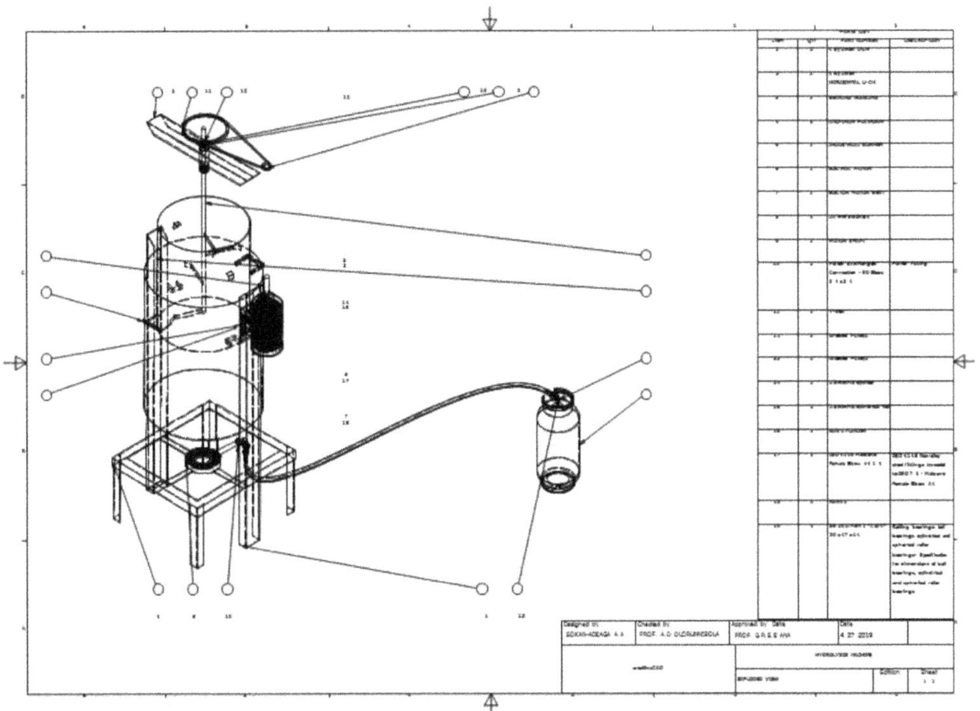

FIGURE 14.A1 Engineering sketch of the mixing machine/hydrolysis chamber.

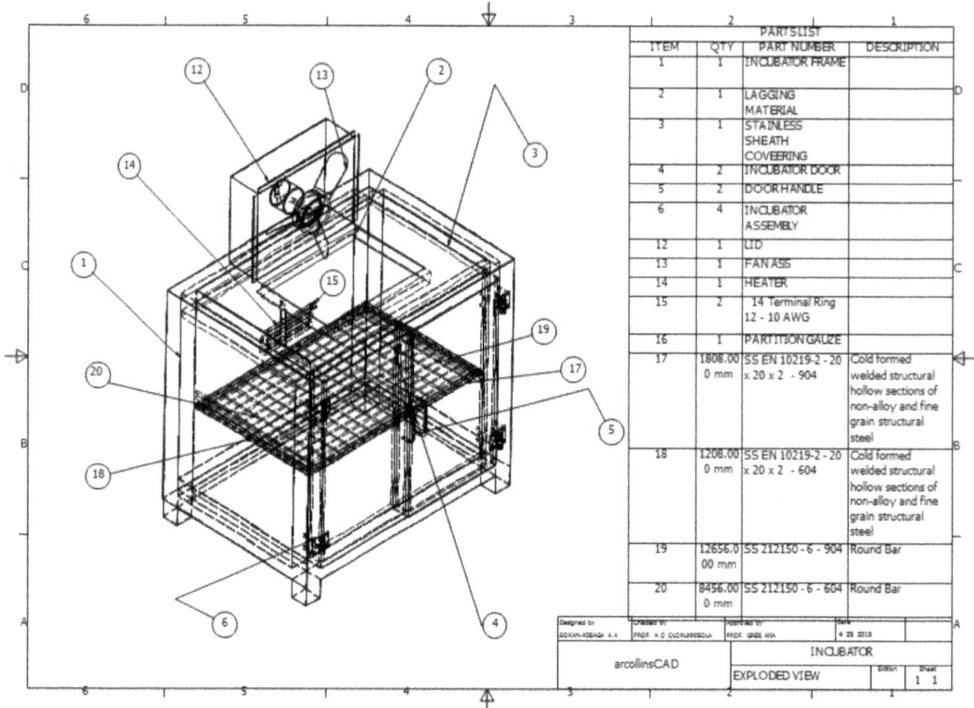

FIGURE 14.A2 Engineering sketch of the incubator.

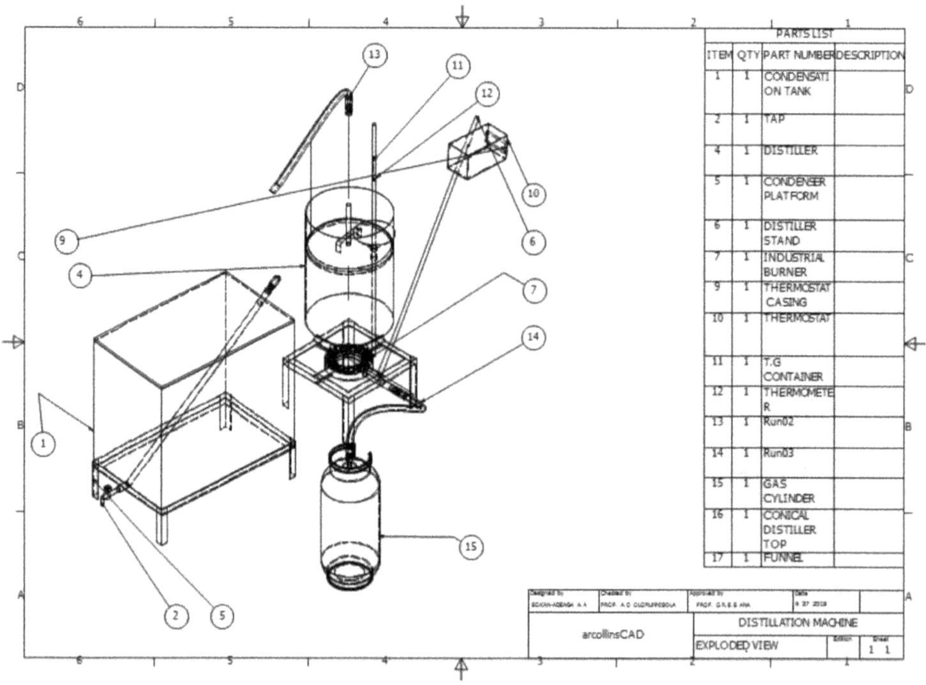

FIGURE 14.A3 Engineering sketch of the distillation machine (distiller and cooling tank).

Appendix: Supplementary Information on the Small-Scale Production Facility (SSPF)

Mixing Machine/Hydrolysis Chamber

S/N	Item Name	Quantity	Material	Dimension
1	Electric motor (2 hp, single phase) (1,500 W)	1		
2	Belt	1	Rubber	B × 56
3	Pulley	1	Stainless steel	Diameter (Φ) = 25 cm
4	Inner cylinder	1	Galvanized steel	Φ = 47 cm; height = 60 cm
4	Outer cylinder	1	Galvanized steel	Φ = 60 cm; height = 79 cm
5	U-channel 4 × 2 in (as supporting frame for the electric motor and resting of the pulley)	2	Mild steel	Length = 152 cm; width = 9 cm
6	U-channel 3 × 2 in (for suspending the stirring rod)	1	Mild steel	Length = 71 cm; width = 9 cm
7	Industrial burner	1	Mild steel	
8	Angle rod (for the base and connection of the burner)	4 (2 length)	Mild steel	
9	Stirrer (stirring rod)	1	Mild steel	Length = 115 cm
10	Bolt	4	Mild steel	12–13 mm
11	Electric wire	1	Copper covered with rubber	3 yards
12	Stopper (16 mm rod)	4	Mild steel	5 cm
13	Bearing housing	2	Mild steel	Height = 17 cm; Φ = 5 cm
14	Washer (to avoid seizure)	1	Steel	Φ = 5 cm
15	Industrial burner	1	Mild steel	Φ = 18 cm
15	Gas cylinder	1	Galvanized steel	12.5 kg
16	Hose	1	Rubber	2 yds
17	Regulator	1		

Incubator

S/N	Item Name	Quantity	Material	Dimension
1	Electric heating filament (2,500 W)	1	Mild steel	Spiral shape
2	Industrial electric fan/blower (10 W)	4 blades	Aluminium	
3	Digital temperature gauge (10 W)	1	Thermoelectric substance cage in a plastic container	
4	Case for temperature gauge	1	Galvanized steel	Length = 20 cm; width = 8 cm; height = 11 cm
5	Sensors	1	Copper coil	
6	Covering case for incubator	1	Galvanized steel	Length = 60 cm; width = 40 cm; height = 26 cm
7	Lag (to prevent heat loss)	—	Fiber	—
8	Incubator door	2	Galvanized steel	Length = 87 cm; width = 50 cm
9	Handle for incubator door (twist rod 11 mm rod)	2	Mild steel	Length = 15 cm; width = 5 cm
10	Incubator body	1	Galvanized steel	Length = 107 cm; width = 76 cm; heigth = 95 cm
11	Handle for incubator body	2	Mild steel	Length = 76 cm; width = 5 cm
12	Stand/supporting rod	4	Galvanized steel	Length = 6 cm; width = 3 cm
13	Electric wire	1	Copper covered with rubber	2 yds
14	Hanger (tray) to create partition inside the incubator where container can rest	1	Galvanized steel	Length = 100 cm; width = 47 cm

Distillation Machine

S/N	Item Name	Quantity	Material	Dimension
1	Distiller	1	Stainless steel (1.5 sheets)	Φ = 95 cm; height = 46 cm
2	Conical top of distiller	1	Stainless steel	Slant height = 16 cm; Φ = 95 cm
3	Distiller funnel	1	Stainless steel	Φ = 12 cm; slant height = 8 cm
4	Tap for distiller	1	Stainless steel	Φ = 2.0 cm
5	Collection pipe for distillate	1	Stainless steel	Length = 50 cm; Φ = 8 cm
6	Adjoining nut (for connecting the collection pipe in distiller to the collection pipe of the cooling tank)	1	Galvanized steel	Φ = 10 cm
7	Supporting frame for distiller Stand Rectangular base	4 1	Galvanized steel Galvanized steel	Height = 46 cm 30 cm × 30 cm
8	Cooling tank	2 sheets	Mild steel	Length = 64 cm; width = 39 cm; height = 54 cm
9	Stand for the cooling tank (Angle rod)	1 length	Mild steel	Length = 62 cm; width = 57 cm; height = 36 cm
10	Collection pipe in cooling tank	1	Stainless steel	Length = 90 cm; Φ = 7 cm
11	Tap for cooling tank	1	Galvanized steel	Φ = 1.5 cm
11	Outlet for discharging water	1	Galvanized steel	Length = 18 cm; Φ = 12 cm
12	Digital temperature gauge	1	Thermoelectric substance cage in a plastic container	
13	Case for temperature guage	1	Galvanized steel	Length = 18 cm; width = 10 cm; height = 10 cm
14	Supporting stand for temperature gauge	1	Mild steel	Length = 150 cm
15	Low-pressure burner	1	Mild steel	Φ = 14 cm
16	Gas cylinder	1	Galvanized steel	12.5 kg
17	Hose	1	Rubber	2 yds
18	Regulator			
19	Electric wire	1	Copper covered with rubber	1 yd

REFERENCES

1. Chandel, A.K., Chan, E.S., Rudravaram, R., Narasu, M.L., Rao, L.V., and Ravindra, P. (2007a). Economics and environmental impact of bioethanol production technologies: an appraisal. *Biotechnology and Molecular Biology Review*, 2(1): 14–32.
2. Sandia National Laboratories Archives (2010). *Energy and Climate (EC) Program Management Unit (PMU): 2010 LDRD Annual Report*. Available at: www.sandia-gov/research/laboratoryess. Accessed on February 2010.
3. Sharma, N., Kalra, K.L., Oberoi, H.S., & Bansal, S. (2007). Optimization of fermentation parameters for production of ethanol from kinnow taste and banana peels by simultaneous saccharification and fermentation. *Indian Journal of Microbiology*, 47: 10–316.
4. Sokan-Adeaga, A.A., & Ana, G.R.E.E. (2015). A comprehensive review of biomass resources and biofuel production in Nigeria: potential and prospects. *Reviews on Environmental Health*, 30(3): 143–162. http://doi.org/10.1515/reveh-2015-0015.
5. Sokan-Adeaga, A.A., Ana, G.R.E.E., & Sokan-Adeaga, E.D. (2015). *Evaluation of bio-Ethanol Potentials of Selected Lignocellulosic Wastes*. Verlag LAMBERT Academic Publishing, Saarland, Germany, pp. 1–150. ISBN: 978-3-659-79582-4.

6. Agarwal, A.K. (2005). Biofuels. In *Wealth from Waste; Trends and Technologies.* Edited by B. Lal and M.R.V.P. Reedy, 2nd edition. New Delhi: The Energy and Resources Instituted (TERI) Press. ISBNSI-7993-067-X.
7. Sokan-Adeaga, A.A. (2022). Nigeria's perennial energy crisis and the way forward. *Academia Letters*, Article 5098. https://doi.org/10.20935/AL5098
8. Demirbas, A. (2008). Biofuels sources, biofuel policy, biofuel economy and global biofuel projections. *Energy Conversion and Management*, 49: 2106–2116.
9. Oseji, E.M., Ana, G.R.E.E., and Sokan-Adeaga, A.A. (2017). Evaluation of biogas yield and microbial species from selected multi-biomass feedstocks in Nigeria. *London Journal Research Science (LJRS)*, 17(1): 1–19. https://doi.org/10.17472/LJRSVOL17IS1PG1.
10. Food and Agriculture Organisation (FAO), (2008). The role of agricultural biotechnologies for production of bioenergy in developing countries. *Electronic Forum on Biotechnology in Food and Agriculture: Conference 15 from 10 Nov. to 14 December, 2008*, 25 August 2009. Available at https://www.fao.org/3/article/al311e.pdf
11. Hossain, A.B.M.S., Saleh, A., Boyce, A.N., Partha, P., and Naqiuddin, M. (2008). Bioethanol production from agricultural waste biomass as a renewable bioenergy resource in biomaterials. *The 4th International Biomedical Engineering Conference Nikko Hotel*, Kuala Lumpur, Malaysia. 26 Jun 2008 to 28 Jun 2008, University of Malaya, Malaysia, Proceeding.
12. Hossain, A.B.M.S., and Fazliny, A.R. (2010). Creation of alternative energy by bio-ethanol production from pineapple waste and the usage of its properties for engine. *African Journal of Microbiology Research*, 4(6): 813–819.
13. Mathewson, S.W. (1980). *The Manual for Home and Farm Production of Alcohol Fuel.* Berkeley, CA: Ten Speed Press, JA. Diaz Publications, pp.1–12.
14. Dhillon, G.S., Bansal, S., and Oberoi, H.S. (2007). Cauliflower waste incorporation into cane molasses improves ethanol production using Saccharomyces cerevisiae MTCC 178. *Indian Journal of Microbiology*, 47: 353–357.
15. Lakhfif, F., Nemouchi, Z., and Mebarek-Oudina, F. (2016). Numerical investigation of the different spray combustion models under diesel condition. *International Journal of Applied Engineering Research*, 11(18): 9393–9399.
16. Balat, M., and Balat, H. (2009). Recent trends in global production and utilisation of bioethanol fuel. *Applied Energy*, 86: 2273–2282.
17. Farrell, A.E., Plevin, R.J., Turner, B.T., Jones, A.D., O'Hare M., and Kammen, D.M. (2006). Ethanol can contribute to energy and environmental goals. *Science*, 311(5760): 506–508.
18. Gourari, S., Mebarek-Oudina, F., Makinde, O.D., and Rabhi, M. (2021). Numerical investigation of gas-liquid two-phase flows in a cylindrical channel. *Defect and Diffusion Forum*, 409: 39–48. https://doi.org/10.4028/www.scientific.net/DDF.409.39.
19. Keeney, D., and Muller, M. (2007). Ethanol production: environmental effects. *Institute for Agriculture and Trade Policy*, Minneapolis, August 15, 2011. Available at: www.iatp.org./iatp.
20. Sielhorst, S., Molenaar, J.W., and Offermans, D. (2008). *Biofuels in Africa: An Assessment of Risks and Benefits for African Wetlands.* Wageningen, The Netherlands: Wetlands International.
21. Sahel and West Africa Club (SWAC)/Organization for Economic Co-operation and Development (OECD). *Green Fuels for Development? Improving Policy Coherence in West Africa.* SWAC Briefing Note 2, September. SWAC/OECD, 2008.
22. Sokan-Adeaga, A.A., Ana, G.R.E.E., Sokan-Adeaga, M.A., and Sokan-Adeaga, E.D. (2016). Lignocelluloses: An economical and ecological resource for bio-ethanol production—A review. *International Journal of Natural Resource Ecology and Management*, 1(3): 128–144. http://doi.org/10.11648/j.ijnrem.20160103.18
23. Mathew, G., Sukumaran, R., Singhania, R., and Pandey, A. (2009). Progress in research on fungal cellulases for lignocellulose degradation. *Journal of Scientific and Industrial Research*, 67: 898–907.
24. Sokan-Adeaga, A.A., and Ana, G.R.E.E. (2018). Bioconversion of some selected lignocelluloses into bioethanol—A form of waste management strategy. *Winning Essay and Extract of the Proceedings of ISWA-SWIS Winter School 2018*, Held at the University of Texas, Arlington, USA, 15–26 January,

2018, Pp 1–10. Published by International Solid Waste Association (ISWA). Available online as ISWA SWIS Winter School 2018 Proceedings Extract at. Available at: www.iswa.org/media/publications/knowledge-base/

25. Sokan-Adeaga, A.A (2019). Waste quantification, characterization and exploration of bioethanol production potentials of some selected lignocelluloses wastes in Ibadan, Nigeria. In *2018 ISWA-SWIS Winter School Proceedings eBook* (pp. 183–192). Edited by M.D. Sahadat Hossain, A. Vance Kemler, and Brenda A. Haney. SWIS Publications. ISBN 978-0-9976542-2-6. (pp. 1–310). Published 29 May, 2019. www.amazon.com/2018-ISWA-SWIS-WINTER-SCHOOL-PROCEEDINGS-ebook/dp/B07SH2P7QB
26. Zhu, M., Li P., Gong, X., and Wang, J. (2012). A comparison of the production of ethanol between simultaneous saccharification and fermentation and separate hydrolysis and fermentation using unpretreated cassava pulp and enzyme cocktail. *Bioscience, Biotechnology, and Biochemistry*, 76(4): 671–678.
27. Wingrel, A., Galbe, M., and Zacchi, G. (2003). Techno-economic evaluation of producing ethanol from softwood—a comparison of SSF and SHF and identification of bottlenecks. *Biotechnology Progress*, 19(4): 1109–1117.
28. Farone, W.A., and Cuzens, J.E. (1996a). *Method of Producing Sugars Using Strong Acid Hydrolysis of Cellulosic and Hemicellulosic Materials* (US Patent No. 5 562 777). Irvine, CA: Arkenol, Inc.
29. Farone, W.A., and Cuzens, J.E. (1996b). *Method of Separating Acids and Sugars Resulting from Strong Acid Hydrolysis* (US Patent No. 5 580 389). Irvine, CA.
30. Ana, G.R.E.E., and Sokan-Adeaga, A.A. (2015). Bio-ethanol yield from selected lignocellulosic wastes. *International Journal of Sustainable and Green Energy*, 4(4): 141–149.
31. Bolade, D.O., Ana, G.R.E.E., Lateef, S.A., and Sokan-Adeaga, A.A. (2019). Exploration of the bioethanol yield of single and multi-substrate biomass from cassava processing wastes. *Journal of Solid Waste Technology and Management (JSWTM)*, 45(3): 305–314(10). https://doi.org/10.5276/jswtm/2019.305.
32. Hossain, A.B.M.S., Ahmed, S.A., Ahmed, M.A., Faris M.A.A., Annuar, M.S.M., Hadeel, M., and Norah, H. (2011). Bioethanol fuel production from rotten banana as an environmental waste management and sustainable energy. *African Journal of Microbiology Research*, 5(6): 586–598.
33. Ezekiel, O.O., & Aworh, O.C. (2013). Solid state fermentation of cassava peel with enrichment. *International Journal of Biological, Biomolecular, Agricultural, Food and Biotechnological Engineering*, 7(3): 202–209.
34. Otache, M.A., Ubwa, S.T., and Godwin, A.K. (2017). Proximate analysis and mineral composition of peels of three sweet cassava cultivar. *Asian Journal of Physical and Chemical Sciences*, 3(4): 1–10.
35. Ubalua, A.O. (2007). Cassava wastes: Treatment options and value addition alternatives. *African Journal of Biotechnology*, 6(18): 2065–2073.
36. Yoonan, K., & Kongkiattikajorn, J. (2004). A study of optimal conditions for reducing sugars production. *Kasetsart Journal (Natural Science)*, 38: 29–35.
37. Adelekan, B.A. (2012). Cassava as a potent energy crop for the production of ethanol and methane in tropical countries. *International Journal of Thermal & Environmental Engineering*, 4(1): 25–32.
38. Sokan-Adeaga, A.A., and Ana, G.R.E.E (2015). Source identification and characterisation of lignocellulosic wastes from selected biomass sources in Ibadan, Nigeria. *Journal of Solid Waste Technology and Management*, 41(3): 262–269.
39. McCauley, A., Jones, C., and Jacobsen, J. (2009). Soil pH and organic matter. *Nutrient Management Module*, 8(2): 1–12.
40. Lal, B., and Reddy, M.R.V.P. (2005). *Wealth from Waste: Trends and Technologies*, 2nd edition. New Delhi, India: The Energy and Resources Institute (TERI) Press.
41. Mshandete, A., Kivaisi, A., Rubindamayugi, M., and Mattiasson, B. (2004). Anaerobic batch co-digestion of sisal pulp and fish wastes. *Bioresource Technology*, 95(1): 19–24.
42. Yen, H.W., and Brune, D.E. (2007). Anaerobic co-digestion of algal sludge and waste paper to produce methane. *Bioresource Technology*, 98(1): 130–134.
43. Wellinger, A., and Linberg, A. (2000). *Task 24: Energy from Biological Conversion of Organic Waste. Biogas Upgrading and Utilisation*. Paris: International Energy Agency.

44. Beatrix, R., Miklos, S., and Gyorgy, F. (2010). Codigestion of organic waste and sewage sludge by dry batch anaerobic treatment. *Biotechnology Bioengineering*, 23: 1591–1610.
45. Malherbe, S., and Cloete, T.E. (2003). Lignocellulose biodegradation: Fundamentals and Applications—A review. *Environmental Science and Biotechnology*, 1: 105–111.
46. Zhu, M., Li, P., Gong, X., and Wang, J. (2012). A comparison of the production of ethanol between simultaneous saccharification and fermentation and separate hydrolysis and fermentation using unpretreated cassava pulp and enzyme cocktail. *Bioscience, Biotechnology, and Biochemistry*, 76(4): 671–678.
47. Sobrinho, V.S., Ferreira da Silva, V.C., and Cereda, M.P. (2011). Fermentation of sugar cane juice (Sacharum officinarum) cultivar RB 7515 by wild yeasts resistant to UVC. *Journal of Biotechnology and Biodiversity*, 2: 3–21.
48. Fakruddin, M., Islam, A., Ahmed, M.M., and Chowdhury, N. (2013). Process optimisation of bioethanol production by stress tolerant yeasts isolated from agro-industrial waste. *International Journal of Renewable and Sustainable Energy*, 2(4): 133–139.
49. Fakruddin, M.D., Abdul Quayum, M.D., Ahmed, M.M., and Choudhury, N. (2012). Analysis of key factors affecting ethanol production by saccharomyces cerevisiae IFST-072011. *Biotechnology*, 11: 248–252.
50. McMeckin, T.A., Olley, J., Ratkwsky, D.A., and Ross, T. (2002). Predictive microbiology: Towards the interface and beyond. *International Journal Food Microbiology*, 73: 395–407.
51. Phisalaphong, M., Srirattana, N., & Tanthapanichakoon, W. (2006). Mathematical modeling to investigate temperature effect on kinetic parameters of ethanol fermentation. *Biochemistry Engineering Journal*, 28: 36–43.
52. Vaidya, H., Prasad, K.V., Khan, M.I., Mebarek-Oudina, F., Tlili, I., Rajashekhare, C., Elattar, S., Khan, M.I., and Al-Gamdi, S.G. (2022). Combined effects of chemical reaction and variable thermal conductivity on MHD peristaltic flow of Phan-Thien-Tanner liquid through inclined channel. *Case Studies in Thermal Engineering*, 36: 102214.
53. Balachandra, H., Choudhari, R., Vaidya, H., Mebarek-Oudina, F., Manjunatha, G., Prasad, K., and Prathiksha, V. (2021). Homogeneous and heterogeneous reactions on the peristalsis of bingham fluid with variable fluid properties through a porous channel. *Journal of Advanced Research in Fluid Mechanics and Thermal Sciences*, 88(3): 1–14. https://doi.org/10.37934/arfmts.88.3.119
54. Abdulkareem, A., Saka, A., Afolabi, S., and Ogochukwu, M.U. (2015). Production and characterisation of bioethanol from sugarcane bagasse as alternative energy sources. *Proceedings of the World Congress on Engineering 2015 Vol II WCE 2015*, July 1–3, 2015, London, UK.
55. Pippo, W.A., and Carlos, L.A. (2013). Sugarcane energy use: Accounting feedstock energy considering current agro-industrial trends and their feasibility. *International Journal of Energy and Environmental Engineering*: 4–10.
56. Suryawati, L., Wilkins, M.R., Bellmer, D.D., Huhnke, R.L., Maness, N.O., and Banat, I.M. (2008). Simultaneous saccharification and fermentation of Kanlow switchgrass pretreated by hydrothermolysis using Kluyveromyces marxianus IMB4. *Biotechnology and Bioengineering*, 101(5): 894–902.
57. Faga, B.A., Wilkins, M.R., and Banat, I.M. (2010). Ethanol production through simultaneous saccharification and fermentation of switchgrass using Saccharomyces cerevisiae D5A and thermotolerant Kluyveromyces marxianus IMB strains. *Bioresource Technology*, 101: 2273–2279.
58. Maity, J.P., Hou, C.P., Majumder, D., Bundschuh, J., Kulp, T.R., Chen, C.Y., Chuang, L.T., Chen, C.N.N., Jean, J., Ojeda, T.K., Sánchez, E., and Kafarov, V. (2011). Sustainable ethanol production from lignocellulosic biomass—application of exergy analysis. *Energy*, 36: 2119–2128.
59. Pimpakan, P., Yongmanitchai, W., and Limtong, S. (2012). Bioethanol production from sugar cane syrup by thermo-tolerant yeast, Kluyveromyces marxianus DMKU3-1042, using fed-batch and repeated-batch fermentation in a nonsterile system. *Kasetsart Journal (Natural Science)*, 46: 582–591.
60. Li, Y., Horsman, M., Wu, N., Lan, C.Q., and Dubois-Calero, N. (2008). Biofuels from microalgae. *Biotechnology Progress*, 24: 815–820.

61. Umamaheswari, M., Jayakumari, M., Maheswari, K., Subashree, K.M., Mala, P., Sevanthi, T., and Manikandan, T. (2011). Bioethanol production from cellulosic materials. *International Journal of Current Research*, 1: 005–011.
62. Ohimain, E.I. (2010). Emerging bio-ethanol projects in Nigeria: Their opportunities and challenges. *Energy Policy*, 38: 7161–7168.

15 Modelling and Control of a New Au/SiO_2 Optical Nano-Robot Using Backstepping Adaptive-Based Strategy

F. Srairi, K. Chara, and K. Mokhtari

15.1 INTRODUCTION

Recently, the locomotion of self-propelled devices in micro/nano-scale has attracted a great interest of researches in different disciplines, such as physics and micro/nanotechnologies. To perform living actions, the self-propelled devices can adopt energy from the external field or chemical reactions. The optical nano-robots play a fundamental role in science and technological development; it can be used in sub-diffraction imaging as well as microsphere lenses basing on meta-material super-lenses and hyper-lenses [1–10].

The thermo-diffusion is a theoretically known aspect, but the molecules explanations under viscous liquid remains under debate. Thermo-diffusion in water might lead to forceful all-optical screening ways for biomolecules and colloids that motivate the theoretical understanding. As well, the thermo-diffusion can move the molecules optically, which allows to complete the methods as optical tweezers or electrophoresis. Furthermore, the thermo-diffusion allows the manipulation on micro-scale of particles and molecules [11, 12].

The manufacture of nano-robot requires understanding of the phenomena of nature. In particular, the autonomous displacement of natural bio-motors is based on the advantage of the spontaneous hydrolysis of biological energy units. It is worth noting that the movements and swimming principle of microorganisms, bacteria, and devices, including spermatozoids, are based on the deformation of their bodies [3, 10].

The nano-robot based on chemical power can generate a thermal gradient or bubbles to propel in a fluid medium by triggering chemical reactions as a reactant and/or a catalyst. Additionally, based on the Mg-water reaction, Mg-based Janus can be propelled efficiently by the chemical reaction that continuously generates hydrogen bubbles [4]. Due to the fateful advancement in this field these last few years, chemical feed devices have been manufactured and examined from the practical point of view. In order to produce a movement of displacement, an external magnetic field has to be applied on nanoparticles. In this context, numerous works are carried out with respect to various aspects [12]. Indeed, ultrasound waves have often been exploited as an external stimulus for the excitation of nanoparticles [11]. The main drawbacks of these works concern the shape of the proposed structures, which is spherical, where the drag force is important.

In this chapter, a mechanical model of nano-robot is developed by taking into account the effect of all forces acting on the structure, including the Brownian force impact. The propped structure is composed of two optical metals, namely, the gold (Au) and the silicon dioxide (SiO_2), in order to boost optical performances. The aim of the modelling way is to reduce refracting light and to improve the absorbance behavior, which permits the increase of the nano-robot thrust force. Thus, the shape of the proposed structure is ellipsoidal, which allows the decrease of both the drag force and the impact of vortices. In addition, a new numerical model based on accurate solution of Navier–Stokes equation is developed, which allows the study of the nano-robot behavior in H_2O_2

DOI: 10.1201/9781003299608-15

fluid, where the absorbance is calculated. This study shows that the ellipsoidal shape has a great impact to improve the overall electromechanical performances in comparison with conventional designs. Furthermore, the robustness and the quality of the tracking along a reference trajectory are improved thanks to a backstepping adaptive-based control strategy.

15.2 MATHEMATICAL MODELLING

Figure 15.1 depicts the ellipsoidal shape of the considered structure composed of Au and SiO_2, where the different acting forces are shown. It is assumed that the movement of the nano-robot is mainly due to three origins: a Brownian, thermo-diffusion, and viscous force that is governed by the Stokes equation.

15.1.1 Study of the Photo-Thermal Mechanism

The photo-thermal materials are able to absorb the electromagnetic waves under light irradiation to produce a photo-thermal effect that can be used to provide directional guidance by light intensity [13].

In the case of mono-crystalline materials, the mobile carriers generated by external irradiation gain energy, and then electron–electron collisions occur and produce hot electrons. Subsequently, these hot electrons cool rapidly due to the aqueous environment that absorbs the thermal energy by phonon–phonon relaxation. This latter causes a rise in temperature of the environment and degrades the thermal equilibrium that is established between electrons and crystalline assembly by electron–phonon interaction.

In terms of measurement, the temperature distribution around the nanoparticle can only be measured beyond 1 μm from the surface of the nanoparticle.

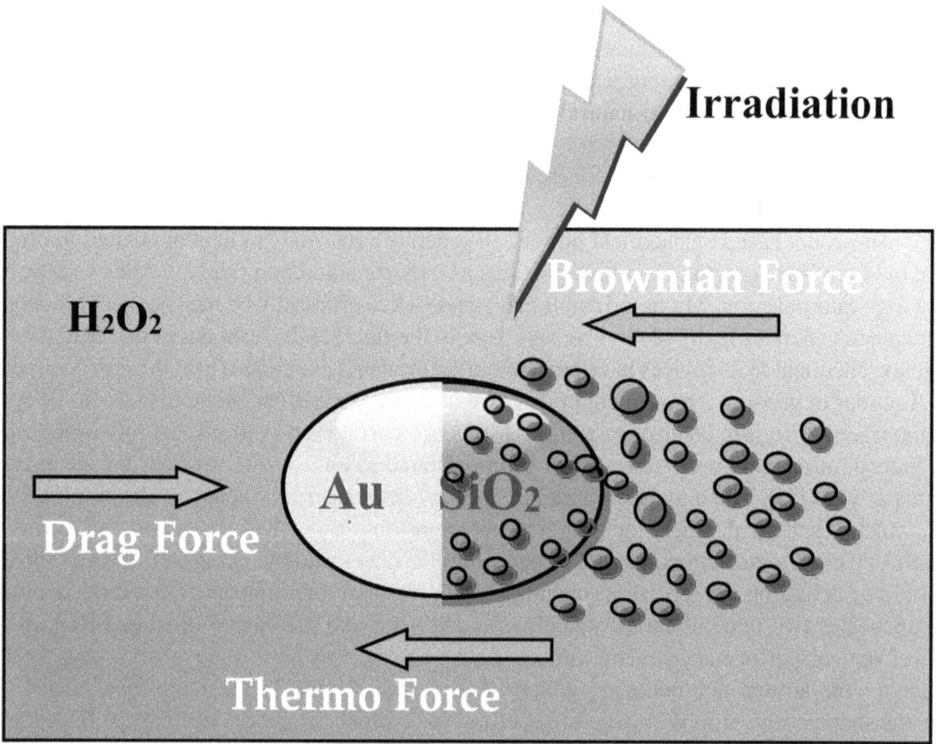

FIGURE 15.1 The proposed optical nano-robot.

In general, the heat transfer and thermal diffusion equations provide the theoretical expression of the thermal distribution in the local neighborhood of the photo-thermal particle, given by [2, 5]:

$$\rho_p(r)c(r)\frac{\partial T(r,t)}{\partial t} = \nabla k(r)\nabla T(r,t) + Q(r,t) \tag{15.1}$$

Where r and t are the spherical coordinate and time, respectively; $\rho(r)$, $c(r)$, and $k(r)$ represent the mass density, the specific heat, and the thermal conductivity, respectively; $T(r, t)$ is the local temperature; and $Q(r, t)$ represents the energy of the light irradiation of the laser stimulus.

More precisely, $Q(r,t) = \langle j(r,t), E(r,t)\rangle_t$

$j(r,t)$ is the current density, and $E(r,t)$ represents the electric field in the system that is calculated from a Maxwell equation system.

In steady state, the approximate temperature of the nanoparticle is described by:

$$\Delta T(r) = \frac{VQ}{4\pi k_0 r} \tag{15.2}$$

In which r is the distance from the center of the particle, with k_0 as the thermal conductivity of the medium and V as the volume of the photo-thermal particle.

The temperature is asymmetric around the nanoparticle due to half-side absorption. In view of this, the temperature difference across the nanoparticle is described as [12]:

$$\Delta T = \frac{3\varepsilon I a}{2(2k_e + k_{NP})} \tag{15.3}$$

Where ε is the photo-thermal-side absorption efficiency, I is the laser intensity, a is the minor radius of the nanoparticle, and k_e and k_{NP} are, respectively, the thermal conductivity of the surrounding fluid and the nanoparticle. The thermo-diffusion rate is described according to the Soret coefficient as [12]:

$$U = -DS_T\frac{\Delta T}{3a} \tag{15.4}$$

Where $S_T = \frac{D_T}{D}$; S_T is the Soret coefficient, D_T represents the coefficient of thermo-diffusion, and D is the diffusivity of Einstein in an infinitely diluted liquid environment.

15.1.2 Thermo-Diffusion Force

The thermo-diffusion force can be considered as deriving from speed. The expression connects the hydrodynamic force F_T exerted by the viscous and thermal stresses on the nano-robot at its speed U_{th} [4].

$$F_T = \frac{U_{th}}{M} \tag{15.5}$$

According to the Stokes law, $M = (6\pi\eta a U_{th} k')^{-1}$.

On the other hand, the gradient can be imposed externally, for which the speed is described by the Soret coefficient [3, 4].

$$U_{th} = -DS_T\nabla T \tag{15.6}$$

15.1.3 Brownian Force

Some symmetrical photo-thermal particles usually exhibit significant Brownian motion under light irradiation. To include such effects in the simulation, the Brownian force F_{Br} is modeled as a random Gaussian white noise process that has zero mean value [12–15]. This force is produced by the impact of adjacent fluid molecules; it is considered equivalent to a force acting on the center of the nano-robot.

$$F_{Br} = \chi\sqrt{\frac{12\pi K_B \mu a T_0}{\Delta t}} \tag{15.7}$$

Where χ is a Gaussian random number of zero mean, K_B is the Boltzmann constant, T_0 represents the thermodynamic temperature, μ is the fluid viscosity, and Δt is the observation time interval that is also the step used in the simulation.

15.1.4 Drag Force

The drag force represents the influence of particles on the fluid flow field. It is calculated by averaging all the influences in the volume of the fluid element. In this case, the drag force applied on the optical nano-robot is approximated by [12]:

$$F_d = -\frac{1}{2}\rho A C_d V_r^2 \tag{15.8}$$

Where ρ is the fluid density, A represents the nano-robot's frontal surface, and C_d is the drag coefficient, which is expressed in function of Reynolds number; for an optical nano-robot immersed in a fluid, the drag coefficient is given by:

$$C_d = \frac{24}{R_e};\ R_e = \frac{2a\rho V}{\mu} \tag{15.9}$$

Where a is the minor radius of ellipsoidal shape of the optical nano-robot.

For an ellipsoid shape, the drag force of the optical nano-robot is given by [16, 17]:

$$F_d = 6\pi\mu a U_r k' \tag{15.10}$$

Where k' represents the corrector shape form. The formula of k' is calculated as:

$$k' = \frac{\frac{3}{4}(\alpha^2 - 1)}{\frac{(2\alpha^2 - 1)}{(\alpha^2 - 1)^{1/2}}\ln[\alpha^2 + (\alpha^2 - 1)^{1/2}] - \alpha} \tag{15.11}$$

Where $\alpha = \frac{b}{a}$; b represents the major radius of the ellipsoidal shape.

In order to implement the control law, it is necessary to establish the state representation of the model defining the motion.

For a sufficiently small nanoparticle, the thermal distribution becomes large enough to generate Brownian motion. The external field effect must also be introduced into the Langevin equation in order to model the motion of the nanoparticle under several physical field effect. The present case of thermal diffusion is the photo-thermal effect.

At the boundary layer, the liquid environment exists as a layer of vapor surrounding the nanoparticle. This layer of vapor gas generates a thermal diffusion force on the Au-coated side, while on the other half, without Au, the vapor layer reduces the viscous force that represents the drag force.

In the present case of thermo-diffusion due to the temperature gradient, Langevin establishes the governing equation of the nanoparticle as [13, 15–21]:

$$m\frac{\partial}{\partial t}\vec{V} = \sum_{i=1}^{3}\vec{F_i} = \overrightarrow{F_D} + \overrightarrow{F_{Br}} + \overrightarrow{F_T} \tag{15.12}$$

(x, y) is the Cartesian coordinates representation of the nano-robot position in a reference devolution plane $\Im\left(0,\vec{i},\vec{j}\right)$. The 2D model is established starting from the differential equations of the system movement, thus defining the dynamic behavior of the nano-robot.

$$\begin{cases} m\dfrac{\partial x}{\partial t} = F_{Dx} + F_{Brx} + F_{Tx} \\ m\dfrac{\partial y}{\partial t} = F_{Dy} + F_{Bry} + F_{Ty} \end{cases} \tag{15.13}$$

15.3 BACKSTEPPING ADAPTIVE CONTROL INVESTIGATION

In the state–space representation, the variables x_1 and x_2 (x_3 and x_4) are, respectively, the nano-robot position and velocity along the projection axis. Assuming that the positions x_1 and x_3 are measurable variables, according to the forces expressions equations (15.6, 15.7, and 15.8) and adequate projections, the optical nano-robot system equation (15.13) can be written in the control form as [15, 18]:

$$\begin{cases} \dot{x}_1 = x_2 \\ \dot{x}_2 = f_2\left(x_2\right) + Cu_1 \\ \dot{x}_3 = x_4 \\ \dot{x}_4 = f_4\left(x_4\right) + Cu_2 \\ y = \left(x_1, x_3\right)^T \end{cases} \tag{15.14}$$

Where $C = -\dfrac{DS_T}{M}$.

The control of the system is ensured by the inputs u_1 and u_2 corresponding to the gradients of temperature distributions given by:

$$\begin{cases} u_1 = \nabla T_x \\ u_2 = \nabla T_y \end{cases} \tag{15.15}$$

The functions f_2 and f_4 are the state functions that represent the normalized forces:

$$\begin{cases} f_2(.) = F_{Dx\_n} + F_{Brx\_n} \\ f_4(.) = F_{Dy\_n} + F_{Bry\_n} \end{cases} \tag{15.16}$$

Before dealing with the control strategy problem of the proposed model equation (15.12), the trajectory reference that minimizes control efforts has to be determined. Based on the robot model, several physical and physiological parameters that are affected by uncertainties require estimation in order to improve the robustness of parametric errors.

The aim of backstepping adaptive controller is twofold. Since the μ is assumed to be unknown, this technique makes it possible to evaluate the various estimated parameters and to update the control law simultaneously. The control law update must ensure the convergence and minimize the error between the real and estimated values of the various parameters. Besides the system control, inputs must stabilize and minimize the tracking error between the real trajectory and the reference one, where a triangular shape is necessary:

$$X = \begin{pmatrix} x_1 \\ x_3 \end{pmatrix}; \quad Z = \begin{pmatrix} x_2 \\ x_4 \end{pmatrix}; \quad U = c\begin{pmatrix} u_1 \\ u_2 \end{pmatrix} \tag{15.17}$$

The new triangular form of the system is obtained from equation (15.14) and by substituting the change of variables:

$$\begin{cases} \dot{X} = Z \\ \dot{Z} = \phi(Z)A + \delta(U)B + \zeta \\ Y = X \end{cases} \tag{15.18}$$

Where ζ is the perturbation.

$$F(X,Z) = \begin{cases} F_{Dx} + F_{Brx} + F_{Tx} \\ F_{Dy} + F_{Bry} + F_{Ty} \end{cases} \tag{15.19}$$

The law control is determined according to the change of variables and the error definition given by:

$$\begin{aligned} E_1 &= X - X_{ref} \\ E_2 &= Z - Z_{ref} \end{aligned} \tag{15.20}$$

and

$$\begin{aligned} \dot{E}_1 &= \dot{X} - \dot{X}_{ref} \\ \dot{E}_2 &= \dot{Z} - \dot{Z}_{ref} \end{aligned} \tag{15.21}$$

Where X, $\dot{X}$, and $\dot{Z}$ are the position, linear velocity, and linear acceleration, respectively.

In this case, the candidate Lyapunov function is taken as:

The first Lyapunov function:

$$V = \frac{1}{2} E_1^T E_1 \tag{15.22}$$

Its derivative is:

$$\dot{V} = -k_1 E_1^T E_1 + E_1^T (k_1 E_1 + Z - \dot{X}_{ref}) \tag{15.23}$$

In order to minimize the error energy, the condition $\dot{V}_1 \leq 0$ has to be satisfied. In this case:

$$V E_1 + Z - \dot{X}_{ref} = 0 \tag{15.24}$$

When Z approaches Z_{ref}, the formula becomes:

$$Z_{ref} = K_1 E_1 - \dot{X}_{ref} \tag{15.25}$$

Where $\dot{Z}_{ref} = -K_1 Z + K_1 \dot{X}_{ref} + \ddot{X}_{ref}$

The second Lyapunov function is given by:

$$V_2 = V_1 + \frac{1}{2} E_2^T E_2 + \tilde{A}^T \Gamma_1^{-1} \tilde{A} + \tilde{B}^T \Gamma_1^{-1} \tilde{B} \tag{15.26}$$

The first derivative of V_2 is given as:

$$\begin{aligned} \dot{V}_2 &= -V_1 E_1^T E_1 - K_2 E_2^T E_2 + E_2^T [K_2 E_2 + \phi(Z)\tilde{A} + \delta(U)\tilde{B} + \xi - \dot{Z}_{ref}] \\ &+ \tilde{A}^T \Gamma_1^{-1} (\dot{\hat{A}} - \Gamma_1 \phi^T (Z) E_2) + \tilde{B}^T \Gamma_2^{-1} (\dot{\hat{B}} - \Gamma_2 \delta^T (U) E_2 + E_1) \end{aligned} \tag{15.27}$$

In order to minimize the energy and to ensure that V_2 is negative definite, we set:

$$K_2 E_2 + \phi(Z)\tilde{A} + \delta(U)\tilde{B} + \xi + K_1 Z - K_1 \dot{X}_{ref} - \ddot{X}_{ref} + E_1 = 0 \tag{15.28}$$

$$\dot{\hat{A}} - \Gamma_1 \phi^T (Z) E_2 = 0 \tag{15.29}$$

$$\dot{\hat{B}} - \Gamma_2 \delta^T (u) E_2 = 0 \tag{15.30}$$

The control law $\delta(U)$ and the estimate parameters update equations are rewritten here:

$$\begin{cases} \delta'(U) = [\dot{\tilde{X}}_{ref} - (K_1 + K_2)(Z - \dot{X}_{ref}) - (1 + K_1 K_2)(X - X_{ref}) - \phi(Z)\tilde{A}]\tilde{B}^{-1} \\ \dot{\hat{A}} = \Gamma_1 \phi^T (Z)[Z - \dot{X}_{ref} + K_1 (X_1 - X_{ref})] \\ \dot{\hat{B}} = \Gamma_2 \delta^T (U)[Z - \dot{X}_{ref} + K_1 (X_1 - X_{ref})] \end{cases} \tag{15.31}$$

15.4 NUMERICAL MODELLING

In fact, to study the impact of external field absorbance on improvement of the optical performances, a numerical model of the nano-robot based on accurate solutions of Maxwell's equations is developed. In this context, a new structure of the optical nano-robot is proposed. Several obstacles can be encountered during modelling to derive the optical performance of the structure composed of Au and SiO$_2$ layers. Therefore, in order to approach behavioral modelling, it is imperative to effectively solve Maxwell's equations, which constitute the main challenge that can hamper the study. In the developed model, periodic boundary conditions are used to describe the periodicity [22–24].

The numerical study is focused essentially on the evaluation of the optical performances of the proposed structure. To compare the optical performances with the conventional ones, the effect of the proposed structure on the total reflection absorbance improvement over the whole wavelength range is carried out. The integral absorbance and reflectance are given by the following formulations:

$$A(\lambda) = \frac{\int_V \frac{1}{2} \left| \vec{E}_z(\vec{r}) \right|^2 \omega \varepsilon_0 \varepsilon''(\lambda) dV}{\int_S \frac{1}{2} \mathrm{Re}\left\{ \vec{E}_z(\vec{r}) \times \vec{H}^*(\vec{r}) \right\} dS} \tag{15.32a}$$

Where ε_0 refers to the vacuum permittivity and ε'' is the imaginary part of the complex material dielectric constant. It is worthy to note that $\vec{H}^*$ expresses the complex magnetic field conjugate.

$$R(\lambda)=\frac{\int_{port1}(E_c-E_1)E_1^*dA_1}{\int_{port1}(E_1E_1^*)dA_1}=\frac{reflected\ power[W]}{Incident\ power[W]} \tag{15.32b}$$

Where E_c is the calculated electric field of the port, consisting of the excitation and reflection fields, and E_l represents the electrical mode of port 1. Details of the calculation of total reflection and average absorption can be found in [24].

15.5 RESULTS AND DISCUSSIONS

The index for evaluating the performance of optical nano-robots is the spectral absorbance over the entire wavelength range. In this case, Figure 15.2 shows the spectral absorption versus light wavelength for conventional and investigated hybrid optical nano-robot structures. As can be seen from this figure, the proposed design provides better absorption performance than the conventional design. Furthermore, it is clear that the proposed design with Au and SiO_2 shows higher spectral absorption behavior, resulting in improved electrical efficiency. These remarkable results of improving the average absorption of the optical nano-robots in the visible range can be explained by the result that the optical confinement occurring in the gold layer in the nano-robot structure leads to the reduction of total internal reflection. This morphology has a major role to enhance the electric field profile of the proposed design compared to the conventional optical nano-robot. It is also noticed that successive concentrated electric field intensities located in the Au layer lead to a high optical confinement, which confirms the effective role of Au layer in the structure; consequently, the superior integral absorbance can be achieved through modulating the electric field behavior.

From Figure 15.3, it is clearly shown that the proposed system is relatively stable, where the stability of the closed-loop system is not affected by the parametric error. Nevertheless, it is observable that in the critical phase, the tracking error is important between t = 0 s and t = 1 s. In this critical phase, the control inputs reach the saturation of actuator in the range $t \in [0\ 0.2]$ s; in this range, all the parameters are not updated. Then, in the range $t\in [0.2\ 1]$s, all the parameters are updated. However, from these results, where the update law stabilizes the parametric error to zero, it is clearly noticeable that both tracking efficiency and estimation of system parameters are improved. The adaptive backstepping strategy presented in this simulation reaches two goals simultaneously.

The estimated parameters in the transient phase have not converged; the aim of the controller is to ensure a good tracking and estimation of degraded velocity. Thus, this fact shows robustness to parameters uncertainties. In case where the estimated parameters have converged to their real values, the controller stabilizes in a perfect way along the reference trajectory.

Figure 15.4 presents both the thermo-diffusion force and the Brownian force as a function of the optical nano-robot radius. It is clearly shown that the harvesting of the thermo-diffusion force is important when the radius increases. This graphical representation allows choosing the best parameters design according to the forces values. The thermo-diffusion force presents the strength of the nano-robot actuator, so it is necessary to boost it in order to increase the controllability performances of the proposed system. From this representation, the variation range of the thermo-diffusion force can be compared to the range of Brownian force variation; it is noticed that the thermo-diffusion force is sufficient to control the nano-robot. Furthermore, in terms of comparison with the conventional design, the thermo-diffusion force of the proposed nano-robot is more important.

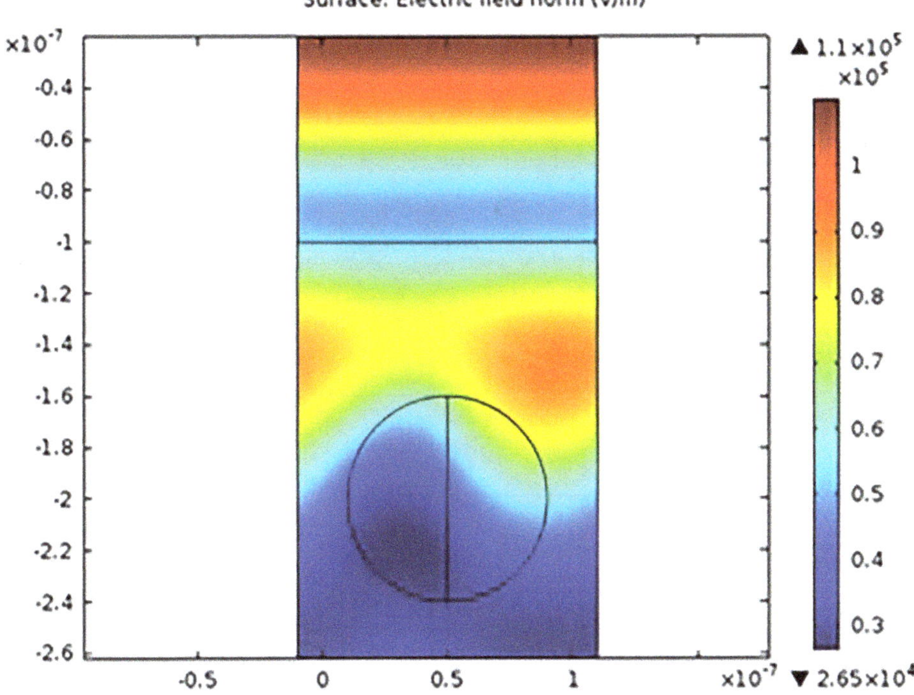

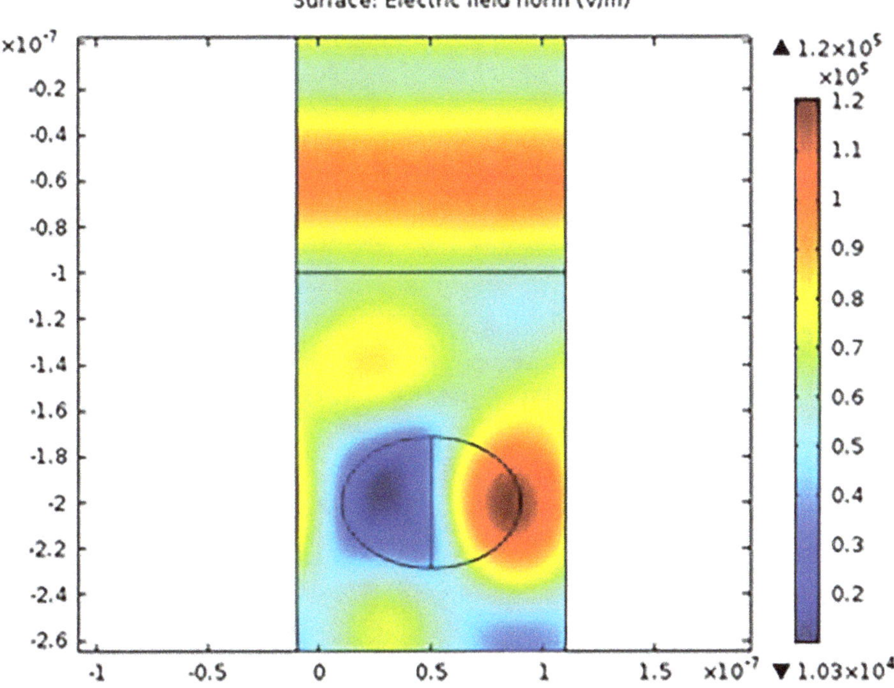

FIGURE 15.2 Electric field profile for both (a) conventional design and (b) proposed structure for specific wavelength (λ = 150 nm), with a = 100 nm, b = 250 nm.

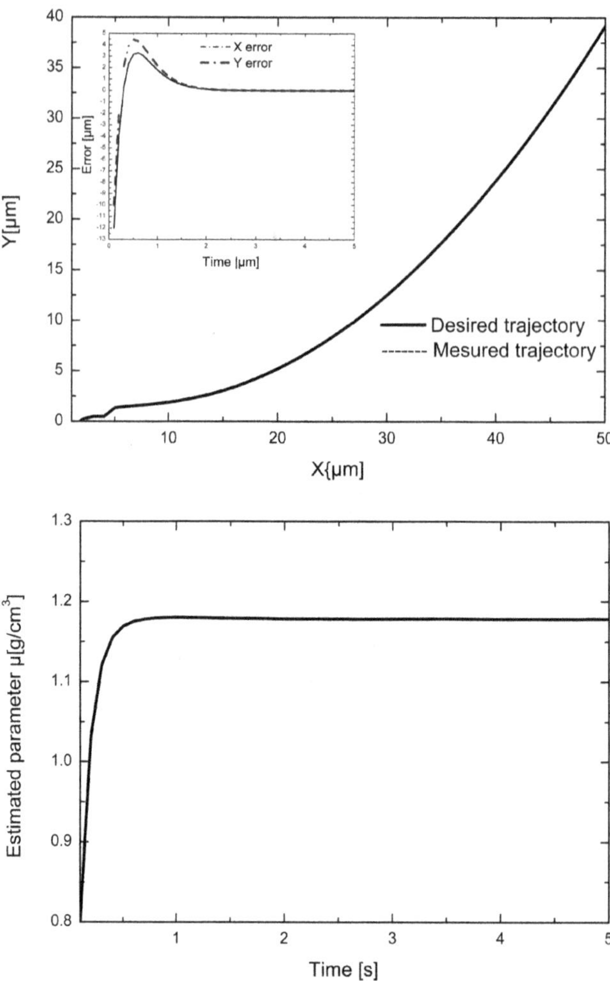

FIGURE 15.3 (a) The control behavior of trajectory tracking in different axis, and (b) the estimated parameter.

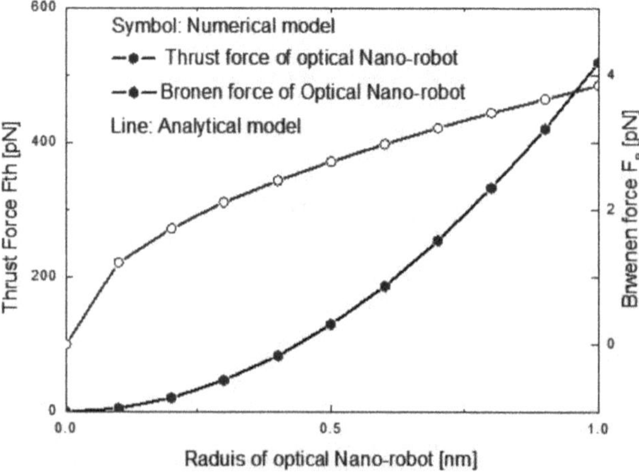

FIGURE 15.4 Variation of the thermo-diffusion force as a function of the optical nano-robot radius.

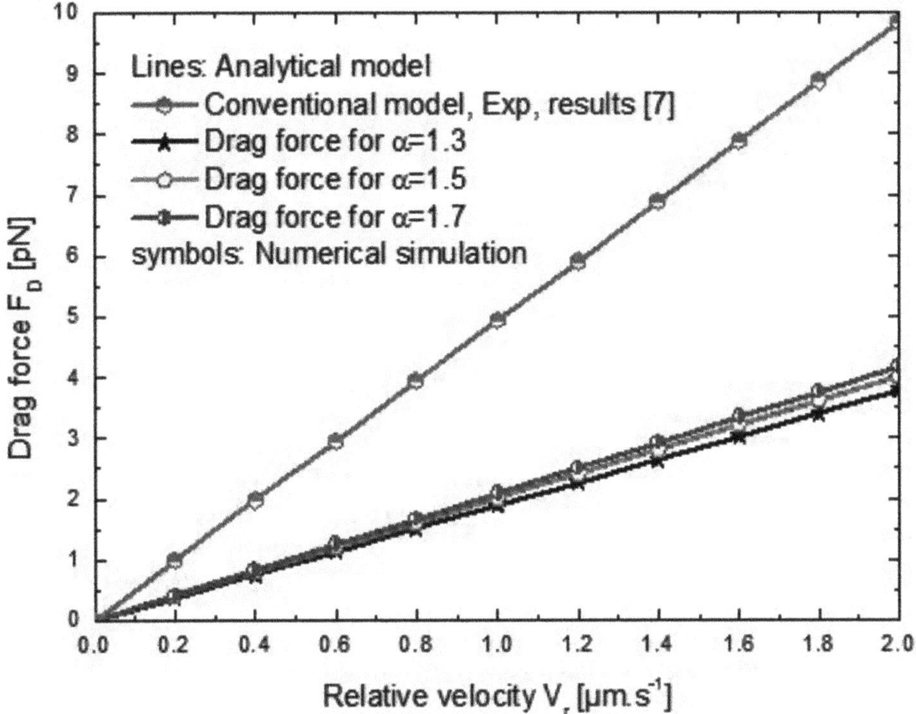

FIGURE 15.5 Variation of the drag force as a function of the relative velocity.

Figure 15.5 shows the drag force as a function of the relative velocity between the nano-robot displacement and the flow fluid. From this representation, it is clear that the ellipsoidal shape of the nano-robot has a very strong influence for decreasing the drag force, which hinders the displacement. However, the design parameters are chosen according to the minimal value of this force. Therefore, it is clear that the drag force increases with the relative velocity in an almost-linear manner with low gradient. Then the drag force has a relative relationship with the optical nano-robot volume. The nano-robot volume is loaded with its different sensors and dragged to reach the target. It is worth noting that the choice of $\alpha = 1.5$ permits a reasonable head's volume. Besides, in terms of comparison, it is clear that the drag force of the proposed design is very small compared to that of the conventional one.

15.6 CONCLUSION

In this work, a new structure of optical nano-robot is proposed for enhancing the optical performance, where the nano-robot behavior has been inspected numerically using both accurate solutions of Maxwell and Navier–Stokes equations. It was found that the proposed method has an important role for enhancing the electrical and optical performances of the nano-robot thanks to its two materials.

Furthermore, in contrast with the conventional nano-robot, the proposed control strategy has shown better performances. This study highlighted the applicability of the backstepping-based strategy for the improvement of the nano-robot control and showed that the proposed method is efficient. Globally, the obtained results constitute a significant contribution, which demonstrate that the proposed design may be promising for many applications.

15.7 ACKNOWLEDGMENTS

The authors would like to thank the CDTA, Centre de Développement des Technologies Avancées, staff for their support and assistance (COMSOL).

Nomenclature

R	spherical coordinate
T	time
ρ_p	mass density of particle
C	specific heat of particle
K	thermal conductivity of particle
T	local temperature
Q	energy of the light irradiation
J	current density
E	electric field
k_0	thermal conductivity of the medium
V	volume of the photo-thermal particle
ε	photo-thermal-side absorption efficiency
I	laser intensity
A	minor radius of the nanoparticle
k_e	thermal conductivity of surrounding fluid
k_{NP}	thermal conductivity of nanoparticle
U_{th}	thermal diffusion rate
S_T	Soret coefficient
D_T	coefficient thermo-diffusion
D	diffusivity of Einstein
F_T	thermo-diffusion
U	nano-robot speed
F_{Br}	Brownian force
K_B	Boltzmann constant
T_0	thermodynamic temperature
μ	fluid viscosity
F_d	drag force
ρ	fluid density
A	nano-robot's frontal surface
C_d	drag coefficient
R_e	Reynolds number
a	minor radius of ellipsoidal shape
m	mass of nano-particle
(x, y)	Cartesian coordinates
ε_0	vacuum permittivity
H	magnetic field
E_c	calculated electric field of the port
E_1	electrical mode of port

REFERENCES

[1] Devasena, U., Brindha, P., & Thiruchelvi, R. (2018). A review on DNA nanobots: A new techniques for cancer treatment. *Asian J Pharm Clin Res*, 11(6), 61–64.

[2] Green, M. A., Zhao, J., & Wang, A. (1998, July). 23% module and other silicon solar cell advances. In *Proc. 2nd World Conf. and Exhibition on Photovoltaic Solar Power Energy Conversion* (pp. 1187–1192). Joint Research Centre European Commission.

[3] Xu, L., Mou, F., Gong, H., Luo, M., & Guan, J. (2017). Light-driven micro/nanomotors: From fundamentals to applications. *Chemical Society Reviews*, 46(22), 6905–6926.
[4] Jiang, H. R., Yoshinaga, N., & Sano, M. (2010). Active motion of a Janus particle by self-thermophoresis in a defocused laser beam. *Physical Review Letters*, 105(26), 268302.
[5] Mebarek-Oudina, F., & Chabani, I. (2023). Review on nano enhanced PCMs: Insight on nePCM application in thermal management/storage systems. *Energies*, 16(3), 1066. https://doi.org/10.3390/en16031066
[6] Chabani, I., Mebarek-Oudina, F., Vaidya, H., & Ismail, A. I. (2022). Numerical analysis of magnetic hybrid nano-fluid natural convective flow in an adjusted porous trapezoidal enclosure. *Journal of Magnetism and Magnetic Materials*, 564, 170142.
[7] Mebarek-Oudina, F. (2019). Convective heat transfer of Titania nanofluids of different base fluids in cylindrical annulus with discrete heat source. *Heat Transfer—Asian Research*, 48(1), 135–147.
[8] Mebarek-Oudina, F., & Chabani, I. (2022). Review on nano-fluids applications and heat transfer enhancement techniques in different enclosures. *Journal of Nanofluids*, 11(2), 155–168.
[9] Govorov, A. O., & Richardson, H. H. (2007). Generating heat with metal nanoparticles. *Nano Today*, 2(1), 30–38.
[10] Lin, X., Si, T., Wu, Z., & He, Q. (2017). Self-thermophoretic motion of controlled assembled micro-/nanomotors. *Physical Chemistry Chemical Physics*, 19(35), 23606–23613.
[11] Golestanian, R., Liverpool, T. B., & Ajdari, A. (2007). Designing phoretic micro-and nano-swimmers. *New Journal of Physics*, 9(5), 126.
[12] Zhao, C., Fu, J., Oztekin, A., & Cheng, X. (2014). Measuring the Soret coefficient of nanoparticles in a dilute suspension. *Journal of Nanoparticle Research*, 16(10), 1–11.
[13] Weinert, F. M., & Braun, D. (2008). Observation of slip flow in thermophoresis. *Physical Review Letters*, 101(16), 168301.
[14] Li, A., & Ahmadi, G. (1992). Dispersion and deposition of spherical particles from point sources in a turbulent channel flow. *Aerosol Science and Technology*, 16(4), 209–226.
[15] Chen, F., Ehlerding, E. B., & Cai, W. (2014). Theranostic nanoparticles. *Journal of Nuclear Medicine*, 55(12), 1919–1922.
[16] Srairi, F., Saidi, L., & Hassam, A. (2018). Modeling control and optimization of a new swimming microrobot using flatness-fuzzy-based approach for medical applications. *Arabian Journal for Science and Engineering*, 43(6), 3249–3258.
[17] Arcese, L., Fruchard, M., & Ferreira, A. (2009, October). Nonlinear modeling and robust controller-observer for a magnetic microrobot in a fluidic environment using MRI gradients. In *2009 IEEE/RSJ International Conference on Intelligent Robots and Systems* (pp. 534–539). IEEE.
[18] Sadelli, L., Fruchard, M., & Ferreira, A. (2016). 2D observer-based control of a vascular microrobot. *IEEE Transactions on Automatic Control*, 62(5), 2194–2206.
[19] Khesrani, S., Hassam, A., Boubezoula, M., & Srairi, F. (2017, May). Modeling and control of mobile platform using flatness-fuzzy based approach with gains adjustment. In *2017 6th International Conference on Systems and Control (ICSC)* (pp. 173–177). IEEE.
[20] de Jesús Rubio, J. (2015). Adaptive least square control in discrete time of robotic arms. *Soft Computing*, 19(12), 3665–3676.
[21] de Jesús Rubio, J., Lopez, J., Pacheco, J., & Encinas, R. (2018). Control of two electrical plants. *Asian Journal of Control*, 20(5), 1–15.
[22] Barroso, Á., Landwerth, S., Woerdemann, M., Alpmann, C., Buscher, T., Becker, M., . . . & Denz, C. (2015). Optical assembly of bio-hybrid micro-robots. *Biomedical Microdevices*, 17(2), 1–8.
[23] Yan, J. L., & Zhang, Z. Y. (2015). Two-grid methods for characteristic finite volume element approximations of semi-linear sobolev equations. *Engineering Letters*, 23(3).
[24] Srairi, F., Djeffal, F., & Ferhati, H. (2017). Efficiency increase of hybrid organic/inorganic solar cells with optimized interface grating morphology for improved light trapping. *Optik*, 130, 1092–1098.

16 LBM Analysis of Magnetohydrodynamic Mixed Convection of Nanoliquid in a Double Lid-Driven Heated Incinerator-Shaped Cavity with Discrete Heating

Bouchmel Mliki, Mohamed Ammar Abbassi, and Fateh Mebarek-Oudina

16.1 INTRODUCTION

The problem of magnetic nanoliquid mixed convection in different geometries has been the object of numerous studies in the past. Many papers have been published in the literature that focused on different applications of this new kind of nanofluid, such as Nandy and Yanuar [1], Sohel et al. [2], and Heris et al. [3].

Ahmed et al. [4] have investigated mixed convection of micropolar nanoliquid in a double lid-driven. They reported an enhancement in Nu_m as the ϕ, while it decreases when the heat source length increases. Teamah et al. [5] reported a problem of mixed convection in a square cavity occupied by nanofuids. It is noticed that Nu_m increases as the Re and Ra, while it decreases when the ϕ increases. Another innovative study, the mixed convection of nanoliquid, has been examined numerically by Garoosi et al. [6]. The influence of Re and Ra on heat transfer, fluid movement, and entropy generation of a nanoliquid in a square cavity was examined by Mirmasoumi and Behzadmehr [7]. It is noticed that Nu_m increase as Re, Ra, and ϕ. Many investigations on mixed convection have been investigated by many researchers [8–13]

The effect of the magnetic field on the thermal configuration in different geometries was extensively investigated by many researchers [14–30]. Mliki et al. [8, 9] used LBM to examine the effect of nanoparticles Brownian motion on fluid movement in different geometries. It is noticed that $|\psi|_{max}$ decreases with an increase in Ha. Also, they found that the heat transfer increased when the role of Brownian motion of nanoparticles is considered. Another innovative study, Alsabery et al. [13] examined the magnetic nanoliquid in a heated square cavity. They considered in their study a two-phase nanoliquid model. Elshehabey and Ahmed [14] have discussed magnetic nanoliquid mixed convection in a cavity. Observation indicates that an inclined magnetic field ceases the fluid movement. In another innovative study, the magnetic mixed convection has been examined numerically by Kefayati [15]. Analysis shows that the heat transfer reduced with the increase of Hartmann number (Ha).

The novelty of this study is the investigation of mixed convection combined with entropy generation in a double lid-driven heated incinerator filled with magnetic nanoliquid. Our numerical results are based on visualization of the flow, thermal fields, and local entropy generation.

DOI: 10.1201/9781003299608-16

16.2 PROBLEM DEFINITION

We consider the magnetic nanoliquid mixed convection in a heated incinerator. The problem configuration of the physical model is shown in Figure 16.1. Two heat sources are situated at the bottom wall at constant temperature. As visible from the graphical view, the left and right walls of the cavity are cold and moving with a constant velocity ($\vec{u}_0$).

The effect of an inclined magnetic field ($\vec{B}$) is depicted in this figure. The incinerator is filled with CuO/water nanofluid (Table 16.1). The viscous dissipation, radiation effects, Joule heating, and induced electric current are neglected.

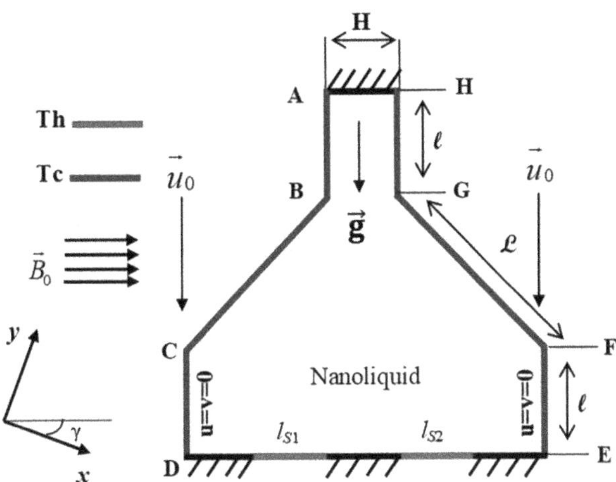

FIGURE 16.1 Geometry of the problem.

TABLE 16.1
Thermophysical Properties of Fluid and Nanoparticles [8]

Thermophysical Properties	H_2O	CuO
C_p (J/kg.K)	4179	540
ρ (kg/m^3)	997.1	6500
k (W/m.K)	0.631	18
$\beta \times 10^{-5}$ (1/K)	21	0.85
σ (Ω/m)$^{-1}$	0.05	2.7 10^{-8}

16.3 GOVERNING EQUATIONS

The governing equations for steady-state two-dimensional heated incinerator-shaped cavity are expressed as [16, 17]:

$$\frac{\partial u}{\partial x}+\frac{\partial v}{\partial y}=0 \tag{16.1}$$

$$\rho_{nf}(u\frac{\partial u}{\partial x}+v\frac{\partial u}{\partial y})=-\frac{\partial p}{\partial x}+\mu_{nf}(\frac{\partial^2 u}{\partial x^2}+\frac{\partial^2 u}{\partial y^2})+(\rho\beta_T)_{nf}\,g(T-T_c)\sin\gamma \tag{16.2}$$

$$\rho_{nf}(u\frac{\partial v}{\partial x}+v\frac{\partial v}{\partial y})=-\frac{\partial p}{\partial y}+\mu_{nf}(\frac{\partial^2 v}{\partial x^2}+\frac{\partial^2 v}{\partial y^2})+(\rho\beta_T)_{nf}g(T-T_c)\cos\gamma-B^2.\sigma_{nf}v \tag{16.3}$$

$$u\frac{\partial T}{\partial x}+v\frac{\partial T}{\partial y}=\alpha_{nf}(\frac{\partial^2 T}{\partial x^2}+\frac{\partial^2 T}{\partial y^2}) \tag{16.4}$$

Where α_{hnf}, β_T, g, and γ are the fluid thermal diffusivity, the thermal expansion coefficient, the gravitational acceleration, and the inclination angle of the cavity, respectively.

In the case of a magnetohydrodynamic nanoliquid mixed convection flow, the local entropy generation S_{gen} can be written as follows [18]:

$$\begin{aligned} S_{gen} &= \frac{k_{hnf}}{T_0^2}\left[\left(\frac{\partial T}{\partial x}\right)^2+\left(\frac{\partial T}{\partial y}\right)^2\right]+\frac{\mu_{hnf}}{T_0}\left[2\left(\frac{\partial u}{\partial x}\right)^2+2\left(\frac{\partial v}{\partial y}\right)^2+\left(\frac{\partial u}{\partial y}+\frac{\partial v}{\partial x}\right)^2\right]+\frac{\sigma_{hnf}B_0^2}{T_0} \\ &= S_{gen,h}+S_{gen,v}+S_{gen,m} \end{aligned} \tag{16.5}$$

Effective density, heat capacitance, thermal expansion coefficient, and thermal diffusivity electrical conductivity of the nanoliquid, respectively, are defined by [19, 20]:

$$\rho_{nf}=(1-\phi)\rho_f+\phi\rho_p \tag{16.6}$$

$$(\rho C_p)_{nf}=(1-\phi)(\rho C_p)_f+\phi(\rho C_p)_p \tag{16.7}$$

$$(\rho\beta)_{nf}=(1-\phi)(\rho\beta)_f+\phi(\rho\beta)_p \tag{16.8}$$

$$\alpha_{nf}=\frac{k_{nf}}{(\rho C_p)_{nf}} \tag{16.9}$$

$$\frac{\sigma_{nf}}{\sigma_f}=1+\frac{3(\frac{\sigma_s}{\sigma_f}-1)\phi}{(\frac{\sigma_s}{\sigma_f}+2)-(\frac{\sigma_s}{\sigma_f}-1)\phi} \tag{16.10}$$

By the help of Maxwell's model [21], the thermal conductivity of nanoliquid is calculated as follows:

$$k_{static}=k_f\frac{k_P+2k_f-2\phi(k_f-k_P)}{k_P+2k_f+\phi(k_f-k_P)} \tag{16.11}$$

The nanofluid effective dynamic viscosity is calculated based on the Brinkman model [22]:

$$\mu_{static}=\frac{\mu_f}{(1-\phi)^{2.5}} \tag{16.12}$$

A numerical study for mixed convection has been conducted using the following dimensionless variables:

$$\begin{aligned} &X=\frac{x}{L},\quad Y=\frac{y}{L},\quad U=\frac{u}{U_0},\quad V=\frac{v}{U_0},\quad \theta=\frac{T-T_c}{T_h-T_c},\quad P=\frac{p}{\rho_{nf}U_0^2},\quad Pr=\frac{v_f}{\alpha_f}, \\ &Re=\frac{U_0L}{v_f},\quad Gr=\frac{g\beta(\mathrm{T}-\mathrm{T}_c)L^3}{v_f^2},Ri=\frac{Gr}{Re^2},\quad Ha=LB\sqrt{\frac{\sigma_f}{\mu_f}},\quad S_T=s\frac{T_0^2L^2}{k_f(\mathrm{T}_h-\mathrm{T}_c)^2} \end{aligned} \tag{16.13}$$

Introducing the preceding dimensionless variables, the non-dimensional forms of the governing equations (16.1–16.4) are reduced as follows:

$$\frac{\partial U}{\partial X}+\frac{\partial V}{\partial Y}=0 \tag{16.14}$$

$$U\frac{\partial U}{\partial X}+V\frac{\partial U}{\partial Y}=-\frac{\partial P}{\partial X}+\frac{1}{Re}\frac{\rho_f}{\rho_{nf}}\frac{1}{(1-\phi)^{2.5}}(\frac{\partial^2 U}{\partial X^2}+\frac{\partial^2 U}{\partial Y^2})+Ri\frac{\rho_f}{\rho_{nf}}\left(1-\phi+\frac{(\rho\beta)_p}{(\rho\beta)_f}\right)\theta\sin\gamma \tag{16.15}$$

$$U\frac{\partial V}{\partial X}+V\frac{\partial V}{\partial Y}=-\frac{\partial P}{\partial X}+\frac{1}{Re}\frac{\rho_f}{\rho_{nf}}\frac{1}{(1-\phi)^{2.5}}(\frac{\partial^2 V}{\partial X^2}+\frac{\partial^2 V}{\partial Y^2})+Ri\frac{\rho_f}{\rho_{nf}}\left(1-\phi+\frac{(\rho\beta)_p}{(\rho\beta)_f}\right)\theta\cos\alpha$$
$$-V\frac{\rho_f}{\rho_{nf}}\frac{\sigma_{nf}}{\sigma_f}\frac{Ha^2}{Re} \tag{16.16}$$

$$U\frac{\partial \theta}{\partial X}+V\frac{\partial \theta}{\partial Y}=\frac{\alpha_{nf}}{\alpha_f}\frac{1}{Re\,Pr}(\frac{\partial^2 \theta}{\partial X^2}+\frac{\partial^2 \theta}{\partial Y^2}) \tag{16.17}$$

Dimensionless entropy generation S_{gen} can be obtained as:

$$S_T=\frac{k_{nf}}{k_f}\left[\left(\frac{\partial\theta}{\partial X}\right)^2+\left(\frac{\partial\theta}{\partial Y}\right)^2\right]+\chi\frac{\mu_{nf}}{\mu_f}\left[2\left(\frac{\partial U}{\partial X}\right)^2+2\left(\frac{\partial V}{\partial Y}\right)^2+\left(\frac{\partial U}{\partial Y}+\frac{\partial V}{\partial X}\right)^2\right]$$
$$+\chi Ha^2\frac{\sigma_{nf}}{\sigma_f}(U\sin\alpha-V\cos\alpha)^2 \tag{16.18}$$

Where χ is the irreversibility factor. It is defined by:

$$\chi=\frac{\mu_f T_0}{k_f}\left(\frac{U_0}{T_h-T_c}\right)^2 \tag{16.19}$$

The average entropy generation is calculated by:

$$S_{avr}=\frac{1}{V}\int_V S_T dV \tag{16.20}$$

Where V is the total volume of the physical domain.

The local and average Nusselt numbers along the two heat sources (l_{S1}, l_{S2}) can be obtained as:

$$Nu=-\frac{k_{nf}}{k_f}\left(\frac{\partial\theta}{\partial Y}\right)\bigg|_{Y=0} \tag{16.21}$$

$$\overline{Nu_{l_{S1}}}=\int_{0.2}^{0.4}Nu\,dX\quad,\quad Nu_{l_{S2}}=\int_{0.6}^{0.8}Nu\,dX \tag{16.22}$$

16.4 NUMERICAL METHOD

The governing equations were solved by using LBM [23]. By Bhatnagar–Gross–Krook approximation, LBM utilized two distribution functions g and f of the temperature and the flow field, respectively [23].

$$f_i\left(x+c_i\Delta t,t+\Delta t\right)=f_i\left(x,t\right)-\frac{1}{\tau_\upsilon}\left(f_i\left(x,t\right)-f_i^{eq}\left(x,t\right)\right)+\Delta tc_iF_i \tag{16.21}$$

$$g_i\left(x+c_i\Delta t,t+\Delta t\right)=g_i\left(x,t\right)-\frac{1}{\tau_a}\left(g_i\left(x,t\right)-g_i^{eq}\left(x,t\right)\right) \tag{16.22}$$

Two local equilibrium distribution functions for the temperature and flow fields g_i^{eq} and f_i^{eq} are calculated with equations (16.23)–(16.24) [23]:

$$f_i^{eq}=w_ir\left[1+\frac{3\left(c_i.u\right)}{c^2}+\frac{9\left(c_i.u\right)^2}{2c^4}-\frac{3u^2}{2c^2}\right] \tag{16.23}$$

$$g_i^{eq}=w_iT\left[1+3\frac{c_i.u}{c^2}\right] \tag{16.24}$$

The nine-velocities (D_2Q_9) lattice model (Figure 16.2) is applied to the present study with uniform grid size of $dx=dy$ for simulating the steady magnetohydrodynamic (MHD) mixed convection of nanoliquid. According to this model, the weighting factors w_i and the discrete particle velocity vectors c_i can be defined as [23]:

$$w_0=\frac{4}{9},w_i=\frac{1}{9}\ for\ i=1,2,3,4\ \ and\ \ w_i=\frac{1}{36}\ for\ i=5,6,7,8 \tag{16.25}$$

$$c_i=\begin{cases}0 & i=0\\ (cos[(i-1)\,p/2],\,sin[(i-1)p/2])c & i=1,2,3,4\\ \sqrt{2}\left(cos[(i-5)\,p/2+p/4],\,sin[(i-5)p/2+p/4]\right)c & i=5,6,7,8\end{cases} \tag{16.26}$$

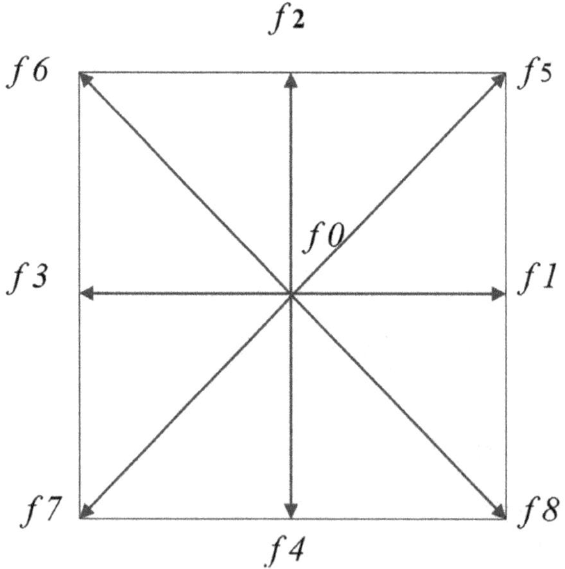

FIGURE 16.2 Direction of streaming velocities, D2Q9.

In the case of inclined cavity, the external force term in equations (16.2) and (16.3) is given by:

$$F_i = 3\omega_i \rho g_y \beta (T - T_c)(\cos\gamma + \sin\gamma) \tag{16.27}$$

Finally, macroscopic variables (ρ, u, and T) are calculated using the following equations:

$$\rho = \sum_i f_i, \quad \rho u = \sum_i f_i c_i, \quad T = \sum_i g_i \tag{16.28}$$

16.5 MESH VERIFICATION AND VALIDATION

For the grid sensitivity, Table 16.2 is performed. Four different mesh sizes were used in this study for the case: nanofluid (CuO-water) with grid independence test for $Re = Ri = 50$; $\phi = 4{\cdot}10^{-2}$; $Ha = 30$; and $\alpha = 45^{\circ}$. From these simulations, the difference between the results of the grid size 100 and that of the grid size 150 is very small. Therefore, the chosen grid is equal to 100 × 100.

For the validation of data, the present results are compared with the numerical results obtained by Lai and Yang [24] for the case of nanoliquid natural convection in a square enclosure (Figure 16.3).

TABLE 16.2
Grid Independence Test for $Re = Ri = 50$; $\phi = 4.10^{-2}$; $Ha = 30$; $\alpha = 45^{\circ}$

Grid Size	$\overline{Nu}_{ls_1}$	$\overline{Nu}_{ls_2}$
50 × 50	3.965541	2.839216
75 × 75	4.212218	3.108770
100 × 100	4.447725	3.322563
150 ×150	4.452828	3.331345

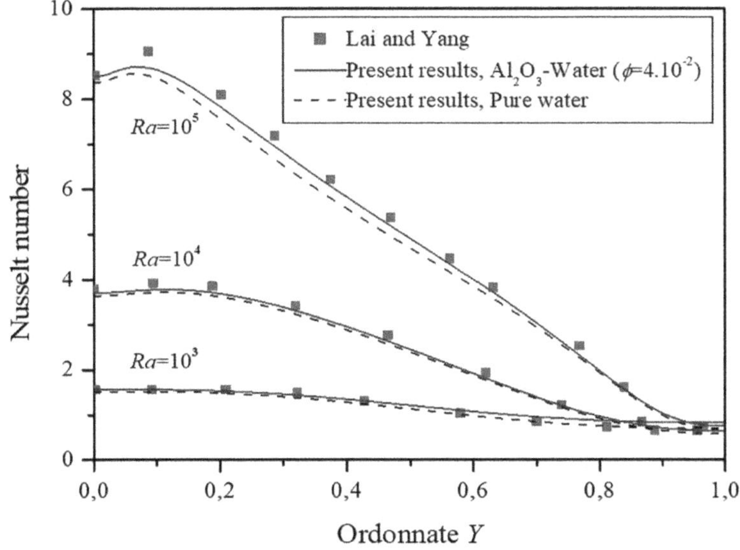

FIGURE 16.3 Comparison of the local Nusselt number along the hot wall between the present results and numerical results by Lai and Yang [24].

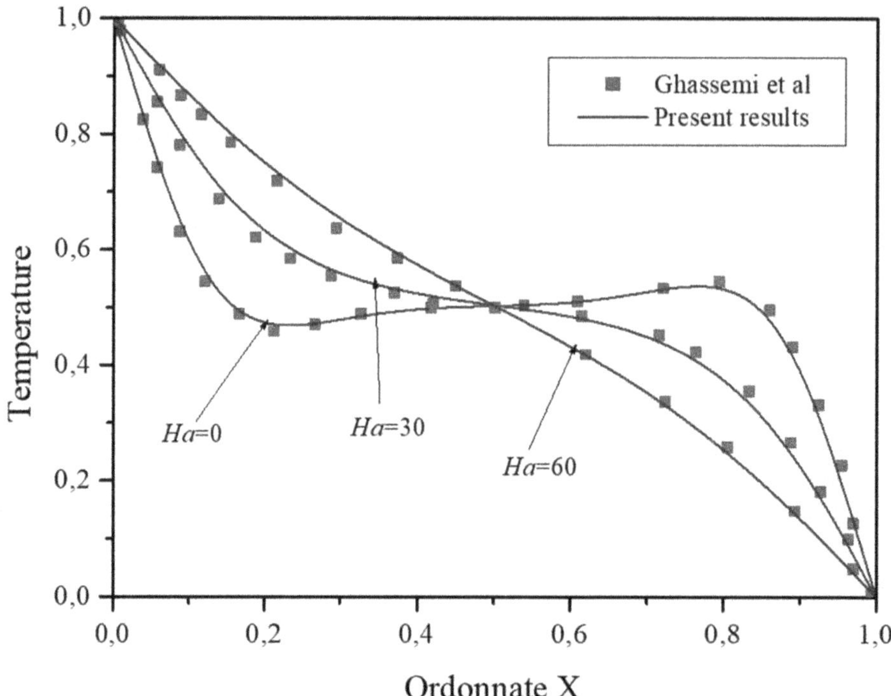

FIGURE 16.4 Comparison of the temperature on the axial midline between the present results and numerical results by Ghassemi et al. [25] ($\phi = 3.10^{-2}$, Ra = 10^5).

In addition, a comparison of the temperature on the axial midline is made between the present results and the numerical results provided by Ghassemi et al. [25] for the case of magnetic nanoliquid mixed convection in a square enclosure (Figure 16.3).

Another test for validation has been performed for the case of mixed convection (Figure 16.5). In this test case, the present numerical results have been compared with those of Talebi et al. [26]. Based on the aforementioned comparisons, the developed code is reliable for studying MHD mixed convection of a nanoliquid confined in a double lid-driven heated incinerator-shaped cavity.

The study of natural convection in vertical incinerator with rectangular hot block for different controlling parameters using lattice Boltzman method is made by Abbassi et al. [27–28]. Numerical results are shown as isotherms, stream function, and iso-contours of entropy generation. In this case, the established code is also validated with reference benchmark results in many cases.

16.6 RESULTS AND DISCUSSION

Numerical analysis of two-dimensional MHD mixed convection is studied for a CuO–water nanoliquid in a double lid-driven heated incinerator.

16.6.1 Effects of Reynolds Number

Th primary objective of this numerical study is to evaluate the effects of *Re* on isotherms, local entropy generation, and streamlines inside the heated incinerator saturated with CuO-water nanoliquid for $Ri = 20$, $\phi = 4.10^{-2}$, $Ha = 30$, $\gamma = 0$. Globally, the flow, local entropy generation, and

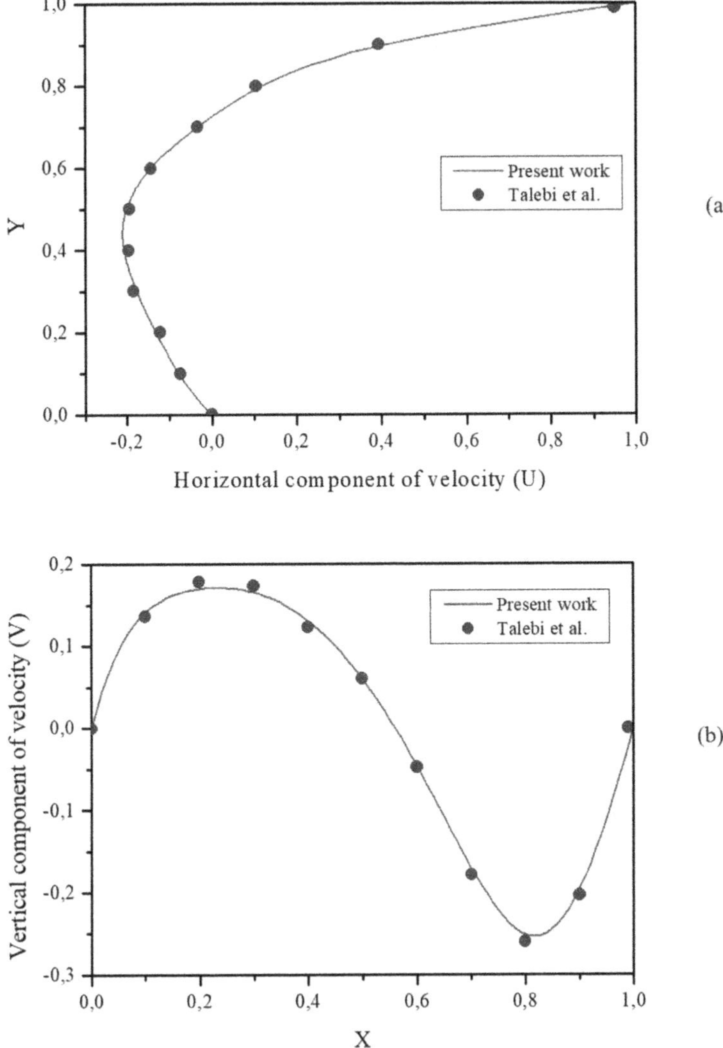

FIGURE 16.5 (a) Horizontal component of velocity, (b) vertical component of velocity with those of de Talebi et al. [23].

temperature contours are symmetrical about the vertical centerline ($X = 0.5$) of the heated incinerator and are concentrated along the two heated sources (l_{S1} and l_{S2}) due to enhanced fluid movement in these regions. For all Reynolds number (Re = 1, 10, 50, and 100), the flow is characterized by the presence of two symmetric anticlockwise vortices, turning in opposite rotating direction and having the same strength ($|\psi|_{max}$ = 7.86 10^{-2}, 3.91 10^{-1}, 5.26 10^{-1}, and 8.1 10^{-1}), respectively (Figure 16.6). Physically, this is true because of the symmetrical boundary conditions about the horizontal X-axis (X = 0.5). However, as Reynolds number increases, the influence of moving walls (CD and EF) becomes more significant and the circulation cell becomes stronger due to increase in buoyancy forces. This is a good reason for the intensification of the maximum stream function magnitude.

Regarding the structure of isotherms presented in this figure, for a low value of Reynolds number (Re = 1 and 10), the heat transfer regime inside the incinerator is dominated by the conduction, where isotherms are approximately quasi-parallel to the isothermal walls. Moreover, enhancement in Reynolds number (Re = 50 and 100) causes to push the cold nanoliquid to the bottom corners

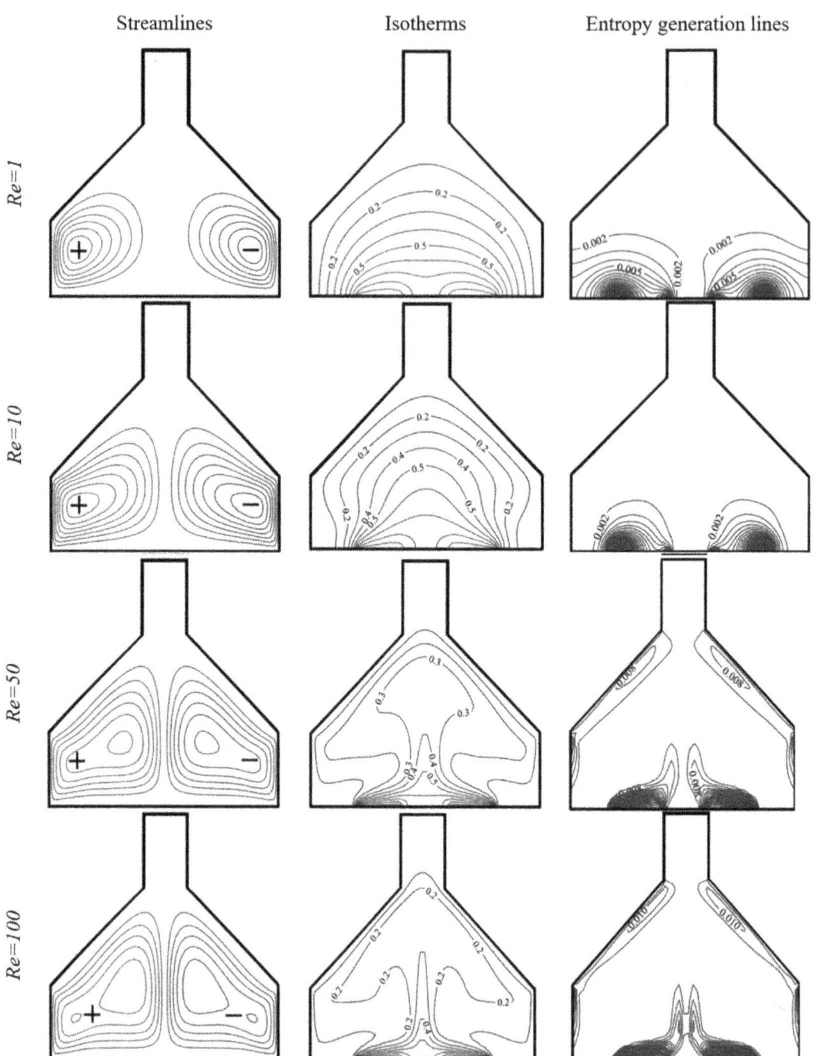

FIGURE 16.6 Streamlines, isotherms, and entropy generation lines for different *Re* at $Ri = 30$, $\gamma = 0$, $Ha = 0$, and $\phi = 4.10^{-2}$.

(D and E), and consequently, the temperature patterns are compressed adjacent to the discrete heat sources (l_{S1} and l_{S2}). The nanoliquid is well circulated in the top part of the heated incinerator. This is an indication of a higher heat transfer rate.

Also, the effect of *Re* on S_{gen} is displayed in Figure 16.6. Increasing Reynolds number, which means an increment in fluid movement, causes more contours of the entropy generation near the hot discrete sources. This is a good reason for the formation of active regions for $S_{gen,T}$, especially in higher Reynolds number ($Re = 50$ and 100).

The effect of solid concentration on Nu_m along the discrete heat sources (l_{S1} and l_{S2}) is shown in Figure 16.7. Examining equation (16.3), the increase in solid concentration leads to enhancing the effective thermal conductivity. It is obvious that when ϕ increases from 0 to $4.10^{-2}\%$ for $Ri = 20$, the Nu_m increases by about, for $Re = 1$, 10.23%, and about 6.86% for $Re = 100$. So at low Reynolds number, the presence of nanoparticle has more effect to enhance the heat transfer rate with respect to the pure fluid.

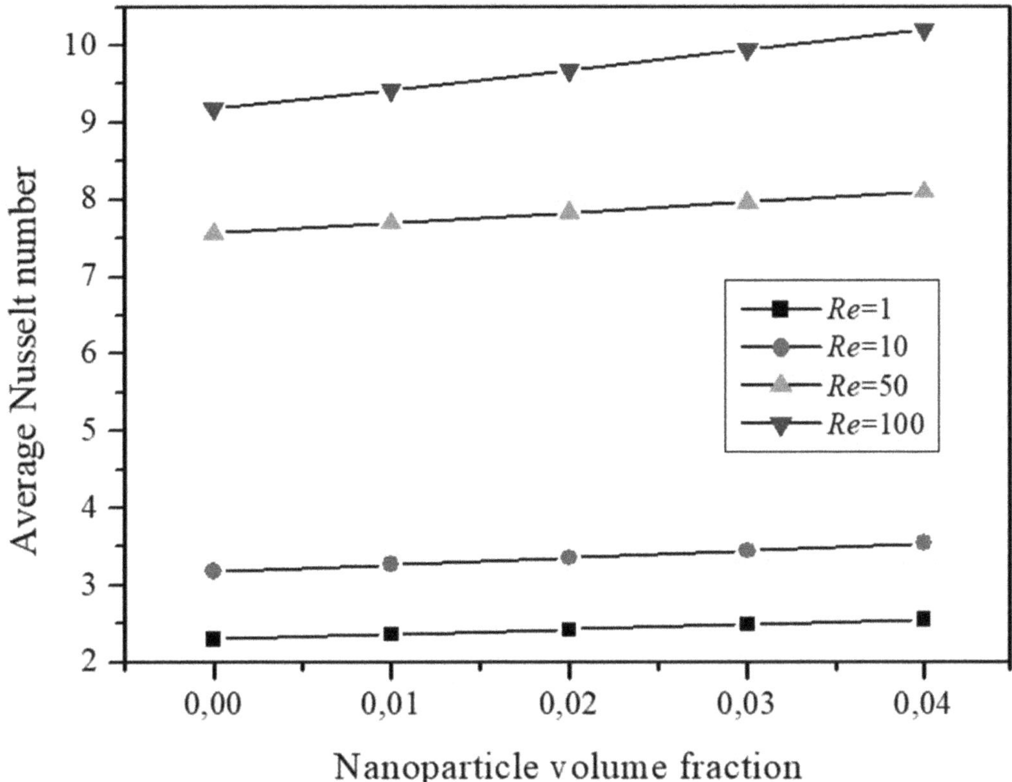

FIGURE 16.7 Average Nusselt number for different values of the Reynolds numbers and volumetric fraction of nanoparticles at $Ri = 20$, $\gamma = 0$, and $Ha = 0$.

The effects of ϕ on S_T for various Reynolds number is displayed in Figure 16.8. It is observed that the S_T increases with an increase in Re and ϕ. This is due to the enhanced fluid movement and energy transferred by the discrete heat sources (l_{S1} and l_{S2}).

Figures 16.9 shows the profile of the dimensionless temperature in the incinerator chimney at $y/L = 0.75$ for different values of Re. It is possible to observe that the temperature in the incinerator chimney increases by an increase in Re due to a domination of the heat transfer regime mixed convection.

16.6.2 Effects of Hartmann Number

The effects of Ha on the streamlines, isotherms, and entropy generation contours for $Re = 100$, $Ri = 20$, $\phi = 4.10^{-2}$, and $\gamma = 0$ are depicted in Figure 16.10. The influence of moving walls (CD and EF) and buoyancy force on its adjacent nanoliquid layers leads to push the hot nanoliquid to the upper parts (incinerator chimney) and causes a favorable enhancement of the heat transfer. The presence of magnetic field leads to a substantial suppression of the mixed convection and causes an unfavorable enhancement of the heat transfer. Physically, this is true, because the effect of Lorentz force is opposite to the buoyancy force.

As can be seen, the increase in Hartmann number leads to reduce the flow intensity, so that the minimum value of $|\psi|_{max}$ is obtained at the maximum Hartmann number. The maximum value of the maximum stream function equal to 5.78 10^{-1} for $Ha = 0$, 4.12 10^{-1} for $Ha = 40$, and 1.99 10^{-1} for $Ha = 80$. This is a good reason for the stagnation of the stream function for $Ha = 80$. However,

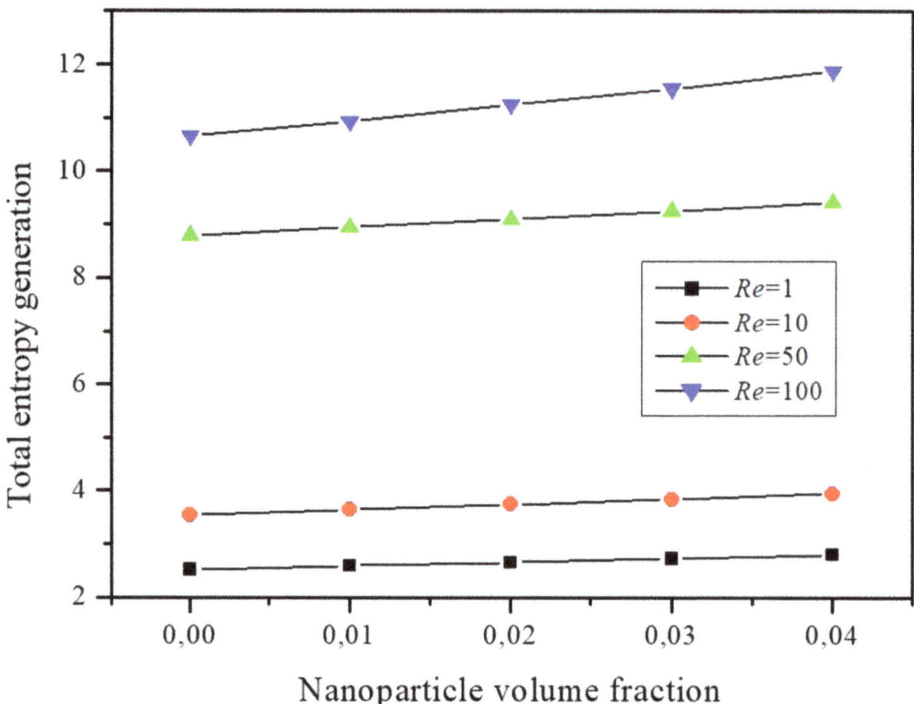

FIGURE 16.8 Total entropy generation for different values of the Reynolds numbers and volumetric fraction of nanoparticles at $Ri = 20$, $\gamma = 0$, and $Ha = 0$.

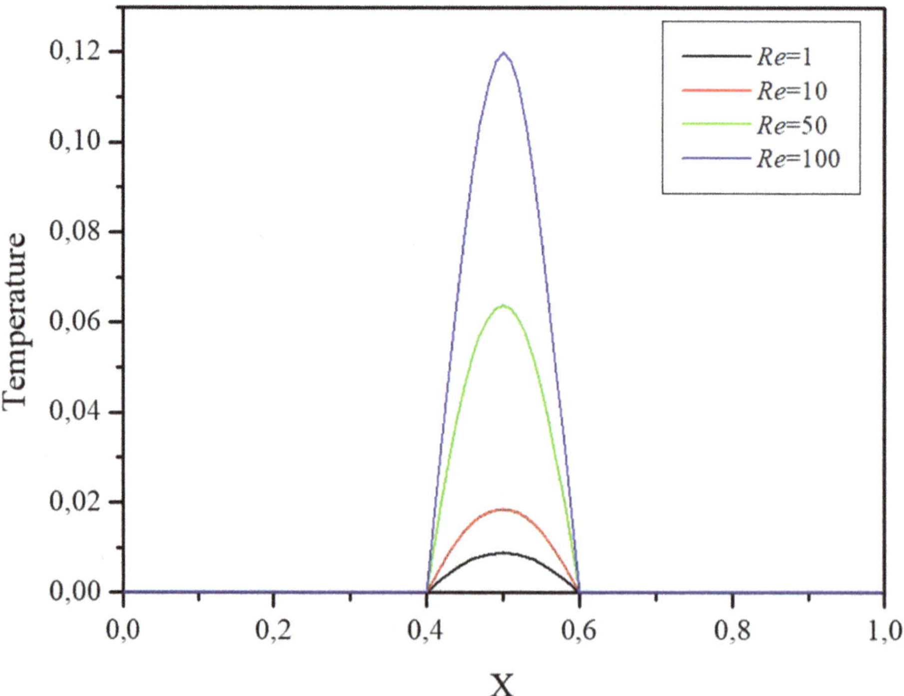

FIGURE 16.9 Profiles of the dimensionless temperature in the middle of the heated incinerator $y/L = 0.75$ for different Reynolds numbers at $Ri = 20$, $\gamma = 0$, and $Ha = 0$.

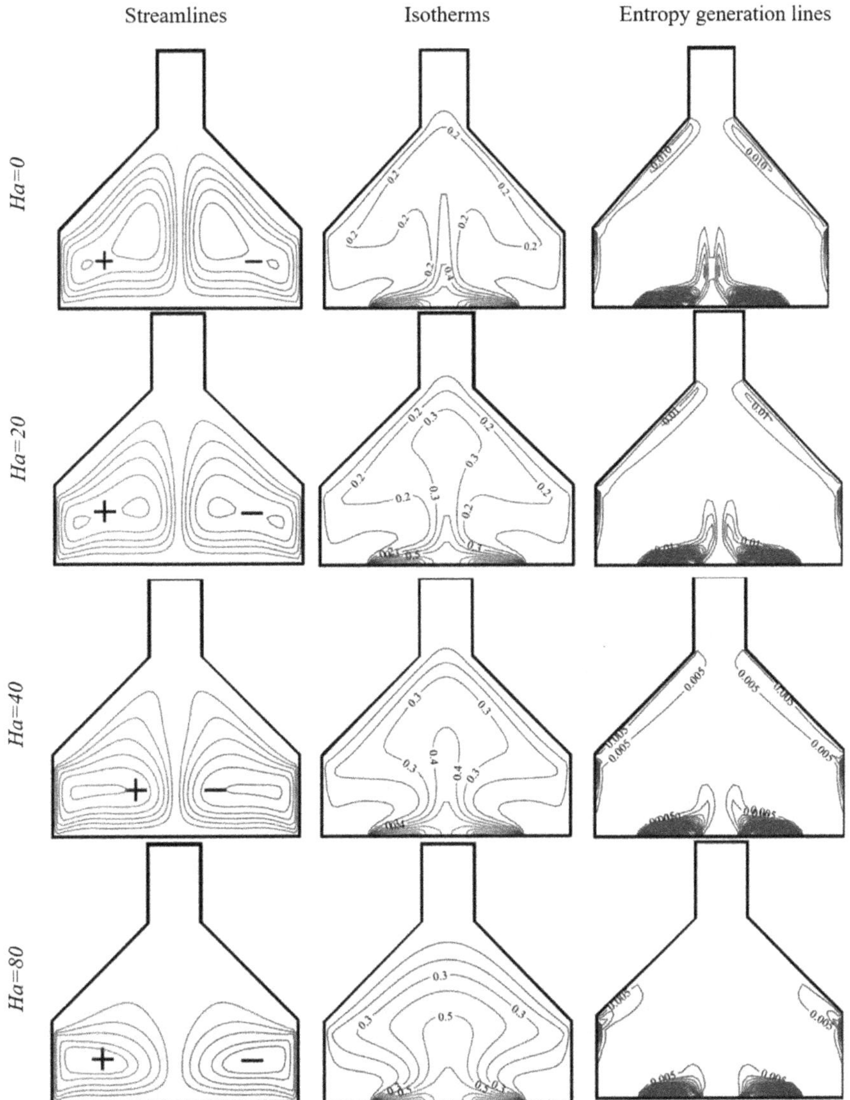

FIGURE 16.10 Streamlines, isotherms, and entropy generation lines for different Ha at Re = 100, Ri = 20, $\gamma = 0$, and $f = 4.10^{-2}$.

as Ha increases, the influence of Lorentz forces becomes more significant, and the density of the temperature distribution contours near the hot discrete sources reduced, which is an indication of a low heat transfer rate.

The effect of Ha on the normalized Nu_m ($Nu_{(Ha)}/Nu_{(Ha=0)}$) for various Reynolds numbers at Ri = 20, $f = 4.10^{-2}$, $\gamma = 0$ is presented in Figure 16.11. From this figure, it is noticed that the effect of Ha on S_T is neglected at low Reynolds numbers (Re = 1). This can be explained due to the fact that the favorable enhancement of the heat transfer by the buoyancy force dominates on the Lorentz force effect. Then, at high Re, the difference between $Nu_{(Ha)}$ and $Nu_{(Ha=0)}$ becomes faster. Consequently, a significant effect in the heat transfer rate with magnetic field can be found for high values of Reynolds number, and the unfavorable enhancement of the heat transfer by Lorentz force dominates on the buoyancy force effect.

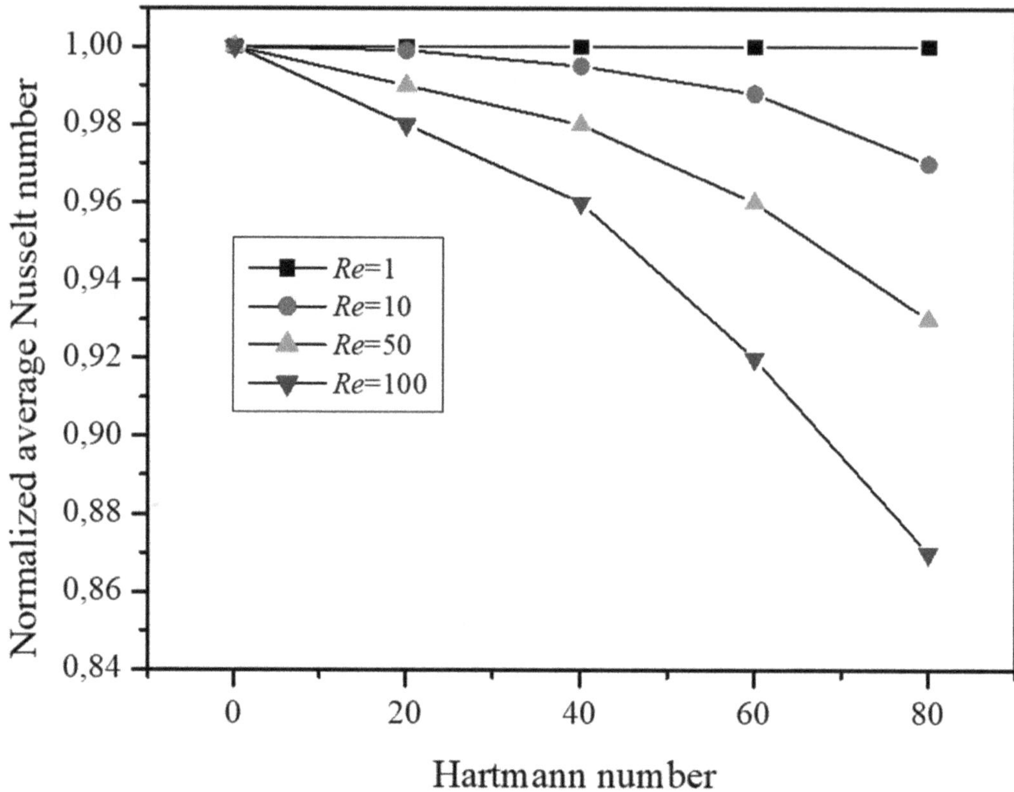

FIGURE 16.11 Effect of Hartmann number on normalized average Nusselt number for different Reynolds numbers at $Ri = 20$, $\phi = 4.10^{-2}$, $\gamma = 0$.

The variation of the S_T is similar to the Nu_m (Figure 16.12). It can be seen from this figure that the effect of magnetic field on S_T is neglected at low Reynolds numbers. Conversely, a significant effect is obtained at high *Re* ($Re > 40$).

The analysis of the profile of the dimensionless temperature in the incinerator chimney at $y/L = 0.75$, presented in Figure 16.13, shows that the temperature in the incinerator chimney decreases by an increase in *Ha*. This is a good reason for the stagnation of the temperature distribution contours near the bottom wall for $Ha = 80$.

16.6.3 Effects of Inclination Angle

The effects of γ on Nu_m are presented in Figure 16.14. From this figure, it is noticed that the heat transfer due to left heat source *S*1 and right heat source *S*2 is the same at $\gamma = 0$. As γ increases from 0 to $\pi/12$, $\pi/6$, $\pi/4$, $2\pi/6$, $5\pi/12$, and $\pi/2$, the heat transfer due to the left heat source *S*1 is enhanced. Conversely, an opposite effect occurs in the heat transfer due to the right heat source *S*2. Generally, increasing γ causes an increase of the Nu_m ($Nu_{S1} + Nu_{S2}$) and, consequently, enhanced heat transfer. The maximum value of the Nu_m is 8.08 and occurs for $\gamma = 0$.

Effect of γ on the S_T is depicted in Figure 16.15. It is shown that S_T slightly decreases with the inclination angle for $\gamma < 30$. If inclination angle is increased to 90, the difference between Nu ($\gamma = \pi/2$) and Nu ($\gamma = 0$) becomes faster. Consequently, a significant effect in the entropy generation with inclination angle can be found for $\gamma > 30$.

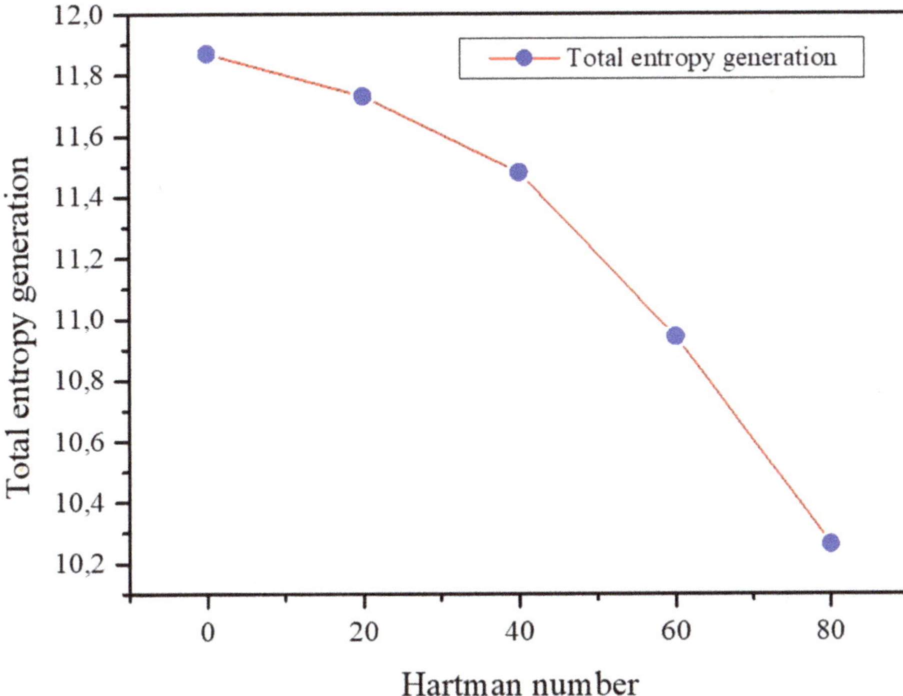

FIGURE 16.12 Effect of Hartmann number on total entropy generation for different Reynolds numbers at $Ri = 20$, $\phi = 4.10^{-2}$, $\gamma = 0$.

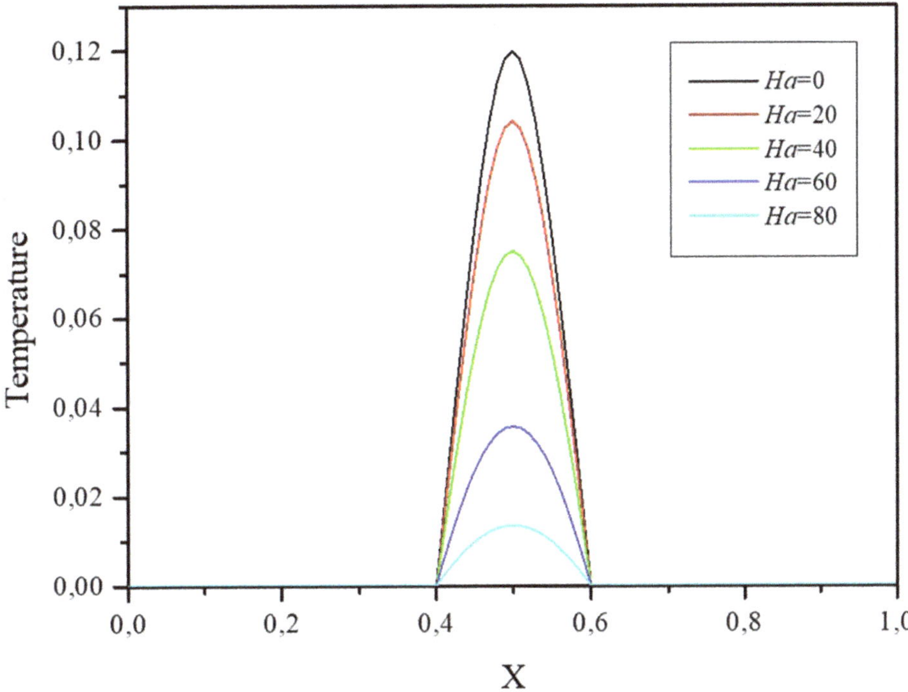

FIGURE 16.13 Profiles of the dimensionless temperature in the middle of the heated incinerator $y/L = 0.75$ for different Hartmann numbers at at $Ri = 20$, $\phi = 4.10^{-2}$, $\gamma = 0$.

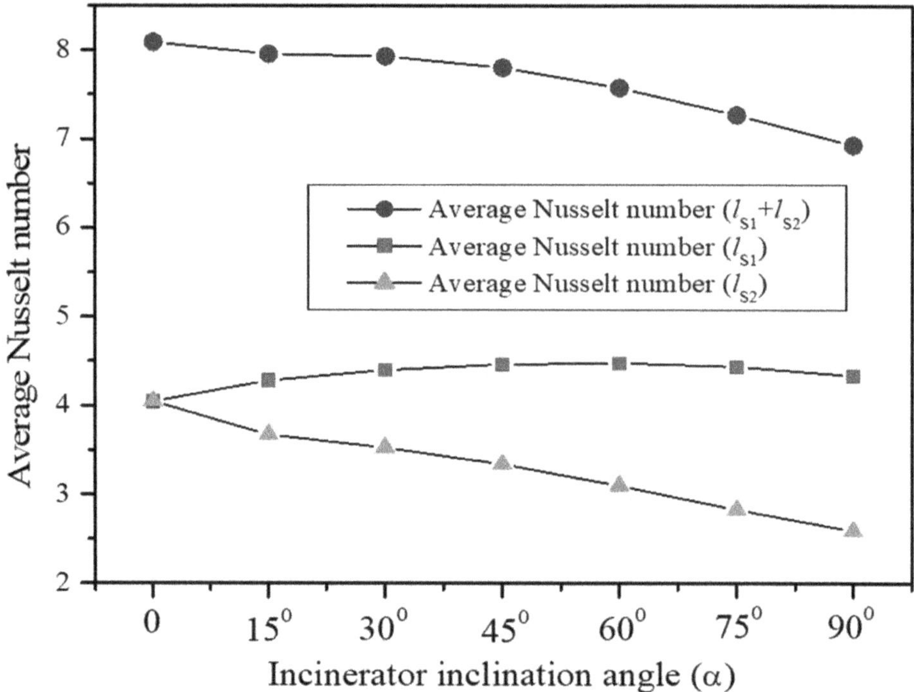

FIGURE 16.14 Average Nusselt number for different γ at Re = 50, Ri =20, Ha = 0, and ϕ = 4.10^{-2}.

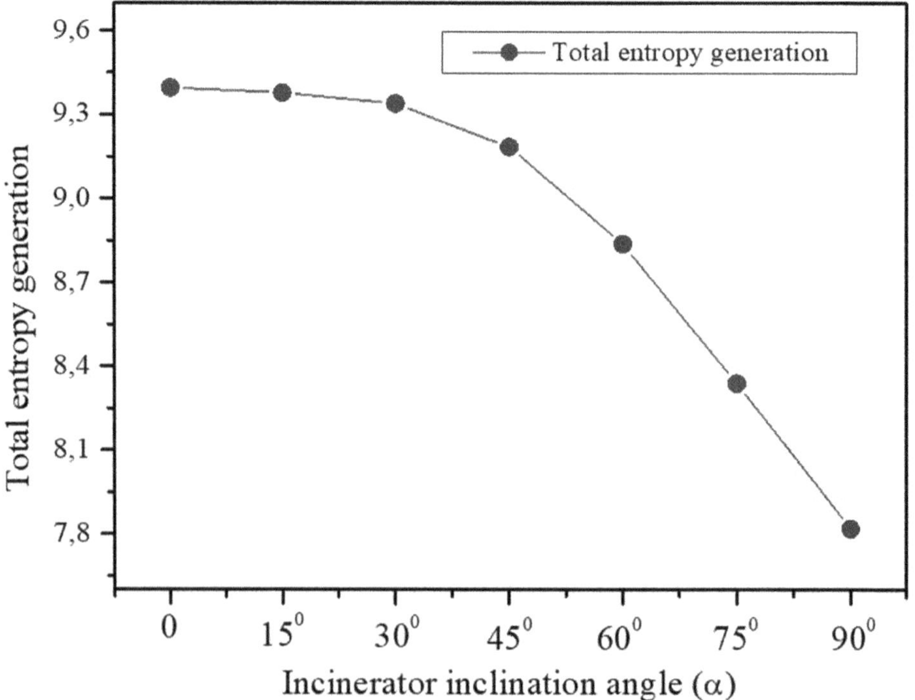

FIGURE 16.15 Total entropy generation for different γ at Re = 50, Ri = 20, Ha = 0, and f = 4.10^{-2}.

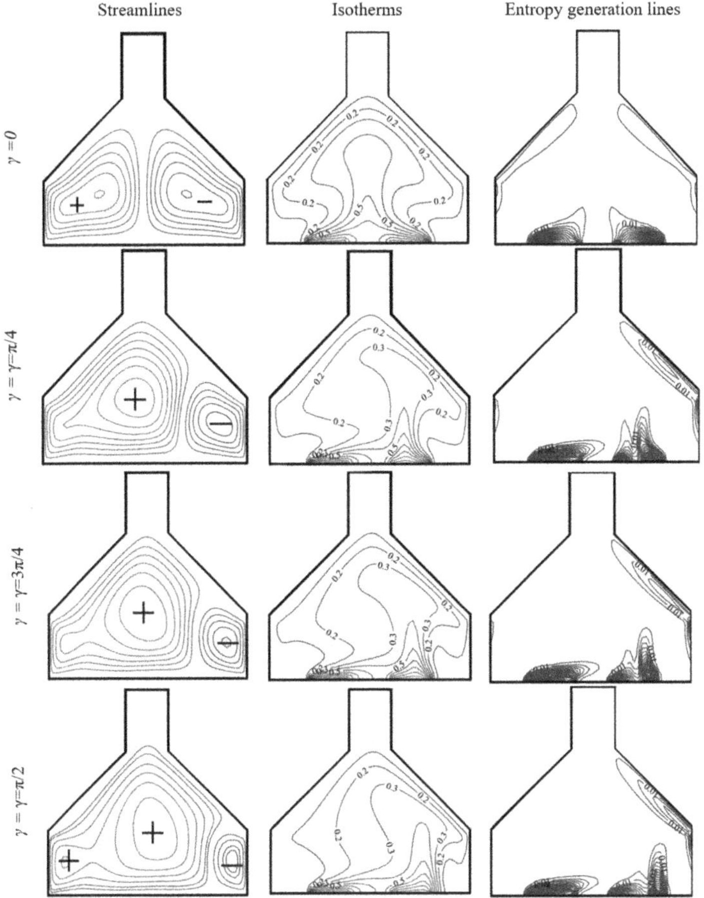

FIGURE 16.16 Streamlines, isotherms, and entropy generation lines for different γ at $Re = 50$, $Ri = 20$, $Ha = 0$, and $\phi = 4.10^{-2}$.

For a selected case $Re = 50$, $Ri = 20$, and $\phi = 4.10^{-2}$, streamlines, temperature contours, and entropy generation are presented at various values of γ (Figure 16.16). For $\gamma = 0$, two symmetrical cells with clockwise and anticlockwise rotations are formed in the bottom part of the heated incinerator due to the symmetrical boundary conditions about the horizontal X-axis ($X = 0.5$). It is noticed that with an increase in γ, the circulation cell that exists inside the left vertical portion of the incinerator becomes stronger, and the thermal boundary layer thicknesses near the left moving walls (EF) increase. Moreover, as inclination angle γ increases, the density of the entropy generation distribution contours in the right part of the incinerator are enhanced. This is an indication of a higher heat transfer rate in this region.

16.7 CONCLUSIONS

In the present study, magnetohydrodynamic (MHD) mixed convection of CuO-water nanoliquid in a double lid-driven heated incinerator-shaped cavity with discrete heating is investigated. Numerical results show that:

- With an increase in Re and ϕ, the values of Nu_m and S_T also increase.
- The maximum value of $|\psi|_{max}$ is obtained for $Re = 100$.

- With applying magnetic field, the velocity of the nanoliquid and, subsequently, the maximum value of $|\psi|_{max}$ decrease.
- Nu_m and S_T are in the indirect relation with *Ha*.
- The temperature in the incinerator chimney increases by an increase in *Re*. Conversely, an opposite effect is obtained by an increase in *Ha*.
- The effect of *Ha* on Nu_m and S_T is more significant at high Reynolds numbers (*Re* = 50 and 100).
- The maximum values of the Nu_m and S_T were obtained at $\gamma = 0$.

16.8 ACKNOWLEDGMENTS

This work was supported by the Tunisian Ministry of Higher Education and Scientific Research under grant 20/PRD-22.

Nomenclature

B	Magnetic field $\left(\text{unit: Tesla} = \text{N}/\left(\text{A} \bullet \text{m}^2\right)\right)$
c	Lattice speed
c_s	Speed of sound
c_i	Discrete particle speed
c_p	Specific heat at constant pressure (unit:$\text{J} \bullet \text{kg}^{-1} \bullet \text{K}^{-1}$)
F_i	External forces (unit : N)
Ha	Hartmann number
k	Thermal conductivity (unit: (W • m¯1)/K)
Nu_m	Average Nusselt number
Nu	Local Nusselt number
P	Pressure (unit:Pa)
Pr	Prandtl number
Ra	Rayleigh number
S	Entropy Generation
Ra	Rayleigh number
T	Temperature (unit:K)
u, v	Velocities (unit:$\text{m} \bullet \text{s}^{-1}$)
x, y	Lattice coordinates (unit:m)
H	Height of cavity (unit:m)

Greek Letters

α	Thermal diffusivity (unit:m^2s^{-1})
β	Coefficient of thermal expansion (unit:K^{-1})
ϕ	Solid volume fraction$\left(\text{unit: Pa}\right)$
μ	Dynamic viscosity (unit:$\text{kgm}^{-1}\text{s}^{-1}$)
ρ	Fluid density (unit:kgm^{-3})
θ	Non-dimensional temperature
v	Kinematic viscosity (unit:m^2s^{-1})
σ	Electrical conductivity (unit : $(\text{Wm})^{-1}$)
ψ	Stream function (unit:m^2s^{-1})

Subscripts

t_h	Cold temperature
t_c	Hot temperature
n_h	Hybrid-nanofluid

REFERENCES

1. Nandy, P., Iskandar, F. N. Application of nanofluid to a heat pipe liquid-block and the thermoelectric cooling of electronic equipment. *Experimental Thermal and Fluid Science*, 2011, 5: 1274–1281.
2. Sohel, M. R., Saidur, R., Mohd, S. M., Kamalisarvestani, M., Elias, M. M., Ijam, A. Investigating the heat transfer performance and thermophysical properties of nanofluid in a circular micro-channel. *International Communications in Heat and Mass Transfer*, 2013, 42: 75–81.
3. Heris, S. Z., Esfahany, M. N., Etemad, S. G. Experimental investigation of convective heat transfer of Al_2O_3/water nanofluid in circular tube. *International Journal of Heat and Fluid Flow*, 2007, 28: 203–210.
4. Ahmed, S. E., Mansour, M. A., Hussein, A. K., Sivasankaran, S. Mixed convection from a discrete heat source. *Engineering Science and Technology, an International Journal*, 2016, 19: 364–376.
5. Teamah, M. A., Sorour, M. M., El-Maghlany, W. M., Afifi, A. Numerical simulation of double diffusive laminar mixed convection. *Alexandria Engineering Journal*, 2013, 52: 227–239.
6. Garoosi, F., Rohani, B., Rashidi, M. M. Two-phase mixture modeling of mixed convection of nanofuids in a square cavity. *Powder Technology*, 2015, 275: 304–321.
7. Mirmasoumi, S., Behzadmehr, A. Effect of nanoparticles mean diameter on mixed convection heat transfer of a nanofluid in a horizontal tube. *International Journal of Heat and Fluid*, 2008, 29: 557–566.
8. Mliki, B., Abbassi, M. A., Omri, A., Zeghmati, B. Lattice Boltzmann analysis of MHD natural convection of CuO-water nanofluid in inclined C-shaped enclosures under the effect of nanoparticles Brownian motion. *Powder Technology*, 2017, 308: 70–83.
9. Mliki, B., Abbassi, M. A., Omri, A. Lattice boltzmann simulation of magnethydrodynamics natural convection in an L-shaped enclosure. *International Journal of Heat and Technology*, 2016, 34(4).
10. Khan, U., Mebarek-Oudina, F., Zaib, A., Ishak, A., Abu Bakar, S., Sherif, E. M., Baleanu, D. An exact solution of a Casson fluid flow induced by dust particles with hybrid nanofluid over a stretching sheet subject to Lorentz forces. *Waves in Random and Complex Media*, 2022, 32: 1–14.
11. Hassan, M., Mebarek-Oudina, F., Faisal, A., Ghafar, A., Ismail, A. I. Thermal energy and mass transport of shear thinning fluid under effects of low to high shear rate viscosity. *International Journal of Thermofluids*, 2022, 15: 100176. https://doi.org/10.1016/j.ijft.2022.100176
12. Swain, K., Mahanthesh, B., Mebarek Oudina, F. Heat transport and stagnation-point flow of magnetized nanoliquid with variable thermal conductivity, Brownian moment, and thermophoresis aspects. *Heat Transfer*, 2021, 50(1): 754–764.
13. Alsabery, A. I., Sheremet, M. A., Chamkha, A. J., Hashim, I. MHD convective heat transfer in a discretely heated square cavity with conductive inner block using two-phase nanofuid model. *Scientific Reports*, 2018, 8: 7410. https://doi.org/10.1038/s41598-018-25749-2
14. Elshehabey, H. M., Ahmed, S. E. MHD mixed convection in a lid-driven cavity filled by a Nanofluid. *International Journal of Heat and Mass Transfer*, 2015, 88: 181–202.
15. Kefayati, G. R. Magnetic field effect on heat and mass transfer of mixed convection of shear-thinning fluids in a lid-driven enclosure with non-uniform boundary conditions. *Journal of the Taiwan Institute of Chemical Engineers*, 2015, 51: 20–33.
16. Marzougui, S., Mebarek Oudina, F., Assia, A., Magherbi, M., Zahir, S., Ramesh, K. Entropy generation on magneto-convective flow of copper–water nanofluid in a cavity with chamfers. *Journal of Thermal Analysis and Calorimetry*, 2021, 143: 2203–2214.
17. Mliki, B., Abbassi, M. A., Omri, A., Zeghmati, B. Augmentation of natural convective heat transfer in linearly heated cavity by utilizing nanofluids in the presence of magnetic field and uniform heat generation/absorption. *Powder Technology*, 2015, 284: 312–325.
18. Mliki, B., Abbassi, M. A., Omri, A. Lattice Boltzmann simulation of MHD double dispersion natural convection in a C-shaped enclosure in the presence of a nanofluid. *Fluid Dynamic and Material Processing*, 2015, 11(1): 87–114.
19. Mliki, B., Abbassi, M. A., Omri, A. Lattice Boltzmann simulation of natural convection in an L-shaped enclosure in the presence of Nanofluid. *Engineering Science and Technology, an International Journal*, 2015, 18: 503–511.
20. Mliki, B., Abbassi, M. A., Omri, A., Zeghmati, B. Effects of nanoparticles Brownian motion in a linearly/sinusoidally heated cavity with MHD natural convection in the presence of uniform heat generation/absorption. *Powder Technology*, 2016, 295: 69–83.
21. Maxwell, J. C. *A treatise on electricity and magnetism*. Oxford University Press, Cambridge, 1873, 2.

22. Brinkman, H. C. The viscosity of concentrated suspensions and solutions. *Journal of Chemical Physics*, 1952, 20: 571–581.
23. Bhatnagar, P. L., Gross, E. P., Krook, M. A model for collision processes in gases, I: small amplitude processes in charged and neutral one-component systems. *Physical Review*, 1954, 94(3): 511.
24. Lai, F. H., Yang, Y. T. Lattice Boltzmann simulation of natural convection heat transfer of Al_2O_3/water nanofluids in a square enclosure. *International Journal of Thermal Sciences*, 2011, 50: 1930–1941.
25. Ghasemi, B., Aminossadati, S. M., Raisi, A. Magnetic field effect on natural convection in a nanofluid-filled square enclosure. *International Journal of Thermal Sciences*, 2011, 50: 1748–1756.
26. Talebi, F., Mahmoudi, A. H., Shahi, M. Numerical study of mixed convection flows in a square lid-driven cavity utilizing Nanofluid. *International Communications in Heat and Mass Transfer*, 2010, 37: 79–90.
27. Abbassi, M. A., Safaei, M. R., Djebali, R., Guedri, K., Zeghmati, B., Alrashed, A. A. LBM simulation of free convection in a nanofuid flled incinerator containing a hot block. *International Journal of Mechanical Sciences*, 2018, 148.
28. Abbassi, M. A., Djebali, R., Guedri, K. Effects of heater dimensions on nanofuid natural convection in a heated incinerator shaped cavity containing a heated block. *Journal of Thermal Engineering*, 2018, 4(3).
29. Warke, A. S., Ramesh, K., Mebarek Oudina, F., Abidi, A. Numerical investigation of the stagnation point flow of radiative magnetomicropolar liquid past a heated porous stretching sheet. *Journal of Thermal Analysis and Calorimetry*, 2022, 147(12): 6901–6912.
30. Djebali, R., Mebarek-Oudina, F., Choudhari, R. Similarity solution analysis of dynamic and thermal boundary layers: Further formulation along a vertical flat plate. *Physica Scripta*, 2021, 96(8): 085206.

17 Influence of Lorentz Force on Cu-Water Nanoliquid Convective Thermal Transfer in a Porous Lid-Driven Cavity with Uniform and Linearly Heated Boundaries

L. Jino, A. Vanav Kumar, M. Berlin, Swapnali Doley, G. Saravanakumar, Ashwin Jacob, and H. A. Kumara Swamy

17.1 INTRODUCTION

The exploration of the fluid flow in a porous matrix has various applications, such as chemical catalytic reactors, solidification of castings, crystal growth, drying of porous solids, dispersion of chemical contaminants, separation process in chemical industries, etc. The detailed pieces of information about the liquid flow in porous media are explored in some works of the literature [1–5]. Djebali et al. [6] focused on the fluid dynamics of a boundary layer during free convection over a heated plate. The significance of nanofluid motion in the medium of porous and its utilization is reviewed by Khanafer and Vafai [7]. The application of nanofluid in daily needs has increased due to its abnormal increase in thermal conductivity. Suspended nanoparticles and their properties play a pivotal role in the heat transfer advancement of nanofluids [8–11]. For instance, Sankar et al. [12,13] analyzed the energy transfer with the help of entropy generation plots for a convective nanofluid (alumina-based) flow within the annulus. It is noted that fluid friction irreversibility dominates for Rayleigh number above 10^5 during convection flow. Pushpa et al. [14] investigated the Cu-water nanofluid flow within a partially separated (using a thin baffle plate) annulus under buoyant convection and heat dissipation. It is found that heat transfer by water is more at a lower Rayleigh number and heat transfer by Cu-water nanofluid is more at a higher Rayleigh number.

Mahapatra et al. [15] discussed the buoyancy ratio effects with constant heat and non-uniform heat boundaries in an LDC under natural convection. Mahmoodi et al. [16] entertained the nanofluids, such as Al2O3/TiO2/CuO/Cu-water, to flow within a cavity due to free convection. It is noted that the heat transfer rate is higher in the case of using Cu-water nanofluid as a running fluid. Also, Ghalambaz et al. [17] addressed the natural convective effect in Ag-MgO-water-filled cavity. An augmentation in heat transfer is carried by conduction during the quantification of hybrid nanofluid at lower Rayleigh numbers, and convection is dominated for higher values of Rayleigh number. Sivasankaran et al. [18] did an investigation of heat transfer due to sinusoidally varying temperature boundary conditions under mixed convective MHD fluid flow in the LDC. Their study concluded that diminishing heat transfer takes place with advancing Hartmann and Richardson numbers. Studies relating to the magnetic field over the motion of fluid inside LDC were done by Kefayati

DOI: 10.1201/9781003299608-17

et al. [19] with linear varying temperature boundaries. The study established that the change in the orientation/direction of the magnetic field alters the stream function behavior and the rate of heat transfer. Kiran et al. [20] inspected the thermal dissipation and convection flow circulation within a cavity due to buoyancy convection. The results illustrate that the amplitude ratio of sinusoidal heating leads to the quantification of buoyant heat transfer rate. Ahmed et al. [21] investigated the MHD mixed convective flow within a two-sided LDC. Here, the flow circulation is noticed to be reduced due to the development of Lorentz force, and thus, the flow is damped with an augmentation of the Hartmann number.

A tool for heatline (heat flow line) visualization for heat transfer by fluid flow was proposed by Kimura and Bejan [22]. Moreover, Costa [23] discussed the significance of streamlines, heatlines, and masslines visualization over the anisotropic media. Subsequently, Basak and Chamkha [24] used the heat function (or heatlines) visualization technique for analyzing the heat flow in a nanofluid-loaded enclosure with respect to various types of thermal boundaries. The study also illustrates that Cu-water nanofluid dispenses good heat transfer. Şahin [25] analyzed the entropy generation and heatlines for exergy and heat transfer due to a linearly heated cavity from the middle of the left wall. Mondal et al. [26] reported the magnetic field effects, such as its strength and inclination on convection flows within LDC. The study involves the highlight of heatline flow directed to the cool boundary from the heated region during the presence of an applied magnetic strength. Similarly, Azizul [27] depicted the heat transfer from the below wavy heated wall/boundary to the upper wall using heatlines inside the mixed convective LDC.

Sathiyamoorthy et al. [28] investigated the free convective effect on linearly varying temperature wall-bounded porous square cavity. Here, the porous medium is framed using the Darcy–Bringman relation. Due to the linearly heated boundary condition, the local Nusselt number exhibits oscillation because of the presence of a secondary streamline circulation. Ghasemi et al. [29] discussed the heat flow, streamlines flow distribution, and entropy generation within a linear heated porous (Darcy–Brinkman–Forchheimer model) square cavity. It is found that, at higher *Ra*, swirling gets stronger due to the convective dominated heat transfer regime within a cavity. Additionally, Malik and Nayak [30] studied the flow due to sinusoidally heated boundary over magnetohydrodynamic convective motion in a porous cavity. It is noticed that by increasing the value of the Grashof number to $Gr = 10^5$, convection domination occurs due to the elevation in fluid velocity, which consequently increases the entropy generation. Hence, the augmentation in the heat transfer occurs while hiking Darcy number. Mebarek-Oudina [31] studied the convection flow of liquid within a vertical annulus, which is heated with a source of different lengths. Later, Mebarek-Oudina et al. [32] extended a similar study by considering nanofluid as a working fluid. The main findings of the study signify that the heat transfer/flow stability can be controlled by selecting the proper length of the heat source. Biswas et al. [33] demonstrated the convective heat flow using heatlines in a hybrid nanofluid crammed porous cavity with a sinusoidally heated wall. The reduction in thermal buoyancy effect and Nusselt number is found with the augmentation in Hartmann number. Similarly, Tilehnoee et al. [34] made the flow analysis (nanofluid streamline flow and heat flow) on MHD natural convection within the incinerator porous cavity. More recently, Swamy et al. [35] and Kumar et al. [36] focused on a Cu-water and MWCNT-water nanofluid flow transport over a convective porous (Darcy's model) cavity with tilted and non-tilted orientation. The study also examines entropy transport. From the studies, it is observed that the shallow annular cavity affords a better heat transfer rate, and entropy generation augments with permeability as well as buoyancy force. Jino and Kumar [37,38] studied the Cu-water nanofluid flow patterns which are generated due to MHD convection in a right-wall heated porous enclosure using the Carman–Kozeny model. Kemparaju et al. [39] studied the source–sink heated flow due to convection within a square porous annulus using the Brinkman-extended Darcy-based model, and Fares et al. [40] considered the Brinkman–Forchheimer-based model for deriving porous matrix. It is noted that the convection flow/heat transfer improves with advancing Darcy number (permeability) in both studies. Chabani et al. [41] considered the EG-Cu-TiO2 nanofluid within a zigzag-edged triangular porous cavity to explore the MHD convective flow. The results show that the Nusselt number increases with increasing Rayleigh number and volume fraction of nanoparticles.

Oztop [42] made a study on the convection heat transmission in a porous LDC by a partially active wall. It is observed that the heater on the left wall gives the best heat transfer rate rather than any other wall. Also, Basak et al. [43] analyzed a mixed convective effect on the LDC porous cavity with the boundary heated linearly. Heat transfer and fluid motion intensity rise with the rise in Darcy number (*Da*), while considering both the vertical walls or only left wall as linearly heated. Sivasankaran and Pan [44] presented a convection study on porous LDC by using the Darcy–Brinkman–Forchheimer model under non-uniform thermal hot wall conditions. It is noticed that the higher heat transfer occurs within a cavity when both walls are heated than the single side vertical wall. Kumar et al. [45] explored heatline trajectories during MHD-free convection in a Cu-water nanofluid-filled porous LDC. Further, Jino and Kumar [46, 47] visualized and spotted the heat flow of Cu-water nanofluid from the heated wall using heat function/heatlines during the MHD quadratic convective porous cavity. The study considers the Brinkman–Darcy model for designing the porous matrix. Later, Jino and Kumar [48] discussed the heatlines for Al_2O_3 water-Cu hybrid nanofluid due to MHD-free convection under porous conditions. It is observed that the Cu nanoparticle addition to the water causes development in the heat transfer process. Marzougui et al. [49] examined flow irreversibility and convection flow of Cu-water nanofluid within a porous LDC. The porous media is mathematically modeled using Darcy–Brinkman–Forchheimer relation. It is noted that the entropy generation reduces with the magnetic field effect and porous medium.

In the preceding literature, it is challenging to find the discussion of fluid movement in a porous media by using the Carman–Kozeny equation model, as this is a special case used in the modelling of phase change problems. The phase change that occurs is defined using the liquid fraction of a liquid solution and is illustrated mathematically using the Carman–Kozeny equation. In this particular study, the Carman–Kozeny-based model is incorporated to define the nanofluid flow over the porous matrix. The current study includes the MHD natural convection effects within the lid-driven porous cavity with uniformly heated and non-evenly heated (linearly varying temperature) left wall. The forced convection effects are neglected in this study, and only the effects due to free convection are considered. Flow patterns are discussed by using the streamlines, heatlines, and, isotherms. Heat transfer from a heated portion of the boundary to a cold boundary can be clearly viewed by heatlines in this study.

17.2 PROBLEM MODEL

The two-dimensional porous LDC of dimension $H \times H$ packed with nanofluid (Cu-water) is shown in Figure 17.1. The LDC is heated to the left wall at a constant temperature (T_h) or linearly varying temperature $(T_h(T_h - T_c)y/H)$ with maximum temperature at the bottom. Both the horizontally oriented walls are thermally protected (adiabatic walls). In addition, only the right wall is maintained at a lower temperature. The top horizontal wall is moved with a constant velocity of $u = U_0, v = 0$, and the other remaining walls are stationary $(u = 0 = v)$. The Cu-water nanofluid packed within the cavity is presumed to be a Newtonian laminar incompressible medium. Other effects, such as Joule heating, radiation, and viscous dissipation, are ignored. Whereas the Boussinesq approximation is included. The gravity, g, and magnetic field strength, B, act parallel to the vertical walls. The properties of nanoparticle (Cu) and base fluid (water) are listed in Table 17.1.

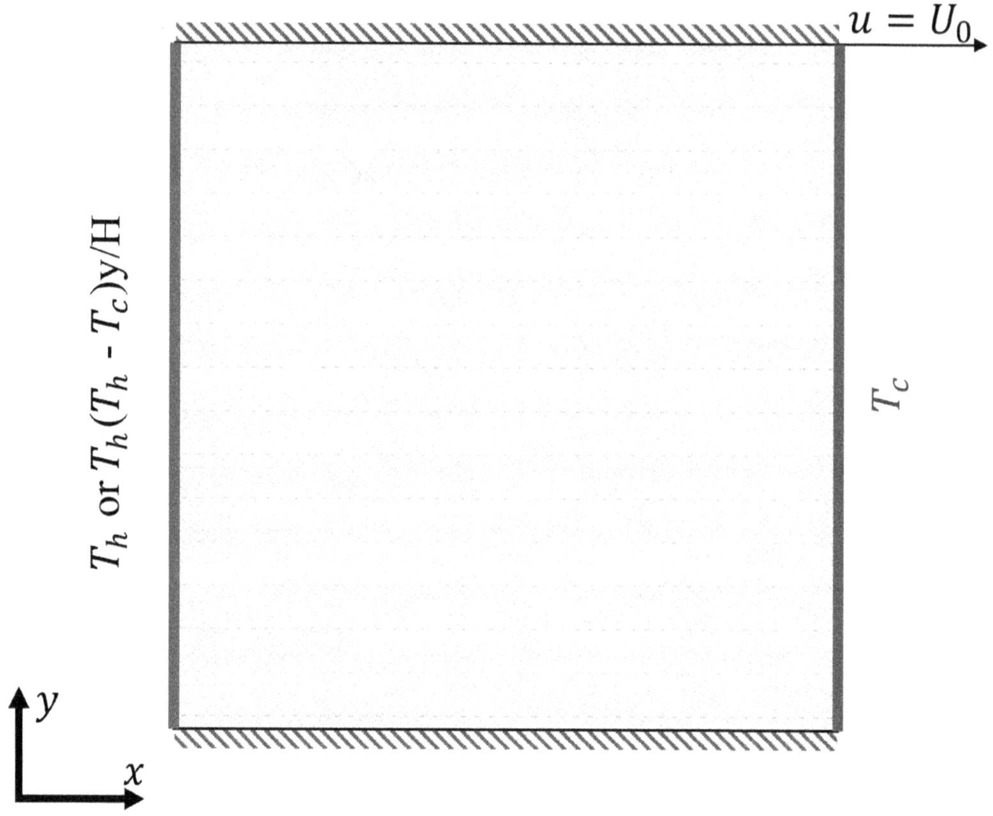

FIGURE 17.1 Physical problem model.

TABLE 17.1
Properties

	Water	Cu
Density (kg/m^3)	997.1	8954
Viscosity (Pas)	8.9×10^{-4}	–
Specific heat (J/kgK)	4179	383
Thermal conductivity (W/mK)	0.6	400
Thermal expansion coefficient ($1/K$)	2.1×10^{-4}	1.67×10^{-5}

17.2.1 Mathematical Description

Following conservation laws, such as mass, momentum, and energy laws, the equation governing the fluid motion in the cavity is constructed with the valid assumptions as discussed earlier, represented by:

$$\frac{\partial u}{\partial x} + \frac{\partial v}{\partial y} = 0 \tag{17.1}$$

$$\rho_n \frac{\partial u}{\partial t} + \rho_n u \frac{\partial u}{\partial x} + \rho_n v \frac{\partial u}{\partial y} = -\frac{\partial p}{\partial x} + \mu_n \left(\frac{\partial^2 u}{\partial x^2} + \frac{\partial^2 u}{\partial y^2} \right) - \rho_n A(\lambda) u \tag{17.2}$$

$$\rho_n \left[\frac{\partial v}{\partial t} + u \frac{\partial v}{\partial x} + v \frac{\partial v}{\partial y} \right] = -\frac{\partial p}{\partial y} + \mu_n \left(\frac{\partial^2 v}{\partial x^2} + \frac{\partial^2 v}{\partial y^2} \right) - \rho_n A(\lambda) v - \sigma_n B^2 v + (\rho\beta)_n g (T - T_0) \tag{17.3}$$

$$(\rho c_p)_n \frac{\partial T}{\partial t} + (\rho c_p)_n u \frac{\partial T}{\partial x} + (\rho c_p)_n v \frac{\partial T}{\partial y} = k_n \left(\frac{\partial^2 T}{\partial x^2} + \frac{\partial^2 T}{\partial y^2} \right) \tag{17.4}$$

Where the velocities u and v correspond to the x- and y-axis respectively. The pressure and temperature are symbolized as P and T, respectively.

The fluid passes through the porous matrix, which is assumed to be homogeneous and isotropic and does not undergo distortion. The porous matrix is controlled by the velocities as a function of the liquid fraction of colloid (λ), and it is given by the Carman–Kozeny equation:

$$A(\lambda) = C_m (1-\lambda)^2 / \lambda^3 \tag{17.5}$$

The thermophysical and transport properties, such as ρ, μ, β, σ, k, and c_p, are given by:

$$\left[\rho_n, \sigma_n, (\rho c_p)_n, (\rho\beta)_n \right] = (1-\phi) \left[\rho_f, \sigma_f, (\rho c_p)_f, (\rho\beta)_f \right] + \phi \left[\rho_p, \sigma_p, (\rho c_p)_p, (\rho\beta)_p \right] \tag{17.6}$$

$$\mu_n = \mu_f / (1-\phi)^{2.5}, \tag{17.7}$$

$$\alpha_n = k_n / (\rho C_p)_n, \tag{17.8}$$

$$k_n = \left(k_p + 2k_f - 2\phi \left(k_f - k_p \right) \right) k_f / k_p + 2k_f - \phi \left(k_f - k_p \right). \tag{17.9}$$

Using the constants and dimensionless quantity, $(X,Y) = \frac{(x,y)}{H}$, $(U,V) = \frac{(u,v)H}{\alpha_f}$, $\tau = \frac{t\alpha_f}{H^2}$, $P = \frac{pH^2}{\rho_n \alpha_f^2}$, $\theta = \frac{T - T_0}{T_h - T_0}$, $Pr = \frac{v_f}{\alpha_f}$, $Da = \frac{\mu_f \lambda^3}{C_m (1-\lambda)^2 H^2}$, $Ra = \frac{g\beta_f H^3 (T_h - T_c)}{v_f \alpha_f}$, $Ha = BH\sqrt{\frac{\sigma_n}{\rho_n v_n}}$, and applying them to equations (17.1)–(17.5), the dimensionless governing equations of the fluid flow are given by:

$$\frac{\partial U}{\partial X} + \frac{\partial V}{\partial Y} = 0 \tag{17.10}$$

$$\frac{\partial U}{\partial \tau} + U \frac{\partial U}{\partial X} + V \frac{\partial U}{\partial Y} = -\frac{\partial P}{\partial X} + \frac{\mu_n}{\rho_n \alpha_f} \left(\frac{\partial^2 U}{\partial X^2} + \frac{\partial^2 U}{\partial Y^2} \right) - \rho_f \frac{Pr}{Da} U \tag{17.11}$$

$$\frac{\partial V}{\partial \tau} + U \frac{\partial V}{\partial X} + V \frac{\partial V}{\partial Y} = -\frac{\partial P}{\partial Y} + \frac{\mu_n}{\rho_n \alpha_f} \left(\frac{\partial^2 V}{\partial X^2} + \frac{\partial^2 V}{\partial Y^2} \right) - \rho_f \frac{Pr}{Da} V - Ha^2 Pr V + \frac{(\rho\beta)_n}{\rho_n \beta_f} Ra Pr \theta \tag{17.12}$$

$$\frac{\partial \theta}{\partial \tau} + U \frac{\partial \theta}{\partial X} + V \frac{\partial \theta}{\partial Y} = \frac{\alpha_n}{\alpha_f} \left(\frac{\partial^2 \theta}{\partial X^2} + \frac{\partial^2 \theta}{\partial Y^2} \right) \tag{17.13}$$

Along with boundary conditions, given by:

$$U(X,0)=V(X,0)=\frac{\partial\theta}{\partial Y}(X,0)=0, \tag{17.14}$$

$$U(X,1)=1,\ V(X,1)=\frac{\partial\theta}{\partial Y}(X,1)=0, \tag{17.15}$$

$$U(0,Y)=V(0,Y)=0,\ \theta(0,Y)=1\ or 1-Y, \tag{17.16}$$

$$U(1,Y)=V(1,Y)=\theta(1,Y)=0. \tag{17.17}$$

To visualize the motions of the fluid, the plots are used for stream function or streamlines. These streamlines are the functions of velocities, given by:

$$\frac{\partial^2\psi}{\partial X^2}+\frac{\partial^2\psi}{\partial Y^2}=\frac{\partial U}{\partial Y}-\frac{\partial V}{\partial X}, \tag{17.18}$$

Where $\partial\psi/\partial Y-U=0$, $\partial\psi/\partial X+V=0$.

The heat flow in a cavity can stay pictured using the heat function or heatlines. These heatlines are discerned with the conductive and convective heat fluxes which are given by:

$$\frac{\partial^2\Pi}{\partial X^2}+\frac{\partial^2\Pi}{\partial Y^2}=\frac{\partial U\theta}{\partial Y}-\frac{\partial V\theta}{\partial X}, \tag{17.19}$$

Where $\partial\Pi/\partial Y+(\alpha_n/\alpha_f)\partial\theta/\partial X=U\theta$, $\partial\Pi/\partial X-(\alpha_n/\alpha_f)\partial\theta/\partial Y=-V\theta$.

The enhancement of heat transfer can be understood by the average Nusselt number, which is specified by Nu as:

$$Nu=\int_{\text{Leftwall}}\left(-\frac{k_n}{k_f}\left(\frac{\partial T}{\partial X}\right)_{\text{Leftwall}}\right)dY. \tag{17.20}$$

17.3 NUMERICAL ASPECTS

The energy equation as well as the vorticity equation formulated from the non-dimensional continuity and momentum equations are solved implicitly with the technique based on finite difference (FD). The FD-based discretization can be found in articles [50–52]. For visualization, the computation of stream functions and heat functions is carried out by using the SOR method in eqaution (17.18) and equation (17.19). The following convergence norm is applied for the simulation $|\xi^{n+1}-\xi^{n}|<10^{-5}$, where ξ is replaced with the velocities U, V or temperature θ, with n representing the time step.

Grid spacing is kept at 122 × 122 for solving the mathematical presiding equations, shown in equations (17.10)–(17.13), which are bounded with boundary conditions, as shown in equations (17.14)–(17.17). The computed outcomes are authorized with the results obtained by Basak et al. [43]. The results obtained are shown in Figure 17.2 and are in good agreement with that of streamlines and isotherms by [43]. Hence, this shows that our computations are authentic and can be implemented for the present study.

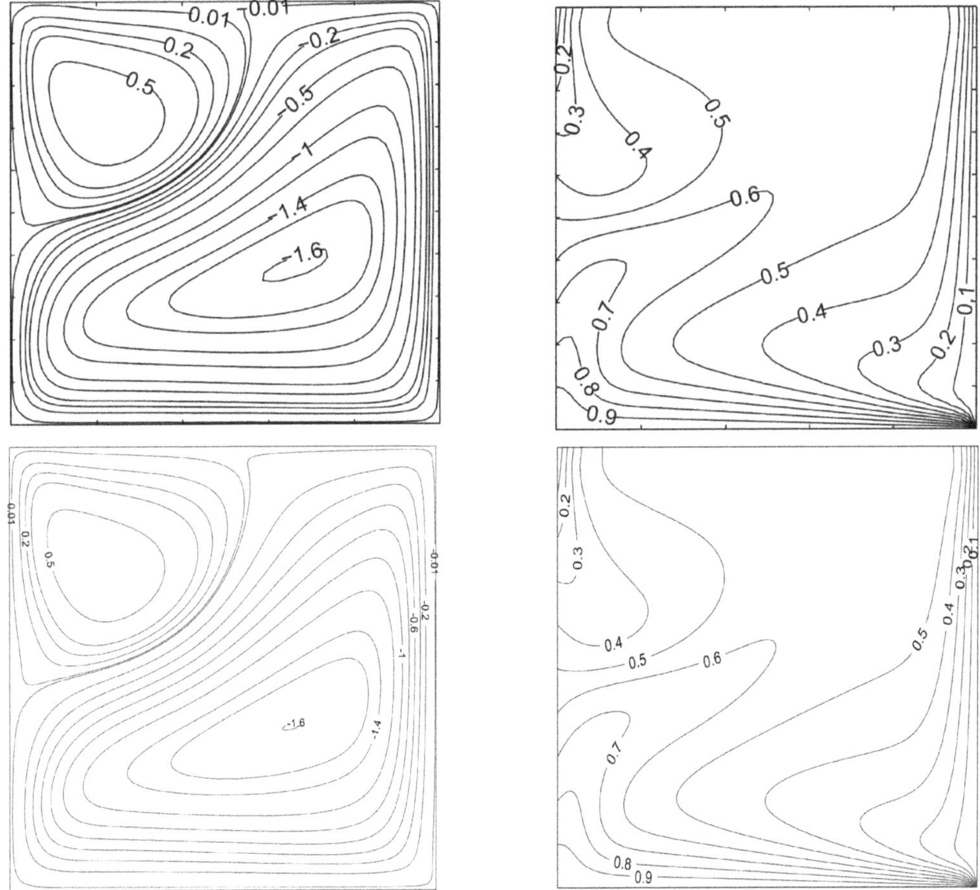

FIGURE 17.2 Validation of streamlines and isotherms against the previous published results.

17.4 RESULTS AND DISCUSSION

The flow analysis of Cu-water nanofluid exposed to the magnetic field in a porous LDC due to natural convection is presented. The numerical results/outcomes are visualized in terms of the contour/curve represented by streamlines, isotherms, heatlines, and Nusselt number (ϕ, θ, Π, Nu) for the steady-state condition. The results are discussed for different sets of values for Darcy number $Da\left(10^{-2}-10^{2}\right)$, Rayleigh number $Ra\left(10^{4}-10^{6}\right)$, solid volume fraction $\phi(0.01-0.1)$ and Hartmann number $Ha(0-100)$ at fixed Prandtl number $(Pr=6.2)$.

17.4.1 Heat/Fluid While Uniform Heating

Figure 17.3 describes the flow field in a cavity for different values of Ha at $Da=10^{-2}$, $\phi=0.05$ and $Ra=10^{6}$. The streamlined circulation covers the whole cavity for every Hartmann number. An increase in Ha from 0 to 50 makes a reduction in the diameter of the innermost circulation $(\psi=-0.53)$. With further increases in Ha to 100, the inner circulation $\psi=-0.53$ vanishes and is replaced by $\psi=-0.51$. The isotherms are almost linearly distributed from the left to the right walls and look similar for various Hartmann numbers. The indication of parallel lines in the heatlines denotes the domination of conduction over the convection. Since there is no circulation with the curvy nature of heatlines, it is concluded that the conduction mode is dominant.

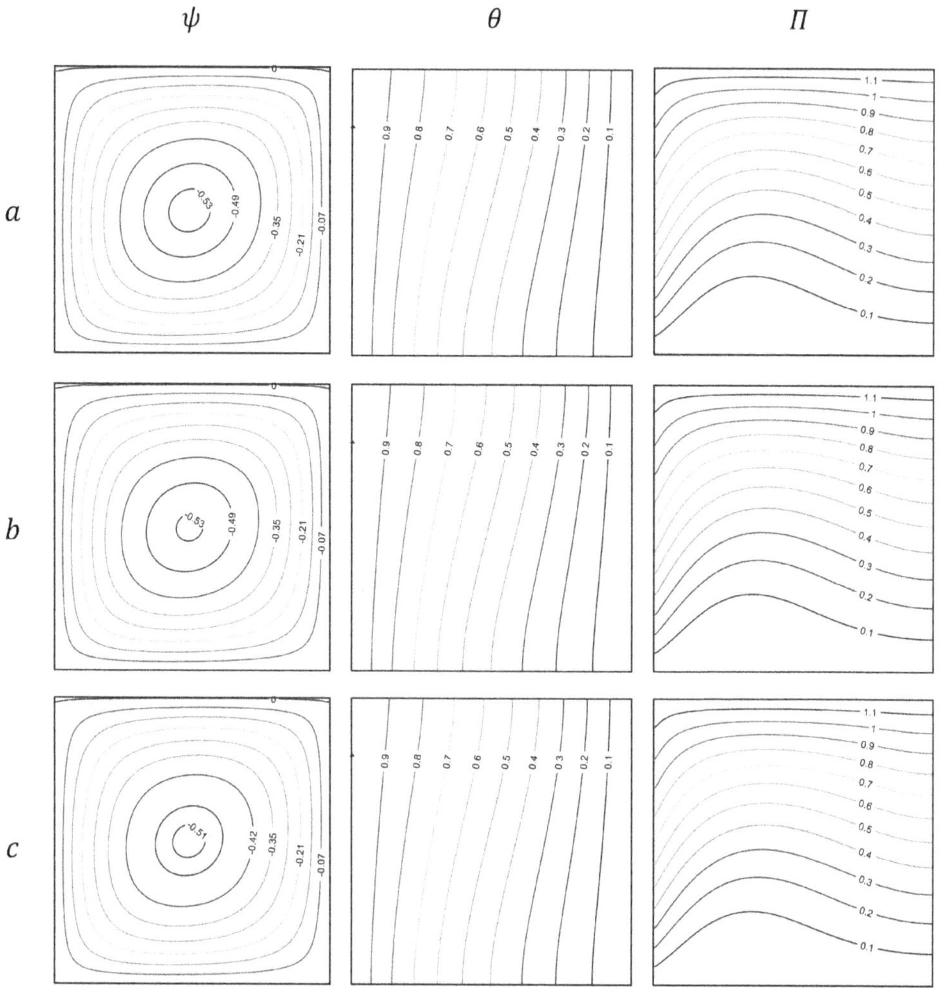

FIGURE 17.3 Streamlines (ψ), Isotherms (θ), and Heatlines (Π) contours when $Da = 10^{-2}$, $\phi = 0.05$, $Ra = 10^6$ at (a). $Ha = 0$, (b). $Ha = 50$, (c). $Ha = 100$ while uniform heating.

Figure 17.4 represents the streamlines, isotherms, and heat function contours for $Da = 10^2$ with various Hartmann numbers. An advancement in Darcy number increases the permeability level and makes the intensity of circulation from $\psi = -0.53$ to -19. The effect of natural convection is increased; as a part of this, two inner circulations are formed in the middle. An increase in Ha reduces the shape of the inner circulations and weakens the swirling. A rise in Hartmann number reduces the sharp bends that are found in the nonappearance of an applied magnetic field in the isotherms contour. The higher circulation in the heatlines and denser lines nearby the left wall indicate the domination of convective effect in a cavity at $Ha = 0$. A rise in Hartmann number causes a diminishing in the intensity of circulation in the heatlines. This signifies the reduction or fall in convection effects.

Flow simulations for varying Ra at $\phi = 0.1$, $Da = 10^{-2}$, and $Ha = 50$ are shown in Figure 17.5. Streamlines for $Da = 10^{-2}$ denotes the involvement of effect from the top lid-driven in the cavity, which makes the secondary circulation adjacent to the top wall. A rise in Ra leads to a rise in buoyancy effect and primary circulation flow intensities. The natural convective effect dominates

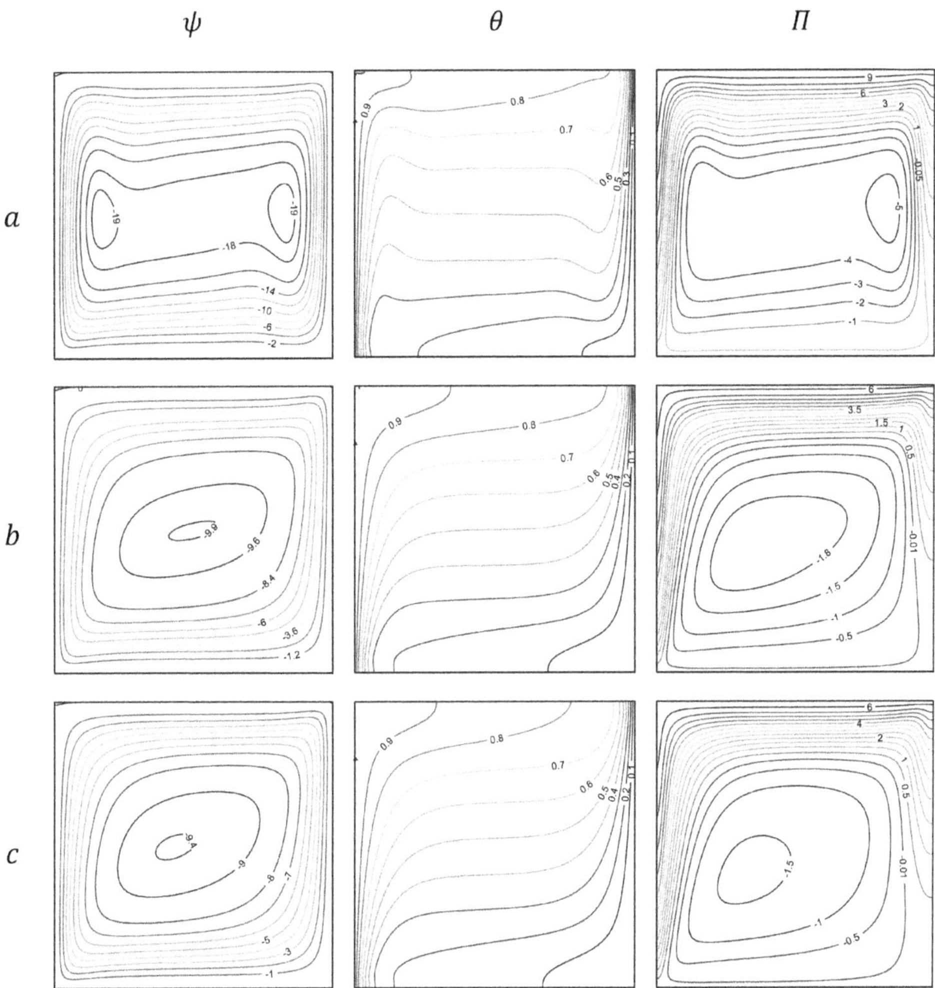

FIGURE 17.4 Streamlines (ψ), Isotherms (θ), and Heatlines (Π) contours when $Da = 10^2$, $\phi = 0.05$, $Ra = 10^6$ at (a). $Ha = 0$, (b). $Ha = 50$, (c). $Ha = 100$ while uniform heating.

the circulation generated by the lid-driven wall and thus reduces the secondary circulations. The conductive mode heat flow in a cavity decreases towards the right wall with quantification in the *Ra*. Parallel heatlines for $Ra = 10^4$ and 10^5 indicate that conduction primarily dominates in the heat transfer. This fact is attributed to less permeability and the viscosity of the fluid caused due to the addition of nanoparticles. Growth in convection effects was observed for an increase in *Ra*, which is visualized as disturbances in the straight heatlines at the bottom of the cavity.

The pores are increased by changing the Darcy number from 10^{-2} to 10^2 and flow for $Ha = 50$ is shown in Figure 17.6. The pores are responsible for fluid to flow freely and make the flow intensity higher. An ascendance in *Ra* leads to ascendance in buoyancy effects, and thereby, flow intensity gets increased. An alteration is observed in isotherms due to an increase in the domination of convection effects. Heatlines for the $Ra = 10^4$ indicate the conduction to be dominant, and circulations grow as an increase in *Ra*. This circulation is a sign of representation for the convection mode of heat transfer. The more denser lines observed near the left wall of a cavity indicate the improved convection effects at $Ra = 10^6$.

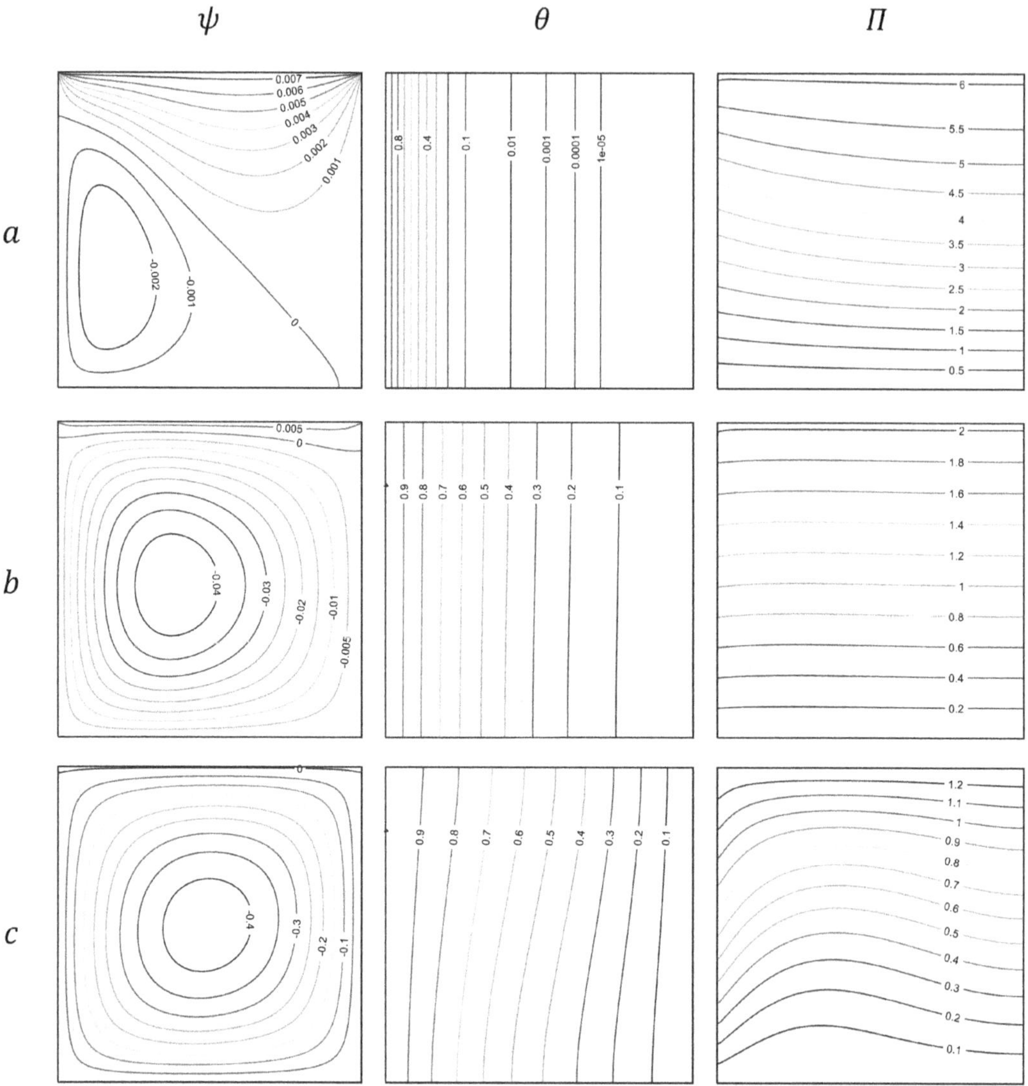

FIGURE 17.5 Streamlines (ψ), Isotherms (θ), and Heatlines (Π) contours when $Da = 10^{-2}$, $\phi = 0.1$, $Ha = 50$ at (a). $Ra = 10^4$, (b). $Ra = 10^5$, (c). $Ra = 10^6$ while uniform heating.

17.4.2 Heat/Fluid Flow Contours While Non-Uniform Heating

Figure 17.7 demonstrates the ϕ, θ, Π for non-uniform boundary conditions by varying the Hartmann number 0 to 100 at $Da = 10^{-2}$, $Ra = 10^6$, and $\phi = 0.05$. The Ha has less effect on isotherms and streamlines with a small limitation of circulations in the middle of a cavity. This behavior is similar to the uniformly heated cavity. A shifting of circulations towards the left lower region of the cavity is due to the stronger heating detected near the bottom of a cavity, and heating intensity decreases linearly towards the positive y-direction. The heatlines for all the Hartmann number indicate no circulation, and it asserts that conduction dominates the heat transfer and also noted that the heat transfer is reduced as similar to the uniform heated boundary.

Figure 17.8 shows the heat and fluid flow patterns for distinct values of Ha, at Darcy number $Da = 10^{-2}$. Secondary circulation emerges in streamlines, and also, primary circulations are hiked with a higher Darcy number. The primary circulation intensity is more for linearly heating compared to uniform heating at $Ha = 0$. For an increase in Ha, the primary flow rush in streamlines

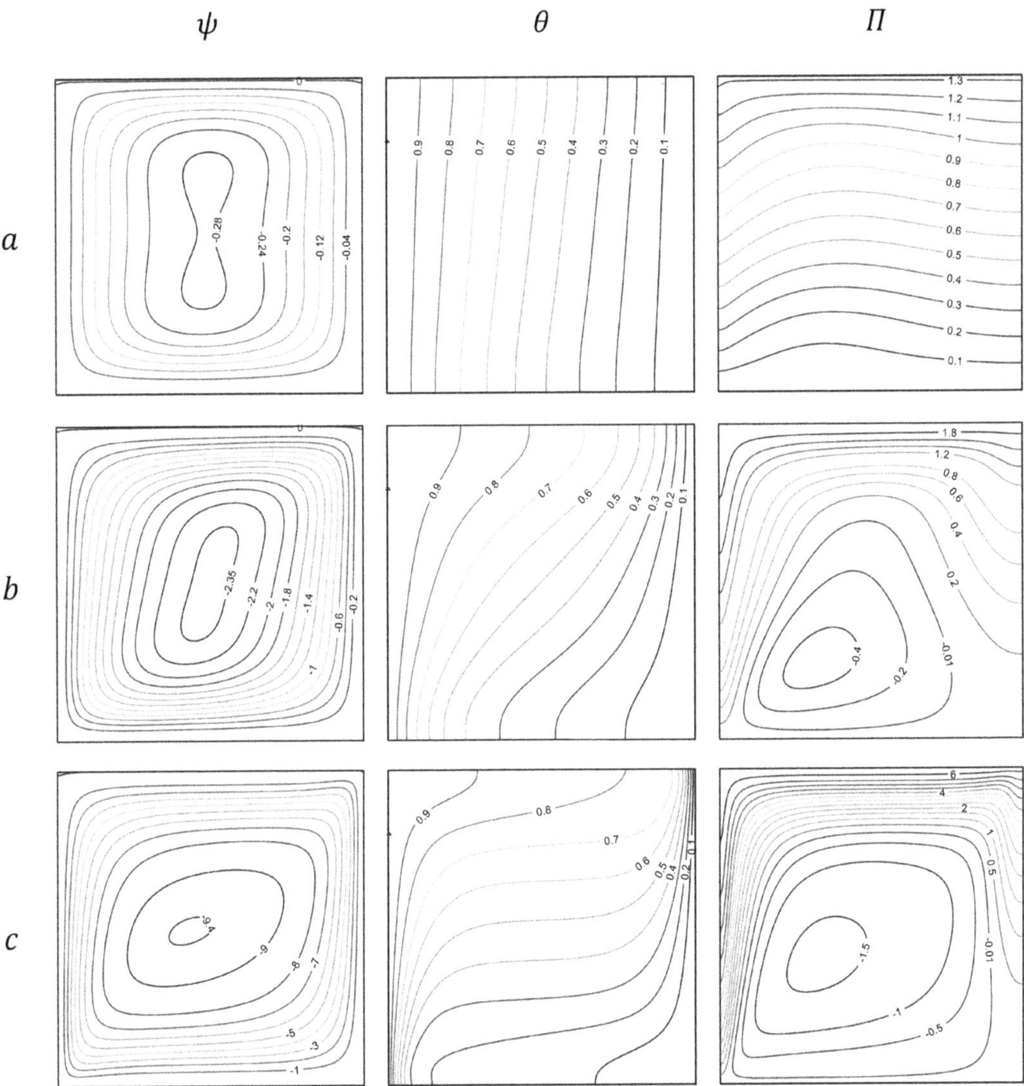

FIGURE 17.6 Streamlines (ψ), Isotherms (θ), and Heatlines (Π) contours when $Da = 10^2$, $\phi = 0.1$, $Ha = 50$ at (a). $Ra = 10^4$, (b). $Ra = 10^5$, (c). $Ra = 10^6$ while uniform heating.

is reduced in comparison to the uniform heating. The temperature contour ($\theta = 0.1, 0.2, 0.3$) starts to straighten up, and sharp bends disappear. Also, the temperature contour $\theta = 0.4$ moves inwards along the inner cavity because of an increase in the magnetic field effect. The secondary facing circulations are formed in heatlines for linear heating and which get vanished due to quantification of Hartmann number to 100. The inferior circulation formed in the heat function because of the secondary circulation generated in streamlines at the top left corner of a cavity. The heatlines denote the domination of the convective effect, which was reduced with an increase in the magnetic field effects.

The effect of lid-driven velocity is found adjacent to the wall for $Ra = 10^4$ when the $Da = 10^{-2}$, and this gets shrunk with an increase in Ra, which is shown in Figure 17.9. This reduction of secondary circulations of streamlines is because of the increase in buoyancy effect. Lesser heat transfer is found in the cavity for the linearly heating wall compared with the uniform heated wall.

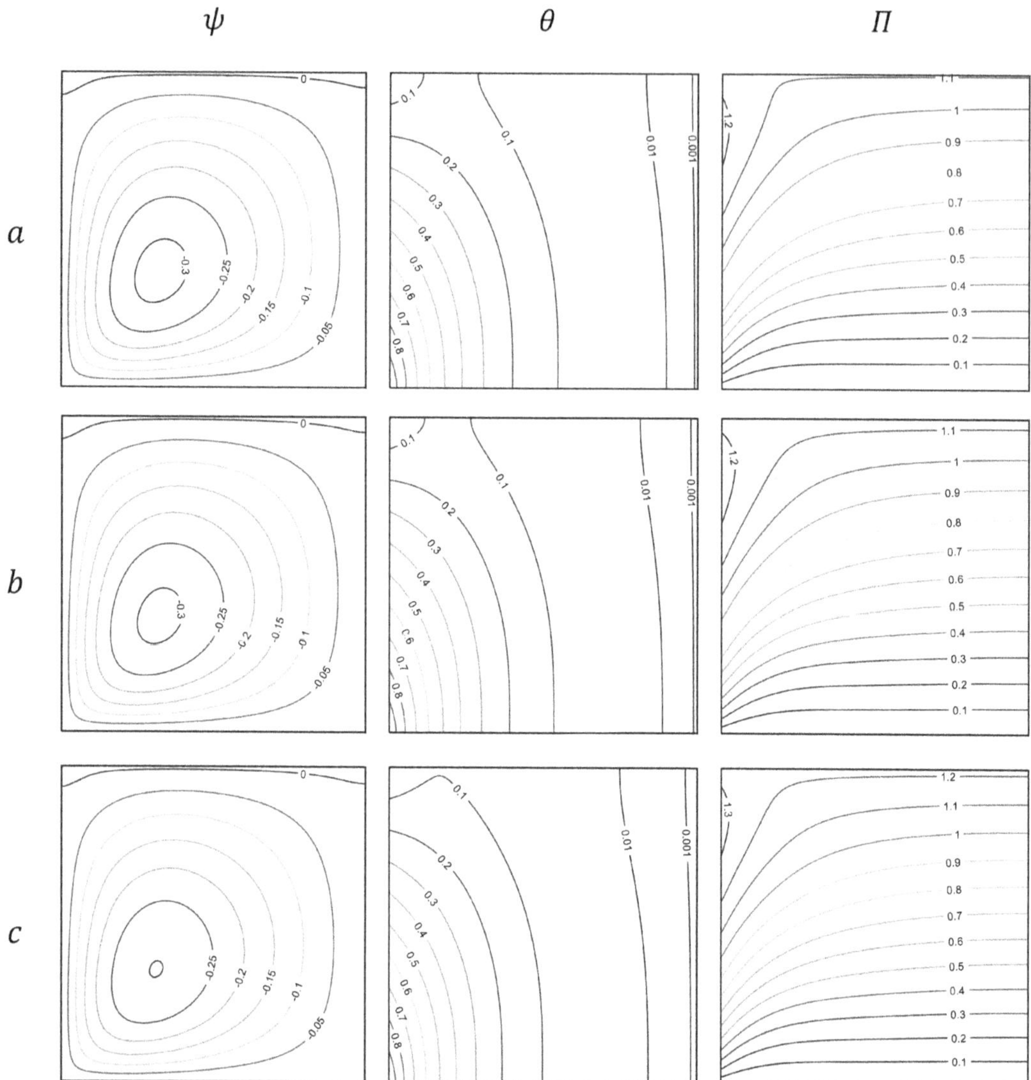

FIGURE 17.7 Streamlines (ψ), Isotherms (θ), and Heatlines (Π) contours when $Da = 10^{-2}$, $\phi = 0.05$, $Ra = 10^6$ at (a). $Ha = 0$, (b). $Ha = 50$, (c). $Ha = 100$ while non-uniform heating.

There is denser swirling found in the streamlines at the bottom, and the middle of the cavity as Da is increased from 10^{-2} to 10^2. Also, the secondary circulation originates from the top at the left corner of the cavity, as shown in Figure 17.10. This secondary circulation begins because of temperature variation; however, there is no extra circulation in the case of the boundary with even heating. The movement of temperature distribution gets increased because of augmentation in the buoyancy-driven effects. The growth of circulations in Π indicates the convection heat transfer and denser heatlines are noticed in the bottom region of the cavity's left wall.

17.4.3 Heat Transfer Curves While Uniform/Non-Uniform Heating

The effect of Ra, Ha, Da, on Nu is shown in Figure 17.11(a–b). A rise in Ha leads to a mild increase in Nu at $Da = 10^{-2}$ and decreases at $Da = 10^2$ for both even and linear heating of walls. The average Nusselt number diminishes for an increase in Ra for a lesser Darcy number ($Da = 10^{-2}$)

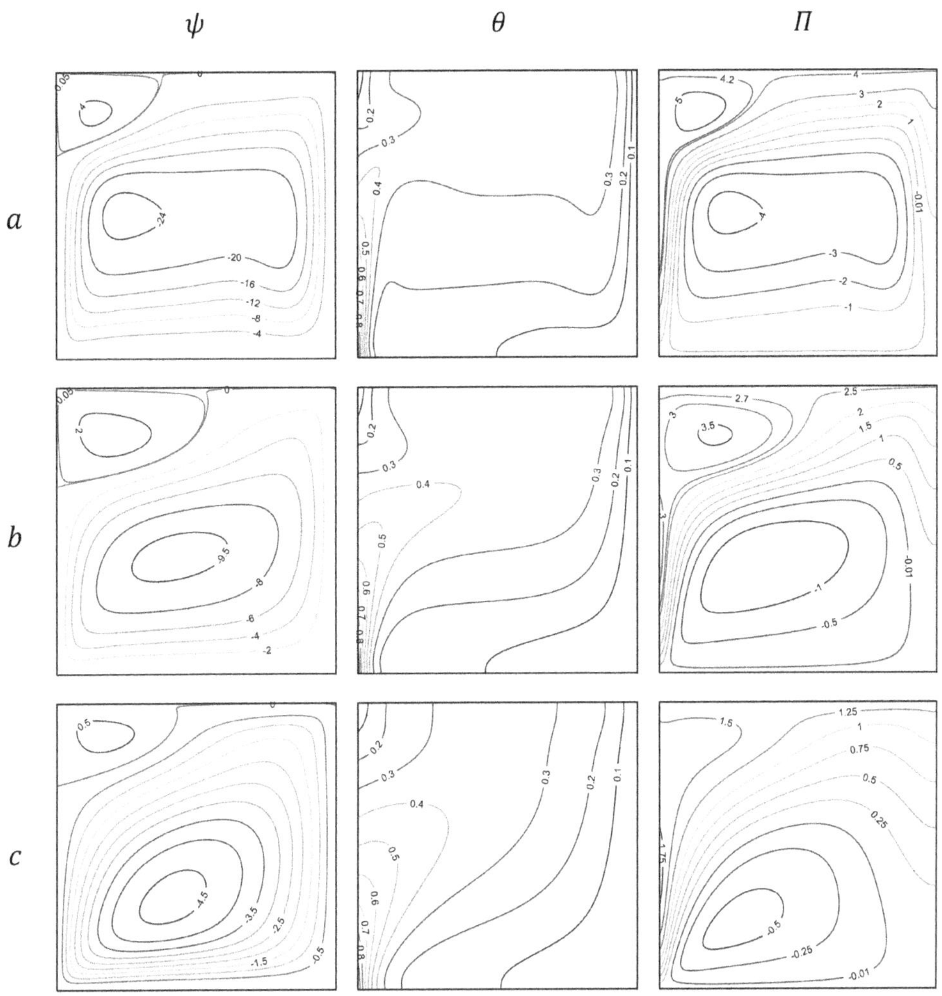

FIGURE 17.8 Streamlines (ψ), Isotherms (θ), and Heatlines (Π) contours when $Da = 10^2$, $\phi = 0.05$, $Ra = 10^6$ at (a). $Ha = 0$, (b). $Ha = 50$, (c). $Ha = 100$ while non-uniform heating.

because the heat transfer that occurs here is mainly based on conduction. The quantification in the Rayleigh number tries to deny the conduction effect, which in turn reduces *Nu*. For $Da = 10^2$, the permeability is increased and causes an increase in flow intensity. This leads to an upshot in the convection effects while increasing *Ra*, which makes an augment in *Nu*. As the intensity of the magnetic field increases, the convection effects and average Nusselt number (due to convection) decrease. At $Ra = 10^4$, the *Nu* increases when the Hartmann number changes from 50 to 100. This is due to the hike in conduction heat transfer domination. In addition, it is noticed that the heat transfer is less for linearly heated walls as compared to the uniform heated walls.

Figure 17.12(a–b) illustrates the change in *Nu* for varying the values of solid volume fraction, $\phi = 0.01$ to 0.1. An increase in ϕ increases the *Nu* for both the linear and uniformly heated boundary conditions at $Da = 10^{-2}$. Less effect is observed for *Nu* at $Da = 10^2$ for various values of ϕ. An increment in the Hartmann number makes a small rise in heat transfer for $Da = 10^{-2}$, and *Nu* decreases for $Da = 10^2$. A rise in ϕ leads to a growth in the density of the fluid and causes a rise in the conduction rate.

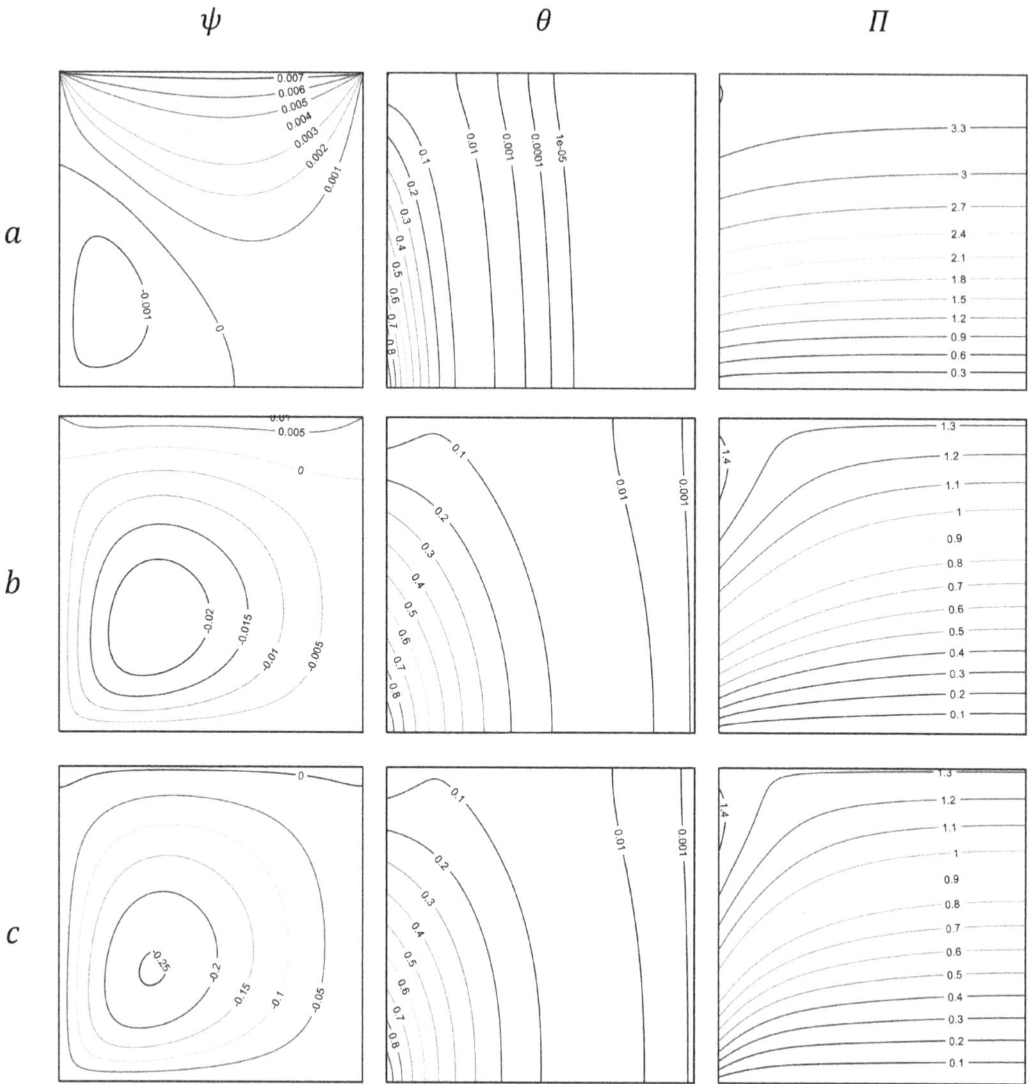

FIGURE 17.9 Streamlines (ψ), Isotherms (θ) and Heatlines (Π) contours when $Da = 10^{-2}$, $\phi = 0.1$, $Ha = 50$ at (a). $Ra = 10^4$, (b). $Ra = 10^5$, (c). $Ra = 10^6$ while non-uniform heating.

17.5 CONCLUSION

The fluid flow and heat transfer in a square porous LDC using the Carman–Kozeny model are discussed. The effect of several parameters, such as Darcy number Da, Hartmann number Ha, Rayleigh number Ra, and solid volume fraction ϕ, are considered. It is observed that the rise in Darcy number makes the rise in fluid flow intensity and thereby enhances the convective heat transfer. Moreover, higher Ra enhances the convection heat transfer effect, and the increase in Ha reduces the convection heat transfer. At lesser $Da = 10^{-2}$, conduction dominates the convection effects, thereby causing a small hike in Nu with an increase in the Hartmann number. For higher Darcy number $Da = 10^2$, the reduction in Nu is noticed with an increase in the magnetic field effect. The secondary circulations are generated for a linearly heated boundary, and the secondary circulations are not found in the case of a uniform heated boundary. Finally, it is quoted that the heat transfer rate is lesser within the cavity during the linearly heated left wall, and more heat transfer rate during the evenly heated wall.

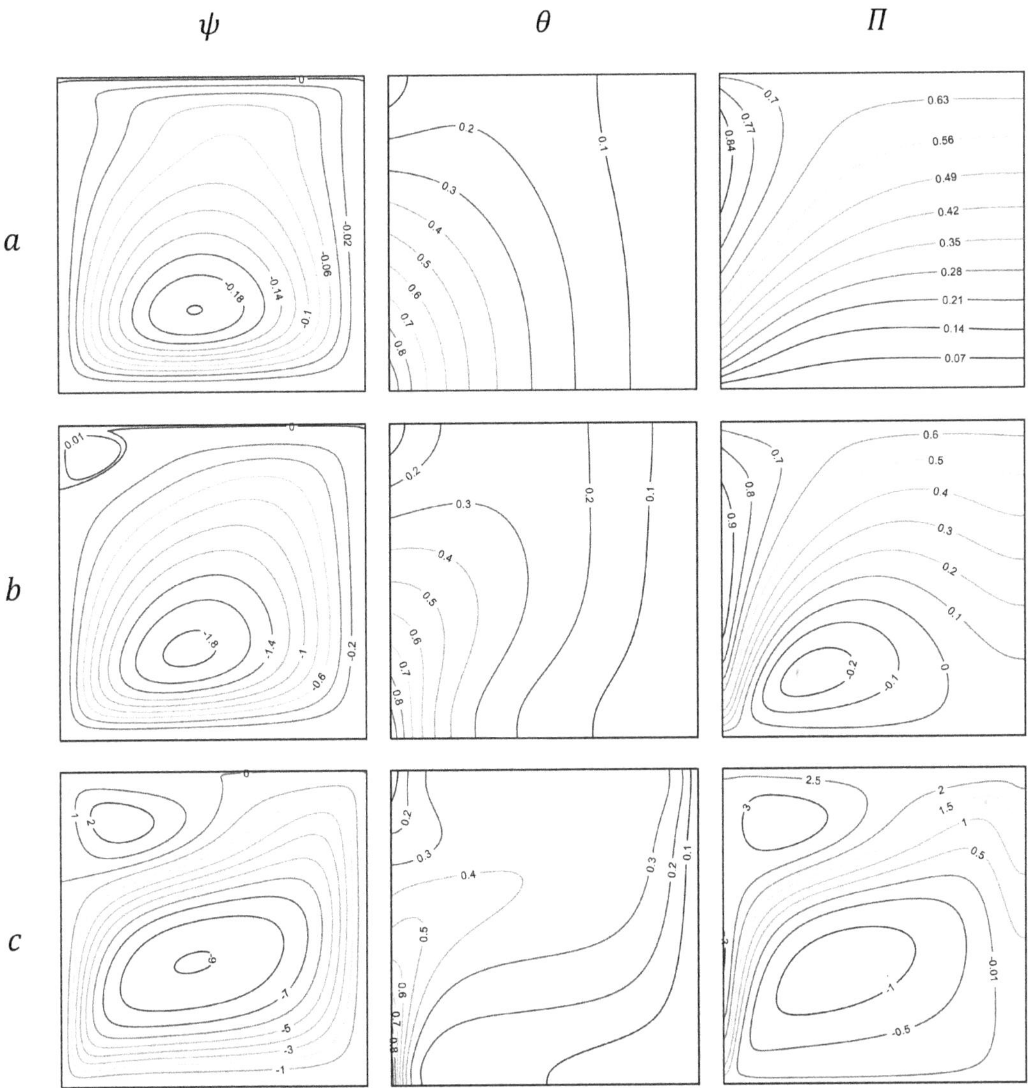

FIGURE 17.10 Streamlines (ψ), Isotherms (θ), and Heatlines (Π) contours when $Da = 10^2$, $\phi = 0.1$, $Ha = 50$ at (a). $Ra = 10^4$, (b). $Ra = 10^5$ (c). $Ra = 10^6$ while non-uniform heating.

Nomenclature

B	magnetic field strength
c_p	specific heat
g	gravity
Da	Darcy number
ρ	density
X,Y	dimensionless coordinates
τ	dimensionless time
θ	dimensionless temperature
U,V	dimensionless velocities
μ	dynamic viscosity
σ	electrical conductivity
Ha	Hartmann number

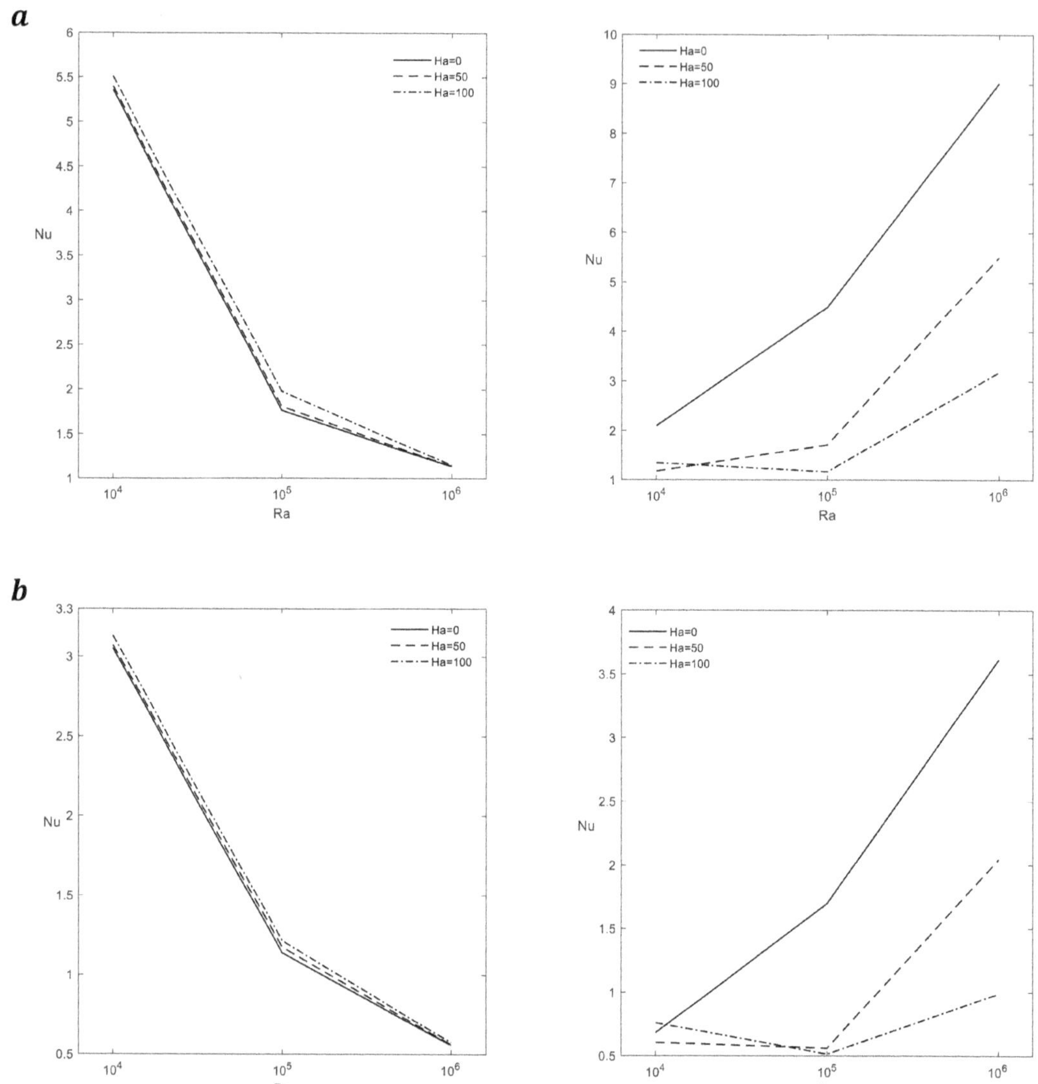

FIGURE 17.11 Average Nusselt number for various values of Ra at $\phi = 0.1$, $Da = 10^{-2}$ (left), and $Da = 10^{2}$ (Right) when (a). uniform heating (b). linear heating.

ν	kinematic viscosity
λ	liquid fraction of colloid
Nu	Nusselt number
C_m	porosity constant
Pr	Prandtl number
p	pressure
Ra	Rayleigh number
k	thermal conductivity
β	thermal expansion coefficient
α	thermal diffusivity
t	time
T	temperature

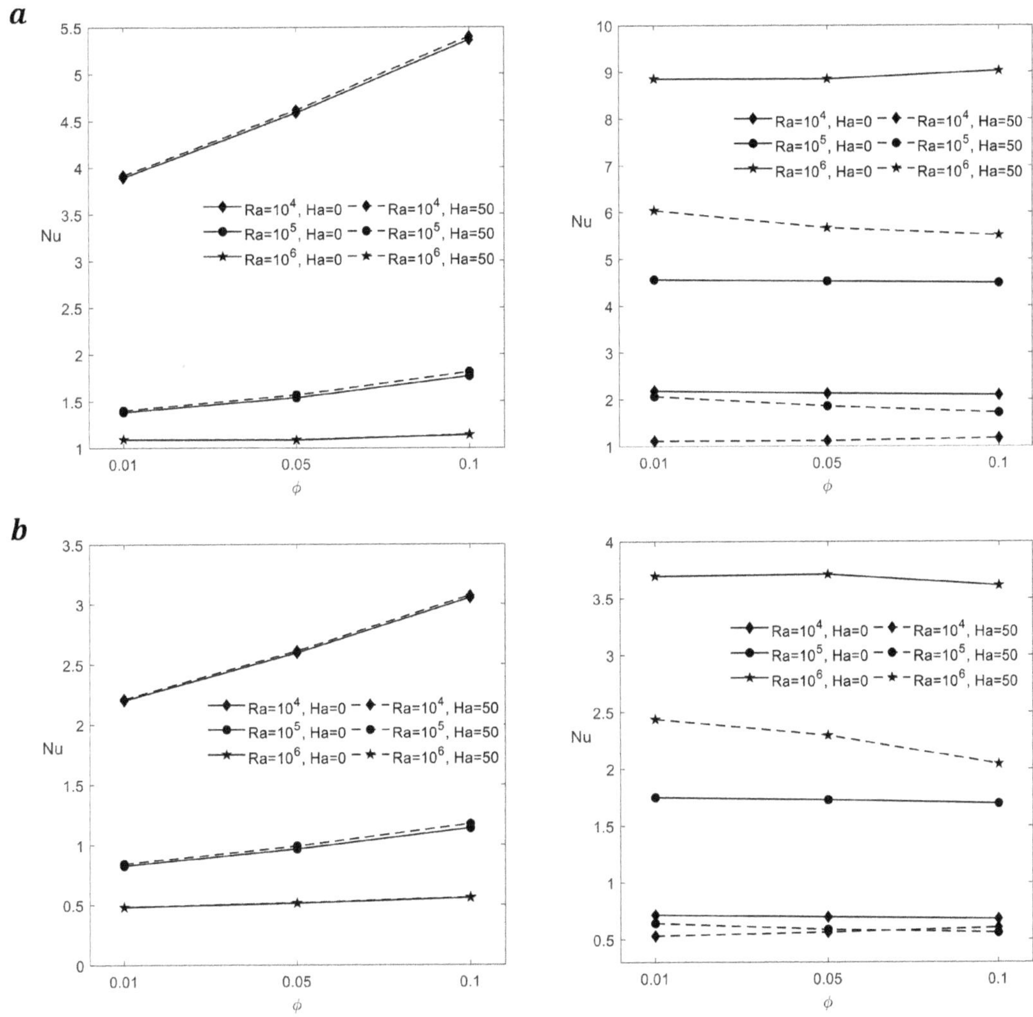

FIGURE 17.12 Average Nusselt number for various values of ϕ and Ra at Da = 10^{-2} (left) and Da = 10^{2} (Right) when (a). uniform heating (b). linear heating.

u,v	velocities
x, y	Cartesian coordinates
ϕ	volume fraction of the nanoparticles

REFERENCES

1. Ingham, D. B., & Pop, I. (1998). *Transport Phenomena in Porous Media.* Elsevier, Pergamon.
2. Nield, D. A., & Bejan, A. (2006). *Convection in Porous Media.* Springer, New York. https://doi.org/10.1007/0-387-33431-9
3. Vafai, K. (2015). Handbook of porous media. In K. Vafai (Ed.), *Handbook of Porous Media.* CRC Press, Boca Raton, FL. https://doi.org/10.1201/b18614
4. Lawrence, J., Mohanadhas, B., Narayanan, N., Kumar, A. V., Mangottiri, V., & Govindarajan, S. K. (2021). Numerical modelling of nitrate transport in fractured porous media under non-isothermal conditions. *Environmental Science and Pollution Research, 29*(57), 85922–85944. https://doi.org/10.1007/s11356-021-15691-8

5. Warke, A. S., Ramesh, K., Mebarek-Oudina, F., & Abidi, A. (2022). Numerical investigation of the stagnation point flow of radiative magnetomicropolar liquid past a heated porous stretching sheet. *Journal of Thermal Analysis and Calorimetry*, *147*(12), 6901–6912. https://doi.org/10.1007/s10973-021-10976-z
6. Djebali, R., Mebarek-Oudina, F., & Rajashekhar, C. (2021). Similarity solution analysis of dynamic and thermal boundary layers: further formulation along a vertical flat plate. *Physica Scripta*, *96*(8), 085206. https://doi.org/10.1088/1402-4896/ABFE31
7. Khanafer, K., & Vafai, K. (2019). Applications of nanofluids in porous medium: a critical review. *Journal of Thermal Analysis and Calorimetry*, *135*(2), 1479–1492. https://doi.org/10.1007/s10973-018-7565-4
8. Buongiorno, J. (2006). Convective transport in nanofluids. *Journal of Heat Transfer*, *128*(3), 240–250. https://doi.org/10.1115/1.2150834
9. Mebarek-Oudina, F., & Chabani, I. (2022). Review on nano-fluids applications and heat transfer enhancement techniques in different enclosures. *Journal of Nanofluids*, *11*(2), 155–168. https://doi.org/10.1166/jon.2022.1834
10. Doley, S., Kumar, A. V., & Jino, L. (2022). Time fractional transient magnetohydrodynamic natural convection of hybrid nanofluid flow over an impulsively started vertical plate. *Computational Thermal Sciences: An International Journal*, *14*(3), 59–82. https://doi.org/10.1615/ComputThermalScien.2022041607
11. Dhif, K., Mebarek-Oudina, F., Chouf, S., Vaidya, H., & Chamkha, A. J. (2021). Thermal analysis of the solar collector cum storage system using a hybrid-nanofluids. *Journal of Nanofluids*, *10*(4), 616–626. https://doi.org/10.1166/jon.2021.1807
12. Sankar, M., Swamy, H. A. K., Do, Y., & Altmeyer, S. (2022). Thermal effects of nonuniform heating in a nanofluid-filled annulus: buoyant transport versus entropy generation. *Heat Transfer*, *51*(1), 1062–1091. https://doi.org/10.1002/htj.22342
13. Swamy, H. A. K., Sankar, M., Reddy, N. K., & Manthari, M. S. Al. (2022). Double diffusive convective transport and entropy generation in an annular space filled with alumina-water nanoliquid. *The European Physical Journal Special Topics*, *231*(1), 2781–2800. https://doi.org/10.1140/epjs/s11734-022-00591-w
14. Pushpa, B. V., Sankar, M., & Mebarek-Oudina, F. (2021). Buoyant convective flow and heat dissipation of Cu–H 2 O nanoliquids in an annulus through a thin baffle. *Journal of Nanofluids*, *10*(2), 292–304. https://doi.org/10.1166/jon.2021.1782
15. Mahapatra, T. R., Pal, D., & Mondal, S. (2013). Effects of buoyancy ratio on double-diffusive natural convection in a lid-driven cavity. *International Journal of Heat and Mass Transfer*, *57*(2), 771–785. https://doi.org/10.1016/j.ijheatmasstransfer.2012.10.028
16. Mahmoodi, M., Abbasian Arani, A. A., Mazrouei Sebdani, S., Nazari, S., & Akbari, M. (2014). Free convection of a nanofluid in a square cavity with a heat source on the bottom wall and partially cooled from sides. *Thermal Science*, *18*(Suppl.2), 283–300. https://doi.org/10.2298/TSCI110406011A
17. Ghalambaz, M., Doostani, A., Izadpanahi, E., & Chamkha, A. J. (2020). Conjugate natural convection flow of Ag–MgO/water hybrid nanofluid in a square cavity. *Journal of Thermal Analysis and Calorimetry*, *139*(3), 2321–2336. https://doi.org/10.1007/s10973-019-08617-7
18. Sivasankaran, S., Ananthan, S. S., & Hakeem, A. K. A. (2016). Mixed convection in a lid-driven cavity with sinusoidal boundary temperature at the bottom wall in the presence of magnetic field. *Scientia Iranica*, *23*(3), 1027–1036. https://doi.org/10.24200/sci.2016.3871
19. Kefayati, G. R., Gorji-Bandpy, M., Sajjadi, H., & Ganji, D. D. (2012). Lattice boltzmann simulation of MHD mixed convection in a lid-driven square cavity with linearly heated wall. *Scientia Iranica*, *19*(4), 1053–1065. https://doi.org/10.1016/j.scient.2012.06.015
20. Kiran, S., Sankar, M., Swamy, H. A. K., & Makinde, O. D. (2022). Unsteady buoyant convective flow and thermal transport analysis in a nonuniformly heated annular geometry. *Computational Thermal Sciences: An International Journal*, *14*(2), 1–17. https://doi.org/10.1615/ComputThermalScien.2021039723
21. Ahmed, S. E., Raizah, Z. A. S., & Aly, A. M. (2021). Magnetohydrodynamic convective flow of nanofluid in double lid-driven cavities under slip conditions. *Thermal Science*, *25*(3), 1703–1717. https://doi.org/10.2298/TSCI190811141A
22. Kimura, S., & Bejan, A. (1983). The "heatline" visualization of convective heat transfer. *Journal of Heat Transfer*, *105*(4), 916–919. ASME. https://doi.org/10.1115/1.3245684
23. Costa, V. A. F. (2003). Unified streamline, heatline and massline methods for the visualization of two-dimensional heat and mass transfer in anisotropic media. *International Journal of Heat and Mass Transfer*, *46*(8), 1309–1320. https://doi.org/10.1016/S0017-9310(02)00404-0

24. Basak, T., & Chamkha, A. J. (2012). Heatline analysis on natural convection for nanofluids confined within square cavities with various thermal boundary conditions. *International Journal of Heat and Mass Transfer*, *55*(21–22), 5526–5543. https://doi.org/10.1016/j.ijheatmasstransfer.2012.05.025
25. Şahin, B. (2020). Effects of the center of linear heating position on natural convection and entropy generation in a linearly heated square cavity. *International Communications in Heat and Mass Transfer*, *117*, 104675. https://doi.org/10.1016/j.icheatmasstransfer.2020.104675
26. Mondal, C., Sarkar, R., Sarkar, S., Biswas, N., & Manna, N. K. (2020). Magneto-thermal convection in lid-driven cavity. *Sādhanā*, *45*(1), 227. https://doi.org/10.1007/s12046-020-01463-6
27. Azizul, F. M., Alsabery, A. I., Hashim, I., & Chamkha, A. J. (2021). Heatline visualization of mixed convection inside double lid-driven cavity having heated wavy wall. *Journal of Thermal Analysis and Calorimetry*, *145*(6), 3159–3176. https://doi.org/10.1007/s10973-020-09806-5
28. Sathiyamoorthy, M., Basak, T., Roy, S., & Pop, I. (2007). Steady natural convection flow in a square cavity filled with a porous medium for linearly heated side wall(s). *International Journal of Heat and Mass Transfer*, *50*(9–10), 1892–1901. https://doi.org/10.1016/j.ijheatmasstransfer.2006.10.010
29. Ghasemi, K., & Siavashi, M. (2017). Lattice Boltzmann numerical simulation and entropy generation analysis of natural convection of nanofluid in a porous cavity with different linear temperature distributions on side walls. *Journal of Molecular Liquids*, *233*, 415–430. https://doi.org/10.1016/j.molliq.2017.03.016
30. Malik, S., & Nayak, A. K. (2017). MHD convection and entropy generation of nanofluid in a porous enclosure with sinusoidal heating. *International Journal of Heat and Mass Transfer*, *111*, 329–345. https://doi.org/10.1016/j.ijheatmasstransfer.2017.03.123
31. Mebarek-Oudina, F. (2017). Numerical modeling of the hydrodynamic stability in vertical annulus with heat source of different lengths. *Engineering Science and Technology, an International Journal*, *20*(4), 1324–1333. https://doi.org/10.1016/j.jestch.2017.08.003
32. Mebarek-Oudina, F., Keerthi Reddy, N., & Sankar, M. (2018). Heat source location effects on buoyant convection of nanofluids in an annulus. In *Lecture Notes in Mechanical Engineering: Advances in Fluid Dynamics* (pp. 923–938). www.springer.com/series/11693
33. Biswas, N., Manna, N. K., & Chamkha, A. J. (2021). Effects of half-sinusoidal nonuniform heating during MHD thermal convection in Cu–Al2O3/water hybrid nanofluid saturated with porous media. *Journal of Thermal Analysis and Calorimetry*, *143*(2), 1665–1688. https://doi.org/10.1007/s10973-020-10109-y
34. Hashemi-Tilehnoee, M., Dogonchi, A. S., Seyyedi, S. M., Chamkha, A. J., & Ganji, D. D. (2020). Magnetohydrodynamic natural convection and entropy generation analyses inside a nanofluid-filled incinerator-shaped porous cavity with wavy heater block. *Journal of Thermal Analysis and Calorimetry*, *141*(5), 2033–2045. https://doi.org/10.1007/s10973-019-09220-6
35. Swamy, H. A. K., Sankar, M., & Reddy, N. K. (2022). Analysis of entropy generation and energy transport of cu-water nanoliquid in a tilted vertical porous annulus. *International Journal of Applied and Computational Mathematics*, *8*(1), 1–23. https://doi.org/10.1007/s40819-021-01207-y
36. Kumar, A. V., Lawrence, J., & Saravanakumar, G. (2022). Fluid friction/heat transfer irreversibility and heat function study on MHD free convection within the MWCNT–water nanofluid-filled porous cavity. *Heat Transfer*, *51*(5), 4247–4267. https://doi.org/10.1002/htj.22498
37. Jino, L., Vanav Kumar, A., Maity, S., Mohanty, P., & Sankar, D. S. (2021). Mathematical modeling of a nanofluid in a porous cavity with side wall temperature in the presence of magnetic field. *AIP Conference Proceedings*, *020007*(March), 020007. https://doi.org/10.1063/5.0046109
38. Jino, L., Vanav Kumar, A., Doley, S., Berlin, M., & Mohanty, P. K. (2022). Numerical modelling of porous square cavity heated on vertical walls in presence of magnetic field. In *Advances in Thermofluids and Renewable Energy* (pp. 127–137). Springer, Singapore. https://doi.org/10.1007/978-981-16-3497-0_10
39. Kemparaju, S., Kumara Swamy, H. A., Sankar, M., & Mebarek-Oudina, F. (2022). Impact of thermal and solute source-sink combination on thermosolutal convection in a partially active porous annulus. *Physica Scripta*, *97*(5), 55206. https://doi.org/10.1088/1402-4896/ac6383
40. Fares, R., Mebarek-Oudina, F., Aissa, A., Bilal, S. M., & Öztop, H. F. (2022). Optimal entropy generation in Darcy-Forchheimer magnetized flow in a square enclosure filled with silver based water nanoliquid. *Journal of Thermal Analysis and Calorimetry*, *147*(2), 1571–1581. https://doi.org/10.1007/s10973-020-10518-z
41. Chabani, I., Mebarek-Oudina, F., & Ismail, A. A. I. (2022). MHD flow of a hybrid nano-fluid in a triangular enclosure with zigzags and an elliptic obstacle. *Micromachines*, *13*(2), 1–17. https://doi.org/10.3390/mi13020224

42. Oztop, H. F. (2006). Combined convection heat transfer in a porous lid-driven enclosure due to heater with finite length. *International Communications in Heat and Mass Transfer*, *33*(6), 772–779. https://doi.org/10.1016/j.icheatmasstransfer.2006.02.003
43. Basak, T., Roy, S., Singh, S. K., & Pop, I. (2010). Analysis of mixed convection in a lid-driven porous square cavity with linearly heated side wall(s). *International Journal of Heat and Mass Transfer*, *53*(9–10), 1819–1840. https://doi.org/10.1016/j.ijheatmasstransfer.2010.01.007
44. Sivasankaran, S., & Pan, K. L. (2012). Numerical simulation on mixed convection in a porous lid-driven cavity with nonuniform heating on both side walls. *Numerical Heat Transfer; Part A: Applications*, *61*(2), 101–121. https://doi.org/10.1080/10407782.2011.643741
45. Kumar, A. V., Jino, L., Doley, S., & Berlin, M. (2021). Magnetic field effect on lid-driven porous cavity heated to the right wall. *Science & Technology Asia*, *26*(December), 27–37. https://doi.org/10.14456/scitechasia.2021.63
46. Jino, L., & Kumar, A. V. (2021). Cu-water nanofluid MHD quadratic natural convection on square porous cavity. *International Journal of Applied and Computational Mathematics*, *7*(4), 164. https://doi.org/10.1007/s40819-021-01103-5
47. Jino, L., & Vanav Kumar, A. (2021). Fluid flow and heat transfer analysis of quadratic free convection in a nanofluid filled porous cavity. *International Journal of Heat and Technology*, *39*(3), 876–884. https://doi.org/10.18280/ijht.390322
48. Jino, L., & Kumar, A. V. (2021). MHD natural convection of hybrid nanofluid in a porous cavity heated with a sinusoidal temperature distribution. *Computational Thermal Sciences: An International Journal*, *13*(5), 83–99. https://doi.org/10.1615/computthermalscien.2021037663
49. Marzougui, S., Mebarek-Oudina, F., Magherbi, M., & Mchirgui, A. (2022). Entropy generation and heat transport of Cu–water nanoliquid in porous lid-driven cavity through magnetic field. *International Journal of Numerical Methods for Heat and Fluid Flow*, *32*(6), 2047–2069. https://doi.org/10.1108/HFF-04-2021-0288/FULL/XML
50. Reddy, N. K., Swamy, H. A. K., & Sankar, M. (2021). Buoyant convective flow of different hybrid nanoliquids in a non-uniformly heated annulus. *European Physical Journal: Special Topics*, *230*(5), 1213–1225. https://doi.org/10.1140/epjs/s11734-021-00034-y
51. Lawrence, J., & Alagarsamy, V. K. (2021). Mathematical modelling of MHD natural convection in a linearly heated porous cavity. *Mathematical Modelling of Engineering Problems*, *8*(1), 149–157. https://doi.org/10.18280/mmep.080119
52. Kiran, S., Sankar, M., Swamy, H. A. K., & Makinde, O. D. (2022). Unsteady buoyant convective flow and thermal transport analysis in a nonuniformly heated annular geometry. *Computational Thermal Sciences*, *14*(2), 1–17. https://doi.org/10.1615/ComputThermalScien.2021039723

18 Magnetohydrodynamics Free Convection in a Wavy Cavity Partially Heated from Below and Saturated with a Nanofluid

Imene Rahmoune and Saadi Bougoul

18.1 INTRODUCTION

In the last years, research on heat transfer improvement has been an important objective in many applications. It has received considerable interest because it tries to satisfy the increasing requirements of thermal efficiency of certain energy systems, as cooling of electronic components, energy production, solar collectors, etc.

The search for new techniques leading to further improve heat transfer has become essential. Among these techniques, we find one that improves the thermal properties of conventional fluids like water, oil, and ethylene glycol. In 1995, Choi [1] discovered a new class of fluids with improved thermal conductivity; this class of fluids is named nanofluids. The nanofluid is an injection of nanometric particles into the conventional fluid.

The analysis of free convection in partially heated cavities interests several researchers. Many numerical studies have been conducted on free convection in a rectangular cavity having warm bottom walls and cold vertical ones. Among these, we find the works of Hasnaoui et al. (1992) [2], Ganzarolli and Milanez (1995) [3], Aydin and Yang (2000) [4], Corcione (2003) [5], Mebarek-Oudina (2017) [6], Pushpa et al. (2021) [7], Merzougui et al. (2021) [8], Mebarek-Oudina and Chabani (2022) [9], Chabani et al. (2022) [10], and Merzougui et al. (2021) [11]. Other studies are available to complete this work, including the following.

Calcagni et al. (2005) [12] developed an experimental and numerical investigation of heat transmission by free convection in a square enclosure saturated with air and equipped with a discrete heating element positioned on the lower wall. Cooling of this cavity is ensured by sides. It was noted that increasing heat source length gives a growth in heat transmission rate, especially for significant Rayleigh numbers. Sharma et al. (2007) [13] established a simulation of turbulent free convection in a square cavity with heating from below and cooling from vertical sidewalls. They observed that Nusselt number rises with growing heated width.

Aminossadati and Ghasemi (2009) [14] developed a numerical investigation of free convection in a cavity with restricted heating at the bottom and saturated with a nanofluid. The results obtained show that a rise in Rayleigh numbers reinforces heat fluxes by free convection, and the growth of nanoparticles solid volume fractions leads to a diminution in the maximum temperature of the heating part. Mansour et al. (2016) [15] simulated the consequence of magnetohydrodynamics (MHD) on free convection in an enclosure with heating element located at the bottom and saturated with a nanofluid. They noticed that a growing Hartmann number gives a reduction in heat transmission rate, and the opposite phenomenon is observed when growing Rayleigh number. Rahmoune et al. (2022) [16] numerically studied free laminar convection of Al_2O_3-water nanofluid in an enclosure

DOI: 10.1201/9781003299608-18

of a shape unlike that of H. They noted that the average Nusselt number inside the cavity rises with nanoparticles volume fraction and Rayleigh number. Bougoul and Rahmoune (2021) [17] presented a numerical investigation of entropy production and free convection flow in a hollow enclosure with a specific shape and saturated with Al_2O_3-H_2O nanofluid. This cavity is submitted to a magnetic field. It is noted that entropy generation and average Nusselt number rise with Rayleigh number and nanofluid concentration, but they decrease with Hartmann number.

Recently, studies of heat transmission in corrugated cavities saturated with a nanofluid have received considerable importance from several researchers, where they have considered the corrugated walls as a means of improving heat transfer. We can cite the following:

Nasrin et al. (2013) [18] investigated a numerical analysis of free convection in the presence of a nanofluid in a solar collector made up of a glass cover and a corrugated absorber. They found that the solar collector is of high performance when it works with water-Al_2O_3 nanofluid than by a conventional fluid. Öğüt et al. (2017) [19] conducted a simulation of nanofluid free convection intilted cavity with corrugated walls. Results highlight that growing undulations, nanofluid concentrations, and Rayleigh number significantly raise heat transmission rate. Raizah et al. (2021) [20] treated the impact of the magnetic field on mixed convection flow in a corrugated cavity saturated with hybrid nanofluid and a porous medium. They observed that length and the position of localized heating act significantly on nanofluid motion and heat transfer rate.

Heat transmission by free convection of nanofluids contained in a cylindrical annulus heated by a discrete heat source with various base fluids was quantitatively presented by Mebarek-Oudina (2019) [21]. Marzougui et al. (2021) [22] developed a numerical study on entropy production associated with magneto-convective of Cu-water nanofluid in a cavity with chamfers. Under the influence of shear rate viscosity, Hassan et al. (2022) [23] investigated the thermal energy and mass transfer of shear thinning fluid. Shafiq et al. (2022) [24] examined numerically the improvement of heat transmission, which corresponds to a sensibility analysis performed using surface strategies.

In this investigation, we present a numerical analysis of free convection coupled with a magnetic field in a cavity with corrugated walls whose main objective is to know flow structure and heat transmission that occurs inside this cavity saturated with a nanofluid, taking into account the impact of Rayleigh number, Hartmann number, concentration of Al_2O_3-water nanofluid, and corrugated walls.

18.2 PHYSICAL MODEL

In this study, the influence of magnetohydrodynamic on free laminar convection of alumina/water nanofluid inside a square wavy cavity using the single-phase model is analyzed. Figure 18.1 gives the geometry that will be examined in this investigation. The two-dimensional enclosure consists of a hot source part located on the bottom wall with a temperature (Tc) and length B, whereas the two vertical wavy walls are kept at cold temperature (Tc) with a width L, and the other walls are adiabatic. For all cavity walls, the non-slip condition is imposed.

Nanofluid flow used in this investigation is supposed to be Newtonian, incompressible, stationary, and laminar. Radiation influence and viscous dissipation are supposed to be negligible. All physical parameters of the nanofluids chosen are supposed to be constant; only density is supposed to vary with temperature according to the Boussinesq approximation in buoyancy terms of momentum equations. In addition, a uniform extern magnetic field β_0 is applied to the enclosure.

The governing equations, which define free convection in the enclosure, are given as:

Continuity equation:

This equation is simplified to:

$$\frac{\partial u}{\partial x}+\frac{\partial v}{\partial y}=0 \tag{18.1}$$

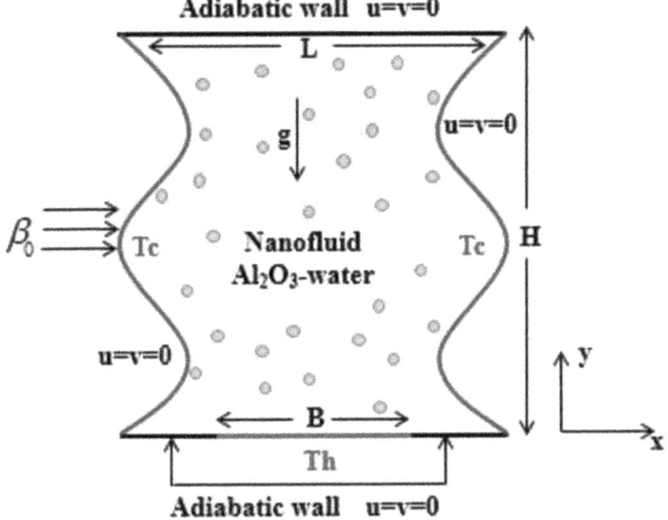

FIGURE 18.1 Geometry of this study with boundary conditions.

Momentum equation:
Momentum equations written in the two directions are:

$$u\frac{\partial u}{\partial x}+v\frac{\partial u}{\partial y}=-\frac{1}{\rho_{nf}}\frac{\partial P}{\partial x}+\upsilon_{nf}\left(\frac{\partial^2 u}{\partial x^2}+\frac{\partial^2 u}{\partial y^2}\right), \tag{18.2}$$

$$u\frac{\partial v}{\partial x}+v\frac{\partial v}{\partial y}=-\frac{1}{\rho_{nf}}\frac{\partial P}{\partial y}+\upsilon_{nf}\left(\frac{\partial^2 v}{\partial x^2}+\frac{\partial^2 v}{\partial y^2}\right)+\beta_{nf}g\left(T-T_C\right)-\frac{\sigma_{nf}}{\rho_{nf}}\beta_0^2 v, \tag{18.3}$$

Energy equation:
The energy equation is reduced to:

$$u\frac{\partial T}{\partial x}+v\frac{\partial T}{\partial y}=\alpha_{nf}\left(\frac{\partial^2 T}{\partial x^2}+\frac{\partial^2 T}{\partial y^2}\right), \tag{18.4}$$

Where ρ_{nf} is density, υ_{nf} the kinematic viscosity, β_{nf} its thermal expansion coefficient, and α_{nf} the thermal diffusivity of the nanofluid; σ_{nf} and β_0 are electrical conductivity and magnetic field.

Expressions of thermal conductivity, viscosity, density, thermal expansion coefficient, and nanofluid specific heat can be evaluated using conventional formulas developed for solid–liquid mixtures [25, 26].

$$K_{nf}=\left[\frac{K_s+2K_f-2\phi\left(K_f-K_s\right)}{K_s+2K_f+\phi\left(K_f-K_s\right)}\right]K_f \tag{18.5}$$

$$\mu_{nf}=\mu_f\left(1-\phi\right)^{-2.5} \tag{18.6}$$

$$\rho_{nf}=\left(1-\phi\right)\rho_f+\phi\rho_s \tag{18.7}$$

$$\left(\rho\beta\right)_{nf}=\left(1-\phi\right)\left(\rho\beta\right)_f+\phi\left(\rho\beta\right)_s \tag{18.8}$$

$$(\rho cp)_{nf} = (1-\phi)(\rho cp)_f + \phi(\rho cp)_s \tag{18.9}$$

The following equation is used to calculate the electrical conductivity [17].

$$\sigma_{nf} = \left[1 + \frac{3\phi(\gamma-1)}{(\gamma+2)-(\gamma-1)\phi}\right]\sigma_f \tag{18.10}$$

Where $\gamma = \frac{\sigma_s}{\sigma_f}$.

The following non-dimensionless parameters considered in this study are defined as:

$$Ra = \frac{g\beta_{nf}(T_h - T_c)H^3}{\nu_{nf}\alpha_{nf}}, \quad Nu_{nf} = \frac{h_{nf}H}{K_f}, \quad Ha = \beta_0 H\sqrt{\frac{\sigma_{nf}}{\mu_{nf}}}$$

Ra, Nu, and Ha are numbers of Rayleigh, Nusselt, and Hartmann, respectively. The Hartmann number is a crucial quantity in the magnetohydrodynamic coupled to free convection. It represents electromagnetic force compared to the viscous one. It characterizes fluid movement having a certain conductivity and exposed to a magnetic field [17].

18.3 NUMERICAL MODEL

In this study of natural convection, the various equations governing this phenomenon were solved numerically using the commercial software Fluent-CFD founded on finite-volume method. In momentum equations, we used the SIMPLE algorithm to ensure pressure–velocity coupling and second-order upwind model for approximation of convective and diffusive expressions of transport equations. The convergence criterion is set to 10^{-6} in the numerical solution of the various transport equations.

For Ra = $7.68.10^4$, three types of meshes were tested (110 × 110), (120 × 120), and (130 × 130) to get the independence of results of the selected mesh (Figure 18.2(b)). In Figure 18.2(b), we show

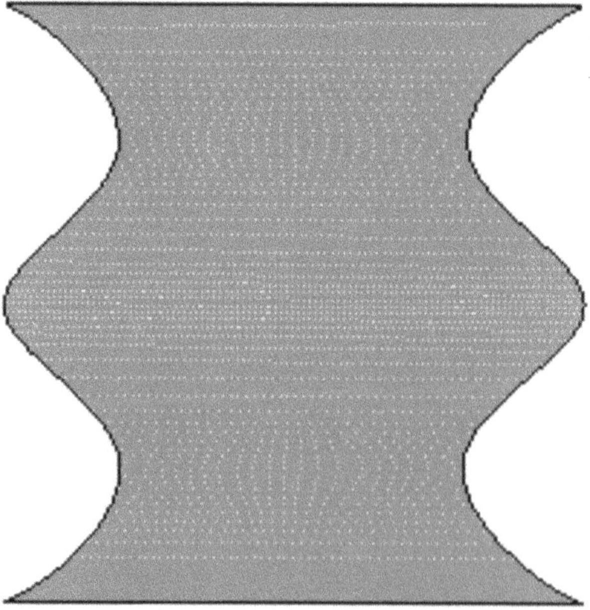

FIGURE 18.2 (a) Mesh used in this investigation; (b) temperature variation for the selected meshes.

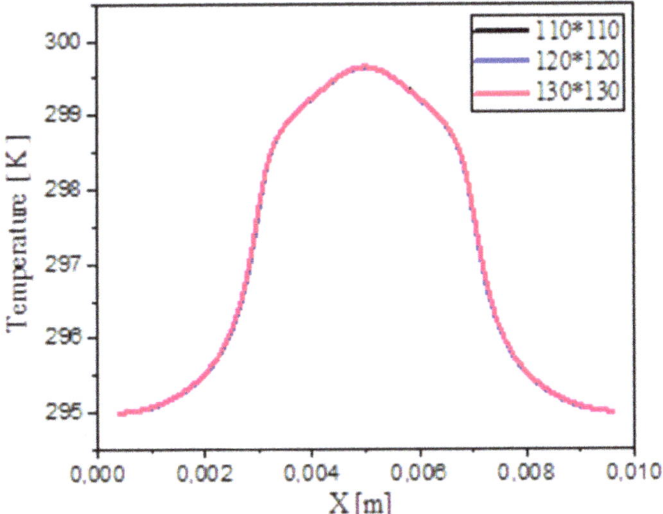

FIGURE 18.2 (Continued)

profiles of temperature near the bottom wall of the enclosure. To reduce calculation time, we have selected the second mesh (120 × 120). Following the dimensions of the cavity studied, the mesh chosen is refined, uniform, and quadratic (Figure 18.2(a)); this makes it possible to capture all the gradients which occur inside the cavity.

18.4 RESULTS AND DISCUSSION

18.4.1 Temperature Variation

Various simulations were conducted, and the results of this analysis of free convection in a corrugated cavity were analyzed. This cavity is under action of a magnetic field oriented in the horizontal direction. Volume fractions chosen take respectively the values 0%, 3%, and 5%. The selected values of the Rayleigh number are $7.68.10^4$, $1.5.10^5$, $2.304.10^5$, and $3.072.10^5$, and the Hartmann number is considered variable between 0 and 75. The results are represented in the following curves in the form of streamlines, isotherms, velocity, and temperature profiles and average Nusselt number.

In Figures 18.3, 18.4, 18.5, and 18.6, we clearly observe the effect of the Rayleigh and volume fractions on the temperature distribution inside the cavity for the chosen Hartmann number.

It is noted that once the Rayleigh number rises, heat transmission by convection becomes dominant (buoyancy force becomes important) and, by conduction, becomes less significant, and growth in the volume fraction increases the temperature without modifying the structure of the isotherms. This is the result of improvement in the conductivity of the conventional fluid.

An increase in Hartmann number has an opposite effect compared to that of the Rayleigh number; for this, the convection becomes less intense, and the conduction becomes more and more dominant. The Hartmann number, which takes into account the action of a magnetic field, expresses the presence of the Lorentz force, which opposes convective motion.

By fixing the Hartmann number and growing the Rayleigh number, thermal boundary layer thickness is reduced, heat transmission by convection intensifies, and isotherms undergo a stratification. Increasing nanoparticles volume fraction and the Rayleigh number promotes heat transmission by convection. Magnetic field limits heat transmission rate by convection without eliminating it, in particular for significant values of Rayleigh number.

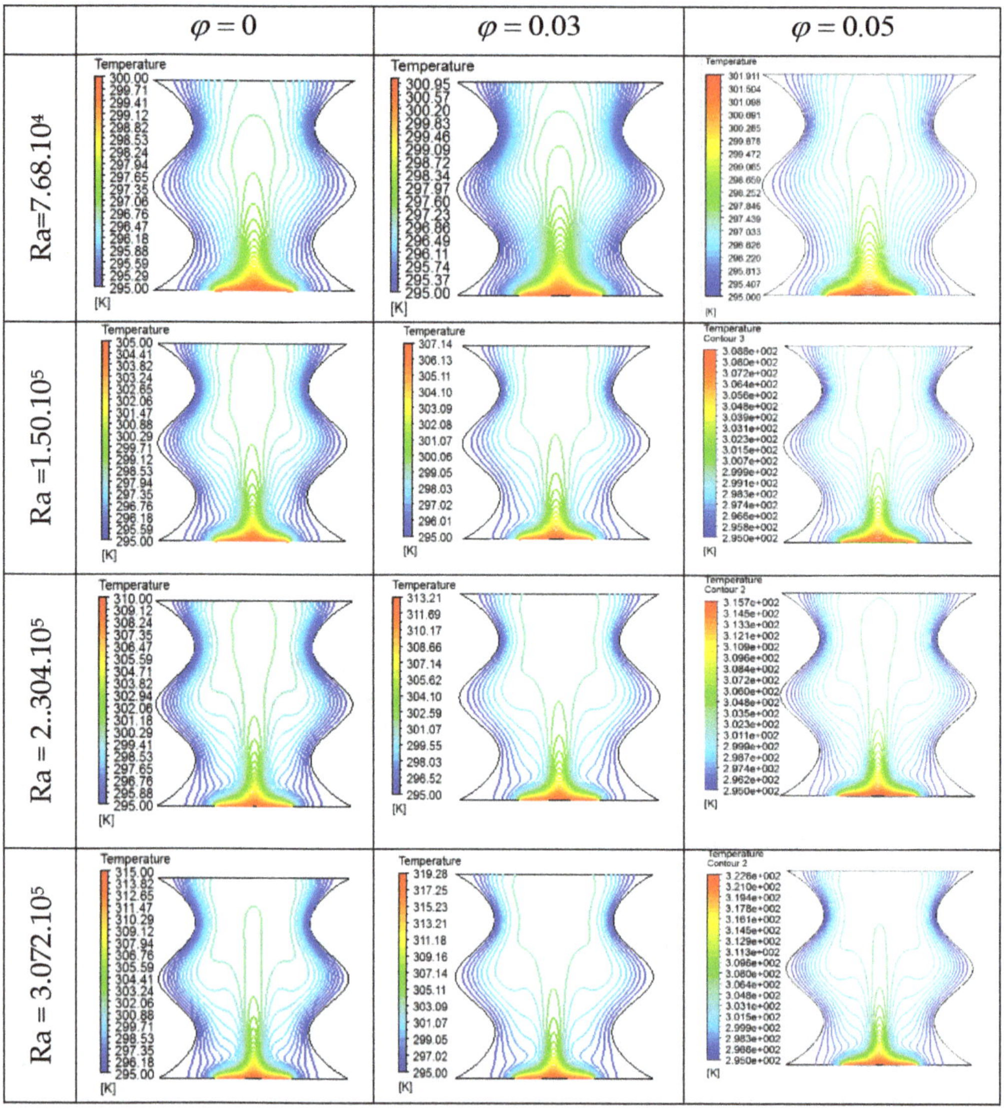

FIGURE 18.3 Isotherms developed by the nanofluid flow for selected Rayleigh numbers (*Ra*) and volume fractions (ϕ) for *Ha = 0.*

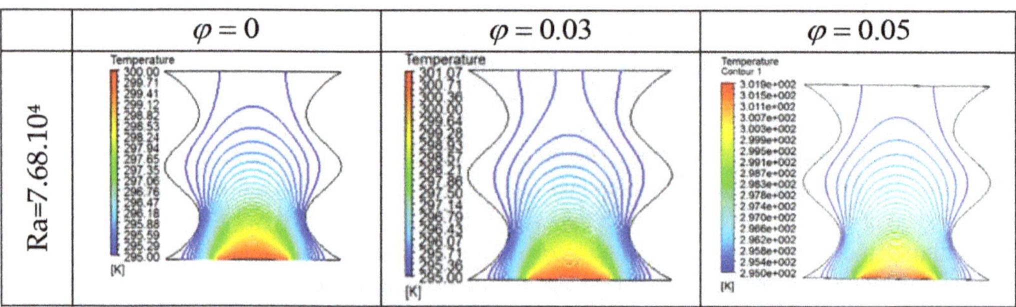

FIGURE 18.4 Isotherms developed by the nanofluid flow for selected Rayleigh numbers (*Ra*) and volume fractions (ϕ) for *Ha = 25.*

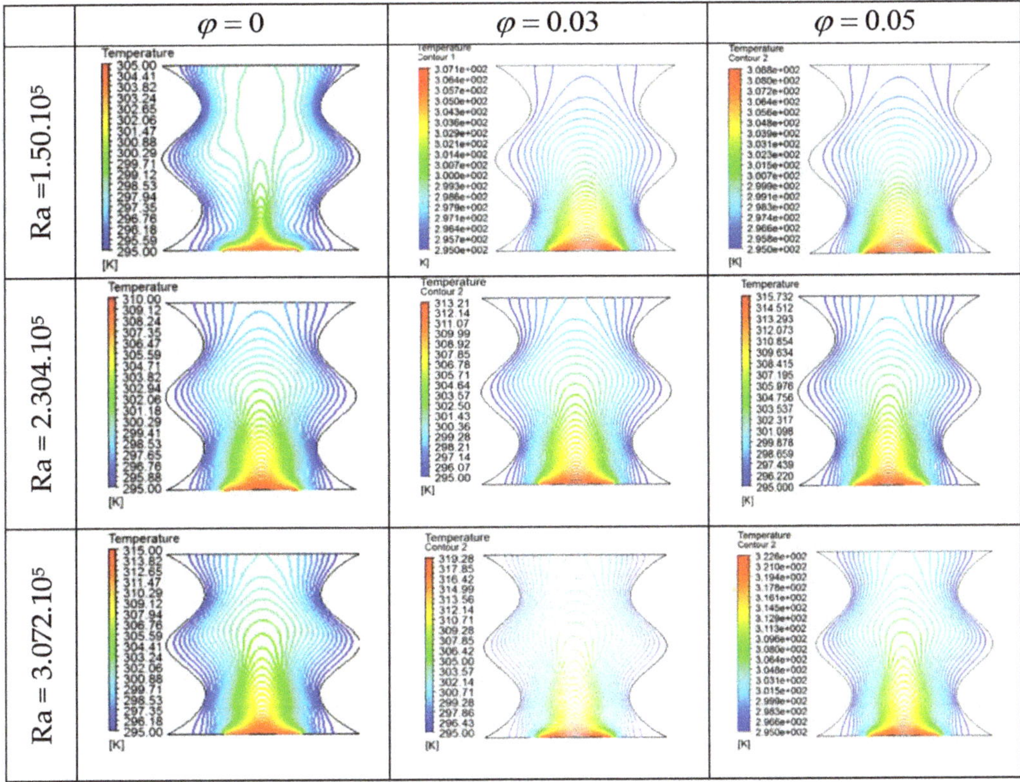

FIGURE 18.4 (Continued)

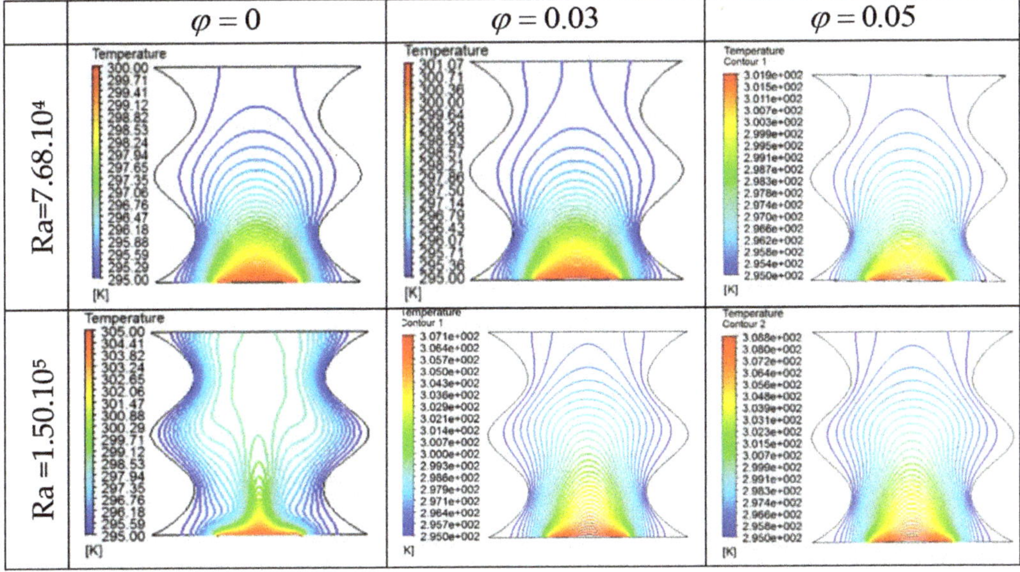

FIGURE 18.5 Isotherms developed by the nanofluid flow for selected Rayleigh numbers (*Ra*) and volume fractions (ϕ) for *Ha = 50.*

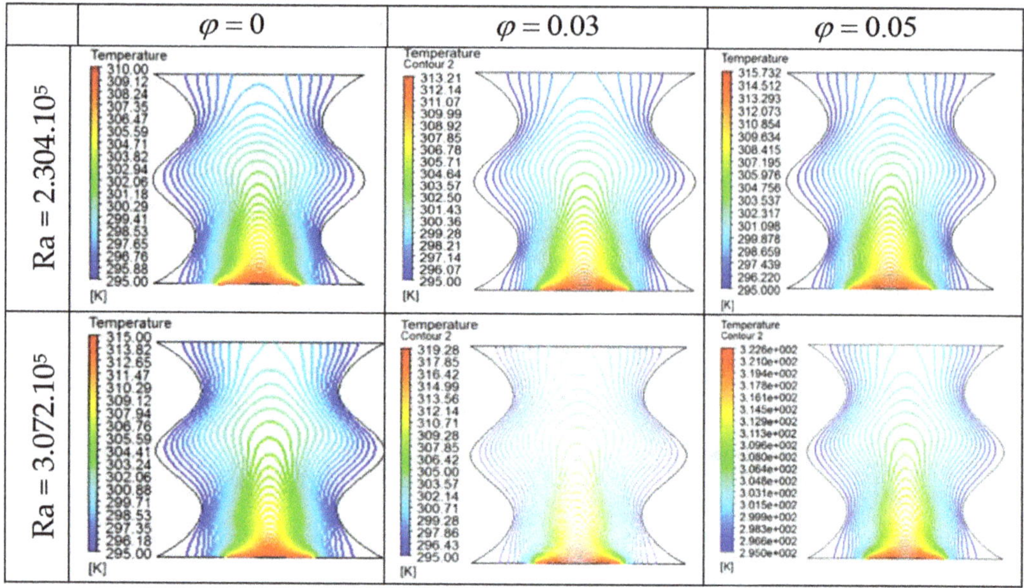

FIGURE 18.5 (Continued)

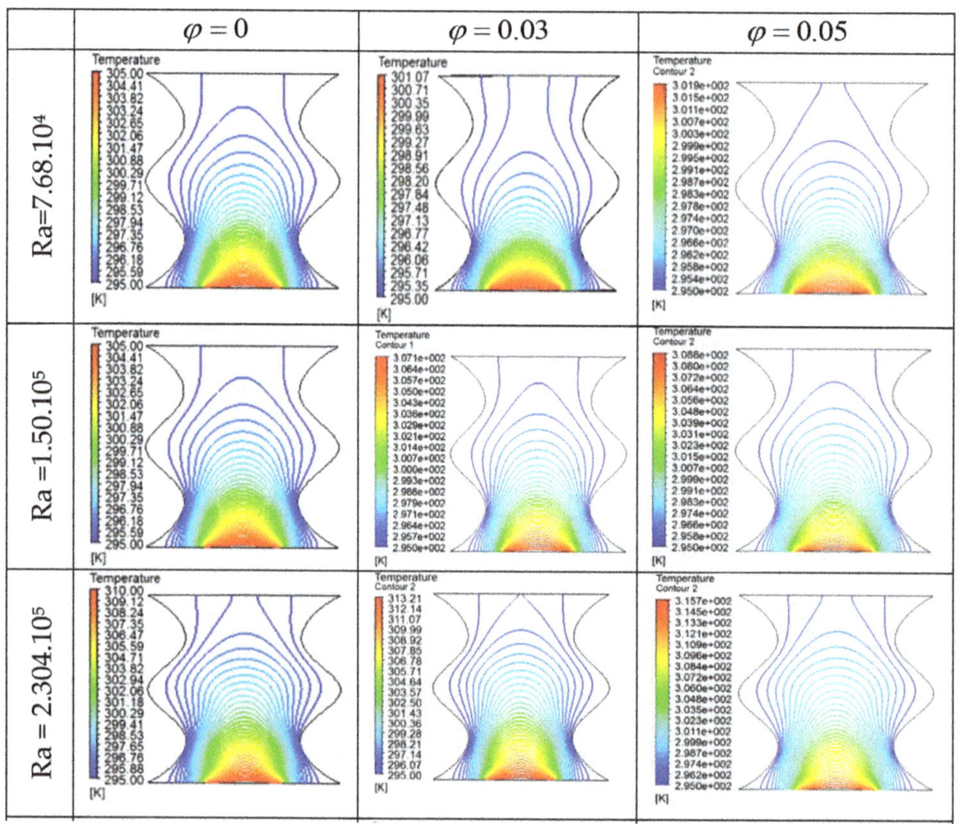

FIGURE 18.6 Isotherms developed by the nanofluid flow for selected Rayleigh numbers (*Ra*) and volume fractions (ϕ) for $Ha = 75$.

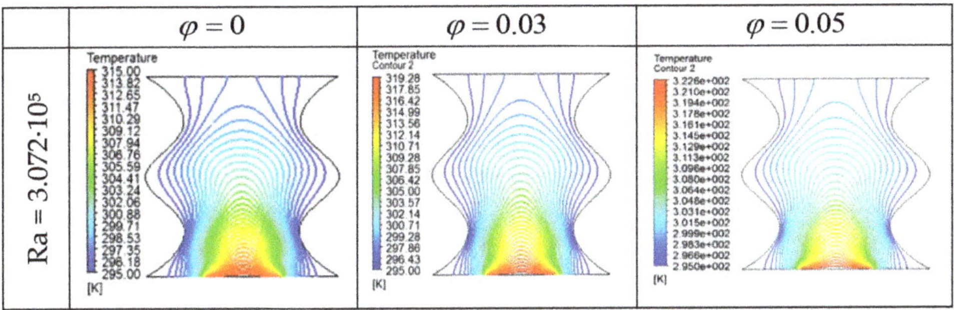

FIGURE 18.6 (Continued)

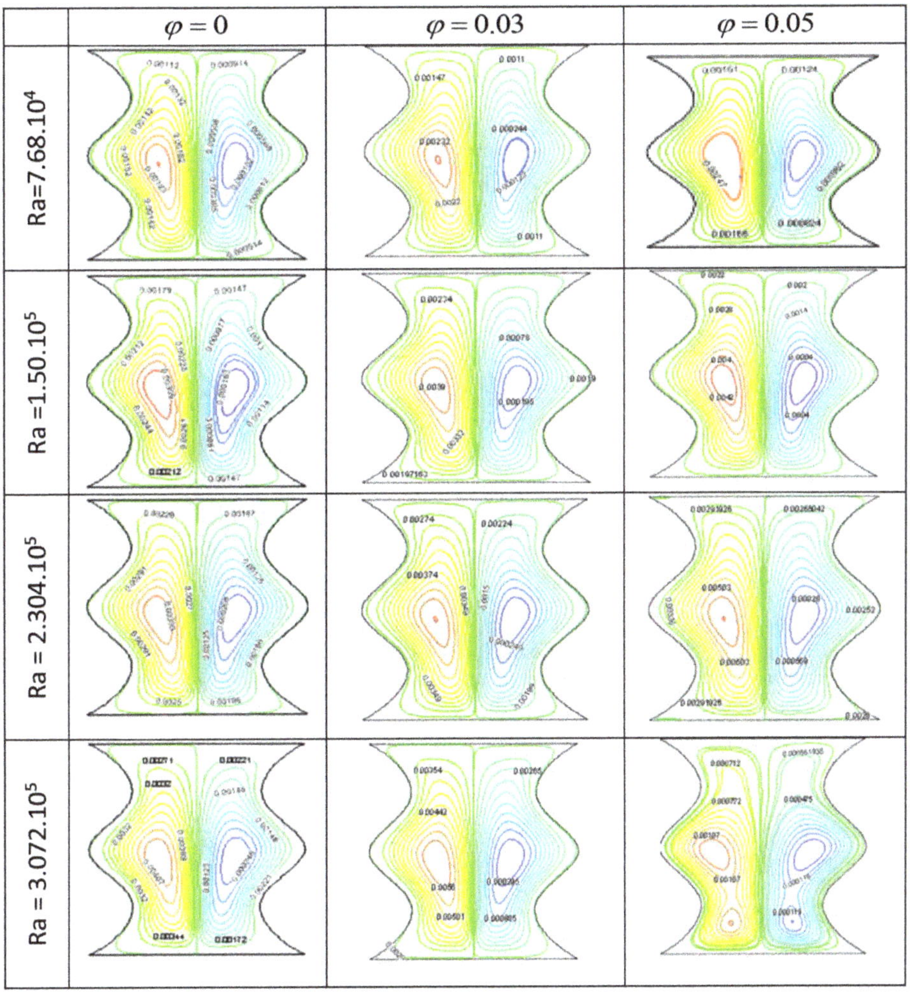

FIGURE 18.7 Streamlines for selected Rayleigh numbers (*Ra*) and volume fractions (ϕ) for *Ha* = 0.

18.4.2 Streamlines

In this part, we will examine the impact of Rayleigh number, volume fraction, and Hartmann number on nanofluid dynamics inside the corrugated cavity. In the case where the Hartmann number is zero and for various volume fractions and the Rayleigh numbers, the streamlines are represented in the following figure (Figure 18.7). For each Rayleigh number and for three chosen volume fractions, two counter-rotating cells arise inside the cavity. Growth in volume fraction does not influence the flow structure; however, it becomes accelerated.

By increasing the Hartmann number to 25, we see that the fluid has almost the same circulation as the case where the Hartmann is zero (Figure 18.8), with a slight deformation of the counter-rotating cells, and the fluid becomes less accelerated compared to the previous case.

For a Hartmann number equal to 25, the flow structure is not too much modified, and a rise in volume fraction allows to accelerate the fluid.

On the other hand, if the Hartmann number exceeds the value 25 (Figures 18.9 and 18.10), we notice the formation of two large convective cells situated on either side of the median axis, and inside each of these two cells, there are two small ones, and the fluid circulation tends to be significant at the bottom of the cavity, and formed cells are compressed close to the bottom wall.

Growth in the Hartmann number modifies the flow structure, and the fluid becomes less accelerated by occupying less space in the cavity.

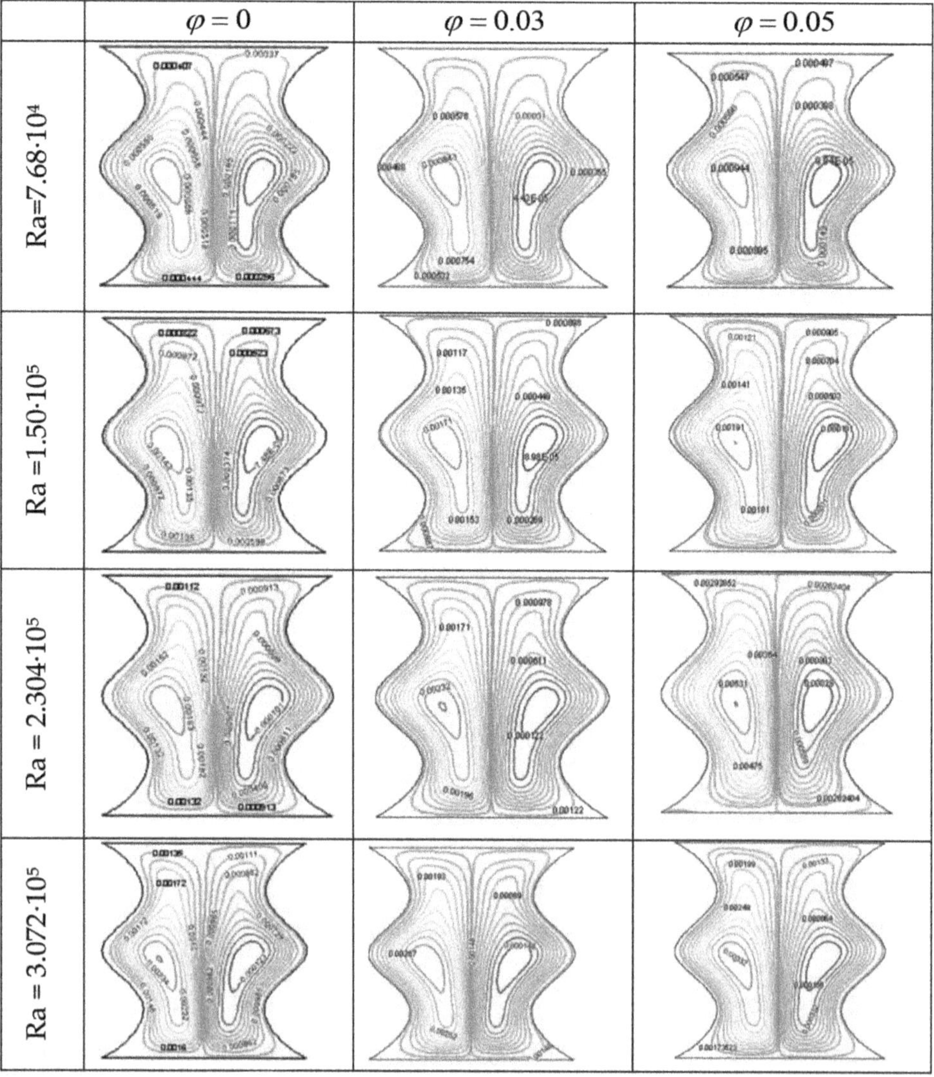

FIGURE 18.8 Streamlines for selected Rayleigh numbers (*Ra*) and volume fractions (ϕ) for *Ha* = 25.

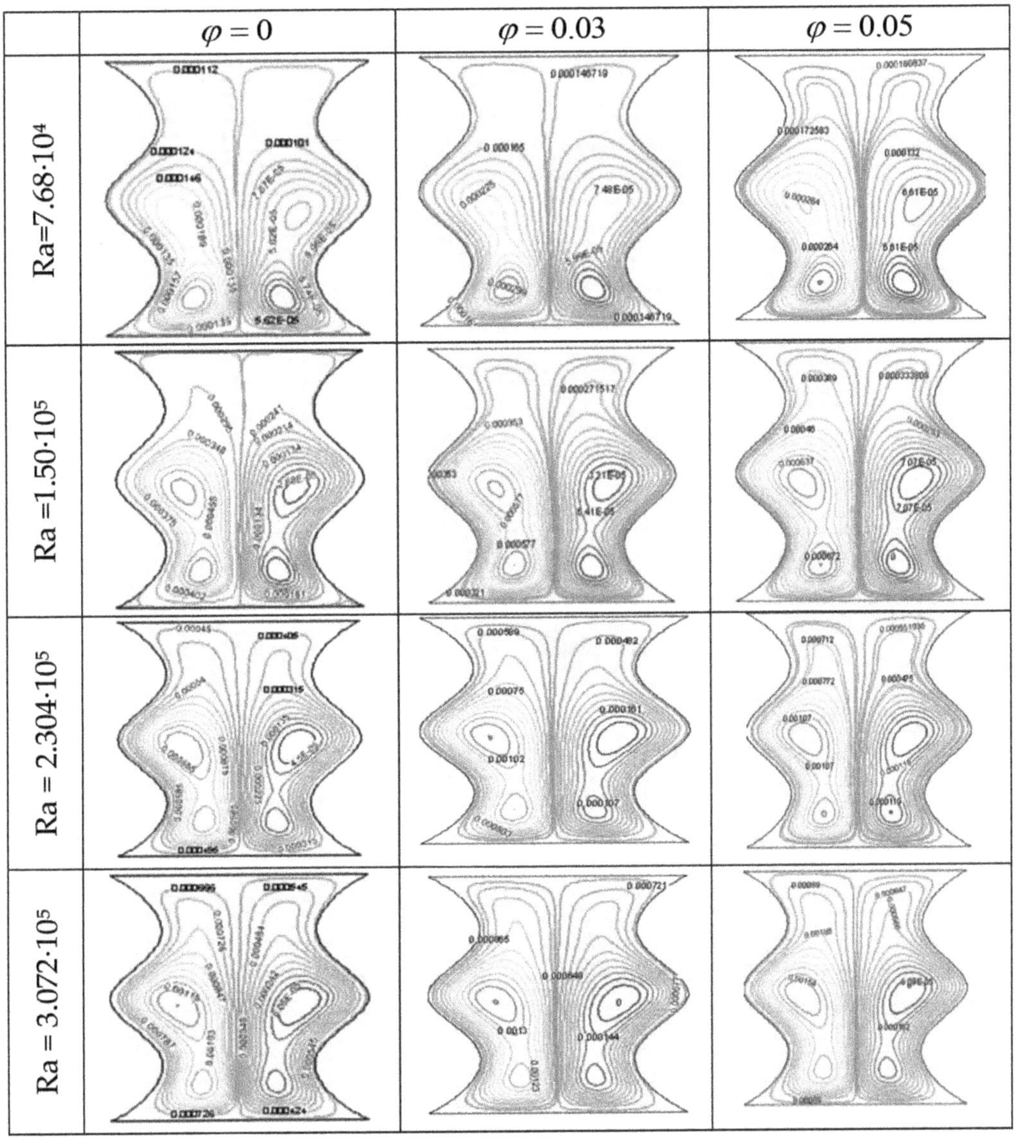

	$\varphi = 0$	$\varphi = 0.03$	$\varphi = 0.05$
$Ra = 7.68 \cdot 10^4$			
$Ra = 1.50 \cdot 10^5$			
$Ra = 2.304 \cdot 10^5$			
$Ra = 3.072 \cdot 10^5$			

FIGURE 18.9 Streamlines for selected Rayleigh numbers (*Ra*) and volume fractions (ϕ) for *Ha* = 50.

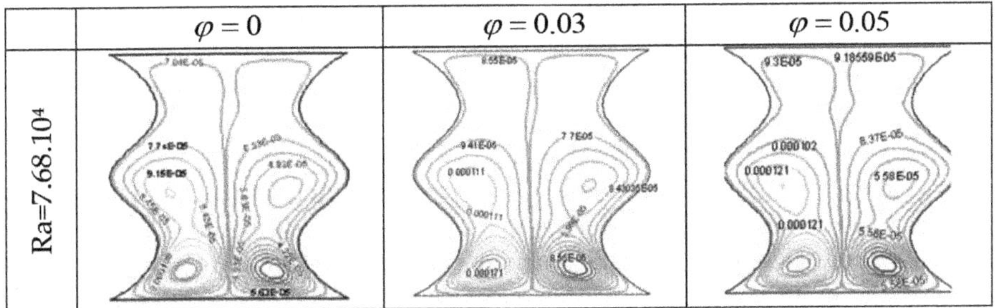

FIGURE 18.10 Streamlines for selected Rayleigh numbers (R*a*) and volume fractions (ϕ) for *Ha* = 75.

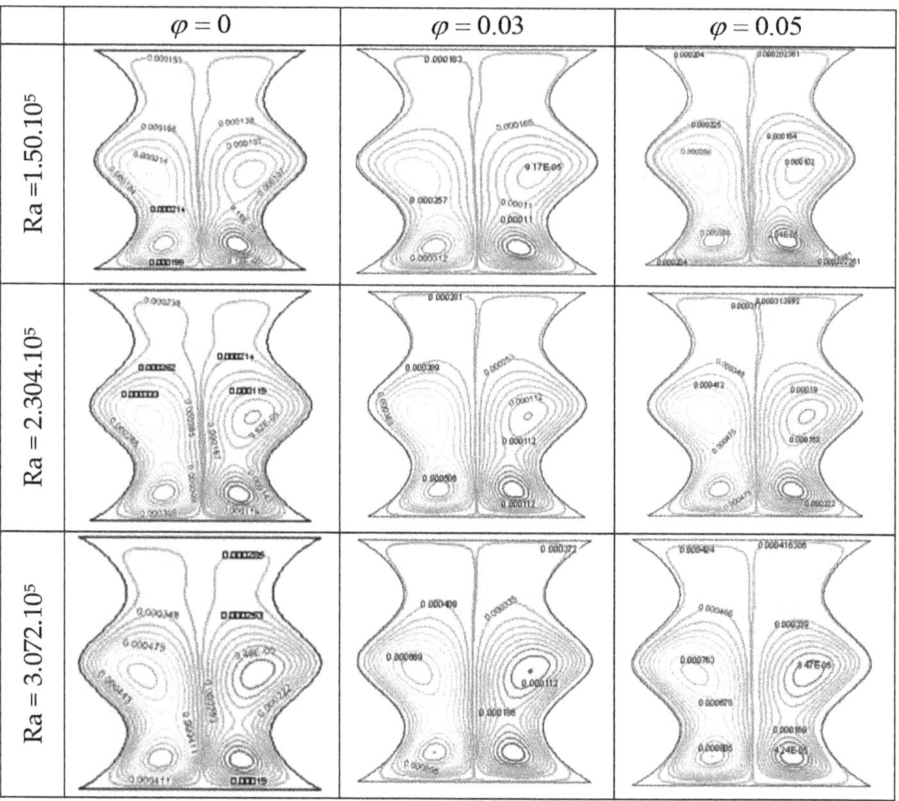

FIGURE 18.10 (Continued)

By growing Rayleigh number, the same physical phenomenon occurs inside the cavity for each fixed Hartmann number. A rise in Rayleigh number accelerates the fluid, and consequently, convection intensity becomes significant.

18.4.3 Velocity Profiles

Vertical component profiles of the velocity are shown in the following figures (Figures 18.11 and 18.12) for two fixed Rayleigh numbers and for three values of the volume fraction (0%, 3%, and 5%).

These profiles are given for three different vertical positions of the cavity (at the bottom close to the horizontal wall, in the middle of the cavity, and at the top close to the adiabatic wall). For the same position, we note that when the Rayleigh number rises, the fluid velocity increases; as a result, the Rayleigh number raises convection intensity.

For positions located at the bottom and at the top of the cavity, speed is low compared to that calculated in the middle of the cavity; this is because close to the horizontal walls, friction is high.

We also note that once Hartmann number rises, velocity decreases, which implies that the convection intensity decreases, and consequently, heat transfer by conduction becomes dominant. For small Rayleigh values and for large Hartmann numbers, the speed can reach zero values. Thus, the Hartmann number limits convection and, in some cases, suppresses it. So the Hartmann number greatly influences heat transfer inside the cavity.

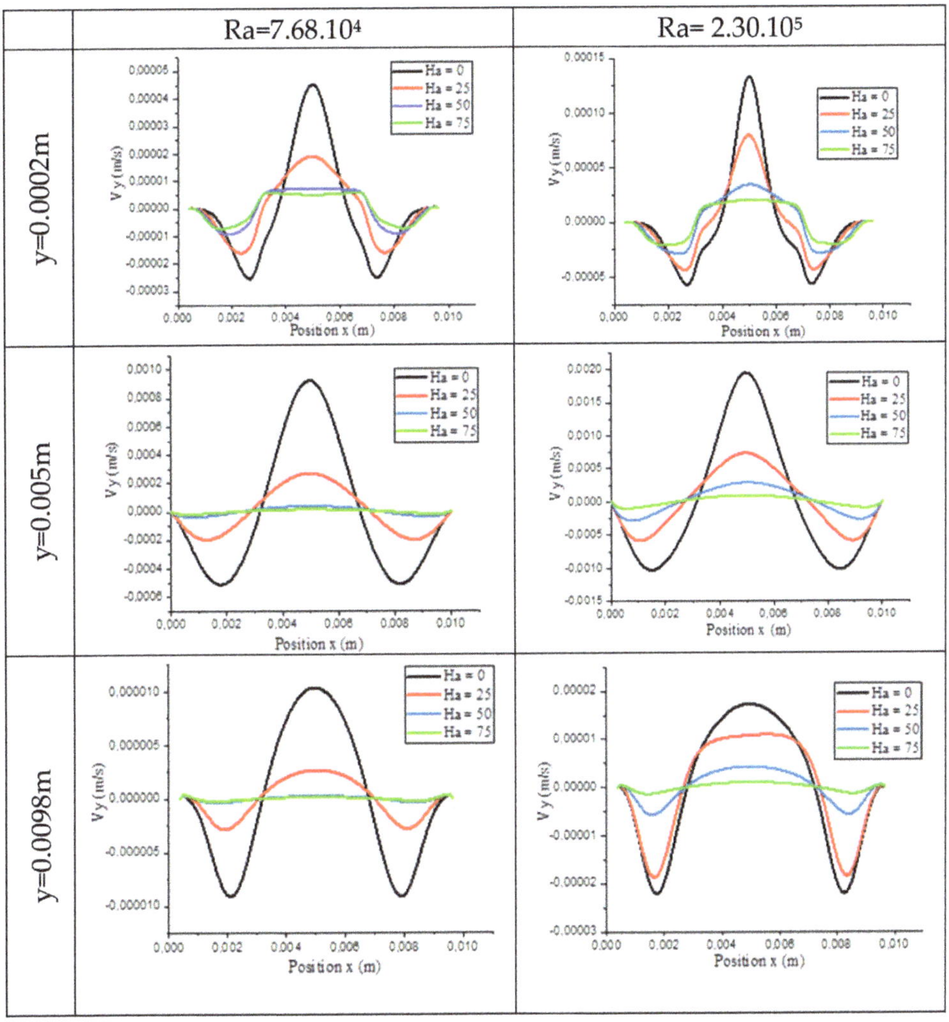

	Ra=7.68.10⁴	Ra= 2.30.10⁵
y=0.0002m		
y=0.005m		
y=0.0098m		

FIGURE 18.11 Velocity vertical component for two Rayleigh numbers and for selected Hartmann numbers ($\phi = 0$).

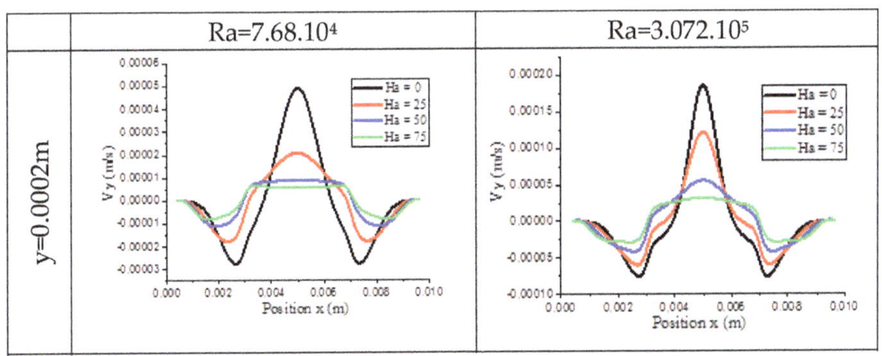

FIGURE 18.12 Velocity vertical component for two Rayleigh numbers and for selected Hartmann numbers ($\varphi = 0.03$).

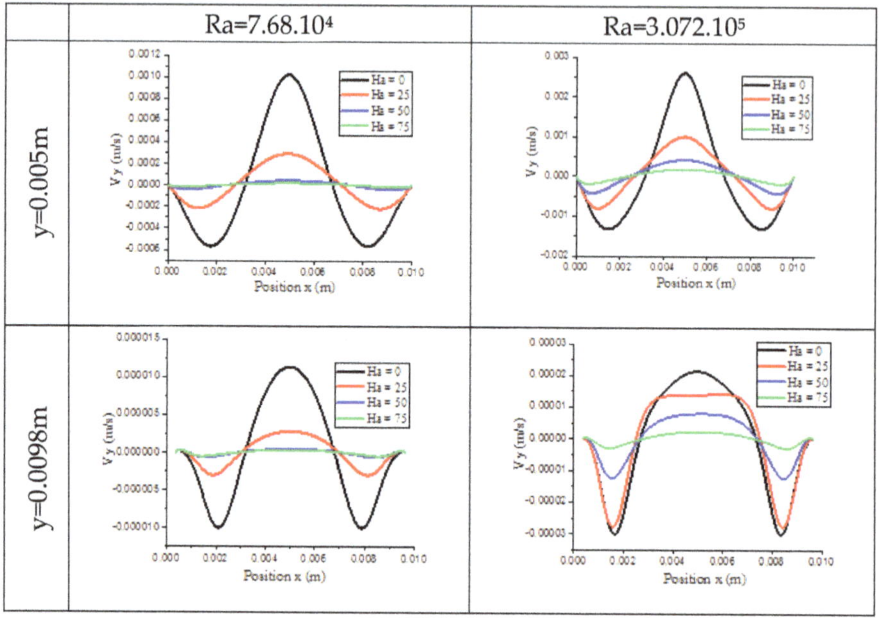

FIGURE 18.12 (Continued)

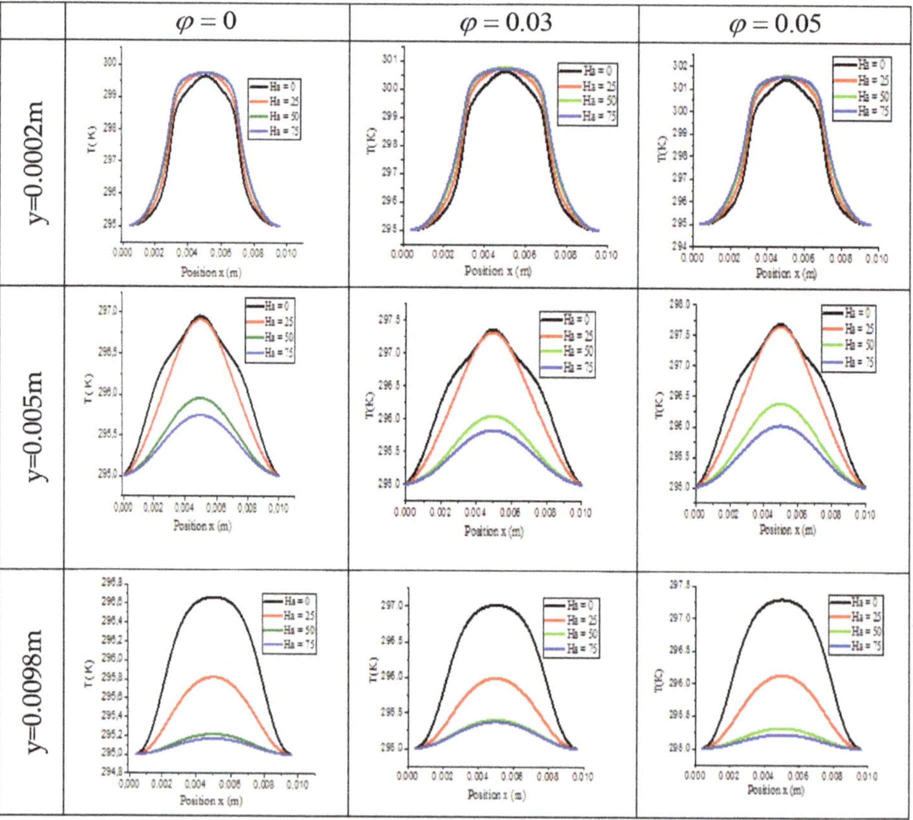

FIGURE 18.13 Temperature variation for the volume fractions chosen and the selected Hartmann numbers for a Rayleigh number $Ra = 7.68.10^4$.

18.4.4 Temperature Profiles

Temperature variation for Rayleigh numbers set at $7.68.10^4$ and $1.50.10^5$ is shown in Figures 18.13 and 18.14. We find that near the bottom wall, the temperature increases with increasing Hartmann. At the center of the cavity, the temperature for Hartmann is equal to 25 and for a weak Rayleigh has a maximum value of the same order as that in the non-existence of a magnetic field. If Rayleigh number rises further, temperature is reduced with growing Hartmann, and in some cases, it becomes higher than that of pure convection.

At the top of the cavity, there is an inverse phenomenon for the temperature variation compared to that at the bottom; the temperature varies inversely with respect to the Hartmann number. As a result, that in the non-existence of a magnetic field, fluid circulation tends to be upwards; however, if a magnetic field is associated with natural convection, the circulation of the fluid tends to take place down, especially when the Hartmann number rises.

By growing the volume fraction, we see that temperature increases.

18.4.5 Nusselt Number Variation

Nusselt number compares convective transfer to conductive one. If heat transfer by convection is significant, the Nusselt number increases and tends towards high values. For low values of Nusselt

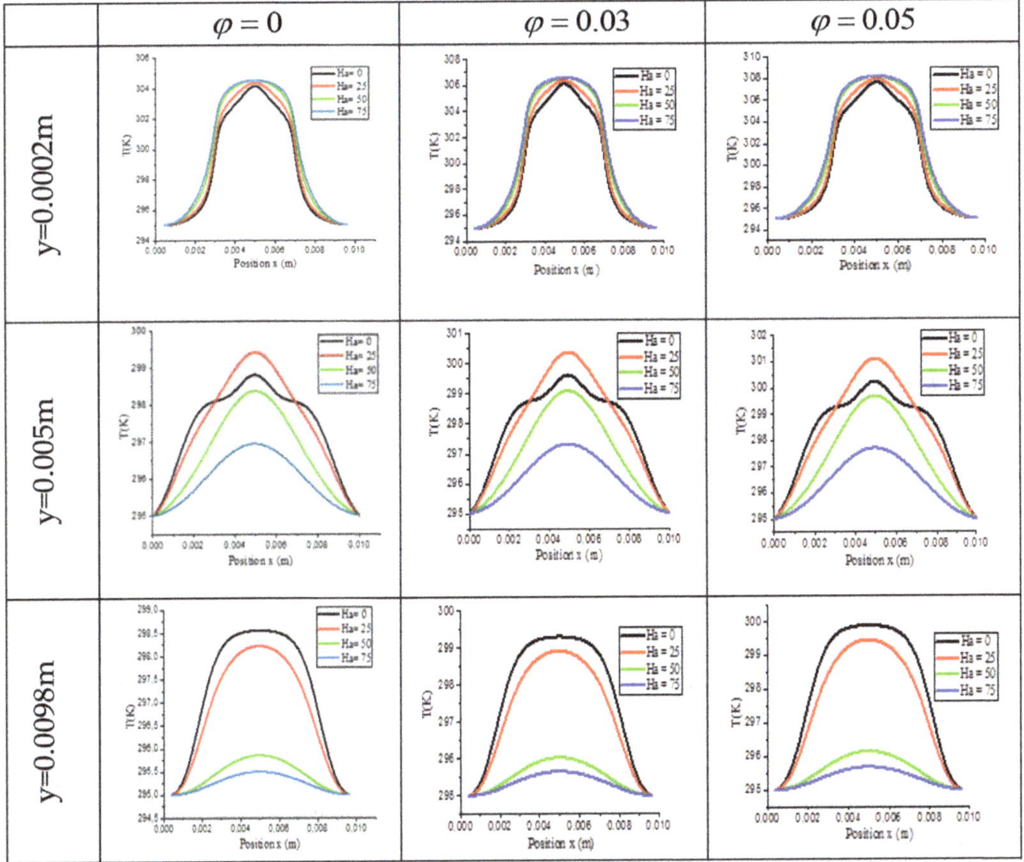

FIGURE 18.14 Temperature variation for the volume fractions chosen and selected Hartmann numbers for a Rayleigh number $Ra = 1.50.10^5$.

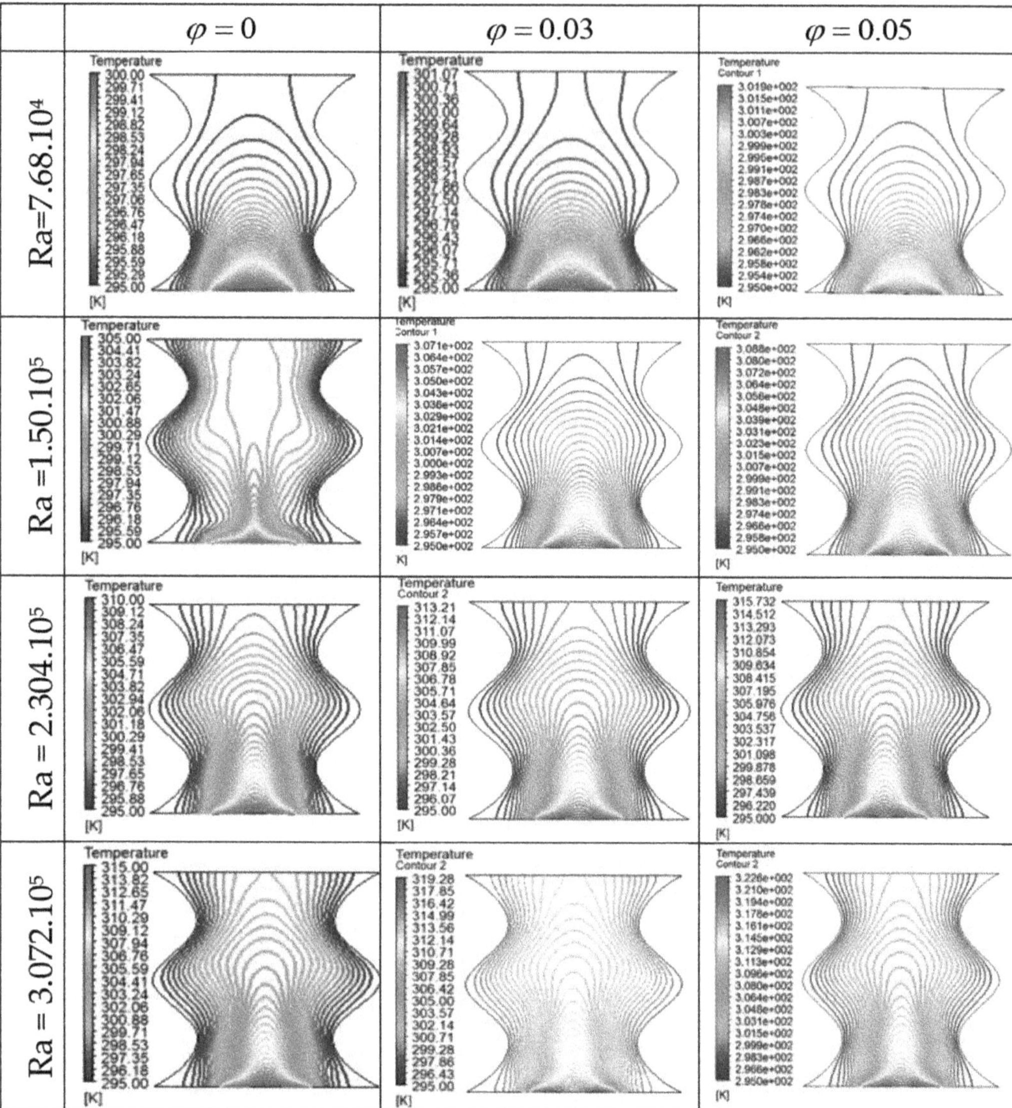

FIGURE 18.15 Nusselt number variation for selected Hartmann and Rayleigh numbers for different volume fractions.

number, conductive transfer becomes important. In this study, we investigated the influence of Hartmann number on Nusselt number for various Rayleigh numbers and different volume fractions of the nanofluid (Figures 18.15 and 18.16). We recall that the Hartmann number compares electromagnetic force and viscous one.

It is distinguished that Nusselt number is reduced with the growth of Hartmann number, which explains that magnetohydrodynamics (MHD) affects the heat exchange inside the cavity. In increasing Hartmann number, we gradually move from convection to conduction heat transfer.

We also notice that the Nusselt number rises with volume fraction of Al_2O_3-water nanofluid and Rayleigh number (Ra). Results obtained are qualitatively validated by those of Raizah et al. (2021) [20].

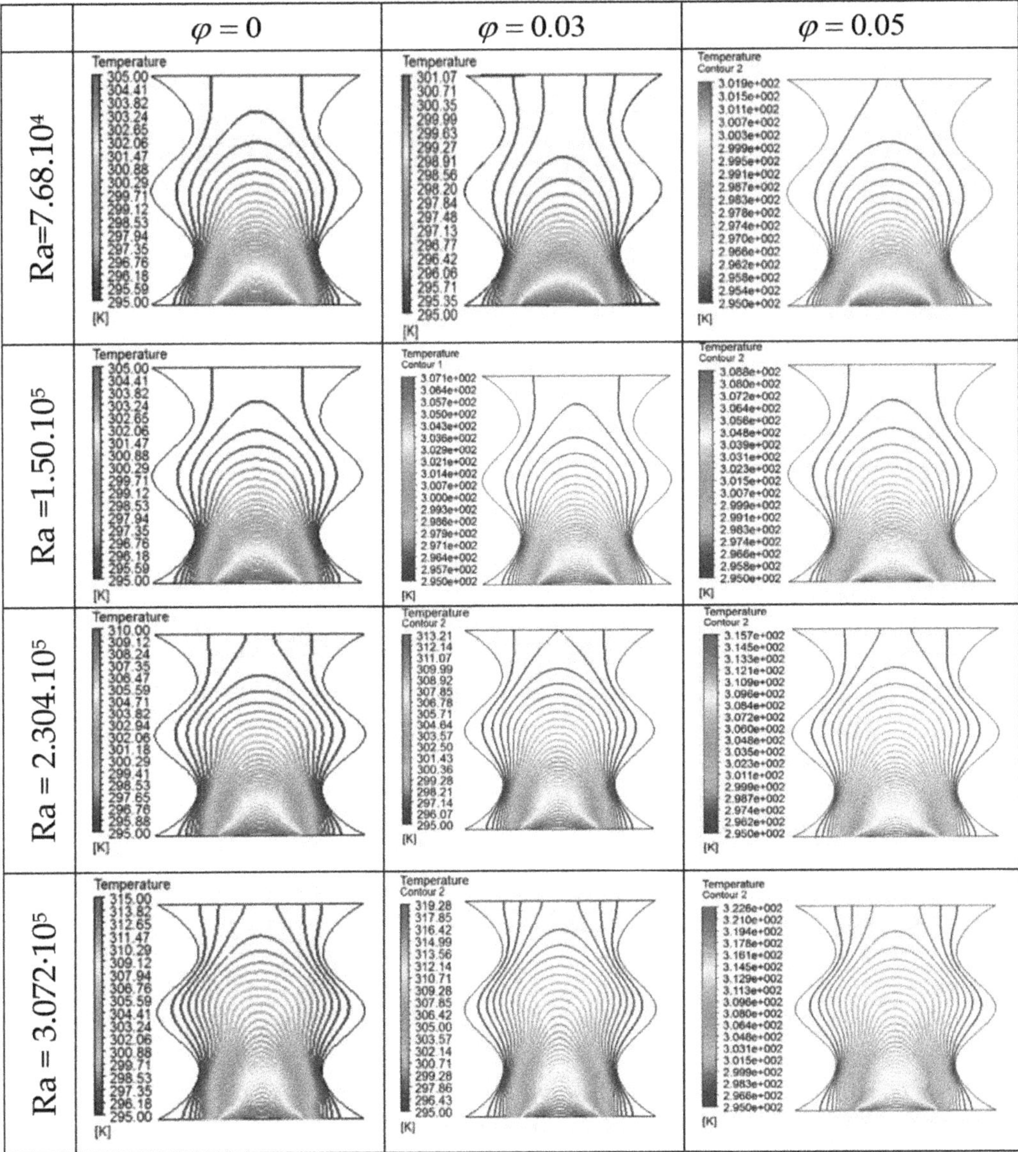

FIGURE 18.16 Nusselt number for selected Hartmann and volume fraction for a Rayleigh number fixed at $3.072.10^5$.

18.5 CONCLUSIONS

In this work, we have developed a numerical analysis of heat transmission by free laminar convection in a corrugated cavity saturated with Al_2O_3-water nanofluid in existence of a constant magnetic field. The effect of Hartmann and Rayleigh numbers and the concentration of nanoparticles on streamlines, isotherms, and mean Nusselt number are analyzed.

The main results obtained in this study are:

- Nusselt number rises with growing Rayleigh number and volume fraction of nanoparticles and decreases with increasing Hartmann.

- The presence of the undulation favors heat transfer, but it presents itself as an obstacle for the movement of the fluid along this wall.
- For high Hartmann number, the magnetic field can limit heat transmission by convection, and heat by conduction becomes dominant.
- Introduction of MHD to free convection considerably affects fluid movement.

Nomenclature

β_0	uniform magnetic field (Tesla)
B	length of hot source part (m)
Cp	specific heat at constant pressure ($J.kg^{-1}K^{-1}$)
g	gravitational acceleration ($m.s^{-2}$)
h	convection heat transfer coefficient ($W.m^{-2}.K^{-1}$)
H	height of the enclosure (m)
Ha	Hartmann number
K	thermal conductivity ($W.m^{-1}.K^{-1}$)
L	length of the enclosure (m)
Nu	average Nusselt number
Ra	Rayleigh number
T	temperature (K)
u, v	velocity components ($m.s^{-1}$)

Greek Symbols

α	thermal diffusivity ($m^2.s^{-1}$)
ρ	density ($kg.m^{-3}$)
β	coefficient of thermal expansion (K^{-1})
σ	electrical conductivity ($\Omega^{-1}.m^{-1}$)
μ	dynamic viscosity ($kg.m^{-1}.s^{-1}$)
v	kinematic viscosity ($m.s^{-2}$)
φ	volume fraction (%)

Subscripts

f	fluid
s	nanoparticle
nf	nanofluid
c	cold wall
h	hot wall

REFERENCES

1. Choi, S. U. S. (1995). Enhancing thermal conductivity of fluids with nanoparticules, developments and applications of Non-Newtonian flows. *ASME Journal of Heat Transfer*, 66, 99–105.
2. Hasnaoui, M., Bilgen, E., Vasseurt, P. (1992). Natural convection heat transfer in rectangular cavities partially heated from below. *Journal of Thermophysics and Heat Transfer*, 6, 255–264.
3. Ganzarolli, M. M., Milanez, L. F. (1995). Natural convection in rectangular enclosures heated from below and symmetrically cooled from the sides. *International Journal of Heat and Mass Transfer*, 38, 1063–1073.
4. Aydin, O., Yang, W. J. (2000). Natural convection in enclosures with localized heating from below and symmetrical cooling from sides. *International Journal of Numerical Methods for Heat & Fluid Flow*, 10, 518–529.

5. Corcione, M. (2003). Effects of the thermal boundary conditions at the sidewalls upon natural convection in rectangular enclosures heated from below and cooled from above. *International Journal of Thermal Sciences*, 42, 199–208.
6. Mebarek-Oudina, F. (2017). Numerical modeling of the hydrodynamic stability in vertical annulus with heat source of different lengths. *Engineering Science and Technology, an International Journal*, 20, 1324–1333.
7. Pushpa, B. V., Sankar, M., Mebarek-Oudina, F. (2021). Buoyant convective flow and heat dissipation of Cu–H_2O nanoliquids in an annulus through a thin baffle. *Journal of Nanofluids*, 10, 292–304.
8. Marzougui, S., Mebarek-Oudina, F., Assia, A. (2021). Entropy generation on magneto-convective flow of copper–water nanofluid in a cavity with chamfers. *Journal of Thermal Analysis and Calorimetry*, 143, 2203–2214.
9. Mebarek-Oudina, F., Chabani, I. (2022). Review on nano-fluids applications and heat transfer enhancement techniques in different enclosures. *Journal of Nanofluids*, 11, 155–168.
10. Chabani, I., Fateh mebarek-Oudina, F., Ismail, A. I. (2022). MHD flow of a hybrid nano-fluid in a triangular enclosure with zigzags and an elliptic obstacle. *Micromachines*, 13, 224.
11. Marzougui, S., Mebarek-Oudina, F., Magherbi, M., Mchirgui, A. (2022). Entropy generation and heat transport of Cu–water nanoliquid in porous lid-driven cavity through magnetic field. *International Journal of Numerical Methods for Heat & Fluid Flow*, 32, 2047–2069.
12. Calcagni, B., Marsili, F., Paroncini, M. (2005). Natural convective heat transfer in square enclosures heated from below. *Applied Thermal Engineering*, 25, 2522–2531.
13. Sharma, A. K., Velusamy, K., Balaji, C. (2007). Turbulent natural convection in an enclosure with localized heating from below. *International Journal of Thermal Sciences*, 46, 1232–1241.
14. Aminossadati, S. M., Ghasemi, B. (2009). Natural convection cooling of a localised heat source at the bottom of a nanofluid-filled enclosure. *European Journal of Mechanics B/Fluids*, 28, 630–640.
15. Mansour, M. A., Ahmed, S. E., Rashad, A. M. (2016). MHD natural convection in a square enclosure using nanofluid with the influence of thermal boundary conditions. *Journal of Applied Fluid Mechanics*, 9, 2515–2525.
16. Rahmoune, I., Bougoul, S., Chamkha, A. J. (2022). Analysis of nanofluid natural convection in a particular shape of a cavity. *The European Physical Journal Special Topics*, 231, 2901–2914.
17. Bougoul, S., Rahmoune, I. (2021). MHD's effects on entropy production and free convection in a specific-shaped enclosure saturated with a nanofluid Al_2O_3—water. *International Journal of Mechanics and Energy*, 9, 1–7.
18. Nasrin, R., Parvin, S., Alim, M. A. (2013). Effect of Prandtl number on free convection in a solar collector filled with nanofluid. *Procedia Engineering*, 56, 54–62.
19. Öğüt, E. B., Akyol, M., Arici, M. (2017). Natural convection of nanofluids in an inclined square cavity with side wavy walls. *Journal of Thermal Science and Technology*, 37, 139–150.
20. Raizah, Z., Aly, A. M., Alsedais, N., Mansour, M. A. (2021). MHD mixed convection of hybrid nanofluid in a wavy porous cavity employing local thermal non-equilibrium condition. *Scientific Reports*, 11, 17151.
21. Mebarek-Oudina, F. (2019). Convective heat transfer of Titania nanofluids of different base fluids in cylindrical annulus with discrete heat source. *Heat Transfer-Asian Research*, 48, 135–147.
22. Marzougui, S., Mebarek-Oudina, F., Aissa, A., Magherbi, M., Shah, Z., Ramesh, K. (2021). Entropy generation on magneto-convective flow of copper-water nanofluid in a cavity with chamfers. *Journal of Thermal Analysis and Calorimetry*, 143, 2203–2214.
23. Hassan, M., Mebarek-Oudina, F., Faisal, A., Ghafar, A., Ismail, A. I. (2022). Thermal energy and mass transport of shear thinning fluid under effects of low to high shear rate viscosity. *International Journal of Thermofluids*, 15, 100176.
24. Shafiq, A., Mebarek-Oudina, F., Sindhu, T. N., Rassoul, G. (2022). Sensitivity analysis for Walters' B nanoliquid flow over a radiative Riga surface by RSM. *Scientia Iranica*, 29, 1236–1249.
25. Rahmoune, I., Bougoul, S., Chamkha, A. J. (2022). Magneto-hydrodynamics natural convection and entropy production in a hollow cavity filled with a nanofluid. *Journal of Nanofluids*, 11, 1–9.
26. Rahmoune, I., Bougoul, S. (2021). Numerical analysis of laminar mixed convection heat transfer of the Al_2O_3–H_2O nanofluid in a square channel. *Journal of Applied Mechanics and Technical Physics*, 62, 920–926.

19 Non-Newtonian Casson Nanoliquid Flowing through a Bi-directional Stretching Device: Physical Impact of Heat Producing and Radiation

MD. Shamshuddin and K. K. Asogwa

19.1 INTRODUCTION

A nanofluid is a liquid comprising nanoparticles, most of which are metal or metal oxide with varying sizes between 1 nm and 100 nm in a base liquid. Nanofluids are advantageous heat transfer liquids for research and production processes, as per statistics suggesting the dramatic increase in the nanofluids-related investigation. The trajectory of thermal transmission in nanofluids is largely dependent on the thermal conductivity of nanoparticles. On the other hand, the Casson fluid or viscoelastic fluid model was established by Casson to understand the behaviour of pigment–oil combinations. Processed fruit beverages and fibers are examples of Casson fluid, according to Khan et al. [1]. Numerous biomedical and economic applications are available for the Casson fluid model. For substances like blood, chocolate, and other rheological attributes, the Casson fluid model performs better than prior viscoelastic models. Casson fluid is an infinitely viscous shear-thinning liquid with zero shear rate. The Casson nanofluid controlled by a moving surface was explored by Faisal et al. [2]. They discovered that for a certain stipulated surface temperature, greater Casson parameter values enhance heat transmission while limiting mass transfer. Rafique et al. [3] used the Keller box approach to explore the effect of Brownian motion on the Casson nanofluid across an inclined stretched sheet. According to their observations, elevating the Casson parameter retarded the velocity distribution and Nusselt number. The Casson and Oldroyd-B fluids within a stratified stretched sheet were addressed by Algehyne et al. [4]. They concluded that the magnetic and buoyancy ratio parameters were found to be more significant for Oldroyd-B fluid than for Cason liquid. Zeeshan et al. [5] inspected the propagation of MHD Casson nanofluid through a cylinder and reported that nanoparticle fraction dispersion decreases as the Casson fluid parameter improves. Ghadikolaei et al. [6] mentioned the role of magnetohydrodynamic Casson nanofluid on a tilting stretched sheet. Nadeem et al. [7] explored magnetohydrodynamic 3D Casson nanofluid across a stretched sheet. Vaidya et al. [8] reported Casson nanofluid across a heated stretched surface. Basha and Sivaraj [9] examined the dual solutions and stability of Casson nanofluid flow via a porous expanding/contracting wedge. Due to two parallel stretchy disks, Nandeppanavar et al. [10] conducted the axisymmetric analysis of Casson nanomaterial. They discovered that the axial momentum field exhibits a rising pattern for the stretching ratio and the Casson parameter. Other research pertinent to this subject is listed in [11–17].

The conveyance of heat is essential to innovations in technology and science. Thermal efficiency, hydrological generation, the cooling of circuit boards, and iron casting are among the numerous applications. The temperature gradient between a body's inside and outside is necessary for heat

 DOI: 10.1201/9781003299608-19

source/generation. Chamkha [18] evaluated the thermal transfer performance created by a heat source in a vertical cavity and reported that the heat source coefficient improves fluid flow. Using finite element method, Ibrahim et al. [19] examined a 2D fluid flow featuring heat source. From their conclusion, they declared that the temperature pattern is boosted as the heat source coefficient increases. Ferdows et al. [20] investigated heat source effect via a moving plate computationally employing various nanoparticles. Masood et al. [21] adopted a homotopy analysis procedure to examine the hybrid nanoparticles flowing triggered by heat-emitting stretched device. They showed that the temperature distribution gets more prominent as the source of heat increases in intensity. Cui et al. [22] employed a Bv4pc computational scheme approach on influence of non-identical simulation on nanoscale formulation. Zainal et al. [23] stated the characteristics of heat source on the flowing hybrid nanoliquid along a progressively stretched device. They concluded that a stronger heat source is associated with a higher thermal transmission rate. Anuradha and Sasikala [24] presented the convective heat characteristics of heat source and MHD nanofluid. Hafeez et al. [25] utilized an optimal homotopy analysis technique to examine the rotating stream of Oldroyd-B fluid long a disk.

In manufacturing output, wind turbines, hydroelectricity, engineering, and medicine, electromagnetic fields at the wall of any object are fundamental. A radiative surface emits heat energy in all directions. Heat is transported through space through conduction or convection, and this process is known as radiation. The impact of radiation on the time-dependent MHD convective flow of Jeffrey fluid was researched by Zin et al. [26]. Employing hall current and radiation, Suba and Muthucumaraswamy [27] explored magnetic field flow subject to a rotating fluid. They showed that elevating the radiation parameter promotes the temperature fluid to move more quickly. Reddy et al. [28] focused on the impact of reactions on MHD close to the stagnation point, when radiation was present. The findings showed that as radiation grows, the fluid temperature rises more quickly. In a time-dependent non-Newtonian flow, Ullah et al. [29] considered radiation and heat generation effect on convective boundary conditions. Blending radiant energy with convective nanoliquid flow across an exponential surface was reported by Gangaiah et al. [30]. Rashidi et al. [31] swotted spontaneous convective heat transport along a permeable extending vertical sheet to account for the existence of radiation. The effect of radiation on a transient magnetic field flow in the presence of slip and a heat sink was observed by Reddy et al. [32] when they were investigating numerical approaches. They observed that as the radiative parameter improves, velocity and temperature profiles diminish. Radiation's effect on magnetohydrodynamics thermal transport on convective nanofluid across a vertical plate was examined by Kumar et al. [33]. Some other studies have been conducted for evaluating magnetohydrodynamics thermal transport in modern industrial and physical applications, which are given in [34–37].

19.2 MODELLED FORMULATION

The schematic diagram of the molding system to be analyzed in this chapter is shown in Figure 19.1. Incompressible, natural convective radiative magnetized Casson nanoliquid flow through bidirectional stretchy device is investigated. Stretching device located at $z = 0$ in a linear mode was discussed also. Boundaries are imposed by convective heat and mass. Casson fluid has characteristics of Brownian movement and thermophoresis. Stretched surface bottom is heated by uni-varying energy and concentration at the wall T_f and C_f that are greater than the ambient heat transfer coefficients h_1 and h_2. The flow of Casson liquid underneath contemplations was subject to the sturdy uniformed transversal magnetic strength B_0 parallel to z-direction. The rheological model illustrating non-Newtonian Casson liquid is considered from the developed models [38–40].

The flow over the deforming sheet obeys the assumptions of the boundary layer, which is governed by the following set of equations (Sulochana et al. [41]).

$$u_x + v_y + w_z = 0, \tag{19.1}$$

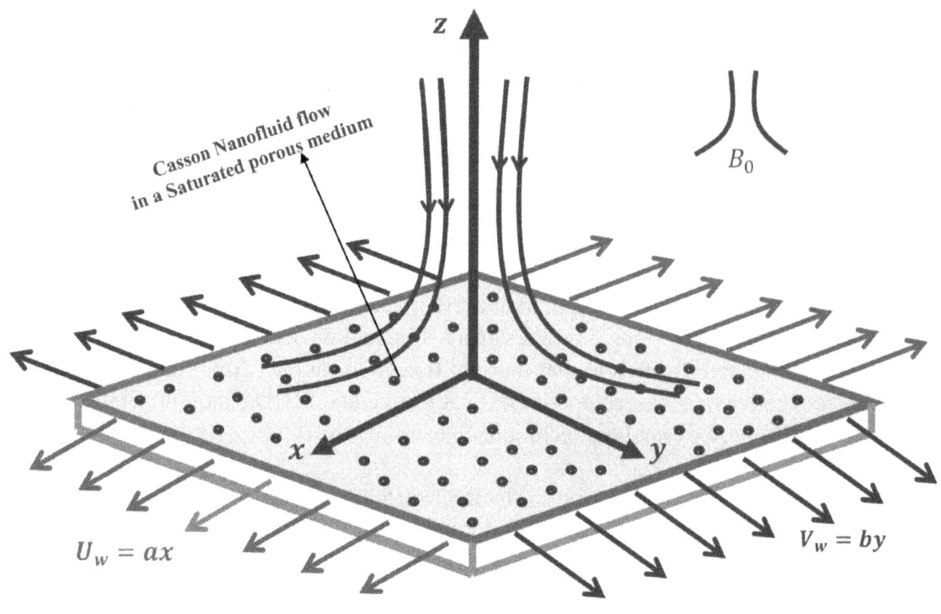

FIGURE 19.1 Geometrical view of problem.

$$u\,u_x + v\,u_y + w\,u_z = \nu\left(1+\frac{1}{\beta}\right)u_{zz} - \frac{\sigma B_0^2}{\rho}u - \frac{\nu}{k_1}u, \tag{19.2}$$

$$u\,v_x + v\,v_y + w\,v_z = \nu\left(1+\frac{1}{\beta}\right)v_{zz} - \frac{\sigma B_0^2}{\rho}v - \frac{\nu}{k_1}v, \tag{19.3}$$

$$u\,T_x + v\,T_y + w\,T_z = \frac{\kappa}{\rho c_p}T_{zz} - \frac{1}{(\rho c)_{nf}}(q_r)_z + + \frac{Q}{(\rho c)_f}(T - T_\infty) + \tau\left[D_B\,C_z\,T_z + \frac{D_T}{T_\infty}(T_y)^2\right], \tag{19.4}$$

$$u\,C_x + v\,C_y + w\,C_z = D_B C_{zz} + \frac{D_T}{T_\infty}(T_x)^2, \tag{19.5}$$

Net thermal radiative heat flux q_r is pursued [42–44] and is presented as:

$$q_r = -\frac{4\sigma^*}{3k^*}\left(T^4\right)_z = -\frac{16\sigma^*}{3k^*}\left(T^3\,T_z\right)_z$$

The conditions at the boundary are:

$$\left.\begin{array}{l} u = U_w(x) = ax, \quad v = V_w(x) = by, \quad w = 0 \\ T_z = -\dfrac{h_f}{\kappa}\left(T_f - T\right), \quad C_z = -\dfrac{h_s}{D_B}(C_s - C) \quad at\ z = 0 \\ u \to 0, \quad v \to 0, \quad w \to 0, \quad T \to T_\infty, \quad C \to C_\infty \quad at\ z \to \infty \end{array}\right\} \tag{19.6}$$

19.2.1 Similarity Transformation

For current model simplification, the defined subsequent similarity quantities are utilized:

$$\eta = z(a/v)^{1/2}, u = a\,x\,f'(\eta),\ v = b\,y\,g'(\eta),\ w = -(av)^{1/2}\left(f(\eta)+c\,g(\eta)\right),$$
$$\theta(\eta) = (T-T_\infty)/(T_f-T_\infty),\ \phi(\eta) = (C-C_\infty)/(C_f-C_\infty) \tag{19.7}$$

Equations (19.1) to (19.6) are reformed with implementation of (19.7) as:

$$\left(1+\frac{1}{\beta}\right)f''' - (f+cg)f'' - (f')^2 - (M+K)f' \;=\; 0 \tag{19.8}$$

$$\left(1+\frac{1}{\beta}\right)g''' - (f+cg)g'' - (g')^2 - (M+K)g' \;=\; 0 \tag{19.9}$$

$$\left(1+Rd\right)\theta'' + \Pr\left\{\left(f+cg\right)\theta' + Nb\,\theta'\phi' + Nt\,\theta'^2 + S\theta\right\} \;=\; 0 \tag{19.10}$$

$$\phi'' + \Pr Le\left(f+cg\right)\phi' + \frac{Nt}{Nb}\theta'' \;=\; 0 \tag{19.11}$$

The transformed boundary conditions are:

$$\left.\begin{array}{l} at\ \eta=0\ \ \left|f=0,\ g=0,\ f'=1,\ g'=c,\ \theta'=-Bi_1(1-\theta(0)),\ \phi'=-Bi_2(1-\phi(0))\right. \\ at\ \eta\to\infty\left|f'\to 0,\ g'\to 0,\ \theta\to 0,\ \phi\to 0\right. \end{array}\right\} \tag{19.12}$$

Where the primes refer to differentiation with regard to similarity variable η. Dimensionless physical parameters are defined as follows:

$$\left.\begin{array}{l} \beta = \mu_B\sqrt{2\pi_c}/p_z, c=\dfrac{b}{a}, M=\dfrac{\sigma B_0^2}{\rho a},\ K=\dfrac{v}{ak_1},\ \Pr=\dfrac{\rho v C_p}{\kappa},\ Rd=\dfrac{16\sigma^* T_\infty^3}{3\kappa k^*}, \\ Nb=\dfrac{\tau D_B\left(C_f-C_\infty\right)}{v},\ Nt=\dfrac{\tau D_T\left(T_f-T_\infty\right)}{vT_\infty}, S=\dfrac{Q}{c(\rho c)_f} Le=\dfrac{v_f}{D_B}, \\ Bi_1=\dfrac{h_f}{\kappa}\sqrt{\dfrac{v}{a}},\ Bi_1=\dfrac{h_s}{D_B}\sqrt{\dfrac{v}{a}} \end{array}\right\} \tag{19.13}$$

19.2.2 Engineering Quantities of Curiosity

To express for skin friction factor, the characteristics of the model of the current fluid movement at sheet $z=0$ in x- and y-directions were calculated by:

$$Cf_x = \frac{\tau_{xz}}{\rho_f U_w^2},\qquad Cf_y = \frac{\tau_{yz}}{\rho_f U_w^2} \tag{19.14}$$

The Nusselt quantity in the expression of rates of temperature transportations at the sheet surface as well as Sherwood quantity in the expression of rates of concentration transportations at the sheet surface are specified by:

$$Nu_x = -\frac{x}{\left(T_f - T_\infty\right)}\left(T_z\right)\Big|_{z=0} + \frac{xq_r}{\kappa\left(T_f - T_\infty\right)}, \quad Sh_x = -\frac{x}{\left(C_f - C_\infty\right)}\left(C_z\right)\Big|_{z=0} \tag{19.15}$$

Invoking the non-dimensional transformations, we get:

$$Cf_x\sqrt{\mathrm{Re}_x} = \left(1+\frac{1}{\beta}\right)f''(0), \qquad Cf_y\sqrt{\mathrm{Re}_x} = \left(1+\frac{1}{\beta}\right)g''(0) \tag{19.16}$$

$$Nu_x\left(\mathrm{Re}_x\right)^{-1/2} = -\left[1+Rd\right]\theta'(0), \quad Sh_x\left(\mathrm{Re}_x\right)^{-1/2} = -\phi'(0) \tag{19.17}$$

In the preceding expression, $\mathrm{Re}_x = ax^2/\nu$ represents the local Reynolds number.

19.3 NUMERICAL SOLUTION

In this section, the coupled highly nonlinear ODEs (19.8 to 19.11) along conditions at the boundary (19.12) are numerically tackled according to the study of [41]. The detailed studies about this numerical technique are also found in [45]. Following these studies, this numerical approach consists of a few important features, such as its being based on the classical finite difference method. According to [46], the momentum equation (19.8) can be written as:

$$\left(1+\frac{1}{\beta}\right)f''' - \left(f+cg\right)f'' - f'^2 - \left(M+K\right)f' = 0, \tag{19.18}$$

In (19.8), assume $y(x) = f'(\eta)$, and therefore, equation (19.8) yields:

$$\left(1+\frac{1}{\beta}\right)f''' - \left(f+cg\right)f'' - \left(f'+M+K\right)f' = 0, \tag{19.19}$$

Which takes the following form:

$$P(x)y''(x) + Q(x)y'(x) + R(x)y(x) = S(x), \tag{19.20}$$

Where:

$$\left.\begin{aligned} P(x) &= \left(1+\frac{1}{\beta}\right); Q(x) = -\left(f+cg\right); \\ R(x) &= -\left(f'+M+K\right); S(x) = 0. \end{aligned}\right\}$$

Now, a common finite difference method applied to solving equation (19.19) is located on central differencing and matrix manipulation. An initial guess needed to be employed before starting the calculation of $f'(\eta)$ and $g(\eta)$ between $\eta = 0$ and $\eta = \eta_\infty$, which automatically satisfies the BCs (19.12) and by assuming:

$$\left.\begin{array}{l} f'(\eta) = 1 - \dfrac{\eta}{\eta_\infty}, \quad g'(\eta) = c\left(1 - \dfrac{\eta}{\eta_\infty}\right), \quad \theta'(\eta) = -Bi_1(1-\theta)\left(1 - \dfrac{\eta}{\eta_\infty}\right), \\ \phi'(\eta) = -Bi_2(1-\phi)\left(1 - \dfrac{\eta}{\eta_\infty}\right) \end{array}\right\}$$

BCs = boundary conditions

The profile of $f(\eta)$ is attained within the integration from the $f'(\eta)$ curve. Now, the next gradation is to consider the $g(\eta)$ known and to ascertain a newish estimation for $f'(\eta)$, f'_{new} by solving equation (19.19), as mentioned in the process of the applied method. Then a new profile, such as f and f', is used as new inputs and so on. In this way, the momentum equation (19.18) is solved iteratively until the required convergence is attained.

A similar process is used to solve equations (19.8)–(19.11) after obtaining the $f(\eta)$ profile.

For solving the energy equation, we may write equation (19.10) in following form:

$$(1+Rd)\theta'' + \Pr(f + cg + Nb\phi' + Nt\theta')\theta' + \Pr S\theta = 0, \tag{19.21}$$

Consider $y(x) = \theta(\eta)$, and therefore, equation (19.21) becomes:

$$P_1(x)y''(x) + Q_1(x)y'(x) + R_1(x)y(x) = S_1(x), \tag{19.22}$$

Where:

$$\left.\begin{array}{l} P_1(x) = (1+Rd); Q_1(x) = \Pr(f + cg + Nb\varphi' + Nt\theta'); \\ R_1(x) = \Pr S; S_1(x) = 0 \end{array}\right\}$$

Like in similar manner, the concentration equation can be written as:

$$\phi'' + \Pr Le(f + cg)\phi' + \frac{Nt}{Nb}\theta'' = 0, \tag{19.23}$$

Substituting the values of θ'' from (19.22) in equation (19.23), and assuming $y(x) = \varphi(\eta)$, we have:

$$P_2(x)y''(x) + Q_2(x)y'(x) + R_2(x)y(x) = S_2(x), \tag{19.24}$$

Where:

$$\left.\begin{array}{l} P_2(x) = 1; Q_2(x) = \Pr Le(f + cg) - \dfrac{\Pr}{(1+Rd)}Nt\theta'; \\ R_2(x) = 0; S_2(x) = \dfrac{Nt}{Nb}\dfrac{\Pr}{(1+Rd)}[f + cg + Nt\theta' + S]\theta' \end{array}\right\}$$

Thus, finally, we found the profiles of $\theta(\eta)$ and $\phi(\eta)$ up to a negligible quantity attained. The whole numerical process is ongoing until the trial convergence of this problem is obtained, and for that we assume in numerical computation step size $h = \Delta\eta = 0.01, \eta_{lowest} = 0$ and $\eta_{heighest} = 8$, with convergence 10^{-3}.

19.3.1 Code Validation

To ensure the authenticity of the applied numerical code, we compare our results with those studied in [41] and find it to be in suitable agreement, and these values are captured in Table 19.1, which ensures we implement this new numerical algorithm in this chapter.

19.4 RESULTS AND DISCUSSION

The major theme of this chapter is to express the bilateral reactions carried over MHD Darcy movement of an incompressible, natural convective radiative magnetized Casson nanoliquid flow through bidirectional stretchy device underneath conditions of convective heat as well as mass effects into account. The finite difference procedure is employed to generate figures. The graphical results that are incorporated in the discussion section (Figures 19.2–19.15) provide a comprehensive

TABLE 19.1
The Computed Results for Comparison of Friction Coefficient for *M, K* and β When *Rd* = *S* = 0

M	K	β	Sulochana *et al.* [41]	Current Results
0	0	∞	1.093252	1.094
0	0	5	1.197425	1.197
0	0.5	1	1.836082	1.844
10	0	∞	3.342020	3.343
10	0	5	3.3660730	3.369
10	0.5	1	4.830596	4.831

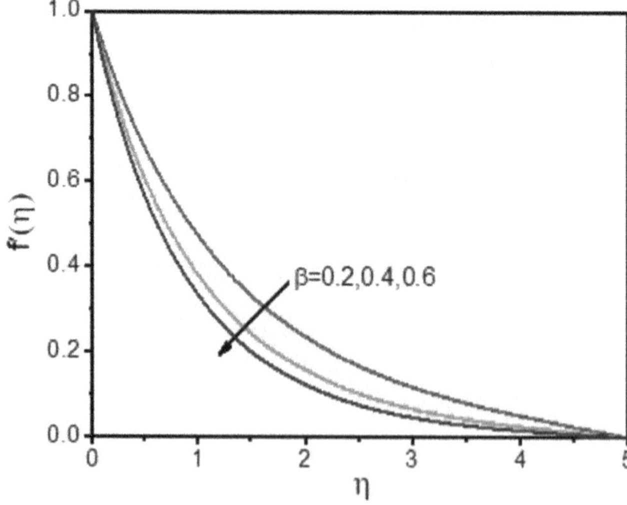

FIGURE 19.2(A, B) Velocity profile variations with β in x- and y-directions.

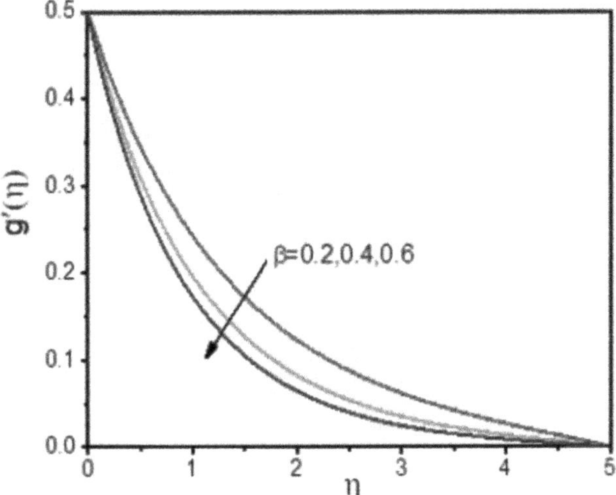

FIGURE 19.2(A, B) (Continued)

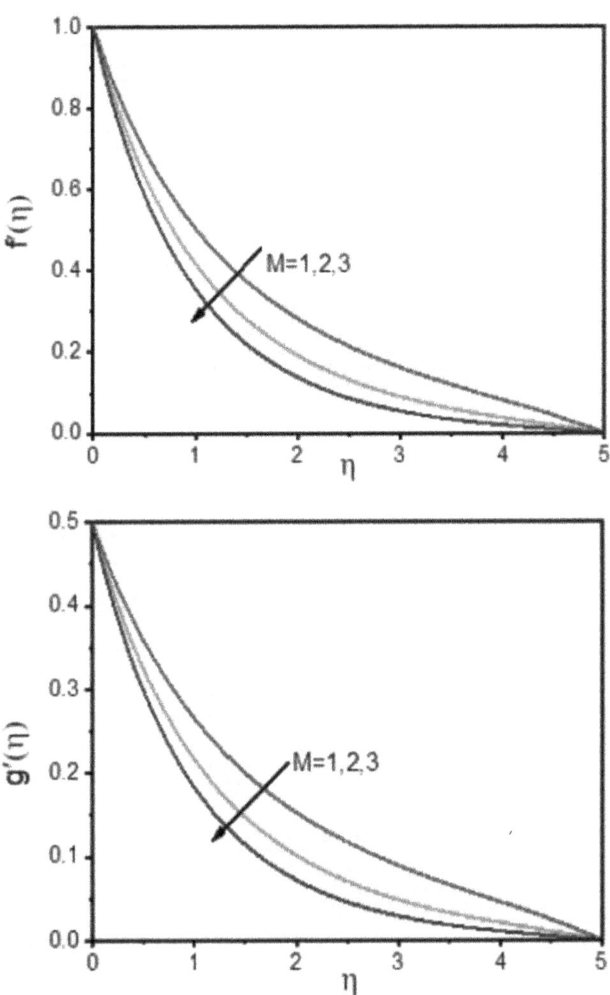

FIGURE 19.3 (A, B) Velocity profile variation with M in x- and y-directions.

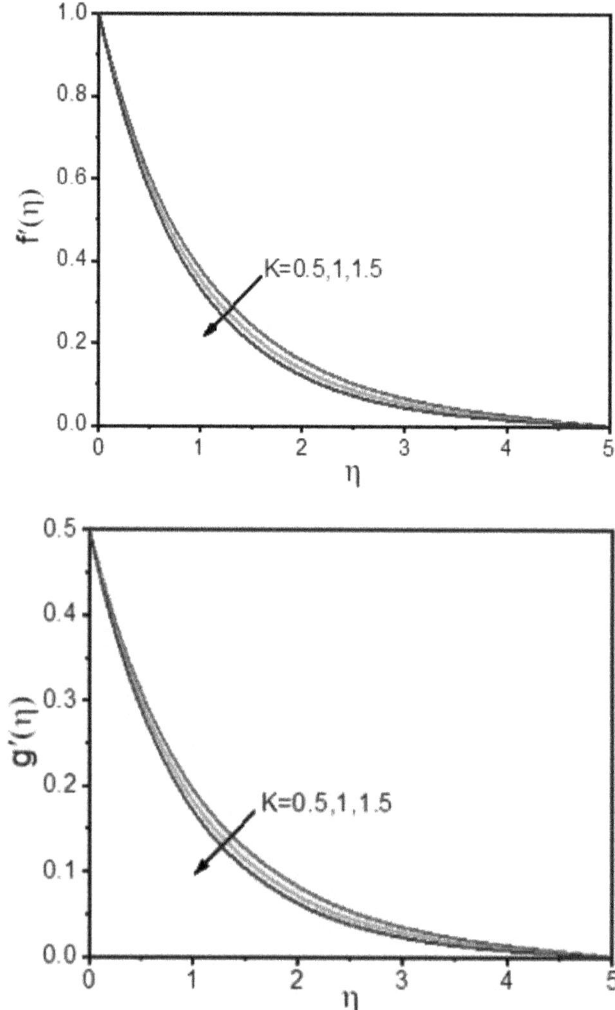

FIGURE 19.4 (A, B) Velocity profile variation with K in x- and y-directions.

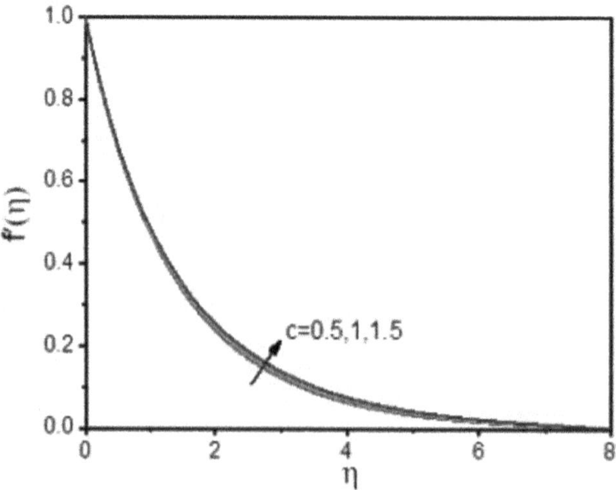

FIGURE 19.5 (A, B) Velocity profile variation with c in x- and y-directions.

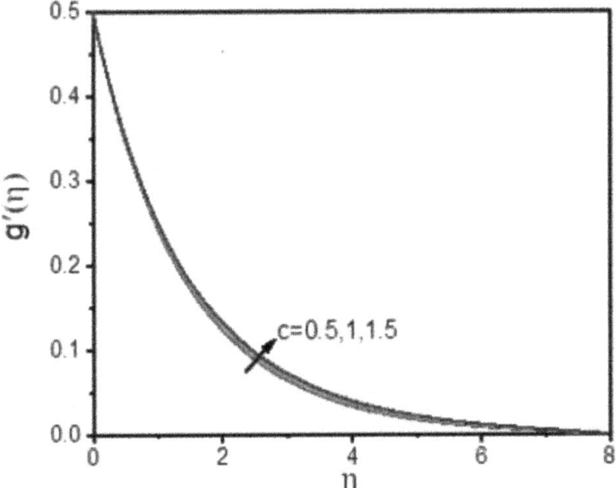

FIGURE 19.5(A, B) (Continued)

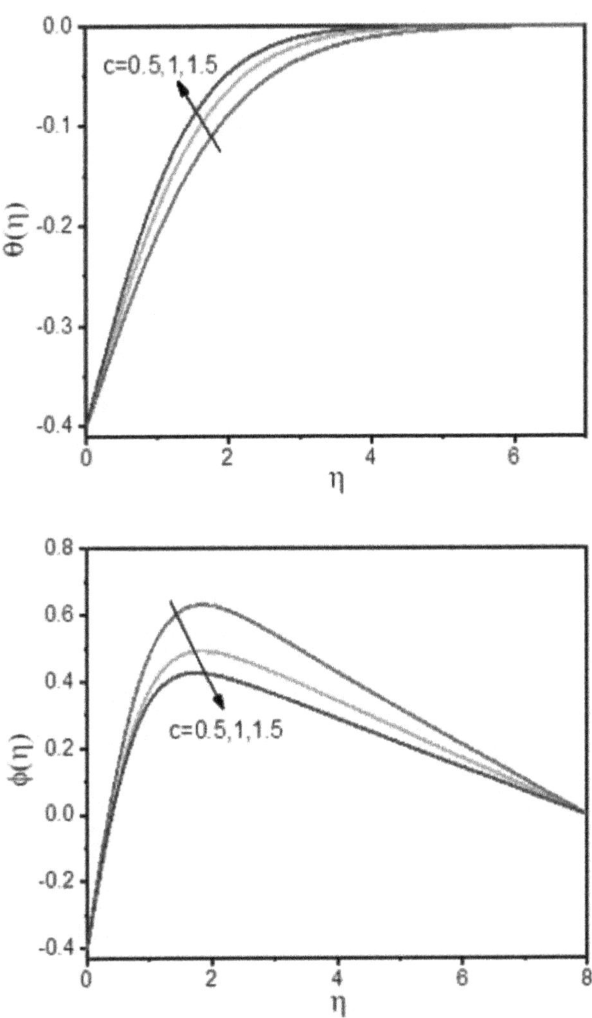

FIGURE 19.6 (A, B) Heat and mass distrubution variations with *c*.

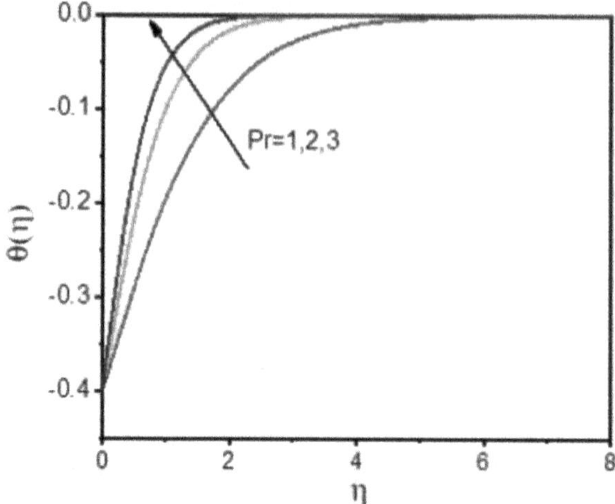

FIGURE 19.7 Heat distribution variations with Pr.

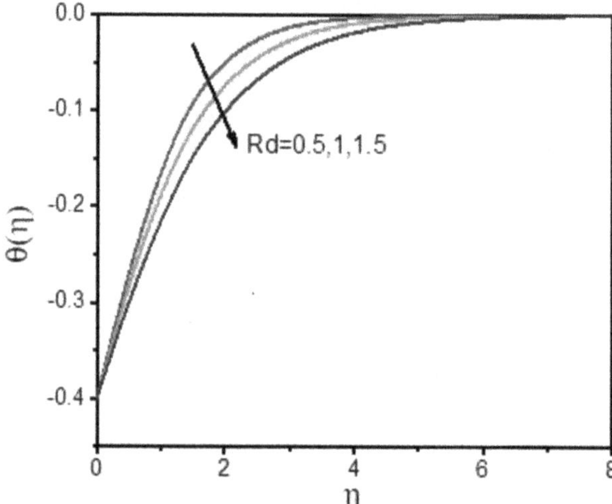

FIGURE 19.8 Heat distribution variations with *Rd*.

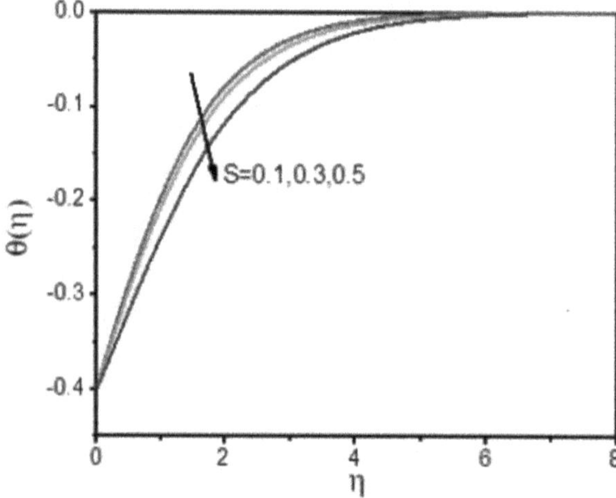

FIGURE 19.9 Heat distribution variations with *S*.

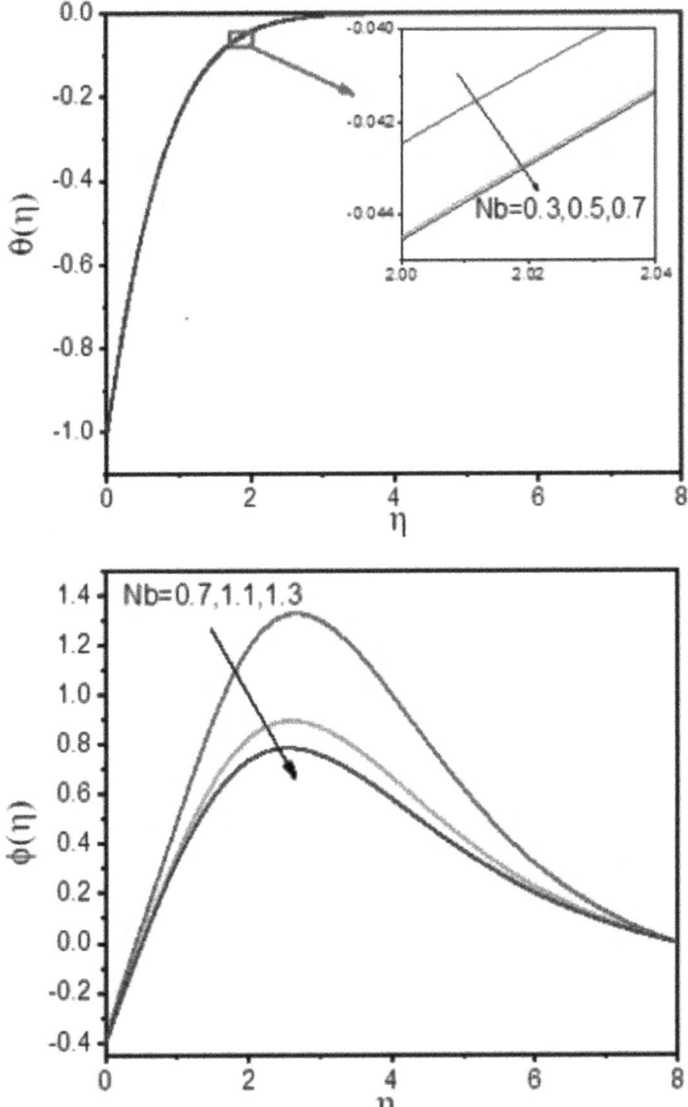

FIGURE 19.10(A, B) Heat and mass distrubution variations with *Nb*.

evaluation and scrutinization of the mathematical computations given in the approximate solution segment by FDM. The thermodynamical parameters considered during the computations are $M = 3$; $\beta = 0.4$; $K = 0.5$; $c = Rd = S = 0.5$; $\Pr = Le = 1$; $Bi_1 = Bi_2 = 0.3$; $Nb = 0.3$; $Nt = 0.1$; except specified on each plot.

19.4.1 Profiles of Velocity in x- and y-Directions

Figures 19.2(a, b) present the impact of Casson parameter (β) on $f'(\eta)$ and $g'(\eta)$ profiles. The horizontal velocity components of x-axis and y-axis paths of the nanofluid transmission characteristics are a declining feature of a progressive Casson parameter, as demonstrated in Figures 19.2(a, b). **Figures 19.3(a, b)** are displayed to illustrate the impact of M on $f'(\eta)$ and $g'(\eta)$ profiles.

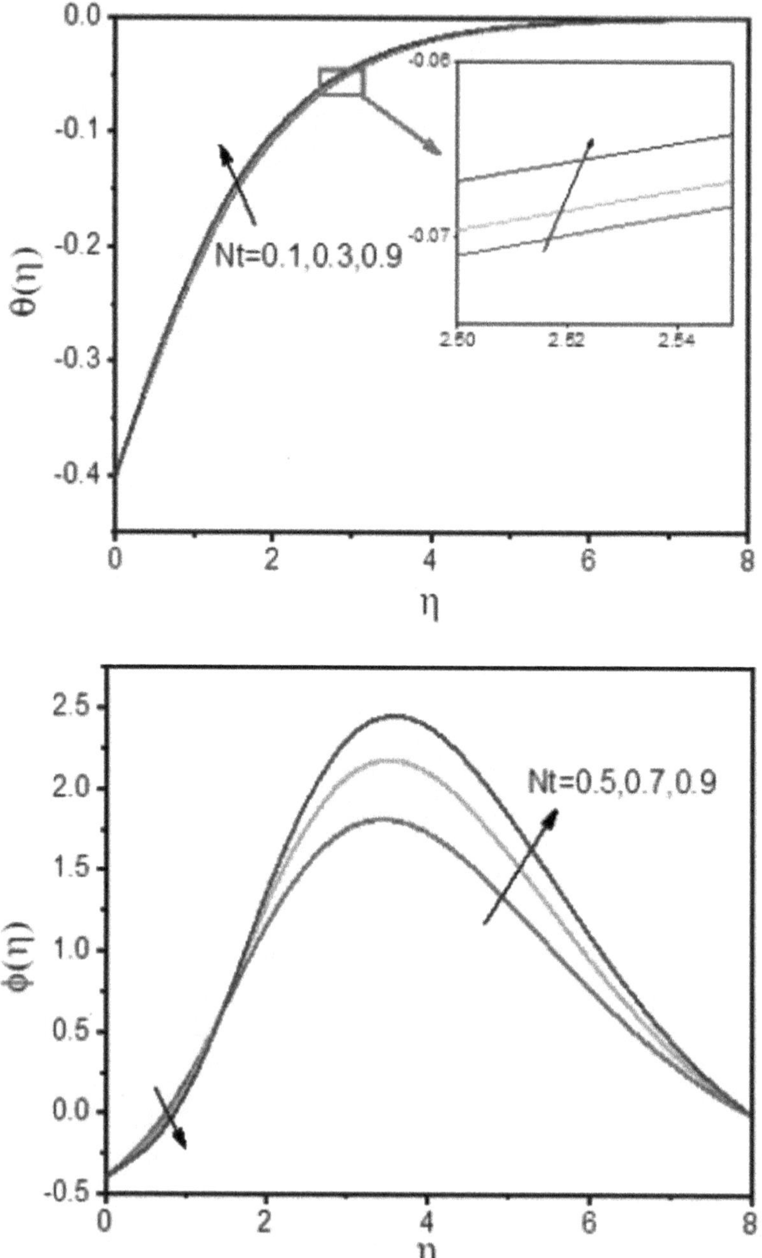

FIGURE 19.11(A, B) Heat and mass distribution variations with *Nt*.

It is essential to demonstrate in Figures 19.3(a, b) that empowered values of M on $f'(\eta)$ and $g'(\eta)$ fields degrade the velocity component of the bidirectional nanofluid transmission processes. The decrease in velocity component can be attributed to the Lorentzian force generated by the rise in magnetic field intensity. Similarly, **Figures 19.4(a, b)** elucidate the implication of permeability parameter (K) on the $f'(\eta)$ and $g'(\eta)$ profiles. From the visual analysis of Figures 19.4, the escalating values of K diminish velocity profiles. This situation occurs because an increase in the permeability parameter promotes a rise in the resistance of the porous layer, resulting in a drop

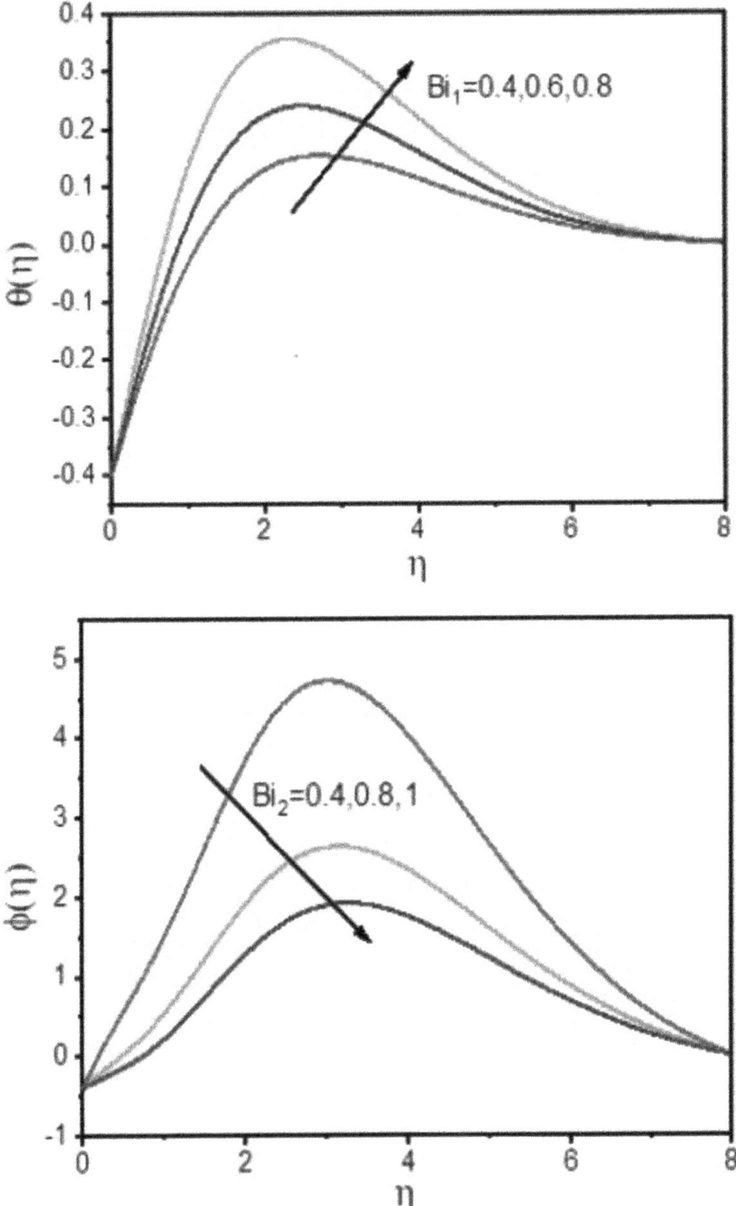

FIGURE 19.12 (A, B) Heat distribution variations with Bi_1 and mass distribution variation with Bi_2.

in the velocity of the nanofluid flow. **Figures 19.5(a, b)** demonstrate that the velocity of nanofluid improves with a stronger stretching parameter c towards the horizontal velocity in both directions.

19.4.2 Temperareture and Concentation Profiles

Figures 19.6(a, b) present the impact of stretching parameter (c) on $\theta(\eta)$ and $\phi(\eta)$ profiles. Temperature diminishes with a stronger stretching parameter c towards the wall, as seen in Figure 19.6(a). In thermal transmission, while velocity is rising, heat energy is drifted away from the

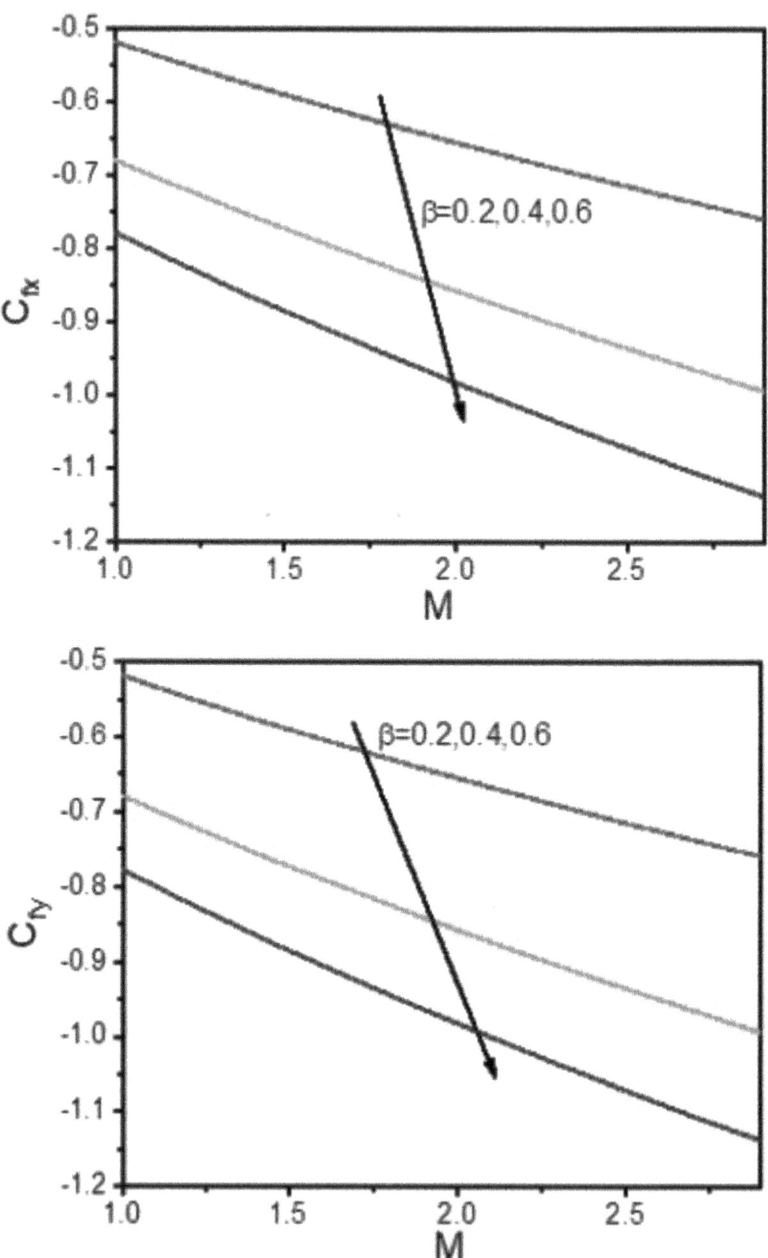

FIGURE 19.13(A, B) Drag friction coefficient profiles variation with M and β in x- and y-directions.

boundary. The obtained result is justified since the detected reduction in the temperature flow is significant. Further, it also indicates that the concentration distribution rises with expansion in stretching parameter (Figure 19.6(b)). Physically, a higher magnitude of stretching parameter $c = b / a$ corresponds to a decreased stretching rate a. Figure 19.7 demonstrates an improvement in Prandtl number which corresponds to a decline in the thermal energy. Ultimately, the thermal field demonstrates a stronger decline in thermal boundary layer thickness with enriched Prandtl number. Given a decline in Prandtl number, nanofluids show a superior property due to their increased thermal

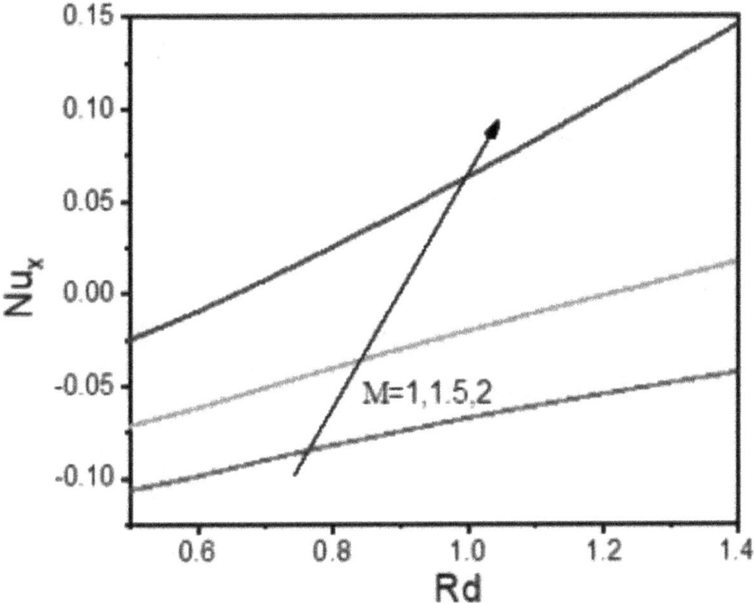

FIGURE 19.14 Temperature gradient coefficient profiles variation with M and Rd.

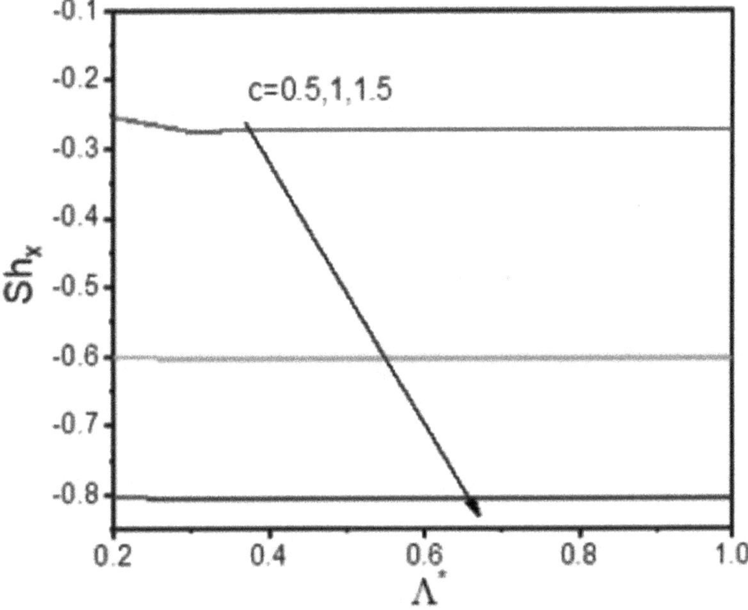

FIGURE 19.15 Concentration gradient coefficient profiles variation with c and Le.

conductivity. The significance of radiative factor on temperature field is depicted in Figure 19.8. The amplification of the radiation component boosts the temperature field, as expected. It is proven in Figure 19.9 that the heat source parameter provides an upsurge in temperature distribution. More so, the heat source parameter shows a spontaneous rise in the boundary layer thickness, realizing entirely well that heat source parameter improves wall temperature. In **Figures 19.10(a, b)**, the

Brownian motion parameter facilitates nanofluid thermal and species variations as the Brownian motion increases. Since the stronger Brownian motion factor empowers the thermal and concentration boundary layer thickness, the fluid nanoparticles migrate from the rapidly expanding surface to a less-nanofluid mobility. Figure 19.11(**a, b**) shows how a growing thermophoresis parameter promotes nanofluid temperature distribution. However, the concentration distribution experienced an initial decline with increasing thermophoresis parameter at $\eta \in [0 \;\; 2]$ and, subsequently, $\eta > 2$. The concentration distribution experienced an upshot with increasing thermophoresis parameter. Physically, the thermophoresis parameter is characterized as the mobility of nanoparticles in the path of a declining thermal gradient. This is attributable to the fact that when there is a large measure of energy provided in the boundary layer, the nanoparticles floating in the liquid tend to migrate to a region with little or no thermal energy, resulting in a drop and upswing in the nanofluid concentration. As the Biot number Bi_1 grows, the nanofluid temperature distribution improves, and the concentration distribution decreases when Bi_2 increases, as displayed in Figure 19.12(**a, b**).

19.4.3 Engineering Quantities of Intrest Profiles

Figures 19.13 depicts the impact of the Casson indicator on friction versus M. The friction coefficients $\left(C_{fx}\right)$ and $\left(C_{fy}\right)$ against M for the chosen values of Casson parameter are displayed graphically. It is evident that the friction coefficients dwindle with escalating Casson parameter values for the specified M values. In addition, it is believed that decreased local skin friction coefficients are consequent of the force exerted by Lorentz force towards the surface as Casson parameter is enhanced. The Nusselt number $\left(Nu_x\right)$ versus N_b for chosen values of N_t as displayed in Figure 19.14. It is remarkable that Nu_x depletes by growing values of N_t. The Nusselt number $\left(Nu_y\right)$ versus R_d for chosen values of M is demonstrated in Figure 19.14. It is conspicious that Nu_y is enhanced by growing values of M. The Sherwood number $\left(Sh_x\right)$ versus Lewis parameter for selected values of stretching parameter is shown in Figure 19.15. It is obvious that Sh_x depletes by growing values of stretching parameter.

19.5 CONCLUDING REMARKS

The examination of highly nonlinear Casson nanofluid flow equations with prescribed thermal transport and surface mass transport is carried out in bidirection stretching surface. Two-phase nanofluid model is utilized in energy equation. The configuration of the flow is restructured and executed via FDM. On the flow momentum, thermal disstrubution, solutal distribution, and engineering quantities distribution, the influence of thermosolutal dynamic parameters is determined and presented quantitatively in graphs. Arising from the chapter, the key findings are:

- Increasing values of M on $f'(\eta)$ and $g'(\eta)$ fields degrade the velocity component of bidirectional nanofluid transmission processes.
- An improvement in yield stress through an enhancement in Casson factor depletes the velocity in both directions.
- An increase in the permeability parameter promotes a rise in the resistance of the porous layer, resulting in a drop in the velocity of the nanofluid.
- The velocity of the nanofluid improves with a stronger stretching parameter c towards the horizontal velocity in x- and y-directions.
- Brownian motion and thermophoresis parameters facilitate nanofluid temperature.
- The presence of heat source has remarkable impact on thermal distribution.
- Friction coefficients dwindle with enriched Casson parameter values for the specified magnetic parameter values.

- It is remarkable that the Nusselt number depletes by growing values of thermophoresis and behaves in opposite manner in case of magnetic field.
- Sherwood number depletes by growing values of Lewis number.

Therefore, the outcomes of this chapter are applicable to improving thermal device performance and nanotechnology advancement. Hence, possible extension of the chapter can be considered for single-phase nanofluid model with hybrid dispersed flow along a porous saturated rotating disk.

Nomenclature

a,b	stetching constants
S	heat source
Nb	Brownian motion prarameter
D_T	thermophoresis coefficient (m^2/s)
v	velocity in y-direction (m/s)
B_0	dimensional magnetic body force
C_{fx}	local skin friction coefficient in x-direction
Bi_1	thermal Biot number
C_{fy}	local skin friction coefficient in y-direction
Bi_2	concentration Biot number
c	stretching parameter
Le	Lewis number
C_f	concentration at the wall
T_∞	temperature at far away from wall (K)
C_p	specific heat $(JKg^{-1}K^{-1})$
C_∞	concentration at far away from wall
D_B	Brownian coefficient (m^2/s)
h_1,h_2	heat and mass transfer coefficients
Nt	thermophoresis prarameter
$\Pr$	Prandtl number
h_s	convective mass transfer
M	magnetic parameter
K	dimensionless permeability parameter
k_1	dimensional permeability
k^*	mean absorption coefficient
Rd	thermal radiation
h_f	convective heat transfer
Nu_x	local Nusselt number
q_r	radiative heat flux (W/m^2)
Re_x	local Reynolds number
T_f	temperature at the wall (K)
U_w	stretching velocity in x-direction (m/s)
V_w	stretching velocity in y-direction (m/s)
u	velocity in x-direction (m/s)
β	component of Casson liquid
κ	thermal conductivity $(Wm^{-1}K^{-1})$
ρ	density of the fluid (Kg/m^3)
τ	ratio of nanoparticle to effective heat capacity
σ^*	Stefan–Boltzmann constant
η	similarity variable

θ	dimensionless temperature
w	velocity in z-direction (m / s)
ϕ	dimensionless concentration
v	viscosity of the fluid (m^2 / s)
T	fluid temperature (K)

REFERENCES

1. Khan, D., Khan, A., Khan, I., Ali, F., Karim, F.U., & Tlili, I. (2019). Effects of relative magnetic field, chemical reaction, heat generation and newtonian heating on convection flow of Casson fluid over a moving vertical plate embedded in a porous medium. *Scientific Reports*, 9, 400. https://doi.org/10.1038/s41598-018-36243-0.
2. Faisal, M., Ahmad, I., & Javed, T. (2020). Significances of prescribed heat sources on magneto Casson nanofluid flow due to unsteady bi-directionally stretchable surface in a porous medium. *SN Applied Sciences*, 2, 1472. https://doi.org/10.1007/s42452-020-03262-4.
3. Rafique, K., Anwar, M.I., Misiran, M., Khan, I., Alharbi, S.O., Thounthong, P., & Nisar, K.S. (2019). Keller-box analysis of buongiorno model with brownian and thermophoretic diffusion for Casson nanofluid over an inclined surface. *Symmetry*, 11, 1370. https://doi.org/10.3390/sym11111370.
4. Algehyne, E.A., Aldhabani, M.S., Saeed, A., Dawar, A., & Kumam, P. (2022). Mixed convective flow of Casson and Oldroyd-B fluids through a stratified stretching sheet with nonlinear thermal radiation and chemical reaction. *Journal of Taibah University of Sciences*, 16(1), 193–203.
5. Zeeshan, A., Mehmood, O.U., Mabood, F., & Alzahrani, F. (2022). Numerical analysis of hydromagnetic transport of Casson nanofluid over permeable linearly stretched cylinder with Arrhenius activation energy. *International Communication in Heat and Mass Transfer*, 130, 105736. https://doi.org/10.1016/j.icheatmasstransfer.2021.105736.
6. Ghadikolaei, S.S., Hosseinzadeh, K., Ganji, D.D., & Jafari, B. (2018). Nonlinear thermal radiation effect on magneto Casson nanofluid flow with Joule heating effect over an inclined porous stretching sheet. *Case Studies in Thermal Engineering*, 12, 176–187.
7. Nadeem, S., Haq, R.U., & Akbar, N.S. (2013). MHD three-dimensional boundary layer flow of Casson nanofluid past a linearly stretching sheet with convective boundary condition. *IEEE Transactions in Nanotechnology*, 13(1), 109–115.
8. Vaidya, H., Prasad, K.V., Vajravelu, K., Wakif, A., Basha, N.Z., Manjunatha, G., & Vishwanatha, U.B. (2020). Effects of variable fluid properties on oblique stagnation point flow of a Casson nanofluid with convective boundary conditions. *Defect and Diffusion Forum*, 401, 183–196.
9. Basha, H.T., & Sivaraj, R. (2022). Stability analysis of Casson nanofluid flow over an extending/contracting wedge and stagnation point. *Journal of Applied Computational Mechanics*, 8(2), 566–579.
10. Nandeppanavar, M.M., Vaishali, S., Kemparaju, M.C., & Raveendra, N. (2020). Theoretical analysis of thermal characteristics of Casson nanofluid flow past an exponential stretching sheet in Darcy porous media. *Case Studies in Thermal Engineering*, 21, 100717. https://doi.org/10.1016/j.csite.2020.100717.
11. Shamshuddin, M.D., Ghaffari, A., & Usman. (2022). Radiative heat energy exploration on Casson-type nanoliquid induced by a convectively heated porous plate in conjunction with thermophoresis and Brownian movements. *InternationalJournal of Ambient Energy*, 43(1), 6329–6340.
12. Sarwar, N., Asjad, M.I., Sitthiwirattham, T., Patanarapeelert, N., & Muhammad, T. (2021). A Prabhakar fractional approach for the convection flow of Casson fluid across an oscillating surface based on the generalized Fourier law. *Symmetry*, 13, 2039. https://doi.org/10.3390/sym13112039.
13. Asogwa, K.K., Bilal, S.M., Animasaun, I.L., & Mebarek-Oudina, F. (2021). Insight into the significance of ramped wall temperature and ramped surface concentration: The case of Casson fluid flow on an inclined Riga plate with heat absorption and chemical reaction. *Nonlinear Engineering*, 10(1), 213–230.
14. Mebarek-Oudina, F. (2019). Convective heat transfer of Titania nanofluids odf different base fluids in cylindrical annulus with discrete heat source. *Heat Transfer*, 48(1), 135–147.
15. Patil, V.S., Shamshuddin, M.D., Ramesh, K., & Rajput, G.R. (2022). Slipperation of thermal and flow speed impacts on natural convective two-phase nanofluid model across Riga surface: computational scrutinization. *International Communication in Heat and Mass Transfer*, 135, 106135. https://doi.org/10.1016/j.icheatmasstransfer.2022.106135.

16. Asogwa, K.K., Uwanta, I.J., Momoh, A.A., & Omokhuale, E. (2013). Heat and mass transfer over a vertical plate with periodic Suction and heat sink. *Research Journal of Applied Sciences, Engineering and Technology*, 5(1), 7–15.
17. Reddy, Y.D., Mebarek-Oudina, F., Goud, B.S., & Ismail, A.I. (2022). Radaition, velocity and thermal slips effect toward MHD boundary layer flow through heat and mass transport of Williamson nanofluid with porous medium. *Arabian Journal of Science and Engineering*, 47, 16355–16369.
18. Chamkha, A.J. (2002). Hydromagnetic combined convection flow in a vertical lid-driven cavity with internal heat source or absorption. *Numerical Heat Transfer, Part A*, 41(5), 529–546.
19. Ibrahim, W., & Gadisa, G. (2020). Finite element solution of nonlinear convective flow of Oldroyd-B fluid with Cattaneo-Christov heat flux model over nonlinear stretching sheet with heat source or absorption. *Propulsion and Power Research*, 9(3), 304–315.
20. Ferdows, M., Shamshuddin, M.D., Salawu, S.O., & Zaimi, K. (2021). Numerical simulation for the steady nanofluid boundary layer flow over a moving plate with suction and heat source. *SN Applied Sciences*, 3, 264. https://doi.org/10.1007/s42452-021-04224-0.
21. Masood, S., Farooq, M., & Anjum, A. (2021). Influence of heat source/absorption and stagnation point on polystyrene–TiO2/H2O hybrid nanofluid flow. *Scientific Reports*, 11, 22381. https://doi.org/10.1038/s41598-021-01747-9.
22. Cui, J., Razzaq, R., Farooq, U., Khan, W.A., Farooq, F.B., & Muhammad, T. (2021). Impact of non-similar modeling for forced convection analysis of nanofluid flow over stretching sheet with chemical reaction and heat source. *Alexiandria Engineering Journal*, 61, 4253–4261.
23. Zainal, N.A., Nazar, R., Naganthran, K., & Pop, I. (2020). Heat generation/absorption effect on MHD flow of hybrid nanofluid over bidirectional exponential stretching/shrinking sheet. *Chinese Journal of Physics*, 69, 118–133.
24. Anuradha, S., & Sasikala, K. (2017). MHD nanofluid flow of a convection slip over a radiating stretching sheet with binary chemical reaction and activation energy. *Global Journal of Pure and Applied Mathematics*, 13(9), 6483–6492.
25. Hafeez, A., & Khan, M. (2021). Flow of Oldroyd-B fluid caused by a rotating disk featuring the Cattaneo-Christov theory with heat source/absorption. *International Communicationin Heat and Mass Trnasfer*, 123, 105179. https://doi.org/10.1016/j.icheatmasstransfer.2021.105179.
26. Zin, N.A.M., Khan, I., & Shafe, S. (2016). Influence of thermal radiation on unsteady MHD free convection flow of Jeffrey fuid over a vertical plate with ramped wall temperature. *Mathematical Problems in Engineering*, 2016, 6257071. https://doi.org/10.1155/2016/6257071.
27. Suba, S.A., & Muthucumaraswamy, R. (2018). Computational manipulation of a radiative MHD flow with Hall current and chemical reaction in the presence of rotating fluid. *IOP Conference Series: Journal of Physics*, 1000, 012144. https://doi.org/10.1088/1742-6596/1000/1/012144.
28. Reddy, S.H., Naidu, K.K., Babu, D.H., SatyaNarayana, P.V., & Raju, M.C. (2020). Significance of chemical reaction on MHD near stagnation point flow towards a stretching sheet with radiation. *SN Applied Sciences*, 2, 1822. https://doi.org/10.1007/s42452-020-03621-1.
29. Ullah, I., Bhattacharyya, K., Shafie, S., & Khan, I. (2016). Unsteady MHD Mixed convection slip flow of Casson fluid over nonlinearly stretching sheet embedded in a porous medium with chemical reaction, thermal radiation, heat generation/absorption and convective boundary conditions. *PLoS ONE*, 11(10), e0165348. https://doi.org/10.1371/journal.pone.0165348.
30. Gangaiah, T., Saidulu, N., & Lakshmi, A.V. (2019). The influence of thermal radiation onmixed convection MHD flow of a Casson nanofluid over an exponentially stretching sheet. *International Journal of Nanoscience and Nanotechnology*, 15(2), 83–98.
31. Rashidi, M.M., Rostami, B., Freidoonimehr, N., & Abbas-bandy, S. (2014). Free convective heat and mass transfer for MHD fluid flow over a permeable vertical stretching sheet in the presence of the radiation and buoyancy effects. *Ain Shams Engineering Journal*, 5(3), 901–912.
32. Reddy, Y.D., Goud, B.S., & Kumar, M.A. (2021). Radiation and heat absorption effects on an unsteady MHD boundary layer flow along an accelerated infinite vertical plate with ramped plate temperature in the existence of slip condition. *Partial Differnatial Equationsin Applied Mathematics*, 4, 100166. https://doi.org/10.1016/j.padiff.2021.100166.
33. Kumar, M.A., Reddy, Y.D., Rao, V.S., & Goud, B.S. (2021). Thermal radiation impact on MHD heat transfer natural convective nano fluid flow over an impulsively started vertical plate. *Case Studies in Thermal Engineering*, 24, 100826. https://doi.org/10.1016/j.csite.2020.100826.

34. Warke, A.S., Ramesh, K., Mebarek-Oudina, F., & Abidi, A. (2022). Numerical investigation of the stagnation point flow of radiative magnetomicropolar liquid past a heated porous stretching sheet. *Journal of Thermal Analysis and Calorimetry*, 147, 6901–6912.
35. Swain, K., Mahanthesh, B., & Mebarek-Oudina, F. (2021). Heat transport and stagnation-point flow of magnetized nanoliquid with variable thermal conductivity, Brownian moment, and thermophoresis aspects. *Heat Transfer*, 50(1), 754–767.
36. Djebali, R., Mebarek-Oudina, F., & Rajashekhar, C. (2021). Similarity solution analysis of dynamic and thermal boundary layers: further formulaion along a vertical flat plate. *Physica Scripta*, 96, 085206. https://doi.org/10.1088/1402-4896/abfe31.
37. Hassan, M., Mebarek-Oudina, F., Faisal, A., Ghafar, A., & Ismmail, A.I. (2022). Thermal energy and mass transport of shear thinning fluid under effects of low to high shear rate viscosity. *International Journal of Thermofluids*, 15, 100176. https://doi.org/10.1016/j.ijft.2022.100176.
38. Casson, N. (1959). A flow equation for pigment oil suspensions of the printing ink type. In: *Rheology of disperse systems*. Mill, C.C. (Ed.). Pergamon Press, Oxford, pp. 84–102.
39. Ibrahim, W., & Makinde, O.D. (2016). Magnetohydrodynamic stagnation point flow and heat transfer of Casson nanofluid past a stretching sheet with slip and convective boundary condition. *Journal of Aerospace Engineering*, 29(2), 04015037-1–04015037-11.
40. Archana, M., Gireesha, B.J., Prasannakumar, B.C., & Gorla, R.S.R. (2018). Influence of nonlinear thermal radiation on rotating flow of Casson nanofluid. *Nonlinear Engneering*, 7(2), 91–101.
41. Sulochana, C., Ashwinkumar, P., & Sandeep, N. (2016). Similarity solution of 3D Casson nanofluid flow over a stretching sheet with convective boundary conditions. *Journal of Nigerian Mathematical Society*, 35(1), 128–141.
42. Sheikholeslami, M., & Rokni, H.B. (2017). Magnetohydrodynamic CuO-water nano-fluid in a porous complex shaped enclosure. *Journal of Thermal Scienceand Engineering Application*, 9(4), 041007. https://doi.org/10.1115/1.4035973.
43. Sheikholeslami, M. (2018). CuO-water nanofluid flow due to magnetic field inside aporous media considering Brownian motion. *Journal of Molecular Liquids*, 249, 429–437.
44. Sheikholeslami, M. (2018). Numerical investigation of nanofluid free convection under the influence of electric field in a porous enclosure. *Journal of Molecular Liquids*, 249, 1212–1221.
45. Kafousias, N.G., & Williams, E.W. (1993). An improved approximation technique to obtain numerical solution of a class of two-point boundary value similarity problems in fluid mechanics. *International Journal of Numerical Methods in Fluids*, 17, 145–162.
46. Murtaza, M.G., Tzirtzilakis, E.E., & Ferdows, M. (2017). Effect of electrical conductivity and magnetization on the bio-magnetic fluid flow over a stretching sheet. *Zeitschrift fur Angewandte Mathematik und Physik (ZAMP)*, 68, 93. https://doi.org/10.1007/s000033-017-0839-z.

20 Insinuation of Radiative Bio-Convective MHD Flow of Casson nanofluid with Activation Energy and Swimming Microorganisms

Muhammad Jawad, Sajjad Hussain, Fateh Mebarek-Oudina, and Khurrem Shehzad

20.1 INTRODUCTION

Recently, the study of Casson fluid has achieved an unbelievable position among numerous mathematicians because of its dynamic thermal efficiency and potential in several heat flow problems without any pressure drop. It behaves like a solid material if the applied yield stress has high values as compared to the shear stress of the fluid, and it acts as a liquid if the value of the applied yield stress is lower than the shear stress of the fluid. Non-Newtonian fluids, on the other hand, are discovered to be very important in commercial and industrial applications, such as slurries, clay coating, shampoo, paints, grease, custard, cosmetic items, and so on. Researchers have studied the use of these fluids in order to learn more about non-Newtonian fluids. Many foodstuff and biomaterials like blood are examples of Casson fluids. Dash et al. [1] found the impact of yield stress on the flow properties of a Casson fluid bounded by a circular tube by using the Brinkman model. They examined the role of yield stress on the porous medium and velocity distributions. Hayat et al. [2] employed thermophoresis and Brownian motion impact in the nanofluid. They modelled nonlinear PDEs to nonlinear ODEs by using useful transformation and proved that the behaviors of thermophoretic parameters and Brownian motion on the nanofluid are quite reverse. Nadeem et al. [3] measured the effect of electrical conducting flow of Casson fluid through a shrinking exponential surface. Immanuel et al. [4] analyzed the Casson fluid and experimentally proved that the thermal conductivity of base fluids (liquid, gas) is lesser as compared to nanofluids in the occurrence of uniform magnetic fields. Choi [5] explains the term "nanofluid" and pronounces a liquid suspension enclosing a diameter of less than 100 nm having ultra-fine atoms. The thermal conductivity of base fluid (liquid/gas) increases due to thermophoresis, and Brownian motion, as pointed by Buongiorno [6], also showed that the smash of nanoparticles is liable for the rise in thermal conductivity of nanofluids. Pal et al. [7] explored the impact of nonlinear shrinking/extending surface on the flow of nanofluid with thermal radiation. Raju and Sandeep [8] calculated the magnetic field phenomenon and found the impact of ferrous on the nanoparticles and considered a Casson flow model as a topic of improvement in recent research. Hassan et al [9] found the role of chemical diffusion and thermal radiative heat transfer on the magnetohydrodynamic flow passing a porous extending plate. Numerous investigators have studied the rheology of nanofluids to design the viscosity of nanofluids, as explored by Hady et al. [10]. Mukhopadhyay et al. [11] investigated radiative Casson fluid flow over extending sheet. Sulochana et al. [12] recognized

DOI: 10.1201/9781003299608-20

the Casson nanofluids flow in three dimensions through a porous extending surface numerically. Readers who are interested might read [13–16] for further information. Microorganisms such as microalgae and bacteria have a higher density than water and swing in the opposing direction of gravity. Microorganisms aggregate on the topmost of the suspension, causing the density of the upper layer to exceed that of the lower layer, resulting in an unstable density distribution between the upper and lower layers. The commencement of convection configurations begins as a result of this convective instability. *Bio-convection* is the term for the random movement of microorganisms. They have imperative utilizations in commercial, ecological, and industrial products, such as fuels, ethanol, and fertilizers. The movement of mobile microorganisms results in macroscopic convective motion, which is referred to as bio-convection. Microorganisms are denser than fluid density. A stratified layer forms on top of the fluid due to the self-derived upward random migration of swimming microorganisms. Different stimulators, such as photosynthesis, chemotaxis, and gyrotactic, replicate the motion of microorganisms. Several scholars [17–19] have worked on this problem, and there have been several recent improvements. Joule heating and thermal radiation impact and electrical conducting flow, which is the non-Newtonian fluids flow over an extending sheet having several practical utilization in cooling of metallic plates, food processing, oil recovery, plastic sheets polymers, coated surface, and optical fibers, were analyzed. Hayat et al. [20] executed the flow of viscous magnetohydrodynamic fluids on the convective boundary over a smooth surface. The term magnetohydrodynamic has enormous application in field of science and technology. In addition, it has enormous aid in procedures resembling the star formation, X-ray radiation, electrolytes, magnetic field of the Earth, solar wind, tumor therapy, plasmas, fusion and cooling of fission reactors, etc. Scientists and engineers are interested in the effects of MHD fluid moving over a stretched surface in a variety of applications containing geothermal reservoirs, increased oil recovery, and catalytic reactors. Because their flow may be synchronized by an external magnetic field, a solution of polyisobutylene and polyethene oxide is used as a cooling liquid for enhancing quality. Taking into consideration these significant appliances, various investigators have focused on these flow difficulties. Because of their multiple applications in various domains, heat and mass flow analysis techniques with extended boundaries have significant importance for investigators. Similarly, because it handles heat and mass transfer, the heat absorption/generation influence on heat transfer has gained significant relevance. Several researchers have looked into the source/sink effects of different flow patterns proposed by Basak and Chamkha [21]. In the recent study, we have evaluated the electrical conducting mixed convection flow of a Casson nanofluid in the occurrence of gyrotactic microorganism over a porous shrinking sheet. Unlike typical studies, the porous shrinking sheet is considered for the proposed flow problem. The associated boundary value problem is executed numerically by applying the shooting approach after transforming it into a first-order initial value problem. It has been successfully employed for different interesting problems [22–29]. Finally, results are presented for discussion purpose. In view of this, the author has inspired to investigate the influence of radiation and activation energy on an MHD nano-Casson fluid with chemical reaction and porous medium. This chapter is organized as follows: in Section 20.2, mathematical modelling as well as closed-form solutions are presented for the velocity and temperature distributions. In Section 20.3, the numerical results are discussed for the velocity and temperature profiles through graphs. In Section 20.4, the important conclusions are presented for the studied problems.

20.2 FORMULATION AND SOLUTION OF THE PROBLEM

Steady two-dimensional flow of Casson nanofluid with thermal radiation, chemical reaction, and activation energy embedded in a porous shrinking sheet is discussed in the occurrence of gyrotactic microorganisms. The components of velocity u and v are assumed in x- and y-direction, respectively. The governing equations of problems in the Cartesian coordinates are given in the following.

The equation of mass conservation:

$$u_x + v_y = 0 \tag{20.1}$$

The equation of momentum:

$$uu_x + vu_y = v\left(1+\frac{1}{\beta}\right)u_{yy} + \frac{g}{\rho_f}\begin{bmatrix}(1-C_\infty)(T-T_\infty)\beta\rho_f - \\ \dot{\gamma}(n-n_\infty)(\rho_m-\rho_r)- \\ (\rho_p-\rho_f)(C-C_\infty)\end{bmatrix}$$

$$-\frac{\sigma_e B_0^2 u}{\rho_f} - \left(1+\frac{1}{\beta}\right)\frac{vu}{K} \tag{20.2}$$

Subjected to:

$$v = 0, u = ax \;\; u \to U = ax$$

The equation of temperature:

$$uT_x + vT_Y = \frac{k}{(\rho c)_f}(T_{yy}) + \frac{(\rho c)_p}{(\rho c)_f}\left\{\frac{D_T}{T_\infty}(T_y)^2 + D_B C_y T_y\right\}$$

$$-\frac{1}{(\rho c)_f}q_{r_y} + \frac{Q_0}{(\rho c)_f}(T-T_\infty) \tag{20.3}$$

Subjected to:

$$T = T_W \;\; as \;\; y \to 0$$
$$T \to T_\infty \;\; as \;\; y \to \infty$$

The equation of concentration:

$$uC_x + vC_y = D_B C_{yy} + \frac{D_T}{T_\infty}T_{yy} - K_0(C-C_\infty)\left(\frac{T}{T_\infty}\right)^n \exp\left(-\frac{E_a}{kT}\right) \tag{20.4}$$

Subjected to:

$$C = C_W \;\; as \;\; y \to 0$$
$$C \to C_\infty \;\; as \;\; y \to \infty$$

The equation of density of microorganism:

$$un_x + vn_y + \frac{bW_c}{C_W - C_\infty}\left[n_y C_{yy}\right] = D_m n_{yy} \tag{20.5}$$

Subjected to:

$$n = n_W \;\; as \;\; y \to 0$$
$$n \to n_\infty \;\; as \;\; y \to \infty,$$

In these governing equations, the components of velocity u and v are assumed in x- and y-direction, respectively; ρ_f is the density of the base fluid, v is the viscosity, σ^* is the electrical intensity, k^* is the absorption constant, β is the volume expansion coefficient, g is gravity, ρ_p is density of micro-organism particles, $\dot{\gamma}$ represents the volume of the microorganism, n is the concentration of the microorganism in the fluid, T is temperature of nanofluid, α is thermal diffusivity, $(\rho c)_f$ is heat capacity of liquid, $(\rho c)_p$ is effective heat capacity of nanoparticles, q_r is radiative heat flux, W_c is the maximum cell swimming speed, D_B is for Brownian diffusivity, D_T is for thermophoretic diffusion coefficient, k_c is chemical reaction parameter, and D_m is the diffusivity of microorganisms.

Where $k_c(C-C_\infty)\left(\frac{T}{T_\infty}\right)^n \exp\left(-\frac{E_a}{kT}\right)$ represents the Arrhenius expression. The temperature-dependent thermal conductivity is expressed as:

$$k = k_\infty\left(1+\varepsilon\frac{T-T_\infty}{T_{w-}T_\infty}\right), \tag{20.6}$$

According to radiative heat flux theory:

$$q_r = \frac{-4\sigma^*}{3k^*}\frac{\partial T^4}{\partial y}, \tag{20.7}$$

Where k^* stands for absorption coefficient and σ^* denotes the Stefan–Boltzmann constant. Using an expansion of Taylor's series about T_∞, we get:

$$T^4 = 4T_\infty{}^3T - 3T_\infty{}^4, \tag{20.8}$$

With the utilization of equations (11)–(12) in (4), we have:

$$\begin{aligned} uT_x + vT_Y &= \frac{k}{(\rho c)_f}\left(T_{yy}\right) + \frac{(\rho c)_p}{(\rho c)_f}\left\{\frac{D_T}{T_\infty}\left(T_y\right)^2 + D_BC_yT_y\right\} + \\ &\frac{16\sigma^* T_\infty^3}{3k^*(\rho c)_f}\frac{\partial^2 T}{\partial y^2} + \frac{Q_0}{(\rho c)_f}\left(T-T_\infty\right) \end{aligned} \tag{20.9}$$

Let us implement the following similarity approaches to transform partial differential equations to ordinary differential equations [30–32]:

$\psi = (av)^{\frac{1}{2}} xf(\eta), \eta = \left(\frac{a}{v}\right)^{\frac{1}{2}} y,$ and $C = C_\infty + \left(C_w - C_\infty\right)\phi(\eta),$ $T = T_\infty + \left(T_w - T_\infty\right)\theta(\eta), n = n_\infty + \left(n_w - n_\infty\right)\chi(\eta),$ where $\psi(x,y)$ is the stream line function, defined as $u=\psi_y$ and $v=-\psi_x$, which tropically fulfill the equation of mass conservation, and η is the similarity variable. Equations (20.2)–(20.5) are reduced to:

$$\begin{aligned} &\left(1+\frac{1}{\beta}\right)f'''[\eta] - f'^2[\eta] + f[\eta]f''[\eta] - M(f'[\eta]-1) + \\ &1 - K_1\left(1+\frac{1}{\beta}\right)f'[\eta] + \lambda'(\theta[\eta] - N_c\chi[\eta] - N_r\phi[\eta]) = 0, \end{aligned} \tag{20.10}$$

$$\left(1+\frac{4Rd}{3}\right)\theta''[\eta]+p_r f[\eta]\theta'[\eta]+p_r Nt\theta'^2[\eta]+p_r Nb\theta'[\eta]\phi'[\eta]+p_r\Delta\theta[\eta]=0, \tag{20.11}$$

$$\phi''[\eta]+\Pr Lef[\eta]\phi'[\eta]+\left(\frac{N_t}{N_b}\right)\theta''[\eta]-\Pr Le\gamma\left(1+\delta\theta[\eta]\right)^n\exp\left(\frac{-E}{1+\delta\theta}\right)\phi[\eta]=0, \tag{20.12}$$

$$\chi''[\eta]-P_e\chi[\eta]\phi'[\eta]-P_e\sigma\phi''[\eta]+P_e\chi'[\eta]\phi''[\eta]+S_c f[\eta]\chi'[\eta]=0, \tag{20.13}$$

The boundary condition becomes:

$$f(\eta)=S,\ f'(\eta)=-\lambda,\ \chi(\eta)=1,\ \theta(\eta)=1,\ \phi(\eta)=1\ as\ \eta\to 0 \tag{20.14}$$

$$f'(\eta)=1,\ \theta(\eta)=0,\ \phi(\eta)=0,\ \chi(\eta)=0\ as\ \eta\to\infty \tag{20.15}$$

Where $\gamma=\frac{K_0}{a}$ is the chemical reaction parameter, $\delta=\frac{T_w-T_\infty}{T_\infty}$ is the temperature difference, $\lambda'=\frac{\beta\dot{\gamma}(1-C_\infty)(T_w-T_\infty)x^3}{aU_w}$ is the mixed convection parameter, $E=\frac{E_a}{kT_\infty}$ is the activation energy, $Rd=\frac{4\sigma^*T_\infty^3}{k^*k_\infty}$ is thermal radiation, $Nr=\frac{(\rho_p-\rho_f)(C_w-C_\infty)}{\beta\rho_f(1-C_\infty)T_\infty\beta}$ is buoyancy ratio parameter, $K_1=\frac{\nu}{aK}$ is porosity parameter, $P_r=\frac{\nu_f}{\alpha}$ is Prandtl number, $Nc=\frac{\dot{\gamma}(\rho_m-\rho_f)(n_w-n_\infty)}{\beta\rho_f(1-C_\infty)T_\infty}$ is Rayleigh number, $\sigma=\frac{N_\infty}{N_w-N_\infty}$ is microorganism concentration difference, $Nb=\frac{D_B\tau(C_w-C_\infty)}{\nu}$ is Brownian motion, $\Delta=\frac{Q_0}{a(\rho_C)_f}$ is heat generation/absorption coefficients, $Nt=\frac{D_T\tau(T_w-T_\infty)}{T_\infty\nu}$ is thermophoresis parameters, $Le=\frac{\nu}{D_B}$ is bio-convection Lewis number, and $Pe=\frac{bW_c}{D_m}$ is Peclet number. The physical quantities of interest are define as:

$$C_f=\frac{\tau_w}{\rho U_w^2},Nu_x=\frac{xq_w}{k(T_w-T_\infty)},$$

$$Sh_x=\frac{xq_m}{D_B(C_w-C_\infty)},Nn_x=\frac{xq_m}{D_n(N_w-N_\infty)} \tag{20.16}$$

$$C_f\,\mathrm{Re}_x^{1/2}=-f''(0),Nu_x\,\mathrm{Re}_x^{1/2}=-\theta'(0),$$
$$Sh_x\,\mathrm{Re}_x^{1/2}=-\phi'(0),Nn_x\,\mathrm{Re}_x^{1/2}=-\chi'(0) \tag{20.17}$$

Where $\mathrm{Re}_x^{1/2}=U_w\,x/\upsilon$ is the local Reynolds number.

20.2.1 Numerical Solution

Nonlinear ordinary differential equations (20.8)–(20.13), with given boundary conditions equations (20.14)–(20.15), are solved numerically, exhausting the similarty approach. To use this numerical method, create a new variable, as shown in the following. Put $f(\eta)$ by $z_1(\eta)$, $\theta(\eta)$ by $z_4(\eta)$, $\phi(\eta)$ by $z_6(\eta)$, and $\chi(\eta)$ by $z_8(\eta)$. The consequent equations are:

$$\begin{aligned}
z_1' &= z_2\\
z_2' &= z_3\\
z_3' &= \begin{pmatrix} z_2 * z_2 - z_1 * z_3 + M * z_2 - \\ \lambda'*\left(z_4 - Nr * z_6 - Nc * z_8\right) \end{pmatrix} / \left(\frac{1}{\beta} + 1\right) + K_1 * z_2\\
z_4' &= z_5\\
z_5' &= -\left(\Pr / \left(1 + \frac{4Rd}{3}\right)\right)\begin{pmatrix} z_1 * z_5 + Nt * \left(z_5 * z_5\right) + \\ Nb * (z_5 * z_6) + \Delta * z_4 \end{pmatrix}\\
z_6' &= z_7\\
z_7' &= -\left(Le * z_1 * z_7 + \left(Nt / Nb\right) * z_5'\right) -\\
&Le * \gamma * \left(1 + \delta\theta[\eta]\right)^n \exp\left(\frac{-E}{1 + \delta\theta}\right) z_6\\
z_8' &= z_9\\
z_9' &= Pe * \left(z_8 * z_7 + z_7' * \left(z_9 + \sigma\right)\right) - Sc * \left(z_1 * z_9\right)
\end{aligned} \tag{20.18}$$

The transformed conditions are:

$$z_1(\eta) = S,\ z_2(\eta) = -\lambda,\ z_8(\eta) = 1,\ z_4(\eta) = 1,\ z_6(\eta) = 1 \ as\ \eta \to 0 \tag{20.19}$$

$$z_2(\eta) = 1,\ z_4(\eta) = 0,\ z_6(\eta) = 0,\ z_8(\eta) = 0 \ as\ \eta \to \infty \tag{20.20}$$

20.3 RESULTS AND DISCUSSION

The results are presented in graphical and tabular formats in this section. Our interests' physical dimensionless quantities are chemical reaction parameter γ, temperature difference δ, mixed convection parameter λ', activation energy E, thermal radiation Rd, buoyancy ratio parameter Nr, porosity parameter K_1, magnetic parameter M, Prandtl number P_r, Rayleigh number Nc, microorganism concentration difference σ, Brownian motion Nb, heat generation/absorption coefficients Δ, bio-convection Lewis number Le, thermophoresis parameters Nt, and Peclet number Pe. We express skin friction coefficient and density number, Sherwood and local Nusselt numbers, at sheet to understand the behavior of the mathematical model of the problem. For this we set the suitable values as $\lambda = -1.24$, $\beta = 0.8$, M = 0.5, Pr = 0.7 d = 0.2, S = 2, Sc = 1.0, Nb = 0.4, K = 0.14, Nt = 0.4, Le = 0.9, Nr = 0.4, Nc = 0.4, λ' = 0.9, δ = 0.3, Rd = 0.2, Pe = 1.0, γ = 0.10, Δ = 0.6. Variations of parameters in the computation scheme are shown in graphs and tables. The results for speed $f'(\eta)$, concentration distribution $\phi(\eta)$, microorganism density profile $\chi(\eta)$, and temperature field $\theta(\eta)$ are attained.

20.3.1 VELOCITY PROFILE

Plot 20.1 sketches the influence of β on velocity function $f'(\eta)$, which is diminished when the Casson parameter is improved. With increasing variations of Hartmann number M and shrinking parameter λ, as well as permeability parameter K, a similar pattern of velocity distribtuin was discovered, as illustrated in Plots 20.2, 20.3, and 20.4. As seen in Plots 20.5 and 20.6, the curve of the velocity field reduces as the bio-convection Rayleigh number *Nc* and the buoyancy ratio parameter *Nr* rise due to the occurrence of buoyancy forces.

20.3.2 Temperature Field

Temperature is enhanced by increasing values of *M* (see Plot 20.7). Lorentz forces that are resistive forces are included in the Hartmann number. As *M* boosts up, the Lorentz force is enhanced, causing a resistance of the flow of liquid as a result of velocities declining and temperature enriching. The heat generation parameter Δ (Δ>0), the nonlinear thermal radiation parameter *Rn*, and *Nt* show a growing influence on the energy field $\theta(\eta)$, as exhibited individually in Plots 20.8–20.10. The inspiration of the Prandtl number on the temperature function $\theta(\eta)$ is illustrated in Plot 20.11. The temperature of the nanoparticles drops because of enhancement in *Pr*. The Prandtl number is termed as the ratio among thermal conductivity of fluid and thermal diffusivity of a fluid. In consequence, the minimum value of Prandtl number is a consequence of the maximum thermal diffusivity, while this causes lesser boundary layer thickness and temperature. However, shrinking parameters λ have reducing influence on energy field, as revealed in Plot 20.12.

20.3.3 Concentration Field

The thermophoresis parameter *Nt* and chemical reaction parameter γ manifest a momentous growth in the nanofluid concentration field ϕ as described separately (in Plots 20.13–20.14). But the shrinking parameter $\lambda > 0$ and Lewis number *Le* have diminishing influence on dimensionless nanofluid concentration field ϕ, as demonstrated in Plots 20.15–20.16 correspondingly.

20.3.4 Microorganism Density Field

The curve of motile microorganism density function $\chi(\eta)$ decreases with an increase of the Schmidt number *Sc*, Lewis number *Le*, microorganism concentration difference σ, shrinking parameter $\lambda > 0$, and bio-convection Peclet number *Pe* exposed in Plots 20.17–20.21, respectively. But thermophoresis illustrations inversely influence the microorganism density field, as presented in Plot 20.22.

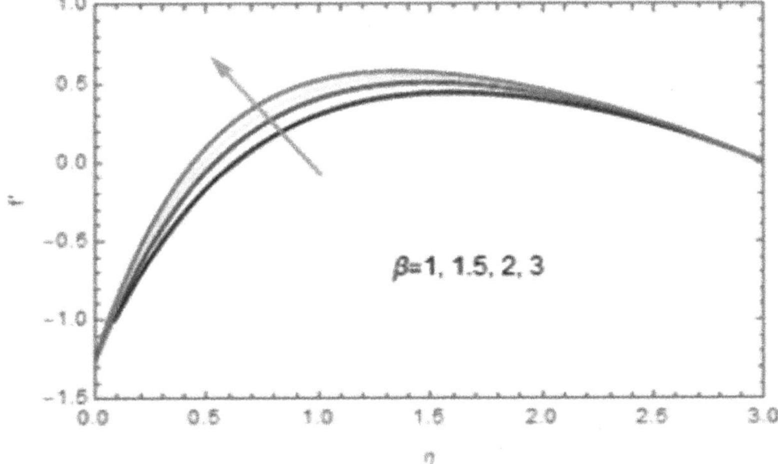

PLOT 20.1 Influence of β on $f'(\eta)$.

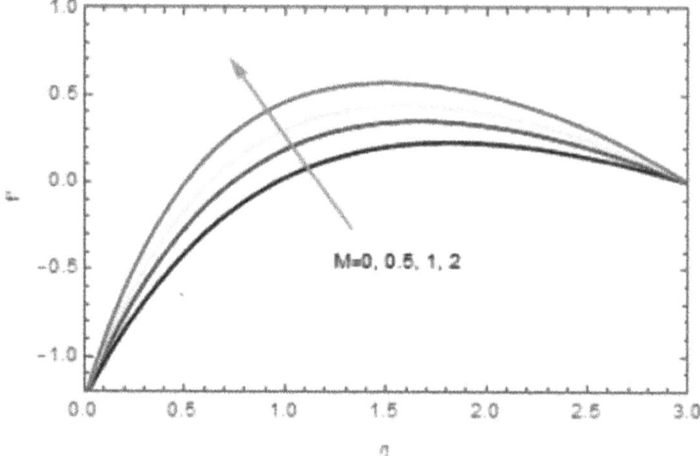

PLOT 20.2 Variation of M on $f'(\eta)$.

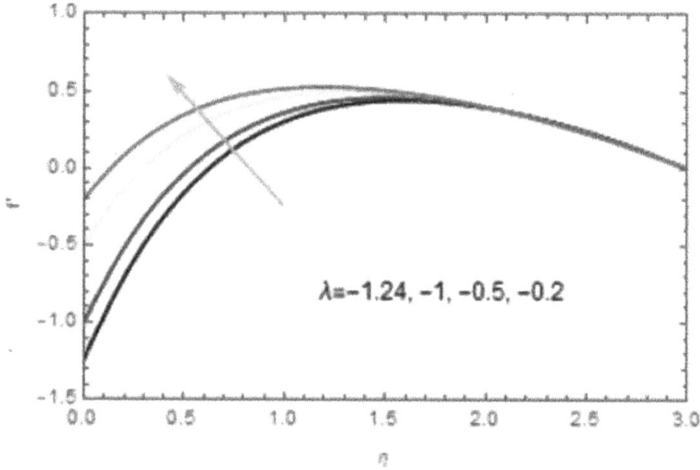

PLOT 20.3 Variation of λ on $f'(\eta)$.

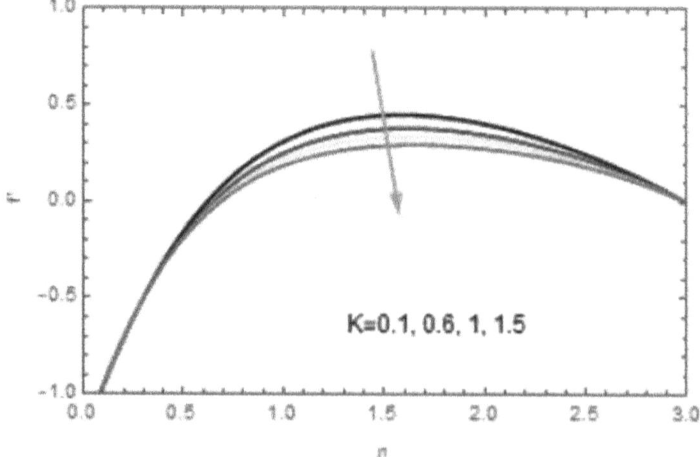

PLOT 20.4 Inspiration of K on $f'(\eta)$.

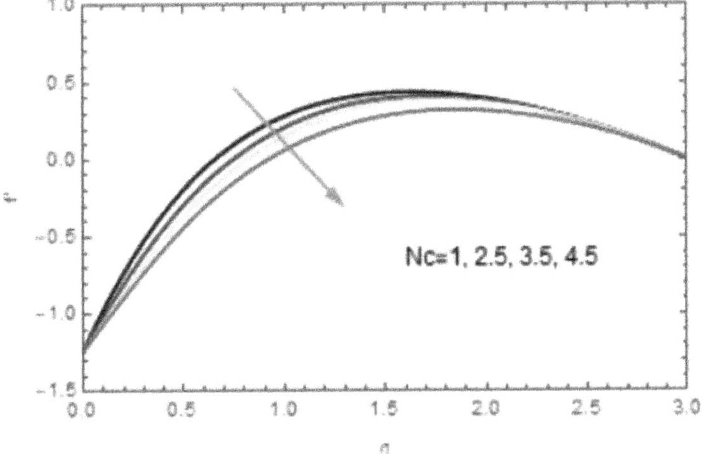

PLOT 20.5 Change of Nc on $f'(\eta)$.

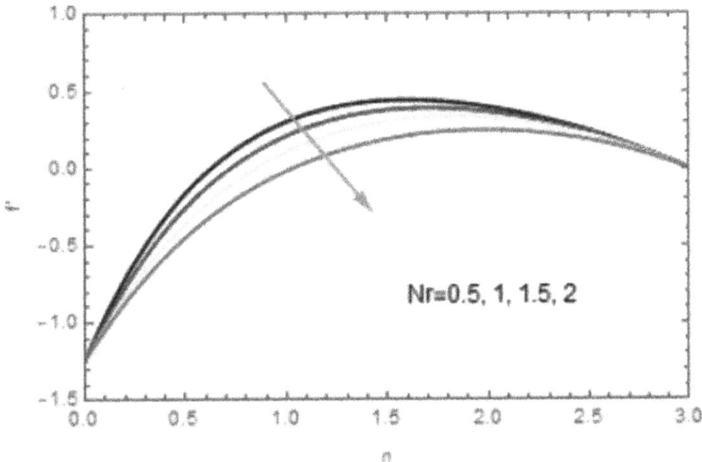

PLOT 20.6 Change of Nr on $f'(\eta)$.

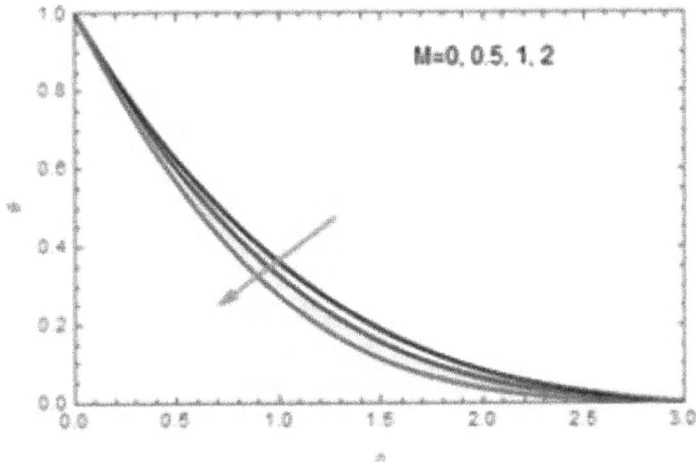

PLOT 20.7 Variation of M on θ.

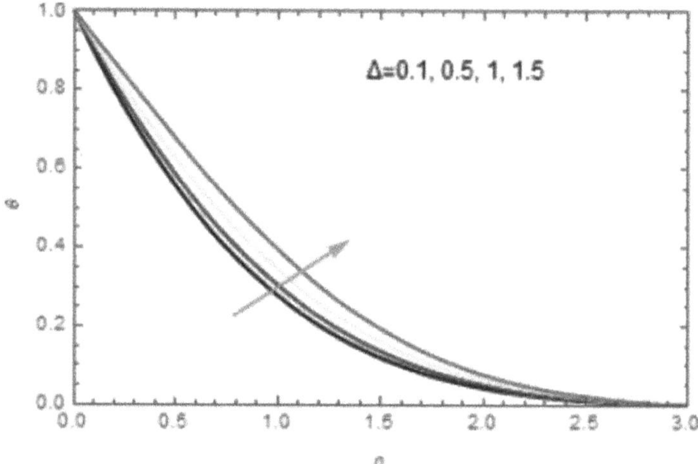

PLOT 20.8 Effect of heat absorption Δ on θ.

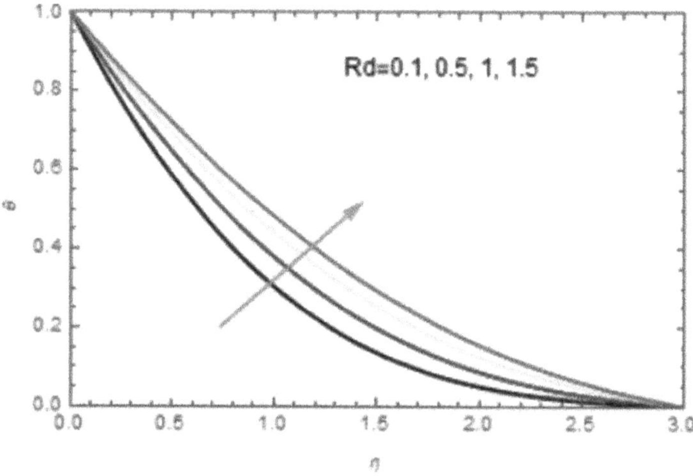

PLOT 20.9 Change of Rd on q.

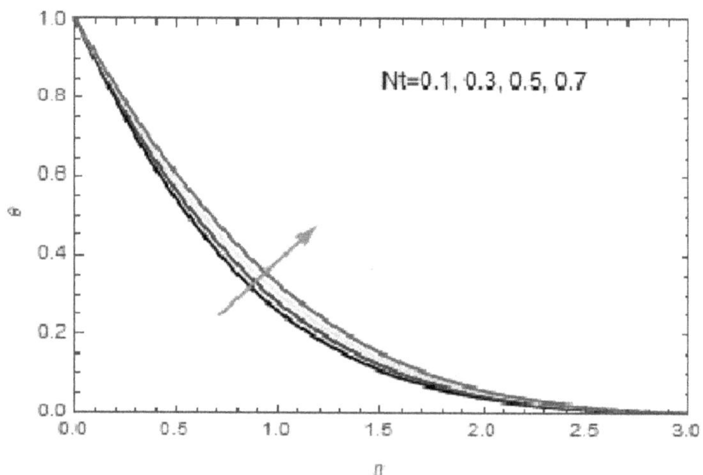

PLOT 20.10 Effect of *thermophoresis* on θ.

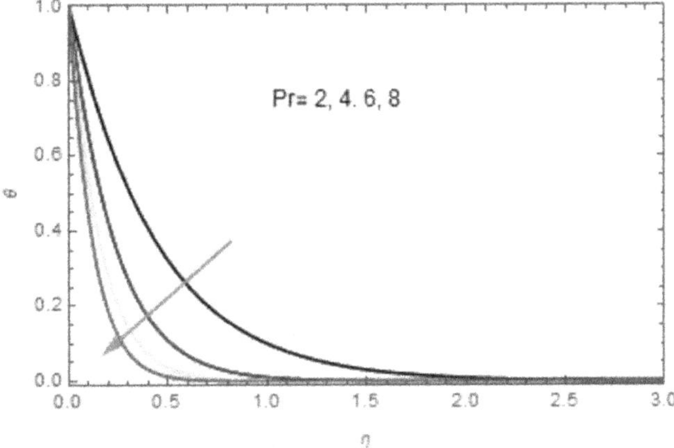

PLOT 20.11 Impact of *Pr* on *q*.

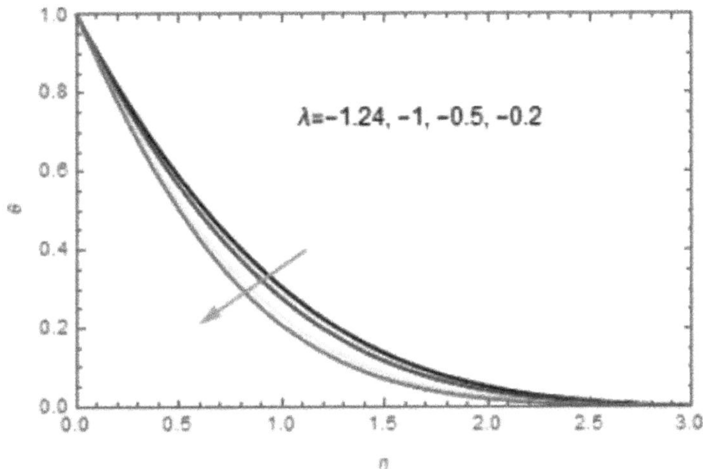

PLOT 20.12 Impact of λ on *q*.

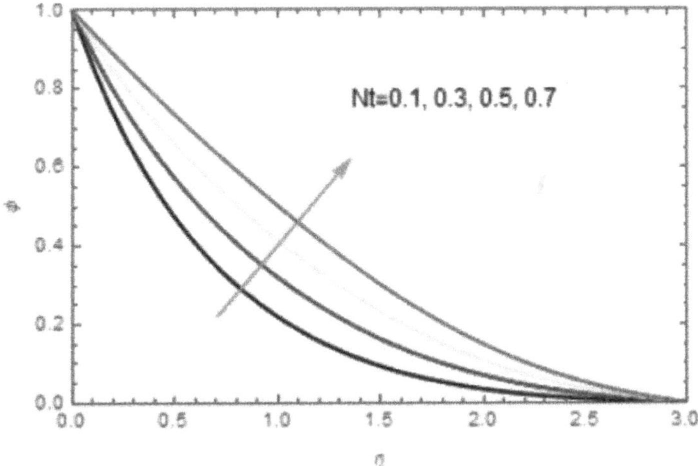

PLOT 20.13 Inspiration of *thermophoresis* on $\varphi(\eta)$.

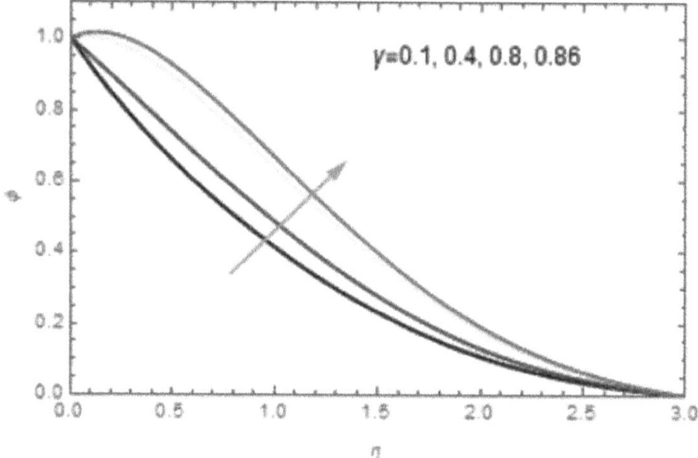

PLOT 20.14 Change of γ on $\varphi(\eta)$.

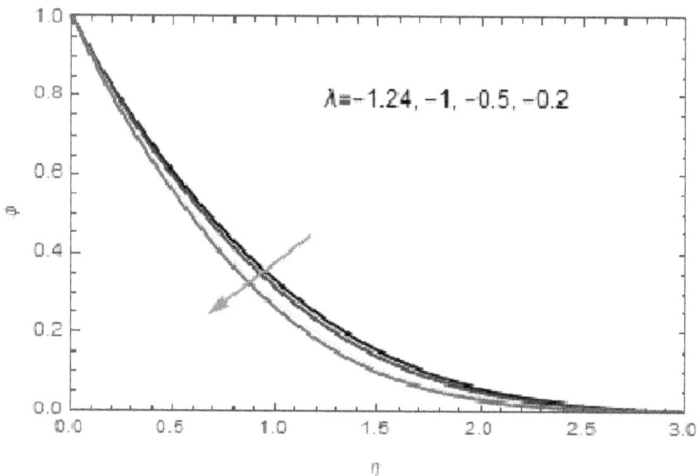

PLOT 20.15 Impact of λ on $\varphi(\eta)$.

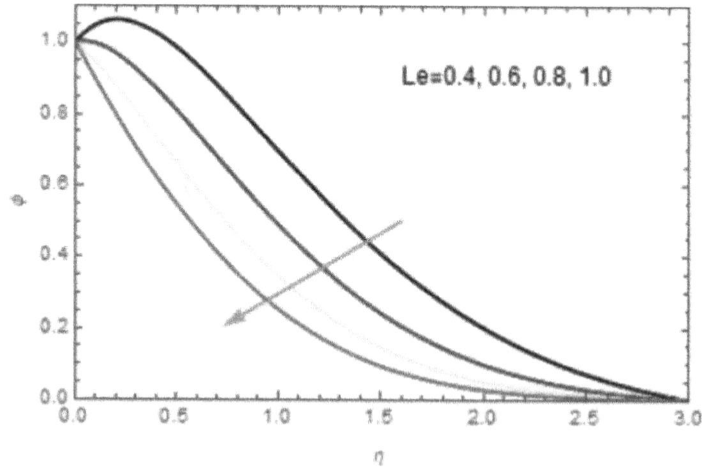

PLOT 20.16 Effect of Le on $\varphi(\eta)$.

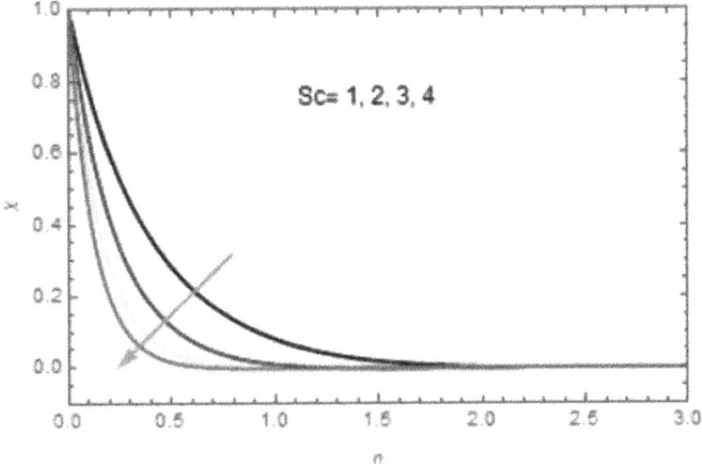

PLOT 20.17 Influence of *Sc* on the density field.

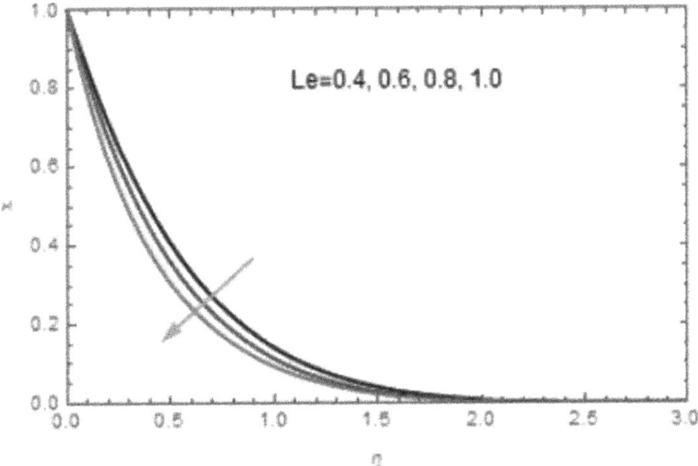

PLOT 20.18 Influence of *Le* on density distribution.

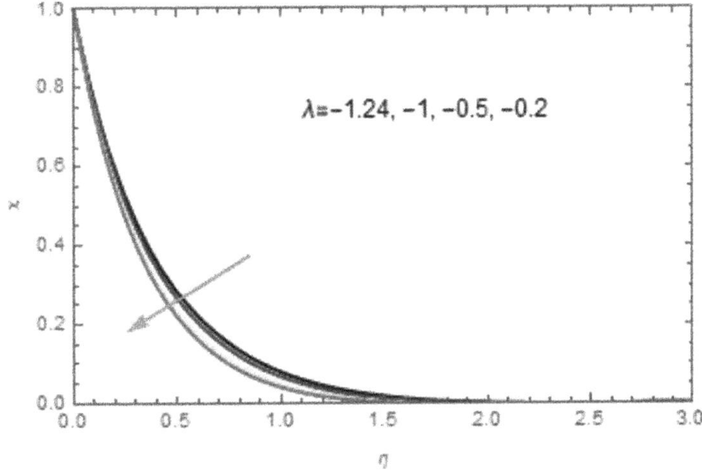

PLOT 20.19 Influence of λ on density profile.

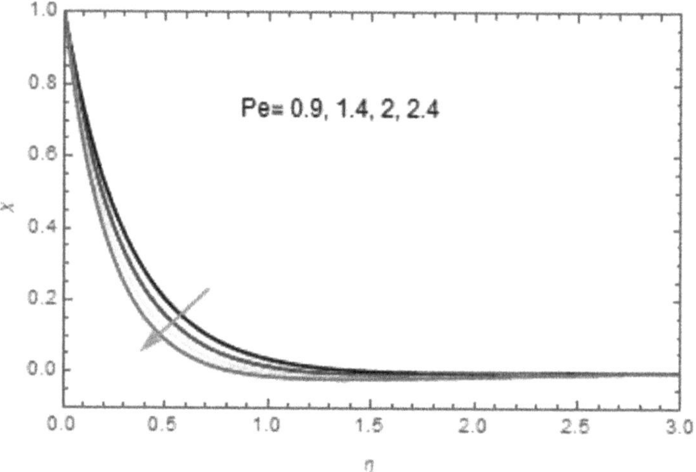

PLOT 20.20 Influence of *Pe* on density function.

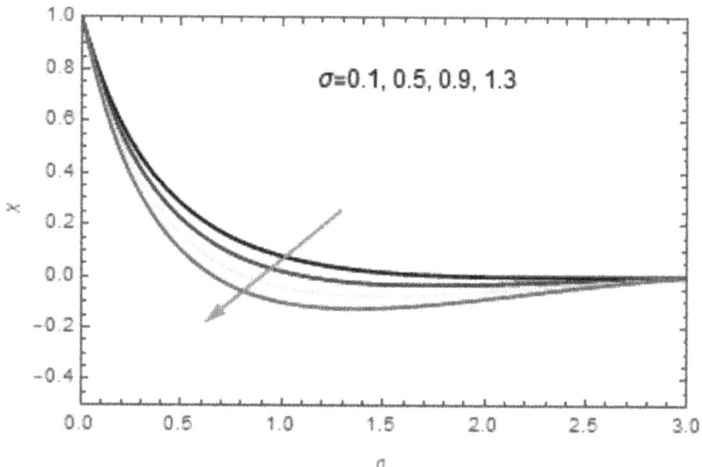

PLOT 20.21 Influence of *sigma* on density field.

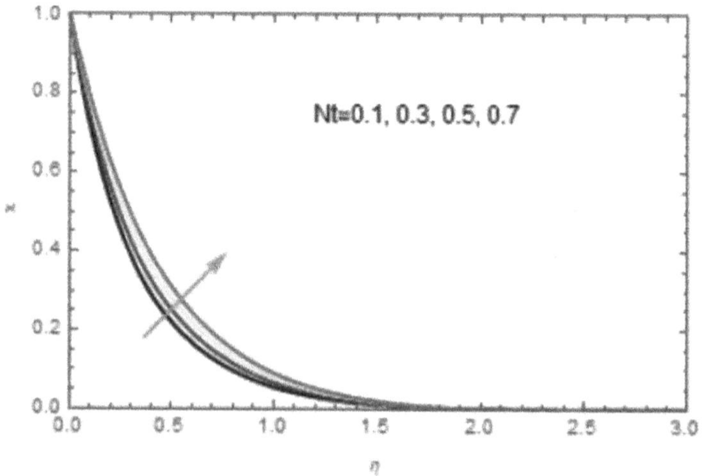

PLOT 20.22 Influence of *Nt* on density profile.

TABLE 20.1
Numerical Computation of $-f''(0)$ for Some Promising Parameters

M	λ'	Nr	Nc	$-f''(0)$
0.5	0.2	0.2	0.2	0.8583
				0.9679
				1.0971
0.2	0.2	0.2	0.2	0.7278
0.6				0.9197
1.0				1.2269
0.5	0.5	0.2	0.2	0.8065
	0.8			1.0101
	1.2			1.3535
0.5	0.2	0.2	0.1	0.8575
		0.5		0.8573
		1.0		0.8571
0.5	0.2	0.2	0.2	0.8638
			0.6	0.8894
			1.0	0.9154

TABLE 20.2
Numerical Computations for Local Nusselt Number for Prominent Parameters

M	λ'	Pr	Nb	Nt	Nr	Nc	Le	Rd	$\theta'(0)$
0.2	0.1	2.0	0.2	0.3	0.2	0.2	2.0	0.5	0.2106
0.6									0.2072
1.0									0.2015
0.5	0.5	2.0	0.2	0.3	0.2	0.2	2.0	0.5	0.1717
	0.8								0.1725
	1.2								0.1737
0.5	0.1	1.2	0.2	0.3	0.2	0.2	2.0	0.5	0.1864
		1.5							0.1956
		1.8							0.2032
0.5	0.1	2.0	0.1	0.3	0.2	0.2	2.0	0.5	0.2074
			0.5						0.2075
			1.0						0.2076
0.5	0.1	2.0	0.2	0.1	0.2	0.2	2.0	0.5	0.2088
				0.5					0.2060
				1.0					0.2022
0.5	0.1	2.0	0.2	0.3	0.2	0.2	2.0	0.5	0.2083
					0.6				0.2081
					1.0				0.2079
0.5	0.1	2.0	0.2	0.3	0.2	0.2	2.0	0.5	0.2082
						0.6			0.2077
						1.0			0.2071
0.5	0.1	2.0	0.2	0.3	0.2	0.2	1	0.5	0.2079
							1.4		0.2077
							1.8		0.2075
0.5	0.1	2.0	0.2	0.3	0.2	0.2	0.2	0.1	0.2196
								0.4	0.2102
								0.8	0.1979

TABLE 20.3
Numerical Computations for Local Sherwood Numbers for Various Values Emerging Parameters

M	λ'	Nr	Nc	Nt	Nb	Le	Pr	Rd	E	$\phi'(0)$
0.2	0.1	0.2	0.2	0.3	0.2	0.2	0.2	0.5	0.1	0.3159
0.6										0.3108
1.0										0.3023
0.5	0.5	0.2	0.2	0.3	0.2	0.2	0.2	0.5	0.1	0.4525
	0.8									0.4569
	1.2									0.4639
0.5	0.1	0.2	0.2	0.3	0.2	0.2	0.2	0.5	0.1	0.3125
		0.6								0.3122
		1.0								0.3120
0.5	0.1	0.2	0.2	0.3	0.2	0.2	0.2	0.5	0.1	0.3123
			0.6							0.3115
			1.0							0.3106
0.5	0.1	0.2	0.2	0.1	0.2	0.2	0.2	0.5	0.1	0.1044
				0.5						0.5151
				1.0						1.10112
0.5	0.1	0.2	0.2	0.3	0.1	0.2	0.2	0.5	0.1	0.6222
					0.5					0.1245
					1.0					0.0622
0.5	0.1	0.2	0.2	0.3	0.2	1.0	0.2	0.5	0.1	0.3118
						1.4				0.3115
						1.8				0.3113
0.5	0.1	0.2	0.2	0.3	0.2	0.2	1.2	0.5	0.1	0.2796
							1.5			0.2935
							1.8			0.3048
0.5	0.1	0.2	0.2	0.3	0.2	0.2	0.2	0.1	0.1	0.3294
								0.4		0.3154
								0.8		0.2999
0.5	0.1	0.2	0.2	0.3	0.2	0.2	0.2	0.5	0.2	0.4936
									0.6	0.3357
									0.9	0.0743

TABLE 20.4
Numerical Computations for Motile Microorganism's Density Number

M	λ'	Nr	Nc	Pe	Le	$-\chi'(0)$
0.2	0.1	0.2	0.2	0.2	1.0	0.6722
0.6						0.6384
1.0						0.5876
0.5	0.5	0.2	0.2	0.2	1.0	0.4677
	0.8					0.4899
	1.2					0.4665
0.5	0.1	0.2	0.2	0.2	1.0	0.6492
		0.6				0.6473
		1.0				0.6460

M	λ'	Nr	Nc	Pe	Le	$-\chi'(0)$
0.5	0.1	0.2	0.2	0.2	1.0	0.6477
			0.6			0.6426
			1.0			0.6372
0.5	0.1	0.2	0.2	0.2	1.0	0.6697
				0.6		0.7868
				1.0		0.9046
0.5	0.1	0.2	0.2	0.2	1.0	0.4667
					1.4	0.3368
					1.8	0.1676

20.4 CONCLUSIONS

On the account of industrial and technological applications, the enhancement of energy by inserting nanoparticles is a hot topic in the present century. Therefore, the current analysis presents a theoretical analysis regarding the flow of electrically conducted nanofluid with variable viscosity over a stretching surface in the presence of the swimming gyrotactic microorganism. In addition, the influence of buoyancy forces, thermal conductivity, chemical diffusion, and Arrhenius activation energy is considered. By using the opposite transformation, the system of contemporary partial differential expressions is first converted into nonlinear ordinary differential system. The set of these transmuted equations is solved with the help of the shooting method. Reliable results are obtained for the velocity profile, temperature, motile microorganism density, and concentration. It is evaluated that by increasing the value of bio-convection Peclet and Lewis numbers, the microorganism distribution exhibited diminishing behavior. These results may be useful in improving the efficiency of heat transfer devices and microbial fuel cells. The main results of this study are summarized as:

- The speediness of flow $f'(\eta)$ diminishes with surpassing values of Hartmann number M, permeability parameter K, and shrinking parameter λ.
- The magnitude of energy distribution $\theta(\eta)$ for shrinking surface diminishes with increments in radiation parameter Rd and Hartmann number M, but P_r reveals an opposite performance.
- Enlarging the value of thermophoresis parameter Nt, activation energy parameter E, and chemical reaction parameter γ on the nanofluid concentration field $\phi(\eta)$ illustrations boosts actions.
- The motile microorganism density function $\chi(\eta)$ progressively reduces with an increase in the value of the Peclet number Pe and the bio-convective Lewis number.

Nomenclature

T	fluid temperature
σ_e	electrical conductivity
α	thermal diffusivity
B_0	magnetic field
D_B	Brownian diffusion coefficient
D_T	thermophoresis diffusion coefficient
ω	kinematic viscosity
$(\rho c)_f$	heat capacity of liquid
$(\rho c)_p$	effective heat capacity of nanoparticles

D_n diffusivity of microorganisms
b chemotaxis constant
W_c maximum cell swimming speed
k thermal conductivity of the fluid
N nanoparticle volume fraction
C concentration of microorganisms
v density of base fluid
K_0 chemical reaction rate coefficient
ρ_f heat capacity ratio
ρ_p density of nanoparticles
Pr Prandtl number
N_b Brownian motion parameter
Sc Schmidt number
Nt thermophoresis parameter
Pe bio-convection Peclet number
Δ heat generation/absorption coefficients
M magnetic parameter number
σ dimensionless constant
Le Lewis number
λ mixed convection parameter
Nr Buoyancy ratio parameter
Nc bio-convection Rayleigh number
γ chemical reaction parameter
Rd thermal radiation parameter
K porous medium parameter
ψ stream function
ρ density of the fluid
η dimensionless similarity variable

REFERENCES

[1] Dash, R. K., Mehta, K. N., & Jayaraman, G. (1996). Effect of yield stress on the flow of a Casson fluid in a homogeneous porous medium bounded by a circular tube. *Applied Scientific Research*, *57*(2), 133–149.

[2] Hayat, T., Khan, M. I., Waqas, M., Yasmeen, T., & Alsaedi, A. (2016). Viscous dissipation effect in flow of magnetonanofluid with variable properties. *Journal of Molecular Liquids*, *222*, 47–54.

[3] Nadeem, S., Haq, R. U., & Lee, C. (2012). MHD flow of a Casson fluid over an exponentially shrinking sheet. *Scientia Iranica*, *19*(6), 1550–1553.

[4] Immanuel, Y., Pullepu, B., & Kirubhashankar, C. K. (2015). Casson flow of MHD fluid moving steadily with constant velocity. *Applied Mathematical Sciences*, *9*(30), 1503–1508.

[5] Choi, S. U., & Eastman, J. A. (1995). *Enhancing thermal conductivity of fluids with nanoparticles* (No. ANL/MSD/CP-84938; CONF-951135-29). Argonne National Lab. (ANL), Argonne, IL.

[6] Buongiorno, J. (2006). Convective transport in nanofluids. *Journal of Heat Transfer*, *128*(3), 240–250. http://doi.org/10.1115/1.2150834.

[7] Pal, D., Mandal, G., & Vajravalu, K. (2016). Soret and Dufour effects on MHD convective–radiative heat and mass transfer of nanofluids over a vertical non-linear stretching/shrinking sheet. *Applied Mathematics and Computation*, *287*, 184–200.

[8] Raju, C. S. K., & Sandeep, N. (2017). Unsteady Casson nanofluid flow over a rotating cone in a rotating frame filled with ferrous nanoparticles: a numerical study. *Journal of Magnetism and Magnetic Materials*, *421*, 216–224.

[9] Hassan Waqas, N. T., Naseem, R., Farooq, S., Khalid, S., & Hussain, S. (2017). The chemical diffusion and bouyancy effects on MHD flow of Casson fluids past a stretching inclined plate with non-uniform heat source. *Journal of Applied Environmental and Biological Sciences*, *7*(6), 135–142.

[11] Hady, F. M., Ibrahim, F. S., Abdel-Gaied, S. M., & Eid, M. R. (2012). Radiation effect on viscous flow of a nanofluid and heat transfer over a nonlinearly stretching sheet. *Nanoscale Research Letters*, *7*(1), 1–13.

[12] Mukhopadhyay, S., De, P. R., Bhattacharyya, K., & Layek, G. C. (2013). Casson fluid flow over an unsteady stretching surface. *Ain Shams Engineering Journal*, *4*(4), 933–938.

[13] Anjali Devi, S. P., & Prakash, M. (2015). Temperature dependent viscosity and thermal conductivity effects on hydromagnetic flow over a slendering stretching sheet. *Journal of the Nigerian Mathematical Society*, *34*(3), 318–330.

[14] Khalique, C. M., Safdar, R., & Tahir, M. (2019). First analytic solution for the oscillatory flow of a Maxwells fluid with annulus. *Open Journal of Mathematical Sciences*, *2*, 1–9.

[15] Imran, M., Ching, D. L. C., Safdar, R., Khan, I., Imran, M. A., & Nisar, K. S. (2019). The solutions of non-integer order burgers' fluid flowing through a round channel with semi analytical technique. *Symmetry*, *11*(8), 962.

[16] Safdar, R., Imran, M., Tahir, M., Sadiq, N., & Imran, M. (2020). MHD flow of burgers' fluid under the effect of pressure gradient through a porous material pipe. *Punjab University Journal of Mathematics*, *50*(4).

[17] Safdar, R., Gulzar, I., Jawad, M., Jamshed, W., Shahzad, F., & Eid, M. R. (2022). Buoyancy force and Arrhenius energy impacts on Buongiorno electromagnetic nanofluid flow containing gyrotactic microorganism. *Proceedings of the Institution of Mechanical Engineers, Part C: Journal of Mechanical Engineering Science*, 09544062221095693.

[18] Safdar, R., Jawad, M., Hussain, S., Imran, M., Akgül, A., & Jamshed, W. (2022). Thermal radiative mixed convection flow of MHD Maxwell nanofluid: Implementation of buongiorno's model. *Chinese Journal of Physics*, *77*, 1465–1478.

[19] Kuznetsov, A. V. (2010). The onset of nanofluid bioconvection in a suspension containing both nanoparticles and gyrotactic microorganisms. *International Communications in Heat and Mass Transfer*, *37*(10), 1421–1425.

[20] Kuznetsov, A. V. (2011). Nanofluid bioconvection in water-based suspensions containing nanoparticles and oxytactic microorganisms: oscillatory instability. *Nanoscale Research Letters*, *6*(1), 1–13.

[21] Nagendramma, V., Raju, C. S. K., Mallikarjuna, B., Shehzad, S. A., & Leelarathnam, A. (2018). 3D Casson nanofluid flow over slendering surface in a suspension of gyrotactic microorganisms with Cattaneo-Christov heat flux. *Applied Mathematics and Mechanics*, *39*(5), 623–638.

[22] Shehzad, S. A., Hayat, T., & Alsaedi, A. (2015). Influence of convective heat and mass conditions in MHD flow of nanofluid. *Bulletin of the Polish Academy of Sciences. Technical Sciences*, *63*(2), 465–474.

[23] Basak, T., & Chamkha, A. J. (2012). Heatline analysis on natural convection for nanofluids confined within square cavities with various thermal boundary conditions. *International Journal of Heat and Mass Transfer*, *55*(21–22), 5526–5543.

[24] Hayat, T., Muhammad, T., Alsaedi, A., & Alhuthali, M. S. (2015). Magnetohydrodynamic three-dimensional flow of viscoelastic nanofluid in the presence of nonlinear thermal radiation. *Journal of Magnetism and Magnetic Materials*, *385*, 222–229.

[25] Mebarek-Oudina, F., Aissa, A., Mahanthesh, B., & Öztop, H. F. (2020). Heat transport of magnetized Newtonian nanoliquids in an annular space between porous vertical cylinders with discrete heat source. *International Communications in Heat and Mass Transfer*, *117*, 104737.

[26] Swain, K., Mebarek-Oudina, F., & Abo-Dahab, S. M. (2022). Influence of MWCNT/Fe3O4 hybrid nanoparticles on an exponentially porous shrinking sheet with chemical reaction and slip boundary conditions. *Journal of Thermal Analysis and Calorimetry*, *147*(2), 1561–1570.

[27] Jawad, M., Mebarek-Oudina, F., Vaidya, H., & Prashar, P. (2022). Influence of bioconvection and thermal radiation on MHD Williamson nano Casson fluid flow with the swimming of gyrotactic microorganisms due to porous stretching sheet. *Journal of Nanofluids*, *11*(4), 500–509.

[28] Marzougui, S., Mebarek-Oudina, F., Magherbi, M., & Mchirgui, A. (2021). Entropy generation and heat transport of Cu–water nanoliquid in porous lid-driven cavity through magnetic field. *International Journal of Numerical Methods for Heat & Fluid Flow*, *32*(6), 2047–2069.

[29] Mebarek-Oudina, F., Fares, R., Aissa, A., Lewis, R. W., & Abu-Hamdeh, N. H. (2021). Entropy and convection effect on magnetized hybrid nano-liquid flow inside a trapezoidal cavity with zigzagged wall. *International Communications in Heat and Mass Transfer*, *125*, 105279.

[30] Hayat, T., Imtiaz, M., Alsaedi, A., & Kutbi, M. A. (2015). MHD three-dimensional flow of nanofluid with velocity slip and nonlinear thermal radiation. *Journal of Magnetism and Magnetic Materials*, *396*, 31–37.

[31] Jawad, M., Shehzad, K., & Safdar, R. (2021). Novel computational study on MHD flow of nanofluid flow with gyrotactic microorganism due to porous stretching sheet. *Punjab University Journal of Mathematics*, *52*(12).

[32] Jawad, M., Hameed, M. K., Majeed, A., & Nisar, K. S. (2022). Arrhenius energy and heat transport activates effect on gyrotactic microorganism flowing in maxwell bio-nanofluid with nield boundary conditions. *Case Studies in Thermal Engineering*, 102574.

21 3D Boundary Layer Flow of Conducting Nanoliquid Over a Stretching Sheet with Homogeneous and Heterogeneous Reactions

B. C. Prasannakumara, K. Ramesh,
R. Naveen Kumar, and R. J. Punith Gowda

21.1 INTRODUCTION

Nanofluids have made industrial operations, heat exchangers, electrical devices, and cooling processes more efficient because of their ability to transmit heat more effectively. Particles of 1–100 nm in size dispersed in base fluids yield nanofluids, which were proposed by Choi. It is common for researchers to view nanoliquid as a distinct homogeneous phase that encompasses the thermophysical properties of nanoliquids to simulate their qualities. Titanium dioxide (TiO_2) is found in nature as the minerals anatase, rutile, and brookite. Titanium dioxide is used in everyday life products such as disposable wrappers, which assist customers in protecting food from contamination. Coffee creamer, confectionery, white sauces, and cake decorations all include titanium dioxide, which has a white tint. It is a food additive in the form of E171 [1–2], which aids in the preservation of food for a long time. Recently, numerous researchers have explored the suspension of TiO_2 in different base liquids [3–5]. The thermal conductivity value measured experimentally differs significantly from the value predicted using a conventional fluid model. The explanation for this discrepancy is that the influence of nanoparticle (NP) aggregation kinematics was overlooked. The fractal model, which depicts the interfacial patterns formed by fluid molecules interacting with NPs, sheds light on the effects of NP aggregation on the thermal conductivity ratio [6–7]. The authors proved the fractal dimension for nanoparticle clusters. They said that, the thermal conductivity, was determined mainly by the cluster formation and NP interaction processes. Recently, Sabu et al. [8] investigated the dynamics of NP aggregation in a convective nanomaterial stream passing over an inclined flat plate. Mahanthesh [9] quizzed the heat transport of NPs in the presence of NP aggregation. By using NPs aggregation, Mackolil and Mahanthesh [10] swotted the magnetic field influence on the Marangoni convective stream of nanoliquid.

The magnetic field was extensively utilized in semiconductor crystal formation as well as casting methods to reduce unwanted heat convection and mass transport oscillations in the melt. Lund et al. [11] investigated the radiation and dissipation impacts on Casson liquid stream passing through a surface and swotted the stability analysis. Kumar et al. [12] investigated the chemically reactive Casson nanoliquid stream passing across a coiled sheet. The effects of radiation on the Marangoni convective stream of nanoliquid above a permeable surface were explained by Jawad et al. [13]. The dissipative flow of a liquid with the suspension of dual nanoparticles across a porous surface with a magnetic effect was quizzed by Lund et al. [14]. By considering the viscous dissipation and magnetic influence, Shah et al. [15] explained the nanoliquid flow between the porous plates.

DOI: 10.1201/9781003299608-21

In the industrial manufacturing, radiant heat transfer flow is critical for designing reliable equipments, nuclear power plants, gas turbines and other propulsion mechanisms for missiles, aeroplanes, satellites, and space vehicles. Inspired by these applications, numerous scholars explored the flow of different liquids with radiation effect. By including bio-convection, Al-Khaled et al. [16] elaborated on the radiative flow of nanoliquid. The radiative stream of Williamson nanoliquid over a stretchable geometry was explained by Hashim et al. [17]. The radiative stream of nanoliquid with heat production/absorption was studied by Ali et al. [18]. The radiative flow of nanoliquid across a melting surface was described by Reddy et al. [19]. The encouragement of radiation and magnetic dipole on the bio-convective flow of Jeffery nanofluid was explained by Ijaz et al. [20].

Chemically reacting systems incorporating homogeneous–heterogeneous (H-H) processes include biochemical frameworks, burning, and catalysis. Particularly perplexing is the link between H-H processes. With the exception of the presence of a catalyst, a proportion of reactions may proceed slowly. Chemical reactions are also employed in food processing, hydrometallurgical processes, fog formation, polymer manufacture, and ceramics, among other applications. The H-H reactions in the Sutterby liquid stream over the spinning disk were swotted by Khan et al. [21]. Rashid et al. [22] quizzed how H-H reactions influenced the Oldroyd-B liquid flow via a permeable medium. The effects of H-H reactions were imparted on a fluid stream embedded with CNTs running over an increasingly expanding surface by Hayat et al. [23]. The features of H-H reactions on a liquid stream on a revolving disk with a heat sink or source was quizzed by Abbas et al. [24]. Ali et al. [25] investigated the flow of a cross liquid through a stretchy surface with H-H responses.

Melting is a phase change that is accompanied by the absorption of thermal energy. The thermal engineering applications of melting heat transport in diverse liquids include oil extraction, geothermal energy recovery, magma solidification, permafrost melting, and semiconductor material preparation. By establishing a systematic and straightforward approach for calculating melting rates in difficult flow scenarios, Epstein cleared the path for melting transfer research. Khan et al. [26] modelled the Falkner–Skan nanofluid flow with the melting process. Amirsom et al. [27] explored the thermophysical characteristics of MHD nanofluid. Farooq et al. [28] inspected the melting heat transmission capacity of an unstable compressed nanofluid travelling through a Darcy porous medium. They also examined the melting potential of the stretched sheet. Radhika et al. [29] quizzed the melting heat transport in a Cu-Fe2SO4-H20 hybrid nanofluid. Using a nanofluid model, Mabood et al. [30] investigated melting heat transfer.

Through the previously cited articles, the current study examines the effect of H-H chemical reactions on an electrically conducting stream of nanoliquid through a stretching surface with NPs aggregation that has not yet been investigated. However, no numerical solution has previously been examined for the impact of H-H chemical reaction effect on the flow of nanoliquid past a stretching sheet. The present chapter's main focus is on numerically examining the earlier-stated flow.

21.2 MATHEMATICAL FORMULATION

Consider a three-dimensional steady boundary layer flow of an incompressible electrically conducting nanoliquid (TiO_2-ethylene glycol-based nanofluid) over a stretching sheet induced due to stretching of the sheet with $u_w(x) = cx$ and $v_w(y) = cy$ velocity in x- and y-directions, correspondingly. The flow takes place in the domain $z > 0$. Heat transfer analysis is taken into account in the presence of thermal radiation. Moreover, the H-H reactions and melting effect are also considered. We assume that the temperature of the melting surface is less than the ambient temperature ($T_m < T_\infty$). Here, we use Chaudhary and Merkin's [31–32] simple boundary layer flow model for the interaction among H-H processes involving dual chemical species A^* and B^*, as follows:

$$A^* + 2B^* \rightarrow 3B^*, rate = k_r ab^2 \tag{21.1}$$

$$A^* \to B^*, rate = k_s a \tag{21.2}$$

After utilizing boundary layer approximations, one can have the following problem statements (see [33–35]):

$$\frac{\partial u}{\partial x} + \frac{\partial v}{\partial y} + \frac{\partial w}{\partial z} = 0 \tag{21.3}$$

$$u\frac{\partial u}{\partial x} + v\frac{\partial u}{\partial y} + w\frac{\partial u}{\partial z} = \nu_{nf}\left(\frac{\partial^2 u}{\partial z^2}\right) - \frac{\nu_{nf}}{K}u - \frac{\sigma_{nf}}{\rho_{nf}}B_0^2 u \tag{21.4}$$

$$u\frac{\partial v}{\partial x} + v\frac{\partial v}{\partial y} + w\frac{\partial v}{\partial z} = \nu_{nf}\left(\frac{\partial^2 v}{\partial z^2}\right) - \frac{\nu_{nf}}{K}v - \frac{\sigma_{nf}}{\rho_{nf}}B_0^2 v \tag{21.5}$$

$$u\frac{\partial T}{\partial x} + v\frac{\partial T}{\partial y} + w\frac{\partial T}{\partial z} = \frac{k_{nf}}{(\rho C_p)_{nf}}\left(\frac{\partial^2 T}{\partial z^2}\right) + \frac{1}{(\rho C_p)_{nf}}\frac{16\sigma^* T_\infty^3}{3k^*}\frac{\partial}{\partial z}\left(\frac{\partial T}{\partial z}\right) \tag{21.6}$$

$$u\frac{\partial a}{\partial x} + v\frac{\partial a}{\partial y} + w\frac{\partial a}{\partial z} = D_A\left[\frac{\partial^2 a}{\partial z^2}\right] - k_r ab^2 \tag{21.7}$$

$$u\frac{\partial b}{\partial x} + v\frac{\partial b}{\partial y} + w\frac{\partial b}{\partial z} = D_B\left[\frac{\partial^2 b}{\partial z^2}\right] + k_r ab^2 \tag{21.8}$$

The boundary constraints for the current flow model are given as:

$$\left.\begin{aligned} &Z = 0 : u = U_w(x) = cx, v = V_w(y) = dy, k_{nf}\left(\frac{\partial T}{\partial z}\right) = \rho_{nf}\left(c_s\left(T_m - T_0\right) + \lambda\right)w, T = T_m, \\ &D_A\frac{\partial a}{\partial z} = k_s a, D_B\frac{\partial b}{\partial z} = -k_s a, \\ &Z \to \infty : u \to 0, v \to 0, T \to T_\infty, a \to a_0, b \to 0. \end{aligned}\right\} \tag{21.9}$$

Transformations are taken as follows:

$$\left.\begin{aligned} &u = cxf'(\eta) = U_w(x)f'(\eta), v = cyg'(\eta), w = -\sqrt{c\nu_f}\left(f(\eta) + g(\eta)\right), \\ &\theta(\eta) = \frac{T - T_m}{T_\infty - T_m}, a = a_0 h(\eta), b = a_0 j(\eta), \eta = z\sqrt{\frac{c}{\nu_f}}. \end{aligned}\right\} \tag{21.10}$$

21.3 THERMOPHYSICAL PROPERTIES FOR AGGREGATION APPROACH

It is commonly known that nanofluids have a high thermal conductivity, based on experimental evidence. Furthermore, the haphazard motion of NPs or aggregation of NPs generating percolation behaviour may be used to improve the thermal property. When compared to aggregation, which increases the mass of aggregates, Brownian randomness deteriorates, but aggregate percolation behaviour may increase thermal conductivity. As a result, the effective viscosity, density, heat capacitance, and thermal conductivity of nanofluid for nanoparticle aggregation are stated as follows (see Refs. [36–38]):

$$\mu_{nf} = \mu_f \left(1 - \frac{\phi_{agg}}{\phi_{\max}}\right)^{-2.5*\phi_{\max}} \tag{21.11}$$

$$\rho_{nf} = \left(1 - \phi_{agg}\right)\rho_f + (\phi\rho)_{agg} \tag{21.12}$$

$$\left(\rho C_p\right)_{nf} = \left(1 - \phi_{agg}\right)\left(\rho C_p\right)_f + \phi_{agg}\left(\rho C_p\right)_{agg} \tag{21.13}$$

$$k_{nf} = k_f \left(\frac{k_{agg} + 2k_f + 2\phi_{agg}\left(k_{agg} - k_f\right)}{k_{agg} + 2k_f - \phi_{agg}\left(k_{agg} - k_f\right)}\right) \tag{21.14}$$

$$\sigma_{nf} = \left(1 + \frac{3\left(\frac{\sigma_s}{\sigma_f} - 1\right)\phi_{agg}}{\left(\frac{\sigma_s}{\sigma_f} + 2\right) - \left(\frac{\sigma_s}{\sigma_f} - 1\right)\phi_{agg}}\right)\sigma_f \tag{21.15}$$

21.4 THERMAL CHARACTERISTICS OF PARTICLES AGGREGATION

The correlations for the effective viscosity and thermal conductivity are taken from the modified Krieger and Dougherty model and the modified Maxwell model, respectively (see refs. [36–38]):

$$\phi_{agg} = \frac{\phi}{\phi_{\text{int}}}, \phi_{\text{int}} = \left(\frac{R_{agg}}{R_p}\right)^{D-3} \tag{21.16}$$

$$\rho_{agg} = \left(1 - \phi_{\text{int}}\right)\rho_f + \phi_{\text{int}}\rho_s \tag{21.17}$$

$$\left(\rho C_p\right)_{agg} = \left(1 - \phi_{\text{int}}\right)\left(\rho C_p\right)_f + \phi_{\text{int}}\left(\rho C_p\right)_s \tag{21.18}$$

$$k_{agg} = \frac{k_f}{4}\left(\begin{array}{l}\left[3\phi_{\text{int}} - 1\right]\frac{k_s}{k_f} + \left[3\left(1 - \phi_{\text{int}}\right) - 1\right] + \\ \left[\left[\left(3\phi_{\text{int}} - 1\right)\frac{k_s}{k_f} + \left(3\left(1 - \phi_{\text{int}}\right) - 1\right)\right]^2 + \frac{8k_s}{k_f}\right]^{\frac{1}{2}}\end{array}\right) \tag{21.19}$$

The maximum particle packing fraction is $\phi_{\max} = 0.605$ for particles of spherical shape. From the theory of fractal, R_p and R_{agg} correspond to primary NPs and radii of aggregates (the value of $\frac{R_{agg}}{R_p}$ is taken as 3.34). D is the fractal index, which takes the value 1.8 in general for spherical particles. The more realistic Maxwell and Bruggeman model can be used to precisely estimate the effective thermal conductivity of TiO_2-ethylene glycol nanoliquid.

Using equation (21.10), equation (21.3) is fulfilled automatically, and the remaining equations (21.4)–(21.8) are reduced to the following forms:

$$\left.\varepsilon_1 f''' - f'^2 + \left(ff'' + gf''\right) - \left(\varepsilon_1 K^* + \varepsilon_2 \frac{\sigma_{nf}}{\sigma_f} M\right) f' = 0\right\} \tag{21.20}$$

$$\left.\varepsilon_1 g''' - g'^2 + \left(fg'' + gg''\right) - \left(\varepsilon_1 K^* + \varepsilon_2 \frac{\sigma_{nf}}{\sigma_f} M\right) g' = 0\right\} \tag{21.21}$$

$$\varepsilon_3 \left(\frac{k_{nf}}{k_f} + R\right)\theta'' + \Pr\left(f\theta' + g\theta'\right) = 0 \tag{21.22}$$

$$\frac{1}{Sc} h'' + (f+g)h' - Hhj^2 = 0, \tag{21.23}$$

$$\frac{\delta}{Sc} j'' + \left(f+g\right) j' + Hhj^2 = 0. \tag{21.24}$$

Where:

$$\varepsilon_1 = \frac{\left(1 - \frac{\phi_{agg}}{\phi_{\max}}\right)^{-2.5*\phi_{\max}}}{\left(1-\phi_{agg}\right) + \phi_{agg}\frac{\rho_{agg}}{\rho_f}}, \; \varepsilon_2 = \frac{1}{\left(1-\phi_{agg}\right) + \phi_{agg}\frac{\rho_{agg}}{\rho_f}}, \; \varepsilon_3 = \frac{1}{\left(1-\phi_{agg}\right) + \phi_{agg}\frac{\left(\rho C_p\right)_{agg}}{\left(\rho C_p\right)_f}}. \tag{21.25}$$

And the reduced boundary constraints take the following form:

$$\left.\begin{aligned} &f'(0) = 1, g'(0) = \alpha, \varepsilon_3 \frac{k_{nf}}{k_f} M_e \theta'(0) + \Pr\left(f(0) + g(0)\right) = 0, \theta(0) = 0, h'(0) = \gamma h(0), \\ &\delta j'(0) = -\gamma h(0), f'(\infty) \to 0, g'(\infty) \to 0, \theta(\infty) \to 1, h(\infty) \to 1, j(\infty) \to 0. \end{aligned}\right\} \tag{21.26}$$

Where:

$$Sc = \frac{\nu_f}{D_A}, \delta = \frac{D_B}{D_A}, \gamma = \frac{k_s}{D_A}\sqrt{\frac{\nu_f}{c}}, H = \frac{k_r a_0^2}{c}, M = \frac{\sigma_f B_0^2}{\rho_f c}, \alpha = \frac{d}{c},$$

$$K^* = \frac{\nu_f}{Kc}, M_e = \frac{C_{pf}(T_\infty - T_m)}{\lambda + C_s(T_m - T_0)}, R = \frac{16\sigma^* T_\infty^3}{3k^* k_f}, \Pr = \frac{\rho_f \nu_f \left(C_p\right)_f}{k_f}.$$

The A^* and B^* are assumed to be of identical magnitude in this case. As a result of this logic, we must assume that the D_A and D_B are equivalent, that is, $\delta = 1$, and thus:

$$h(\eta) + j(\eta) = 1 \tag{21.27}$$

Now, by using equation (21.27) in equations (21.23) and (21.24), we obtain:

$$\frac{1}{Sc} h'' + (f+g)h' - Hh\left(1-h\right)^2 = 0. \tag{21.28}$$

Along with the boundary conditions:

$$h'(0) = \gamma h(0), h(\infty) \to 1. \tag{21.29}$$

The skin friction coefficients and local Nusselt number, which are specified by the following relations, are the most relevant physical variables for the problem from an environmental perspective:

$$C_{fx} \mathrm{Re}_x^{\frac{1}{2}} = \left(1 - \frac{\phi_{agg}}{\phi_{\max}}\right)^{-2.5*\phi_{\max}} f''(0), \tag{21.30}$$

$$C_{fy} \mathrm{Re}_y^{\frac{1}{2}} = \left(1 - \frac{\phi_{agg}}{\phi_{\max}}\right)^{-2.5*\phi_{\max}} g''(0), \tag{21.31}$$

$$Nu \mathrm{Re}_x^{-\frac{1}{2}} = -\left(\frac{k_{nf}}{k_f} + R\right)\theta'(0). \tag{21.32}$$

Where $\mathrm{Re}_x = \frac{U_w(x+y)}{\nu_f}$ and $\mathrm{Re}_y = \frac{V_w(x+y)}{\nu_f}$ are the local Reynolds number along the x- and y-directions, respectively.

21.5 RESULTS AND DISCUSSION

This section aims to go through the physical characteristics of the figures and explain the real mechanism behind the flow, heat, and mass transport changes generated by the major dimensionless parameters. The RKF-45 procedure and the shooting approach are used to solve the reduced ODEs. In this study, two alternative scenarios are considered: with and without NPs aggregation. In the graphs, dashed curves represent flow without NP aggregation, and solid lines represent flow with NP aggregation. A wide range of parameter values are chosen to replicate the process to test the suggested model's insight thoroughly. Changes in velocity, heat, and mass transfer with skin friction and Nusselt number may all be seen with the proper graphs. Furthermore, the present study's numerical solutions are compared to previous research, and the results are found to be quite similar (see Tables 21.1 and 21.2).

Figures 21.1 and 21.2 show the effect of M on the velocities $f'(\eta)$ and $g'(\eta)$, respectively, for two different cases, namely, with and without NPs aggregation. The $f'(\eta)$ and $g'(\eta)$ decrease for both cases as the value of M increases. On both velocity distribution functions, the M has a diminishing influence. M is a dimensionless parameter that is used to control the velocity of the liquid. The Lorentz force generated by the applied magnetic field opposes the fluid flow, causing both velocity functions to decrease. Furthermore, the $f'(\eta)$ and $g'(\eta)$ decline faster in the absence of nanoparticles aggregation for improved values of M. The influence of K^* on the $f'(\eta)$ and $g'(\eta)$ for two different cases is displayed in Figures 21.3 and 21.4, respectively. The rise in values of K^* declines both $f'(\eta)$ and $g'(\eta)$ for both cases. This is because the porous zone generates resistive power. When the porousness is vast, the liquid has more area to flow. The velocity decreases due to increased viscous force generated by augmented values of K^*. From both figures, we conclude that the velocity of the liquid declines faster in the absence of nanoparticles aggregation for improved values of K^*. Figure 21.5 depicts the effects of the α on non-dimensional velocity profile in response to various stretching ratio values. The stretching ratio is the ratio of the stretched sheet's transverse and axial velocity. As the stretching ratio increases, the transverse velocity gets bigger than the axial velocity. As shown in Figure 21.5, a growth in the α value causes the $g'(\eta)$ to decrease. Here,

TABLE 21.1
Comparision of $-f''(0)$ and $-g''(0)$ Values for Some Reduced Cases

		$-f''(0)$	$-g''(0)$
$\alpha = 0$	Wang [39]	1	0
	Shehzad et al. [40]	1	0
	Present results	1	0
$\alpha = 0.25$	Wang [39]	1.048813	0.194564
	Shehzad et al. [40]	1.04881	0.19457
	Present results	1.048815	0.194565
$\alpha = 0.50$	Wang [39]	1.093097	0.465205
	Shehzad et al. [40]	1.09309	0.46522
	Present results	1.093098	0.465204
$\alpha = 0.75$	Wang [39]	1.134485	0.794622
	Shehzad et al. [40]	1.13450	0.79462
	Present results	1.134486	0.794623
$\alpha = 1.0$	Wang [39]	1.173720	1.173720
	Shehzad et al. [40]	1.17372	1.17372
	Present results	1.173721	1.173721

TABLE 21.2
Comparison of $\theta'(0)$ Values for Some Reduced Cases

Pr	1	5	10
Liu et al. [41]	−0.54964375	−1.5212390	−2.25742372
Nayak et al. [42]	−0.5496446407	−1.5212389587	−2.2574236707
Swain and Mahanthesh [43]	−0.54964207	−1.52130241	−2.25741543
Present results	−0.549642	−1.521395	−2.257419

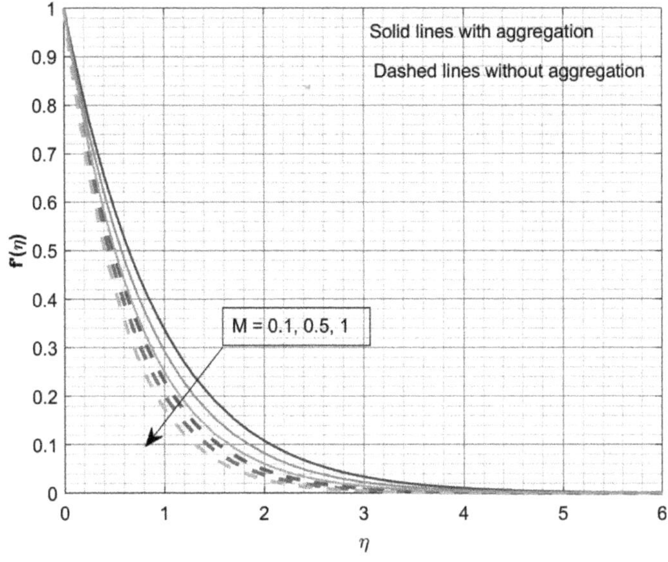

FIGURE 21.1 The impact of M on $f'(\eta)$.

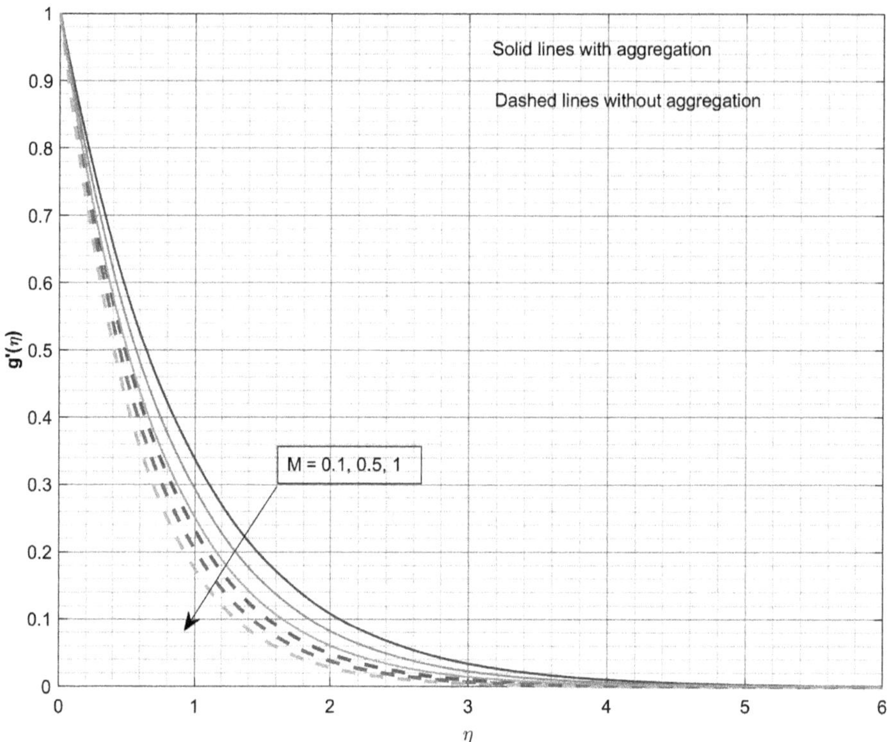

FIGURE 21.2 The impact of M on $g'(\eta)$.

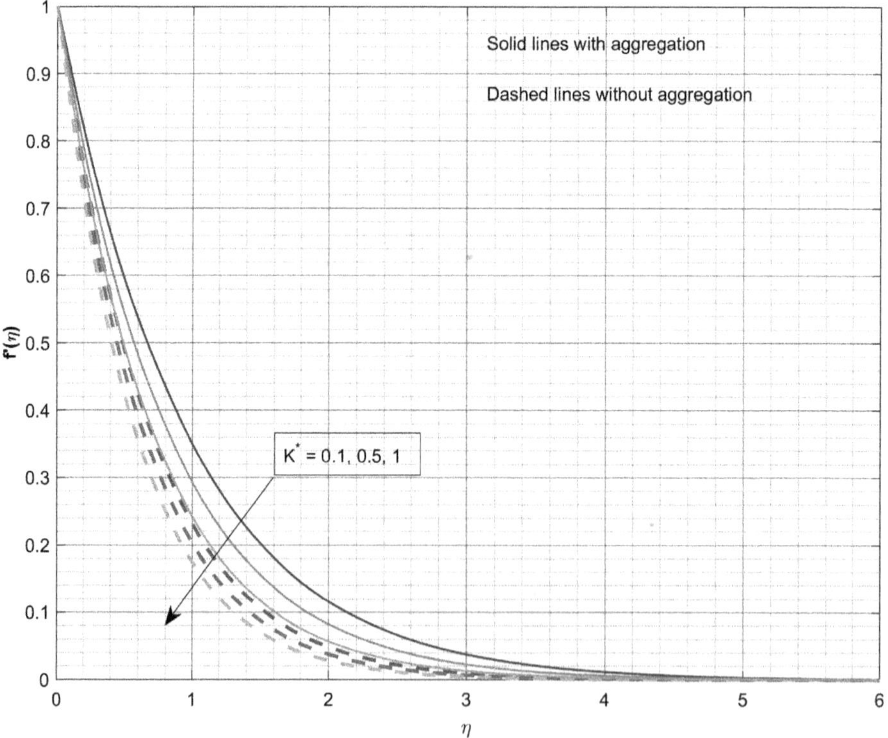

FIGURE 21.3 The impact of K^{*} on $f'(\eta)$.

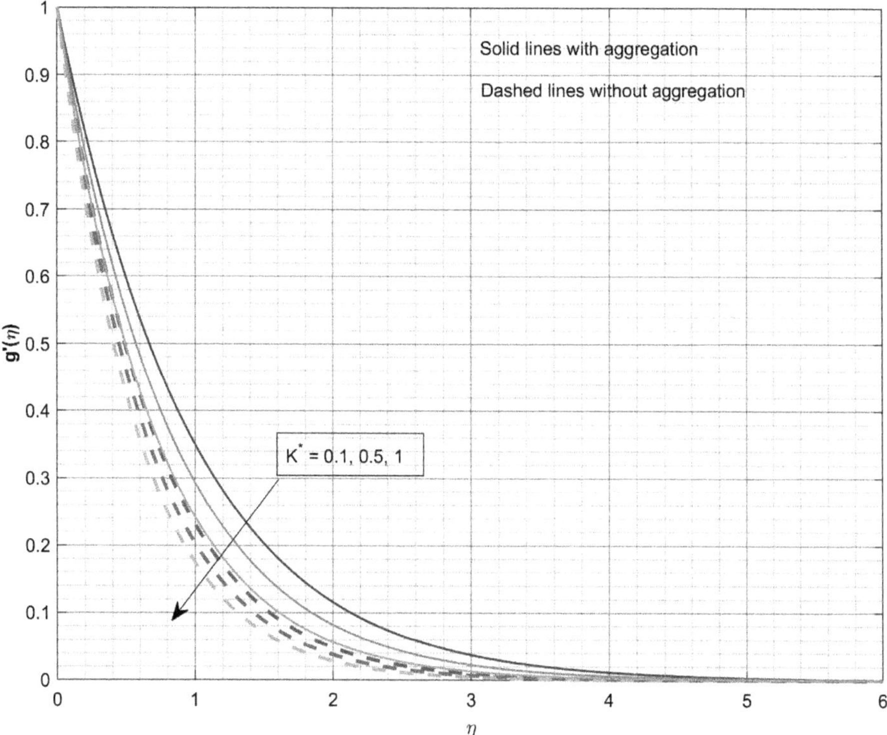

FIGURE 21.4 The impact of K^* on $g'(\eta)$.

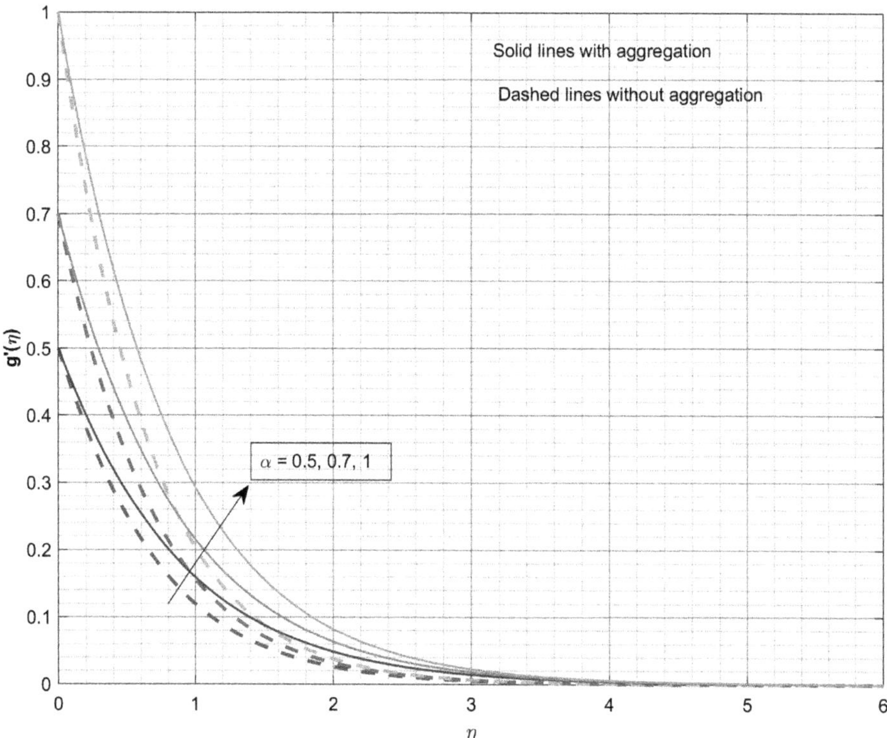

FIGURE 21.5 The impact of α on $g'(\eta)$.

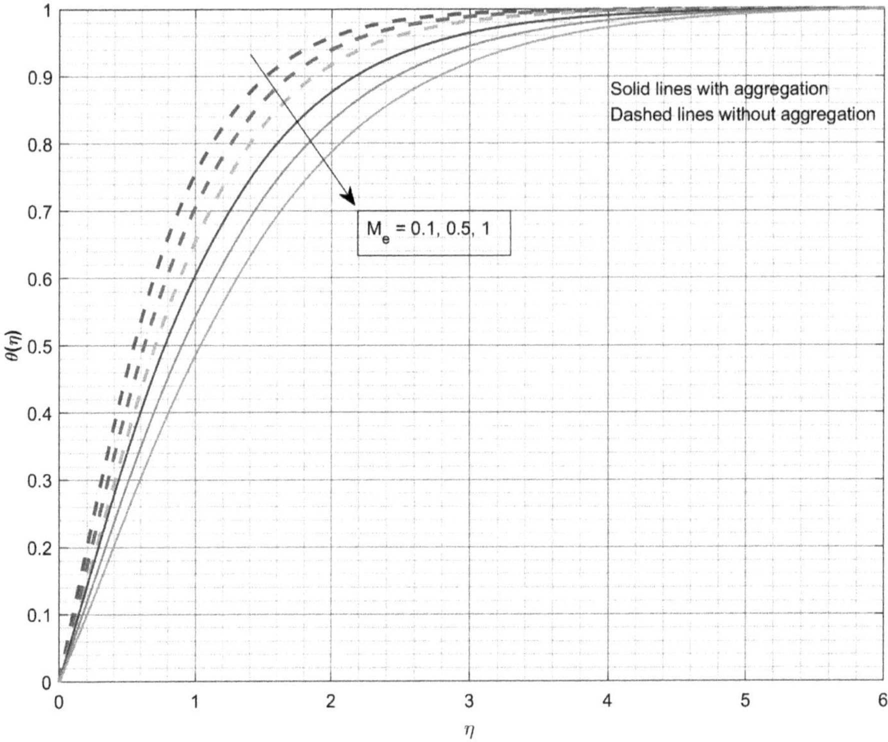

FIGURE 21.6 The impact of M_e on $\theta(\eta)$.

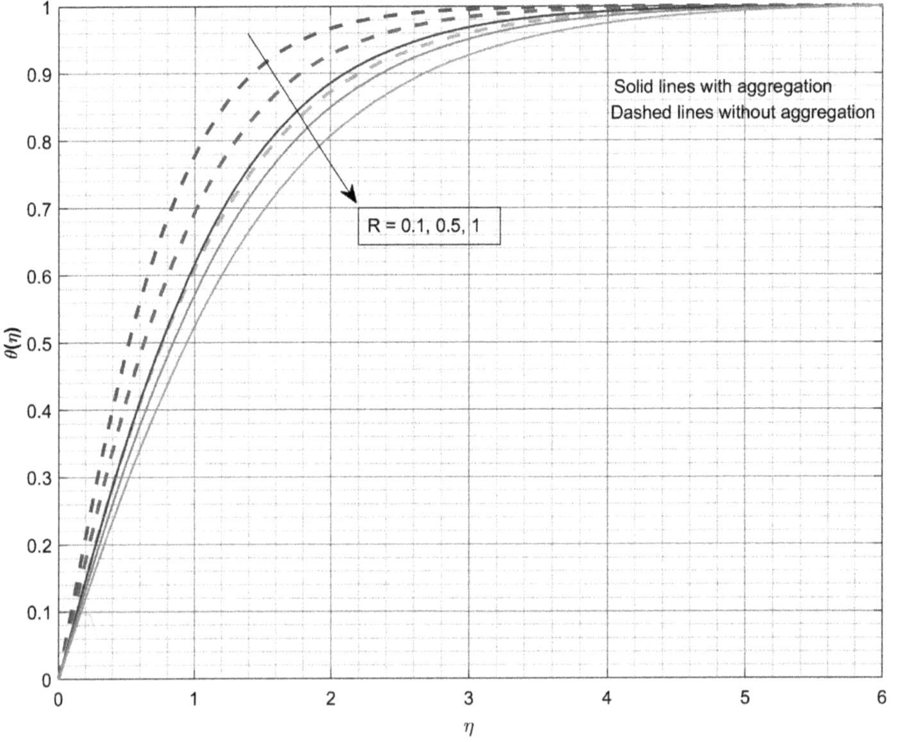

FIGURE 21.7 The impact of R on $\theta(\eta)$.

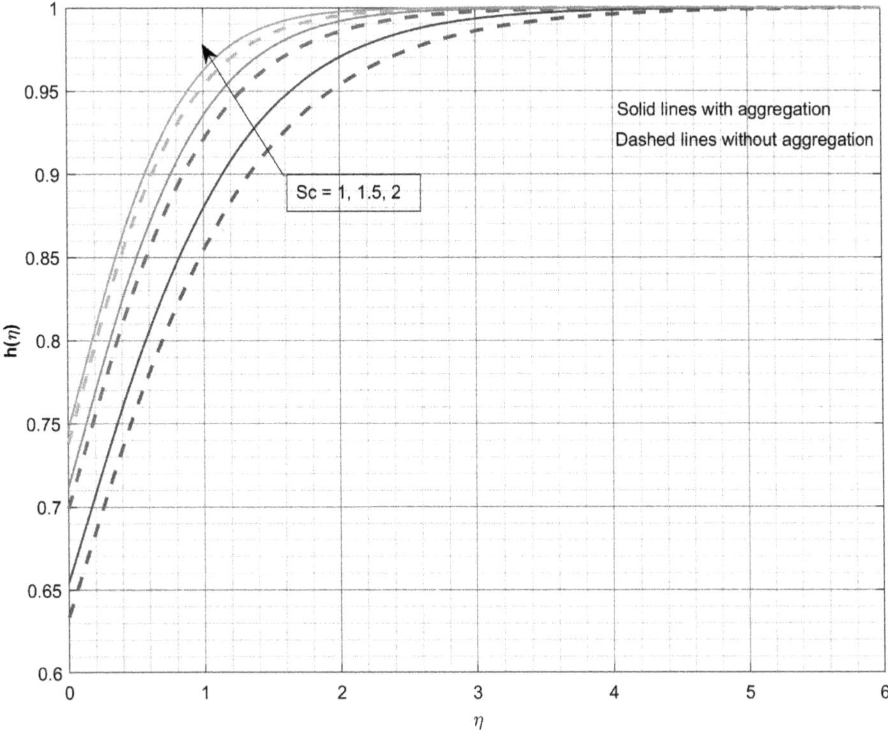

FIGURE 21.8 The impact of Sc on $h(\eta)$.

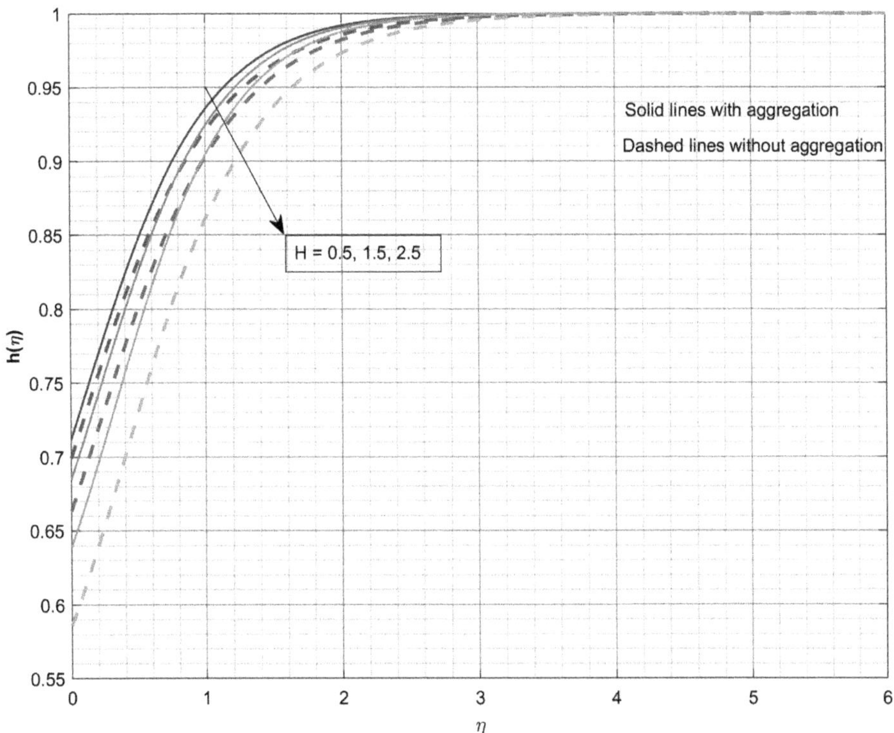

FIGURE 21.9 The impact of H on $h(\eta)$.

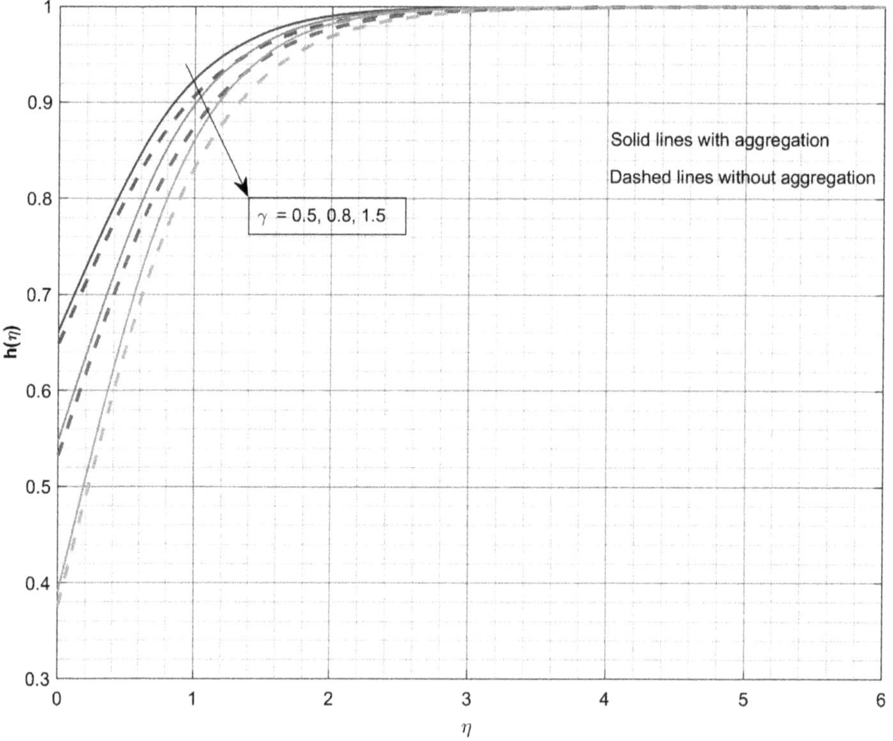

FIGURE 21.10 The impact of γ on $h(\eta)$.

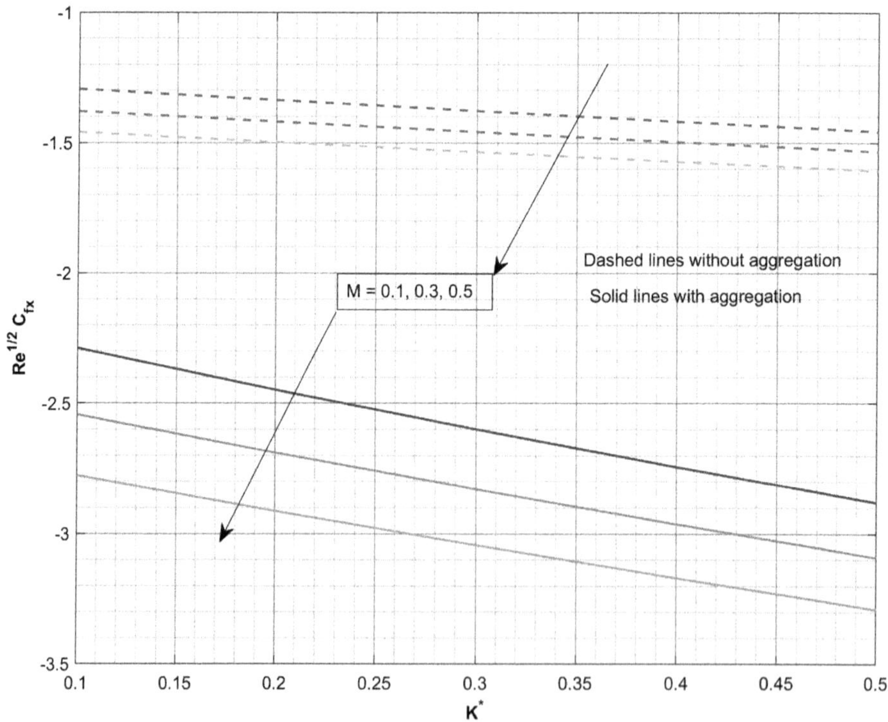

FIGURE 21.11 The impact of M on $\mathrm{Re}^{1/2} C_{fx}$ versus K^{*}.

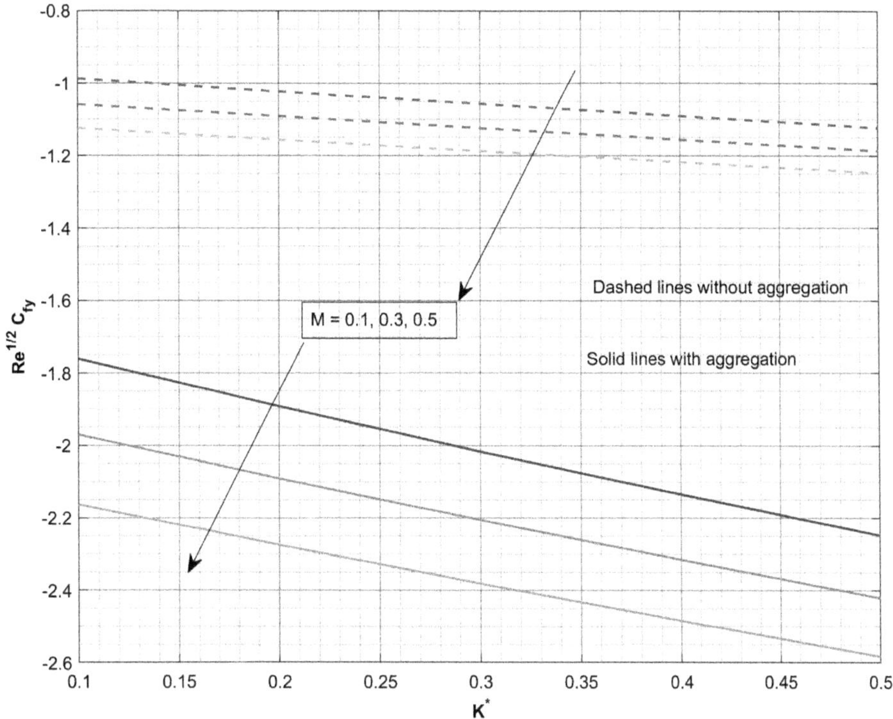

FIGURE 21.12 The impact of K^* on $\mathrm{Re}^{1/2} C_{fy}$ versus M.

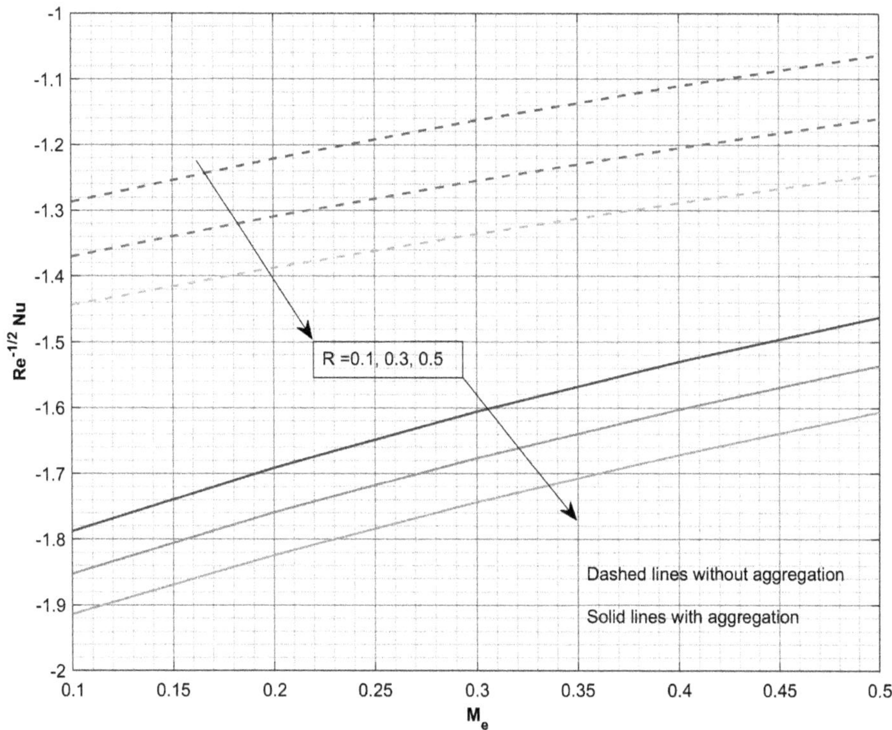

FIGURE 21.13 The impact of R on $\mathrm{Re}^{-1/2} Nu$ versus M_e.

the velocity of the liquid improves faster in the presence of nanoparticles aggregation for improved values of α.

Figure 21.6 shows the significance of M_e on $\theta(\eta)$ for both cases. The gain in the value of M_e declines the $\theta(\eta)$ for both cases. Here, the larger M_e values correspond to an increased convective stream from the heated fluid to the cold surface, resulting in a reduction in heat transfer. Furthermore, the fluid flow in the absence of NPs aggregation exhibits better heat transmission than the other cases. With NPs aggregation, we see the least heat transfer for fluid flow. Figure 21.7 demonstrates the consequence of R on $\theta(\eta)$ for both cases. The upsurge in the value of R declines the $\theta(\eta)$ for both cases. Moreover, the fluid flow in the absence of NPs aggregation case shows better-quality heat transport for intensified values of R than the remaining case. Here, we observe the least heat transfer for fluid flow with NPs aggregation.

Figure 21.8 displays the effect of Sc on $h(\eta)$ for both cases. The upsurge in the value of Sc inclines the $h(\eta)$ for both cases. Momentum diffusivity rises as Sc rises, causing the mass transfer to the incline. Furthermore, fluid flow in the presence of NP aggregation displays better mass transfer than other fluid flow. We see the least mass transfer for assumed fluid flow without NP aggregation. The impact of H on $h(\eta)$ is presented in Figure 21.9 for both flow cases. The rise in the value of H deteriorations the $h(\eta)$ for both flow cases. The mass transfer depreciates as the reactants are consumed throughout the homogeneous reaction. Furthermore, for increased values of H, $h(\eta)$ for fluid flow in the presence of aggregation drops slower than for the other. In the presence of aggregation, we see increased mass transfer for fluid flow. The encouragement of γ on $h(\eta)$ is shown in Figure 21.10 for both flow cases. The increase in value of γ declines the $h(\eta)$ for both flow cases. An increase in the γ is helpful in increasing the chemical species concentration in this case. More chemical species are likely to be involved in a chemical reaction when the rate of change in velocity for a heterogeneous reaction speeds up. Furthermore, for increased values of γ, $h(\eta)$ for fluid flow in the presence of aggregation drops slower than for the other. In the presence of aggregation, we see increased mass transfer for fluid flow.

The influence of M on $\mathrm{Re}^{1/2}\, C_{fx}$ with respect to varied values of K^* is portrayed in Figure 21.11 for two different flow cases, as mentioned earlier. The rise in M and K^* values declines the $\mathrm{Re}^{1/2}\, C_{fx}$ for both cases. Furthermore, the fluid flow in the absence of NPs aggregation shows improved skin friction for augmented values of M and K^*. The influence of K^* on $\mathrm{Re}^{1/2}\, C_{fy}$ with respect to varied values of M is portrayed in Figure 21.12 for two different cases. The rise in M and K^* values declines the $\mathrm{Re}^{1/2}\, C_{fy}$ for both cases. Also, the fluid flow in the absence of NPs aggregation shows improved skin friction for augmented values of M and K^*. The effect of R on $\mathrm{Re}^{-1/2}\, Nu$ with respect to varied values of M_e is portrayed in Figure 21.13 for two different flow cases. The rise in values of R declines the $\mathrm{Re}^{-1/2}\, Nu$ for both cases, but a contrary trend is seen for enhanced M_e values. Moreover, the fluid flow in the absence of NPs aggregation shows an improved heat transport rate.

21.6 CONCLUSIONS

A three-dimensional laminar flow of an electrically conducting nanoliquid (TiO_2-EG based nanoliquid) over a stretching sheet is explored in this study. Also, two alternative scenarios are considered: the presence of NPs aggregation and the absence of NPs aggregation. The equations representing the specified flow are transformed to ODEs by choosing relevant similarity variables. The behaviour of flow profiles is clearly understood using a suitable numerical method (RKF-45 by adopting a shooting scheme). Graphs are used to investigate the impact of dimensionless factors on the involved profiles. The following are the study's most significant results:

- The velocity of the liquid declines faster in the absence of nanoparticles aggregation for improved values of M.

- The velocity of the liquid declines faster in the absence of nanoparticles aggregation for improved values of K^*.
- The fluid flow in the absence of NPs aggregation exhibits better heat transmission for increased values of M_e than the other case.
- The fluid flow in the absence of NPs aggregation case shows better-quality heat transfer for intensified values of R than the remaining case.
- The $g'(\eta)$ in the presence of aggregation condition declines slower for upward values of γ.
- The $g'(\eta)$ in the presence of aggregation condition declines slower for increasing values of H.
- The rise in values of R deteriorates the $\mathrm{Re}^{-1/2}\,Nu$ for both cases, but the inverse trend is seen for enhanced values of M_e. Additionally, the fluid flow in the absence of NPs aggregation shows an improved heat transport rate.

Nomenclature

K	porous medium permeability
α	stretching ratio parameter
T_∞	ambient temperature
k	thermal conductivity
k_r *and* k_s	rate constant
λ	latent heat of the fluid
T_0	solid temperature
K^*	porosity parameter
R	radiation parameter
k^*	mean absorption coefficient
M_e	melting parameter
(u, v, w)	constituents of velocity
C_{fx}, C_{fy}	skin frictions
T	temperature
T_m	temperature of the melting surface
Nu	Nusselt number
Pr	Prandtl number
ρ	density
U_w	stretching velocity
(x, y, z)	directions
ρC_p	heat capacitance
A^* and B^*	chemical species
D_A and D_B	diffusion coefficients
Sc	Schmidt number
v	dynamic viscosity
C_s	heat capacity of the solid surface
σ^*	Stefan–Boltzmann constant
$f'(\eta), g'(\eta)$	dimensionless velocity profiles
η	dimensionless variable
ψ	stream function

$\theta(\eta)$	dimensionless thermal profile
ϕ	volume fraction
Re	local Reynolds number
c,d	constants
μ	dynamic viscosity
H	strength of homogeneous reaction parameter
γ	strength of heterogeneous reaction parameter

Subscripts

f	fluid
agg	aggregate
nf	nanofluid
s	solid nanoparticle

REFERENCES

[1] M.-H. Ropers, H. Terrisse, M. Mercier-Bonin, and B. Humbert, *Titanium dioxide as food additive*. Intech Rijeka, 2017.

[2] A. Zeeshan, N. Shehzad, T. Abbas, and R. Ellahi, "Effects of radiative electro-magnetohydrodynamics diminishing internal energy of pressure-driven flow of titanium dioxide-water nanofluid due to entropy generation," *Entropy*, vol. 21, no. 3, Art. no. 3, Mar. 2019, http://doi.org/10.3390/e21030236.

[3] R. Naveen Kumar, R. J. P. Gowda, B. J. Gireesha, and B. C. Prasannakumara, "Non-Newtonian hybrid nanofluid flow over vertically upward/downward moving rotating disk in a Darcy–Forchheimer porous medium," *Eur. Phys. J. Spec. Top.*, vol. 230, no. 5, pp. 1227–1237, Jul. 2021, http://doi.org/10.1140/epjs/s11734-021-00054-8.

[4] R. J. Punith Gowda, R. Naveen Kumar, and B. C. Prasannakumara, "Two-phase darcy-forchheimer flow of dusty hybrid nanofluid with viscous dissipation over a cylinder," *Int. J. Appl. Comput. Math.*, vol. 7, no. 3, p. 95, May 2021, http://doi.org/10.1007/s40819-021-01033-2.

[5] M. R. Zangooee, Kh. Hosseinzadeh, and D. D. Ganji, "Hydrothermal analysis of MHD nanofluid (TiO2-GO) flow between two radiative stretchable rotating disks using AGM," *Case Stud. Therm. Eng.*, vol. 14, p. 100460, Sep. 2019, http://doi.org/10.1016/j.csite.2019.100460.

[6] B.-X. Wang, L.-P. Zhou, and X.-F. Peng, "A fractal model for predicting the effective thermal conductivity of liquid with suspension of nanoparticles," *Int. J. Heat Mass Transf.*, vol. 46, no. 14, pp. 2665–2672, Jul. 2003, http://doi.org/10.1016/S0017-9310(03)00016-4.

[7] H. Xie, M. Fujii, and X. Zhang, "Effect of interfacial nanolayer on the effective thermal conductivity of nanoparticle-fluid mixture," *Int. J. Heat Mass Transf.*, vol. 48, no. 14, pp. 2926–2932, Jul. 2005, http://doi.org/10.1016/j.ijheatmasstransfer.2004.10.040.

[8] A. S. Sabu, J. Mackolil, B. Mahanthesh, and A. Mathew, "Nanoparticle aggregation kinematics on the quadratic convective magnetohydrodynamic flow of nanomaterial past an inclined flat plate with sensitivity analysis," *Proc. Inst. Mech. Eng. Part E J. Process Mech. Eng.*, vol. 36, no. 3, Dec. 2021, http://doi.org/10.1177/09544089211056235.

[9] B. Mahanthesh, "Flow and heat transport of nanomaterial with quadratic radiative heat flux and aggregation kinematics of nanoparticles," *Int. Commun. Heat Mass Transf.*, vol. 127, p. 105521, Oct. 2021, http://doi.org/10.1016/j.icheatmasstransfer.2021.105521.

[10] J. Mackolil and B. Mahanthesh, "Inclined magnetic field and nanoparticle aggregation effects on thermal Marangoni convection in nanoliquid: A sensitivity analysis," *Chin. J. Phys.*, vol. 69, pp. 24–37, Feb. 2021, http://doi.org/10.1016/j.cjph.2020.11.006.

[11] L. A. Lund, Z. Omar, I. Khan, D. Baleanu, and K. S. Nisar, "Dual similarity solutions of MHD stagnation point flow of Casson fluid with effect of thermal radiation and viscous dissipation: Stability analysis," *Sci. Rep.*, vol. 10, no. 1, Art. no. 1, Sep. 2020, http://doi.org/10.1038/s41598-020-72266-2.

[12] R. S. V. Kumar, P. G. Dhananjaya, R. Naveen Kumar, R. J. P. Gowda, and B. C. Prasannakumara, "Modeling and theoretical investigation on Casson nanofluid flow over a curved stretching surface with the influence of magnetic field and chemical reaction," *Int. J. Comput. Methods Eng. Sci. Mech.*, pp. 1–8, Mar. 2021, http://doi.org/10.1080/15502287.2021.1900451.

[13] M. Jawad, A. Saeed, P. Kumam, Z. Shah, and A. Khan, "Analysis of boundary layer MHD Darcy-Forchheimer radiative nanofluid flow with soret and dufour effects by means of marangoni convection," *Case Stud. Therm. Eng.*, vol. 23, p. 100792, Feb. 2021, http://doi.org/10.1016/j.csite.2020.100792.

[14] L. A. Lund, Z. Omar, J. Raza, and I. Khan, "Magnetohydrodynamic flow of Cu–Fe3O4/H2O hybrid nanofluid with effect of viscous dissipation: Dual similarity solutions," *J. Therm. Anal. Calorim.*, vol. 143, no. 2, pp. 915–927, Jan. 2021, http://doi.org/10.1007/s10973-020-09602-1.

[15] Z. Shah, E. O. Alzahrani, W. Alghamdi, and M. Z. Ullah, "Influences of electrical MHD and Hall current on squeezing nanofluid flow inside rotating porous plates with viscous and joule dissipation effects," *J. Therm. Anal. Calorim.*, vol. 140, no. 3, pp. 1215–1227, May 2020, http://doi.org/10.1007/s10973-019-09176-7.

[16] K. Al-Khaled, S. U. Khan, and I. Khan, "Chemically reactive bioconvection flow of tangent hyperbolic nanoliquid with gyrotactic microorganisms and nonlinear thermal radiation," *Heliyon*, vol. 6, no. 1, p. e03117, Jan. 2020, http://doi.org/10.1016/j.heliyon.2019.e03117.

[17] Hashim, A. Hamid, M. Khan, and U. Khan, "Thermal radiation effects on Williamson fluid flow due to an expanding/contracting cylinder with nanomaterials: Dual solutions," *Phys. Lett. A*, vol. 382, no. 30, pp. 1982–1991, Aug. 2018, http://doi.org/10.1016/j.physleta.2018.04.057.

[18] U. Ali, M. Y. Malik, A. A. Alderremy, S. Aly, and K. U. Rehman, "A generalized findings on thermal radiation and heat generation/absorption in nanofluid flow regime," *Phys. Stat. Mech. Its Appl.*, vol. 553, p. 124026, Sep. 2020, http://doi.org/10.1016/j.physa.2019.124026.

[19] M. Gnaneswara Reddy, R. Punith Gowda, R. Naveen Kumar, B. Prasannakumara, and K. Ganesh Kumar, "Analysis of modified Fourier law and melting heat transfer in a flow involving carbon nanotubes," *Proc. Inst. Mech. Eng. Part E J. Process Mech. Eng.*, p. 09544089211001353, Mar. 2021, http://doi.org/10.1177/09544089211001353.

[20] M. Ijaz, S. Nadeem, M. Ayub, and S. Mansoor, "Simulation of magnetic dipole on gyrotactic ferromagnetic fluid flow with nonlinear thermal radiation," *J. Therm. Anal. Calorim.*, vol. 143, no. 3, pp. 2053–2067, Feb. 2021, http://doi.org/10.1007/s10973-020-09856-9.

[21] M. I. Khan, M. W. A. Khan, S. Ahmad, T. Hayat, and A. Alsaedi, "Transportation of homogeneous–heterogeneous reactions in flow of Sutterby fluid confined between two co-axially rotating disks," *Phys. Scr.*, vol. 95, no. 5, p. 055211, Feb. 2020, http://doi.org/10.1088/1402-4896/ab4627.

[22] S. Rashid, M. I. Khan, T. Hayat, M. Ayub, and A. Alsaedi, "Theoretical and analytical analysis of shear rheology of Oldroyd-B fluid with homogeneous–heterogeneous reactions," *Appl. Nanosci.*, vol. 10, no. 8, pp. 3035–3043, Aug. 2020, http://doi.org/10.1007/s13204-019-01037-x.

[23] T. Hayat, A. Aziz, T. Muhammad, and A. Alsaedi, "Significance of homogeneous–heterogeneous reactions in Darcy–Forchheimer three-dimensional rotating flow of carbon nanotubes," *J. Therm. Anal. Calorim.*, vol. 139, no. 1, pp. 183–195, Jan. 2020, http://doi.org/10.1007/s10973-019-08316-3.

[24] S. Z. Abbas, W. A. Khan, M. Waqas, M. Irfan, and Z. Asghar, "Exploring the features for flow of Oldroyd-B liquid film subjected to rotating disk with homogeneous/heterogeneous processes," *Comput. Methods Programs Biomed.*, vol. 189, p. 105323, Jun. 2020, http://doi.org/10.1016/j.cmpb.2020.105323.

[25] M. Ali, F. Sultan, M. Shahzad, and W. A. Khan, "Influence of homogeneous-heterogeneous reaction model for 3D Cross fluid flow: A comparative study," *Indian J. Phys.*, vol. 95, no. 2, pp. 315–323, Feb. 2021, http://doi.org/10.1007/s12648-020-01706-6.

[26] W. A. Khan, M. Ali, M. Irfan, M. Khan, M. Shahzad, and F. Sultan, "A rheological analysis of nanofluid subjected to melting heat transport characteristics," *Appl. Nanosci.*, vol. 10, no. 8, pp. 3161–3170, Aug. 2020, http://doi.org/10.1007/s13204-019-01067-5.

[27] N. A. Amirsom, M. J. Uddin, M. F. Md Basir, A. Kadir, O. A. Bég, and A. I. Md. Ismail, "Computation of melting dissipative magnetohydrodynamic nanofluid bioconvection with second-order slip and variable thermophysical properties," *Appl. Sci.*, vol. 9, no. 12, Art. no. 12, Jan. 2019, http://doi.org/10.3390/app9122493.

[28] M. Farooq, S. Ahmad, M. Javed, and A. Anjum, "Melting heat transfer in squeezed nanofluid flow through darcy forchheimer medium," *J. Heat Transf.*, vol. 141, no. 1, Oct. 2018, http://doi.org/10.1115/1.4041497.

[29] M. Radhika, R. J. P. Gowda, R. Naveenkumar, Siddabasappa, and B. C. Prasannakumara, "Heat transfer in dusty fluid with suspended hybrid nanoparticles over a melting surface," *Heat Transf.*, vol. 50, no. 3, pp. 2150–2167, 2021, https://doi.org/10.1002/htj.21972.

[30] F. Mabood, T. A. Yusuf, and W. A. Khan, "Cu–Al2O3–H2O hybrid nanofluid flow with melting heat transfer, irreversibility analysis and nonlinear thermal radiation," *J. Therm. Anal. Calorim.*, vol. 143, no. 2, pp. 973–984, Jan. 2021, http://doi.org/10.1007/s10973-020-09720-w.

[31] M. A. Chaudhary and J. H. Merkin, "A simple isothermal model for homogeneous-heterogeneous reactions in boundary-layer flow. I Equal diffusivities," *Fluid Dyn. Res.*, vol. 16, no. 6, p. 311, Nov. 1995, http://doi.org/10.1016/0169-5983(95)00015-6.

[32] M. A. Chaudhary and J. H. Merkin, "Homogeneous-heterogeneous reactions in boundary-layer flow: Effects of loss of reactant," *Math. Comput. Model.*, vol. 24, no. 3, pp. 21–28, Aug. 1996, http://doi.org/10.1016/0895-7177(96)00097-0.

[33] T. Hayat, M. Rashid, and A. Alsaedi, "Three-dimensional radiative flow of magnetite-nanofluid with homogeneous-heterogeneous reactions," *Results Phys.*, vol. 8, pp. 268–275, Mar. 2018, http://doi.org/10.1016/j.rinp.2017.11.038.

[34] T. Hayat, K. Muhammad, A. Alsaedi, and S. Asghar, "Numerical study for melting heat transfer and homogeneous-heterogeneous reactions in flow involving carbon nanotubes," *Results Phys.*, vol. 8, pp. 415–421, Mar. 2018, http://doi.org/10.1016/j.rinp.2017.12.023.

[35] B. Mahanthesh, B. J. Gireesha, and R. S. R. Gorla, "Nonlinear radiative heat transfer in MHD three-dimensional flow of water based nanofluid over a non-linearly stretching sheet with convective boundary condition," *J. Niger. Math. Soc.*, vol. 35, no. 1, pp. 178–198, Apr. 2016, http://doi.org/10.1016/j.jnnms.2016.02.003.

[36] R. Ellahi, M. Hassan, and A. Zeeshan, "Aggregation effects on water base Al2O3—nanofluid over permeable wedge in mixed convection," *Asia-Pac. J. Chem. Eng.*, vol. 11, no. 2, pp. 179–186, 2016, http://doi.org/10.1002/apj.1954.

[37] N. Acharya, K. Das, and P. K. Kundu, "Effects of aggregation kinetics on nanoscale colloidal solution inside a rotating channel," *J. Therm. Anal. Calorim.*, vol. 138, no. 1, pp. 461–477, Oct. 2019, http://doi.org/10.1007/s10973-019-08126-7.

[38] L. Th. Benos, E. G. Karvelas, and I. E. Sarris, "Crucial effect of aggregations in CNT-water nanofluid magnetohydrodynamic natural convection," *Therm. Sci. Eng. Prog.*, vol. 11, pp. 263–271, Jun. 2019, http://doi.org/10.1016/j.tsep.2019.04.007.

[39] C. Y. Wang, "The three-dimensional flow due to a stretching flat surface," *Phys. Fluids*, vol. 27, no. 8, pp. 1915–1917, Aug. 1984, http://doi.org/10.1063/1.864868.

[40] S. A. Shehzad, A. Alsaedi, T. Hayat, and M. S. Alhuthali, "Thermophoresis particle deposition in mixed convection three-dimensional radiative flow of an Oldroyd-B fluid," *J. Taiwan Inst. Chem. Eng.*, vol. 45, no. 3, pp. 787–794, May 2014, http://doi.org/10.1016/j.jtice.2013.08.022.

[41] I.-C. Liu, H.-H. Wang, and Y.-F. Peng, "Flow and heat transfer for three-dimensional flow over an exponentially stretching surface," *Chem. Eng. Commun.*, vol. 200, no. 2, pp. 253–268, Jan. 2013, http://doi.org/10.1080/00986445.2012.703148.

[42] M. K. Nayak, S. Shaw, and A. J. Chamkha, "3D MHD free convective stretched flow of a radiative nanofluid inspired by variable magnetic field," *Arab. J. Sci. Eng.*, vol. 44, no. 2, pp. 1269–1282, Feb. 2019, http://doi.org/10.1007/s13369-018-3473-y.

[43] K. Swain and B. Mahanthesh, "Thermal enhancement of radiating magneto-nanoliquid with nanoparticles aggregation and joule heating: A three-dimensional flow," *Arab. J. Sci. Eng.*, vol. 46, no. 6, pp. 5865–5873, Jun. 2021, http://doi.org/10.1007/s13369-020-04979-5.

22 Analysis of Chemically Reactive Casson Nanofluid Flow Over a Stretched Sheet with Joule Heating and Viscous Dissipation

Abha Kumari, Kharabela Swain, and Renuprava Dalai

22.1 INTRODUCTION

Casson fluid is a non-Newtonian fluid because of its rheological properties with respect to the shear stress–strain relationship. At low shear and strain, and above a critical stress value, it behaves like an elastic solid; otherwise, it behaves like a Newtonian fluid. Mukhopadhyay et al. [1] investigated Casson fluid flow over an unsteady stretching surface. Pramanik [2] investigated the heat transfer analysis towards an exponentially stretching surface in the presence of suction or injection. Ramesh et al. [3] studied the Casson fluid flow near the stagnation point over a stretching sheet in the presence of thermal radiation. El-Aziz and Afify [4] examined the entropy generation of Casson fluid flow over an extending sheet. Das et al. [5] investigated heat and mass transfer in a porous medium using an unsteady-flow Casson fluid past a flat plate. Several researchers [6–17] have investigated Newtonian and non-Newtonian fluid flow over a stretching sheet. The flow of an incompressible Newtonian/non-Newtonian fluid over a stretching surface has a significant impact on several technological processes, such as polymer extraction in a melt spinning process, glass blowing, and continuous metal casting, etc. Crane [18] found a closed-form similarity solution to a flow due to stretching surface. Mahapatra and Gupta [19] studied the viscous stagnation point flow towards a stretching sheet. In recent years, the study of non-Newtonian fluid has gained more importance due to its industrial applications. Further, Misra and Sinha [20] studied the biological applications of flow on stretching surface.

In view of the aforementioned literature survey, the objective of the present analysis is laid down as follows:

- The momentum transport equation of Casson nanofluid has been modified due to temperature- as well as space-dependent free stream (potential flow) stretching.
- Further, the inclusion of two body forces, one of electromagnetic force, and another is the permeability of the saturated porous medium embedding the stretching sheet through which flow occurs.
- The inclusion of Brownian motion, thermophoresis, thermal radiation, heat source, Joulian, and viscous dissipation in energy equation.
- The presence of chemical reaction in concentration equation.
- Further, the present analysis brings to its fold the viscous flow by letting $\gamma \to \infty$ and constant surface condition by letting the coefficients $T_0, C_0 \to 0$.

The inclusion of the aforementioned criteria makes the analysis susceptible to a variety of technological and/or therapeutic applications. Suitable similarity transforms convert partial differential

DOI: 10.1201/9781003299608-22

equations (PDEs) into nonlinear ordinary differential equations (ODEs) and solved numerically by MATLAB software using bvp4c code with reasonable accuracy. The validity of the current results is supported by work reported in text/published papers.

22.2 PROBLEM DESCRIPTION

Consider an unsteady two-dimensional flow of an electrically conducting and chemically reactive Casson nanofluid flow over a stretched sheet entrenched in a porous matrix. The plate placed along the x-axis and y-axis is normal to it (Figure 22.1). The flow confined to the plane $y > 0$ is due to elongated bounding surface and free stream. We assume that the magnetic Reynolds number of the fluid is very small so that the induced magnetic field effects are neglected as compared to applied magnetic field. Furthermore, the buoyancy forces are also neglected.

The rheological equation of state for an isotropic and incompressible flow of a Casson fluid [21] is expressed as:

$$\tau_{ij} = \begin{cases} 2\left(\mu_B + \dfrac{p_y}{\sqrt{2\pi}}\right)e_{ij}, \pi > \pi_c \\ 2\left(\mu_B + \dfrac{p_y}{\sqrt{2\pi_c}}\right)e_{ij}, \pi < \pi_c \end{cases} \tag{22.1}$$

Where $e_{ij} = \dfrac{1}{2}\left(\dfrac{\partial u_i}{\partial x_j} + \dfrac{\partial u_j}{\partial x_i}\right)$ is the rate of strain tensor, τ_{ij} is the component of stress tensor, μ_B is the Casson coefficient of viscosity, $\pi = e_{ij}e_{ij}$ is the product of the rate of strain tensor with itself, π_c is the critical value of the product of the rate of strain tensor with itself, and p_y is the yield stress of the fluid.

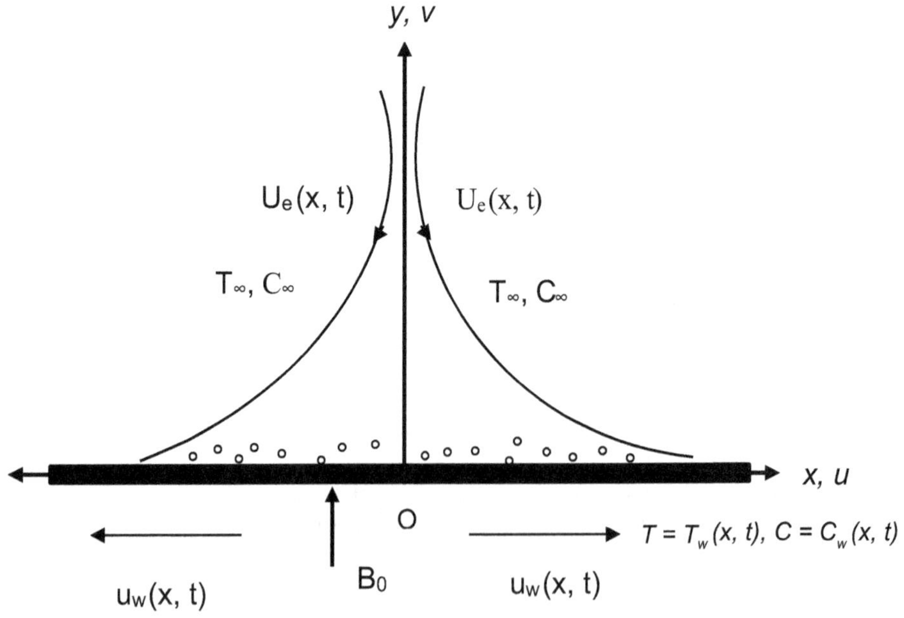

FIGURE 22.1 Flow geometry.

The continuity, momentum, energy, and concentration boundary layer equations with prescribed boundary conditions of the unsteady incompressible Casson nanofluid flow following [22] are given by:

$$\frac{\partial u}{\partial x}+\frac{\partial v}{\partial y}=0 \tag{22.2}$$

$$\frac{\partial u}{\partial t}+u\frac{\partial u}{\partial x}+v\frac{\partial u}{\partial y}=\frac{\partial U_e}{\partial t}+U_e\frac{\partial U_e}{\partial x}+\upsilon_f\left(1+\frac{1}{\gamma}\right)\frac{\partial^2 u}{\partial y^2}-\frac{\sigma B_0^2}{\rho_f}(u-U_e)-\frac{\upsilon_f}{K_p^*}(u-U_e) \tag{22.3}$$

$$\begin{aligned}\frac{\partial T}{\partial t}+u\frac{\partial T}{\partial x}+v\frac{\partial T}{\partial y}&=\left(\alpha+\frac{16\sigma^* T_\infty^3}{3k^*}\right)\frac{\partial^2 T}{\partial y^2}+\tau\left[D_B\frac{\partial T}{\partial y}\frac{\partial C}{\partial y}+\frac{D_T}{T_\infty}\left(\frac{\partial T}{\partial y}\right)^2\right]\\&+\frac{\sigma B_0^2 u^2}{(\rho c)_f}+\frac{\mu_f}{(\rho c)_f}\left(1+\frac{1}{\gamma}\right)\left(\frac{\partial u}{\partial y}\right)^2+\frac{Q}{(\rho c)_f}(T-T_\infty)\end{aligned} \tag{22.4}$$

$$\frac{\partial C}{\partial t}+u\frac{\partial C}{\partial x}+v\frac{\partial C}{\partial y}=D_B\frac{\partial^2 C}{\partial y^2}+\left(\frac{D_T}{T_\infty}\right)\frac{\partial^2 T}{\partial y^2}-Kc^*\left(C-C_\infty\right) \tag{22.5}$$

$$\left.\begin{aligned}&u=u_w(x,t)=\frac{ax}{1-\lambda t},v=0,T=T_w(x,t),C=C_w(x,t),\ at\ y=0\\&u=U_e(x,t)=\frac{bx}{1-\lambda t},T\to T_\infty,C\to C_\infty,\qquad as\ y\to\infty\end{aligned}\right\} \tag{22.6}$$

Here, the stretching velocity $u_w(x,t)=\frac{ax}{1-\lambda t}$, the free stream velocity $U_e(x,t)=\frac{bx}{1-\lambda t}$, and the wall temperature $\left(T_W\right)$ and nanoparticle volume fraction $\left(C_W\right)$ are respectively given by $T_w(x,t)=T_\infty+T_0\left[\frac{ax^2}{(1-\lambda t)}\right]$ and $C_w(x,t)=C_\infty+C_0\left[\frac{ax^2}{(1-\lambda t)}\right]$, respectively.

Introducing the following similarity variables, transformations, and parameters in equations (22.2)–(22.6), we get:

$$\left.\begin{aligned}&\eta=\sqrt{\frac{a}{\upsilon_f(1-\lambda t)}}y,\psi=\sqrt{\frac{a\upsilon_f}{(1-\lambda t)}}xf(\eta),u=\frac{\partial\psi}{\partial y},v=-\frac{\partial\psi}{\partial x},\beta=\frac{b}{a},A=\frac{\lambda}{a},M=\frac{\sigma B_0^2}{\rho_f a},\\&T=T_\infty+T_0\left[\frac{ax^2}{(1-\lambda t)^2}\right]\theta(\eta),C=C_\infty+C_0\left[\frac{ax^2}{(1-\lambda t)^2}\right]\phi(\eta),\Pr=\frac{\upsilon_f}{\alpha},R=\frac{16T_\infty^3\sigma^*}{3k^*k},Sc=\frac{\upsilon_f}{D_B},\\&K_p=\frac{\upsilon_f}{K_p^*a},Kc=\frac{Kc^*}{a},Nb=\frac{\tau D_B(C_w-C_\infty)}{\upsilon_f},Nt=\frac{\tau D_T(T_w-T_\infty)}{\upsilon_f T_\infty},Ec=\frac{u_w^2}{\left(c_p\right)_f\left(T_w-T_\infty\right)}\end{aligned}\right\} \tag{22.7}$$

Equation (22.2) is identically satisfied, and equations (22.3)–(22.6) become:

$$\left(1+\frac{1}{\gamma}\right)f'''+ff''-f'^2-\left(M+K_p\right)\left(f'-\beta\right)+\beta^2-A\left(\frac{1}{2}\eta f''+f'-\beta\right)=0 \tag{22.8}$$

$$\frac{1}{\Pr}(1+R)\theta''+f\theta'-2f'\theta+Nb\theta'\phi'+Nt\theta'^2$$
$$+MEcf'^2+Ec\left(1+\frac{1}{\gamma}\right)f''^2+S\theta-A\left(\frac{1}{2}\eta\theta'+2\theta\right)=0 \tag{22.9}$$

$$\phi''+Sc\left[f\phi'-2f'\phi-Kc\phi-A\left(\frac{1}{2}\eta\phi'+2\phi\right)\right]+\frac{Nt}{Nb}\theta''=0 \tag{22.10}$$

$$\left.\begin{array}{ll} f'(\eta)=1, f(\eta)=0, \theta(\eta)=1, \phi(\eta)=1 & at\ \eta=0 \\ f'(\eta)=\beta, \theta(\eta)=0, \phi(\eta)=0 & as\ \eta\rightarrow\infty \end{array}\right\} \tag{22.11}$$

The quantities of physical interest in this problem are the skin friction coefficient C_f, the local Nusselt number Nu_x, and the local Sherwood number Sh_x, which are defined as follows:

$$C_f=\frac{\mu}{\rho_f U_w^2}\left(\frac{\partial u}{\partial y}\right)_{y=0}\Rightarrow \mathrm{Re}_x^{0.5}C_{fx}=f''(0) \tag{22.12}$$

$$Nu_x=\frac{-x}{(T_w-T_\infty)}\left[\frac{\partial T}{\partial y}-\frac{4\sigma^*}{3k'^*}\left(\frac{\partial T^4}{\partial y}\right)\right]_{y=0}\Rightarrow \mathrm{Re}_x^{-0.5}Nu_x=-(1+R)\theta'(0) \tag{22.13}$$

$$Sh_x=\frac{-x}{(C_w-C_\infty)}\left(\frac{\partial C}{\partial y}\right)_{y=0}\Rightarrow \mathrm{Re}_x^{-0.5}Sh_x=-\phi'(0) \tag{22.14}$$

Where $\mathrm{Re}_x=\frac{u_w x}{\upsilon_f}$ is the local Reynolds number.

The present problem reduces to steady-state flow when $A=0$.

22.3 RESULTS AND DISCUSSION

The set of nonlinear coupled ordinary differential equations (22.8)–(22.10) with boundary conditions (22.11) are solved numerically by MATLAB software using bvp4c code. Due to the stretching ratio parameter $\beta\ (\beta<1)$, the inverted boundary layer is formed, and hence, the effects of parameters are reversed. During numerical simulations, the default values of the parameters are taken as $M=K_p=\gamma=0.5, Nb=Nt=R=Ec=S=0.1,\ A=Kc=0.3, \Pr=5, n=2$, and $Sc=1$, unless otherwise specified. To check the validity of the integration procedure, the values of $f''(0)$ have been computed for various values of β when $M=Kp=A=0, \gamma\rightarrow\infty$ in Table 22.1, and it is remarked that our results are in good agreement with the results presented by [19, 23–24].

Figures 22.2 and 22.3 show the effects of magnetic parameter and stretching ratio parameters on velocity distribution. The boundary layer is formed when $\beta>1$, that is, with higher rate of free stream stretching than bounding surface. The inverted boundary layer is formed for $\beta<1$. Due to resistive Lorentz force and the presence of a porous matrix, the magnetic and porosity parameters reduce velocity, respectively, and the effect is reversed in the case of an inverted boundary layer. The Casson parameter reduces velocity, resulting in a thinning of the boundary layer.

Figures 22.5–22.8 show the temperature distribution for various parameter values. It can be seen that increasing the Eckert number and radiation parameter raises the temperature as they contribute to more thermal energy (Figures 22.5 and 22.6). An increase in Ec contributes to higher temperature. Since Ec is the measure of heat energy addition due to viscous dissipation, it can also be seen that the temperature rises with higher values of the heat source parameter and falls in the presence of a sink (Figure 22.7). According to Figure 22.8, increasing the Brownian motion parameter increases the thermal energy, and thus the temperature, but the reverse effect is well marked in

TABLE 22.1
Comparison of $f''(0)$ for Different Values of β When $M = Kp = A = 0$, $\gamma \to \infty$

	$f''(0)$			
β	Mustafa et al. [23]	Mahapatra and Gupta [19]	Oyelakin et al. [24]	Present Study
0.1	−0.96939	−0.9694	−0.96937	−0.96951
0.2	−0.918107	−0.9181	−0.918111	−0.91816
0.5	−0.66735	−0.6673	−0.66726	−0.66733
2	2.01757	2.0175	2.01750	2.01758
3	4.72964	4.7294	4.72928	4.72931

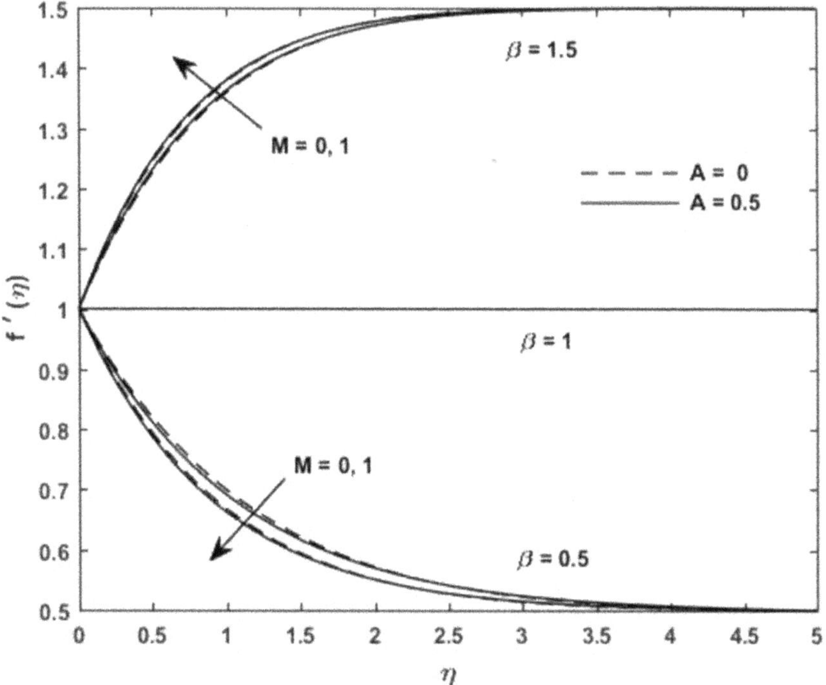

FIGURE 22.2 Velocity profiles for various values of M and β .

the case of solutal concentration/volume fraction. In case of thermophoresis parameter (Nt), both temperature and volume fraction get accelerated with thermophoretic processes.

Figures 22.9 and 22.10 portray the solutal concentration of nanoparticle. It is seen that higher stretching as well as higher Schmidt number (Sc) deplete the concentration level. Because both higher-rate stretching and Schmidt number (heavier species) slow mass diffusion, this results in a thinner solutal boundary layer. The same effects are observed for Kc (Figure 22.10). Upon careful observation, it is further seen that slight instability is marked in concentration distribution for low Sc, that is, for lighter species and Casson parameter (γ).

Table 22.2 shows the variations of $-f''(0)$, $-\theta'(0)$, and $-\phi'(0)$ for different values of parameters. It is seen that for fixed values of other parameters, the wall shear stress $\{-f''(0)\}$ increases with the increase in the values of M and S, whereas it decreases with increase in the value of β. Therefore, it is suggested that the magnetic intensity is to be reduced to decrease the shear at the bounding surface. Further, it is concluded that the higher the unsteadiness, the greater the shearing stress at the bounding

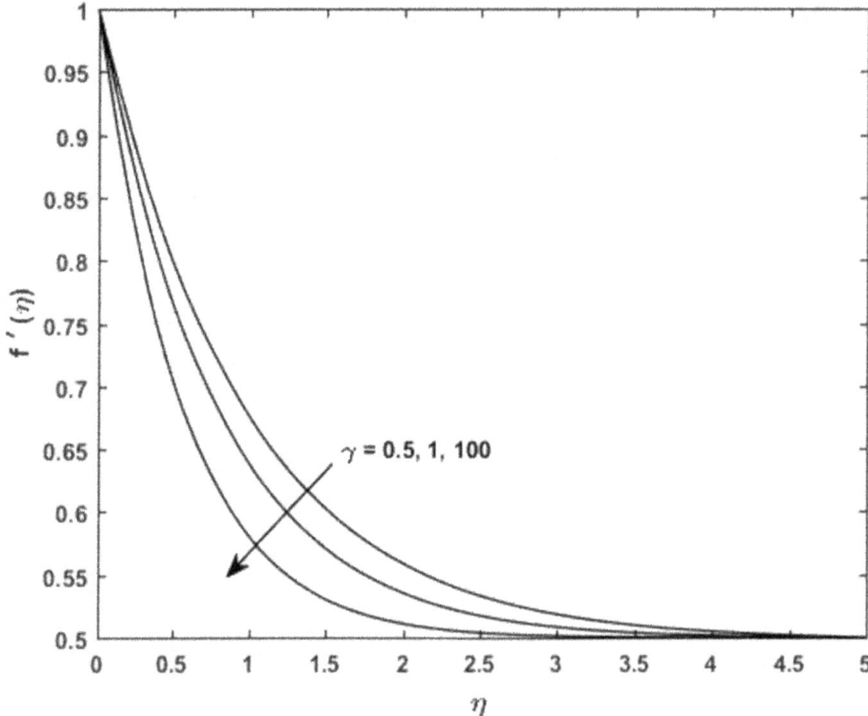

FIGURE 22.3 Velocity profiles for various values of γ when $A = 0.5$.

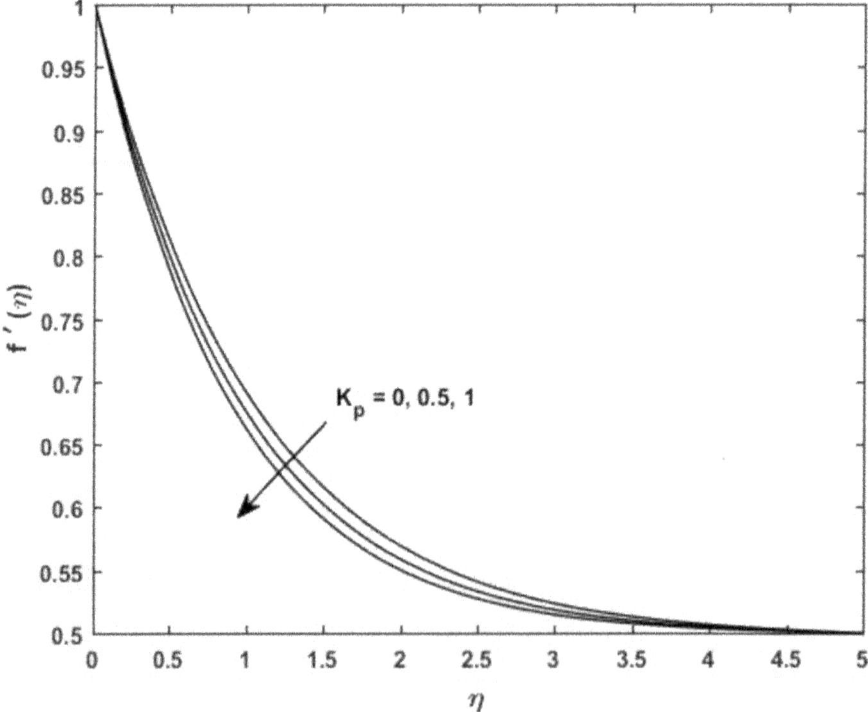

FIGURE 22.4 Velocity profiles for various values of K_p.

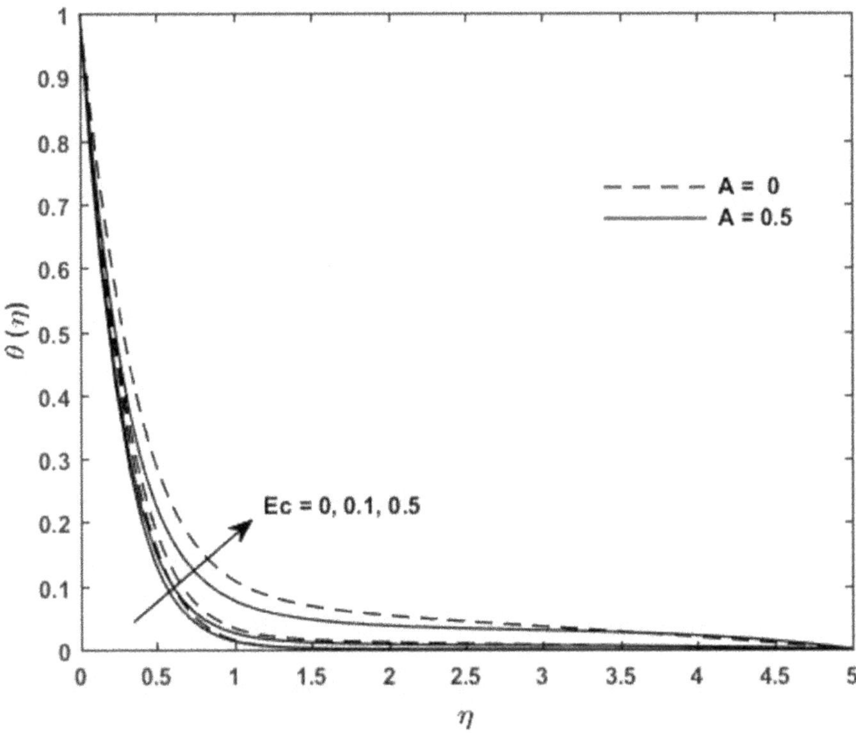

FIGURE 22.5 Temperature profiles for various values of *Ec*.

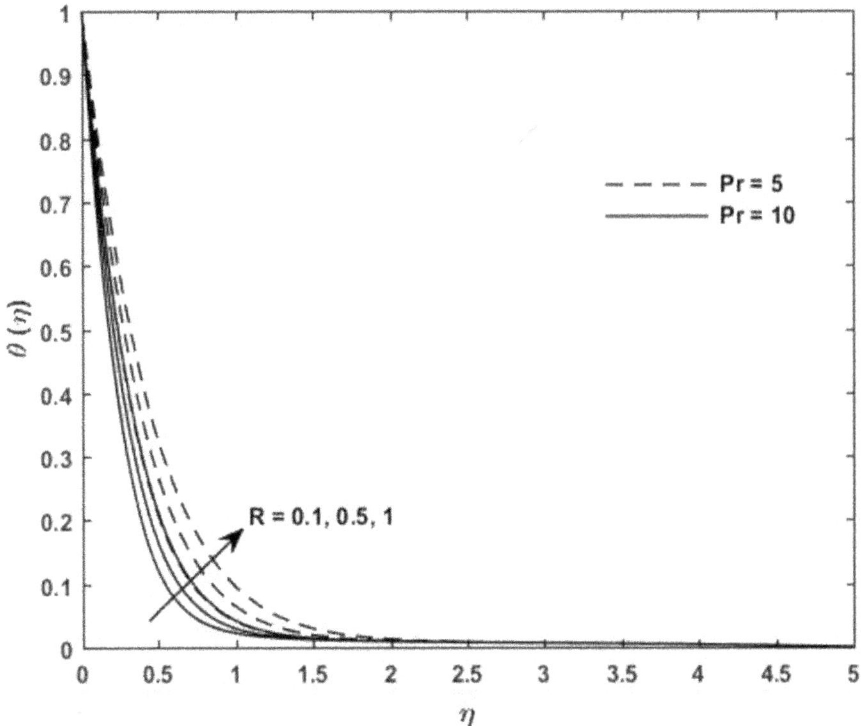

FIGURE 22.6 Temperature profiles for various values of *R* and Pr.

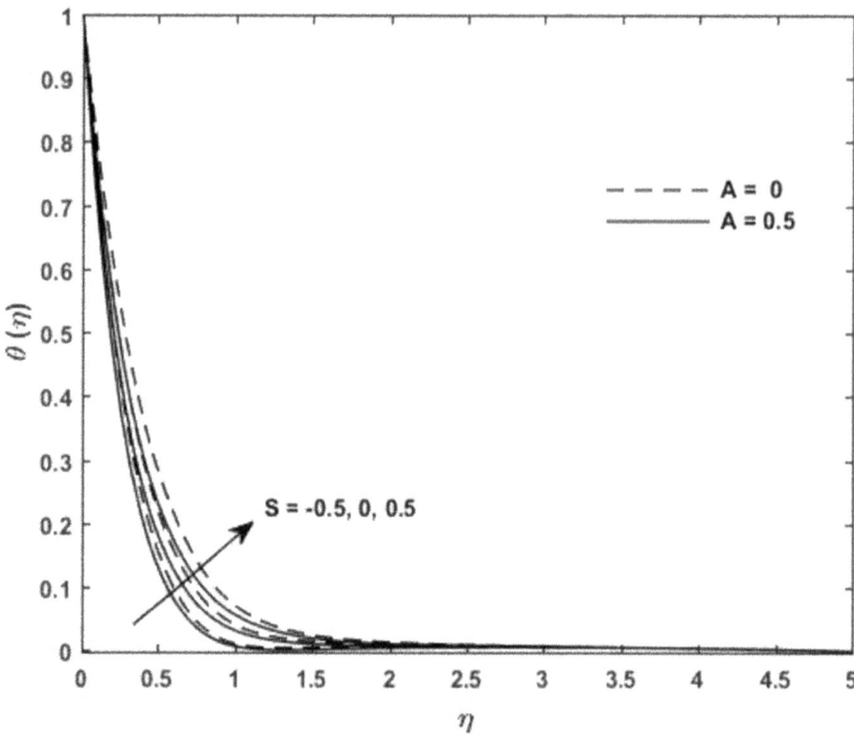

FIGURE 22.7 Temperature profiles for various values of S.

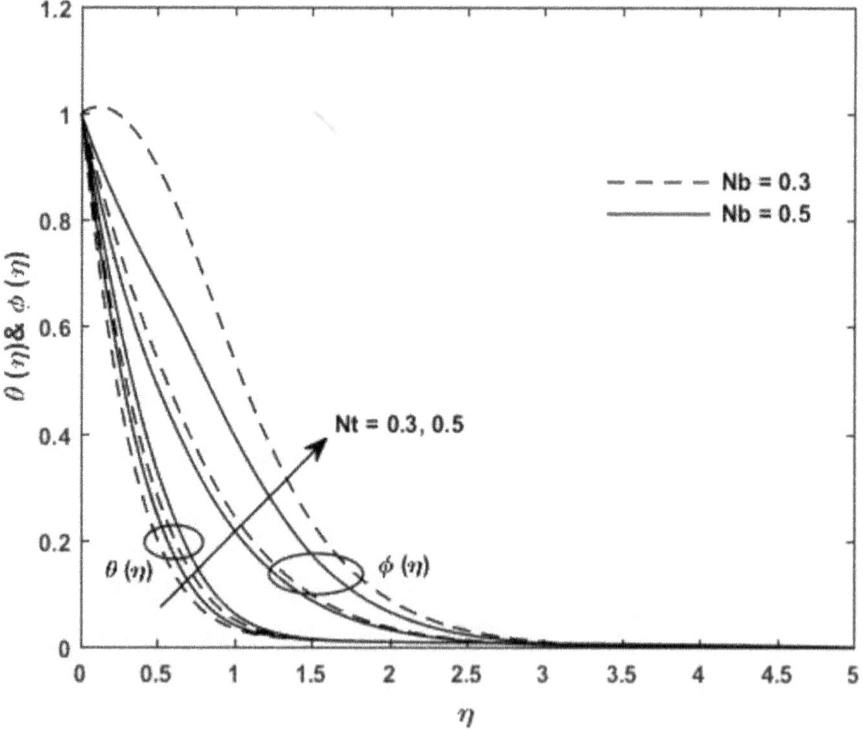

FIGURE 22.8 Temperature and concentration profiles for various values of Nb and Nt.

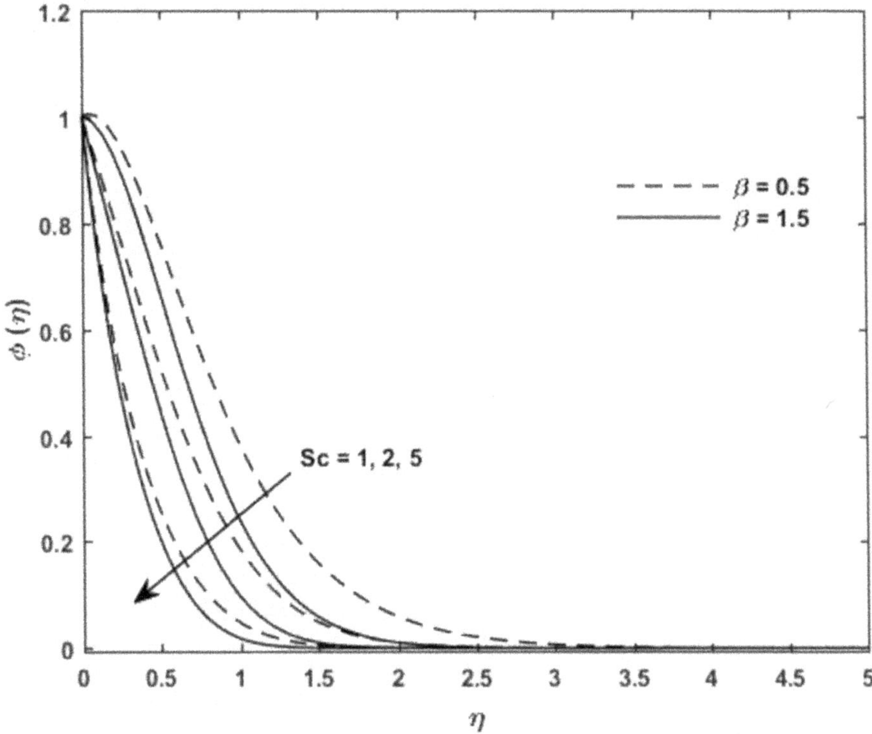

FIGURE 22.9 Concentration profiles for various values of Sc and β.

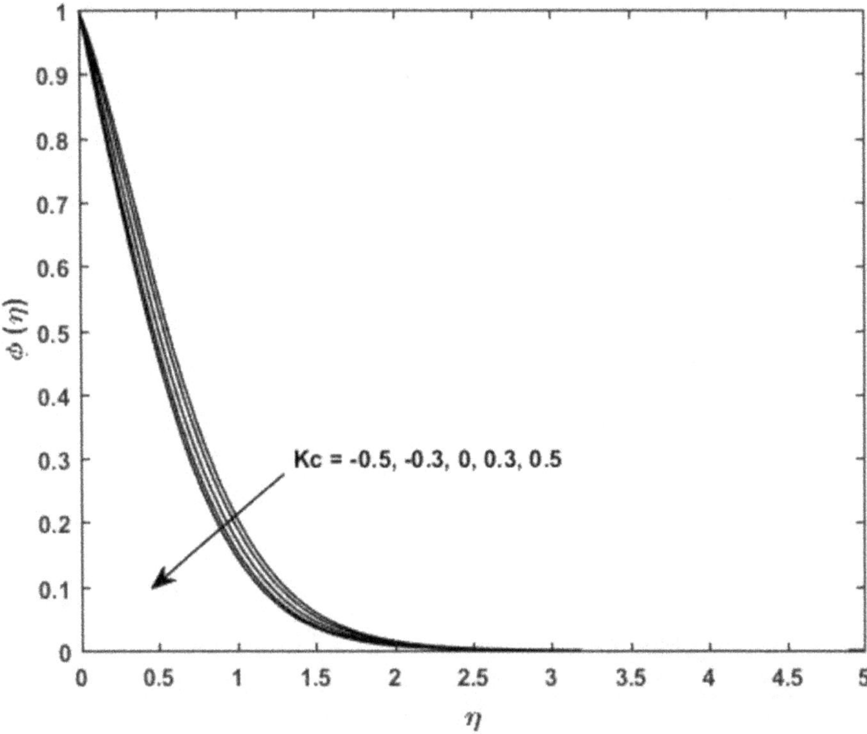

FIGURE 22.10 Concentration profiles for various values of Kc when $A = 0.5$.

TABLE 22.2
Computation of $f''(0)$, $-\theta'(0)$ and $-\phi'(0)$ When $K_p = \gamma = 0.5$, $Nb = Kc = 0.3$, $n = 2$

M	B	A	Pr	R	Ec	Sc	S	$-f''(0)$	$-\theta'(0)$	$-\phi'(0)$
0.1	0.1	0	2	0.1	0.1	1	0.1	0.690075	1.593641	0.662945
0.5								0.763479	1.548217	0.663050
1								0.846787	1.490263	0.668581
	0.3							0.695486	1.550537	0.722526
	0.5							0.521781	1.606041	0.767754
		0.5						0.550046	1.959737	0.885626
		1						0.577368	2.259591	0.990695
			3					0.577368	2.578165	0.779118
			5					0.577368	2.971211	0.508111
				0.3				0.577368	3.473128	0.619323
				0.5				0.577368	3.935834	0.712331
					0.3			0.577368	3.639983	0.839830
					0.5			0.577368	3.341396	0.968801
						2		0.577368	3.080330	2.138669
						5		0.577368	2.745269	4.207764
							0.5	0.577368	2.511262	4.277873
							1	0.577368	2.215085	4.367104
							−0.5	0.577368	3.091518	4.104659
							−1	0.577368	3.375783	4.020584

surface. Moreover, the rates of heat transfer and solutal concentration at the bounding surface increase with β and A, but in case of M, the rate of heat transfer decreases, but the rate of solutal concentration increases. It is observed that $-\theta'(0)$ increases with increase in Pr as well as strength of the heat sink ($S < 0$), whereas $-\phi'(0)$ decreases. It is remarked that Ec, Sc and $S > 0$ affect $-\theta'(0)$ and $-\phi'(0)$ adversely as compared to that of Pr and $S < 0$. Further, fluid with lower Pr will possess higher thermal conductivity so that heat can diffuse from the sheet faster than higher-Pr fluids. These results warrant the presence of nanoparticles. But from the work of Koo and Kleinstreuer [25], it is reported that the presence of 20 nm copper nanoparticles at low volume fraction (1 to 4%) to high Prandtl number liquids significantly increases the heat transfer performance of a microchannel heat sink [26]. The effect of the increase in viscous dissipation parameter (Ec) is to reduce the wall temperature gradient $-\theta'(0)$, as an increase in Ec increases the temperature, since more heat energy is stored up in the fluid due to frictional heating. Further, it is to note that higher Sc (heavier species of diffusion) and heat source $(S > 0)$ also reduce $-\phi'(0)$. But most interestingly, the rate of solutal concentration at the wall shows the opposite effect compared to the rate of heat transfer. This may be attributed to the fact that higher thermal energy enhances the solutal diffusion, causing the fall of concentration and, hence, the flux at the wall. Thus, it is concluded that the presence of nanoparticles in the base fluid reduces the shearing effect at the plate surface so that it may impose stability or avoid backflow in the downstream.

22.4 CONCLUSIONS

From the present study, the major findings are as follows:

- The unsteadiness of the flow reduces the momentum transport but enhances the thermal energy irrespective of the effects of other parameters.
- The stretching ratio of free stream and plate surface plays a vital role in the formation of the boundary layer and the inverted boundary layer, causing the flow reversal.
- The higher stretching rate decreases the temperature of the fluid.
- Thermophoresis favors the rise in temperature and volume fraction of the nanofluid.

- Unsteadiness flow decreases the level of concentration, but Casson fluidity enhances it.
- Use of high-Prandtl-number base fluid and nanoparticle of high thermal conductivity could be of practical use to increase the rate of heat transfer and to avoid nanoparticle accumulation.
- The presence of nanoparticles in the base fluid reduces the shearing stress at the plate surface so as to avoid backflow.

Nomenclature

u,v	velocities along x- and y-directions, respectively
a	stretching rate
b	strength of stagnation flow
t	time
B_0	magnetic field strength
M	magnetic parameter
K_p	porosity parameter
n	exponential index
Pr	Prandtl number
Nb	Brownian motion parameter
Nt	thermophoresis parameter
Sc	Schmidt number
S	heat source/sink parameter
Kc	chemical reaction parameter
U_e	ambient fluid velocity
T_w	temperature of the wall
T_∞	ambient temperature
C_∞	ambient concentration
R	radiation parameter
A	unsteadiness parameter
D_B	Brownian diffusion coefficient
D_T	thermophoresis diffusion coefficient
c_p	specific heat at constant temperature
Kc^*	chemical reaction coefficient
k	thermal conductivity coefficient
Kp^*	permeability of the medium
Q	heat source/sink coefficient
η	similarity variable
σ	electrical conductivity
ψ	stream function
λ	positive constant
γ	Casson parameter
β	stretching ratio parameter
α	thermal diffusivity
τ	ratio of the nanoparticle heat capacity to the base fluid heat capacity
$(\rho c)_f$	heat parameter of base fluid
$(\rho c)_p$	heat parameter of nanoparticle
μ_f	dynamic viscosity of base fluid
υ_f	kinematic viscosity of base fluid
ρ_f	density of base fluid

REFERENCES

[1] Mukhopadhyay, S., De, PR., Bhattacharyya, K., Layek, GC. Casson fluid flow over an unsteady stretching surface. *Ain Shams Engineering Journal*, 2013, 4: 933–938.

[2] Pramanik, S. Casson fluid flow and heat transfer past an exponentially porous stretching surface in presence of thermal radiation. *Ain Shams Engineering Journal*, 2014, 5: 205–212.

[3] Ramesh, GK., Prasannakumara, BC., Gireesha, BJ., Rashidi, MM. Casson fluid flow near the stagnation point over a stretching sheet with variable thickness and radiation. *Journal of Applied Fluid Mechanics*, 2016, 9: 1115–1122.

[4] El-Aziz, MA., Afify, AA. MHD Casson fluid flow over a stretching sheet with entropy generation analysis and hall influence. *Entropy*, 2019, 21: 592.

[5] Das, M., Mahanta, G., Shaw, S., Parida, SB. Unsteady MHD chemically reactive double-diffusive Casson fluid past a flat plate in porous medium with heat and mass transfer. *Heat Transfer – Asian Research*, 2019, 48: 1761–1777.

[6] Reddy, YD., Mebarek-Oudina, F., Goud, BS., Ismail, AI. Radiation, velocity and thermal slips effect toward MHD boundary layer flow through heat and mass transport of Williamson nanofluid with porous medium. *Arabian Journal for Science and Engineering*, 2022, 47: 16355–16369.

[7] Hassan, M., Mebarek-Oudina, F., Faisal, A., Ghafar, A., Ismail, AI. Thermal energy and mass transport of shear thinning fluid under effects of low to high shear rate viscosity. *International Journal of Thermofluids*, 2022, 15: 100176.

[8] Chabani, I., Mebarek-Oudina, F., Vaidya, H., Ismail, AI. Numerical analysis of magnetic hybrid Nanofluid natural convective flow in an adjusted porous trapezoidal enclosure. *Journal of Magnetism and Magnetic Materials*, 2022, 564: 170142.

[9] Khan, U., Mebarek-Oudina, F., Zaib, A., Ishak, A., Abu Bakar, S., Sherif, EM., Baleanu, D. An exact solution of a Casson fluid flow induced by dust particles with hybrid nanofluid over a stretching sheet subject to Lorentz forces. *Waves in Random and Complex Media*, 2022. https://doi.org/10.1080/17455030.2022.2102689

[10] Shafiq, A., Mebarek-Oudina, F., Sindhu, TN., Rassoul, G. Sensitivity analysis for Walters' B nanoliquid flow over a radiative Riga surface by RSM. *Scientia Iranica*, 2022, 29: 1236–1249.

[11] Raza, J., Mebarek-Oudina, F., Ali Lund, L. The flow of magnetised convective Casson liquid via a porous channel with shrinking and stationary walls. *Pramana—Journal of Physics*, 2022, 96: 229.

[12] Mebarek-Oudina, F. Convective heat transfer of Titania nanofluids of different base fluids in cylindrical annulus with discrete heat source. *Heat Transfer-Asian Research*, 2019, 48: 135–147.

[13] Swain, K., Mebarek-Oudina, F., Abo-Dahab, SM. Influence of MWCNT/Fe3O4 hybrid-nanoparticles on an exponentially porous shrinking sheet with chemical reaction and slip boundary conditions. *Journal of Thermal Analysis and Calorimetry*, 2022, 147: 1561–1570.

[14] Mebarek-Oudina, F. Numerical modeling of the hydrodynamic stability in vertical annulus with heat source of different lengths. *Engineering Science and Technology*, 2017, 20: 1324–1333.

[15] Rout, BC., Mishra, SR. Thermal energy transport on MHD nanofluid flow over a stretching surface: A comparative study. *Engineering Science and Technology, an International Journal*, 2018, 21: 60–69.

[16] Swain, K., Parida, SK., Dash, GC. Effects of non-uniform heat source/sink and viscous dissipation on MHD boundary layer flow of Williamson nanofluid through porous medium. *Defect and Diffusion Forum*. 2018, 389: 110–127.

[17] Swain, K., Parida, SK., Dash, GC. MHD heat and mass transfer on stretching sheet with variable fluid properties in porous medium. *Modelling Measurement and Control B*. 2017, 86: 706–726.

[18] Crane, LJ. Flow past a stretching plate. *Zeit Angew Math Phys*, 1970, 21: 645–647.

[19] Mahapatra, TR., Gupta, AS. Heat transfer in stagnation-point flow towards a stretching sheet. *Heat and Mass Transfer*, 2002, 38: 517–521.

[20] Misra, JC., Sinha, A. Effect of thermal radiation on MHD flow of blood and heat transfer in a permeable capillary in stretching motion. *Heat Mass Transfer*, 2013, 49: 617–628.

[21] Senapati, M., Swain, K., Parida, SK. Numerical analysis of three-dimensional MHD flow of Casson nanofluid past an exponentially stretching sheet. *Karbala International Journal of Modern Science*, 2020, 6: 93–102.

[22] Das, K., Duari, PR., Kundu, PK. Nanofluid flow over an unsteady stretching surface in presence of thermal radiation. *Alexandria Engineering Journal*, 2014, 53: 737–745.

[23] Mustafa, M., Hayat, T., Pop, I., Hendi, A. Stagnation-point flow and heat transfer of a Casson fluid towards a stretching sheet. *Zeitschrift für Naturforschung*, 2011, 67a: 70–76.

[24] Oyelakin, IS., Mondal, S., Sibanda, P. Unsteady Casson nanofluid flow over a stretching sheet with thermal radiation, convective and slip boundary conditions. *Alexandria Engineering Journal*, 2016, 55: 1025–1035.

[25] Koo, J., Kleinstreuer, C. A new thermal conductivity model for nanofluids. *Journal of Nanoparticle Research*, 2004, 6: 577–588.

[26] Das, SK., Choi, SS., Yu, W., Pradeep, T. *Nanofluids Science and Technology*. John Wiley & Sons INC. 2007: 343.

23 Study on Falkner–Skan Flow of MWCNT-MgO/EG Hybrid Nanofluid

Mahanthesh Basavarajappa and Dambaru Bhatta

23.1 INTRODUCTION

Nanotechnology is a multidisciplinary area of research due to its broad applications in electronics, thermal systems, agriculture, pharmaceutical medicine, the chemical industry, and many others. Nanoliquids are engineered liquids prepared by accruing nanoparticles into base liquids to exaggerate the thermophysical capability of base liquids. Solid nanoparticles have advanced thermal conductivity when compared to ordinary liquids. The idea of adding solid nanoparticles into a base liquid was first introduced by Choi [1] to intensify the heat transport coefficient of the base fluid. Nanoliquids are utilized inside the absorber and serve as heat transfer liquids. Nanoliquids possess superior thermal features and enhance the performance of the systems. Nanoliquids are used in many applications, such as heat exchangers, solar collectors, chemical progression, cancer psychoanalysis, and biomedicine. Das et al. [2] and Eastman et al. [3] confirmed through their remarkable enrichment in the thermal conductivity of working liquids H_2O and $C_2H_6O_2$ due to alumina Al_2O_3 and copper Cu nanoparticles, respectively. Khanafer et al. [4] proved the enrichment of heat transfer conditions theoretically in an enclosure owing to the dispersion of copper Cu nanoparticles in H_2O. They modeled the problem of nanoliquids by considering the thermophysical features of nanoparticles with base liquid to analyze the heat transport performance. This model of nanoliquid is well-known as the Khanafer–Vafai–Lightstone model (KVLM). Few very recent studies have since emerged in the literature [5–11] to study heat transport in nanoliquids with distinct physical aspects. These authors concluded that the thermal performance of base liquid enhances significantly due to the inclusion of nanoparticles.

Buongiorno [12] developed a model to divulge thermal conductivity advancement in nanoliquids, as the KVL model may not be adequate in cases where Brownian motion, liquid–solid particles interaction, gravity, thermophoresis, dispersion, and sedimentation are imperative. Among the slip mechanisms, the Brownian motion and thermophoretic effects are significant. Some recent studies [13–16] have considered nanoliquid heat transport in various geometries by employing the Buongiorno model. The Buongiorno nanoliquid model no doubt is an advancement over the KVL model even though the thermophysical properties are not incorporated in the model. Therefore, Noor et al. [17] have incorporated the Buongiorno model and effective nanofluid properties to study the dynamics and heat transport of nanoliquids over a moving surface. They addressed the characteristics of the thermophysical properties of nanoparticles, Brownian motion, and thermophoresis in the analysis. This model can be referred to as modified Buongiorno nanofluid model (MBNM), and essentially, KVLM is a limiting case of MBNM. A theoretical answer to the dramatic enhancement in heat transport in nanofluid using MBNM is provided by Yang et al. [18]. Siddheshwar and Kanchana [19] investigated the transient Rayleigh–Bénard convection in nanomaterials by employing MBNM along with linear and local nonlinear stability analysis.

Nanoliquids are promising working liquids in many practical applications because they executed well and disclosed acceptable outcomes that encouraged the researchers to sense the suspension of a diverse amalgamation of nanoparticles in the base liquid, which were developed and

DOI: 10.1201/9781003299608-23

referred to as "hybrid nanoliquids." Makishima [20] pointed out that "when two or more materials are mixed so that their combination has a different chemical bond entitled hybrid metals." Indeed, when two or more types of nanoparticles delivered the homogeneous phase with concurrent mixing, it is named "hybrid nanofluid." This superior class of innovative nanoliquids demonstrated significant augmentation in heat transport features and thermophysical, chemical, and hydrodynamic properties in comparison with mono nanofluids. According to Suresh et al. [21], mixing metallic and non-metallic nanoparticles in base liquid confirms advanced thermal and chemical features. They established that Al_2O_3-Cu-H_2O hybrid nanoliquids (HNFs) acquire higher thermal conductivity than mono nanoliquid, followed by H_2O. Hayat and Nadeem [22] also proved that the hybrid nanoliquid achieved fine with elevated heat transport rate than mono nanofluid in the investigation of the non-transient rotating flow of Ag-CuO-H_2O nanofluid. Studies related to the flow and heat transport of hybrid nanofluids are limited in the literature (Devi and Devi [23], Chamkha et al. [24], Amala and Mahanthesh [25], Ashlin and Mahanthesh [26], Shruthy and Mahanthesh [27], Dadheech et al. [28], Animasaun et al. [29], Hayat and Nadeem [30], Hassan et al. [31], Khan et al. [32], Mourad et al. [33], Abdel-Nour et al. [34], and Soltani and Akbari [35]). So far, the flow and heat transport of novel $\mathrm{MWCNT-MgO/EG}$ hybrid nanofluid by utilizing the MBNM is not yet considered.

On the other hand, the problem of fluid flowing through a wedge can be seen in many thermal engineering applications, such as heat exchangers, thermal insulation, solar collectors, crude oil extraction, nuclear reactors, and geothermal systems. A model of non-transient laminar fluid transport on a wedge has been developed by Falkner and Skan [36] by introducing a new kind of similarity transformation (known as the Falkner–Skan transformation). Later, this model has been used by many authors to study the heat transport behavior of nanofluids by considering various physical aspects (see Kandasamy et al. [37], Khan et al. [38], Khan and Pop [39], and Ibrahim and Tulu [40]). However, the Falkner–Skan flow of MWCNT-MgO/EG hybrid nanofluid over a wedge has not been studied yet.

Therefore, the main objective of this manuscript is to study the Falkner–Skan flow and heat transport of a Newtonian hybrid liquid in a wedge using the modified Buongiorno nanofluid model (MBNM), which includes the mechanisms of Brownian motion and thermophoresis together with the thermophysical properties of nanoparticles. Passive control of nanoparticles and thermal boundary conditions at the boundary are considered. Experimental data of the thermal conductivity and dynamic viscosity of nanofluid are used in the simulations to avoid possible errors. The effects of the Rosseland thermal radiation, magnetic field, and Joule heating are also accounted. The boundary layer theory is also used. The nonlinear governing problem is solved by using MATLAB solver to analyze the behavior of solutions. The following section presents a mathematical model of the problem.

23.2 MATHEMATICAL FORMULATION

The steady laminar flow of an $\mathrm{MWCNT-MgO/EG}$ hybrid nanoliquid over a wedge is considered. The hybrid nanoliquid is assumed to be Newtonian and incompressible. The flow is due to free-stream velocity passing through a wedge with an angle $\Omega = \beta\pi$, where $\beta = 2m/(m+1)$. Here, m is the Falkner–Skan power-law parameter, $m < 0$ corresponds to an adverse pressure gradient (often resulting in boundary layer separation), while $m > 0$ represents a favorable pressure gradient. In 1937, Douglas Hartree showed that physical solutions to the Falkner–Skan equation exist only in the range $-0.090429 \le m \le 2$. The rectangular coordinate system is considered in such a way that the y-axis is normal to the wall of the wedge and the x-axis alongside of the wedge (see Figure 23.1). The variable transverse magnetic field is considered along the y-axis. The modified Buongiorno nanofluid model (MBNM) is employed, which includes the effective thermophysical properties of their constituents along with the Brownian motion and thermophoresis effects.

Under the boundary layer and Boussinesq approximations, the governing equations are stated here:

$$\frac{\partial u}{\partial x}+\frac{\partial v}{\partial y}=0, \tag{23.1}$$

$$u\frac{\partial u}{\partial x}+v\frac{\partial u}{\partial y}=u_e\frac{du_e}{dx}+\frac{\mu_{hnl}}{\rho_{hnl}}\frac{\partial^2 u}{\partial y^2}-\frac{\sigma_{hnl}}{\rho_{hnl}}B^2\left(u-u_e\right), \tag{23.2}$$

$$\begin{aligned} u\frac{\partial T}{\partial x}+v\frac{\partial T}{\partial y}=&\frac{1}{(\rho C_p)_{hnl}}\left(k_{hnl}\frac{\partial^2 T}{\partial y^2}-\frac{\partial q_r}{\partial y}\right)+\frac{\sigma_{hnl}B^2}{(\rho C_p)_{hnl}}u^2 \\ &+\frac{(\rho C_p)_{np}}{(\rho C_p)_{hnl}}\left(D_B\frac{\partial C}{\partial y}\frac{\partial T}{\partial y}+\frac{D_T}{T_\infty}\left(\frac{\partial T}{\partial y}\right)^2\right), \end{aligned} \tag{23.3}$$

$$u\frac{\partial C}{\partial x}+v\frac{\partial C}{\partial y}=D_B\frac{\partial^2 C}{\partial y^2}+\frac{D_T}{T_\infty}\frac{\partial^2 T}{\partial y^2}, \tag{23.4}$$

The boundary conditions (BCs) that include passive control of nanoparticles and temperature jump at the wedge surface are:

$$\begin{cases} u(x,0)=v(x,0)=0 \;, & T(x,0)=T_w+L_1\left.\frac{\partial T}{\partial y}\right|_{y=0} \;, & D_B\left.\frac{\partial C}{\partial y}\right|_{y=0}+\frac{D_T}{T_\infty}\left.\frac{\partial T}{\partial y}\right|_{y=0}=0 \;, \\ u(x,\infty)=u_e(x)=ax^m \;, & T(x,\infty)=T_\infty \;, & C(x,\infty)=C_\infty \;. \end{cases} \tag{23.5}$$

Here, u and v are the components of velocity in the x- and y-directions, respectively; T and C are the temperature and volume fraction of nanoparticles; T_w, T_∞, C_∞ are the wall temperature, ambient temperature, and ambient volume fraction of nanoparticles, respectively; u_e is the free-stream velocity; $a>0$ is the constant; ρ is the density; μ is the dynamic viscosity; ν is the kinematic viscosity; C_p is the specific heat; k is the thermal conductivity; σ is the electrical conductivity; ρC_p is the heat capacity; $q_r\left(=-(4/3)\left(\sigma^*/k^*\right)\left(\partial T^4/\partial y\right)\right)$ is the radiative heat flux [37]; k^* is the mean absorption coefficient; σ^* is the Stefan–Boltzmann constant; D_B and D_T are two coefficients characterizing the diffusion of nanoparticles; $L_1=L_1^* x^{\frac{1-m}{2}}$ is the coefficient of temperature jump; $B\left(=B_0 x^{(m-1)/2}\right)$ is the variable magnetic field strength; B_0 and L_1^* are constants; and the subscripts hnl, np, and l represent the hybrid nanofluid (MWCNT-MgO), nanoparticles, and base fluid (EG), respectively.

After introducing the stream function $\psi(x,y)$ in the mathematical formulation of the physical model, the continuity equation is satisfied by the following relation:

$$(u,v)=\left(\frac{\partial \psi}{\partial y},-\frac{\partial \psi}{\partial x}\right). \tag{23.6}$$

The mathematical analysis of the problem can be simplified by considering the following feasible similarity transformations:

$$\begin{aligned} \eta=y\left[\frac{a(m+1)}{2\nu_l}\right]^{\frac{1}{2}} x^{\frac{(m-1)}{2}}, \psi=\left(\frac{2a\nu_l}{m+1}\right)^{\frac{1}{2}} x^{\frac{(m+1)}{2}} f(\eta), \\ \theta(\eta)=\frac{T-T_\infty}{T_w-T_\infty}, \phi(\eta)=\frac{C-C_\infty}{C_\infty}, \end{aligned} \tag{23.7}$$

Where η is the similarity variable, f is the dimensionless velocity, ψ is the dimensionless stream function, θ is the dimensionless temperature, and ϕ is the scaled nanoparticles volume fraction.

Accordingly, equations (23.2)–(23.4) with their corresponding BCs (23.5) are transformed to:

$$\frac{\Xi_1}{\Xi_2} f''' + ff'' + \beta\left(1 - f'^2\right) - \frac{\Xi_3}{\Xi_2} Ha\left(f' - 1\right) = 0, \tag{23.8}$$

$$\left(\Xi_4 + Rd\right)\theta'' + \Xi_5 \Pr f\theta' + \Pr Nb\theta'\phi' + PrNt\theta'^2 + \Xi_3 \Pr Ec \mathrm{Ha} f'^2 = 0 \tag{23.9}$$

$$\phi'' + PrLef\phi' + \frac{Nt}{Nb}\theta'' = 0, \tag{23.10}$$

$$f(0) = f'(0) = \theta(0) - 1 - Ts\theta'(0) = Nb\phi'(0) + Nt\theta'(0) = 0$$
$$\theta(\infty) = f'(\infty) - 1 = \phi(\infty) = 0 \tag{23.11}$$

The physical parameters appearing earlier are given by:

$$Ec = \frac{\rho_l u_e^2}{\left(\rho C_p\right)_l \left(T_w - T_\infty\right)}, Nb = \frac{\tau D_B C_\infty}{v_l}, Nt = \frac{\tau D_T\left(T_w - T_\infty\right)}{v_l T_\infty}, Rd = \frac{16\sigma^* T_\infty^3}{3k^* k_l},$$

$$Ts = L^* \sqrt{\frac{a(m+1)}{2v_l}}, Le = \frac{k_l}{\left(\rho C_p\right)_l D_B}, \Pr = \frac{v_l (\rho C_p)_l}{k_l}, Ha = \frac{2\sigma_l B_0^2}{\rho_l a(m+1)}, Re_x = \frac{u_e x}{v_l}. \tag{23.12}$$

$$\Xi_1 = \frac{\mu_{hnl}}{\mu_l}, \Xi_2 = \frac{\rho_{hnl}}{\rho_l}, \Xi_3 = \frac{\sigma_{hnl}}{\sigma_l}, \Xi_4 = \frac{k_{hnl}}{k_l}, \Xi_5 = \frac{\left(\rho C_p\right)_{hnl}}{\left(\rho C_p\right)_l}, \tau = \frac{\left(\rho C_p\right)_{np}}{\left(\rho C_p\right)_l}. \tag{23.13}$$

Here, the symbols $Ec, Nb, Nt, Rd, Ts, Le, Pr, Ha$, and Re_x stand for the Eckert number, Brownian motion parameter, thermophoresis parameter, thermal radiation parameter, thermal slip parameter, Lewis number, Prandtl number, modified Hartmann number, and local Reynolds number. Equation (23.9) is obtained by following [37] to simplify the Rosseland radiative heat flux term.

In the present study, MgO-MWCNT/EG hybrid nanofluid is chosen. The effective thermal conductivity and viscosity for various amounts of hybrid nanoparticles are recorded in Table 23.1. The experimental data is chosen from Soltani and Akbari [35]. The total volume fraction ($\chi = \chi_{MWCNT} + \chi_{MgO}$) is chosen as 0.2%, 0.4%, and 0.6%, consisting of the same amount of MWCNT and MgO, where χ_{MWCNT} is the volume fraction of $MWCNT$ and χ_{MgO} is the volume fraction of MgO.

The effective medium theory is implemented to estimate the density, electrical conductivity, and specific heat. The following are the correlations:

$$\Xi_2 = \frac{\rho_{hnl}}{\rho_l} = \left(1 - \chi\right) + \chi_{MWCNT}\frac{\rho_{MWCNT}}{\rho_l} + \chi_{MgO}\frac{\rho_{MgO}}{\rho_l}, \tag{23.14}$$

TABLE 23.1
Experiment Data of Viscosity Ratio ($\Xi_1 = \frac{\mu_{hnl}}{\mu_l}$) and Thermal Conductivity Ratio ($\Xi_4 = \frac{k_{hnl}}{k_l}$)

$\chi = \chi_{MWCNT} + \chi_{MgO}$	Ξ_1	Ξ_4
0.2% (0.1% of MWCNT + 0.1% of MgO)	1.10346	1.11388
0.4% (0.2% of MWCNT + 0.2% of MgO)	1.23450	1.15481
0.6% (0.3% of MWCNT + 0.3% of MgO)	1.57070	1.19750

Source: Retrieved from Soltani and Akbari [35].

$$\Xi_3 = \frac{\sigma_{hnl}}{\sigma_l} = \left(\frac{\frac{\chi_{MWCNT}\sigma_{MWCNT} + \chi_{MgO}\sigma_{MgO}}{\chi} + 2\sigma_l + 2(\chi_{MWCNT}\sigma_{MWCNT} + \chi_{MgO}\sigma_{MgO}) - 2\sigma_l\chi}{\frac{\chi_{MWCNT}\sigma_{MWCNT} + \chi_{MgO}\sigma_{MgO}}{\chi} + 2\sigma_l - 2(\chi_{MWCNT}\sigma_{MWCNT} + \chi_{MgO}\sigma_{MgO}) - 2\sigma_l\chi} \right), \tag{23.15}$$

$$\Xi_4 = \frac{(\rho C_p)_{hnl}}{(\rho C_p)_l} = (1-\chi) + \chi_{MWCNT}\frac{(\rho C_p)_{MWCNT}}{(\rho C_p)_l} + \chi_{MGO}\frac{(\rho C_p)_{MgO}}{(\rho C_p)_l} \tag{23.16}$$

Where the subscripts hnl, l, $MWCNT$, and MgO represent hybrid nanoliquid, ethyne glycol, multiwalled carbon nanotubes, and magnesium oxide nanoparticles, respectively. The values of thermophysical properties at 300 K are $\rho_{MWCNT} = 2100$, $\rho_{MgO} = 3580$, $\rho_l = 1110$, $(\sigma)_{MWCNT} = 10\times 10^{-15}$, $(\sigma)_{MgO} = 5.392\times 10^{-7}$, $(\sigma)_l = 10.7\times 10^{-5}$ $(C_p)_{MWCNT} = 711$, $(C_p)_{MgO} = 879$, and $(C_p)_l = 2415$ (see [38]).

The quantities of engineering interest are expressed in terms of the local skin friction factor, Nusselt, and Sherwood numbers as follows:

$$Cf = \frac{2\mu_{hnl}\left.\frac{\partial u}{\partial y}\right|_{y=0}}{\rho_l u_e^2} \tag{23.17}$$

$$Nu = \frac{-x(k_{nl}\frac{\partial T}{\partial y} + q_r)\Big|_{y=0}}{k_l(T_w - T_\infty)} \tag{23.18}$$

$$Sh = \frac{-xD_B\left.\frac{\partial C}{\partial y}\right|_{y=0}}{D_B C_\infty} \tag{23.19}$$

Because of the similarity variables, the preceding equations were reduced to the following:

$$Sfr = (2m+2)^{-1/2} Re_x^{1/2} C_f = \Xi_1 f''(0), \tag{23.20}$$

$$Nur = \left(\frac{m+1}{2}\right)^{-1/2} Re_x^{-1/2} Nu = -(\Xi_4 + Rd)\theta'(0), \tag{23.21}$$

$$Shr = \left(\frac{m+1}{2}\right)^{-1/2} Re_x^{-1/2} Sh = -\phi'(0). \tag{23.22}$$

23.3 NUMERICAL METHOD AND VALIDATION

The nonlinear differential equations (23.8)–(23.10), along with boundary conditions (23.11), are solved by MATLAB solver. As a first step, the following conversion is carried out:

$$f = \mathcal{Z}_1, f' = \mathcal{Z}_2, f'' = \mathcal{Z}_3 \tag{23.23}$$

$$f''' = \left(-\frac{\Xi_1}{\Xi_2}\right)^{-1}\left(\mathcal{Z}_1\mathcal{Z}_3 + \beta(1-\mathcal{Z}_2^2) - \frac{\Xi_3}{\Xi_2}Ha(\mathcal{Z}_2 - 1)\right), \tag{23.24}$$

$$\theta = Z_4, \theta' = Z_5, \tag{23.25}$$

$$\theta'' = \left(\frac{-Pr}{\Xi_4 + Rd}\right)\left(\Xi_5 Z_1 Z_5 + Nb Z_5 Z_7 + Nt Z_5^2 + \Xi_3 Ec\mathrm{Ha} Z_2^2\right) \tag{23.26}$$

$$\phi = Z_6, \phi' = Z_7 \tag{23.27}$$

$$\phi'' = -PrLe Z_1 Z_7 + \frac{Nt}{Nb}\left(\frac{Pr}{\Xi_4 + Rd}\right)\left(\Xi_5 Z_1 Z_5 + Nb Z_5 Z_7 + Nt Z_5^2 + \Xi_3 Ec\mathrm{Ha} Z_2^2\right) \tag{23.28}$$

$$\text{with } Z_1 = 0, Z_2 = 0, Z_3, Z_4 = 1 + Ts Z_5, Z_5, Nb Z_7 + Nt Z_5 = 0, Z_7 \tag{23.29}$$

The preceding equations are then solved using the MATLAB solver. The far-field conditions ($\eta \to \infty$) are defined at $\eta = 6$ such that:

$$f'(6) - 1 = \theta(6) = \phi(6) = 0 \tag{23.30}$$

The results obtained from MATLAB solver are compared with published data in a limiting case, and comparative results are presented in Tables 23.2 and 23.3 and found in good agreement.

TABLE 23.2
Comparison of $f''(0)$ Values with Those of [40] When $Ha = \chi = 0$

m	Ibrahim and Tulu [40]	Present Results
0.0000	0.46960	0.46959999
0.0141	0.50461	0.50461432
0.0435	0.56898	0.56897776
0.0909	0.65498	0.65497884
0.1429	0.73200	0.73199855
0.2000	0.80213	0.80212560
0.3333	0.92765	0.92765360
1.0000	1.23258	1.23258766

TABLE 23.3
Comparison of $-\theta'(0)$ Values with Those of [40] When Pr = –0.73 and $Rd = Ha = Ec = Ts = 0$

m	Ibrahim and Tulu [40]	Present Results
0.0000	0.42016	0.42015066
0.0141	0.42578	0.42577627
0.0435	0.43548	0.43547576
0.0909	0.44730	0.44729832
0.1429	0.45694	0.45693492
0.2000	0.46503	0.46503089
0.3333	0.47814	0.47813607
1.0000	—	0.50418013

23.4 RESULTS AND DISCUSSION

The main emphasis of this section is to analyze the features of dimensionless velocity $f'(\eta)$, dimensionless temperature $\theta(\eta)$, and nanoparticles volume fraction profile $\phi(\eta)$ for various values of the governing parameters, namely, the Falkner–Skan power-law parameter m, Hartmann number Ha, Brownian motion parameter Nb, thermophoresis parameter Nt, Lewis number Le, radiation parameter Rd, thermal jump parameter Ts, and Eckert number Ec, as illustrated graphically in Figures 23.2–23.17. The significance of Sfr, Nur, and Shr are also analyzed in Tables 23.4 and 23.5.

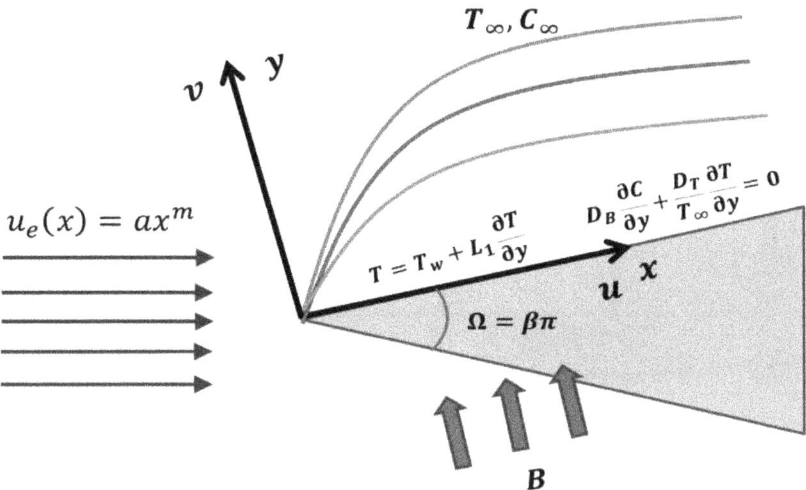

FIGURE 23.1 The geometry of the problem.

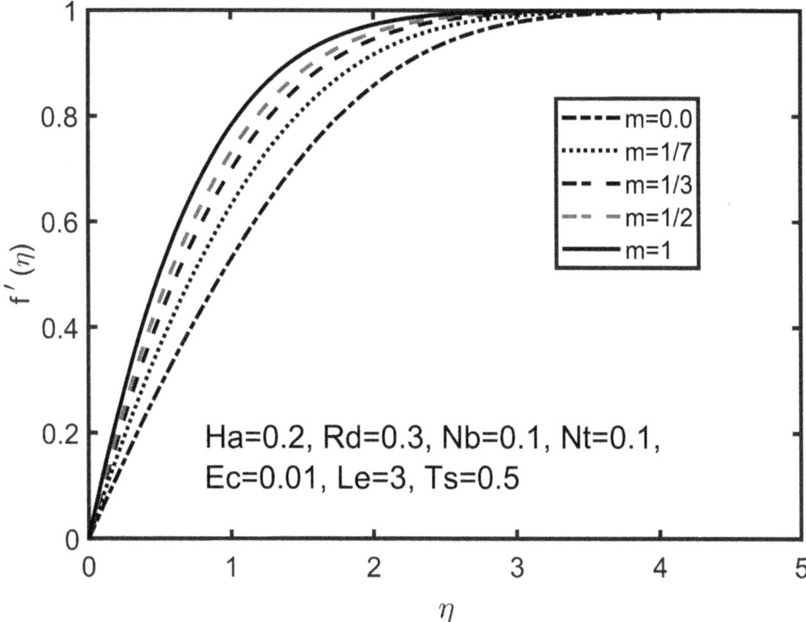

FIGURE 23.2 Velocity profile $f'(\eta)$ for distinct values of m.

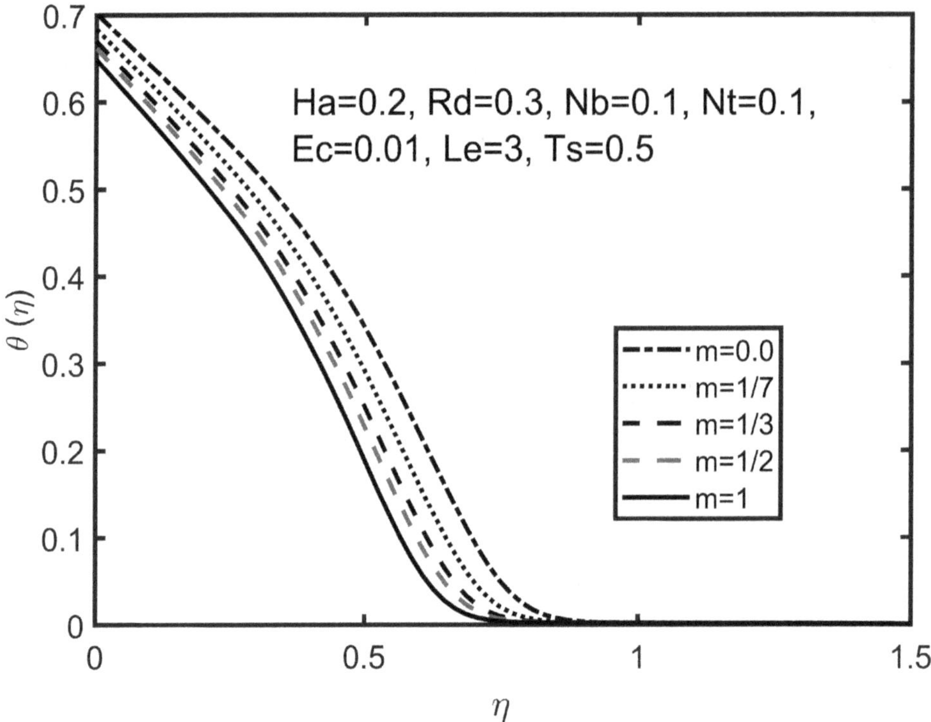

FIGURE 23.3 Temperature profile $\theta(\eta)$ for distinct values of m.

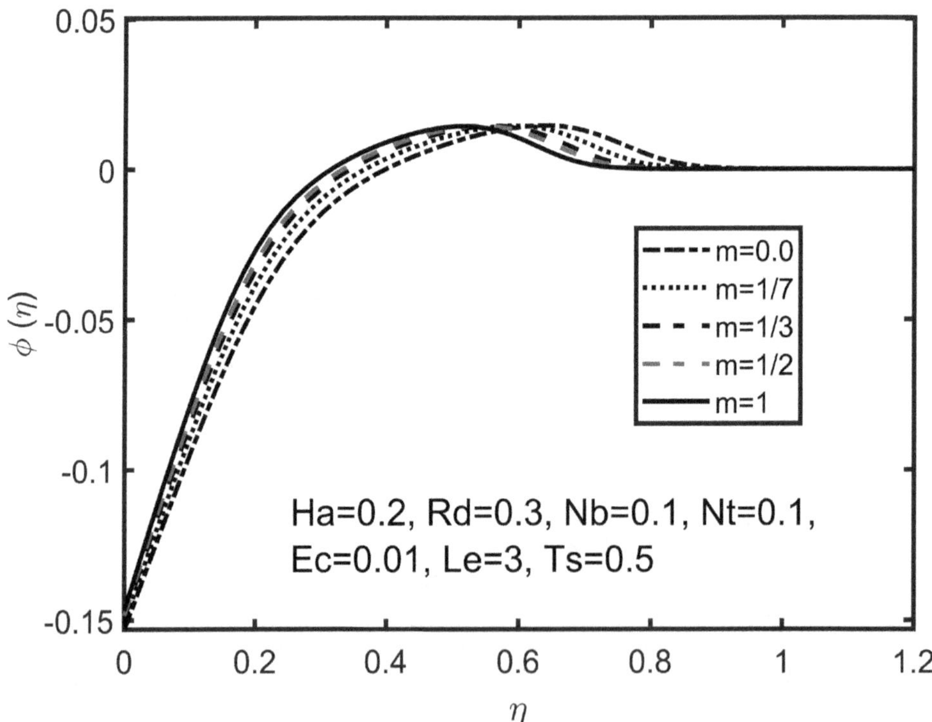

FIGURE 23.4 Nanoparticles volume fraction profile $\phi(\eta)$ for distinct values of m.

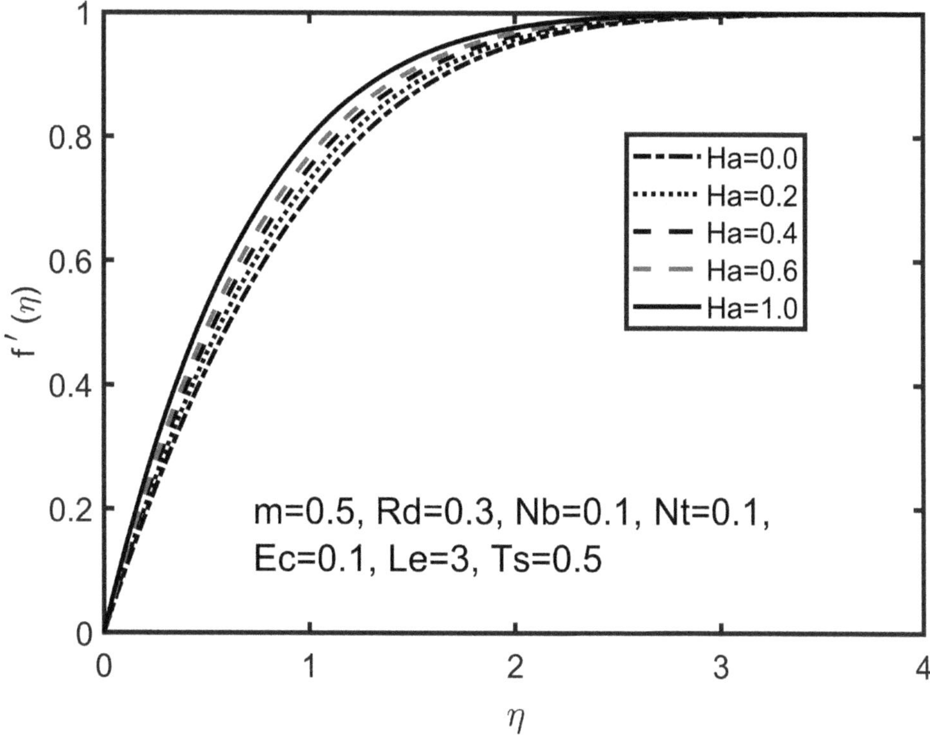

FIGURE 23.5 Velocity profile $f'(\eta)$ for distinct values of *Ha*.

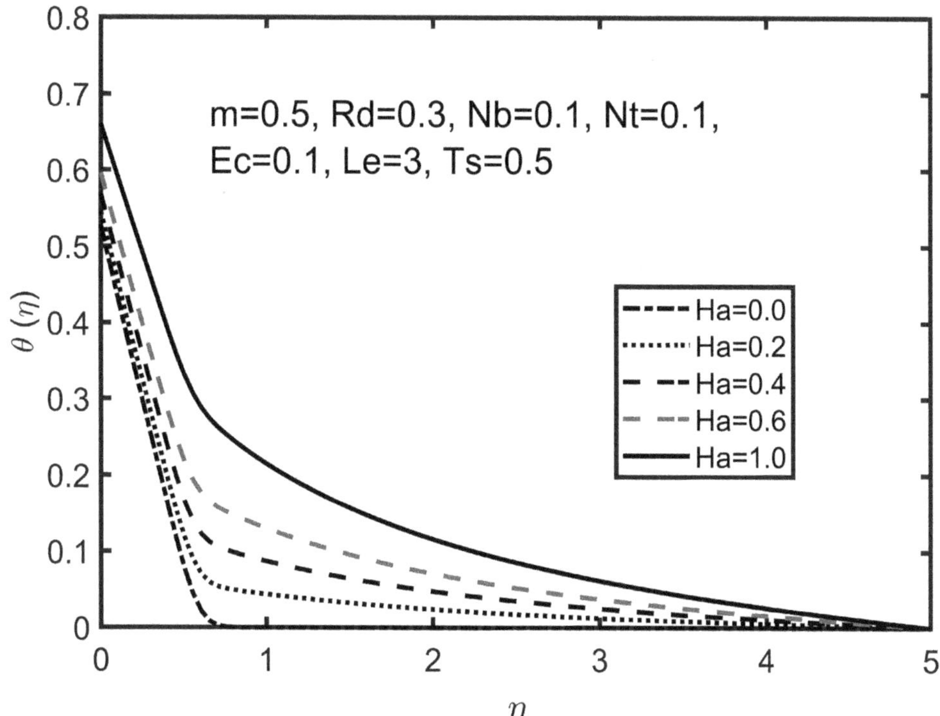

FIGURE 23.6 Velocity profile $\theta(\eta)$ for distinct values of *Ha*.

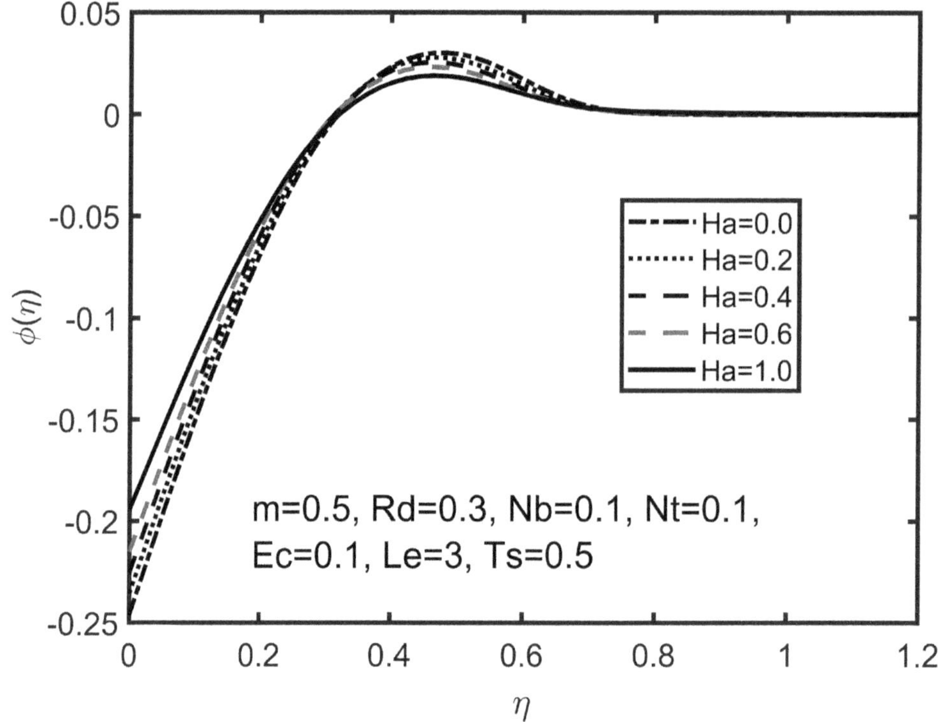

FIGURE 23.7 Nanoparticles volume fraction profile $\phi(\eta)$ for distinct values of Ha.

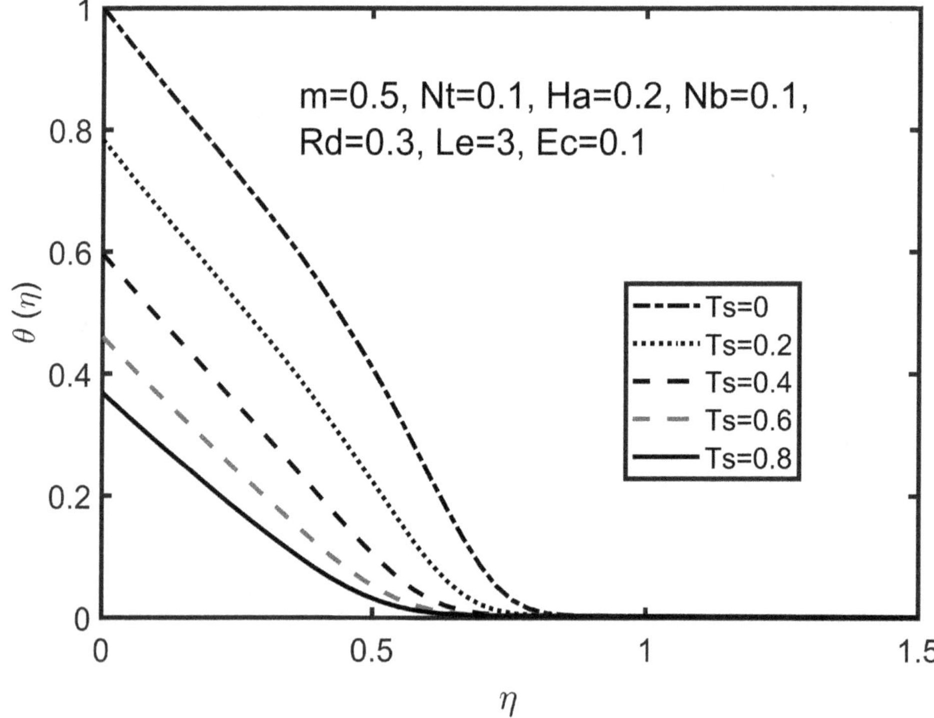

FIGURE 23.8 Velocity profile $\theta(\eta)$ for distinct values of Ts.

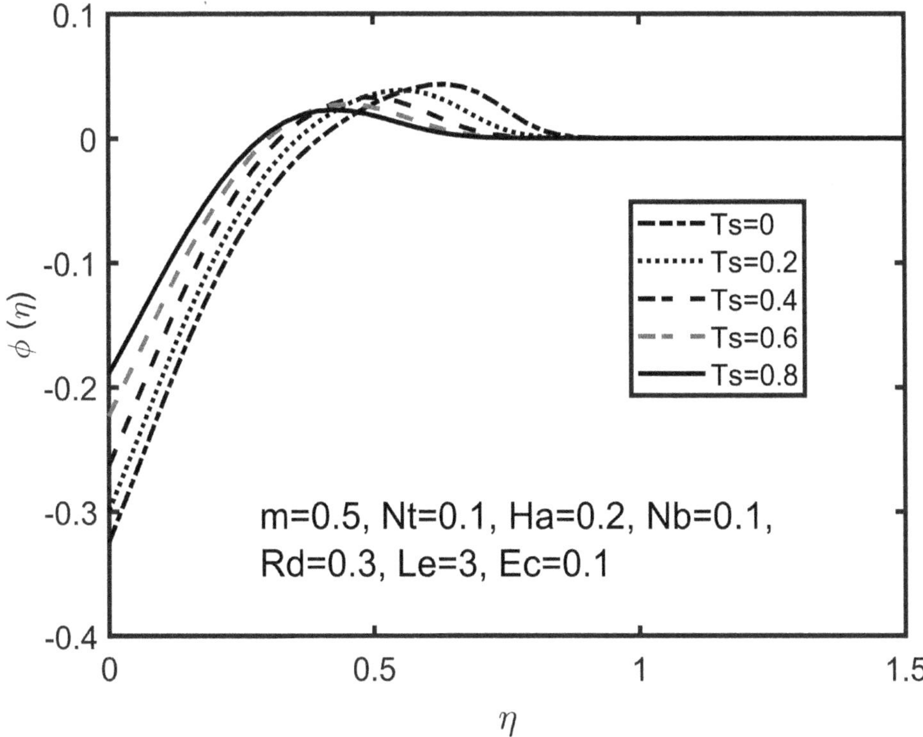

FIGURE 23.9 Temperature profile $\phi(\eta)$ for distinct values of Ts.

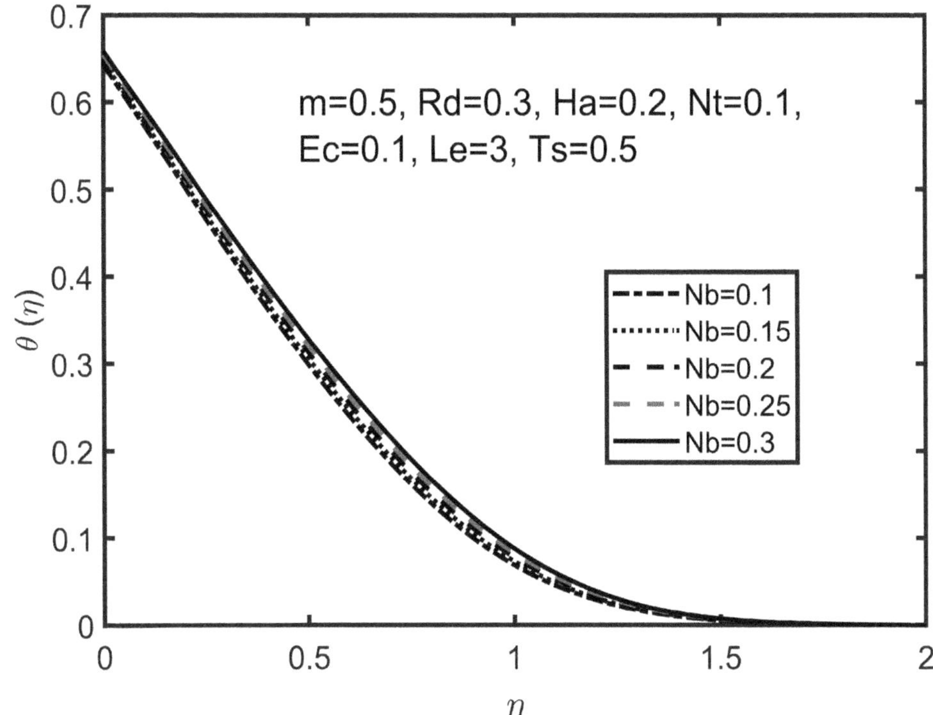

FIGURE 23.10 Temperature profile $\theta(\eta)$ for distinct values of Nb.

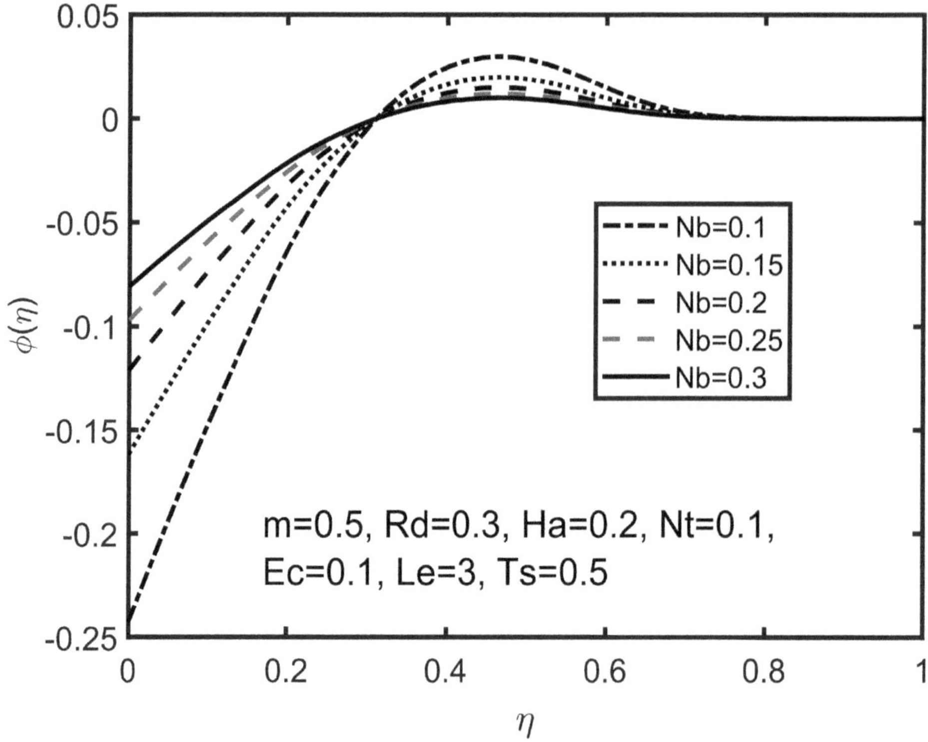

FIGURE 23.11 Nanoparticles volume fraction profile $\phi(\eta)$ for distinct values of Nb.

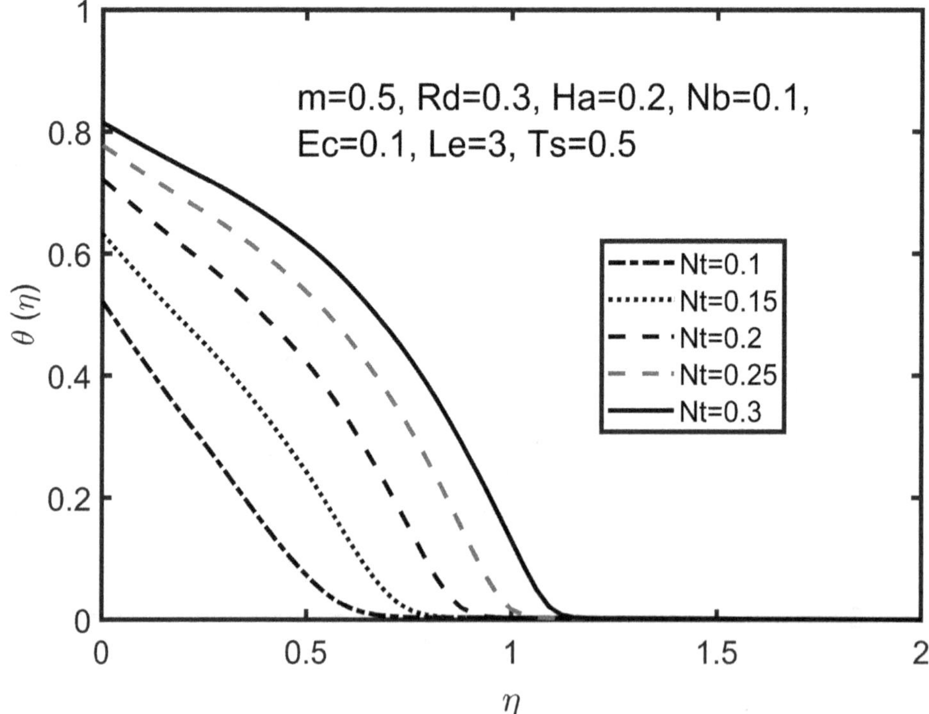

FIGURE 23.12 Temperature profile $\theta(\eta)$ for distinct values of Nt.

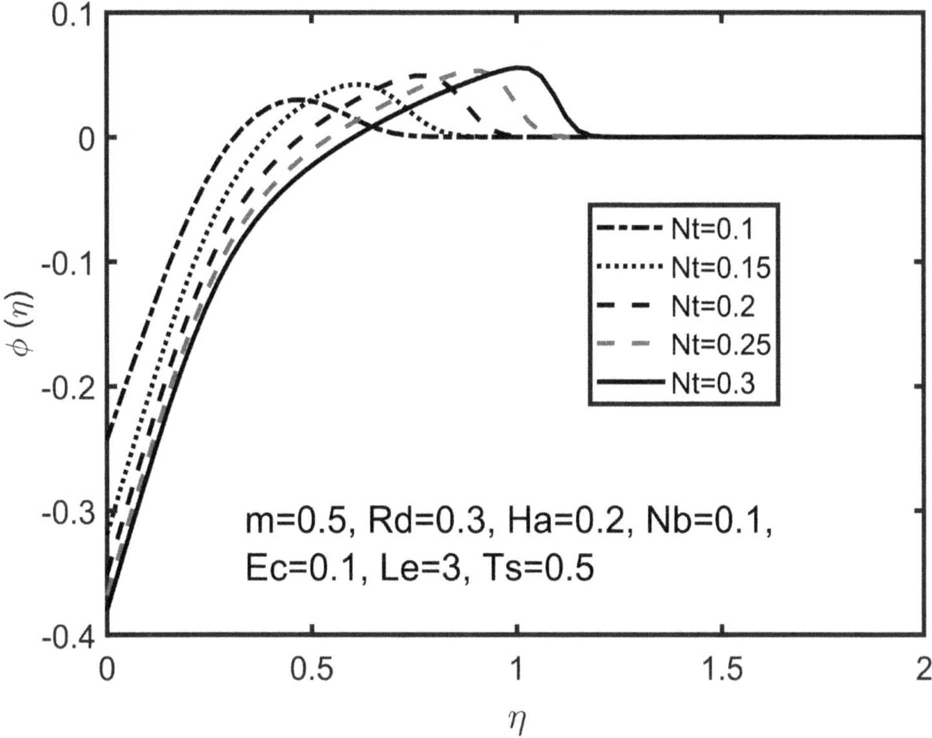

FIGURE 23.13 Nanoparticles volume fraction profile $\phi(\eta)$ for distinct values of Nt.

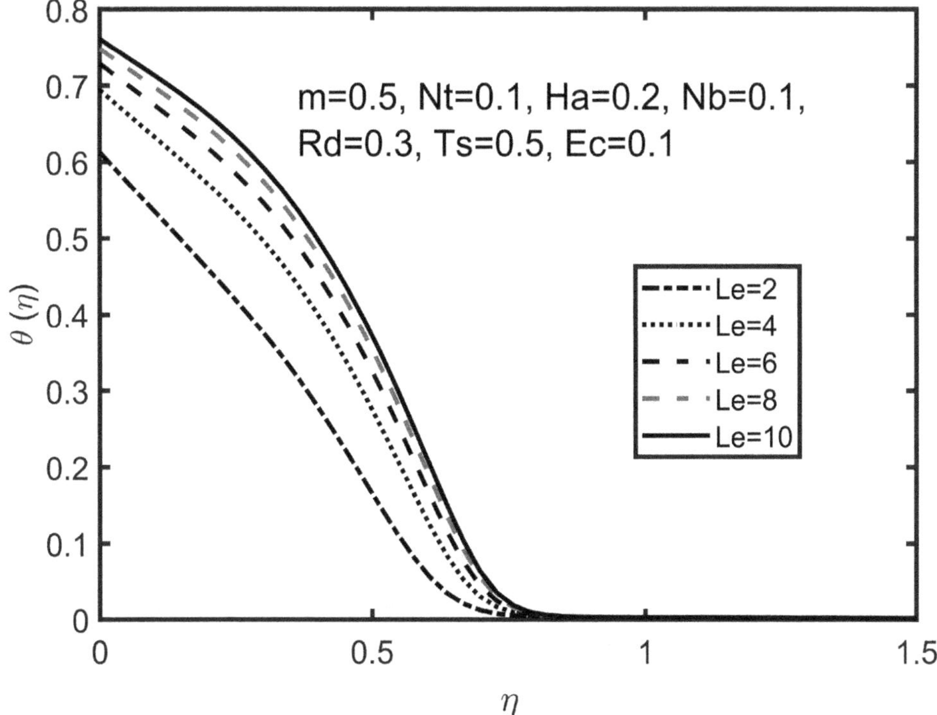

FIGURE 23.14 Temperature profile $\theta(\eta)$ for distinct values of Le.

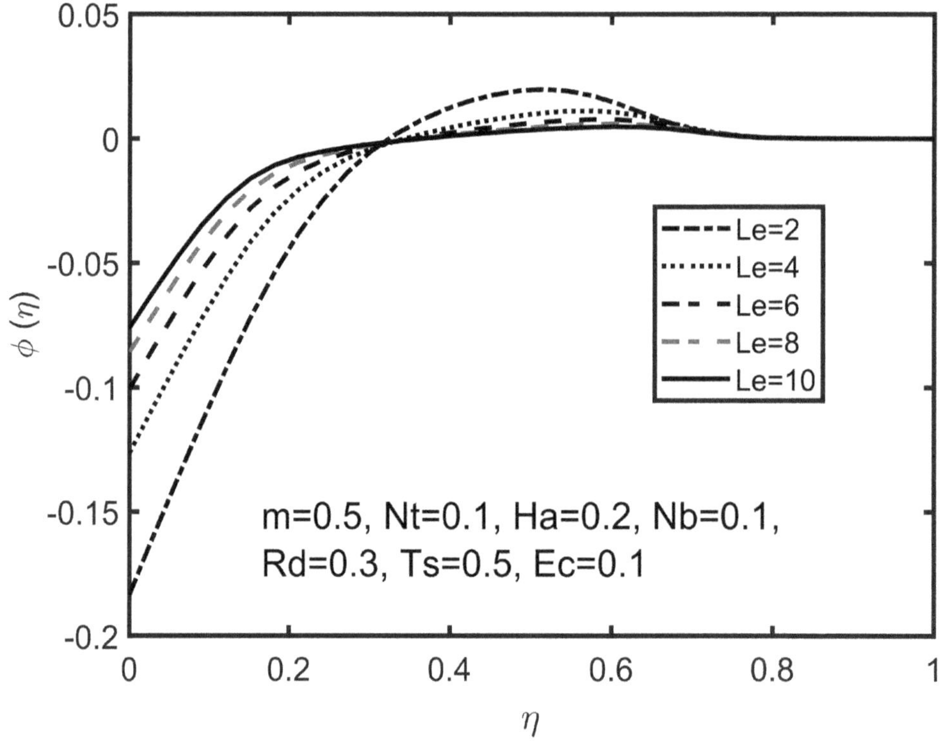

FIGURE 23.15 Nanoparticles volume fraction profile $\phi(\eta)$ for distinct values of *Le*.

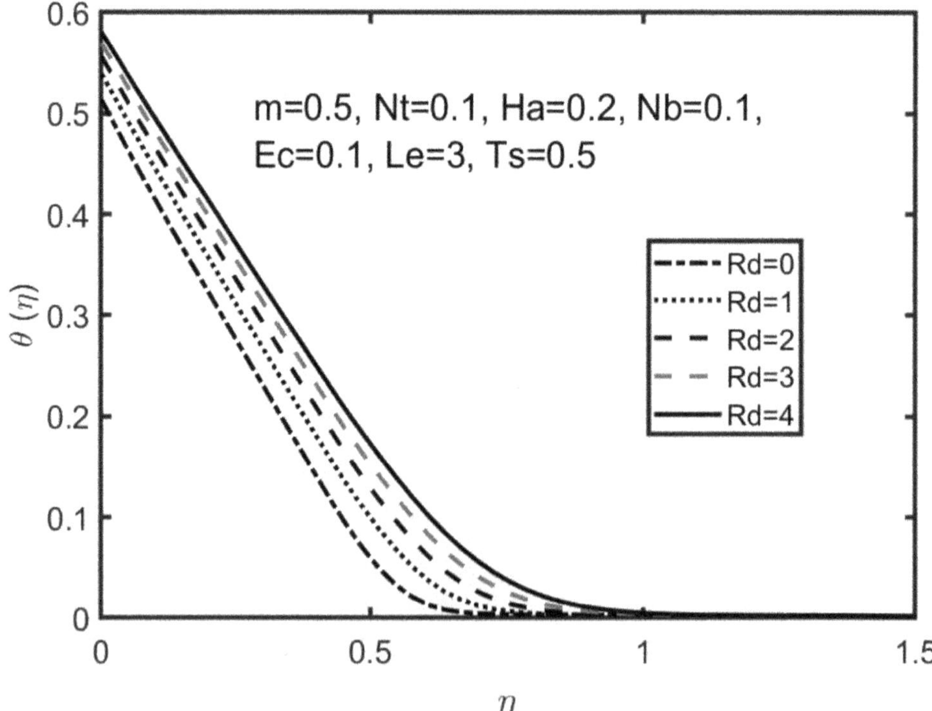

FIGURE 23.16 Temperature profile $\theta(\eta)$ for distinct values of *Rd*.

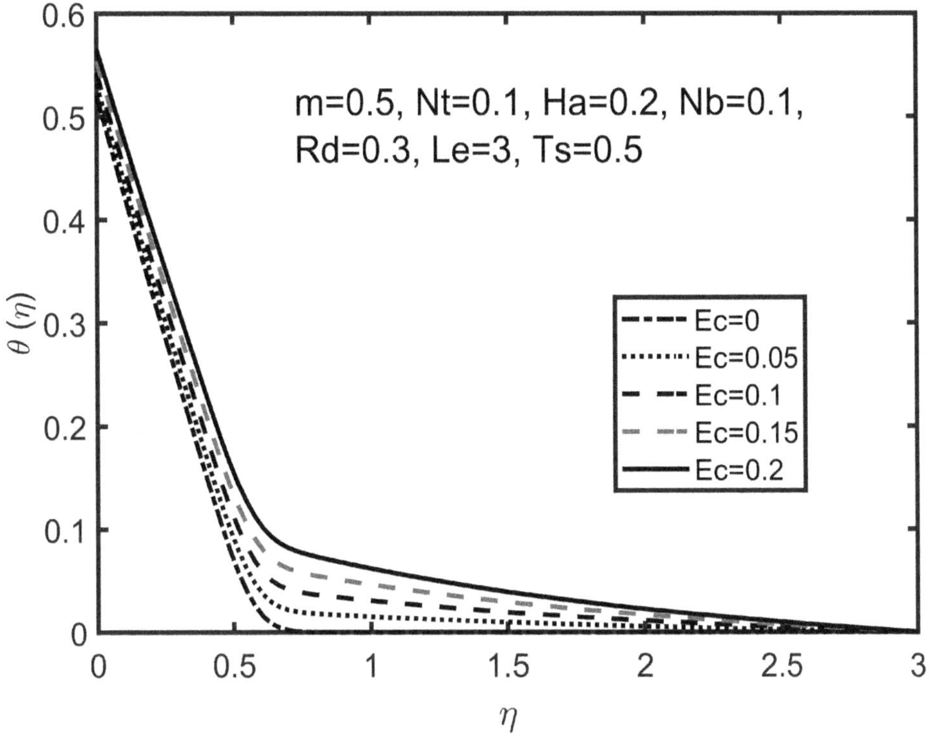

FIGURE 23.17 Temperature profile $\theta(\eta)$ for distinct values of *Ec*.

TABLE 23.4
Impact of χ on *Sfr* and *Nur* When $Rd = 0.3, Ec = 0.1, Nb = 0.1, Nt = 0.1, Ts = 0.5, m = 0.5, Le = 3$ and $Le = 3$

χ	*Sfr*	*Nur*
0	1.12996234	0.86728302
0.2%	1.20334011	0.94987580
0.4%	1.28990251	0.96974029
0.6%	1.47408482	0.97182101
Slope	**5.594649**	**1.667392**

Figures 23.2–23.4 reveal the impact of m on $f'(\eta), \theta(\eta)$ and $\phi(\eta)$ profiles, correspondingly. The Falkner–Skan power-law parameter m with $0 \le m \le 1$, $m = \frac{\beta}{2-\beta}, 0 \le \beta \le 1$. Here, β is the Hartree pressure gradient that signifies $\beta = \Omega / \pi$ for total wedge angle Ω. The zero value of m indicates the horizontal plate case, and the unit value of m corresponds to the vertical plate case. Here, increasing the values of m causes an increment in the dimensionless velocity profile, while the temperature profile decreases for an increase in the values of m. Physically, an increase in the pressure gradient factor induces an impulse in the movement of the fluid, and consequently, the flow gets accelerated. Figure 23.4 reveals that the nanoparticle volume fraction profile increases near the wedge surface while it decreases away from the surface. The features of Ha on $f'(\eta), \theta(\eta)$ and

TABLE 23.5
Effects of *m, Ha, Rd, Ts, Nt, Le,* and *Ec* on *Sfr* and *Nur* When $\chi_{MWCNT} = \chi_{MgO} = 0.3\%$

m	*Ha*	*Rd*	*Ts*	*Nt*	*Le*	*Ec*	*Sfr*	*Nur*
0	2	0.1	0.1	0.1	1	0.1	1.88199369	0.52993063
1/7							2.02787086	0.50322163
1/3							2.16406549	0.47829406
1/2							2.27425067	0.49472846
1							2.41371469	0.43279553
1/3	**0.0**	2	2	0.1	1	0.1	1.21578596	1.24719785
	1.0						1.75368623	0.84143466
	2.0						2.16406549	0.35731068
	3.0						2.50889301	−0.19133687
1/3	0.5	**0.0**	0.1	0.1	1	0.1	1.50815021	1.14258067
		1.0					1.50815021	2.13965514
		2.0					1.50815021	3.09556797
		3.0					1.50815021	4.00451151
1/3	0.1	0.1	**0**	0.1	1	0.1	1.27939656	1.28849299
			0.2				1.27939656	1.26256531
			0.4				1.27939656	1.19361359
			0.6				1.27939656	1.0816217
1/3	0.1	0.1	0.1	**0.1**	1	0.1	1.27939656	1.27931791
				0.15			1.27939656	0.85971355
				0.2			1.27939656	0.64117321
				0.25			1.27939656	0.5098071
1/3	0.1	0.1	0.1	0.2	**1**	0.1	1.27939656	0.64117321
					2		1.27939656	0.45360911
					4		1.27939656	0.34575227
					8		1.27939656	0.28118186
1/3	0.1	0.1	0.1	0.1	1	**0.0**	1.27939656	0.65659556
						0.01	1.27939656	0.65507211
						0.05	1.27939656	0.64893712
						1	1.27939656	0.64117321

$\phi(\eta)$ profiles are presented in Figures 23.5–23.7. Here, both $f'(\eta)$ and $\theta(\eta)$ are increasing functions of Ha. Besides, the nanoparticle volume fraction $\phi(\eta)$ increases near the wedge surface and then reduces for increasing values of Ha. The magnitude of the magnetic parameter signifies the intensity of the Lorentz force. The Lorentz force is a kind of retardation force, due to the decelerated motion of the hybrid nanoliquid. However, the velocity profile increases with the increasing values of the magnetic parameter Ha because of the dominance of free-stream velocity over the Lorentz force. The results of the magnetic field on the velocity are similar to those of Ibrahim and Tulu [40]. It is also noted that the thermal boundary layer is thinner in the absence of Lorentz force ($Ha = 0$).

The impact of the thermal jump parameter Ts on $\theta(\eta)$ and $\phi(\eta)$ is demonstrated in Figures 23.8–23.9 correspondingly. The temperature profile $\theta(\eta)$ decreases with an increase in the temperature jump parameter Ts. $Ts = 0$ corresponds to the isothermal boundary condition. The thermal layer structure is thicker for isothermal boundary conditions. Physically, the thermal jump condition reduces the thermal diffusivity of MWCNT-MgO hybrid nanofluid; as a result, the thermal field diminishes for higher values of Ts. The parameter Ts has a positive effect on $\phi(\eta)$

near the surface of the wedge, as seen in Figure 23.9; however, it declines before reaching the asymptotic value.

Figures 23.10 and 23.11 depict the salient features of Brownian moment *Nb* on $\theta(\eta)$ and $\phi(\eta)$, respectively. The Brownian motion parameter is an energy-generating element in the system due to the random motion of tiny nanoparticles in the fluidic system. Therefore, $\theta(\eta)$ augments when *Nb* is incremented, while $\phi(\eta)$ increases near the wedge surface with Nb and then decreases (see Figure 23.11). Figures 23.12 and 23.13 portray the effect of *Nt* on $\theta(\eta)$ and $\phi(\eta)$, respectively. Larger values of *Nt* yield higher $\theta(\eta)$, because a small number of nanoparticles is pulled away from the hot region to the cold region via the thermophoresis mechanism. Hence, a large number of solid nanomaterials is moved away from the heated region, which inflates the MWCNT-MgO/EG hybrid nanoliquid temperature. Besides, $\phi(\eta)$ decays near the hot wedge surface region when *Nt* is incremented (see Figure 23.13).

The behavior of *Le* on $\theta(\eta)$ and $\phi(\eta)$ is illustrated in Figures 23.14 and 23.15, correspondingly. It is observed that $\theta(\eta)$ significantly increases for enlarging values of *Le*. The Lewis number *Le* is the ratio of the thermal diffusivity to the mass diffusivity. Hence, the thermal diffusivity of MWCNT-MgO/EG increases with *Le*. Consequently, the thermal field and thickness of the thermal layer enhance for upsurging values of *Le*. Besides, the $\phi(\eta)$ increases and then decreases for larger values of *Le* in the boundary layer region. The Brownian diffusion coefficient decays for larger *Le*; as a result, the nanoparticles volume fraction layer thickness is compressed. The role of Rosseland radiation parameter *Rd* on $\theta(\eta)$ is explored in Figure 23.16. An increase in *Rd* augments $\theta(\eta)$. This is due to additional heat supplied by the electromagnetic waves to the system. Therefore, the thermal radiation mechanism could be used as a key factor in the heating processes.

The product of *Ec* and *Ha* represents the impact of the Joule heating phenomena. As has been noticed in Figure 23.6, $\theta(\eta)$ improves with an increase in *Ha* values. Hence, it is expected that $\theta(\eta)$ is augmented when *Ec* is incremented (see Figure 23.17). The relation between the flow of enthalpy difference and kinetic energy is called the Eckert number. It elaborates on the change of kinetic energy into internal energy by work done versus the viscous liquid stresses. The larger Eckert number *Ec* causes a loss of heat from the wedge surface to the hybrid nanoliquid (i.e., cooling of the wedge surface). In other words, larger energy dissipation produces higher hybrid nanoliquid temperature.

The role of the total volume fraction of nanoparticles $(\text{MWCNT} + \text{MgO} : 50\% - 50\%)$ on *Sfr* and *Nur* is presented in Table 23.4. It is revealed that the substitution of solid nanomaterials in ethylene glycol enhances the Nusselt number, hence the enhanced heat transport situation in the system. Increasing the total volume fraction for 0 to 0.6% causes an enhancement in the *Sfr* at the rate of 5.594649 and *Nur* at the rate of 1.667392. From Table 23.4, the following inequities can be drawn:

$$Sfr^{EG} < Sfr^{MWCNT-MgO-EG} \text{ and}$$
$$Nur^{EG} < Nur^{MWCNT-MgO-EG}$$

Table 23.5 depicts the effects of $m, Ha, Rd, Ts\ Nt, Le,$ and Ec on *Sfr* and *Nur* when $\chi_{MWCNT} = \chi_{MgO} = 0.3\%$. Here, the friction factor at the surface, *Sfr*, is enhanced for larger values of m and Ha. The heat transfer rate, *Nur*, is increased for the larger values of *Rd*, while it decreases for increasing values of $m, Ha, Ts, Nt, Le,$ and Ec.

23.5 CONCLUDING REMARKS

A theoretical study of the Falkner–Skan flow of hybrid nanofluids in the presence of linear thermal radiation, Joule heating, and magnetic field is performed. Experimental data on thermal conductivity and dynamic viscosity were used in the simulations. The governing boundary layer equations were solved using a self-similar approach. This analysis leads to the following main observations:

- Higher values of the Hartree pressure gradient increase the momentum of the fluid; consequently, the thickness of the momentum boundary layer increases, while the opposite circumstance is observed for the thermal field.
- Increasing values of the Hartmann number show an improvement in temperature and velocity.
- The impact of thermal jump conditions is unfavorable to the development of the thermal layer structure.
- A larger Lewis number has a higher temperature and a lower nanoparticle volume fraction.
- The temperature and the volume fraction of the nanoparticles through the Brownian motion have an inverse trend.
- The impact of the temperature jump condition is unfavorable to the growth of the thermal boundary layer thickness.
- The effects of Joule heating and thermal radiation are qualitatively the same on the temperature profile.
- Radiative heat flux improves the Nusselt number.
- The friction factor on the wedge surface is improved by increasing the values of the magnetic field parameter and the Hartree pressure gradient parameter.
- $Nur^{EG} < Nur^{MWCNT-MgO-EG}$, and $Sfr^{EG} < Sfr^{MWCNT-MgO-EG}$.

Nomenclature

u, v	components of velocity
$x,\ y$	Cartesian coordinates
T, C	temperature and volume fraction of nanoparticles
T_w	wall temperature of nanoparticles
T_∞	ambient temperature of nanoparticles
C_w	wall volume fraction of nanoparticles
C_∞	ambient volume fraction of nanoparticles
$a, B_0, L_1^*, \tau, \Xi_1, \Xi_2, \Xi_3, \Xi_4, \Xi_5$	constants
m	Falkner–Skan power-law parameter
ρ	density
μ	dynamic viscosity
ν	kinematic viscosity
C_p	specific heat
k	thermal conductivity
σ	electrical conductivity
ρC_p	heat capacity
q_r	radiative heat flux
k^*	mean absorption coefficient
σ^*	Stefan–Boltzmann constant
$D_B,\ D_T$	coefficients characterizing the diffusion of nanoparticles
L_1	coefficient of temperature jump
B	variable magnetic field strength
Ec	Eckert number

Nb	Brownian motion parameter
Nt	thermophoresis parameter
Rd	thermal radiation parameter
Ts	thermal slip parameter
Le	Lewis number
Pr	Prandtl number
Ha	modified Hartmann number
Re_x	local Reynolds number
Cf	local skin friction factor
Nu	Nusselt number
Sh	Sherwood number
ψ	stream function
Ω	wedge angle
β	Hartree pressure gradient

Subscripts

EG	ethylene glycol
MgO	magnesium oxide
MWCNT	multiwalled carbon nanotubes
hnl	hybrid nanofluid
l	base fluid

REFERENCES

1. Choi, S. U. S. (1998). *Nanofluid technology: Current status and future research* (No. ANL/ET/CP-97466) Argonne National Lab. (ANL), Argonne, IL.
2. Das, S. K., Putra, N., Thiesen, P., & Roetzel, W. (2003). Temperature dependence of thermal conductivity enhancement for nanofluids. *Journal of Heat Transfer*, 125(4), 567–574.
3. Eastman, J. A., Choi, S. U. S., Li, S., Yu, W., & Thompson, L. J. (2001). Anomalously increased effective thermal conductivities of ethylene glycol-based nanofluids containing copper nanoparticles. *Applied Physics Letters*, 78(6), 718–720.
4. Khanafer, K., Vafai, K., & Lightstone, M. (2003). Buoyancy-driven heat transfer enhancement in a two-dimensional enclosure utilizing nanofluids. *International Journal of Heat and Mass Transfer*, 46(19), 3639–3653.
5. Animasaun, I. L., Koriko, O. K., Adegbie, K. S., Babatunde, H. A., Ibraheem, R. O., Sandeep, N., & Mahanthesh, B. (2019). Comparative analysis between 36 nm and 47 nm alumina–water nanofluid flows in the presence of Hall effect. *Journal of Thermal Analysis and Calorimetry*, 135(2), 873–886.
6. Mebarek-Oudina, F., Aissa, A., Mahanthesh, B., & Öztop, H. F. (2020). Heat transport of magnetized Newtonian nanoliquids in an annular space between porous vertical cylinders with discrete heat source. *International Communications in Heat and Mass Transfer*, 117, 104737.
7. Koriko, O. K., Animasaun, I. L., Mahanthesh, B., Saleem, S., Sarojamma, G., & Sivaraj, R. (2018). Heat transfer in the flow of blood-gold Carreau nanofluid induced by partial slip and buoyancy. *Heat Transfer-Asian Research*, 47(6), 806–823.
8. Marzougui, S., Mebarek-Oudina, F., Assia, A., Magherbi, M., Shah, Z., & Ramesh, K. (2021). Entropy generation on magneto-convective flow of copper-water nanofluid in a cavity with chamfers. *Journal of Thermal Analysis and Calorimetry*, 143(3), 2203–2214.
9. Liu, H., Animasaun, I. L., Shah, N. A., Koriko, O. K., & Mahanthesh, B. (2020). Further discussion on the significance of quartic autocatalysis on the dynamics of water conveying 47 nm alumina and 29 nm cupric nanoparticles. *Arabian Journal for Science and Engineering*, 45(7), 5977–6004.

10. Zaim, A., Aissa, A., Mebarek-Oudina, F., Mahanthesh, B., Lorenzini, G., Sahnoun, M., & El Ganaoui, M. (2020). Galerkin finite element analysis of magneto-hydrodynamic natural convection of Cu-water nanoliquid in a baffled U-shaped enclosure. *Propulsion and Power Research*, 9(4), 383–393.
11. Marzougui, S., Mebarek-Oudina, F., Magherbi, M., & Mchirgui, A. (2022). Entropy generation and heat transport of Cu–water nanoliquid in porous lid-driven cavity through magnetic field. *International Journal of Numerical Methods for Heat & Fluid Flow*, 32(6), 2047–2069. https://doi.org/10.1108/HFF-04-2021-0288
12. Buongiorno, J. (2006). Convective transport in nanofluids. *Journal of Heat Transfer*, 128, 240–250.
13. Mahanthesh, B., Gireesha, B. J., Sheikholeslami, M., Shehzad, S. A., & Kumar, P. B. S. (2018). Nonlinear radiative flow of Casson nanoliquid past a cone and wedge with magnetic dipole: Mathematical model of renewable energy. *Journal of Nanofluids*, 7(6), 1089–1100.
14. Mahanthesh, B., Gireesha, B. J., & Animasaun, I. L. (2018). Exploration of non-linear thermal radiation and suspended nanoparticles effects on mixed convection boundary layer flow of nanoliquids on a melting vertical surface. *Journal of Nanofluids*, 7(5), 833–843.
15. Waqas, M., Khan, M. I., Hayat, T., Alsaedi, A., & Khan, M. I. (2017). Nonlinear thermal radiation in flow induced by a slendering surface accounting thermophoresis and Brownian diffusion. *The European Physical Journal Plus*, 132(6), 1–13.
16. Swain, K., Mahanthesh, B., & Mebarek-Oudina, F. (2021). Heat transport and stagnation-point flow of magnetized nanoliquid with variable thermal conductivity, Brownian moment, and thermophoresis aspects. *Heat Transfer*, 50(1), 754–767.
17. Noor, A., Nazar, R., Jafar, K., & Pop, I. (2014). Boundary-layer flow and heat transfer of nanofluids over a permeable moving surface in the presence of a coflowing fluid. *Advances in Mechanical Engineering*, 6, 521236.
18. Yang, C., Li, W., Sano, Y., Mochizuki, M., & Nakayama, A. (2013). On the anomalous convective heat transfer enhancement in nanofluids: A theoretical answer to the nanofluids controversy. *Journal of Heat Transfer*, 135(5).
19. Siddheshwar, P. G., Kanchana, C., Kakimoto, Y., & Nakayama, A. (2017). Steady finite-amplitude Rayleigh–Bénard convection in nanoliquids using a two-phase model: Theoretical answer to the phenomenon of enhanced heat transfer. *Journal of Heat Transfer*, 139(1).
20. Makishima, A. (2004). Possibility of hybrid materials, *Ceramic Japan*, 39(2), 90–91.
21. Suresh, S., Venkitaraj, K. P., Selvakumar, P., & Chandrasekar, M. (2011). Synthesis of Al2O3–Cu/water hybrid nanofluids using two step method and its thermo physical properties. *Colloids and Surfaces A: Physicochemical and Engineering Aspects*, 388(1–3), 41–48.
22. Hayat, T., & Nadeem, S. (2017). Heat transfer enhancement with Ag–CuO/water hybrid nanofluid. *Results in Physics*, 7, 2317–2324.
23. Devi, S. S. U., & Devi, S. A. (2016). Numerical investigation of three-dimensional hybrid Cu–Al2O3/water nanofluid flow over a stretching sheet with effecting Lorentz force subject to Newtonian heating. *Canadian Journal of Physics*, 94(5), 490–496.
24. Chamkha, A. J., Doostanidezfuli, A., Izadpanahi, E., & Ghalambaz, M. J. A. P. T. (2017). Phase-change heat transfer of single/hybrid nanoparticles-enhanced phase-change materials over a heated horizontal cylinder confined in a square cavity. *Advanced Powder Technology*, 28(2), 385–397.
25. Amala, S., & Mahanthesh, B. (2018). Hybrid nanofluid flow over a vertical rotating plate in the presence of hall current, nonlinear convection and heat absorption. *Journal of Nanofluids*, 7(6), 1138–1148.
26. Ashlin, T. S., & Mahanthesh, B. (2019). Exact solution of non-coaxial rotating and non-linear convective flow of Cu–Al_2O_3–H_2O hybrid nanofluids over an infinite vertical plate subjected to heat source and radiative heat. *Journal of Nanofluids*, 8(4), 781–794.
27. Shruthy, M., & Mahanthesh, B. (2019). Rayleigh-Bénard convection in Casson and hybrid nanofluids: An analytical investigation. *Journal of Nanofluids*, 8(1), 222–229.
28. Dadheech, P. K., Agrawal, P., Mebarek-Oudina, F., Abu-Hamdeh, N. H., & Sharma, A. (2020). Comparative heat transfer analysis of $MoS_2/C_2H_6O_2$ and SiO_2-$MoS_2/C_2H_6O_2$ nanofluids with natural convection and inclined magnetic field. *Journal of Nanofluids*, 9(3), 161–167.
29. Animasaun, I. L., Yook, S. J., Muhammad, T., & Mathew, A. (2022). Dynamics of ternary-hybrid nanofluid subject to magnetic flux density and heat source or sink on a convectively heated surface. *Surfaces and Interfaces*, 28, 101654.

30. Hayat, T., & Nadeem, S. (2018). An improvement in heat transfer for rotating flow of hybrid nanofluid: A numerical study. *Canadian Journal of Physics*, 96(12), 1420–1430.
31. Hassan, M., Marin, M., Ellahi, R., & Alamri, S. Z. (2018). Exploration of convective heat transfer and flow characteristics synthesis by Cu–Ag/water hybrid-nanofluids. *Heat Transfer Research*, 49(18).
32. Khan, U., Zaib, A., & Mebarek-Oudina, F. (2020). Mixed convective magneto flow of SiO 2–MoS 2/ C2H6O2 hybrid nanoliquids through a vertical stretching/shrinking wedge: Stability analysis. *Arabian Journal for Science and Engineering*, 45(11), 9061–9073.
33. Mourad, A., Aissa, A., Mebarek-Oudina, F., Jamshed, W., Ahmed, W., Ali, H. M., & Rashad, A. M. (2021). Galerkin finite element analysis of thermal aspects of Fe3O4-MWCNT/water hybrid nanofluid filled in wavy enclosure with uniform magnetic field effect. *International Communications in Heat and Mass Transfer*, 126, 105461.
34. Abdel-Nour, Z., Aissa, A., Mebarek-Oudina, F., Rashad, A. M., Ali, H. M., Sahnoun, M., & El Ganaoui, M. (2020). Magnetohydrodynamic natural convection of hybrid nanofluid in a porous enclosure: Numerical analysis of the entropy generation. *Journal of Thermal Analysis and Calorimetry*, 141(5), 1981–1992.
35. Soltani, O., & Akbari, M. (2016). Effects of temperature and particles concentration on the dynamic viscosity of MgO-MWCNT/ethylene glycol hybrid nanofluid: Experimental study. *Physica E: Low-dimensional Systems and Nanostructures*, 84, 564–570.
36. Falkneb, V. M., & Skan, S. W. (1931). Solutions of the boundary-layer equations. *The London, Edinburgh, and Dublin Philosophical Magazine and Journal of Science*, 12(80), 865–896.
37. Kandasamy, R., Muhaimin, I., Khamis, A. B., & bin Roslan, R. (2013). Unsteady Hiemenz flow of Cu-nanofluid over a porous wedge in the presence of thermal stratification due to solar energy radiation: Lie group transformation. *International Journal of Thermal Sciences*, 65, 196–205.
38. Khan, W. A., Hamad, M. A., & Ferdows, M. (2013). Heat transfer analysis for Falkner–Skan boundary layer nanofluid flow past a wedge with convective boundary condition considering temperature-dependent viscosity. *Proceedings of the Institution of Mechanical Engineers, Part N: Journal of Nanoengineering and Nanosystems*, 227(1), 19–27.
39. Khan, W., & A., Pop, I. (2013). Boundary layer flow past a wedge moving in a nanofluid. *Mathematical Problems in Engineering*, 2013. https://doi.org/10.1155/2013/637285.
40. Ibrahim, W., & Tulu, A. (2019). Magnetohydrodynamic (MHD) boundary layer flow past a wedge with heat transfer and viscous effects of nanofluid embedded in porous media. *Mathematical Problems in Engineering*, 2019. https://doi.org/10.1155/2019/4507852.

24 Optimization of Buoyant Nanofluid Flow and Heat Transport in a Porous Annular Domain with a Thin Baffle

Maimouna Al Manthri, M. Sankar, Carlton Azeez, and F. Mebarek-Oudina

24.1 INTRODUCTION

Convective thermal transport driven by buoyant forces in differentially heated finite-shaped geometries, such as rectangular and annular geometries, has been widely studied due to its relevance in heat exchangers, geothermal, nuclear, and solar applications. In many of the applications involving heat exchangers, the control of flow and, hence, thermal transport is vital in transferring thermal energy between two or three different systems effectively. This could be efficiently achieved, among other techniques, with a baffle of finite or negligible thickness. Realizing the importance of the positional and dimensional effects on buoyant flow and thermal transfer in finite-shaped domains, several investigations have been attempted to explore and develop a better understanding of the role of baffle in altering the flow phenomena and associated thermal processes [1–4]. These investigations explored the various impacts of baffle locations and dimensions on buoyant-driven thermal flows in rectangular and annular regions and suggested an optimum dimension and location for the baffle to achieve maximum benefit in terms of higher thermal transport rates.

However, the utilization of conventional fluids in high-cooling-rate applications may lead to thermal inefficiencies of the system. This difficulty leads to the development of a new variety of fluids, by the suspension of nano-sized particles to the conventional fluids, known as nanofluids, pioneered by Choi [5]. Nanofluids exhibited advantageous features in terms of higher heat transport enhancement in the practical applications involving electronic chip cooling, heat exchangers, engine cooling, drug delivery, to name a few. Considering the wide applications of annular geometry, many experimental and numerical studies examined the buoyant convective thermal transfer in an annular geometry. One of the pioneering investigations detailing natural convective flow phenomena in an annular region filled with nanofluids has been performed by Abouali and Falahatpisheh [6] and developed correlations for thermal transport. Recently, Oudina [7, 8] addressed the impacts of discrete heating on buoyant-driven motion of nanofluids in a cylindrical annular domain to control the thermal transport. Pordanjani et al. [9] made an extensive review on different investigations with an emphasis on the impacts of baffle in heat exchanger applications.

Buoyant motion of different fluids in finite-shaped porous domains have also received a great deal of attention among researchers, and this is due to the important role of porous materials in various thermal applications. Among the various finite-shaped domains, the annular domain mimics the physical structure of many vital applications. One of the early and detailed studies in a porous annular domain is by Prasad [10]. The detailed results address the impacts of geometrical and

DOI: 10.1201/9781003299608-24

physical parameters on the buoyant motion in porous domain. Later, Shivakumara et al. [11] made an attempt to explore the non-Darcy impacts on buoyant motion in an upright annular space. Great amount of research has been carried out on convective heat transfer in a porous annular domain subjected to partial heating and cooling due to its importance in cooling applications [12–13] and reported the optimum location of the heater to achieve maximum heat dissipation.

The combined impacts of porosity and nanofluids on buoyant-thermal transport in a porous enclosure have also received wider attention, citing its involvement in various applications. The buoyant-transport analysis in porous cavities filled with nanofluids has been focused on in literature [14, 15] to cater to the need for high-cooling-rate applications. A comprehensive review of various studies on buoyant convection of nanofluid-saturated porous media has been addressed by Mahadi et al. [16] and Menni et al. [17]. Marzougui et al. [18] made an attempt to explore the entropy production and thermal dissipation in a lid-driven porous domain with magnetic force. To augment or suppress the buoyant flow of nanofluid or hybrid nanofluid in a finite-shaped geometry, different mechanisms have been utilized, and successful predictions have been brought out from their analysis [19–21]. The impacts of single or multiple porous baffles on buoyant-transport of nanofluids inside a finite geometry have been investigated, and their effects are compared with a non-porous baffle [22, 23]. Ghalambaz and co-workers [24, 25] utilized a baffle with flexibility to study the buoyant motion of nanofluid inside a square domain and found an optimum baffle elasticity to maximize the heat transport. In recent years, focus towards the choice of nanoparticle to boost thermal dissipation rate, boundary layer formation with nanofluid, and different non-Newtonian fluids has been increased due to their potential applications [26–28].

The detailed review of existing research indicates that attempt has not been made to understand the impact of a circular baffle on the flow behavior and corresponding thermal transport in a differentially heated porous annulus saturated with nanofluid. In the current analysis, different baffle sizes positioned at various locations are considered to achieve the best combination for extracting maximum thermal dissipation.

24.2 MATHEMATICAL FORMULATION

The physical structure chosen for the analysis, as displayed in Figure 24.1, is the porous annular domain formed by two coaxial cylinders. A circular baffle with negligible thickness has been attached with an inner cylinder such that its length and position could be varied over the domain. The inner and outer cylinders are supplied to different uniform heating such that the inner and outer cylindrical surfaces are maintained at higher (T_h) and lower (T_c) temperatures, respectively. The upper and lower regions of the domain are considered to be adiabatic, while the thermal condition of the baffle is chosen as that of hot cylinder. The annular gap is filled with a *Cu*-water nanofluid saturated with porous medium. The thermophysical properties of *Cu* nanoparticle and base fluid are provided in Table 24.1 and taken from [14, 17]. Further, incompressibility and Newtonian fluid with laminar and axisymmetric flow has been assumed. The thermophysical properties are considered to be constant, except the density, which follows the Boussinesq approximation. The nanofluid-saturated porous material is considered to be isotropic as well as homogeneous, while the porous material and nanofluid are assumed to follow the local thermal equilibrium. By assuming the aforementioned conditions and using the Darcy model for the porous media saturated with nanofluid, the dimensional governing equations are:

$$\frac{\partial u}{\partial r} + \frac{\partial w}{\partial z} = 0, \tag{24.1}$$

$$\frac{\partial u}{\partial z} - \frac{\partial w}{\partial r} = -\frac{gK(\rho\beta)_{nf}}{\mu_{nf}} \frac{\partial \theta}{\partial r}, \tag{24.2}$$

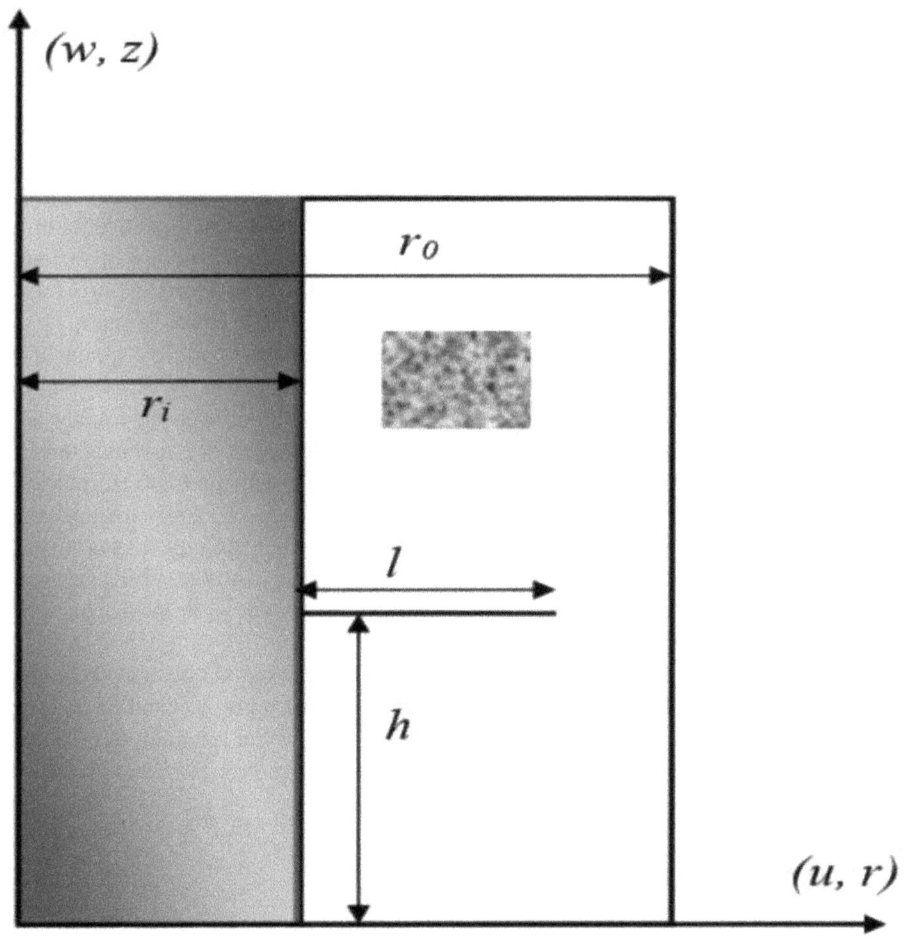

FIGURE 24.1 Axisymmetric structure of porous annulus with geometrical notations.

$$\frac{\partial \theta}{\partial t^*}+u\frac{\partial \theta}{\partial r}+w\frac{\partial \theta}{\partial z}=\alpha_{nf}\left(\frac{\partial^2 \theta}{\partial r^2}+\frac{1}{r}\frac{\partial \theta}{\partial r}+\frac{\partial^2 \theta}{\partial z^2}\right), \tag{24.3}$$

Introducing the stream function $u=\frac{1}{r}\frac{\partial \psi^*}{\partial z}$ and $w=-\frac{1}{r}\frac{\partial \psi^*}{\partial r}$, and the following non-dimensional variables, $(R,Z)=\frac{(r,z)}{D}$, $\psi=\frac{\psi^*}{\alpha_f}$, $(U,W)=\frac{(u,w)}{\alpha_f}D$, $t=\frac{\alpha_f t^*}{D^2}$, $T=\frac{(\theta-\theta_c)}{(\theta_h-\theta_c)}$, the dimensional equations (24.1)–(24.3) can be recast into the following new set of dimensionless governing equations:

$$\frac{\partial T}{\partial t}+U\frac{\partial T}{\partial R}+W\frac{\partial T}{\partial Z}=\frac{\alpha_{nf}}{\alpha_f}\left(\frac{\partial^2 T}{\partial R^2}+\frac{1}{R}\frac{\partial T}{\partial R}+\frac{\partial^2 T}{\partial Z^2}\right), \tag{24.4}$$

$$\frac{\partial^2 \psi}{\partial R^2}-\frac{1}{R}\frac{\partial \psi}{\partial R}+\frac{\partial^2 \psi}{\partial Z^2}=-R(1-\phi)^{2.5}\,Ra_D\,\frac{(\rho\beta)_{nf}}{\rho_{nf}\beta_f}\frac{\partial T}{\partial R}, \tag{24.5}$$

where:

$$U = \frac{1}{R}\frac{\partial \psi}{\partial Z},\ W = -\frac{1}{R}\frac{\partial \psi}{\partial R} \text{ and } Ra_D = \frac{gK\beta_f \Delta\theta D}{\nu_f \alpha_f}.$$

The dimensionless thermal conditions along the inner and outer surfaces of the annular structure are, respectively, $T = 1$ and $T = 0$. The temperature of the baffle is also maintained at $T = 1$. The hydrodynamic conditions along the solid surfaces are $U = W = \frac{\partial \psi}{\partial R} = \frac{\partial \psi}{\partial Z} = 0$. In the current analysis, the global thermal transport is estimated from the Nusselt number and is given by:

$$\overline{Nu} = -\frac{k_{nf}}{k_f}\frac{1}{A}\int_0^A \frac{\partial T}{\partial R} dZ$$

24.3 NUMERICAL PROCEDURE

The coupled governing model equations are numerically solved by utilizing the time-splitting techniques based on an implicit FDM (finite difference method) with second-order accuracy. We have chosen various grid sizes to check the grid independency with $\overline{Nu}$ along the hot and cold cylinders as sensitivity measures for grid independence. After successful tests, the grid size of 161×161 has been chosen to carry out further simulations. The details of grid independency are not provided in the paper for the sake of brevity and can be found in our earlier works [4, 11, 12]. Further, the code used in the current simulations has been validated, in our earlier studies, with several benchmark studies for square and annular domains with and without baffle and are also not provided here for brevity.

24.4 RESULTS AND DISCUSSION

In this section, a detailed analysis of the simulations results is discussed with an objective to attain maximized heat transport for an appropriate choice of geometrical parameters, such as baffle length and position. This is achieved by estimating the global transport rate alongside the hot cylindrical surfaces. Further, the nanofluid flow behavior with respect to these critical parameters are also discussed in the form of flow and thermal contours. To achieve these objectives, we have chosen the Darcy–Rayleigh numbers in the range of $10^1 \le Ra_D \le 10^3$; baffle lengths are varied from 20% to 80% of annular width, baffle positions are considered in the range of 20% to 80% of annular height, and nanoparticle concentrations are chosen as 0%–5% for the case of unit aspect ratio.

24.4.1 Impacts on Flow and Thermal Fields

Figure 24.2 illustrates the impacts of Rayleigh number and *Cu*-particle concentration on the flow and thermal contours for a fixed dimension of baffle size and position. For lower magnitude of Ra, the buoyant motion of nanofluid strength is moderate due to expected lower thermal buoyancy forces, and the flow extends to the entire annular domain with the main eddy towards the right cylinder. However, as the buoyant forces are enhanced with Ra_D, a stronger nanofluid motion is exhibited with threefold increase in the extreme stream function value along with the thick formation of hydrodynamic boundary layers along the inner and outer cylinders as well around the baffle. The flow structure for base and nanofluids are akin for both values of Ra_D. The thermal contours also exhibit similar variation, and the transition from conduction-based to convection-based could be observed vibrantly from thermal contours. The thermal lines reveal a strong symmetrical, saturated structure from merely aligned structure, as the buoyant strength is elevated to its maximum.

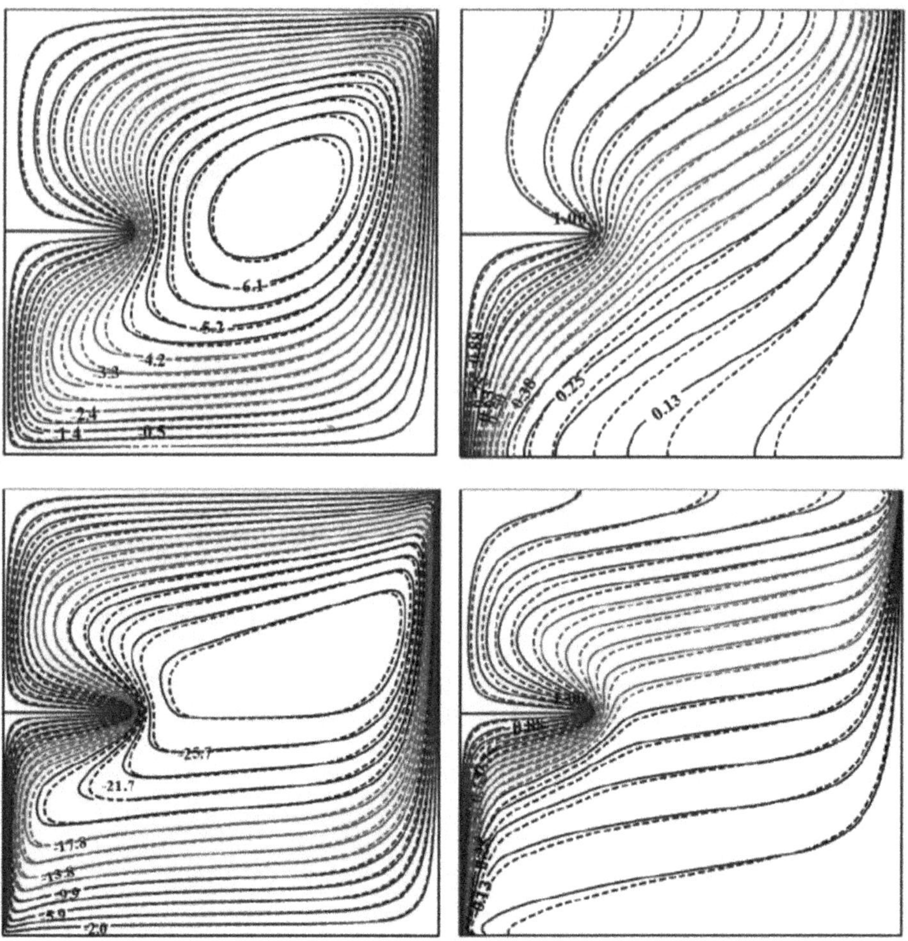

FIGURE 24.2 Flow and thermal contours for different values of Ra_D. Here, $\varphi = 0$ (solid line) and $\varphi = 0.05$ (dotted line). $Ra_D = 10^2$ (top), and $Ra_D = 10^3$ (bottom).

Also, the baffle influences the flow and thermal structure in a significant manner and can be observed from Figure 24.2. The baffle dimension on the flow structure is presented in Figure 24.3 by considering lower, moderate, and higher lengths and by fixing the baffle position at the middle. These lengths correspond to the 20%, 50%, and 80% of the total width of annular domain. The strongest buoyant flow could be observed for longer dimension of baffle, and this could be possible due to the additional buoyant strength added by the hot baffle. As the baffle dimension is decreased, the flow strength also reduces with an increase in the size of the main vortex. As regards to the isotherm contours with baffle dimension, thermal contours occupy the entire annular region for smaller baffle dimension. However, as the baffle size is enhanced, devoid, or stagnant, thermal locations could be observed above the baffle.

The flow regimes could also be effectively modified with the proper identified position of baffle, and hence, the impact of baffle placement on flow structure is exemplified in Figure 24.4 for fixed values of Ra_D and ε. To identify the location influences of baffle, we have chosen three representative positions, namely, lower, middle, and upper locations of baffle, along the inner cylinder. From the streamline contours, we could identify that the baffle position impacts the flow structure and strength in a significant way. First, the main vortex structure has been significantly altered with the

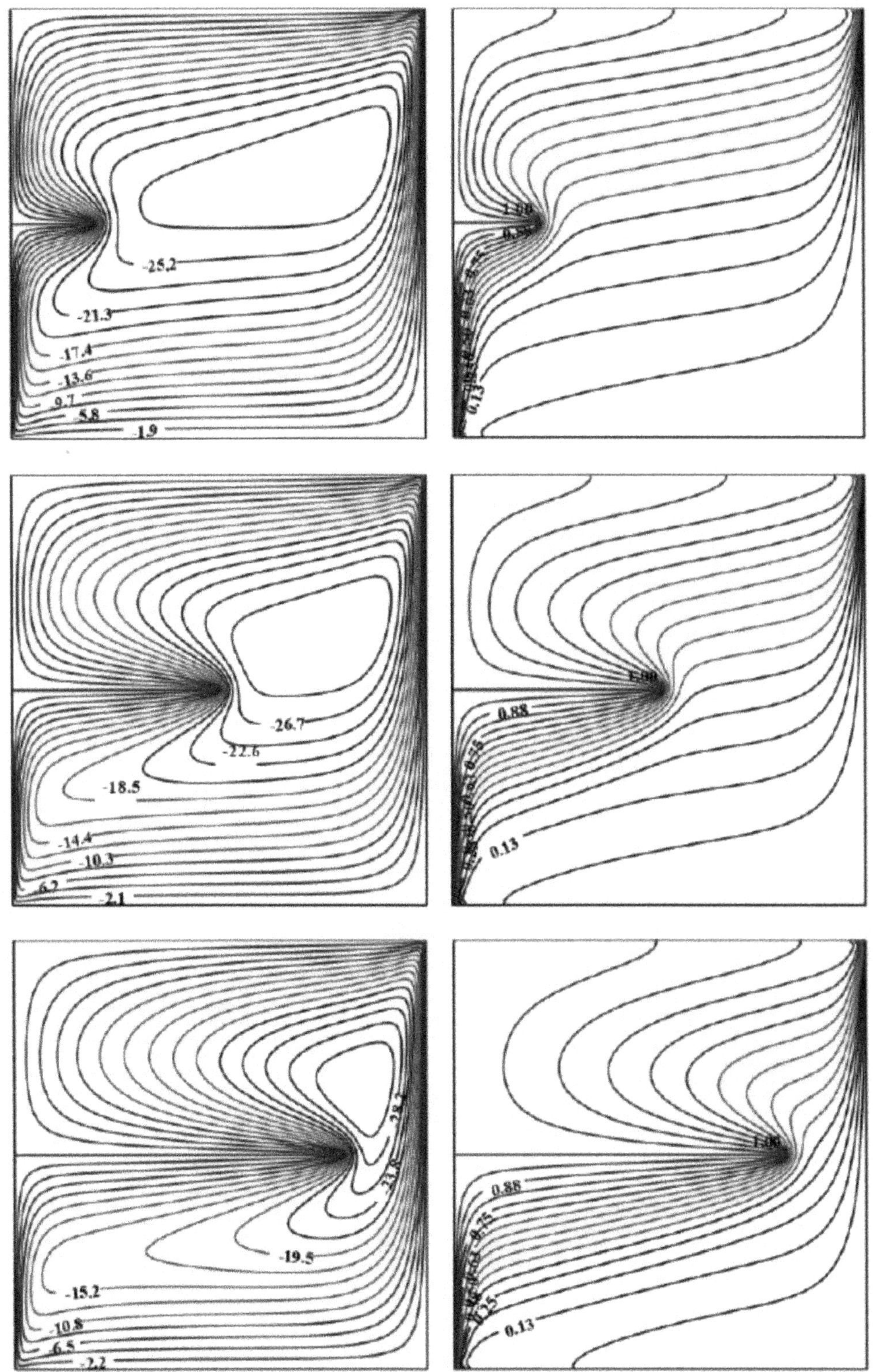

FIGURE 24.3 Impact of baffle dimension on flow and thermal contours for $Ra_D = 10^3$ and $L = 0.5$. The baffle lengths are $\varepsilon = 0.2$ (top), $\varepsilon = 0.5$ (middle), and $\varepsilon = 0.8$ (bottom).

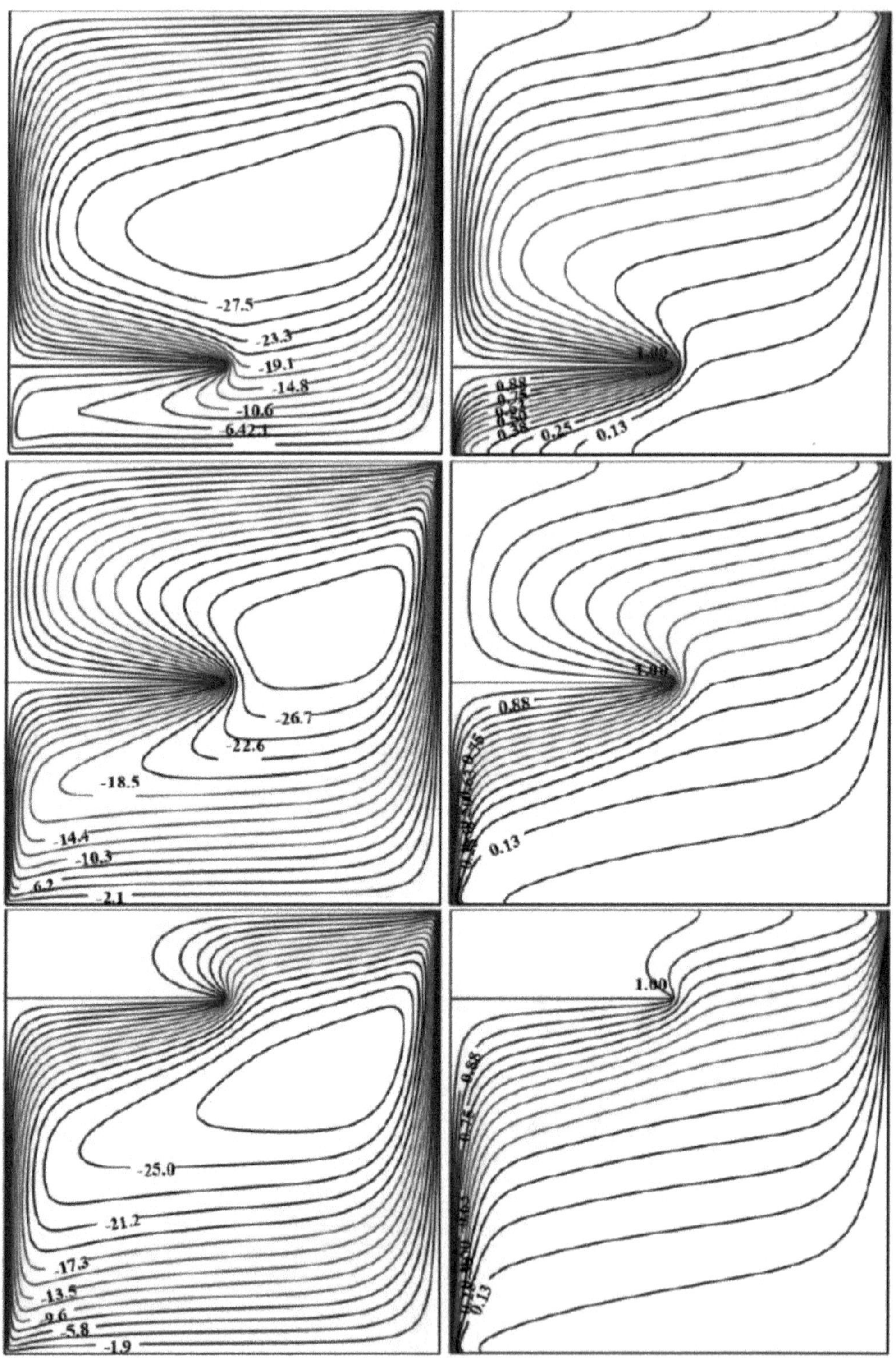

FIGURE 24.4 Impact of baffle position on flow and thermal contours for $Ra_D = 10^3$ and $\varepsilon = 0.5$.

location of baffle. When the baffle is positioned near the lower boundary, a large-sized eddy above the baffle could be seen with a stronger strength. However, as the baffle is elevated to a higher location, near the upper boundary, the position of the main eddy moves below the baffle with reduced strength. Also, for the positioning of baffle near the upper insulated surface, a stagnant or no-flow movement region has been formed above the baffle, and the lowest flow strength is observed for this arrangement. Figure 24.4 also demonstrates the isothermal contours for three different positional arrangements of baffle. These locations are chosen as the representative cases of the entire set of five positions considered in the analysis. The thermal contours reveal a strong stratified thermal structure with a thick formation of the thermal boundary layer along the regions below the inner cylinder, baffle, and upper regions of outer cylinder. This discontinuity in the boundary layer formation is expected due to the presence of baffle and is consistent with the existing literature investigations. Also, as observed in the flow structure, for the positioning of baffle near the upper boundary, a stagnant region without much thermal variation could be observed in Figure 24.4. It is worth mentioning that these contour plots are made for *Cu*-water nanofluid.

24.4.2 Impacts on Heat Dissipation Rates

For the design of any thermal equipment or heat exchangers, the quantitative information is very much essential so that better products could be made. This quantitative measure in any heat transport analysis could be measured in terms of local and global Nusselt numbers. In our simulations, this vital quantitative measure has been estimated, and the impact of various parameters has been portrayed in Figures 24.5–24.8. In thermal transport of nanofluid, the percentage of nanoparticle dispersion is an important quantity to be identified so that maximum thermal transport can be achieved and to avoid sedimentation of the chosen nanoparticles. Figure 24.5 depicts the combined impacts of Ra and *Cu*-concentration on the overall thermal dissipation rates for fixed baffle dimensions. From the simulation results presented in Figure 24.5, we could keenly observe two phenomena. The

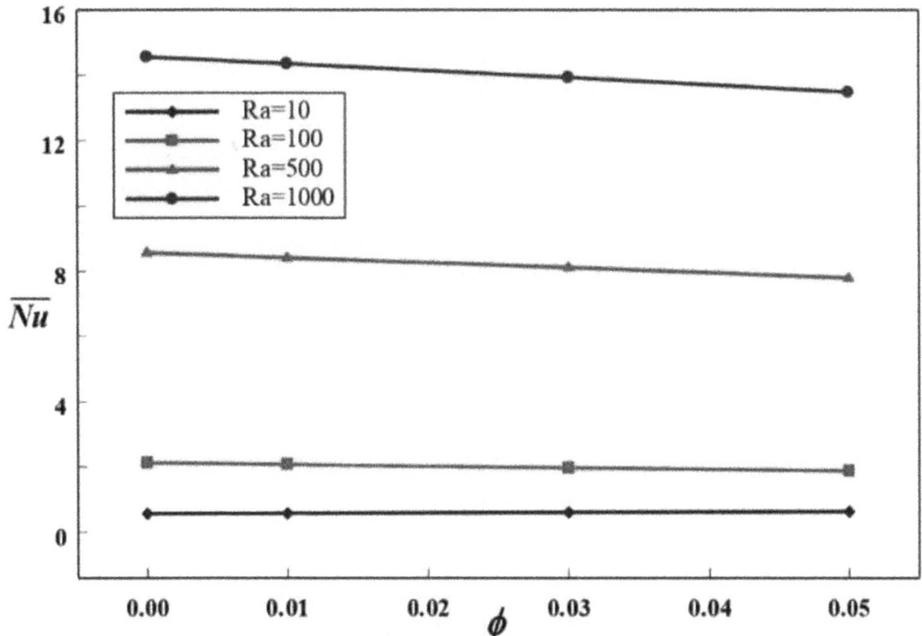

FIGURE 24.5 Impact of Ra_D and φ on average Nu for $L = 0.5$ and $\varepsilon = 0.5$.

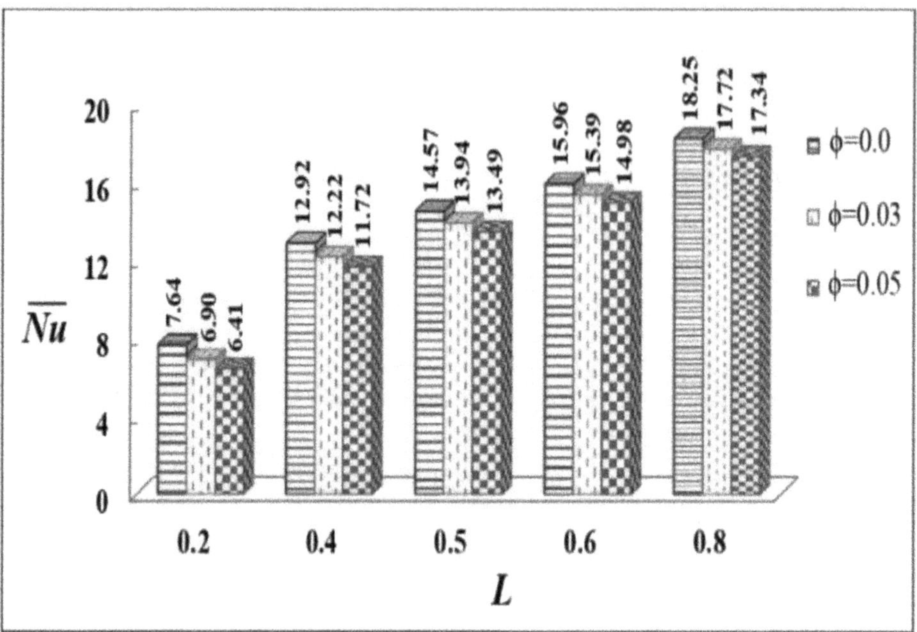

FIGURE 24.6 Impact of φ and L on average Nusselt number for $Ra_D = 10^3$ and $\varepsilon = 0.5$.

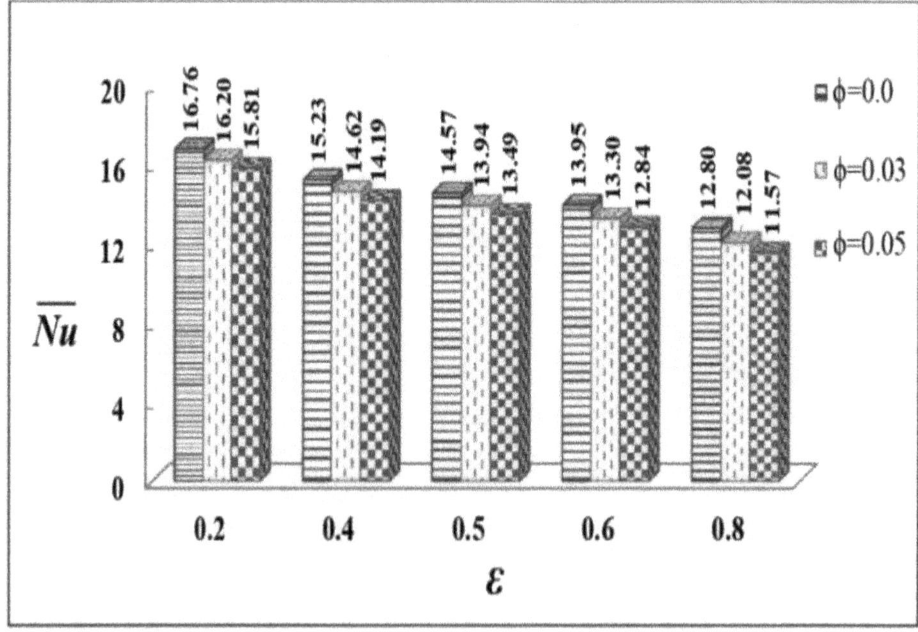

FIGURE 24.7 Impact of φ and ε on average Nusselt number for $Ra_D = 10^3$ and $L = 0.5$.

first one being thermal transport could be improved with an enhancement in Rayleigh number and is very much apparent, as increasing Ra increases the buoyant forces, which in turn intensify the convective motion in the annular domain. The second observation, important and unique for porous domains, is that thermal dissipation rate has been suppressed with an increase in the *Cu* nanoparticle

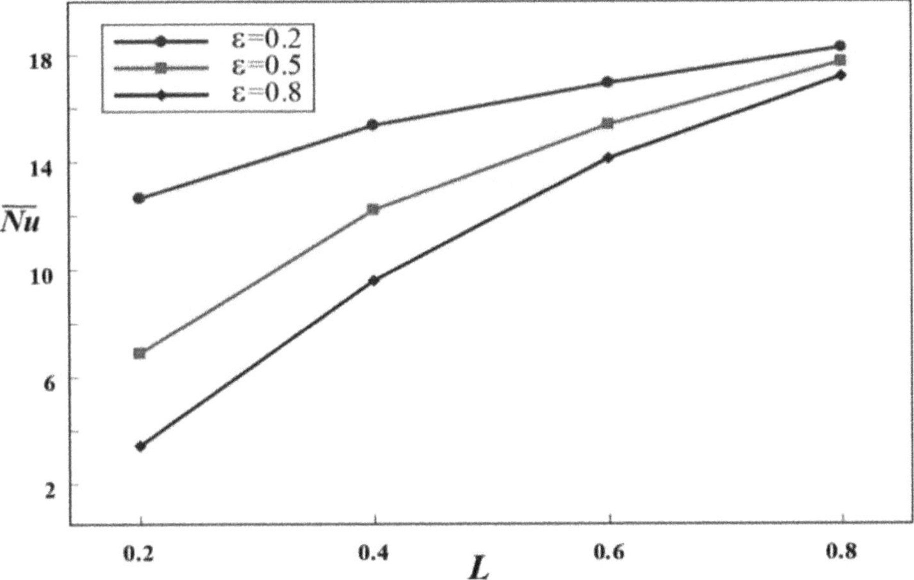

FIGURE 24.8 Impact of L and ε on average Nusselt number for $Ra_D = 10^3$ and $\varphi = 0.03$.

concentration, particularly at higher *Ra*. This could be related to the thermophysical properties of nanofluid, namely, the density and viscosity. It is a known fact that these two physical quantities are higher for nanofluids as compared to base fluid, water. Hence, during higher convective flows, the resistance generated from these two properties of nanofluid acts as a drag force and hence reduces the thermal dissipation rate. This particular phenomenon has been recorded in the literature for similar studies in various finite-shaped domains and hence consistent with literature findings.

Figure 24.6 demonstrates the combined impacts of nanoparticle concentration and baffle positioning on the global thermal dissipation rate by fixing Ra and ε. As discussed in Figure 24.5, the thermal transport rate declines with an increase in *Cu* nanoparticle concentration, and this is because the thermophysical properties of nanofluid produce an extra drag at higher convective flow regimes. That is, for a fixed baffle position, increase in *Cu* nanoparticle leads to suppression in the thermal transport and is true at all baffle locations. As regards baffle positioning, higher thermal dissipation rate could be expected by positioning the baffle at an elevated location near the upper boundary. However, it should be noted that higher flow circulation rates are observed for positioning the baffle near the lower surface. This indicates an important fact that higher circulation flows need not produce higher thermal dissipation rates. This could be attributed to the fact that the baffle positioned at a higher elevation could provide an additional source of energy to dissipate more thermal dissipation to the surrounding nanofluid. This particular finding could be very useful for identifying an optimal positioning of baffle to extract higher thermal dissipation in the porous domain.

In heat exchanger applications, due to space constraints, the dimension of baffle is also an important and critical parameter to be properly identified so that maximum temperature can be dissipated at a desirable space. Figure 24.7 portrays the collective influences of baffle dimension and *Cu* nanoparticle concentration on thermal extraction rate from the hot cylinder. It is interesting to observe that with a baffle of shorter dimension, maximum thermal dissipation could be extracted from the hot boundary. Further, this observation is very much consistent with the findings of Shi and Khodadadi [1] and Tasnim and Collins [2] in a square geometry for a conventional fluid, air. They concluded that a baffle of shorter length is more effective in dissipating the heat rather than a larger baffle. In addition, the variation of thermal dissipation with φ is similar to the observation predicted in earlier results, which indicates the base fluid is more effective in thermal dissipation than the nanofluid.

The individual impacts of baffle dimension and position on thermal extraction have been portrayed and discussed in Figures 24.6 and 24.7. However, an optimum baffle dimension along with proper positioning must be identified so that the impacts of baffle on thermal extraction could be used by the design engineer for better product design. This important aspect has been discussed in Figure 24.8 by choosing different baffle dimensions and locations. The findings from this simulation are very much consistent with the observations made in Figures 24.6 and 24.7, where the individual impacts of baffle have been documented. As observed earlier, thermal dissipation rate increases with an elevated baffle positioning and lowering of the baffle dimension. The estimated result reveals that a baffle with shorter dimension ($\varepsilon = 0.2$) and positioned at a higher elevation on inner cylinder ($L = 0.8$) produces higher thermal dissipation rate compared to other baffle dimensions and positions. It can be concluded, based on the current analysis and with the conditions chosen in this investigation, that a shorter baffle fixed at a location nearer the upper surface could enhance the thermal transport rates among the other dimensions and locations of the baffle considered in this investigation. In other words, this observation could also be used to minimize the thermal transport with other choice of baffle position and dimension, depending on the need of the design engineer with the limitations considered in this analysis.

24.5 CONCLUSIONS

In this theoretical investigation, buoyant convective motion and heat transport of $Cu\text{-}H_2O$ nanofluid in a porous annulus with a baffle of negligible thickness attached to an inner cylinder have been numerically analyzed. The influence of critical parameters along with baffle positions and dimensions on flow and thermal patterns and $\overline{Nu}$ with a fixed value of aspect ratio has been widely analyzed utilizing Darcy's model. The key findings of the present analysis are summarized as follows:

1. The baffle positioning has a strong impact on promoting the buoyant nanofluid convective flow phenomena in the enclosure. The observations could be used to control or modify the flow regimes effectively in the porous domain.
2. Thermal extraction could be maximized for lower baffle dimension and elevated baffle positioning at all magnitudes of Ra_D. The utilization of nanofluid in porous media exhibits reduced thermal dissipation rate as compared to base fluid due to the additional drag force generated by nanofluid.
3. From the detailed analysis, an optimum baffle location along with baffle dimension has been identified to maximize as well as minimize the thermal transport rates.

24.6 ACKNOWLEDGMENTS

The authors, MAL, MS, and CA, acknowledge the financial assistance from UTAS, Ibri, Oman, under the Internal Research Funding via project number DSR-IRPS-2021-22-PROP-1.

TABLE 24.1
Thermophysical Properties of Water and *Cu* [14, 17]

Property	H_2O	Cu
$\rho(kg/m^3)$	997.1	8933
$C_p(J/kg\ K)$	4179	385
$k(W/mK)$	0.613	400
$\beta(K^{-1})$	21×10^{-5}	1.67×10^{-5}

Nomenclature

A	aspect ratio	
H	height of the annular enclosure	(m)
D	annular width	(m)
l	dimensional baffle position	(m)
L	dimensionless baffle position (l/D)	
h	dimensional baffle length	(m)
ε	dimensionless baffle length (h/H)	
Ra_D	Darcy–Rayleigh number	
t	dimensionless time	
θ	dimensional temperature	
(R,Z)	radial, axial coordinates	
(U,W)	velocity components	
α	thermal diffusivity	(m^2/s)
β	thermal expansion coefficient	(K^{-1})
T	dimensionless temperature	
λ	radius ratio	
φ	nanoparticle volume fraction	
nf	nanofluid	
f	fluid	
ψ	dimensionless stream function	

REFERENCES

[1] Shi, X., & Khodadadi, J.M. (2003). Laminar natural convection heat transfer in differentially heated square cavity due to a thin fin on the hot wall. *Transactions of the ASME, Journal of Heat Transfer*, 125, 624–634.

[2] Tasnim, S.H., & Collins, M.R. (2004) Numerical analysis of heat transfer in a square cavity with a baffle on the hot wall. *International Communications in Heat and Mass Transfer*, 31, 639–650.

[3] Wang, L., Wang, W.-W., Cai, Y., Liu, D., & Zhao F.-Y. (2020). Mixed convection and heat flow characteristics in a lid-driven enclosure with porous fins: Full numerical modeling and parametric investigations. *Numerical Heat Transfer, Part A: Applications*, 77(4), 361–390.

[4] Pushpa, B.V., Sankar, M., & Makinde, O.D. (2020). Optimization of thermosolutal convection in vertical porous annulus with a circular baffle. *Thermal Science and Engineering Progress*, 20, 100735.

[5] Choi, S.U.S., & Eastman, J.A. (1995). Enhancing thermal conductivity of fluids with nanoparticles, *ASME International Mechanical Engineering Congress and Exposition*, November 12–17, San Francisco, CA.

[6] Abouali, O., & Falahatpisheh, A. (2009). Numerical investigation of natural convection of Al_2O_3 nanofluid in vertical annuli. *Heat Mass Transfer*, 46, 15–23.

[7] Mebarek-Oudina, F. (2017). Numerical modeling of the hydrodynamic stability in vertical annulus with heat source of different lengths. *International Journal of Engineering Science Technologies*, 20, 1324–1333.

[8] Mebarek-Oudina, F. (2019). Convective heat transfer of Titania nanofluids of different base fluids in cylindrical annulus with discrete heat source. *Heat transfer-Asian Research*, 48, 135–147.

[9] Pordanjani, A.H., et al. (2019). An updated review on application of nanofluids in heat exchangers for saving energy. *Energy Conversion and Management*, 198, 111886.

[10] Prasad, V. (1986). Numerical study of natural convection in a vertical, porous annulus with constant heat flux on the inner wall. *International Journal of Heat and Mass Transfer*, 29, 841–853.

[11] Shivakumara, I.S., Prasanna, B.M.R., Rudraiah, N., & Venkatachalappa, M. (2003). Numerical study of natural convection in vertical cylindrical annulus using a non-Darcy equation. *Journal of Porous Media*, 5, 187–102.

[12] Sankar, M., Park, Y., Lopez, J.M., & Do, Y. (2011). Numerical study of natural convection in a vertical porous annulus with discrete heating. *International Journal of Heat and Mass Transfer*, 54, 1493–1505.

[13] Sankar, M., Park, Y., Lopez, J.M., & Do, Y. (2012). Double-diffusive convection from a discrete heat and solute source in a vertical porous annulus. *Transport in Porous Media*, 91, 753–775.

[14] Bourantas, G.C., Skouras, E.D., Loukopoulos, V.C., & Burganos, V.N. (2014). Heat transfer and natural convection of nanofluids in porous media. *European Journal of Mechanics—B/Fluids*, 43, 45–56.

[15] Alsabery, A.I., Chamkha, A.J., Saleh, H., & Hashim, I. (2017). Natural convection flow of a nanofluid in an inclined square enclosure partially filled with a porous medium. *Scientific Reports*, 7, 2357.

[16] Mahdi, R.A., Mohammed, H.A., Munisamy, K.M., & Saeid, N.H. (2015). Review of convection heat transfer and fluid flow in porous media with nanofluid. *Renewable and Sustainable Energy Reviews*, 41, 715–734.

[17] Menni, Y., Chamkha, A.J., & Azzi, A. (2018). Nanofluid transport in porous media: A review. *Special Topics & Reviews in Porous Media-An International Journal*, 9(4), 1–16.

[18] Marzougui, S., Mebarek-Oudina, F., Mchirgui, A., & Magherbi, M. (2021). Entropy generation and heat transport of cu-water nanoliquid in porous lid-driven cavity through magnetic field. *International Journal of Numerical Methods for Heat & Fluid Flow*, 32(6), 2047–2069.

[19] Chabani, I., Mebarek Oudina, F., & Ismail, A.I. (2022). MHD flow of a hybrid nano-fluid in a triangular enclosure with zigzags and an elliptic obstacle. *Micromachines*, 13(2), 224.

[20] Pushpa, B.V., Sankar, M., & Mebarek-Oudina, F. (2021). Buoyant convective flow and heat dissipation of Cu-H_2O nanoliquids in an annulus through a thin baffle. *Journal of Nanofluids*,10(2), 292–304.

[21] Dhif, K., Mebarek-Oudina, F., Chouf, S., Vaidya, H., & Chamkha, A.J. (2021). Thermal analysis of the solar collector cum storage system using a hybrid-nanofluids. *Journal of Nanofluids*,10(4), 634–644.

[22] Siavashi, M., Yousofvand, R., & Rezanejad, S. (2018). Nanofluid and porous fins effect on natural convection and entropy generation of flow inside a cavity. *Advanced Powder Technology*, 29(1), 142–156.

[23] Wang, L., Liu, R.-Z., Liu, D., Zhao, F.-Y., & Wang, H.-Q. (2020). Thermal buoyancy driven flows inside a differentially heated enclosure with porous fins of multiple morphologies attached to the hot wall. *International Journal of Thermal Sciences*, 147, 106138.

[24] Saleh, H., Hashim, I., Jamesahar, E., & Ghalambaz, M. (2020). Effects of flexible fin on natural convection in enclosure partially-filled with porous medium. *Alexandria Engineering Journal*, 59, 3515–3529.

[25] Mehryan, S.A.M., Alsabery, A., Modir, A., Izadpanahi, E., & Ghalambaz, M. (2020). Fluid-structure interaction of a hot flexible thin plate inside an enclosure. *International Journal of Thermal Sciences*, 153, 106340.

[26] Asogwa, K., Mebarek-Oudina, F., & Animasaun, I. (2021). Comparative investigation of water-based Al_2O_3 nanoparticles through water-based CuO nanoparticles over an exponentially accelerated radiative Riga plate surface via heat transport. *Arabian Journal for Science and Engineering*, 47, 8721–8738.

[27] Djebali, R., Mebarek-Oudina, F., & Choudhari, R. (2021). Similarity solution analysis of dynamic and thermal boundary layers: Further formulation along a vertical flat plate. *Physica Scripta*, 96(8), 085206.

[28] Rajashekhar, C., Mebarek-Oudina, F., Vaidya, H., Prasad, K.V., Manjunatha, G., & Balachandra, H. (2021). Mass and heat transport impact on the peristaltic flow of Ree-Eyring liquid with variable properties for hemodynamic flow. *Heat Transfer*, 50(5), 5106–5122.

25 Second Law Analysis of Magneto-Thermosolutal Convection and Energy Transport in Nanoliquid-Filled Annulus

M. Sankar, H. A. Kumara Swamy, F. Mebarek-Oudina, and N. Keerthi Reddy

25.1 INTRODUCTION

The flow occurs due to the impact of dual density gradients, such as heat and mass, in a system, which is known as the double-diffusive convection or thermosolutal convection. Double-diffusive convection in different finite-sized enclosures has been the area of intensive research over the past few decades due to its priority in natural, industrial, and engineering applications. An annular enclosure between two upright cylinders has been one of the most sought geometries among the finite geometries. The impact of discrete isoflux source length on heat transfer stability in a cylindrical annulus has been numerically analyzed by Mebarek-Oudina [1] and reported that the heat transfer rate is greater with smaller source length. Recently, Husain et al. [2] made an extensive review on experimental and numerical investigations on thermal dissipation rate in an annulus. In addition to energy transport rates, the study on irreversibility distribution measured in terms of entropy generation (EG) has become most popular since the system efficiency could be enhanced by minimizing the entropy generation. Also, through theoretical simulations and experimental visualizations, it has been proven that the application of magnetic field shows a dramatic change in the flow strength, which thereby impacts the energy dissipation rate. The impact of curvature effects on fluid movement in the annulus subjected to aiding cases of thermal and solute concentration has been analyzed by Retiel et al. [3]. Chen et al. [4] investigated thermosolutal convection in an annulus using LBM to analyze the influences of dual buoyancy forces ratio, aspect, and radius ratios. Venkatachalappa et al. [5] analyzed the effect of both aided and opposing cases, the influence of Lorentz force on double-diffusive convection in a vertical annulus, and found that the magnetic field has great impact on double-diffusive flow when it is applied normal to flow direction. Nikbakhti and Rahimi [6] employed the finite difference technique to examine the heat and mass dissipation rate in a partially active enclosure for nine different cases and concluded that the rate of thermal and solutal transport is greater for the bottom-top thermally active section and lower transport rate for the top-bottom thermally active section.

The thermal transport efficiency of traditionally used fluids, such as ethylene glycol, water, oil, and many others, has been found to owe lower thermal conductivity. Through several theoretical and experimental analysis, it has been found that the thermal transport can be enhanced by the inclusion of nano-sized particles having high thermal conductivity [7, 8]. Due to promising prospects of nanoliquids, several research works have been performed by using nanoliquid as a working medium. Abouali and Falahatpisheh [9] investigated the buoyant motion of alumina-water nanoliquid in an annular space and derived correlations for thermal transport rates. A computational study has been

DOI: 10.1201/9781003299608-25

performed on convective flow, irreversibility distribution, and heat transport by considering nanoliquid as working medium in an annular enclosure under various boundary conditions to identify an appropriate thermal condition to increase the thermal efficiency of the system [10–16]. Esfahani and Bordbar [17] conducted a numerical experiment on double-diffusive convection of different nanoliquids filled in a cavity and concluded that the type of nanoparticle plays a key role in increasing/decreasing the heat and mass transport rates. Mejri et al. [18] studied the influence of Lorentz force and non-uniform thermal profile on nanoliquid movement and concluded that the set of parametric values plays a key role to enhance the thermal efficiency of the system. The impact of buoyancy ratio and concentration of SiO_2 nanoparticle on thermal and solutal transport rates has been analyzed by Chen et al. [19] and developed a correlation for average heat and mass transfer rates between Rayleigh number, buoyancy ratio, and nanoparticle concentration. Reddy et al. [20] analyzed the influence of magnetic force on fluid movement induced by dual buoyancy forces raised due to heat and mass gradients. As discussed earlier, in several industrial applications, the fluid flow occurs due to heat and mass gradients. In this regard, irreversibilities from heat transfer, viscous, and diffusive effects are responsible for the production of entropy during double-diffusive convection. Along with heat and mass dissipation rates, the irreversibility distribution due to various components in a partially active tall geometry has been analyzed by Oueslati et al. [21]. Chen et al. [22] numerically studied the influence of nanoparticle concentration and buoyancy ratio on entropy production for laminar and turbulent regimes and reported that the EG enhances more intensively in turbulent regimes and reaches minimum with unit buoyancy ratio. Arun et al. [23] conducted a numerical experiment to analyze the influence of Lorentz force on thermosolutal convection and EG of liquid metal in a cavity containing adiabatic block and concluded that irreversibility in the enclosure could be minimized with greater magnetic field strength. Parveen and Mahapatra [24] reported that an increase in buoyancy ratio, undulation, and Hartmann number leads to a decrease in EG of nanofluid in a cavity with a wavy top wall. Marzougui et al. [25] adopted the finite element method to investigate the irreversibility distribution in a nanoliquid-saturated porous lid-driven cavity under the influence of magnetic force. Recently, the effect of dual buoyancy forces on fluid flow and energy dissipation rate in an annular chamber under different constraints in the absence of magnetic field has been numerically investigated and provided the set of parametric values that enhances the system performance [26, 27].

The nanoliquid movement driven by buoyancy forces, energy dissipation rate, and irreversibility distribution is very sensitive to the structure of the enclosure. In several industries, the shape of an enclosure may be non-regular, which includes inclined square/rectangular geometries, triangular, parallelogram, stretching sheet, horizontal annulus, riga plate, and many other odd shapes. By considering different constraints, numerical studies have been performed on energy transport rates in various geometries [28–34]. To consolidate nanoliquid application and techniques to enhance the thermal transport in various geometries, a detailed review has been recently made by Mebarek-Oudina and Chabani [35].

Motivated by the aforementioned investigations, the novelty of the current investigation lies in analyzing the impact of magnetic field along with thermal and solutal gradients (both aided and opposing) on heat as well as mass transport processes and entropy generation optimization in the annular domain filled with nanoliquid. The current investigation aims to illustrate the impact of pertinent parameters, such as buoyancy ratio, Lewis number, Hartmann number, and nanoparticle concentration on heat and mass transport rates along with entropy production. To the authors' best knowledge, it is believed that the results obtained from the present investigation can contribute to designing an equipment with improvised system efficiency.

25.2 MATHEMATICAL MODELLING

The physical system considered in the current investigation is a two-dimensional annular enclosure obtained by two coaxial concentric cylinders with r_i and r_o as radii of interior and exterior cylinders, respectively, as illustrated schematically in Figure 25.1. The annular gap is occupied with alumina-water nanoliquid. The inner and outer cylinders are respectively maintained at uniform but different temperatures and concentrations (θ_h, S_h) and (θ_c, S_c), while the horizontal are impermeable

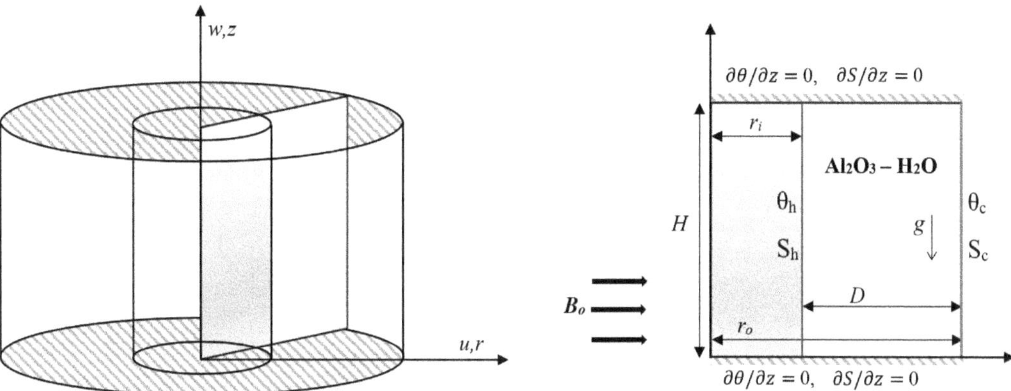

FIGURE 25.1 Schematic representation along with axisymmetric view.

TABLE 25.1
Thermophysical Properties of H_2O and Al_2O_3

	$\rho(Kg\,m^{-3})$	$C_p\,(J\,Kg^{-1}K^{-1})$	$k(W\,m^{-1}K^{-1})$	$\beta\times10^{-5}(K^{-1})$	$\sigma(S\,m^{-1})$
H_2O	997.1	4179	0.613	20.7	0.05
Al_2O_3	3970	765	40	0.85	10^{-10}

and adiabatic. The properties of water and alumina nanoparticle are shown in Table 25.1. Also, the expression for density, thermal conductivity, viscosity, specific heat, thermal expansion coefficient, thermal diffusivity, and electrical conductivity of nanoliquid are given in Table 25.2.

Assumptions:

- The fluid is Newtonian and incompressible.
- Flow is laminar, unsteady, and axisymmetric.
- The magnetic field generated due to the motion of nanoliquid is neglected.
- The Boussinesq approximation, which accounts for the density variation only in the body force term of the momentum equation, is utilized in the current analysis.

25.2.1 Dimensional Equations

For the aforementioned assumptions, the dimensional governing equations obtained by the conservative laws are as follows:

$$\nabla.\vec{q} = 0 \tag{25.1}$$

$$\rho_{nf}\left[\frac{\partial\vec{q}}{\partial t^*} + (\vec{q}.\nabla)\vec{q}\right] = -\nabla p + \mu_{nf}\nabla^2\vec{q} + (\rho\beta_T)_{nf}\,g(\theta-\theta_0) - (\rho\beta_C)_{nf}\,g(S-S_0) + (\vec{J}\times\vec{B}) \tag{25.2}$$

$$\frac{\partial\theta}{\partial t^*} + (\vec{q}.\nabla)\theta = \alpha_{nf}\nabla^2\theta \tag{25.3}$$

$$\frac{\partial S}{\partial t^*} + (\vec{q}.\nabla)S = \alpha_s\nabla^2 S \tag{25.4}$$

Here, $\vec{q} = (u, w)$ is the velocity components, and $\vec{J}$ and $\vec{B}$ are electric current and magnetic field intensities, respectively.

TABLE 25.2
Expression for Properties of Nanoliquid

ρ_{nf}	=	$(1-\phi)\rho_f+\phi\rho_p$
$(\rho C_p)_{nf}$	=	$(1-\phi)(\rho C_p)_f+\phi(\rho C_p)_p$
$(\rho\beta)_{nf}$	=	$(1-\phi)(\rho\beta)_f+\phi(\rho\beta)_p$
μ_{nf}	=	$\dfrac{\mu_f}{(1-\phi)^{2.5}}$
k_{nf}	=	$\left[\dfrac{(k_p+2k_f)-2\phi(k_f-2k_p)}{(k_p+2k_f)+\phi(k_f-2k_p)}\right]k_f$
α_{nf}	=	$\dfrac{k_{nf}}{(\rho C_p)_{nf}}$
σ_{nf}	=	$\left[1+\dfrac{3\left(\dfrac{\sigma_p}{\sigma_f}-1\right)\phi}{\left(\dfrac{\sigma_p}{\sigma_f}+2\right)-\left(\dfrac{\sigma_p}{\sigma_f}-1\right)\phi}\right]\sigma_f$

25.2.2 Non-Dimensional Equations

The preceding continuity, momentum, temperature, and concentration equations are reduced to a dimensionless form by adopting the following transformations:

$$(R,Z)=\frac{(r,z)}{D},\ (U,W)=\frac{D(u,w)}{\alpha_f},\ t=\frac{t^*\alpha_f}{D^2},\ T=\frac{(\theta-\theta_0)}{\Delta\theta},\ C=\frac{(S-S_0)}{\Delta S},\ P=\frac{pD^2}{\rho_{nf}\alpha_f^{\ 2}}.$$

The non-dimensional thermal, concentration, and vorticity-stream function equations are:

$$\frac{\partial T}{\partial t}+U\frac{\partial T}{\partial R}+W\frac{\partial T}{\partial Z}=\frac{\alpha_{nf}}{\alpha_f}\nabla^2{}_1T, \tag{25.5}$$

$$\frac{\partial C}{\partial t}+U\frac{\partial C}{\partial R}+W\frac{\partial C}{\partial Z}=\frac{1}{Le}\nabla^2{}_1C, \tag{25.6}$$

$$\begin{aligned}\frac{\partial \zeta}{\partial t}+U\frac{\partial \zeta}{\partial R}+W\frac{\partial \zeta}{\partial Z}-\frac{U\zeta}{R}=\frac{\mu_{nf}}{\rho_{nf}\alpha_f}\left[\nabla^2{}_1\zeta-\frac{\zeta}{R^2}\right]-\frac{(\rho\beta_T)_{nf}}{\rho_{nf}(\beta_T)_f}Ra\,Pr\left[\frac{\partial T}{\partial R}-N\frac{\partial C}{\partial R}\right]\\+\left(\frac{\rho_f}{\rho_{nf}}\right)\left(\frac{\sigma_{nf}}{\sigma_f}\right)Ha^2\,Pr\frac{\partial W}{\partial R}\end{aligned} \tag{25.7}$$

$$\zeta=\frac{1}{R}\left[\frac{\partial^2\psi}{\partial R^2}-\frac{1}{R}\frac{\partial\psi}{\partial R}+\frac{\partial^2\psi}{\partial Z^2}\right] \tag{25.8}$$

Here:

$$U = \frac{1}{R}\frac{\partial \psi}{\partial Z},\ W = -\frac{1}{R}\frac{\partial \psi}{\partial R},\ \nabla^2_1 = \frac{\partial^2}{\partial R^2} + \frac{1}{R}\frac{\partial}{\partial R} + \frac{\partial^2}{\partial Z^2},$$

$$Ra = \frac{g\beta\Delta\theta D^3}{\upsilon_f \alpha_f},\ Pr = \frac{\upsilon_f}{\alpha_f},\ Ha = B_0 D\sqrt{\frac{\sigma_f}{\mu_f}},\ N = \frac{(\rho\beta_C)_{nf}\,\Delta S}{(\rho\beta_T)_{nf}\,\Delta\theta},\ Le = \frac{\alpha_f}{\alpha_s}.$$

25.2.3 Boundary Conditions

The dimensionless velocity, thermal, and solutal boundary conditions imposed on the motion field are as follows:

At $t = 0$: $U = W = T = C = 0, \psi = \zeta = 0;\quad \frac{1}{\lambda - 1} \le R \le \frac{\lambda}{\lambda - 1},\ 0 \le Z \le Ar$

At $t > 0$;

Along the inner cylinder: $\psi = \frac{\partial \psi}{\partial R} = 0,\ T = C = 1,$

Along the outer cylinder: $\psi = \frac{\partial \psi}{\partial R} = 0,\ T = C = 0,$

Along horizontal surfaces: $\psi = \frac{\partial \psi}{\partial Z} = 0,\ \frac{\partial T}{\partial Z} = \frac{\partial C}{\partial Z} = 0$

25.2.4 Average Nusselt and Sherwood Numbers

The total thermal and solutal dissipation rates along the interior vertical surface of the geometry are respectively represented by the average Nusselt and Sherwood numbers, which are defined by:

$$\overline{Nu} = -\left(\frac{k_{nf}}{k_f}\right)\frac{1}{Ar}\int_0^{Ar}\frac{\partial T}{\partial R}dZ \quad \text{and} \quad \overline{Sh} = -\frac{1}{Ar}\int_0^{Ar}\frac{\partial C}{\partial R}dZ$$

25.2.5 Entropy Generation Equation

In the process of double-diffusive convection, the local entropy generation is owing to the irreversibility nature of thermal transfer, liquid friction, magnetic field impact, and mass transfer. According to local thermodynamic equilibrium with linear transport theory and the assumptions made, and by using the non-dimensional variables, the dimensionless entropy generation due to heat transfer ($S_{l.T}$), liquid friction ($S_{l.\psi}$), magnetic field ($S_{l.M}$), and mass transfer ($S_{l.C}$) is defined as follows:

$$S_{l.T} = \frac{k_{nf}}{k_f}\left[\left(\frac{\partial T}{\partial R}\right)^2 + \left(\frac{\partial T}{\partial Z}\right)^2\right], \tag{25.9}$$

$$S_{l.\psi} = \Phi_1 \frac{\mu_{nf}}{\mu_f}\left[2\left\{\left(\frac{\partial U}{\partial R}\right)^2 + \left(\frac{U}{R}\right)^2 + \left(\frac{\partial W}{\partial Z}\right)^2\right\} + \left(\frac{\partial U}{\partial Z} + \frac{\partial W}{\partial R}\right)^2\right], \tag{25.10}$$

$$S_{l.M} = \Phi_1 W^2 Ha^2 \left(\frac{\sigma_{nf}}{\sigma_f}\right), \tag{25.11}$$

$$S_{l.C} = \Phi_2 \left[\left(\frac{\partial C}{\partial R} \right)^2 + \left(\frac{\partial C}{\partial Z} \right)^2 \right] + \Phi_3 \left[\left(\frac{\partial T}{\partial R} \right) \left(\frac{\partial C}{\partial R} \right) + \left(\frac{\partial T}{\partial Z} \right) \left(\frac{\partial C}{\partial Z} \right) \right]. \tag{25.12}$$

Here, Φ_i $(1 \leq i \leq 3)$ denotes the irreversibility coefficient ratio defined by the following expressions:

$$\Phi_1 = \frac{\mu_f}{k_f} \theta_o \left(\frac{\alpha_f}{D \Delta\theta} \right)^2, \quad \Phi_2 = \frac{R D \theta_o}{k_f C_o} \left(\frac{\Delta S}{\Delta\theta} \right)^2, \quad \Phi_3 = \frac{R D}{k_f} \left(\frac{\Delta S}{\Delta\theta} \right)$$

Which are taken as constant at $\Phi_1 = 10^{-4}$, $\Phi_2 = 0.5$, and $\Phi_3 = 0.01$. The sum of equations (25.9)–(25.12) gives the total local entropy generation in the annulus, which can be written as:

$$S_{GEN} = S_{l.T} + S_{l.\psi} + S_{l.M} + S_{l.C} \tag{25.13}$$

On integrating equation (25.13) across the enclosure, global entropy production inside the geometry can be estimated:

$$S_{tot} = \frac{1}{A} \iint_A S_{GEN}\, dA = \frac{1}{A} \iint_A (S_{l.T} + S_{l.\psi} + S_{l.M} + S_{l.C})\, dA \tag{25.14}$$

Equation (25.14) can be written as $S_{tot} = S_T + S_\psi + S_M + S_C$. Another important non-dimensional number used in the current investigation is the Bejan number, which is defined as the ratio of sum of entropy due to heat and mass to the total entropy production.

$$Be = \frac{S_T + S_C}{S_{tot}} \tag{25.15}$$

25.3 METHODOLOGY, GRID SENSITIVITY, AND VALIDATION

The governing partial differential equations (25.5)–(25.8) are solved by implementing an implicit FDM together with initial and boundary conditions. In particular, the vorticity, temperature, and concentration equations are solved by employing ADI technique, and the SLOR iterative method is adopted to solve equation (25.8). The tridiagonal system of equations that arises is solved by utilizing the tridiagonal matrix algorithm. Local EG due to individual components (equation (25.9)–(25.12)) are discretized and solved by adopting second-order central difference approximation. The detailed discretization of energy equation could be seen in our previous work [15, 16]. Finally, to calculate overall thermal and solutal dissipation rates in the enclosure, the Simpson's rule is established. To estimate the total entropy production, the trapezoidal rule has been utilized. To authenticate the exactness of the developed FORTRAN code, we have compared our results with existing benchmark problems in our previous works [16, 26], and they are not repeated for brevity. The choice of mesh size has been considered by performing a grid independence study, and it has been found that a mesh size of 161×161 is suitable for this investigation.

25.4 RESULTS AND DISCUSSION

25.4.1 Streamlines, Isotherms, Isoconcentrations, Entropy Generation

The effects of *N*, *Le*, and *Ha* on streamlines, isotherms, isoconcentrations, and entropy production are illustrated in Figures 25.2–25.4. The influence of buoyancy ratio has been depicted in Figure 25.2 by fixing *Le* = 2, *Ha* = 10. For *N* = −5 and 5, it can be vividly seen that the magnitude of

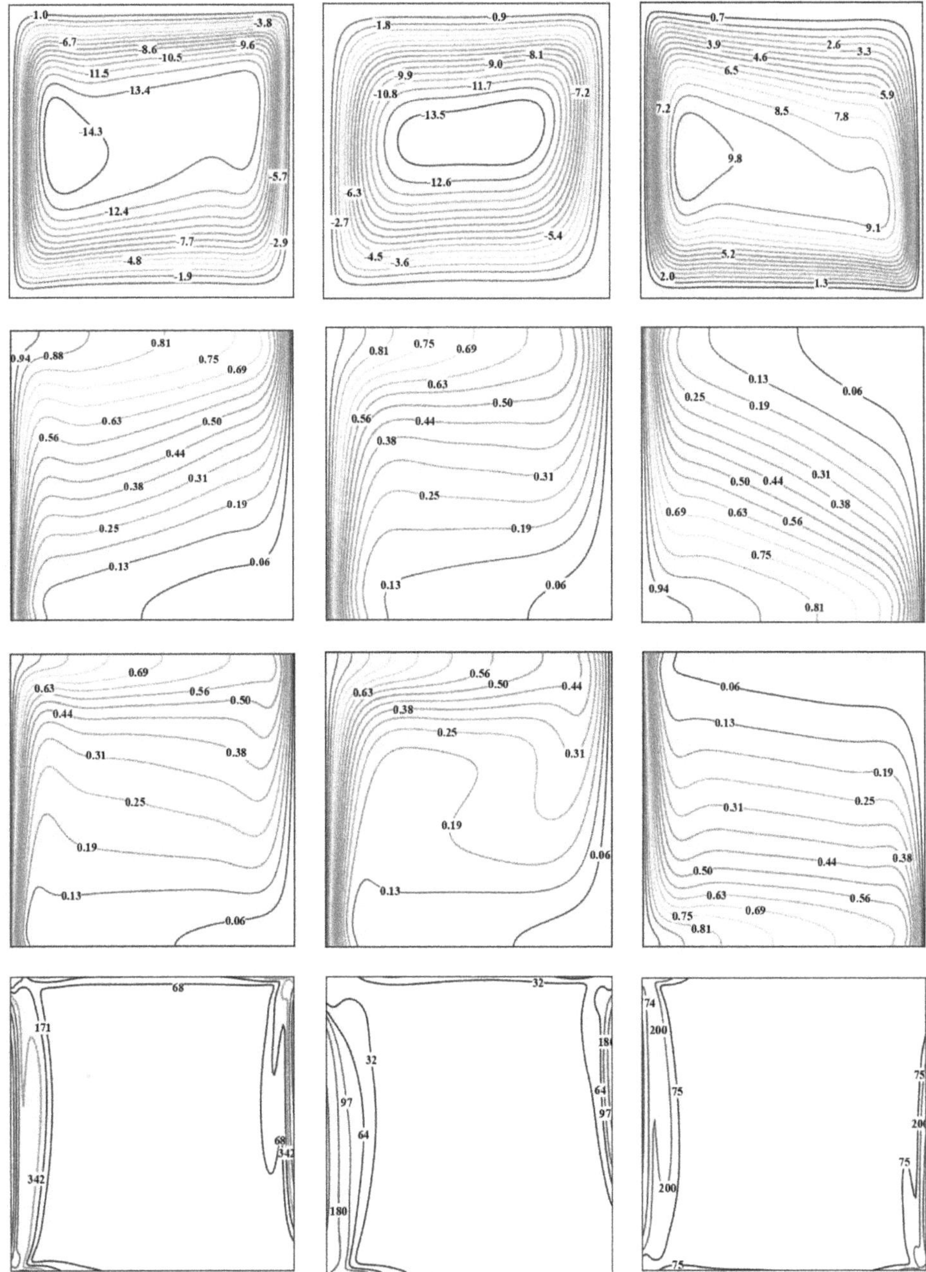

FIGURE 25.2 Streamlines, isotherms, isoconcentrations, entropy generation for N = –5 (left), N = 0 (middle), and N = 5 (right) at Ha = 10, Le = 2, and $\phi = 0.04$.

solutal buoyancy force is greater compared to thermal buoyancy. Therefore, the nanoliquid movement occurs due to the concentration gradient in clockwise or anticlockwise direction, depending on the direction of solutal buoyancy force. At $N = -5$, both the buoyancy forces act in the same direction, due to which the fluid flow occurs in a clockwise direction with greater magnitude, and at $N = 5$, thermal and solutal buoyancy forces act in opposite directions, thus leading to an anticlockwise flow circulations with reduced flow strength. For $N = 0$, the liquid movement takes place due to pure

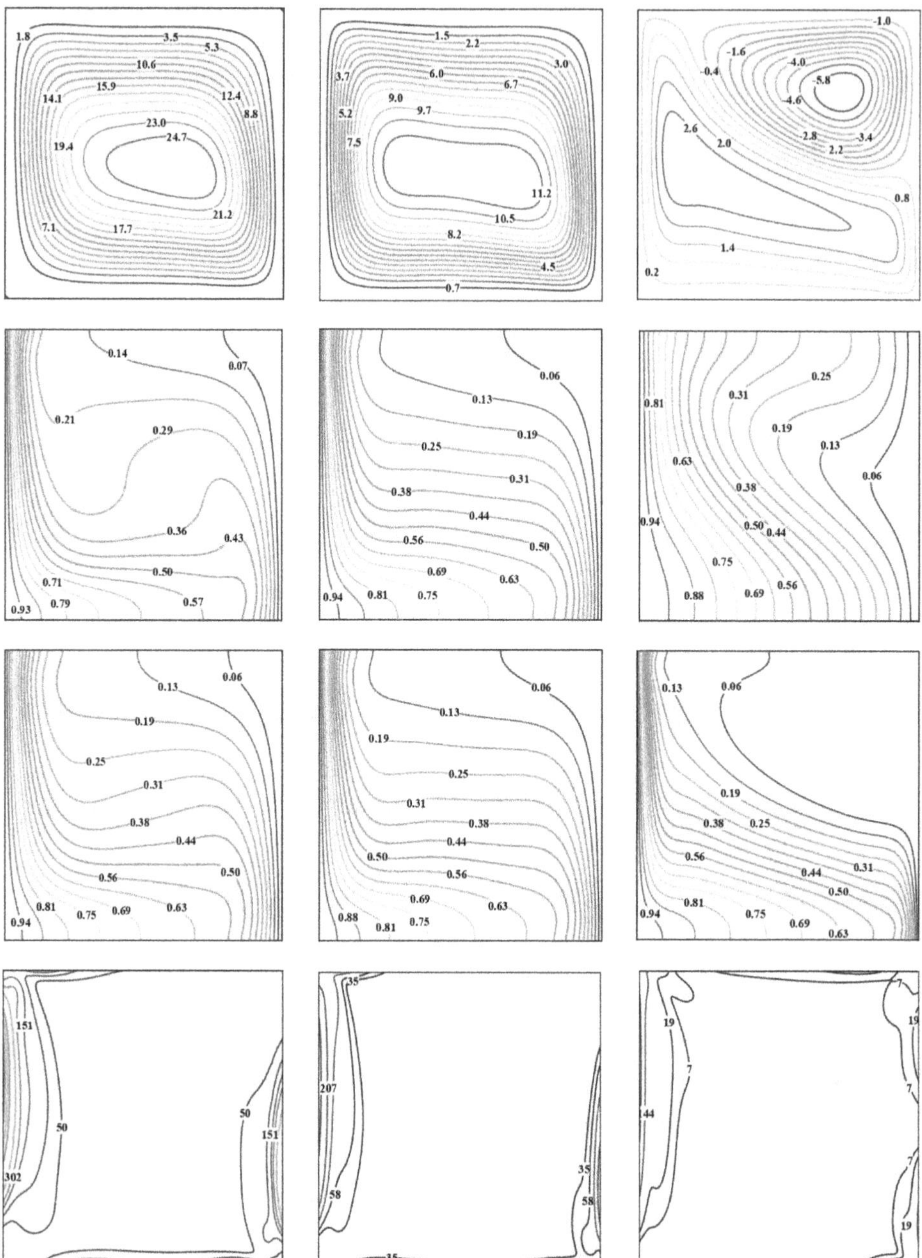

FIGURE 25.3 Streamlines, isotherms, isoconcentrations, entropy generation for Le = 0.5 (left), Le = 1 (middle), and Le = 5 (right) at Ha = 10, N = 2, and $\phi = 0.04$.

thermal convection. As N varies, the isopleths of thermal and concentration have been varied. This indicates that the change in buoyancy ratio results in a change in the thermal and concentration distribution. Also, the prediction shows that, regardless of N value, maximum entropy generates exclusively in the vicinity of vertical surfaces. The irreversibility distribution pattern for $N = -5$ and $N = 5$ seems to be similar, but in opposite direction, due to a change in the direction of nanoliquid flow. On varying the direction of solutal buoyancy force (aided to oppose) by maintaining the same

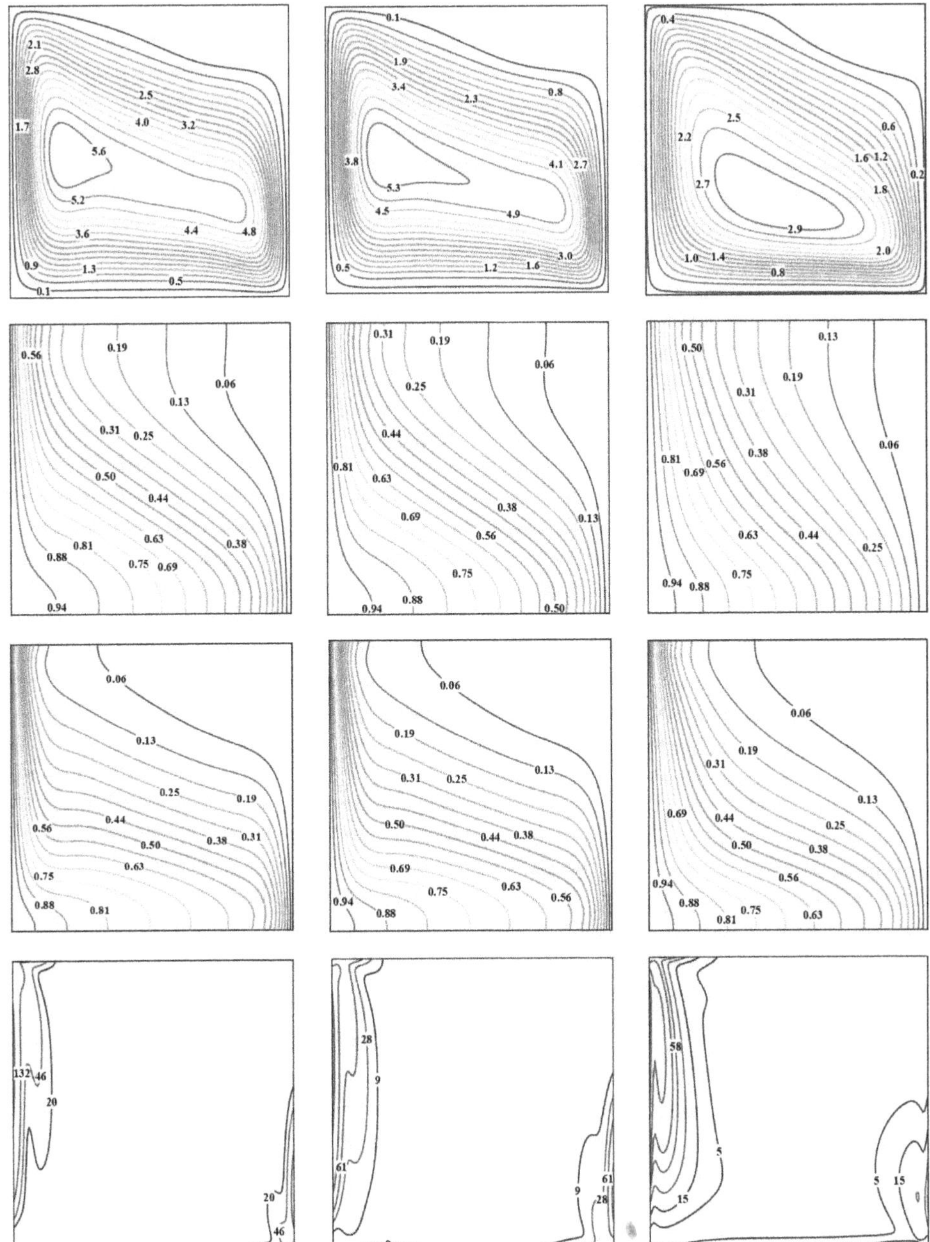

FIGURE 25.4 Streamlines, isotherms, isoconcentrations, entropy generation for Ha = 0 (left), Ha = 10 (middle), and Ha = 50 (right) at Le = 2, N = 2, and $\phi = 0.04$.

strength, S_{max} has been reduced by 48.87%, and in the absence of solutal buoyancy force, 90.58% reduction in entropy generation has been noticed.

Figure 25.3 depicts the impact of the Lewis number on flow pattern, thermal, concentration, and entropy generation contours by fixing other parameters. Since, $N > 1$, the solutal buoyancy force dominates thermal buoyancy and acts in the opposite direction, due to which the flow takes place in a counterclockwise direction. As per the definition of Le, the enhancement of Lewis number declines the solutal resistance, due to which the mass transport could be greater than the thermal transfer in the geometry. In the ongoing analysis, *Le* is altered from 0.5 to 5, which indicates that the study has been carried out by decreasing the solutal resistance. It has been noticed that for *Le* = 0.5 and 1, nanoliquid movement occurs in a single vortex, while enhancing the Lewis number to 5, a bi-cellular flow has been observed with decreased flow strength. A significant change in contour pattern of thermal and concentration fields reveals that as *Le* is varied, the heat and mass transport also vary. Also, due to similar diffusion characteristics, the isopleths of thermal and concentration appear to be the same at unit *Le*. Since the enhancement of *Le* declines the flow strength, this causes a decrease in the irreversibility due to friction and results in a minimum entropy production. In particular, S_{max} has been reduced by 52.21%, with an increase in *Le* from 0.5 to 5.

The effect of Hartmann number on flow, thermal, concentration, and entropy generation contours has been illustrated in Figure 25.4 for $Le = 2$, $N = 2$, and $\phi = 0.04$. As discussed earlier, because of greater buoyancy force acting opposite to the thermal buoyancy direction, regardless of *Ha*, the nanoliquid movement takes place in the anticlockwise direction. Through streamlines, it has been found that the change in movement pattern and flow magnitude is not appreciable on enhancing the magnitude of *Ha* from 0 to 10; however, as *Ha* is enhanced to its maximum value ($Ha = 50$), due to higher magnetic force drag, a significant change in contour structure is noticed, and the flow strength has been declined by 48.65% as compared to the flow strength in the absence of magnetic force. As noticed, the streamlines, isotherms, and isoconcentrations are invariant on the enhancement of *Ha* from 0 to 10; however, a marginal change has been noticed at $Ha = 50$. As regards entropy generation, it has been noticed that for all values of Hartmann number, higher entropy has been generated near left- and right-bottom surfaces of the annulus. As S_{GEN} is proportional to flow strength, enhancement of Hartmann number declines the flow strength, which leads to a decrease in $S_{l,\psi}$ and results in the reduction of S_{GEN} by 66.14% compared to the case of $Ha = 0$.

25.4.2 Heat and Mass Transport

The combined effect of *N* and *Le* on global heat and mass dissipation rates has been depicted in Figure 25.5. As per the definition of *Le*, enhancement of *Le* declines solutal diffusivity, which increases the thermal resistance, and this leads to a decrease in heat dissipation and causes an increase in mass transport rate. For all considered *Le* values, on decreasing the magnitude of *N*, a steep decrease in thermal and solutal dissipation has been noticed and reaches minimum at unit buoyancy ratio. This is because, at unit *N* value, the magnitude of both the buoyancy forces is the

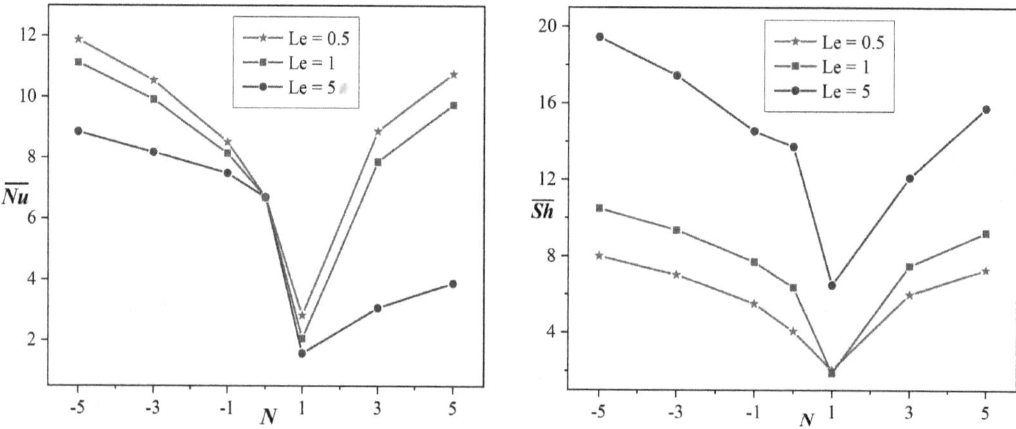

FIGURE 25.5 Impact of N and Le on $\overline{Nu}$ and $\overline{Sh}$ for Ha = 10, $\phi = 0.05$.

same and will be acting in the opposite direction, which declines flow strength and hence heat and mass transport. The same mechanism has been noticed in Figure 25.6 for different values of *Ha* by fixing *Le* and ϕ. Since $Le > 1$, as discussed earlier, the solutal resistance will be less and results in greater mass dissipation compared to thermal, and this holds good for all magnitudes of *Ha*. As regards *Ha*, when the magnitude of *Ha* is increased, the reduction in heat and mass transfer has been observed, and this is due to the fact that the Lorentz force declines the flow strength. Also, it has been found that for a particular choice of *N* value, maximum heat and mass transport rates could be gained during the aiding flow (-*N*) compared to opposing flow (+*N*), and this holds good for all *Le* and *Ha* values. Also, the combined effects of *Ha* and *Le* on average *Nu* and *Sh* have been examined in Figure 25.7 with *N* and ϕ as fixed constants. With reference to the previously discussed physical reasons, an increment in *Ha* decreases the average *Nu* and *Sh*, and this obeys for all Lewis numbers. As regards *Le*, maximum heat transport is noticed at lower *Le,* while greater mass dissipation has been achieved at greater values of *Le*. This is due to the variation in thermal and solutal diffusivity of the nanofluid inside the enclosure.

The impact of ϕ on average *Nu* and *Sh* for different buoyancy ratios and Hartmann numbers is illustrated in Figures 25.8 and 25.9, respectively. Since *Le* has higher magnitude ($Le = 2$), for a

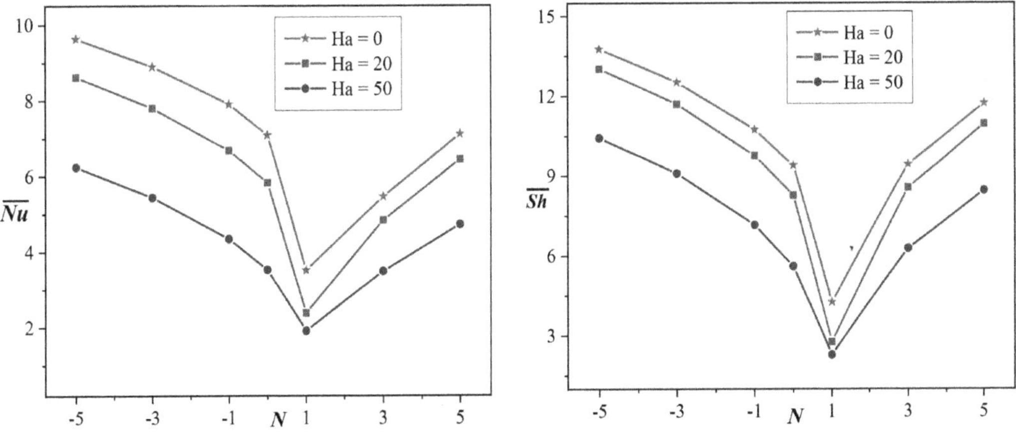

FIGURE 25.6 Impact of N and Ha on $\overline{Nu}$ and $\overline{Sh}$ for Le = 2, $\phi = 0.05$.

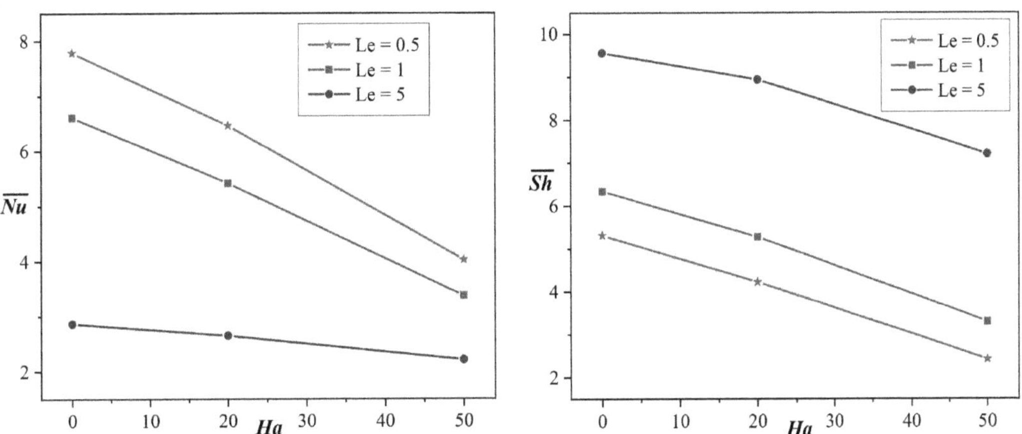

FIGURE 25.7 Effect of Le and Ha on $\overline{Nu}$ and $\overline{Sh}$ for N = 2, $\phi = 0.05$.

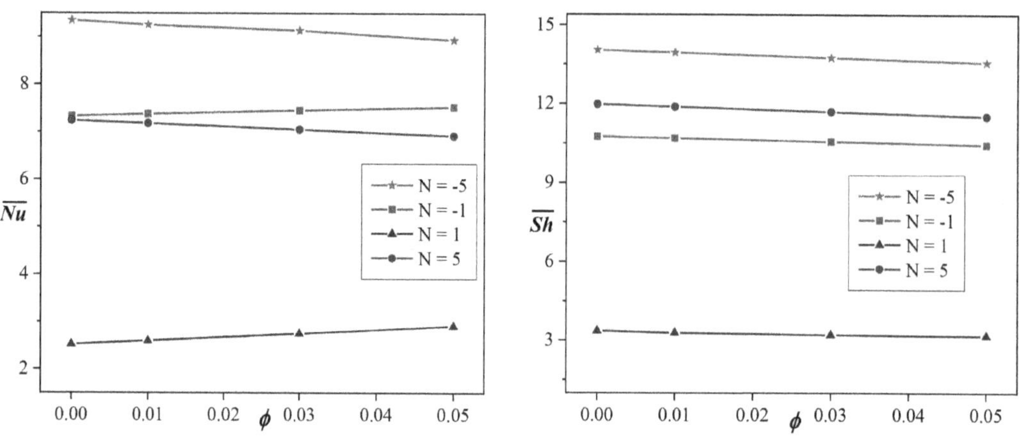

FIGURE 25.8 Impact of N and ϕ on $\overline{Nu}$ and $\overline{Sh}$ for Ha = 10, Le = 2.

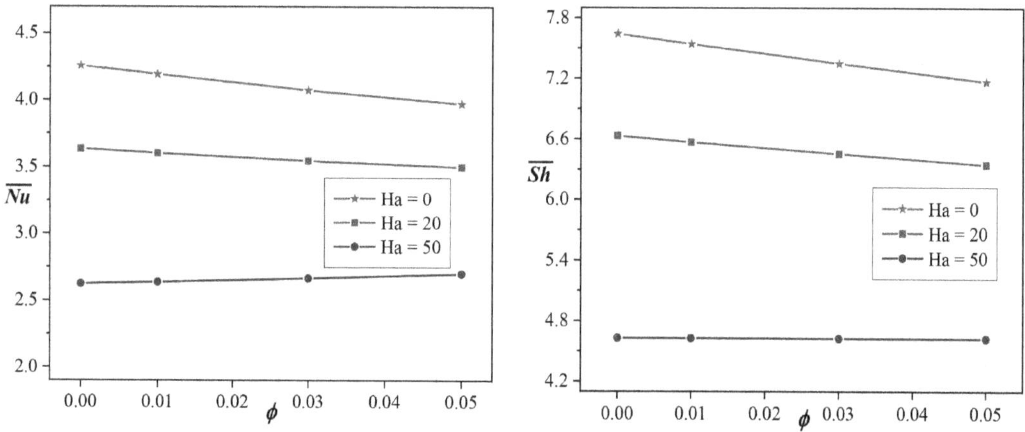

FIGURE 25.9 Effect of Ha and ϕ on and for Le = 2, N = 2.

particular ϕ, regardless of the magnitude of N and Ha, the dissipation of mass takes place in greater amount compared to thermal transport. As regards the fact that increase in ϕ leads to a decrease in flow circulation strength and increase thermal conductivity, in Figure 25.8 it has been noticed that as ϕ increases, heat and mass dissipation rates decline during stronger convection ($N \neq \pm 1$). This is because the percentage of increase in the thermal conductivity would be dominated by the decline in flow strength. However, for $N = \pm 1$, reverse mechanism takes place. As for the variation of N, maximum energy transport rate takes place at $N = -5$, since both buoyancy forces act in the same direction. The variation of thermal and solutal transport rates with the nanoparticle concentration, portrayed in Figure 25.9, reveals that the magnetic force plays a prime role in controlling the energy dissipation rates. In particular, it has been found that average Nu and Sh marginally decreases for Ha = 0 and 20. However, interestingly, for Ha = 50, the average Nu and Sh tends to increase with nanoparticle concentration. This may be expected due to the choice of higher buoyancy ratio.

25.4.3 Total Entropy Generation and Bejan Number

In every energy transfer processes, occurring due to finite temperature/solutal gradients, the production of irreversibility factor could not be evaded, which results in the system's energy degradation.

This could be estimated by means of EG, which reveals the system efficiency. In order to identify the contribution of a particular component of entropy to total EG, the Bejan number (*Be*) is also determined. Figure 25.10 illustrates the combined impact of buoyancy ratio and *Le* on S_{tot} and *Be* for $\phi = 0.05$. Regardless of the magnitude of *Le*, during both aided ($N < 1$) and opposing flows ($N > 1$), a decrease in the buoyancy ratio declines the magnitude of total EG and reaches its minimum at $N = 1$. This could be due to a decrease in the magnitude of flow strength, causing a decline in S_ψ and, hence, S_{tot}. As regards *Le*, except for $N \neq 1$, the higher the Lewis number, the lower the total EG. However, for $N = 1$, a minimum entropy production has been observed at $Le = 1$, because in this case, the movement of nanoliquid is almost stationary, due to which the magnitude of friction irreversibility would be minimal and, hence, minimizes S_{tot} in the enclosure. The influence of the magnetic field on S_{tot} and *Be* for different buoyancy ratios (Figure 25.11) and Lewis numbers (Figure 25.12) has been examined by fixing the other parameters. Irrespective of buoyancy ratio and Lewis number, an increase in the Hartmann number leads to a decrease in total EG. This is due to the degradation in EG due to fluid friction, which results in the reduction of S_{tot}. In similar arguments, decrease in the strength of buoyancy ratio declines total EG and reaches minimum at unit value of *N*, where the fluid is almost stationary (conduction dominant). Through Figure 25.12,

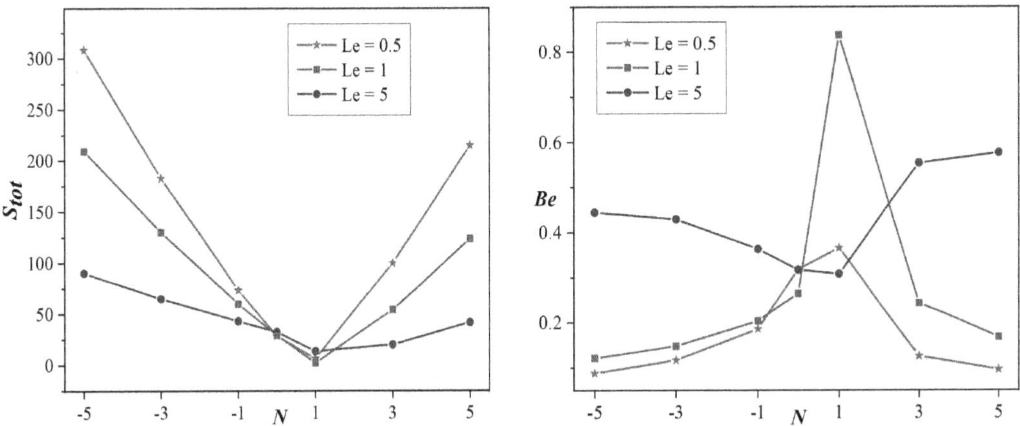

FIGURE 25.10 Effect of N and Le on S_{tot} and Be for Ha = 10, $\phi = 0.05$.

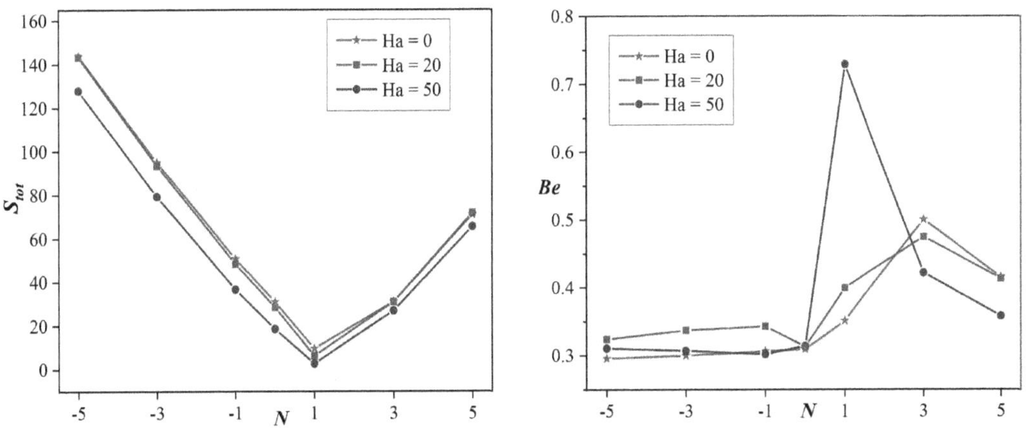

FIGURE 25.11 Effect of N and Ha on S_{tot} and Be for Le = 2, $\phi = 0.05$.

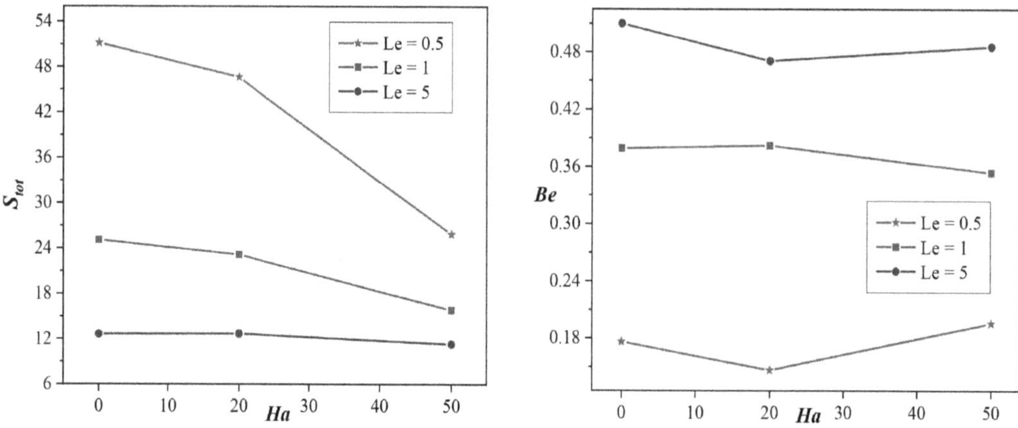

FIGURE 25.12 Effect of Le and Ha on S_{tot} and Be for N = 2, $\phi = 0.05$.

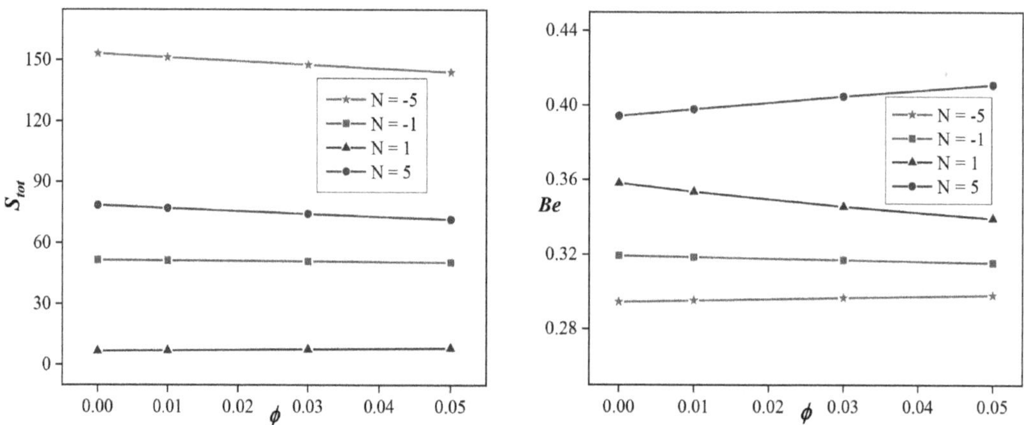

FIGURE 25.13 Effect of N and ϕ on S_{tot} and Be for Ha = 10, Le = 2.

it has been noticed that regardless of *Ha*, minimum EG could be achieved with maximum thermal resistance (*Le* = 5).

Figures 25.13 and 25.14 illustrate the impact of nanoparticle volume fraction on total EG and *Be* for different buoyancy ratios and Hartmann numbers. We noticed that, for all magnitudes of *N* in both aided and opposing cases, S_{tot} declines against ϕ. This could be because the enhancement of ϕ reduces the convective flow strength, which leads to minimum friction irreversibility and causes minimal EG. Similarly, with respect to *N*, maximum and minimum entropy is generated at $N = -5$ and $N = 1$, respectively. Physically, this could be because for -*N*, the dual buoyancy forces act in the same direction, leading to stronger convective flow, which results to maximum S_{ψ} and, hence, S_{tot}. However, for $N = 1$, the thermal buoyancy force is outweighed by its solutal counterpart, reduces the strength of nanoliquid motion, and leads to minimum S_{tot}. Figure 25.14 examines how due to a decline in flow strength, total EG declines against *Ha*. It is very interesting to note that for any values of *Ha* and ϕ, the total EG is minimum with the set of parametric values $N = 2$, $Le = 2$ when compared to other set of *N* and *Le* values. In general, the *Be* plots from Figure 25.10 to 25.14 indicate that during conduction dominant mode, $Be > 0.5$, while $Be < 0.5$ for all other situations.

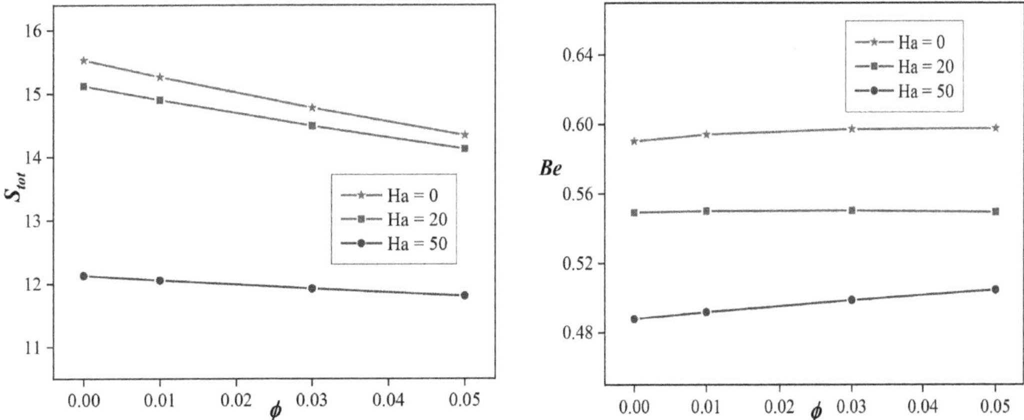

FIGURE 25.14 Effect of Ha and ϕ on S_{tot} and Be for Le = 2, N = 2.

25.5 CONCLUSIONS

1. The magnitude of flow strength, thermal, and solutal dissipation rates along with irreversibility distribution declines with a decrease in the magnitude of $|N|$, and this holds good for all values of *Ha*, *Le*, and ϕ.
2. As the magnitude of Lewis number is enhanced, heat and mass transport rates decrease and increase respectively, and minimum generation of entropy in the annulus is achieved with higher *Le*.
3. The magnetic force which affects the strength of the nanoliquid flow can be utilized to control the convection and, thereby, irreversibility distribution in the enclosure. On enhancement of Ha from 0 to 50, on an average, S_{tot} reduces by 26.98%.
4. The enhancement of nanoparticle concentration increases the thermal conductivity and declines the flow magnitude. So proper care should be taken while choosing the volume fraction in order to enhance the system efficiency.
5. The main objective of this investigation is to identify the choice of parametric values which enhance the efficiency of the system. Since all the variable parameters considered in this investigation play significant role in controlling the irreversibility, a proper choice of parametric values should be considered in order to enhance the efficiency of the system.

Nomenclature

Ar	aspect ratio
Be	Bejan number
B_0	magnetic field strength
C	dimensionless concentration
D	width of the annulus (m)
g	acceleration due to gravity (m/s^2)
H	height of the annulus (m)
Ha	Hartmann number
k	thermal conductivity (W/m.K)
Le	Lewis number
N	buoyancy ratio
Pr	Prandtl number
Ra	Rayleigh number

S_{tot} total entropy generation
T dimensionless temperature
(R,Z) radial, axial coordinates
(U,W) velocity components

Greek letters

α thermal diffusivity (m^2/s)
β thermal expansion coefficient (K^{-1})
θ dimensional temperature (K)
ν kinematic viscosity (m^2/s)
ϕ nanoparticle volume fraction

Subscripts

f base fluid
nf nanofluid
p nanoparticle

REFERENCES

1. Maberak-Oudina, F. (2017). Numerical modeling of the hydrodynamic stability in vertical annulus with heat source of different lengths. *Engineering Science and Technology, an International Journal*, 20, 1324–333.
2. Husain, S., Md Adil, Arqam, M., & Shabani, B. (2021). A review on the thermal performance of natural convection in vertical annulus and its applications. *Renewable and Sustainable Energy Reviews*, 150, 111463.
3. Retiel, N., Bouguerra, E., & Aichouni, M. (2006). Effect of curvature ratio on cooperating double-diffusive convection in vertical annular cavities. *Journal of Applied Sciences*, 6, 2541–2548.
4. Chen, S., Tolke, J., & Krafczyk, M. (2010). Numerical investigation of double-diffusive (natural) convection in vertical annuluses with opposing temperature and concentration gradients. *International Journal of Heat and Fluid Flow*, 31, 217–226.
5. Venkatachalappa, M., Do, Y., & Sankar, M. (2011). Effect of magnetic field on the heat and mass transfer in a vertical annulus. *International Journal of Engineering Science*, 49, 262–278.
6. Nikbakhti, R., & Rahimi, A.B. (2012). Double-diffusive natural convection in a rectangular cavity with partially thermally active side walls. *Journal of the Taiwan Institute of Chemical Engineers*, 43, 535–541.
7. Choi, S.U.S., & Eastman, J.A. (1995). Enhancing thermal conductivity of fluids with nanoparticles. *ASME International Mechanical Engineering Congress and Exposition*, 1995.
8. Putra, N., Roetzel, W., & Das, S.K. (2003). Natural convection of nano-fluids. *Heat and Mass Transfer*, 39, 775–784.
9. Abouali, O., & Falahatpisheh, A. (2009). Numerical investigation of natural convection of Al_2O_3 nanofluid invertical annuli. *Heat and Mass Transfer*, 46, 15–23.
10. Maberak-Oudina, F. (2019). Convective heat transfer of Titania nanofluids of different base fluids in cylindrical annulus with discrete heat source. *Heat Transfer-Asian Research*, 48, 135–147.
11. Maberak-Oudina, F., Reddy, N.K., & Sankar, M. (2021). Heat source location effects on buoyant convection of nanofluids in an annulus. *Lecture Notes in Mechanical Engineering*, 923–937.
12. Sankar, M., Reddy, N.K., & Do, Y. (2021). Conjugate buoyant convective transport of nanofluids in an enclosed annular geometry. *Scientific Reports*, 11, 17122.
13. Reddy, N.K., Swamy, H.A.K., & Sankar, M. (2021). Buoyant convective flow of different hybrid nanoliquids in a non-uniformly heated annulus. *European Physical Journal Special Topics*, 230, 1213–1225.
14. Sankar, M., Swamy, H.A.K., Do, Y., & Altmeyer, S. (2022). Thermal effects of nonuniform heating in a nanofluid-filled annulus: Buoyant transport versus entropy generation. *Heat Transfer*, 51, 1062–1091.
15. Swamy, H.A.K., Sankar, M., & Reddy, N.K. (2022). Analysis of entropy generation and energy transport of Cu-water nanoliquid in a tilted vertical porous annulus. *International Journal of Applied and Computational Mathematics*, 8, 10.

16. Swamy, H.A.K., Sankar, M., & Do, Y. (2022). Entropy and energy analysis of MHD nanofluid thermal transport in a non-uniformly heated annulus. *Waves in Random and Complex Media*, 1–37. https://doi.org/10.1080/17455030.2022.2145522
17. Esfahani, J.A., & Bordbar, V. (2011). Double diffusive natural convection heat transfer enhancement in a square enclosure using nanofluids. *Journal of Nanotechnology in Engineering and Medicine*, 2, 021002.
18. Mejri, I., Mahmoudi, A., Abbassi, M.A., & Omri, A. (2014). Magnetic field effect on entropy generation in a nanofluid filled enclosure with sinusoidal heating on both side walls. *Powder Technology*, 266, 340–353.
19. Chen, S., Yang, B., Luo, K.H., Xiong, X., & Zheng, C. (2016). Double diffusion natural convection in a square cavity filled with nanofluid. *International Journal of Heat and Mass Transfer*, 95, 1070–1083.
20. Reddy, P.N., Murugesan, K., & Koushik, V. (2018). Numerical analysis of MHD double diffusive nanofluid convection in a cavity using FEM. *Annales de Chimie—Science des Matériaux*, 42(4), 589–612.
21. Oueslati, F., Ben-Beya, B., & Lili, T. (2013). Double-diffusive natural convection and entropy generation in an enclosure of aspect ratio 4 with partial vertical heating and salting sources. *Alexandria Engineering Journal*, 52, 605–625.
22. Chen, S., Yang, B., Xiao, X., & Zheng, C. (2015). Analysis of entropy generation in double-diffusive natural convection of nanofluid. *International Journal of Heat and Mass Transfer*, 87, 447–463.
23. Arun, S., & Satheesh, A. (2019). Mesoscopic analysis of MHD double diffusive natural convection and entropy generation in an enclosure filled with liquid metal. *Journal of the Taiwan Institute of Chemical Engineers*, 95, 155–173.
24. Parveen, R., & Mahapatra, T.R. (2019). Numerical simulation of MHD double diffusive natural convection and entropy generation in a wavy enclosure filled with nanofluid with discrete heating. *Heliyon*, 5, e02496.
25. Marzougui, S., Mebarek-Oudina, F., Magherbi, M., & Mchurgui, A. (2022). Entropy generation and heat transport of Cu–water nanoliquid in porous lid-driven cavity through magnetic field. *International Journal of Numerical Methods for Heat & Fluid Flow*, 32, 2047–2069.
26. Swamy, H.A.K., Sankar, M., Reddy, N.K., & Al Manthari, M.S. (2022). Double diffusive convective transport and entropy generation in an annular space filled with alumina-water nanoliquid. *European Physical Journal Special Topics*, 231, 2781–2800.
27. Kemparaju, S., Swamy, H.A.K., Sankar, M., & Mebarek-Oudina, F. (2022). Impact of thermal and solute source-sink combination on thermosolutal convection in a partially active porous annulus. *Physica Scripta*, 97(5), 055206.
28. Dadheech, P.K., Agrawal, P., Mebarek-Oudina, F., Abu-Hamdeh, N.H., & Sharma, A. (2020). Comparative heat transfer analysis of $MoS_2/C_2H_6O_2$ and SiO_2-$MoS_2/C_2H_6O_2$ nanofluids with natural convection and inclined magnetic field. *Journal of Nanofluids*, 9(3), 161–167.
29. Swain, K., Mahanthesh, B., & Mebarek-Oudina, F. (2021). Heat transport and stagnation-point flow of magnetized nanoliquid with variable thermal conductivity, Brownian moment, and thermophoresis aspects. *Heat Transfer*, 50(1), 754–767.
30. Dhif, K., Mebarek-Oudina, F., Chouf, S., Vaidya, H., & Chamkha, A.J. (2021). Thermal analysis of the solar collector cum storage system using a hybrid-nanofluids. *Journal of Nanofluids*, 10, 616–626.
31. Chabani, I., Mebarek-Oudina, F., & Ismail, A.A.I. (2022). MHD flow of a hybrid nano-fluid in a triangular enclosure with zigzags and an elliptic obstacle. *Micromachines*, 13, 224.
32. Asogwa, K.K., Mebarek-Oudina, F., & Animasaun, I.L. (2022). Comparative investigation of water-based Al_2O_3 nanoparticles through water-based CuO nanoparticles over an exponentially accelerated radiative Riga plate surface via heat transport. *Arabian Journal of Science and Engineering*, 47, 8721–8738.
33. Chabani, I., Mebarek-Oudina, F., Vaidya, H., & Ismail, A.I. (2022). Numerical analysis of magnetic hybrid Nano-fluid natural convective flow in an adjusted porous trapezoidal enclosure. *Journal of Magnetism and Magnetic Materials*, 564, 170142.
34. Hassan, M., Mebarek-Oudina, F., Faisal, A., Ghafar, A., & Ismail, A.I. (2022). Thermal energy and mass transport of shear thinning fluid under effects of low to high shear rate viscosity. *International Journal of Thermofluids*, 15, 100176.
35. Mebarek Oudina, F., & Chabani, I. (2022). Review on nano-fluids applications and heat transfer enhancement techniques in different enclosures. *Journal of Nanofluids*, 11, 155–168.

26 Laminar Forced Nanofluids Convection for Studying the Shape and Type of Nanoparticles Effects in Annular Duct

Mohammed Benkhedda and Toufik Boufendi

26.1 INTRODUCTION

Nanotechnology is one of the most important challenges of the modern era. The addition of nano-sized particles has led to a very impressive improvement of thermal-physical properties, especially thermal conductivity, which is the major factor for enhancement in heat transfer. The new fluid with considerable thermophysical properties is called a nanofluid; this term was first induced by Choi [1] at the Argonne National Laboratory.

Nanofluids, or hybrid nanofluids, are a mixture of nanoparticles homogeneously dispersed in a base fluid, which gives a nanofluid, or a mixture of two or more nanoparticles with a base fluid, which gives a hybrid nanofluid. The nanomaterials are metallic, such as Ag, Cu, and Al; metallic oxides, such as TiO_2, CuO, and Al_2O_3; or semiconductors, such as MWCNT and CNT. The base liquids include water, oil, and ethylene glycol. Several factors affect the thermophysical properties, including the shape and diameter of the nanoparticles [2–5]. The preparation of nanofluids in order to obtain the best of them in terms of contribution to heat transfer has aroused great interest from the researchers [6–9]. Today, it is important to find nanofluids in many different applications, such as solar thermal collectors, which convert radiant energy from the sun into thermal or electrical energy [10], thermal storage systems [11, 12], automotive radiators [13, 14], thermal exchangers [15, 16], and electronic cooling [17–19]. Numerous researches have been carried out using metal oxide nanoparticles [20–23]. The heat transfer through the horizontal annulus filled with the nanofluid is a particularly interesting problem, this is due to its wide use in many applications. Bnekhedda et al. [24] carried out a numerically three-dimensional study for horizontal concentric annular duct filled with nanofluids comprising clove-treated multiwalled carbon nanotubes. In their studies, they developed two new, original correlations for the friction factor and the mean Nusselt number. Mebarek-Oudina [25] numerically studied thermal convective and fluid through a concentric annulus filled by titanium oxide–based nanofluid. He discovered that as nanoparticle concentration and Rayleigh number increase, so does the Nusselt number. Alawi et al. [26] studied numerically forced convection turbulent flow using three-dimensional horizontal concentric annulus using various nanoparticle types to estimate the heat transfer and hydrodynamic properties. They show that maximum heat transfer enhancement was found in SiO_2, followed by ZnO, CuO, and Al_2O_3, in that order. Furthermore, nano-platelet particles showed the best enhancements in heat transfer properties, followed by nano-cylinders, nano-bricks, nano-blades, and nano-spheres. Xinglong et al. [27] numerically investigated the laminar forced convection of water–copper nanofluid between two porous horizontal concentric cylinders for two

DOI: 10.1201/9781003299608-26

different geometries for low Reynolds number from 10 to 100, copper nanoparticle volume fraction from 0 to 5%, and the porous medium porosity of 0.5, 0.9. The obtained results show that the increase in the heat transfer coefficient in the second geometry is greater than the first geometry, and in porosity, 0.9 is greater than porosity 0.5. Wang et al. [28] studied experimentally and numerically heat transfer for forced convection for three-dimensional CuO/water nanofluid flow in a horizontal annulus. They showed that in the comparison between pure water, the nanofluids, especially with lower solid volume fraction and smaller nanoparticle diameter, show a superior potential for improving heat transfer. Benkhedda et al. [29] have numerically realized the heat transfer enhancement for a laminar mixed convection flow through a horizontal ring heated by a uniform heat flux imposed on the outer cylinder. The results show that in the comparison between the nanofluid and its hybrid nanofluid, the heat transfer rate of the hybrid nanofluid is very high compared to the nanofluid. In addition, two original correlations were developed for the average Nusselt number, for the nanofluid and the hybrid nanofluid.

In this chapter, we investigate the forced convective heat transfer and laminar forced fluid flow of Al_2O_3/water or TiO_2/water nanofluid between two horizontal concentric cylinders with heated outer cylinder by imposed uniform heat flux. The given Reynolds number, the volume fraction of nanoparticles, and the nanoparticle shape on heat transfer have been reviewed. The investigation problem considering two different nanoparticles types, like A_2O_3 and TiO_2, with four shapes, like blades, platelets, cylindrical, and bricks, which are the most important objective in this study, and nanoparticle volume fraction of 0, 2, 4, and 6% and Reynolds numbers of 800.

26.2 MATHEMATICAL FORMULATION

26.2.1 Problem Description and Governing Equations

The physical domain of the studied problem is illustrated in Figure 26.1. It is a horizontal annulus with the outside diameter D_o two times the inside diameter D_i (D_o = 2 D_i) and the aspect ratio of L / D_h = 200. The inner and outer cylinders are taken as thermally insulated and heated by a constant heat flux, respectively. The annulus is filled with Al_2O_3/water or TiO_2/

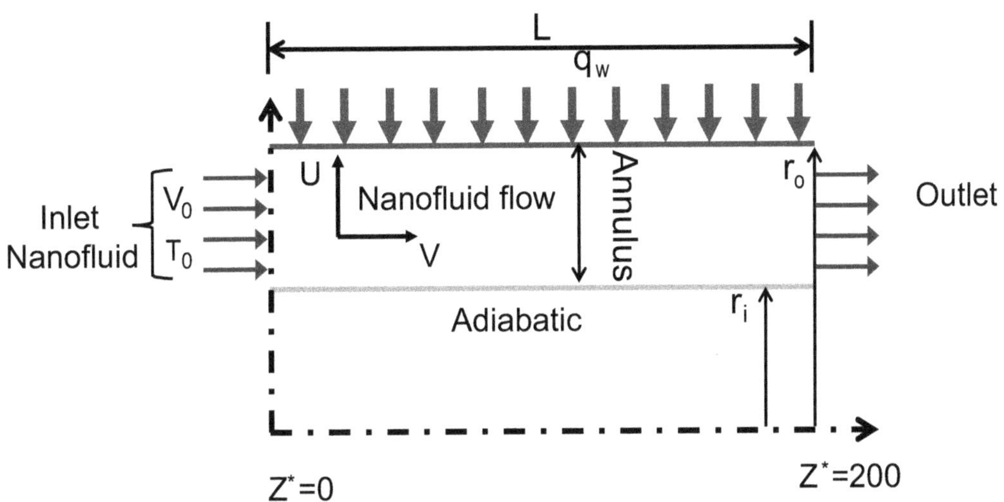

FIGURE 26.1 Physical domain.

water nanofluids. To establish the mathematical model, the following assumptions have been taken into account:

- The problem is considered to be three-dimensional laminar flow along the annular duct.
- A single-phase approach is applied, and the spherical shape of nanoparticles is considered.
- The base fluid and the solid nanoparticles are in thermal equilibrium.
- The nanofluid is Newtonian and incompressible, with negligible viscous dissipation.

The governing equations are written in the dimensionless form according to dimensionless variables deducted from the reference values D_h, v_0:

$$r^* = \frac{r}{D_h},\ z^* = \frac{z}{D_h},\ u^* = \frac{u}{v_0},\ v^* = \frac{v}{v_0},\ w^* = \frac{w}{v_0},\ t^* = \frac{v_0 t}{D_h},\ p^* = \frac{p}{\rho_{nf} v_0^2},\ T^* = \frac{(T - T_0)}{q_w D_h / k_{nf}} \tag{26.1}$$

- Continuity:

$$\frac{\partial u^*}{\partial r^*} + \frac{u^*}{r^*} + \frac{1}{r^*}\frac{\partial w^*}{\partial \theta} + \frac{\partial v^*}{\partial z^*} = 0 \tag{26.2}$$

- Radial momentum:

$$\begin{aligned} &\frac{\partial u^*}{\partial t^*} + u^*\frac{\partial u^*}{\partial r^*} + \frac{w^*}{r^*}\frac{\partial u^*}{\partial \theta} + v^*\frac{\partial u^*}{\partial z^*} - \frac{w^{*2}}{r^*} = -\frac{\partial p^*}{\partial r^*} \\ &\left(\frac{\mu_{nf}/\mu_f}{\rho_{nf}/\rho_f}\right)\frac{1}{\mathrm{Re}_f}\left(\frac{\partial^2 u^*}{\partial r^{*2}} + \frac{1}{r^*}\frac{\partial u^*}{\partial r^*} - \frac{u^*}{r^{*2}} + \frac{1}{r^{*2}}\frac{\partial^2 u^*}{\partial \theta^2} + \frac{\partial^2 u^*}{\partial z^{*2}} - \frac{2}{r^{*2}}\frac{\partial w^*}{\partial \theta}\right) \end{aligned} \tag{26.3}$$

- Angular momentum:

$$\begin{aligned} &\frac{\partial w^*}{\partial t^*} + u^*\frac{\partial w^*}{\partial r^*} + \frac{w^*}{r^*}\frac{\partial w^*}{\partial \theta} + v^*\frac{\partial w^*}{\partial z^*} - \frac{u^* w^*}{r^*} = -\frac{1}{r^*}\frac{\partial p^*}{\partial \theta} \\ &+\left(\frac{\mu_{nf}/\mu_f}{\rho_{nf}/\rho_f}\right)\frac{1}{\mathrm{Re}_f}\left(\frac{\partial^2 w^*}{\partial r^{*2}} + \frac{1}{r^*}\frac{\partial w^*}{\partial r^*} - \frac{w^*}{r^{*2}} + \frac{1}{r^{*2}}\frac{\partial w^*}{\partial \theta} + \frac{\partial}{\partial z^*}\left(\frac{\partial w^*}{\partial z^*}\right) + \frac{2}{r^{*2}}\frac{\partial u^*}{\partial \theta}\right) \end{aligned} \tag{26.4}$$

- Axial momentum:

$$\begin{aligned} &\frac{\partial v^*}{\partial t^*} + u^*\frac{\partial v^*}{\partial r^*} + \frac{w^*}{r^*}\frac{\partial v^*}{\partial \theta} + v^*\frac{\partial v^*}{\partial z^*} = -\frac{\partial p^*}{\partial z^*} + \\ &\frac{\mu_{nf}/\mu_f}{(\rho_{nf}/\rho_f)}\frac{1}{\mathrm{Re}_f}\left(\frac{\partial^2 v^*}{\partial r^{*2}} + \frac{1}{r^*}\frac{\partial v^*}{\partial r^*} + \frac{1}{r^{*2}}\frac{\partial^2 v^*}{\partial \theta^2} + \frac{\partial^2 v^*}{\partial z^{*2}}\right) \end{aligned} \tag{26.5}$$

- Energy:

$$\begin{aligned} &\frac{\partial T^*}{\partial t^*} + u^*\frac{\partial T^*}{\partial r^*} + \frac{w^*}{r^*}\frac{\partial T^*}{\partial \theta} + v^*\frac{\partial T^*}{\partial z^*} = \\ &\frac{(\mathrm{k_{nf}}/\mathrm{k_f})}{(\rho \mathrm{C_p})_{\mathrm{nf}}/(\rho \mathrm{C_p})_f}\frac{1}{\mathrm{Re}_f.\mathrm{Pr}_f}\left(\frac{\partial^2 T^*}{\partial r^{*2}} + \frac{1}{r^*}\frac{\partial T^*}{\partial r^*} + \frac{1}{r^{*2}}\frac{\partial T^*}{\partial \theta^2} + \frac{\partial^2 T^*}{\partial z^{*2}}\right) \end{aligned} \tag{26.6}$$

The governing equations are subject to the following dimensionless initial and boundary conditions:

$$t^* = 0: \quad u^* = v^* = w^* = 0, T^* = 0 \qquad t^* \succ 0 \tag{26.7}$$

At the inlet of the annular duct: $z^* = 0$

$$0.5 \le r^* \le 1 \quad \text{and } 0 \le \theta \le 2\pi: u^* = w^* = T^* = 0, \; v^* = 1 \tag{26.8}$$

At the exit of the annular duct: $z^* = 100$

$$0.5 \le r^* \le 1 \text{ and } 0 \le \theta \le 2\pi: \; \frac{\partial u^*}{\partial z^*} = \frac{\partial w^*}{\partial z^*} = \frac{\partial v^*}{\partial z^*} = \frac{\partial^2 T^*}{\partial z^{*2}} = 0 \tag{26.9}$$

At the inner cylinder: $r_i^* = 0.5$

$$0.5 \le r^* \le 1, \; 0 \le \theta \le 2\pi, \; 0 \le Z^* \le 100, \text{ and } u^* = w^* = v^* = 0, \; \frac{\partial T^*}{\partial r^*} = 0 \tag{26.10}$$

At the outer cylinder: $r_o^* = 1$

$$0.5 \le r^* \le 1, \; 0 \le \theta \le 2\pi, \; 0 \le Z^* \le 100, \text{ and } u^* = w^* = v^* = 0, \; \frac{\partial T^*}{\partial r^*} = -1 \tag{26.11}$$

The local Nusselt number (Nu) at the outer walls is calculated as follows:

$$Nu = \frac{h_o . D_h}{k_{nf}} \tag{26.12}$$

Where h_o is the convective heat transfer coefficient at the outer wall, which is defined by:

$$h_o = \frac{q_w}{(T_w - T_b)} \tag{26.13}$$

$$Nu(\theta, z^*) = \frac{h(\theta, z)_{nf} D_h}{k_f} = \frac{k_{nf}}{k_f}\left[\frac{\left(\partial T^* / \partial r^*\right)\Big|_{r^* = r_o^*}}{T^*(1, \theta, z^*) - T_m^*(z^*)}\right] \tag{26.14}$$

$$Nu(z^*) = \frac{1}{2\pi}\int_0^{2\pi} Nu(\theta, z^*) d\theta = \frac{1}{2\pi}\int_0^{2\pi} \frac{k_{nf}}{k_f}\left[\frac{\left(\partial T^* / \partial r^*\right)\Big|_{r^* = r_o^*}}{T^*(iL, \theta, z^*) - T_m^*(z^*)}\right] d\theta \tag{26.15}$$

$$Nu_A = \frac{1}{(2\pi)(200)}\int_0^{2\pi}\int_0^{200} Nu(\theta, z^*) \mathrm{d}z^* \, d\theta \tag{26.16}$$

At a cross section of the annular duct, the average temperature of the base fluid and nanoparticle mixture is defined by the following relationship:

$$T_m^*(z^*) = \frac{\int_{0.5}^{1}\int_{0}^{2\pi} v^*(r^*,\theta,z^*)T^*(r^*,\theta,z^*)r^*dr^*d\theta}{\int_{0.5}^{1}\int_{0}^{2\pi} v^*(r^*,\theta,z^*)r^*dr^*d\theta} \tag{26.17}$$

26.2.2 Thermophysical Properties of Al_2O_3/Water, TiO_2/Water Nanofuids

The thermophysical properties formulas of both nanofluids (Al_2O_3/water), (TiO_2/water) as, density ρ_{nf}, specific heat $C_{p,nf}$, and thermal conductivity k_{nf} and viscosity μ_{nf} for four nanoparticle shape, blades, cylindrical, platelets, and bricks, are listed in Table 26.1. The thermophysical properties of the base fluid and both nanoparticles Al_2O_3 and TiO_2 are defined in Table 26.2.

TABLE 26.1
Thermal Properties of Al_2O_3, TiO_2 and Water-Based Fluid at T_0 = 300K

Properties	Formulas
Density	$\rho_{nf} = \phi_p \rho_p + (1-\phi)\rho_f$
Specific heat	$(\rho Cp)_{nf} = \phi_p(\rho Cp)_p + (1-\phi)(\rho Cp)_f$
Thermal conductivity	Hamilton and Crosser [30] $\frac{k_{nf}}{k_f} = \frac{k_p + (m-1)k_f - (m-1)\phi(k_f - k_p)}{k_p + (m-1)k_f + \phi(k_f - k_p)}$ $m = 8.6 \rightarrow Baldes$ $m = 4.9 \rightarrow Cylindres$ $m = 5.7 \rightarrow Platelets$ $m = 3.7 \rightarrow Bricks$
Viscosity	Timofeeva et al. [31] $\frac{\mu_{nf}}{\mu_f} = 1 + A\phi + B\phi^2$ $Blades \longrightarrow \begin{cases} A = 14.6 \\ B = 123.3 \end{cases}$ $Cylinders \longrightarrow \begin{cases} A = 13.5 \\ B = 904.4 \end{cases}$ $Platelets \longrightarrow \begin{cases} A = 37.1 \\ B = 612.6 \end{cases}$ $Bricks \longrightarrow \begin{cases} A = 1.9 \\ B = 471.4 \end{cases}$

TABLE 26.2
Thermal Properties of Al_2O_3-TiO_2 and Water-Based Fluid at T_0 = 300K

Thermophysical Propetries	Symbol	Unit	Water	Al_2O_3	TiO_2
Heat capacitance	Cp	$J.kg^{-1}K^{-1}$	4179	765	686.2
Density	ρ	$kg.m^{-3}$	997.1	3970	4250
Thermal conductivity	k	$W m^{-1}K^{-1}$	0.613	40	8.953

TABLE 26.3
Average Nusselt Number

Mesh (r, Θ, Z)	26 × 34 × 142	52 × 44 × 162	52 × 64 × 162
Average Nusselt number	7.05534	7.08315	7.08668

26.3 NUMERICAL PROCEDURE

The finite volume method developed by Patankar [32] is used to discretize the partial differential equations (PDEs). The spatiotemporal numerical discretization is second-order, with errors of the order $(\Delta r^*)^2, (\Delta\theta)^2, (\Delta z^*)^2, (\Delta t^*)^2$. The SIMPLER algorithm has been used for the velocity–pressure coupling using the tridiagonal matrix algorithm, and a sweeping procedure is used for the sequential resolution of the linearized algebraic equations systems.

26.3.1 Grid Generation and Independence Test

In order to get an appropriate mesh for numerical simulations with the FORTRAN-developed code, to ensure a better solution independent of the mesh. Table 26.3 shows the average Nusselt number values for various mesh (23 × 34 × 142), (52 × 44 × 162), and (52 × 64 × 162), in a radial, angular, and axial directions, respectively. The selected mesh for this study is (52 × 44 × 162).

26.3.2 Code Validation

Figure 26.2 (a, b) includes two comparisons for axial and average Nussselt number. Figure 26.2(a) shows the comparison of the local Nusselt number profile variation along the dimensionless length (x/D) with the experimental data from Kim et al. [33]. Figure 26.2(b) shows the comparison between our numerical results on average Nusselt number with those obtained from the Shah equation [34]. The validation confirms that our numerical results are in a good agreement with the results reported previously by Kim et al. [33] and Shah [34].

26.4 RESULTS AND DISCUSSION

26.4.1 Mean and Wall Temperature

We have chosen to represent the two average temperatures of nanofluid (Al_2O_3/water) and the wall temperature for four nanoparticle shapes, which were illustrated in Figure 26.3(a–d). From these figures, we can observe the effect of the nanoparticles for four nanoparticle shapes—blades, cylindrical, platelets, and bricks. The mean temperature of the nanofluid and the wall temperature

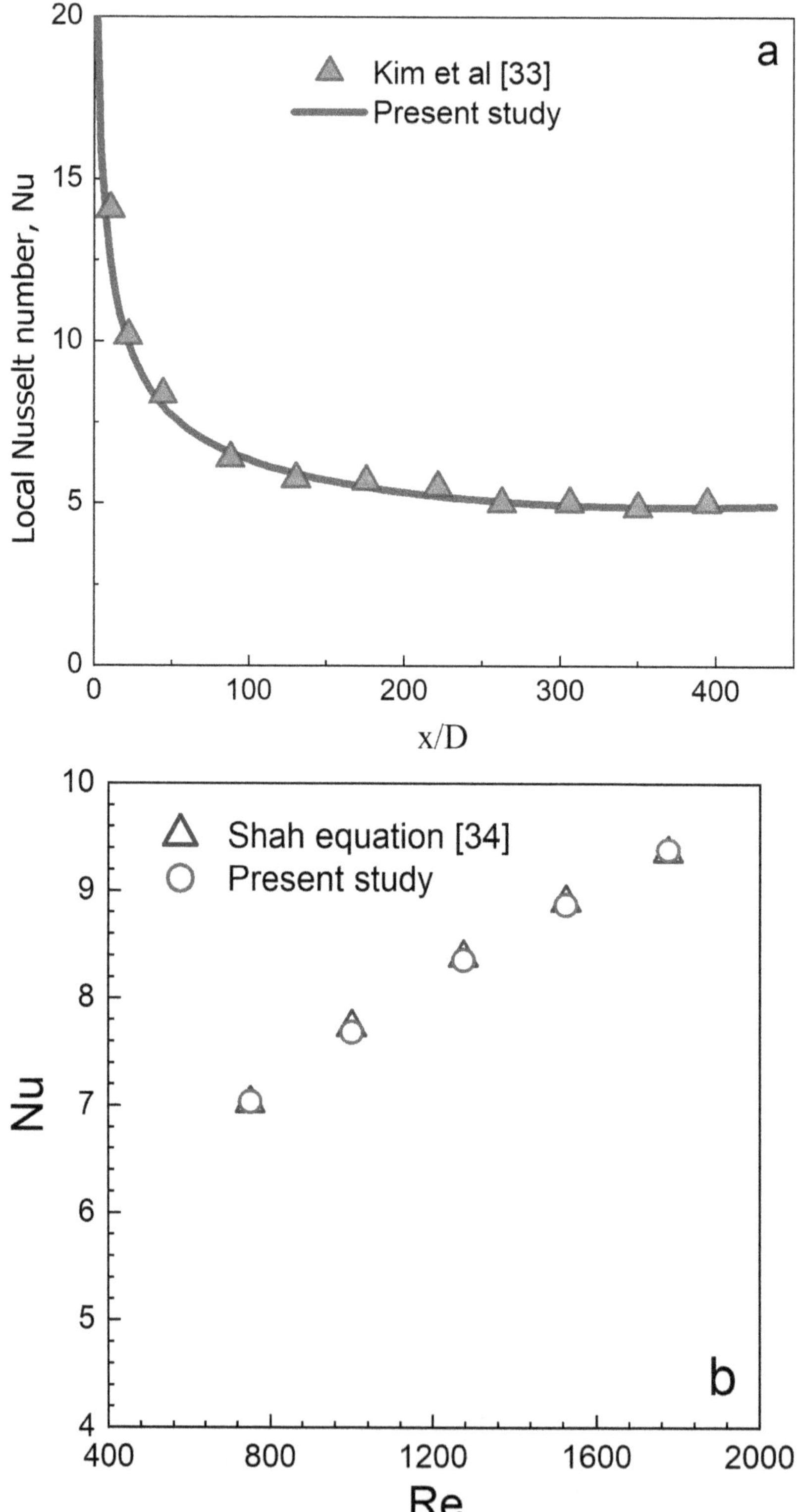

FIGURE 26.2 Comparison of the obtained numerical results between the local Nusselt number (a) along the tube [33] and the average Nusselt number (b) with the Shah equation [34].

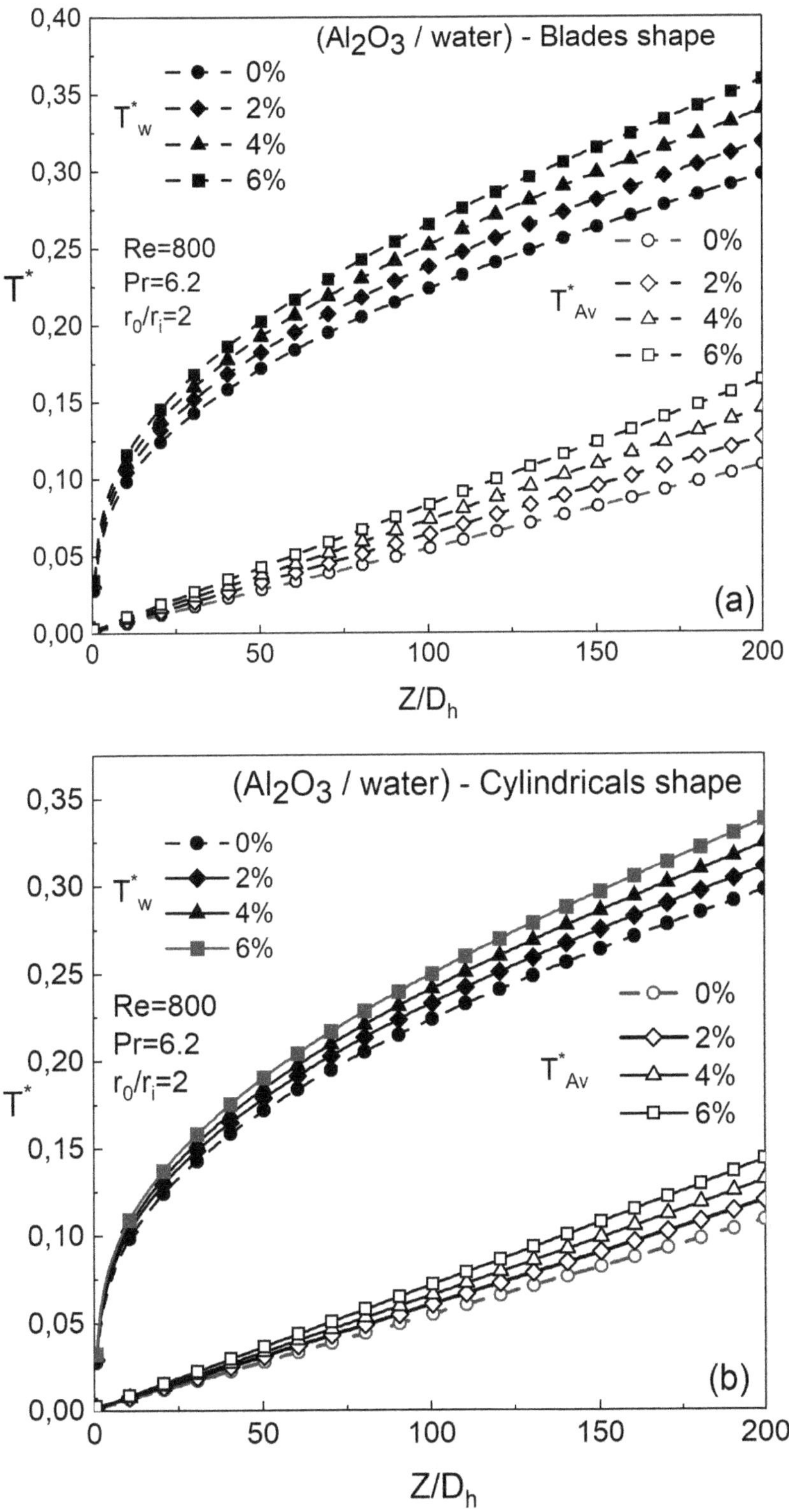

FIGURE 26.3 Mean and wall temperature evolution along the horizontal annulus for four nanoparticle shapes in the order: (a) blades, (b) cylindricals, (c) platelets, and (d) bricks.

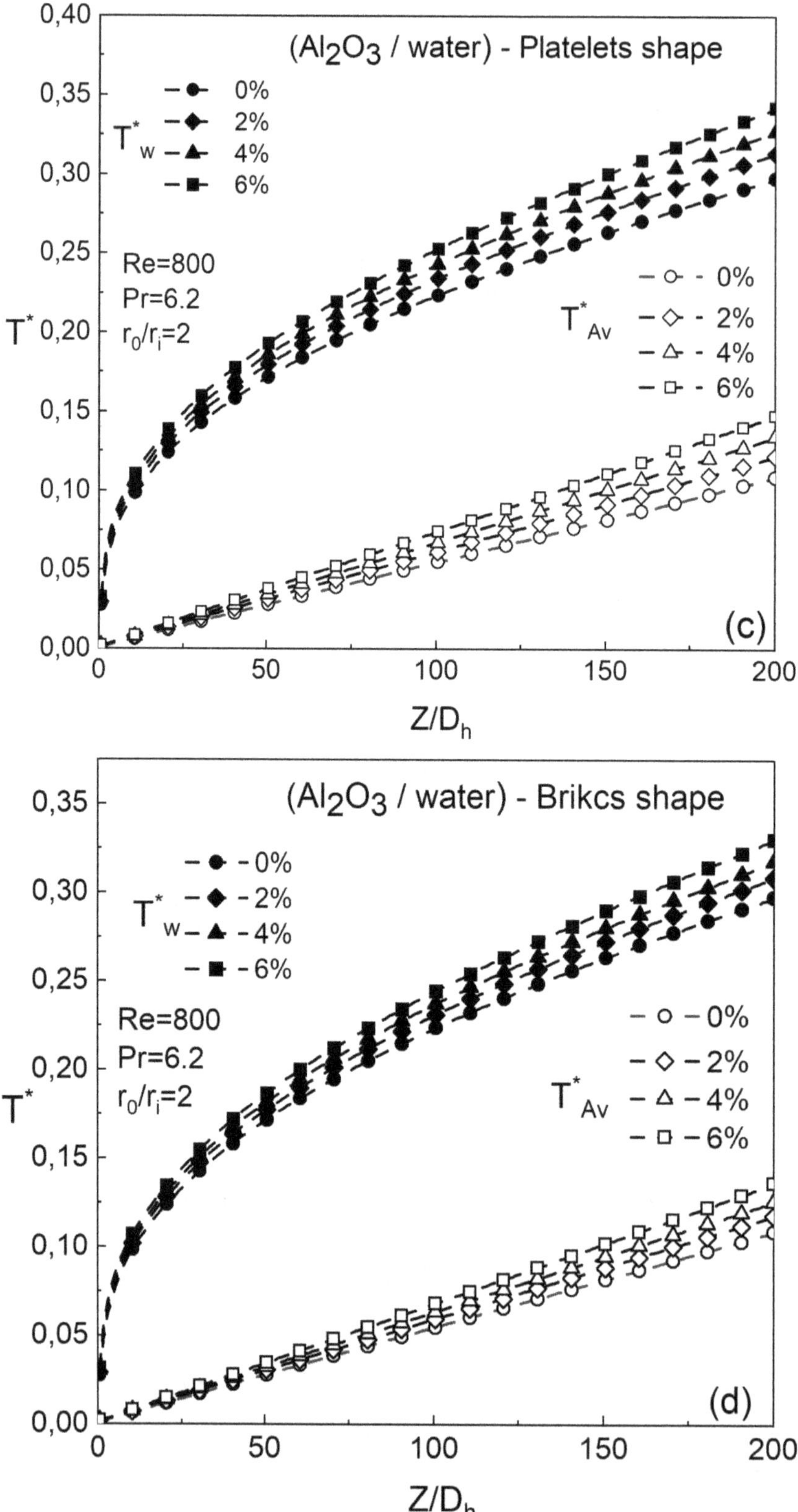

FIGURE 26.3 (Continued)

increase with the increasing volume fraction. Moreover, the profile of mean temperature has a linear increase with increasing volume fraction from the inlet of the annulus to the exit. Also, we can see that the mean temperature and wall temperature obtained using various shapes of nanoparticles are dependent on the shape of nanoparticles.

Figure 26.4(a, b) shows a comparison of average and wall temperatures between two nanoparticle types, Al_2O_3 and TiO_2, for four nanoparticle shapes—blades, cylindricals, platelets, and bricks. It can be seen that for nanoparticle types, the nanoparticle blade shape shows the highest mean temperature of the nanofluid and the wall temperature, followed by the platelet shape, cylinder shape,

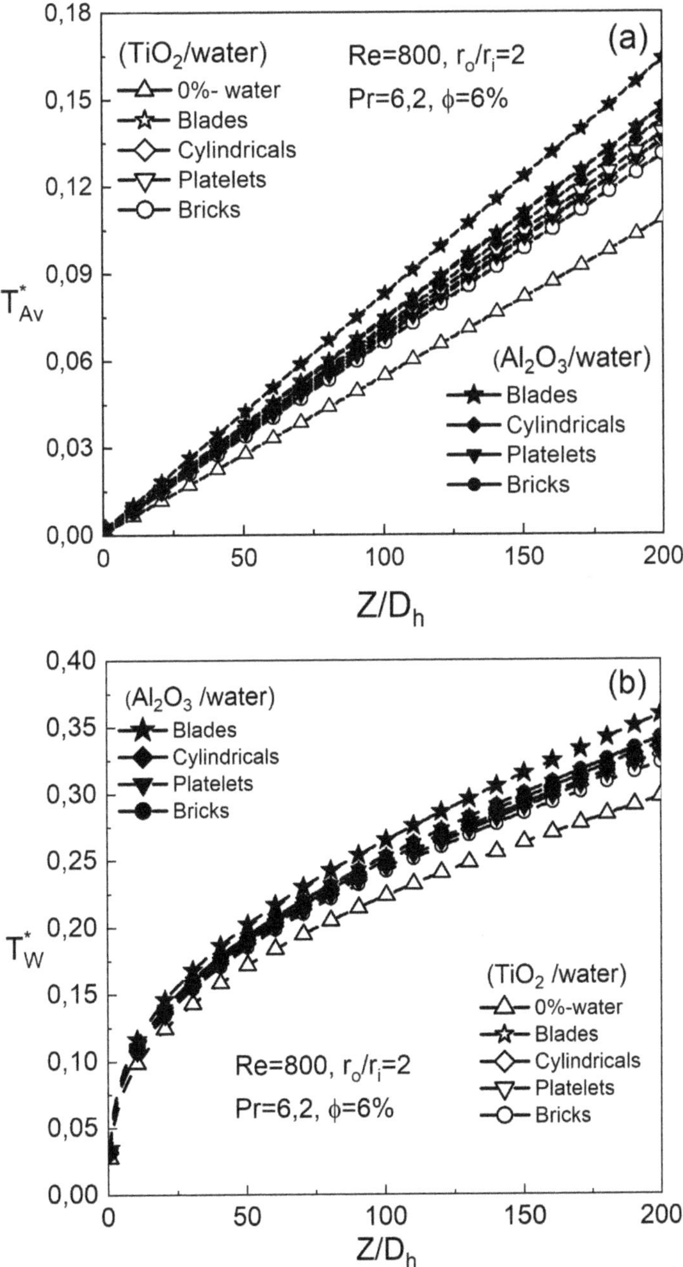

FIGURE 26.4 Comparison between nanoparticle types Al_2O_3 and TiO_2 for blades, cylindricals, platelets, and bricks nanoparticle shape and 6% volume fraction: (a) mean temperature, (b) wall temperature.

and brick shape. Also, in the comparison between both nanoparticle type Al_2O_3 and TiO_2, with the mean temperature and wall temperature for the same nanoparticle shapes, for the Al_2O_3 nanoparticle type, the mean and wall temperature are higher compared with the TiO_2 nanoparticle type.

26.4.2 Nusselt Number

Figure 26.5 presents the local Nusselt number along the annulus hot wall outer cylinder for different values of volume fraction and each nanoparticle's shape following the order: blades, cylindrical,

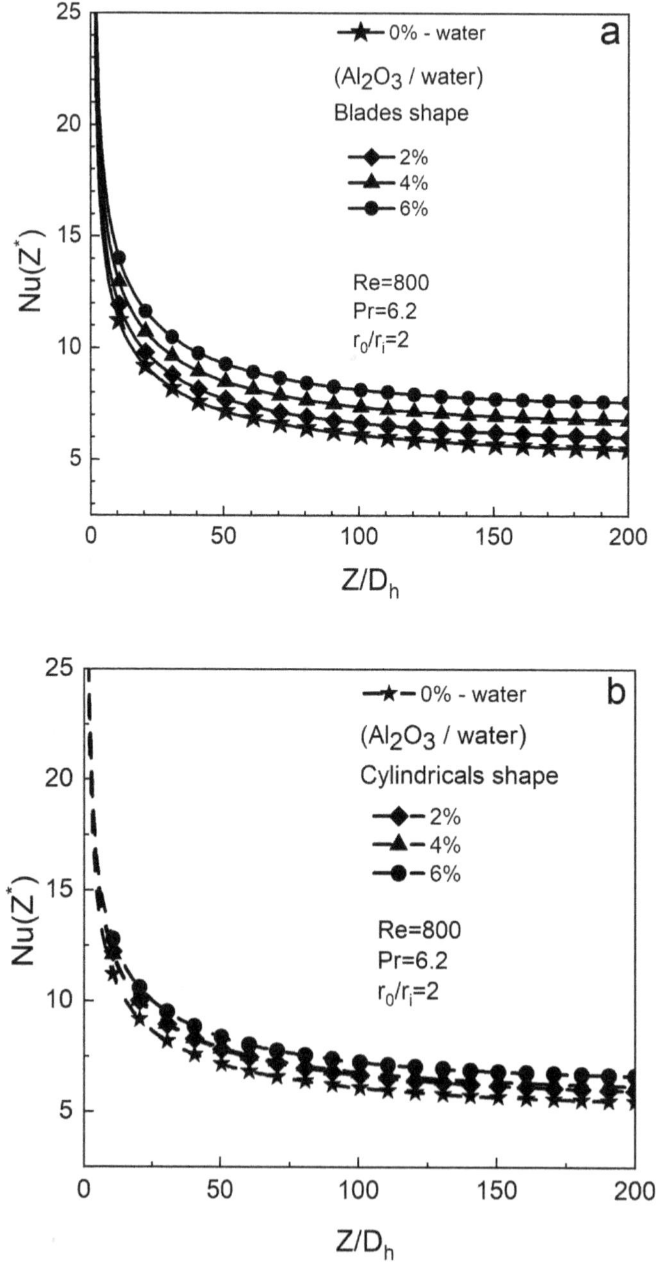

FIGURE 26.5 Variation of axial local Nusselt number of Al_2O_3/water nanofluid for four nanoparticle shapes at volume fraction 6% following the order: (a) blades, (b) cylindrical, (c) platelets, and (d) bricks.

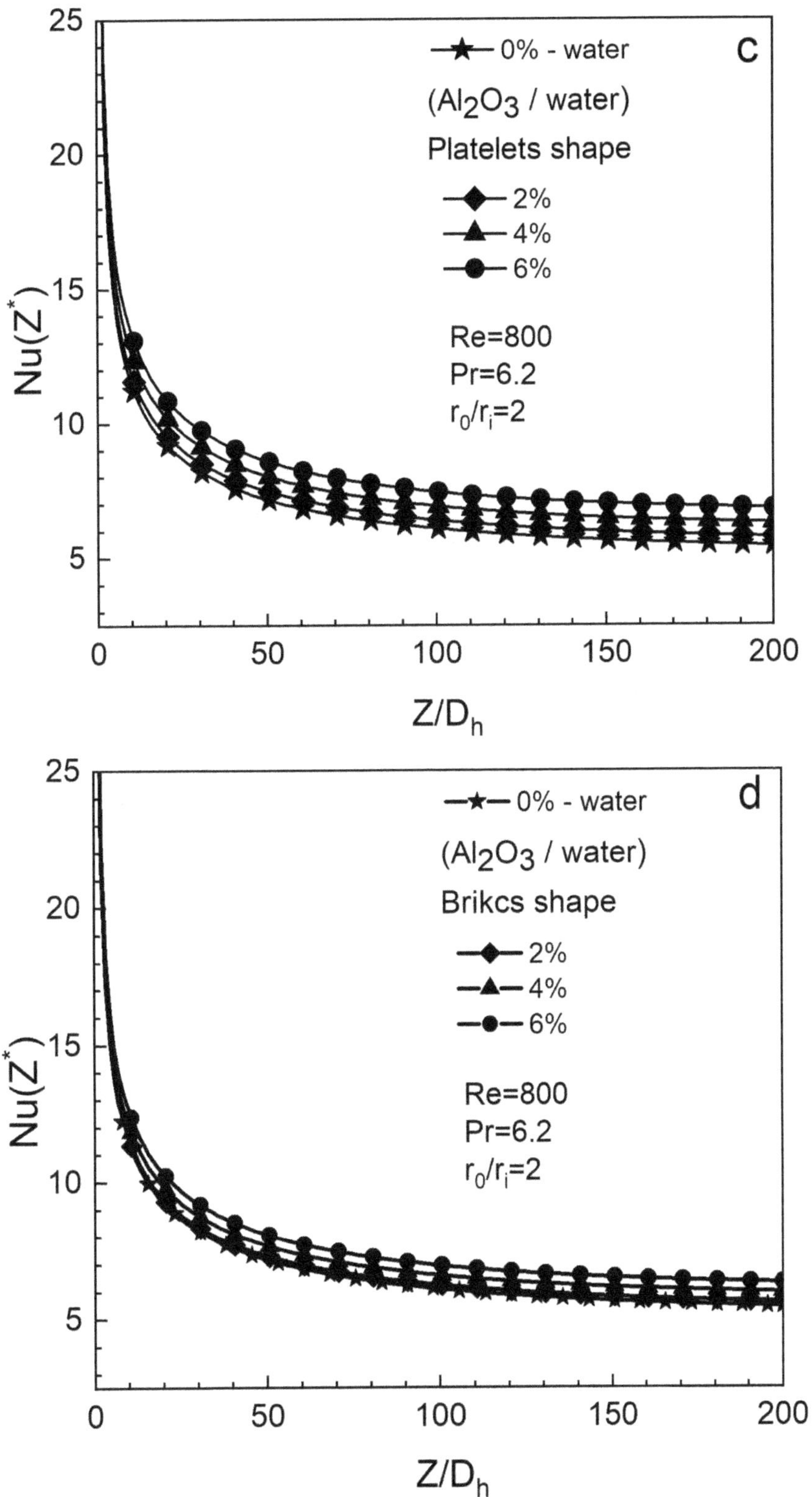

FIGURE 26.5 (Continued)

platelets, and bricks. The local Nusselt number increases by increasing the nanoparticle volume fraction. We illustrate this feature for four nanoparticle shapes and for nanofluids (Al_2O_3/water) studied. Furthermore, the local Nusselt number shows the highest values at the inlet of the annulus due to the thermal boundary layer and the secondary flow, until it becomes stable with a minimum value at the outlet of the annulus, when the flow regime becomes fully developed.

Figure 26.6 shows the comparison between all studied shapes for both nanofluids Al_2O_3/water and TiO_2/water. It can be seen that for both nanofluids, the form by blade-shaped nanoparticles gives higher local Nusselt numbers, followed by the nanofluids formed by cylindrical, platelets, and lastly, bricks nanoparticles shape, with respect to the thermal conductivity order.

Figure 26.7 illustrates the variation of the average Nusselt number via volume fraction for four nanoparticle shapes and both nanofluids and regular water. From this figure, it is observed that the average Nusselt number increases with increasing volume fraction for the two working nanofluids. Furthermore, we can see that the average Nusselt number of Al_2O_3/water is higher compared to the TiO_2/water nanofluid for all nanoparticle shapes; this is because the thermal conductivity of Al_2O_3 is relatively higher compared with TiO_2.

The enhancement of heat transfer (EHT) of both Al_2O_3/water and TiO_2/water nanofluids is represented in Figure 26.8. The obtained results showed clearly an increase in the average Nusselt number for all the considered nanoparticle shapes, with an advantage among them, where the escalation

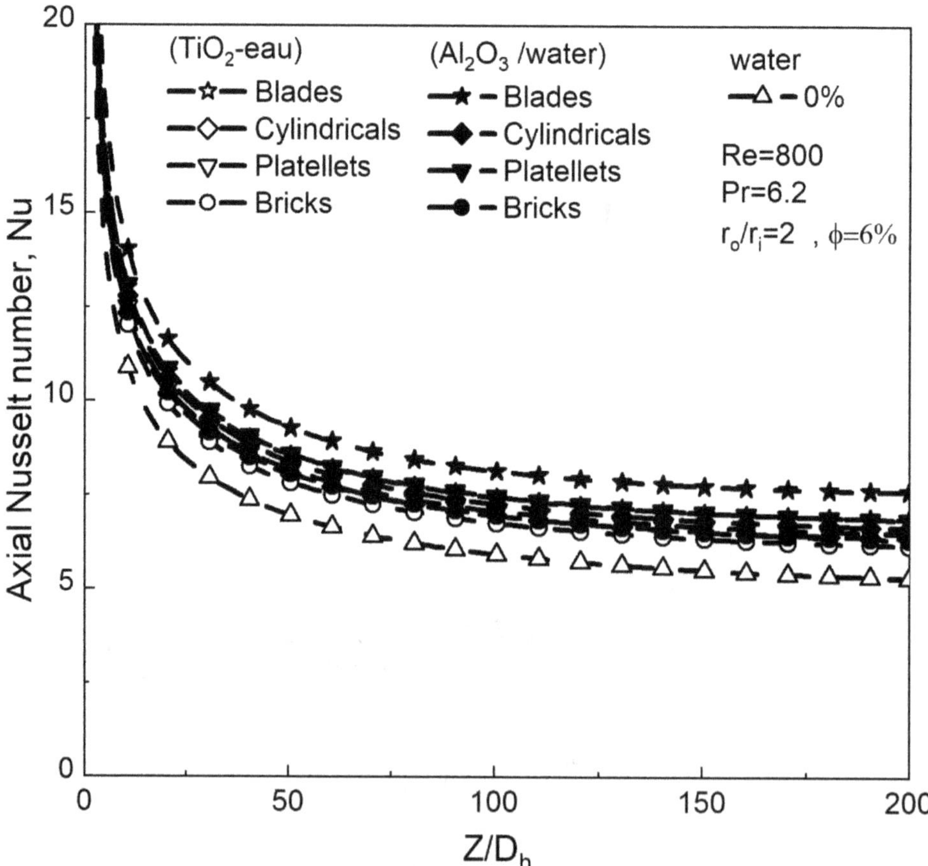

FIGURE 26.6 Comparison of the local Nusselt number between nanofluids compared with water for four nanoparticle shapes at volume fraction 6%.

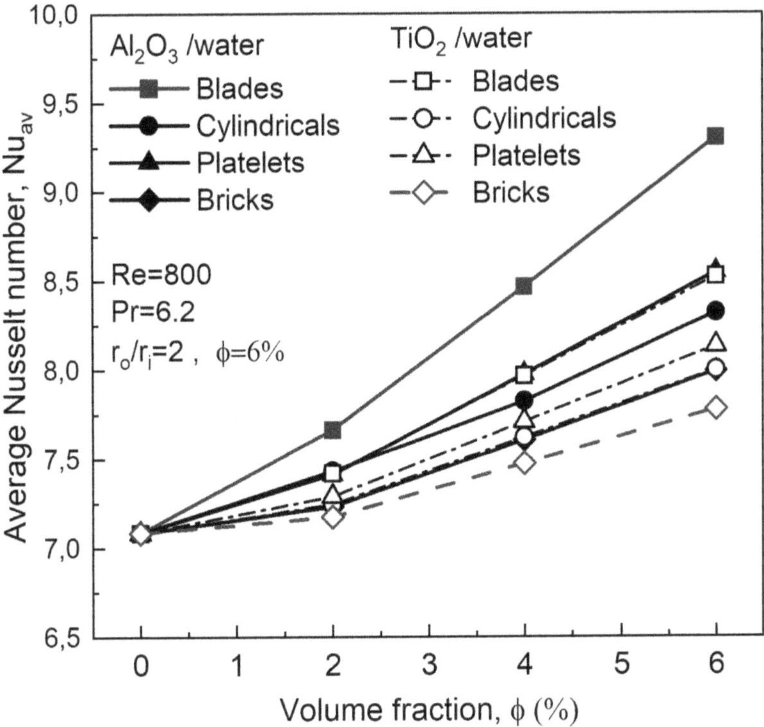

FIGURE 26.7 Average Nusselt number of nanofluids versus volume fraction for four nanoparticle shapes.

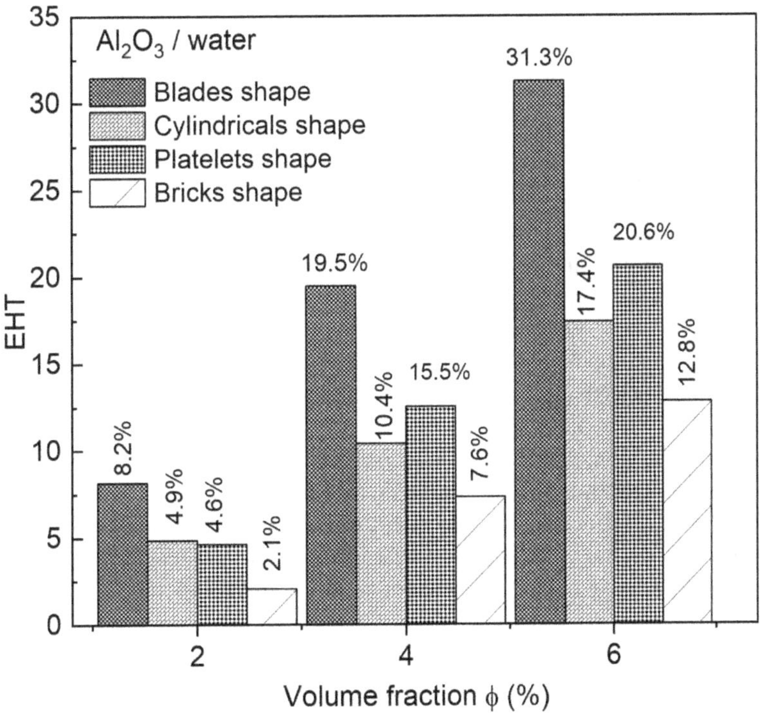

FIGURE 26.8 Enhancement of heat transfer (EHT) of nanofluids according to volume fraction for four nanoparticle shapes.

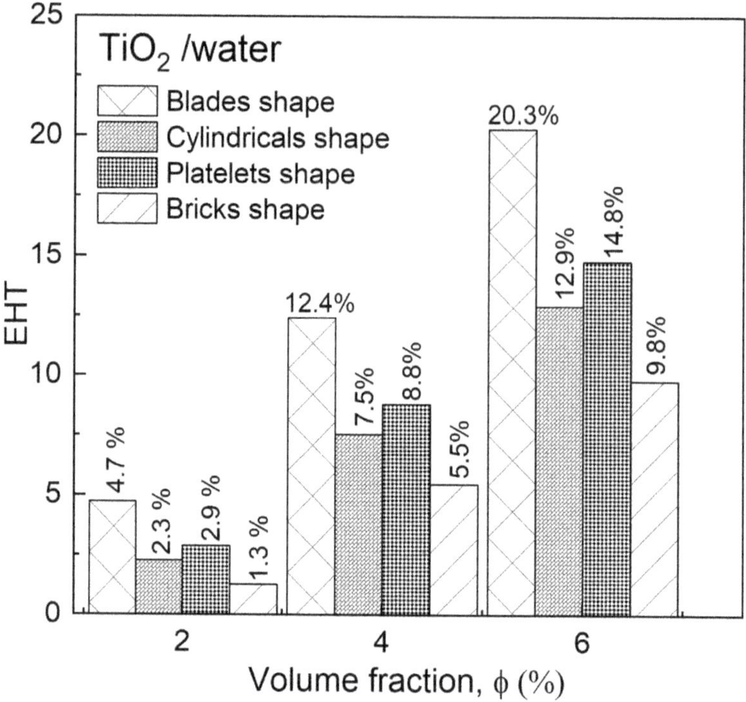

FIGURE 26.8 (Continued)

for 6% volume fraction is almost 31.3%, 17.4%, 20%, and 12.8% for blades, cylindrical, platelets, and bricks for Al_2O_3/water nanofluid. While for the nanofluid TiO_2/water, it is almost 20.3%, 12.9%, 14.8%, and 9.8% for blades, cylindrical, platelets, and bricks, respectively (see Figure 26.7).

26.5 CONCLUSIONS

In this chapter, a numerical study of 3D laminar forced convective heat transfer and fluid flow of two nanofluids (Al_2O_3/water) and (TiO_2/water) in horizontal annulus is carried out. The important findings from this investigation are listed as follows:

1. The important increase in bulk temperature is noticed for the Al_2O_3/water nanofluid formed by blade nanoparticle shape, followed by cylindrical, platelet, and bricks nanoparticles, in this order.
2. The local Nusselt number increases with increasing volume fraction for both nanofluids Al_2O_3/water and TiO_2/water for all studied nanoparticle shapes.
3. The enhancement of heat transfer Nusselt number terms is more considerable for nanofluids Al_2O_3/water or TiO_2/water with blade shape nanoparticles, followed by cylindrical, platelets, and bricks, where it reached the greatest value of 31.3% and 20.3% for Al_2O_3 and TiO_2.

List of Symbols

Cp	heat capacity ($J\ kg^{-1}\ K^{-1}$)
D_h	hydraulic tube diameter (= $D_o - D_i$), (m)
D_i	diameter of the inner tube (m)
D_o	diameter of the outer tube (m)

g gravitational acceleration (m/s^2)
h convective heat transfer coefficients (W m^{-2}K^{-1})
k thermal conductivity ($W\,m^{-1}\,K^{-1}$)
L length (m)
Nu Nusselt number (= $h.D_h/k$)
Pr Prandtl number ($= Cp_f\,.\mu_f\,/\,k_f$)
P pressure (Pa)
P^* dimensionless pressure ($= P\,/\,\rho_{nf}\mathrm{v}_0^2$)
q_w uniform heat flux ($W\,m^{-2}$)
Re Reynolds number ($= \rho_f\,\mathrm{v}_0\mathrm{D}_h/\,\mu_f$)
r_i inner radius (m)
r_o outer radius (m)
r^* dimensionless radius ($= r\,/\,D_h$)
T temperature (K)
T^* dimensionless temperature ($= (T - T_0)/(q_w D_h\,/\,k_{nf})$)
t time (s)
t^* dimensionless time, equal to ($= \mathrm{v}_0\,t\,/\,\mathrm{D}_h$)
u radial velocity component ($m\,s^{-1}$)
u^* dimensionless radial velocity ($= u\,/\,\mathrm{v}_0$)
v axial velocity component ($m\,s^{-1}$)
v^* dimensionless axial velocity ($= v\,/\,\mathrm{v}_0$)
w tangential velocity ($m\,s^{-1}$)
w^* dimensionless tangential velocity ($= w\,/\,\mathrm{v}_0$)
z axial direction (m)
z^* dimensionless axial direction ($= z\,/\,D_h$)

Greek Symbols

β volumetric expansion coefficient (K^{-1})
θ angular coordinate
μ dynamic viscosity ($kg\,m^{-1}\,s^{-1}$)

Subscripts

b bulk
nf nanofluid
f base fluid
0 inlet condition
* dimensionless parameters

Abbreviations

SIMPLER semi-implicit method for pressure-linked equations revised
FVM finite volume method
EHT enhancement of heat transfer

REFERENCES

1. Choi, S. U., & Eastman, J. A. (1995). *Enhancing Thermal Conductivity of Fluids with Nanoparticles.* Argonne National Lab. (ANL), Argonne, IL.
2. Garud, K. S., Hwang, S. G., Lim, T. K., Kim, N., & Lee, M. Y. (2021). First and second law thermodynamic analyses of hybrid nanofluid with different particle shapes in a microplate heat exchanger. *Symmetry*, 13(8), 1466.

3. Hosseinzadeh, K., Asadi, A., Mogharrebi, A. R., Ermia Azari, M., & Ganji, D. D. (2021). Investigation of mixture fluid suspended by hybrid nanoparticles over vertical cylinder by considering shape factor effect. *Journal of Thermal Analysis and Calorimetry*, 143(2), 1081–1095.
4. Benkhedda, M., Boufendi, T., Tayebi, T., & Chamkha, A. J. (2020). Convective heat transfer performance of hybrid nanofluid in a horizontal pipe considering nanoparticles shapes effect. *Journal of Thermal analysis and Calorimetry*, 140(1), 411–425.
5. Reddy, P. S., & Chamkha, A. J. (2016). Influence of size, shape, type of nanoparticles, type and temperature of the base fluid on natural convection MHD of nanofluids. *Alexandria Engineering Journal*, 55(1), 331–341.
6. Afrand, M. (2017). Experimental study on thermal conductivity of ethylene glycol containing hybrid nano-additives and development of a new correlation. *Applied Thermal Engineering*, 110, 1111–1119.
7. Hosseini, M., Sadri, R., Kazi, S. N., Bagheri, S., Zubir, N., Bee Teng, C., & Zaharinie, T. (2017). Experimental study on heat transfer and thermo-physical properties of covalently functionalized carbon nanotubes nanofluids in an annular heat exchanger: A green and novel synthesis. *Energy & Fuels*, 31(5), 5635–5644.
8. Kumar, P. G., Sakthivadivel, D., Meikandan, M., Vigneswaran, V. S., & Velraj, R. (2019). Experimental study on thermal properties and electrical conductivity of stabilized H_2O-solar glycol mixture based multi-walled carbon nanotube nanofluids: Developing a new correlation. *Heliyon*, 5(8), e02385.
9. Bindu, M. V., & Herbert, G. J. (2022). Experimental investigation of stability, optical property and thermal conductivity of water based MWCNT-Al_2O_3-ZnO mono, binary and ternary nanofluid. *Synthetic Metals*, 287, 117058.
10. Nagarajan, P. K., Subramani, J., Suyambazhahan, S., & Sathyamurthy, R. (2014). Nanofluids for solar collector applications: a review. *Energy Procedia*, 61, 2416–2434.
11. Ding, M., Liu, C., & Rao, Z. (2019). Experimental investigation on heat transfer characteristic of TiO_2-H_2O nanofluid in microchannel for thermal energy storage. *Applied Thermal Engineering*, 160, 114024.
12. Brown, T., Schlachtberger, D., Kies, A., Schramm, S., & Greiner, M. (2018). Synergies of sector coupling and transmission reinforcement in a cost-optimised, highly renewable European energy system. *Energy*, 160, 720–739.
13. Rafi, A. A., Haque, R., Sikandar, F., & Chowdhury, N. A. (2019). Experimental analysis of heat transfer with CuO, Al_2O_3/water-ethylene glycol nanofluids in automobile radiator. In *AIP Conference Proceedings* (Vol. 2121, No. 1, p. 040007). AIP Publishing LLC, United States.
14. Elsaid, A. M. (2019). Experimental study on the heat transfer performance and friction factor characteristics of Co3O4 and Al2O3 based H2O/(CH2OH) 2 nanofluids in a vehicle engine radiator. *International Communications in Heat and Mass Transfer*, 108, 104263.
15. Moradi, A., Toghraie, D., Isfahani, A. H. M., & Hosseinian, A. (2019). An experimental study on MWCNT–water nanofluids flow and heat transfer in double-pipe heat exchanger using porous media. *Journal of Thermal Analysis and Calorimetry*, 137(5), 1797–1807.
16. Mohankumar, T., Rajan, K., Sivakumar, K., & Gopal, V. (2019, July). Experimental analysis of heat transfer characteristics of heat exchanger using nano fluids. In *IOP Conference Series: Materials Science and Engineering* (Vol. 574, No. 1, p. 012011). IOP Publishing, Philadelphia, PA.
17. Balaji, T., Selvam, C., & Lal, D. M. (2021, April). A review on electronics cooling using nanofluids. In *IOP Conference Series: Materials Science and Engineering* (Vol. 1130, No. 1, p. 012007). IOP Publishing, Philadelphia, PA.
18. Balaji, T., Selvam, C., & Lal, D. M. (2021, April). A review on electronics cooling using nanofluids. In *IOP Conference Series: Materials Science and Engineering* (Vol. 1130, No. 1, p. 012007). IOP Publishing, Philadelphia, PA.
19. Khaleduzzaman, S. S., Sohel, M. R., Saidur, R., & Selvaraj, J. (2015). Stability of Al2O3-water nanofluid for electronics cooling system. *Procedia Engineering*, 105, 406–411.
20. Adun, H., Kavaz, D., Dagbasi, M., Umar, H., & Wole-Osho, I. (2021). An experimental investigation of thermal conductivity and dynamic viscosity of Al2O3-ZnO-Fe3O4 ternary hybrid nanofluid and development of machine learning model. *Powder Technology*, 394, 1121–1140.
21. Moghari, R. M., Mujumdar, A. S., Shariat, M., Talebi, F., Sajjadi, S. M., & Akbarinia, A. (2013). Investigation effect of nanoparticle mean diameter on mixed convection Al2O3-water nanofluid flow in an annulus by two phase mixture model. *International Communications in Heat and Mass Transfer*, 49, 25–35.

22. Izadi, M. M. S. M., Shahmardan, M. M., Maghrebi, M. J., & Behzadmehr, A. (2013). Numerical study of developed laminar mixed convection of Al2O3/water nanofluid in an annulus. *Chemical Engineering Communications*, 200(7), 878–894.
23. Urmi, W. T., Rahman, M. M., Hamzah, W. A. W., Kadirgama, K., Ramasamy, D., & Maleque, M. A. (2020). Experimental investigation on the stability of 40% ethylene glycol based TiO2-Al2O3 hybrid nanofluids. *Journal of Advanced Research in Fluid Mechanics and thermal Sciences*, 69(1), 110–121.
24. Benkhedda, M., Tayebi, T., & Chamkha, A. J. (2022). Toward the thermohydrodynamic behavior of a nanofluid containing C-MWCNTs flowing through a 3D annulus channel under constant imposed heat flux. *Heat Transfer*, 51(3), 2524–2545.
25. Mebarek-Oudina, F. (2019). Convective heat transfer of Titania nanofluids of different base fluids in cylindrical annulus with discrete heat source. *Heat Transfer—Asian Research*, 48(1), 135–147.
26. Alawi, O. A., Abdelrazek, A. H., Aldlemy, M. S., Ahmed, W., Hussein, O. A., Ghafel, S. T., . . . & Yaseen, Z. M. (2021). Heat transfer and hydrodynamic properties using different metal-oxide nanostructures in horizontal concentric annular tube: An optimization study. *Nanomaterials*, 11(8), 1979.
27. Liu, X., Toghraie, D., Hekmatifar, M., Akbari, O. A., Karimipour, A., & Afrand, M. (2020). Numerical investigation of nanofluid laminar forced convection heat transfer between two horizontal concentric cylinders in the presence of porous medium. *Journal of Thermal Analysis and Calorimetry*, 141(5), 2095–2108.
28. Wang, W., Liu, G., Li, B. W., Rao, Z. H., Wang, H., & Liao, S. M. (2020). Momentum and heat transfer characteristics of three-dimensional CuO/water nanofluid flow in a horizontal annulus: influences of nanoparticle volume fraction and its mean diameter. *Journal of Thermal Analysis and Calorimetry*, 1–16.
29. Benkhedda, M., Boufendi, T., & Touahri, S. (2018). Laminar mixed convective heat transfer enhancement by using Ag-TiO2-water hybrid Nanofluid in a heated horizontal annulus. *Heat and Mass Transfer*, 54(9), 2799–2814.
30. Hamilton, R. L., & Crosser, O. K. (1962). Thermal conductivity of heterogeneous two-component systems. *Industrial & Engineering Chemistry Fundamentals*, 1(3), 187–191.
31. Timofeeva, E. V., Routbort, J. L., & Singh, D. (2009). Particle shape effects on thermophysical properties of alumina nanofluids. *Journal of Applied Physics*, 106(1), 014304.
32. Patankar, S. V. (1980). *Numerical Heat Transfer and Fluid Flow*. CRC Press, New York.
33. Kim, D., Kwon, Y., Cho, Y., Li, C., Cheong, S., Hwang, Y., . . . & Moon, S. (2009). Convective heat transfer characteristics of nanofluids under laminar and turbulent flow conditions. *Current Applied Physics*, 9(2), e119–e123.
34. Shah, R. K. (1972). *Laminar Flow Forced Convection Heat Transfer and Flow Friction in Straight and Curved Ducts—A Summary of Analytical Solutions*. Stanford University Press, Office of Naval Research.

27 Mathematical Modelling on MHD Couple Stress Nanofluid Flow through a Porous Medium with Thermal Radiation

M. S. Tejashwini, R. Mahesh, U. S. Mahabaleshwar, and Basma Souayeh

27.1 INTRODUCTION

The challenge of boundary layer flow and heat transmission through a porous medium has been the subject of several investigations. A lot of work has been done in this field recently because of its importance in manufacturing and industrial processes. Like electrical devices, polyethylene extraction, steel manufacture, nuclear reactors, and many more technical and technological procedures, they use specific flows. Crane [1] proposed the steady flow of fluid on a linearly extended surface, which Wang [2] later extended.

In a very broad range of anticipated applications, nanofluids can be used, such as in heat exchangers, radiators, crystal growth, home refrigerators and electric cooling systems (such as flat plate systems), paper production, and improving diesel generator efficiency, among others. In 1995, Choi [3] generated the term "nanofluid" in an effort to provide extremely advanced thermal conductivity in response to the requirement for increased thermal conductivity in conventional fluids. Nanofluids are a new type of nanotechnology heat transfer medium that are a colloidal mixture of nanomaterials (1–100 nm) with a base fluid. Due to the significance of the occurrence and enhancement of thermal characteristics in real-world applications, many researchers were drawn to this novel form of fluid after Choi's discovery [3]. It has many uses in engineering.

Mahabaleshwar et al. [4, 5] investigated the effect of MHD on heat transfer in a stretching and shrinking sheet using a graphene Casson nanofluid and CNTs. Maranna et al. [6] examined the consequences of MHD–Newtonian fluid on stretching and shrinking sheets. Mahabaleshwar et al. [7] investigated the effect of the viscosity ratio on a two-dimensional nanofluid on an accelerated plate. Anusha et al. [8] investigated the Casson hybrid nanofluids on a porous surface. Umavathi et al. [9] constructed a model of a non-Newtonian nanofluid with a thermosolutal effect.

The couple stress fluid has significant uses in a variety of processes known in the industry, including colloidal solutions, liquid crystal solidification, and extrusion of polymer liquid, rheological complex fluids, and many more. Stokes [10] was a pioneer in developing the Boussinesq–Stokes suspension theory. It presents constitutive equations for couple stress liquids and illustrates the consequences of couple stress (1966).

An MHD couple stress liquid caused by radiation on a stretched sheet is investigated by Mahabaleshwar et al. [11, 12]. By considering the couple stress fluid on the porous layer, Dhananjay et al. [13] investigated the impact of temperature-dependent internal heat generation and the viscosity of thermal instability. Qamar Afzal et al. [14] investigated the non-uniform channel thermal

 DOI: 10.1201/9781003299608-27

convection and nanofluid concentration for the peristaltic flow of the magnetic coupling stress fluid. For the incompressible couple stress fluid, Devakar et al. [15] examined the three classical flow models of Couette, Poiseuille, and generalized Couette flow with the help of slip boundary conditions. Anusha et al. [16] and Sithole et al. [17] examined the nanofluid model of MHD on a porous surface.

There are numerous engineering and medical uses for fluid flow in the presence of magnetism. Electrically conducting fluids are subjected to a magnetic force by MHD boundary layer fluxes. A Lorentz force that opposes the fields and currents is produced by this field. MHD has been seen in a range of fluid flow scenarios, with applications in the pharmaceutical industry, energy generation, and so on. Especially in high-temperature processes, heat radiation has a substantial effect on boundary layer flow. The role of heat radiation is crucial for ensuring product quality because it has an impact on cooling rates. Bognar et al. [18] compared similarity and CFD solutions for the Blasius flow of nanofluid.

The radiation effect has been demonstrated to be a major occurrence in high-temperature industrial operations, such as solar power technologies, metal thinning, and glass blowing. Using the Rosseland approximation, the heat flux energy is distributed by Siddheshwar and Mahabaleshwar [19] to investigate the outcome of the thermal radiation on convection fluid flow through a stretched sheet. Recently, Mahabaleshwar et al. [20–25] studied the role of Newtonian and non-Newtonian fluids by explaining the effect of thermal radiation on shrinking and stretching sheets. The heat sources and sinks in the flow model have a major impact on the heat transfer characteristics because there is a significant temperature differential between the fluid and the surface. This is applicable in industries like electronic chip manufacturing and nuclear reactors. Several studies have recently shown the outcomes of radiation parameters [26–32].

A material is referred to as porous if it has pores that can hold liquid. Porosity and permeability are two characteristics that often set them apart. The porosity of a material affects how much liquid it can store. Only a few of the numerous applications include groundwater hydrology, drainage, the collection of crude oil from storage rock pores, and many other uses [33–37].

The work of Mahabaleshwar et al. [11] served as inspiration for the current work. To justify our contribution to the field, Al_2O_3 nanoparticles were employed on the MHD couple stress with thermal radiation over a porous medium in the present work, where the analytical solution is obtained by using LT in terms of an incomplete gamma function. A number of parameters affecting the flow are further discussed with the help of graphs.

27.2 FORMULATION AND SOLUTION OF THE PROBLEM

Consider a study of MHD couple stress fluid flow in a porous medium with thermal radiation. The Al_2O_3 nanoparticle with H_2O as a base fluid is implemented in this model. The physical flow of the model of two-dimensional (x–y) space is represented in Figure 27.1. The magnetic field B_0 is applied along the y-axis, while the velocity and temperature profiles are drawn across the y-axis.

The governing equation is as follows (see [11]):

$$\frac{\partial u}{\partial x}+\frac{\partial v}{\partial y}=0, \tag{27.1}$$

$$u\frac{\partial u}{\partial x}+v\frac{\partial u}{\partial y}=\frac{\mu_{nf}}{\rho_{nf}}\frac{\partial^2 u}{\partial y^2}-\frac{\eta_0}{\rho_{nf}}\frac{\partial^4 u}{\partial y^4}-\frac{\sigma_{nf}}{\rho_{nf}}B_0^{\,2}u-\frac{\mu_{nf}}{\rho_{nf}k}u, \tag{27.2}$$

$$u\frac{\partial T}{\partial x}+v\frac{\partial T}{\partial y}=\frac{k_{nf}}{\left(\rho C_p\right)_{nf}}\frac{\partial^2 T}{\partial y^2}-\frac{1}{\left(\rho C_p\right)_{nf}}\frac{\partial q_r}{\partial y}, \tag{27.3}$$

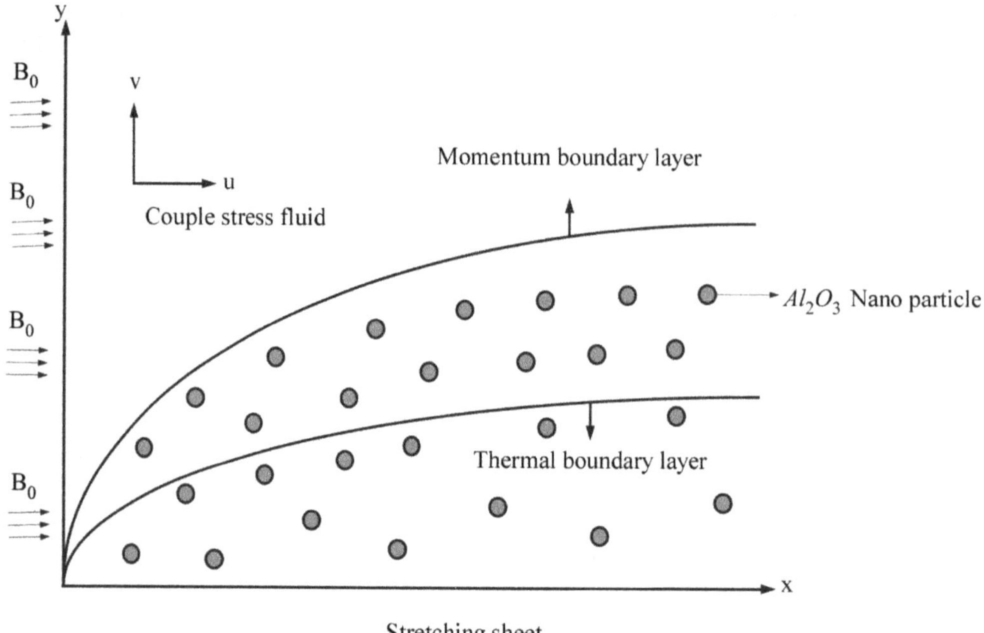

FIGURE 27.1 Physical flow model.

The BCs for the proposed model are:

$$\begin{cases} v=0, & u=u_w=cx, \quad T=T_w, & \text{at} & y=0, \\ u=0, & T\to T_\infty, & \text{as} & y\to\infty, \end{cases} \tag{27.4}$$

The radioactive heat flex is estimated using Rosseland's radiation approximation, as shown in the following:

$$q_r=-\frac{4\sigma^*}{3k^*}\frac{\partial T^4}{\partial y}, \tag{27.5}$$

Where q_r, σ^*, and k^* are known as radiative heat flux, Stefan–Boltzmann constant, and absorption coefficient, respectively. The term with a value T^4 is expanded using Taylor's series as follows (see [32]):

$$T^4\cong 4T_\infty^{\ 3}T-3T^4, \tag{27.6}$$

Equations (27.5) and (27.6) are used to calculate the q_r with respect to y:

$$\frac{\partial q_r}{\partial y}=\frac{16\sigma^*T_\infty^3}{3k^*}\frac{\partial^2 T}{\partial y^2} \tag{27.7}$$

By subsuming equation (27.7) to equation (27.3), we get:

$$u\frac{\partial T}{\partial x}+v\frac{\partial T}{\partial y}=\left(\frac{\kappa_{hnf}}{(\rho Cp)_{hnf}}+\frac{16\sigma^*T_\infty^3}{3(\rho Cp)_{hnf}k^*}\right)\frac{\partial^2 T}{\partial y^2}, \tag{27.8}$$

27.2.1 Thermophysical Properties and Expression of Nanofluid

Physical Properties	Fluid Phase (Water)	Al_2O_3
ρ (kg m-3)	997	3,970
C_p (J kg-1 K-1)	4,180	765
k (W m-1 K-1)	0.6071	40
$\sigma(\Omega/m)^{-1}$	0.05	35×10^6

$$\begin{cases} A_1 = \dfrac{1}{(1-\phi)^{2.5}}, \\ A_2 = (1-\phi)+\phi\dfrac{\rho_s}{\rho_f}, \\ A_3 = (1-\phi)+\phi\dfrac{(\rho Cp)_s}{(\rho Cp)_f}, \\ A_4 = \dfrac{\kappa_s+(n-1)\kappa_f-\varphi(\kappa_f-\kappa_s)}{\kappa_s+(n-1)k_f+\varphi(\kappa_f-\kappa_s)}, \\ A_5 = 1+\dfrac{3(\sigma-1)\varphi}{(\sigma+2)-(\sigma-1)\varphi}, \text{ where } \quad \sigma=\dfrac{\sigma_s}{\sigma_f}, \end{cases} \tag{27.9}$$

27.2.2 Similarity Transformations

We insert the following dimensionless and similarity variables into equations (27.2) and (27.3) in order to convert the governing PDEs into a system of nonlinear ODEs.

$$\begin{cases} \psi = x\sqrt{c\upsilon_f}\,f(\eta), \qquad \eta=\sqrt{\left(\dfrac{c}{\upsilon_f}\right)}y, \qquad \theta(\eta)=\dfrac{T-T_\infty}{T_w-T_\infty}, \\ u=\dfrac{\partial\varphi}{\partial x}=cxf'(\eta),\ v=-\dfrac{\partial\varphi}{\partial x}=-\sqrt{c\upsilon_f}\,f(\eta). \end{cases} \tag{27.10}$$

Using the transformation, equations (27.2) and (27.8) become:

$$Cf_{\eta\eta\eta\eta\eta}-A_1f_{\eta\eta\eta}+A_2(f_\eta^{\ 2}-ff_{\eta\eta})+(A_5Q+A_1Da^{-1})f_\eta=0 \tag{27.11}$$

$$(A_4+N_r)\theta_{\eta\eta}(\eta)+A_3\Pr f(\eta)\theta_\eta(\eta)=0. \tag{27.12}$$

The BCs are as follows:

$$f(0)=0,\quad f_{\eta\eta}(0)=1,\quad f_\eta(\infty)=0, \tag{27.13}$$

$$\theta(0)=1,\quad \theta(\infty)=0, \tag{27.14}$$

Where:

$$\begin{cases} A_1=\dfrac{\rho_{nf}}{\rho_f}, \quad A_2=\dfrac{\mu_{nf}}{\mu_f}, \quad A_3=\dfrac{(\rho C_p)_{nf}}{(\rho C_p)_f}, \quad A_4=\dfrac{k_{nf}}{k_f}, \quad A_5=\dfrac{\sigma_{nf}}{\sigma_f}, \\ Nr=\dfrac{16\sigma^*T_\infty^3}{3k_fk}, \qquad Pr=\dfrac{(\rho C_p)_f v_f}{k_f},\ \text{C}=\dfrac{n_0c}{v^2{}_fp} \end{cases} \tag{27.15}$$

27.2.3 Exact Solution of Momentum Equation

Exact analytical solutions for equation (27.11) are obtained by utilizing equation (27.10) with related boundary condition equation (27.13), as follows:

$$f(\eta)=\frac{(1-\exp[-\alpha\eta])}{\alpha}, \tag{27.16}$$

Where $\alpha>0$ is to determine the use of boundary conditions in equation (27.13) As a result, when using equation (27.16) in equation (27.11):

$$c\beta^4 - A_1\beta^2 + (A_2 + A_5 Q + A_1 Da^{-1}) = 0, \tag{27.17}$$

And the roots are:

$$\beta_1 = -\sqrt{\frac{A_1 - \sqrt{-4CDa^{-1}A_1 + A_1^{\,2} - 4CA_2 - 4CMA_5}}{2C}},$$

$$\beta_2 = \sqrt{\frac{A_1 - \sqrt{-4CDa^{-1}A_1 + A_1^{\,2} - 4CA_2 - 4CMA_5}}{2C}},$$

$$\beta_3 = -\sqrt{\frac{A_1 + \sqrt{-4CDa^{-1}A_1 + A_1^{\,2} - 4CA_2 - 4CMA_5}}{2C}},$$

$$\beta_4 = \sqrt{\frac{A_1 + \sqrt{-4CDa^{-1}A_1 + A_1^{\,2} - 4CA_2 - 4CMA_5}}{2C}},$$

27.2.4 Exact Solution for Temperature Equation

The temperature equation in equation (27.12) can be modified with the aid of boundary conditions in equation (27.14), and by introducing a new variable $t=-\exp[-a\eta]$, it is transformed as follows:

$$t\frac{\partial^2\theta}{\partial t^2} + (m - nt)\frac{\partial\theta}{\partial t} = 0, \tag{27.18}$$

The transformed boundary condition is as follows:

$$\theta(-1)=1, \qquad \theta(0)=0, \tag{27.19}$$

The outcome of equation (27.18) in terms of complete gamma functions is as follows:

$$\theta(\eta)=\frac{\Gamma[1-m,0,-nt]}{\Gamma[1-m,0,n]}, \tag{27.20}$$

Where:

$$p=\frac{\Pr}{\alpha^2}, \qquad m = 1-\frac{A_3 p}{(A_4+N_r)}, \qquad n=\frac{A_3 p}{(A_4+N_r)},$$

27.3 RESULTS AND DISCUSSION

In this analysis, we developed the MHD couple stress fluid model in the presence of thermal radiation in a porous medium. In this mathematical model, we begin converting nonlinear PDEs into nonlinear ODEs by using a similarity variable. The conclusions that came out of these processes instantly are analytical solutions to boundary value problems. This conclusion is reached methodically in terms of the complete gamma function. This work makes a contribution by highlighting the effect of couple stress fluid with Al_2O_3 nanoparticles with water as the base fluid, with Pr set to 6.2 and held constant for the current analysis.

27.3.1 Velocity Profile

Figure 27.2 shows the effect Da^{-1} on the velocity profile. Because rising Darcy represents the rise in the hole of the porous structure, as Da^{-1} rises, the velocity of Al_2O_3 and H_2O fluid decreases. Figure 27.3 shows the effect of the magnetic parameter (M) on the velocity profile. The Lorentz force, a resistive-type force, is produced when a magnetic field is introduced. This force prevents the fluid from flowing, which lowers the fluid's velocity. Furthermore, the velocity of $Al_2O_3 + H_2O$ fluid is decreased with increase in M. The effect of the couple stress (C) on the velocity profile is shown in Figure 27.4. As the value of C increases, the $Al_2O_3 + H_2O$ fluid velocity decreases. Figure 27.5 depicts the effect of the volume fraction on the velocity profile as the value of the $Al_2O_3 + H_2O$ fluid velocity is increased.

27.3.2 Temperature Field

Figure 27.6 shows the effect Da^{-1} on the temperature profile. By increasing the Da^{-1} temperature of $Al_2O_3 + H_2O$, fluid also increases. Figure 27.7 shows the effect of the magnetic parameter (M) on the temperature profile. The thermal boundary of the $Al_2O_3 + H_2O$ fluid increases with an increase in M. Figure 27.8 depicts the effect of the couple stress (C) on the temperature profile by increasing the value of C as the temperature of the $Al_2O_3 + H_2O$ fluid increases. Figures 27.9 shows the effect of the volume fraction on the temperature profile; as the volume fraction increases, so does the temperature of the $Al_2O_3 + H_2$O fluid. Figure 27.10 demonstrates the effect of thermal radiation on

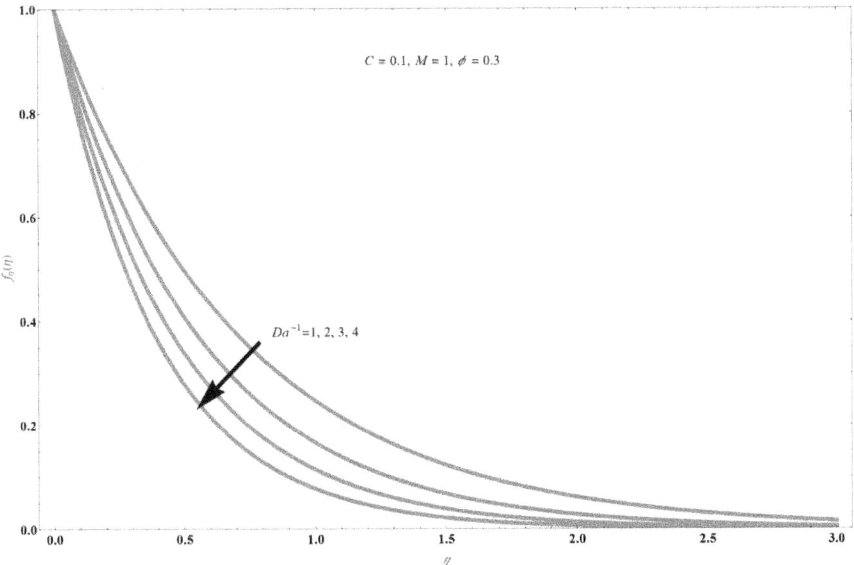

FIGURE 27.2 Effect of the inverse Darcy number on velocity profile for fixed values of $C = 0.1, M = 0.1, \phi = 0.3$.

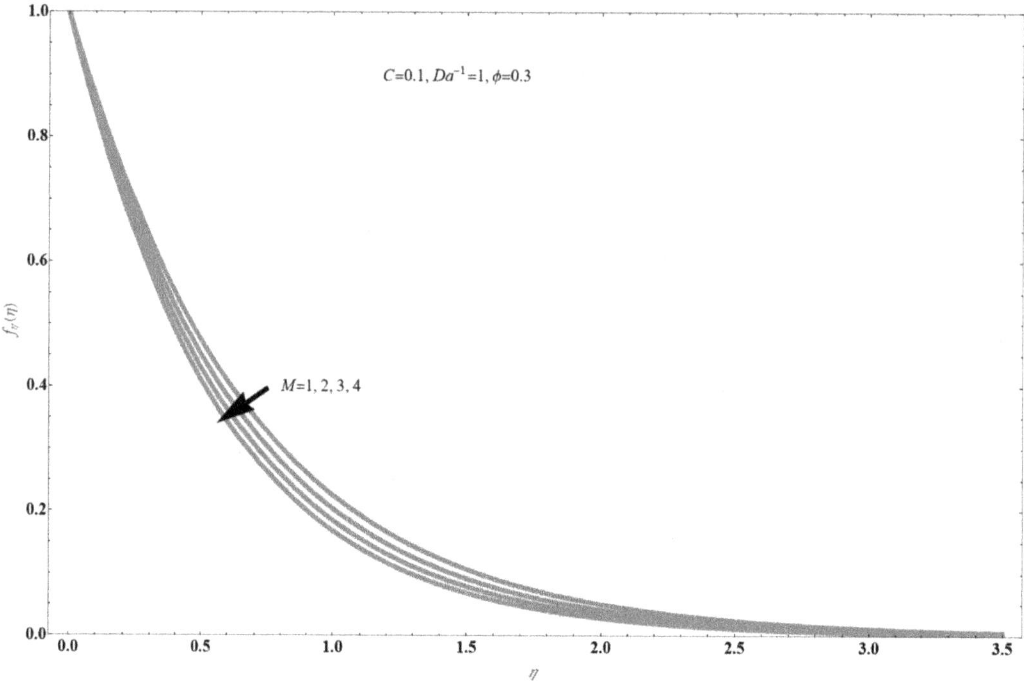

FIGURE 27.3 Effect of magnetic field on a velocity profile for fixed values of $C = 0.1, Da^{-1} = 1, \phi = 0.3$.

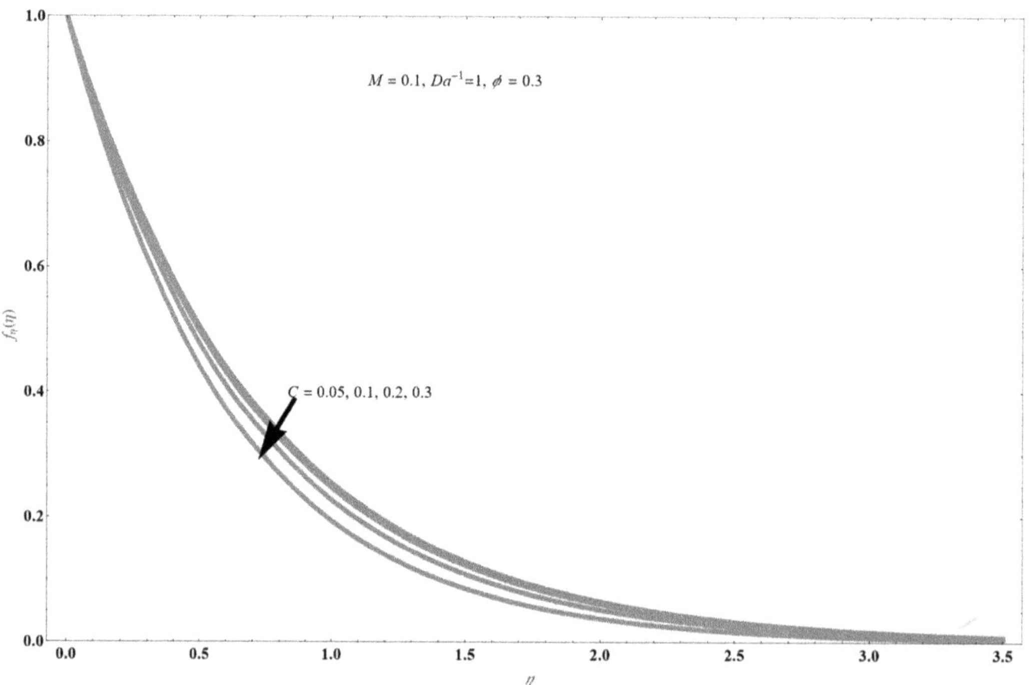

FIGURE 27.4 Effect of couple stress on a velocity profile for fixed values of $Da^{-1} = 1, M = 0.1, \phi = 0.3$.

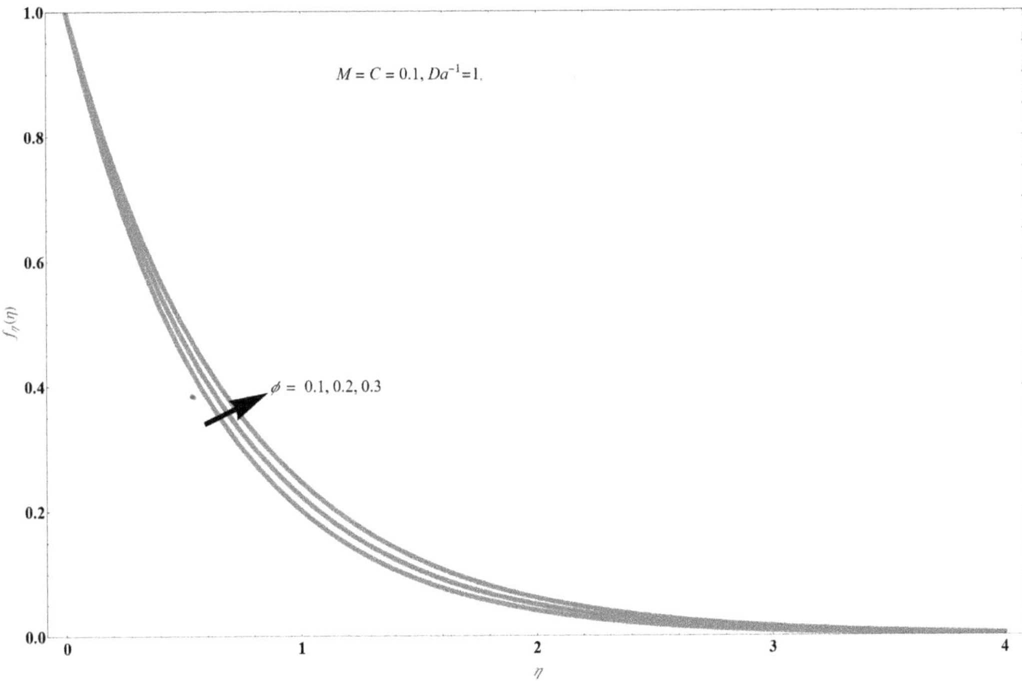

FIGURE 27.5 Effect of volume fraction on a temperature profile for fixed values of $C = M = 0.1, Da^{-1} = 1$.

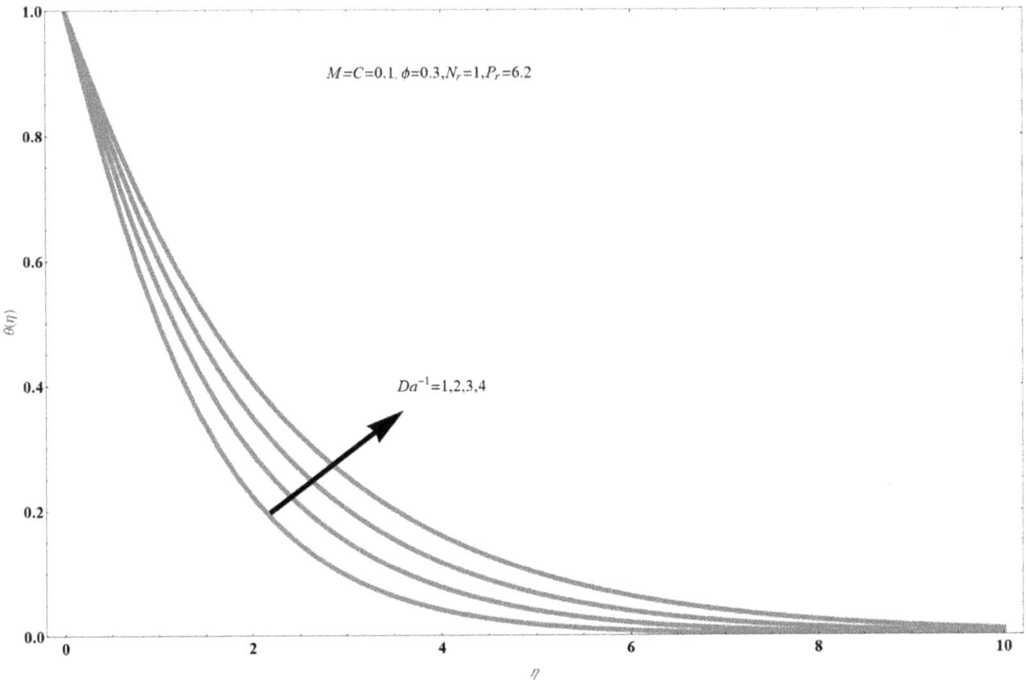

FIGURE 27.6 Effect of the inverse Darcy number on temperature profile for fixed values of $M = C = Nr = 0.1$, $\phi = 0.3, P_r = 6.2$.

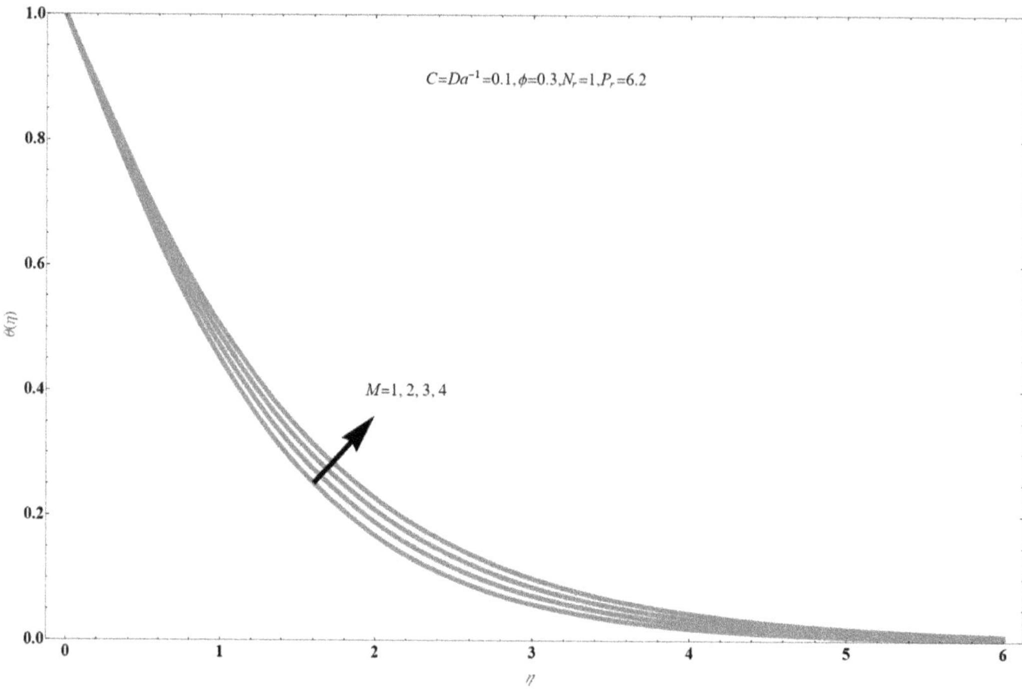

FIGURE 27.7 Effect of the magnetic field on a temperature profile for fixed values of $Da^{-1} = C = Nr = 0.1$, $\phi = 0.3, P_r = 6.2$.

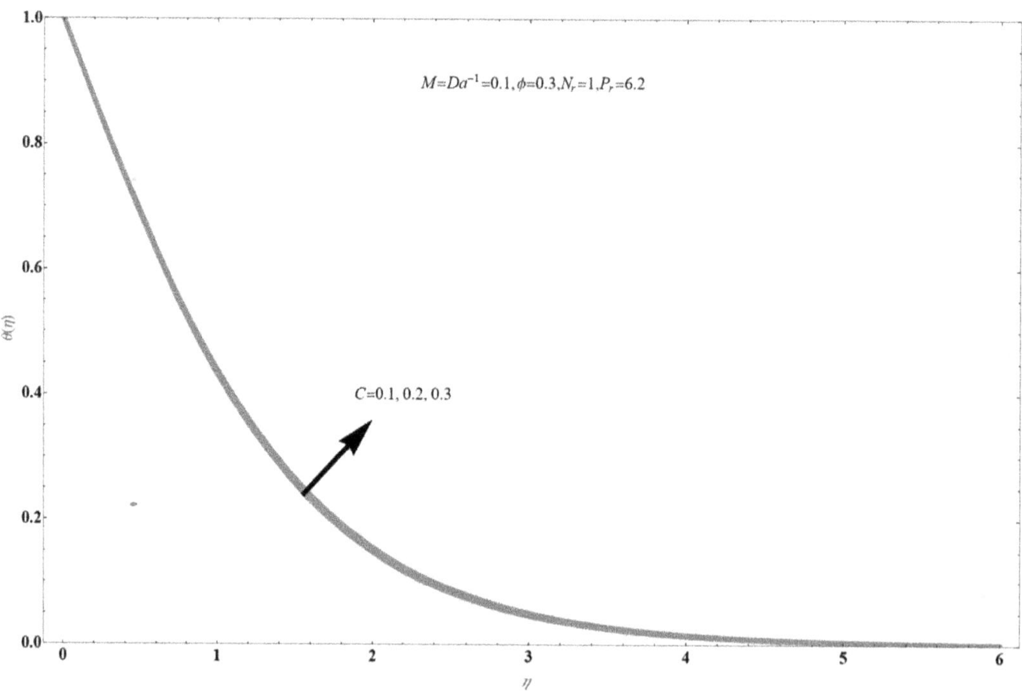

FIGURE 27.8 Effect of couple stress on a temperature profile for fixed values of $Da^{-1} = M = Nr = 0.1$, $\phi = 0.3, P_r = 6.2$.

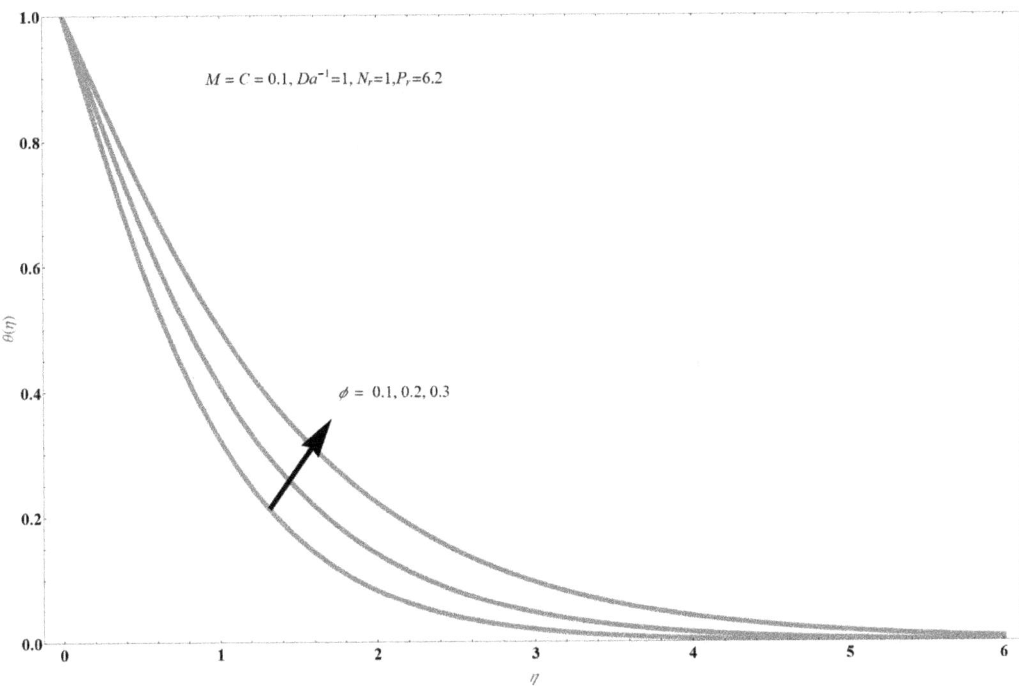

FIGURE 27.9 Effect of volume fraction on a temperature profile for fixed values of $Da^{-1} = C = Nr = 0.1,\ P_r = 6.2$.

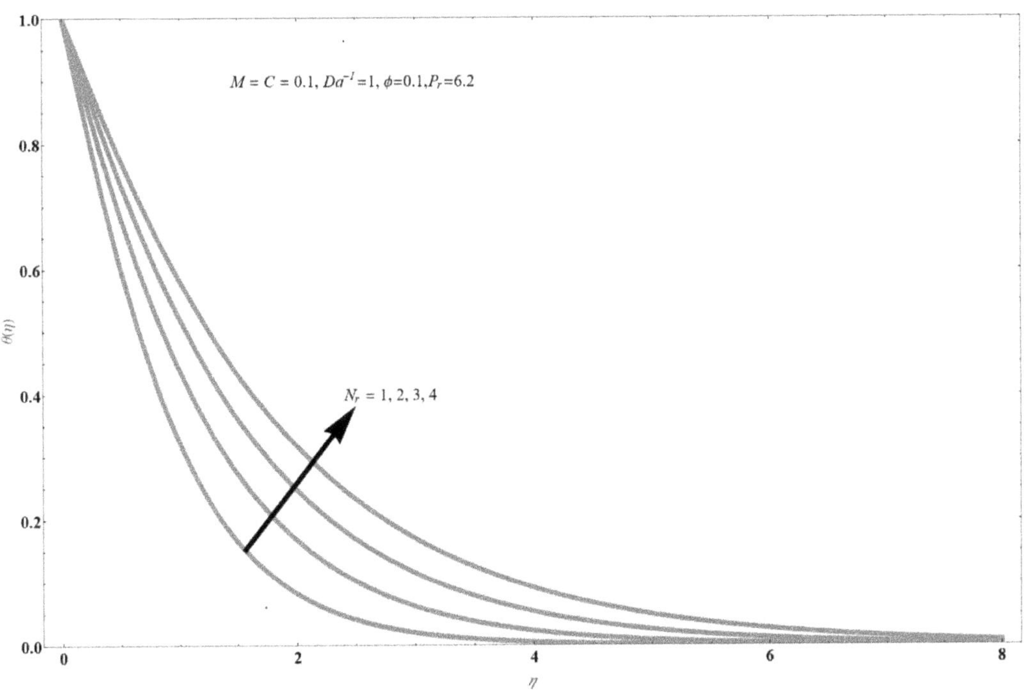

FIGURE 27.10 Effect of thermal radiation and temperature profile for fixed values of $Da^{-1} = C = M = 0.1$, $\phi = 0.3, P_r = 6.2$.

a temperature profile. By increasing the radiation parameter, as a result, thermal radiation shows dominance over conduction in the radiation. It may be utilized to efficiently manage the thermal boundary layer. From the result, it is observed that with the increase in N_r in the $Al_2O_3 + H_2O$ fluid, temperature also increases.

27.4 CONCLUDING REMARKS

In this chapter, we obtained the solution for couple stress fluid flow with radiation, which is represented in the nonlinear ODE equations in equation (27.11) and equation (27.12) with boundary conditions in equations (27.13)–(27.14), which are solved exactly by introducing a closed-form solution for the velocity field and using an incomplete gamma function for the temperature distribution.

- The $Al_2O_3 + H_2O$ fluid flow decreases the velocity boundary and increases the temperature boundary when the physical parameter of the inverse Darcy number $\left(Da^{-1}\right)$, the magnetic field (M), and couples stress (C) are increased.
- The temperature of $Al_2O_3 + H_2O$ fluid flow increases by increasing the thermal radiation (N_r).
- With increasing volume fraction of Al_2O_3 nanoparticles, the velocity and temperature boundaries of $Al_2O_3 + H_2O$ fluid flow increase.

Nomenclature

Symbol	**Explanation**	**SI Unit**
Latin Symbols		
A_1, A_2, A_3, A_4, A_5	constants	$(-)$
B_0	applied magnetic field	(wm^{-2})
C	couple stress fluid parameter	$(-)$
C_P	specific heat at constant pressure	$(JKg^{-1}K^{-1})$
Da^{-1}	inverse Darcy number	$(-)$
k^*	mean absorption coefficient	(m^{-2})
k	permeability of porous medium	(m^{-2})
M	magnetic field	$(-)$
N_r	radiation parameter	$(-)$
Pr	Prandtl number	$(-)$
q_W	local heat flux at the wall	$(-)$
q_r	radiative heat flux	(Wm^{-2})
T	temperature	(K)
(x, y)	coordinate axes	(m)
(u, v)	velocities along x- and y-directions	(ms^{-1})
Greek Symbols		
η	similarity variable	$(-)$
κ	thermal conductivity of fluid	$(WKg^{-1}K^{-1})$
μ	dynamic viscosity of fluid	$(\mathrm{kgm^{-1}s^{-1}})$
ρ	density	(Kgm^{-3})
σ	surface tension	$(-)$
σ^*	Stefan–Boltzmann constant	$\mathrm{Wm^{-2}K^{-4}}$
υ	kinematic viscosity	$(\mathrm{m^2s^{-1}})$
Γ	gamma function	$(-)$

φ	volume fraction Al_2O_3	(–)
ψ	stream function	(–)

Subscripts

f	base fluid	(–)
nf	nanofluid	(–)

Abbreviations

BCs	boundary conditions	(–)
MHD	magnetohydrodynamics	(–)
Al_2O_3	aluminum oxide	(–)
PDEs	partial differential equations	(–)
LT	Laplace transformation	(–)
ODEs	ordinary differential equations	(–)

REFERENCES

1. Crane, L.J. (1970). Flow past a stretching plate. *Journal of Applied Mathematics and Physics (ZAMP)*, 21, 645–647.
2. Wang, C.Y. (1984). The three-dimensional flow due to a stretching flat surface. *Physics of Fluids*, 27(8), 19 15.
3. Choi, S.U.S. (1995). Enhancing thermal conductivity of fluids with nanoparticles. In *Proceedings of the 1995 ASME International Mechanical Engineering Congress and Exposition, San Francisco, CA, USA*, 12–17 November, ASME FED 231/MD 66.
4. Mahabaleshwar, U.S., Aly, E.H., & Vishalakshi, A.B. (2022). MHD and thermal radiation flow of graphene Casson nanofluid stretching/shrinking sheet. *International Journal of Applied Computational and Mathematics*, 8, 113.
5. Mahabaleshwar, U.S., Sneha, K.N., Chan, A., & Zeidan, D. (2022). An effect of MHD fluid flow heat transfer using CNTs with thermal radiation and heat source/sink across a stretching/shrinking sheet. *International Communications in Heat and Mass Transfer*, 135.
6. Maranna, T., Sneha, K.N., Mahabaleshwar, U.S., Sarris, I.E., & Karakasidis, T.E. (2022). An effect of radiation and MHD newtonian fluid over a stretching/shrinking sheet with CNTs and mass transpiration. *Applied Sciences*, 12(11), 5466.
7. Mahabaleshwar, U.S., Bognár, G., Baleanu, D., & Vishalakshi, A.B. (2022). Two-dimensional nanofluid due to an accelerated plate with viscosity ratio. *International Journal of Applied Computational and Mathematics*, 8, 111.
8. Anusha, T., Huang, H.-N., & Mahabaleshwar, U.S. (2021). Two dimensional unsteady stagnation point flow of Casson hybrid nanofluid over a permeable flat surface and heat transfer analysis with radiation. *Journal of the Taiwan Institute of Chemical Engineers*, 127, 79–91.
9. Umavathi, J.C., & Anwar Bég, O. (2020). Modeling the onset of thermosolutal convective instability in a non-Newtonian nanofluid-saturated porous medium layer. *Chinese Journal of Physics*, 68, 147–167.
10. Stokes, V.K. (1966). Couple stress in fluids. *Physics of Fluids*, 9, 1709–1715.
11. Mahabaleshwar, U.S., Sarris, I.E., Hill, A.A., Lorenzini, G., & Pop, I. (2017). An MHD couple stress fluid due to a perforated sheet undergoing linear stretching with heat transfer. *International Journal of Heat and Mass Transfer*, 105.
12. Mahabaleshwar, U.S., Vishalakshi, A.B., Bognar, G.V., & Mallikarjunaiah, S.M. (2022). Effect of thermal radiation on the flow of a boussinesq couple stress nanofluid over a porous nonlinear stretching sheet. *International Journal of Applied and Computational Mathematics*, 8, 169.
13. Dhananjay, Y., Mahabaleshwar, U.S., Wakif, A., & Chand, R. (2021). Significance of the inconstant viscosity and internal heat generation on the occurrence of Darcy-Brinkman convective motion in a couple-stress fluid saturated porous medium: An analytical solution. *International Communications in Heat and Mass Transfer*, 122, 105165.
14. Afzal, Q., Akram, S., Ellahi, R., Sait, S.M., & Chaudhry, F. (2021). Thermal and concentration convection in nanofluids for peristaltic flow of magneto couple stress fluid in a nonuniform channel. *Journal of Thermal Analysis and Calorimetry*, 144, 2203–2218.

15. Devakar, M., Sreenivasu, D., & Shankar, B. (2014). Analytical solutions of couple stress fluid flows with slip boundary conditions. *Alexandria Engineering Journal*, 53(3), 723–730.
16. Anusha, T., Mahabaleshwar, U.S., & Sheikhnejad, Y. (2022). An MHD of nanofluid flow over a porous stretching/shrinking plate with mass transpiration and Brinkman ratio. *Transport in Porous Media*,142, 333–352.
17. Sithole, H., Mondal, H., Goqo, S., Sibanda, P., & Motsa, S. (2018). Numerical simulation of couple stress nanofluid flow in magneto-porous medium with thermal radiation and a chemical reaction. *Applied Mathematics and Computation*, 339, 820–836.
18. Bognár, G., Klazly, M., Mahabaleshwar, U.S., Lorenzini G., & Hriczó, K. (2021). Comparison of similarity and computational fluid dynamics solutions for Blasius flow of nanofluid. *Journal of Engineering and Thermophysics*, 30, 461–475.
19. Siddheshwar, P.G., & Mahabaleshwar, U.S. (2005). Effects of radiation and heat source on MHD flow of a viscoelastic liquid and heat transfer over a stretching sheet. *International Journal of Non-Linear Mechanics*, 40, 807–820.
20. Mahabaleshwar, U.S., Vishalakshi, A.B., & Hatami, M. (2022), MHD micropolar fluid flow over a stretching/shrinking sheet with dissipation of energy and stress work considering mass transpiration and thermal radiation. *International Communications in Heat and Mass Transfer*, 133.
21. Mahabaleshwar, U.S., Vishalakshi, A.B., & Azese, M.N. (2022). The role of Brinkmann ratio on non-Newtonian fluid flow due to a porous shrinking/stretching sheet with heat transfer. *European Journal of Mechanics—B/Fluids*, 92, 153–165.
22. Mahabaleshwar, U.S., Nagaraju, K.R., Vinay Kumar, P.N., & Azese, M.N. (2020). Effect of radiation on thermosolutal Marangoni convection in a porous medium with chemical reaction and heat source/sink. *Physics of Fluids*, 32(11).
23. Mahabaleshwar, U.S., Nagaraju, K.R., Vinay Kumar, P.N., Nadagoud, M.N., Bennacer, R., & Baleanu, D. (2019). An mhd viscous liquid stagnation point flow and heat transfer with thermal radiation and transpiration. *Thermal Science and Engineering Progress*, 16, 100379,
24. Mahabaleshwar, U.S., Anusha, T., Laroze, D., Said, N.M., & Sharifpur, M. (2022). An MHD flow of non-newtonian fluid due to a porous stretching/shrinking sheet with mass transfer. *Sustainability*, 14, 7020.
25. Mahabaleshwar, U.S., Aly, E.H., & Anusha, T. (2022). MHD slip flow of a Casson hybrid nanofluid over a stretching/shrinking sheet with thermal radiation. *Chinese Journal of Physics*, 80, 74–106,
26. Mallikarjun, P., Murthy, R.V., Mahabaleshwar, U.S., & Lorenzini, G. (2019). Numerical study of mixed convective flow of a couple stress fluid in a vertical channel with first order chemical reaction and heat generation/absorption. *Mathematical Modelling of Engineering Problems*, 6(2), 175–182.
27. Benos, L.T., Mahabaleshwar, U.S., Sakanaka, P.H., & Sarris, I.E. (2019). Thermal analysis of the unsteady sheet stretching subject to slip and magnetohydrodynamic effects. *Thermal Science and Engineering Progress*, 13.
28. Kumar, P.N.V., Mahabaleshwar, U.S., Sakanaka, P.H., & Lorenzini, G. (2018). An MHD effect on a newtonian fluid flow due to a superlinear stretching sheet. *Journal of Engineering Thermophysics*, 27(4), 501–506.
29. Singh, B., & Nisar, K.S. (2020). Thermal instability of magnetohydrodynamic couple stress nanofluid in rotating porous medium. *Numerical Methods for Partial Differential Equations*, 1–14.
30. Kumar, K.A., Sugunamma, V., Sandeep, N., & Ramana Reddy, J.V. (2019). Numerical examination of MHD nonlinear radiative slip motion of non-Newtonian fluid across a stretching sheet in the presence of a porous medium. *Heat Transfer Research*, 50, 1163–1181.
31. Krishna, M.V., & Chamkha, A.J. (2020). Hall and ion slip effects on MHD rotating flow of elastico-viscous fluid through porous medium. *International Communications in Heat and Mass Transfer*, 113.
32. Abou-zeid, M.Y. (2018). Homotopy perturbation method for couple stresses effect on MHD peristaltic flow of a non-Newtonian nanofluid. *Microsystem Technologies*, 24, 4839–4846.
33. Hamdan, M.H. (1998) An alternative approach to exact solutions of a special class of Navier-stokes flows. *Applied Mathematics and Computation*, 93(1), 83–90.
34. Hamdan, M.H., & Allan, F.M. (2006). A note on the generalized Beltrami flow through porous media. *International Journal of Pure Applied Mathematics*, 27, 491–500.
35. Vafai, K., & Tien, C.L. (1981). Boundary and inertia effects on flow and heat transfer in porous media. *International Journal of Heat Mass Transfer*, 24(2), 195–203.
36. Islam, S., Muhammad, R.M., & Zhou, C. (2008). Few exact solutions of non-Newtonian fluid in porous medium with hall effect. *Journal of Porous Media*, 11(7), 669–680.
37. Islam, S., & Chaoying, Z. (2007). Certain inverse solutions of a second-grade magnetohydrodynamic aligned fluid flow in a porous medium. *Journal of Porous Media*, 10(4), 401–408.

28 Nonlinear Radiative Falkner–Skan Flow of Hydromagnetic Nanofluid Over a Wedge with Arrhenius Activation Energy

M. Gnaneswara Reddy and K. Ramesh

28.1 INTRODUCTION

The concept of two-dimensional flow for static wedge phenomena was initially pioneered by Falkner and Skan [1] in the year 1931; by using self-similar transformations F-S, they reduced the PDE into ODE with stream-wise pressure gradient with wedge angle for momentum equation, which, in fact, is the generalization of the Blasius boundary layer equation. The fluid flow through a wedge-shaped surface has extensive applications in the field of aero-thermodynamics, geothermal systems, bio-hydrodynamics, and oil recovery. The well-known Falkner–Skan two-point boundary value problem is the third-order nonlinear ODE obtained from PDEs which exhibit the 2D incompressible laminar boundary layer flow model. The Falkner–Skan equation was solved by semi-analytical and numerical techniques by mathematicians and engineers. In the year 1937, Hartree [2] showed that the family of unique solutions for which the limiting point of the slope of tangent is unity for the F-S equations exists under some special conditions. The solution of second-grade incompressible flow is described by the F-S equation, in which the point was kept symmetrically in the direction of the stream, which was investigated by Rajagopalet al. [3]. Series solutions of FSWF were explored by Liao [4] for flat-plate boundary conditions, and for the streCOHing boundary conditions by Yao and Chen [5]. Elnady et al. [6] produced the Chebyshev series method by converting the nonlinear differential equations into a system of algebraic equations. In the year 2009, Alizadeh et al. [7] adopted ADM to obtain the semi-analytical solutions for the Falkner–Skan two-point BVP for wedge flow. Allan and Al Mdallal [8] obtained a series of solutions of the modified Falkner–Skan equation. Temimi and Ben-Romdhane [9] adopted an iterative finite difference scheme based on the Newton Kantorovich quasi-linearization technique to solve the two-point BVP. Khan et al. [10] numerically analyzed the significance of moving-static transient FSWF of MHD nanofluid with nonlinear thermal radiation and convective wall temperature. Khan et al. [11] considered bvp4c to explore the non-Fourier heat flux and two-phase non-homogenous model of MHD Falkner–Skan Sutterby nanofluid past a static permeable wedge.

Activation energy is an essential aspect in the course of chemical reaction and is defined as the minimum obligatory quantity of thermal energy for molecules/atoms enduring a chemical reaction. The Arrhenius activation energy is vital in geothermal engineering, oil emulsions, and mechanics of water. During the 1990s, Bestman [12] explored the analytical study of convective flow over a permeable medium with binary chemical reaction. The non-Newtonian liquid flow on activation energy through a rotating frame has been examined by Shafique et al. [13]. It is noted that the solutal concentration declines for higher activation energy. Khan et al. [14] have scrutinized the impact of activation energy on the hydromagnetic nanoparticle across a cross flow. They resolved the final equations via numerically bvp4c scheme in MATLAB. The heat source effects on the

DOI: 10.1201/9781003299608-28

Casson nanofluid flow over a streCOHing surface studied were by Gireesha et al. [15]. The nonlinear convection transport on non-Newtonian Maxwell nanofluid has been explored by Ijaz and Ayub [16]. They employed the homotopy technique for finding the numerical results. Some of the considerable studies were for the activation energy impact through different configurations and restrictions by the researchers [17–27].

In the transportation of heat energy, the influence of thermal radiation is very much significant and plays a vital role in space technological, industrial, and engineering applications, namely, propulsion devices of an aircraft, satellite vehicles, designing atomic IC engines, solar radiant buildings, etc. Whenever the temperature gradient is practically small, the linear radiation is substantial, while for temperature gradient that is appropriately large, the nonlinear thermal radiation is decisive. Due to its practical prominence in rocket propulsion, several researchers have studied nonlinear thermal radiation effects in several geometrical and material conditions. Tarakaramu and Satya Narayana [28] analyzed the thermal radiation on MHD non-Newtonian viscoelastic nanofluid. Reddy et al. [29] investigated nonlinear thermal radiation effects on Eyring–Powell nanofluid multi-slip flow in the presence of Arrhenius activation energy above the slendering sheet. Archana et al. [30] explored the triple-diffusive nanofluid flow over a horizontal plate along with buoyancy effects. Koriko et al. [31] investigated the combined effects of thermal stratification, quartic autocatalytic chemical reaction, and nonlinear radiation on 3D Eyring–Powell nanofluid motion past a streCOHable surface. Kumar et al. [32] considered the Joule heating and entropy generation effects on Williamson nanofluid motion past a streCOHing sheet with nonlinear radiation. Mahanthesh et al. [33] scrutinized the temperature-dependent heat source and nonlinear radiation effects on Casson and Carreau fluid flows for both dust and fluid phases subjected to convective temperature conditions. Raju and Sandeep [34] addressed the impact of nonlinear radiation on MHD Casson nanofluid flow subjected to convective constraints by employing shooting technique. Waqas et al. [35] characterized the Powell-Eyring nanofluid flow over an elongating sheet in the regime of magnetic dipole with nonlinear radiation. Some more researchers have been reported by incorporating nonlinear thermal radiation effects in various geometries [36–39].

In view of the aforementioned literature, the nonlinear thermal radiation impact through a Falkner–Skan wedge model has not been considered. In the current study, we therefore examine the convective nonlinear radiative magnetohydrodynamic boundary layer Falkner–Skan wedge flow of nanofluid in the presence of Arrhenius activation energy. The transformed conservation boundary layer equations are solved numerically. The numerical solutions are verified with special cases from the literature. The impact of emerging parameters on velocity, temperature, and concentration distributions is evaluated in detail physically. Further, the engineering quantities, such as friction factor, rate of energy, and species at the wall of wedge surface, are reported. The present study is relevant to the rocket propulsion, combustion hybrid IC engines, and plastic industries.

28.2 MATHEMATICAL FORMULATION

A steady two-dimensional (2D) non-Newtonian Falkner–Skan nanofluid flow across a wedge surface is considered. Here, the impacts of applied magnetic field, heat source, and nonlinear radiation are incorporated in the present modelled formulation. In addition, the Newtonian energy and species restrictions have been emphasized. The 2D Cartesian coordinate system is considered by choosing the x-axis along the wedge surface, while the other y-axis is selected vertical to the wedge surface. The geometrical physical flow configuration is displayed in Figure 28.1. The nanofluid passes along the surface of the wedge with the free-stream flow velocity as $u_e(x) = ax^n$, in which a and n are the positive real constants with the property that $0 \leq n \leq 1$. In addition, the angle of wedge is to be considered as $\Omega = \beta\pi$; here, β is the Hartree pressure gradient variable. It is further assumed that the uniform magnetic field of strength (B_0) is exerted vertical to the wedge surface. Furthermore, the induced magnetic field is omitted, owing to the small magnetic Reynolds number.

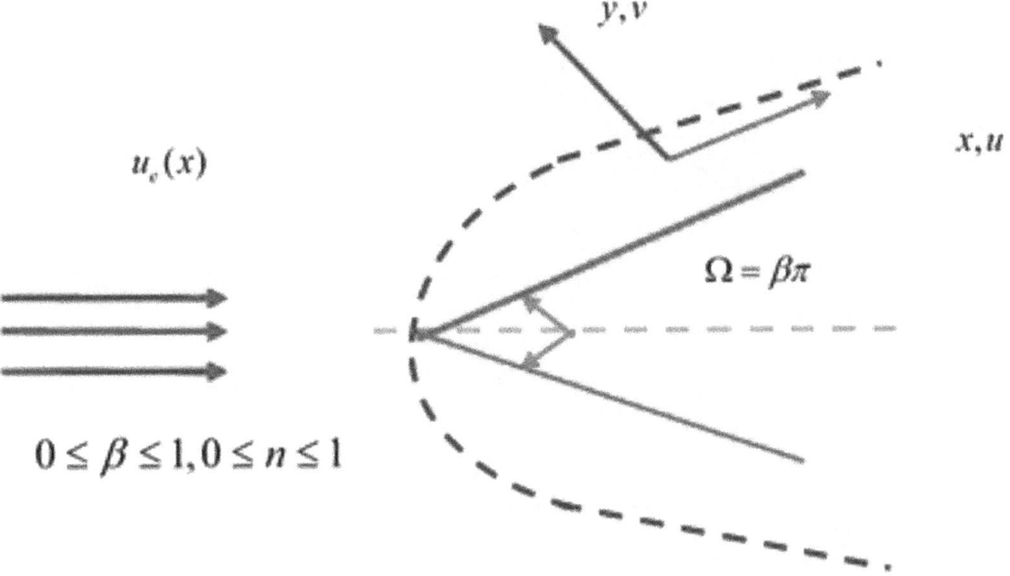

FIGURE 28.1 Flow physical model.

Based on the description analysis and assumptions, the governing equations for the fundamental laws of continuity, momentum, energy, and concentration [34, 40–42] are:

$$\frac{\partial u}{\partial x}+\frac{\partial v}{\partial y}=0 \tag{28.1}$$

$$u\frac{\partial u}{\partial x}+v\frac{\partial u}{\partial y}=u_e\frac{\partial u_e}{\partial x}+\nu\frac{\partial}{\partial y}\left[\frac{\dfrac{\partial u}{\partial y}}{1+\left\{\Gamma\left(\dfrac{\partial u}{\partial y}\right)\right\}^m}\right]$$
$$+\frac{1}{\rho_f}\left[\sigma B_0^2\left(u_e-u\right)+g\beta\left(T-T_\infty\right)+g\beta^*\left(C-C_\infty\right)\right] \tag{28.2}$$

$$u\frac{\partial T}{\partial x}+v\frac{\partial T}{\partial y}=\left[\alpha_m+\frac{16T_\infty^3\sigma^{**}}{3\left(\rho c_p\right)_f k_\infty k^{**}}\right]\frac{\partial^2 T}{\partial y^2}+\tau\left[D_B\frac{\partial C}{\partial y}\frac{\partial T}{\partial y}+\frac{D_T}{T_\infty}\left(\frac{\partial T}{\partial y}\right)^2\right]$$
$$+\frac{Q_0}{\left(\rho c_p\right)_f}\left(T-T_\infty\right) \tag{28.3}$$

$$u\frac{\partial C}{\partial x}+v\frac{\partial C}{\partial y}=D_B\frac{\partial^2 C}{\partial y^2}+\frac{D_T}{T_\infty}\frac{\partial^2 T}{\partial y^2}-k_r\left(C-C_\infty\right)\left(\frac{T}{T_\infty}\right)^m\exp\left[\frac{-E_a}{\kappa T}\right] \tag{28.4}$$

In which u and v are the velocity components along the x- and y-directions; m and Γ are the material parameters; ρ_f is the fluid density; σ is the electrical conductivity; k_f is the fluid thermal conductivity; τ is the ratio of effective heat capacity of the nanofluid to that of the regular fluid; D_B and D_T

represent the Brownian diffusion coefficient and thermophoresis diffusion coefficient; Q_0 is the heat generation/absorption coefficient; and K_r is the chemical reaction rate constant. E_a is the activation energy, and κ is the Boltzmann constant.

The relevant boundary restrictions are [34, 41, 42]:

$$u = 0, \quad v = 0, \quad \frac{\partial T}{\partial y} = -h_s T, \frac{\partial C}{\partial y} = -h_c C \text{ at } y = 0 \ y = 0$$

$$u = u_e(x) = ax^n, T \to T_\infty, C \to C_\infty \text{ as } y \to \infty \ y \to \infty \tag{28.5}$$

Here, the constants h_s and h_c denote the heat and solutal transfer coefficients.

Let us introduce the following usual similarity transformations:

$$\eta = y\sqrt{\frac{a(n+1)}{2\nu}}x^{\frac{n-1}{2}}, \psi = \sqrt{\frac{2\nu a}{n+1}}x^{\frac{n+1}{2}} f(\eta), u = \frac{\partial \psi}{\partial y}, \ v = -\frac{\partial \psi}{\partial x}, We = \left(\frac{(n+1)\Gamma^2 u_e^3}{2\nu x}\right)^{\frac{1}{2}},$$

$$M^2 = \frac{2\sigma B_0^2 x}{\rho_f (n+1) U_e}, \beta = \frac{2n}{n+1}, \lambda = \frac{g\beta^*(T_w - T_\infty)x}{(n+1)u_e^2}, \ \theta = \frac{T - T_\infty}{T_\infty}, \phi = \frac{C - C_\infty}{C_\infty},$$

$$N = \frac{(C_w - C_\infty)x}{\rho(T_w - T_\infty)\beta^*}, \Pr = \frac{\mu c_p}{\alpha_m}, \theta_w = \frac{T_f}{T_\infty}, N_b = \frac{C_\infty D_B \tau}{\nu}, N_t = \frac{D_T \tau}{\nu},$$

$$\lambda_1 = \frac{2Q_0 x}{(\rho c)_f (1+n) u_e}, Sc = \frac{\nu}{D_B}, \sigma_r = \frac{k_r^2}{a}, \ \delta = \frac{T_w - T_\infty}{T_\infty}, \ E = \frac{E_a}{\kappa T_\infty}, \tag{28.6}$$

Here, η is the dimensionless variable, ψ is the dimensional stream function, and θ and ϕ are the dimensionless temperature and concentration functions.

It is clear that the equation of mass (28.1) is trivially satisfied. By utilizing equation (28.6) in the fluid transport, energy, and concentration equations (28.2)–(28.4), we get:

$$f'''\left[\frac{1+(1-m)(Wef'')^m}{\left(1+(Wef'')^m\right)^2}\right] + ff'' + +\beta\left[1-(f')^2\right] + M^2(1-f') + \lambda\left[\theta + N\phi\right] = 0 \tag{28.7}$$

$$\left(1+(N_R(1+(\theta_w - 1)\theta)^3)\theta'\right)' \theta'' + \Pr f\theta' + \Pr Nb\theta'\phi' + \Pr Nt\theta'^2 + \Pr \lambda_1 \theta = 0 \tag{28.8}$$

$$\phi'' + \frac{Nb}{Nt}\theta'' + Sc\left(f\phi' - -\sigma_r(1+\delta\theta)^m \phi \exp\left(-\frac{E}{1+\delta\theta}\right)\right) = 0 \tag{28.9}$$

In which We is the local Weissenberg number, β is the wedge angle parameter, λ is the thermal buoyancy variable, M is the magnetic parameter, N_b is the Brownian motion variable, N_t is the thermophoresis parameter, λ_1 is the local heat generation parameter, and Sc is the Schmidt number.

The reduced dimensionless boundary restrictions (28.5) can be written as:

$$f = 0, \ f' = 0, \ \theta' = -\gamma_1(1+\theta(0)), \ \phi' = -\gamma_2(1+\phi(0)), \text{ at } \eta = 0$$

$$f' \to 1, \ \theta \to 0, \ \phi \to 0, \text{ as } \eta \to \infty \tag{28.10}$$

Here, $\gamma_1 = h_1\sqrt{\dfrac{2\nu x}{(s+1)U_e}}$ and $\gamma_2 = h_2\sqrt{\dfrac{2\nu x}{(s+1)U_e}}$ are the thermal and species conjugate variables, accordingly.

28.3 ENGINEERING PHYSICAL QUANTITIES

The most prominent relations of physical concern in the present modelled analysis are friction factor, rate of heat transfer, and rate of mass transfer at the wall of wedge.

28.3.1 Skin Friction Coefficient

The friction factor (surface drag) can be defined as:

$$C_{fx} = \frac{\tau_w}{\frac{1}{2}\rho u_e^2} \tag{28.11}$$

Where the local wall shear stress τ_w is:

$$\tau_w = \eta_0 \left. \frac{\frac{\partial u}{\partial y}}{1+\left\{\Gamma\left(\frac{\partial u}{\partial y}\right)\right\}^m} \right]_{y=0} \tag{28.12}$$

By employing equation (28.6) in equation (28.11), the reduced skin friction coefficient is:

$$\mathrm{Re}^{\frac{1}{2}} C_{fx} = \frac{2}{\sqrt{2-\beta}}\left[\frac{f''}{1+(Wef'')^m}\right]_{\eta=0}. \tag{28.13}$$

Here, $\mathrm{Re} = \dfrac{xu_e}{\nu}$ is the local Reynolds number.

28.3.2 Nusselt Number

The heat transfer rate at the wall of wedge is given by:

$$Nu_x = \frac{xq_w}{k(T-T_\infty)} \tag{28.14}$$

Where the heat flux at the wall of wedge q_w is:

$$q_w = -k\left(\frac{\partial T}{\partial y}\right)_{y=0} \tag{28.15}$$

With the help of equation (28.6) in equation (28.14), the dimensionless local Nusselt number is:

$$\mathrm{Re}^{-\frac{1}{2}} Nu_x = \frac{\gamma_1}{\sqrt{2-\beta}}\left(1+\frac{1}{\theta(0)}\right). \tag{28.16}$$

28.3.3 Sherwood Number

The solutal transfer rate at the wall of wedge is defined as:

$$Sh_x = \frac{xq_m}{D_B\left(C - C_\infty\right)} \tag{28.17}$$

In which the mass flux at the wall of wedge q_m is given by:

$$q_m = -D_B\left(\frac{\partial C}{\partial y}\right)_{y=0} \tag{28.18}$$

Hence, the non-dimensional Sherwood number with the aid of equation (28.6) is:

$$\mathrm{Re}^{-\frac{1}{2}} Sh_x = \frac{\gamma_1}{\sqrt{2-\beta}}\left(1 + \frac{1}{\phi(0)}\right). \tag{28.19}$$

28.4 RESULTS AND DISCUSSION

The transformed governing flow boundary layer equations (28.7)–(28.9) along with the boundary conditions (28.10) are solved by employing classical Runge–Kutta method in combination with the shooting numerical approach. All the computations were carried out in fixing the parameter values, unless it is specified. The simulated values of the flow behavior for relating parameters are addressed with the aid of plots. To get the validation of the precision of the numerical solution methodology, a comparison with available studies in special cases is carried out, and these are reported in Table 28.1. It is revealed from Table 28.1 that the obtained numerical results are in good correlation with the results examined by Ishak et al. [40], Baoheng Yao [5], and Ullah et al. [42] for some limiting cases ($M = m = \lambda = 0$).

28.4.1 Velocity Field $f'(\eta)$

The momentum distribution $f'(\eta)$ profiles for diverse values of magnetic field (M), Weissenberg number (We), mixed convection (λ), buoyancy ratio (N), and wedge angle (β)parameters are presented in Figures 28.2–28.6. Figure 28.2 displays the enhancement of the $f'(\eta)$ profiles for increasing values of M. It might be attributed to the contribution of dominant pressure gradient force over the Lorentz force to enhance the flow momentum. Thus, to facilitate the momentum of the fluid over

TABLE 28.1

Comparison of Dimensionless Friction Factor $f''(0)$ for Distinct Values of β with $M = m = \lambda = 0$

n	$\beta = \frac{2n}{n+1}$	Ishak et al. [40]	Baoheng Yao [5]	Ullah et al. [42]	Present results
0	0	0.46960	0.46965	0.46963	0.469605
0.1	2/11	0.71498	0.71496	0.71497	0.714986
0.2	1/3	0.80213	0.80214	0.80215	0.802139
0.5	2/3	1.03890	1.03889	1.03890	1.038904
1	1	1.23263	1.23261	1.23262	1.232631

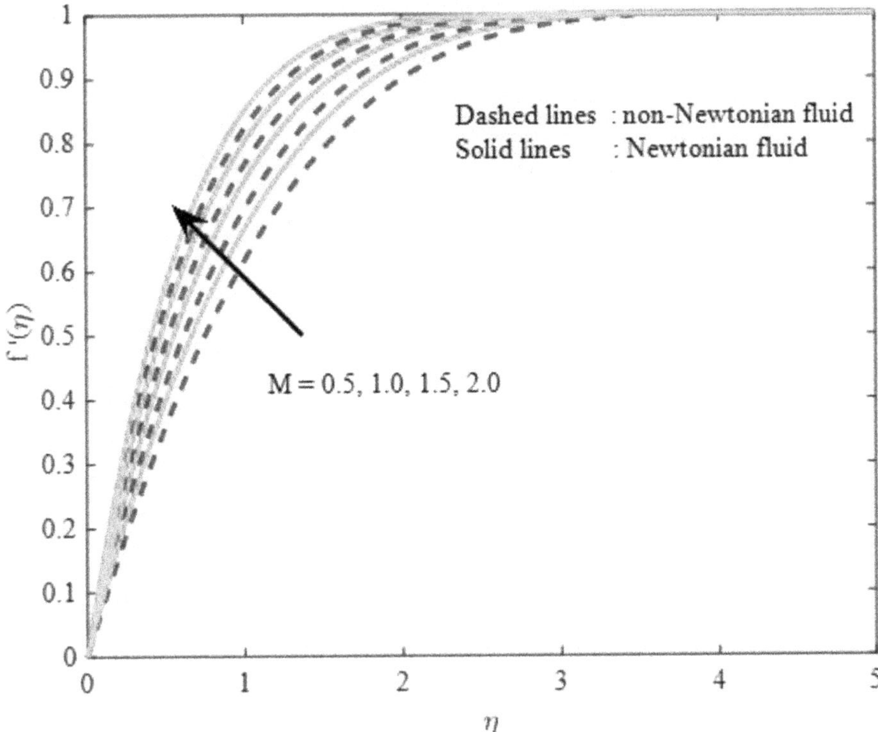

FIGURE 28.2 Impact of M on $f'(\eta)$.

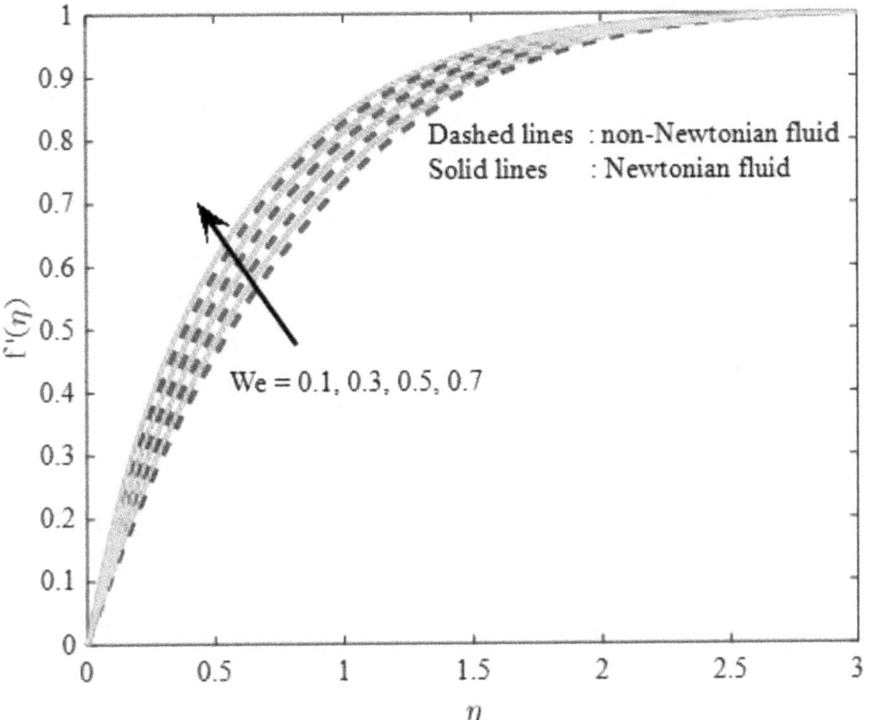

FIGURE 28.3 Impact of We on $f'(\eta)$.

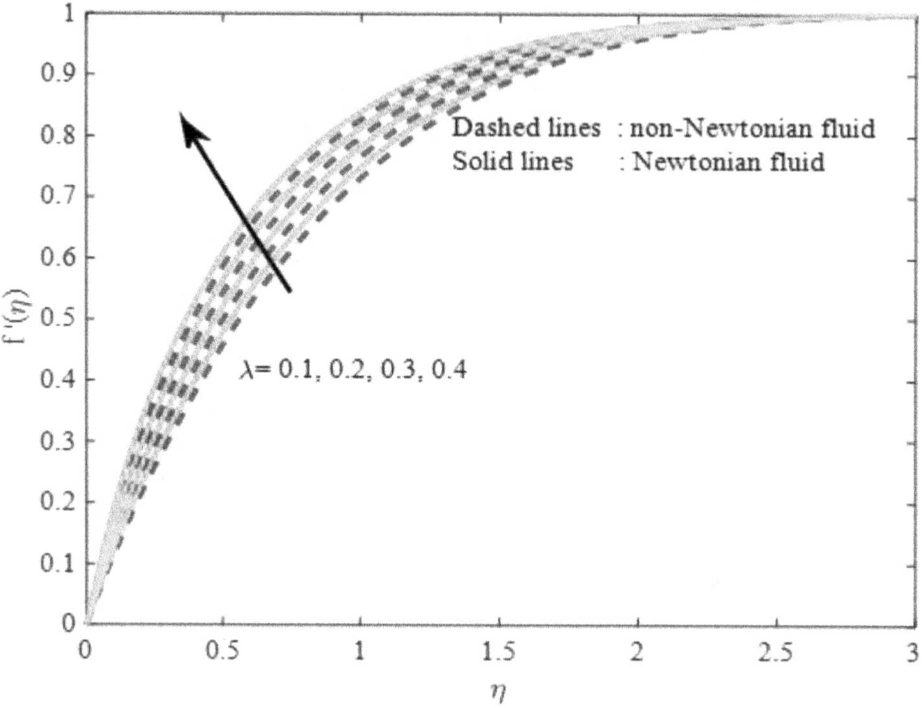

FIGURE 28.4 Impact of λ on $f'(\eta)$.

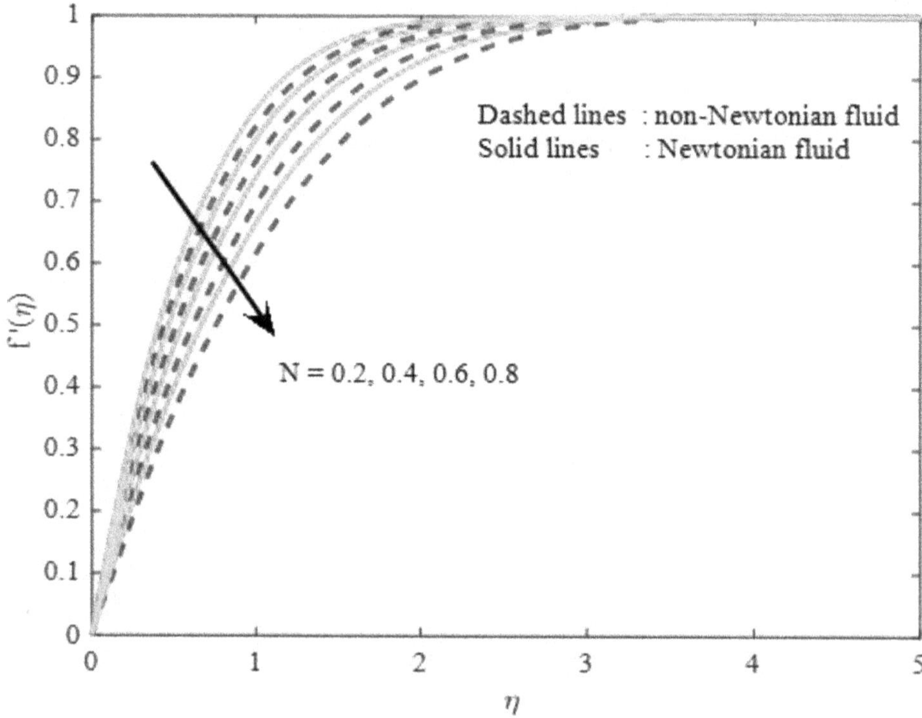

FIGURE 28.5 Impact of N on $f'(\eta)$.

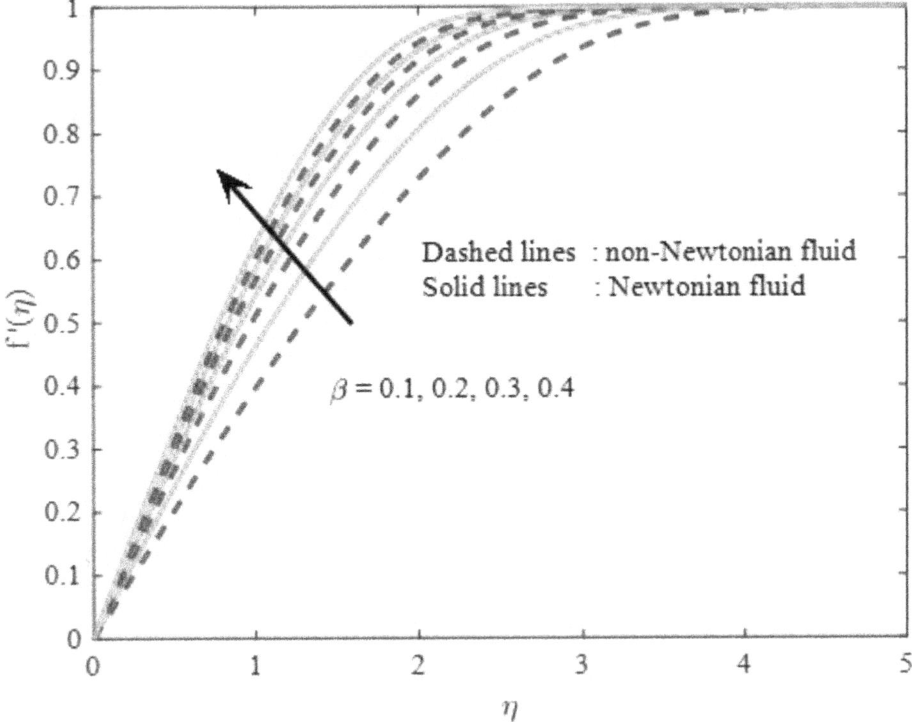

FIGURE 28.6 Impact of β on $f'(\eta)$.

wedge surface, the applied external force plays a vital role. The momentum function $f'(\eta)$ for several values of We is portrayed in Figure 28.3. It is evident from Figure 28.3 that the nanofluid velocity boosts for growing values of We. It is due to the fact that the elastic forces are dominant over the viscous forces; consequently, the thickness of the viscous boundary layer in enhanced. Whenever thermal buoyancy parameter is enhanced, the buoyancy forces are more compared to the frictional forces on the wedge surface; consequently, the thickness of the boundary layer is enhanced, as shown in Figure 28.4. Therefore, for increasing values of λ, this causes an increase in the momentum of the fluid, depicted in Figures 28.4. The application of N on $f'(\eta)$ results in the deceleration of the thickness of the hydrodynamical boundary layer near the wedge surface, which is exhibited in Figure 28.5. This is due to the supremacy of the molecular forces above the thermal buoyancy forces. However, for increasing $\beta(=2n/(n+1))$, where n is the pressure gradient, the momentum of the fluid $f'(\eta)$ increases, as shown in Figure 28.6. Furthermore, it is worthy to note from Figures 28.2–28.6 that the Newtonian nanofluid flow is more in contrast to that of non-Newtonian nanofluid.

28.4.2 Temperature Field $\theta(\eta)$

The features of temperature function $\theta(\eta)$ near the wedge surface for several vital variables $Nt, Nb, \gamma_1, N_R, \theta_W$ and λ_1 are portrayed in Figures 28.7–28.12. Figures 28.7 and 28.8 portrayed that the temperature distributions diminish for increasing values of thermophoresis (Nt) and Brownian motion parameters (Nb), respectively. The thickness of the energy boundary layer escalates for enhancing γ_1, as observed in Figure 28.9. The influence of the nonlinear radiation parameter on fluid temperature function $\theta(\eta)$ is presented in Figure 28.10. The impact of temperature ratio parameter θ_W on temperature is shown in Figure 28.11. The distribution of temperature profiles increases along with θ_W. Thus, the thickness of the thermal boundary layer is enhanced by the contribution of the

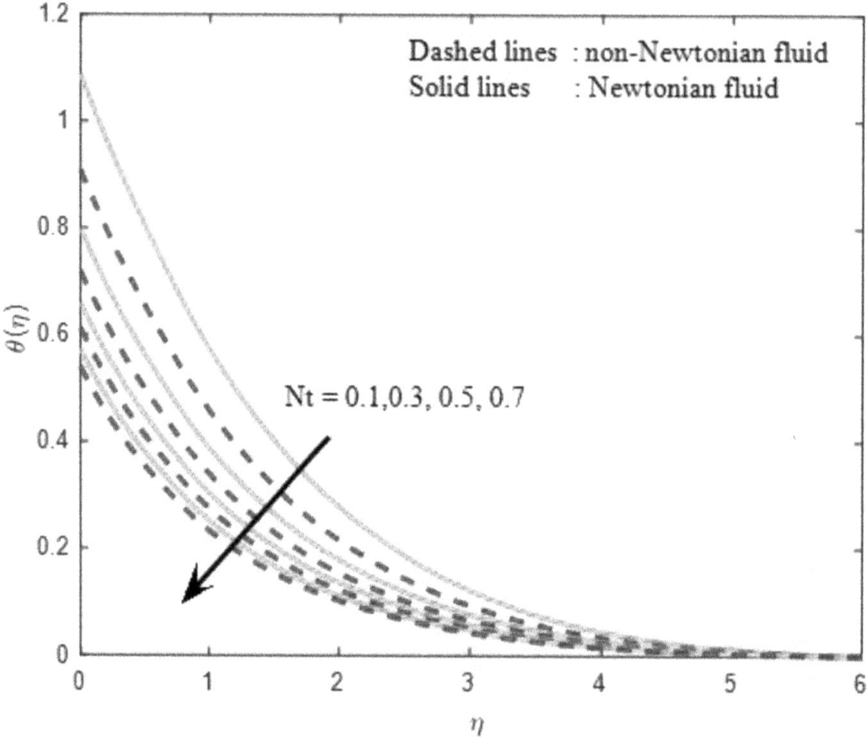

FIGURE 28.7 Impact of *Nt* on $\theta(\eta)$.

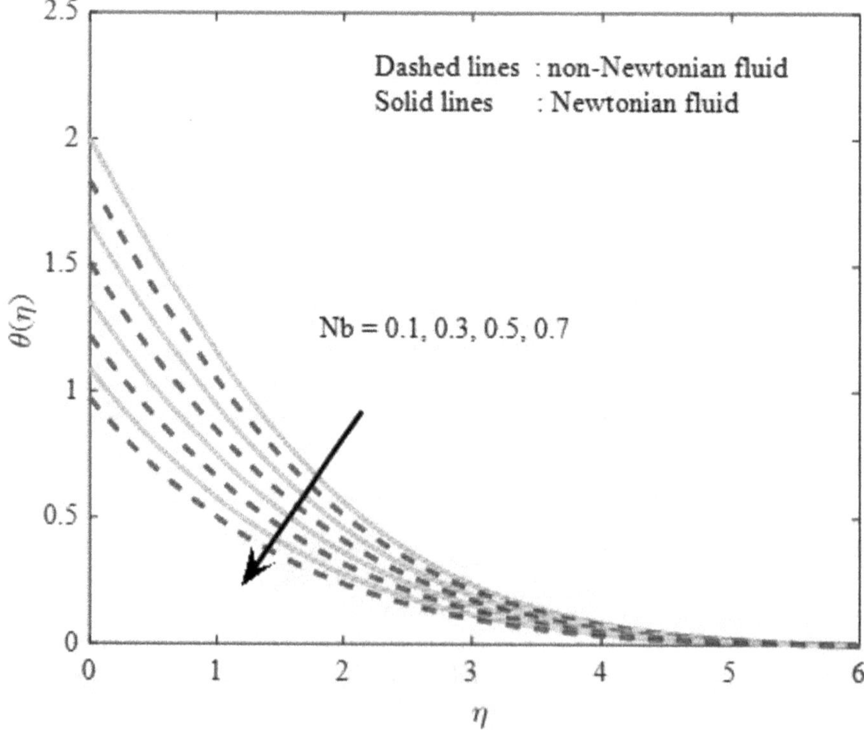

FIGURE 28.8 Impact of *Nb* on $\theta(\eta)$.

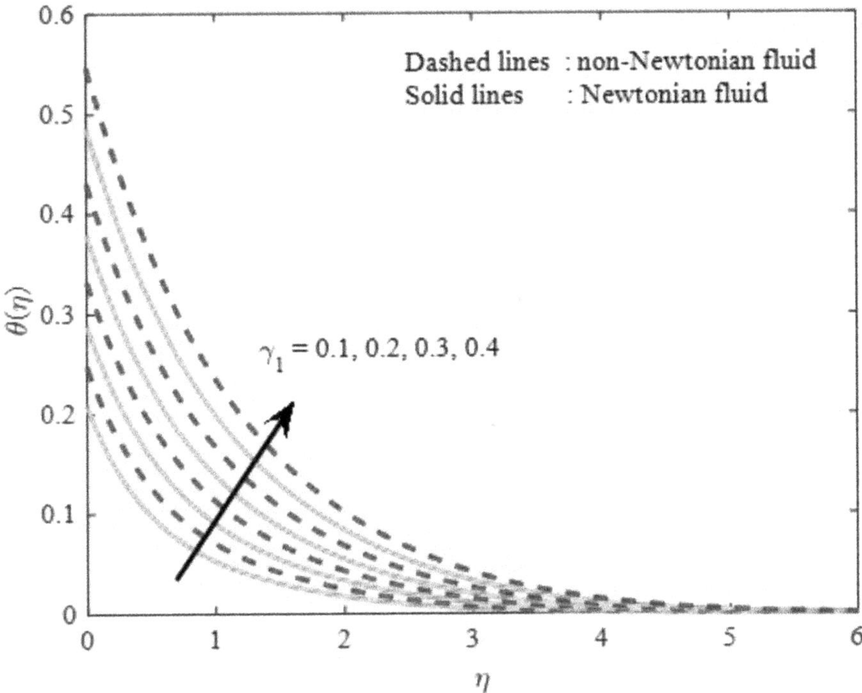

FIGURE 28.9 Impact of γ_1 on $\theta(\eta)$.

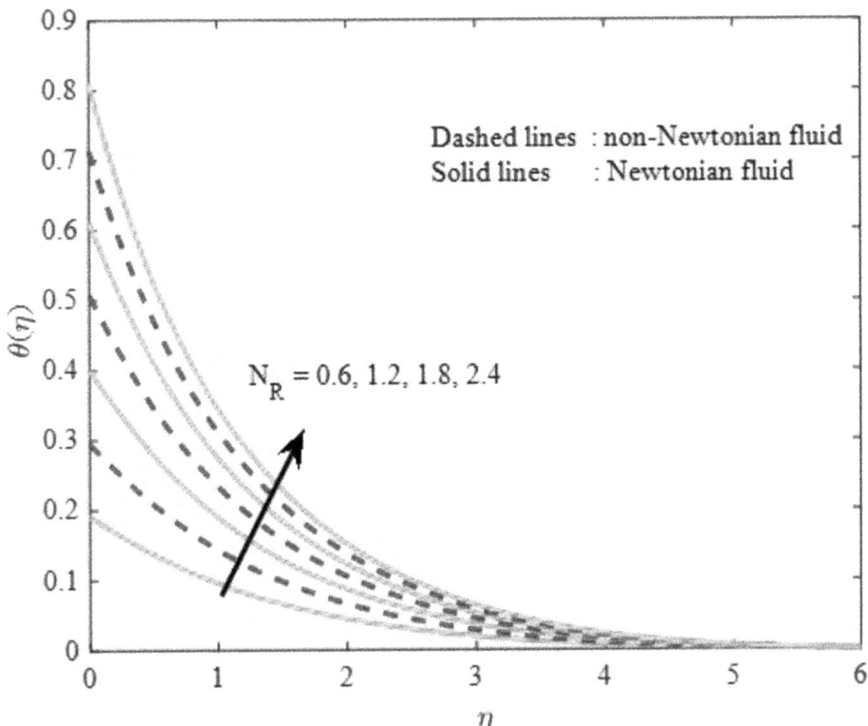

FIGURE 28.10 Impact of N_R on $\theta(\eta)$.

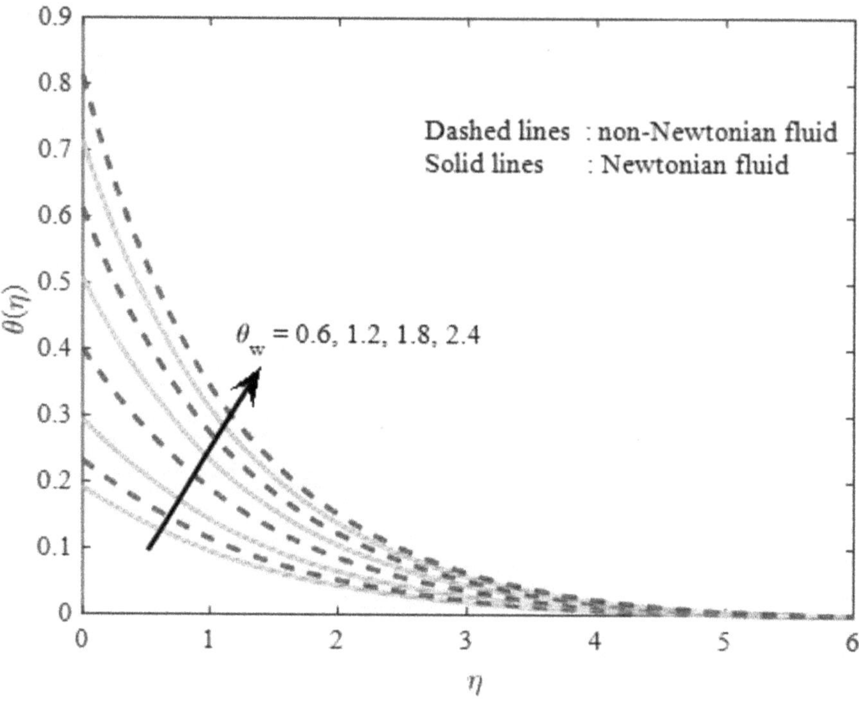

FIGURE 28.11 Impact of θ_w on $\theta(\eta)$.

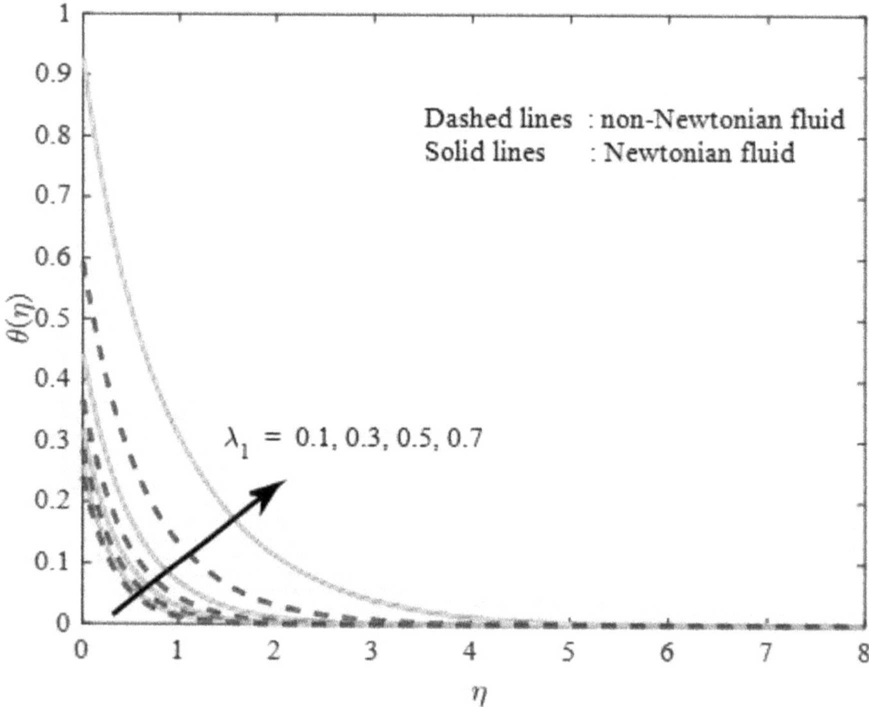

FIGURE 28.12 Impact of λ_1 on $\theta(\eta)$.

internal heat energy. The effect of heat source/sink parameter on the temperature is shown in Figure 28.12. In addition, it is also noticed from Figures 28.7–28.12 that the temperature distribution for non-Newtonian nanofluid is higher than that of Newtonian nanofluid for the parameters γ_1, θ_W, while an opposite trend is observed for Nt, Nb, N_R and λ_1.

28.4.3 Concentration Field $\phi(\eta)$

Figures 28.13–28.17 explore the dispersal of concentration function $\phi(\eta)$ to the impact of thermophoresis (Nt), Brownian motion (Nb), solutal Biot number (γ_2), chemical reaction (σ_r), and dimensionless activation energy (E). As seen from Figure 28.13, the concentration distribution decreases for increasing values of Nt. The enhancement of the boundary layer of concentration is witnessed in Figure 28.14 for boosting Nb. Figure 28.15 exhibits the enhancement of the species boundary layer along with γ_2. The concentration function decreases for rising values of σ_r, as shown in Figure 28.16. Figure 28.17 exhibits the enhancement of the species concentration profiles for increasing values of activation energy E. It is also found that the concentration distribution profiles for non-Newtonian nanofluid are more prominent than that of Newtonian nanofluid, except for the parameter σ_r.

28.4.4 Engineering Quantities

The dimensionless friction factor ($Cf_x \, \mathrm{Re}_x^{1/2}$) near the wedge surface for the range of $M(0:0.2:2)$ is represented in Figure 28.18 at $\beta(0.1:0.1:0.3)$. Figure 28.18 shows that for increasing $M(0 \to 2)$, the distribution $Cf_x \, \mathrm{Re}_x^{1/2}$ decreases gradually, whereas for enhancing β, a significant boosting in $Cf_x \, \mathrm{Re}_x^{1/2}$ is observed near the wedge surface. It is also revealed from Figure 28.18 that the friction factor for the case of non-Newtonian nanofluid is more than the case of Newtonian nanofluid. A significant growth in $Nu_x \, \mathrm{Re}_x^{-1/2}$ is observed near the wedge surface for growing values of $\Pr(0.2:0.2:2)$, as depicted in Figure 28.19. However, the reverse trend is witnessed for N_R. It is also

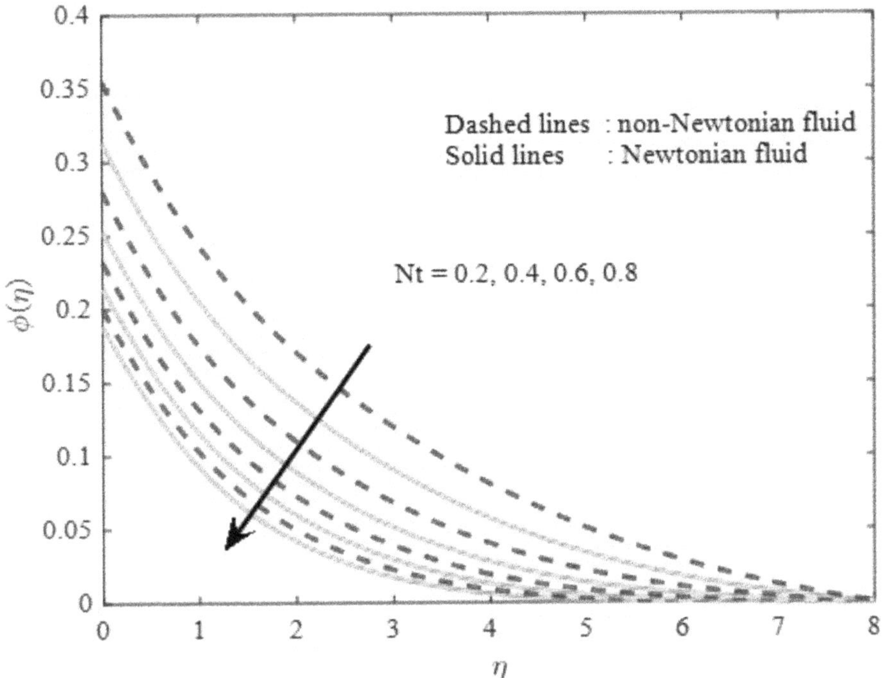

FIGURE 28.13 Impact of Nt on $\phi(\eta)$.

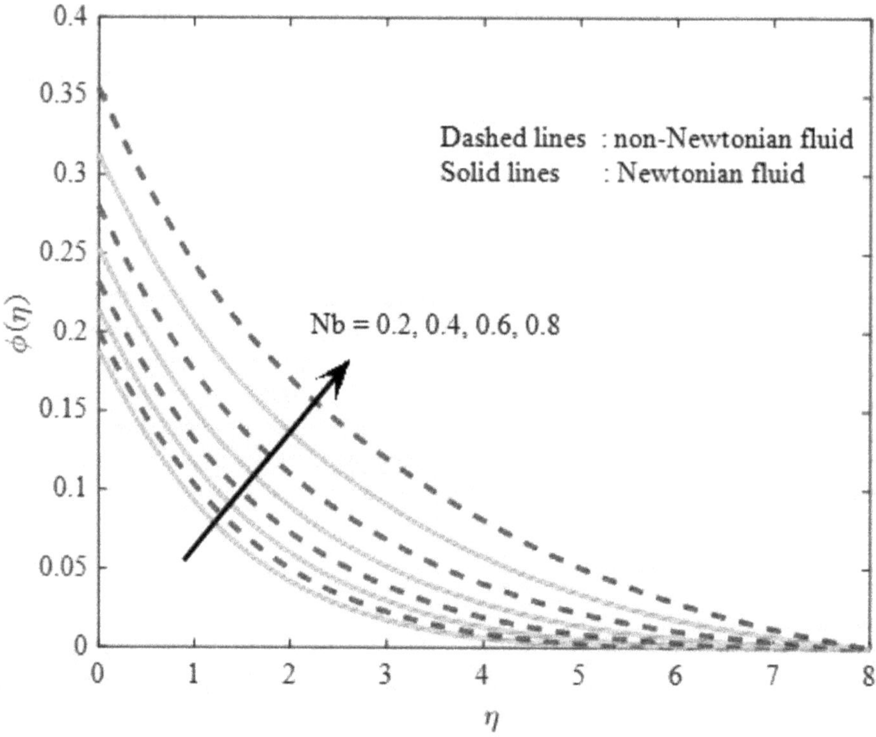

FIGURE 28.14 Impact of Nb on $\phi(\eta)$.

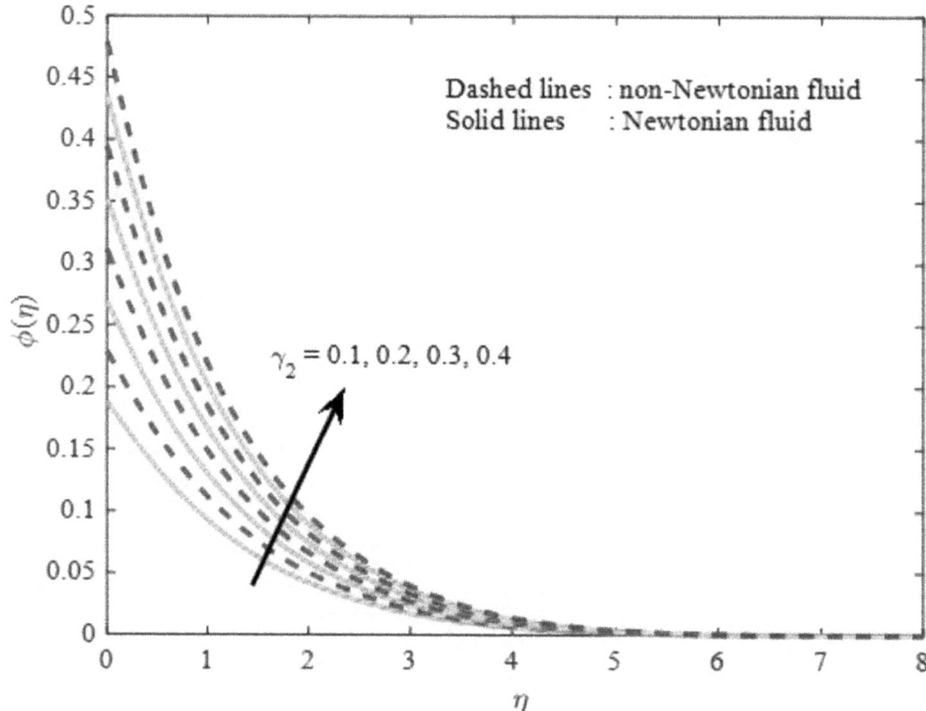

FIGURE 28.15 Impact of γ_2 on $\phi(\eta)$.

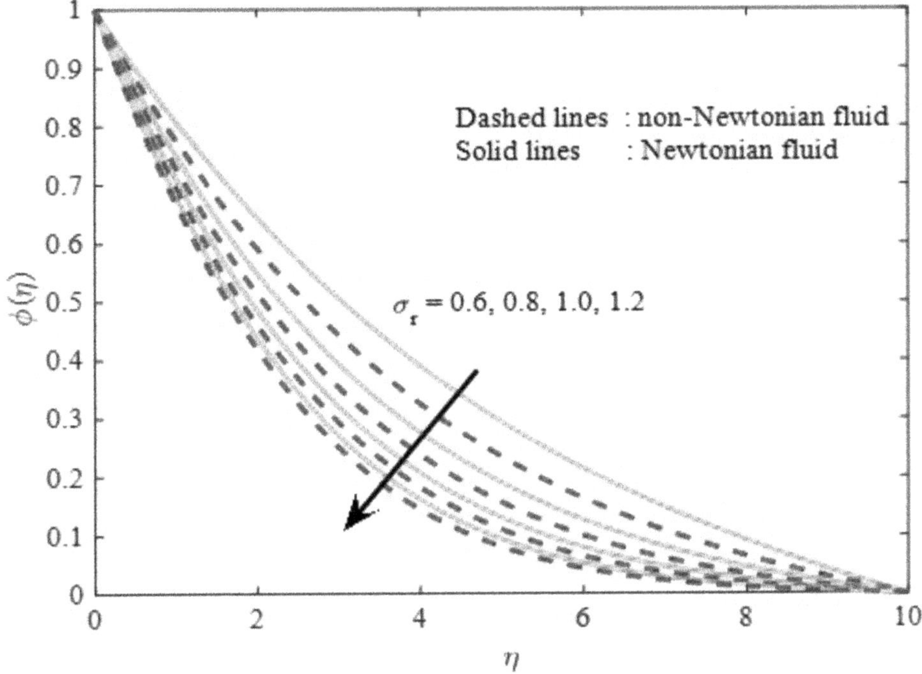

FIGURE 28.16 Impact of σ_r on $\phi(\eta)$.

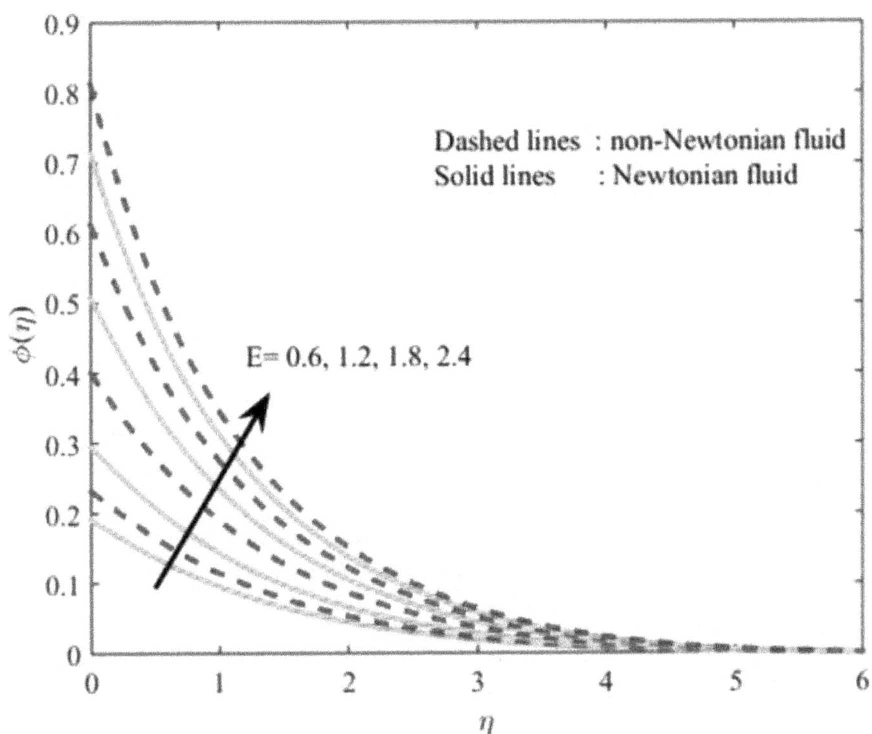

FIGURE 28.17 Impact of E on $\phi(\eta)$.

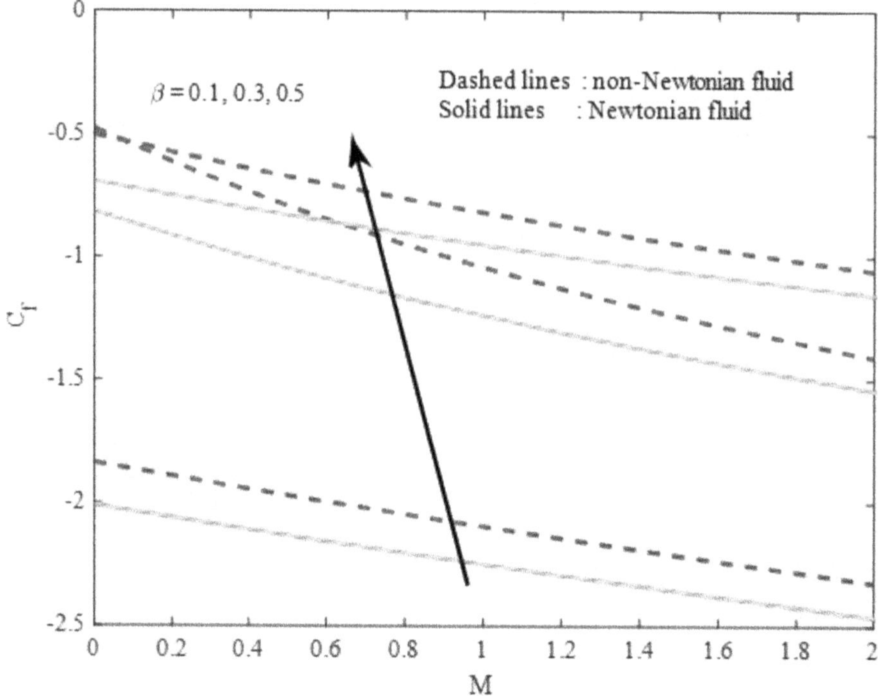

FIGURE 28.18 Impact of M versus β on C_f.

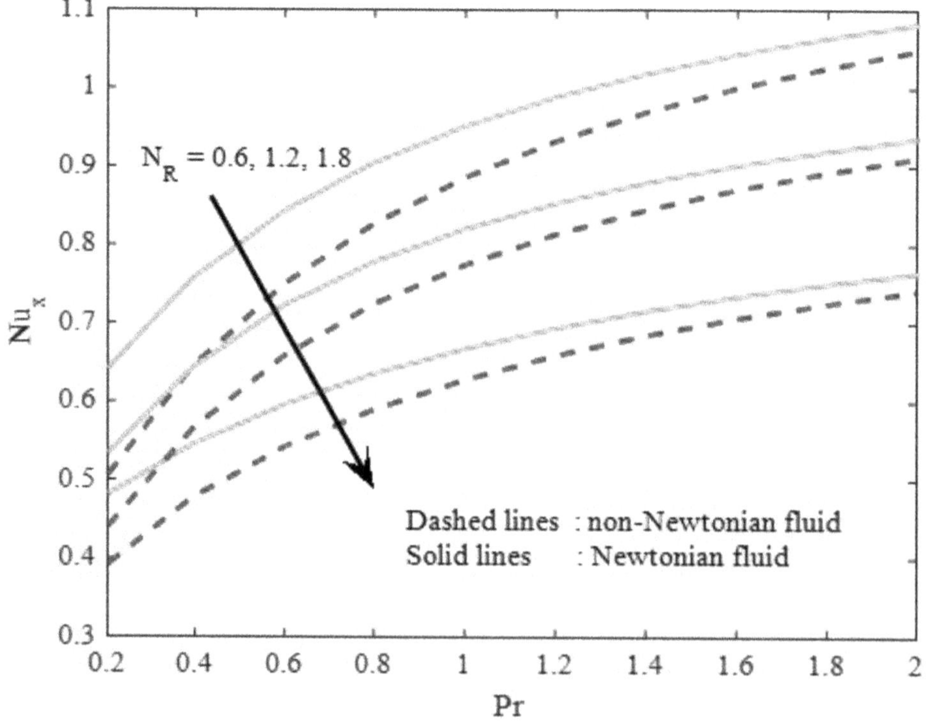

FIGURE 28.19 Impact of Pr versus N_R on Nu_x.

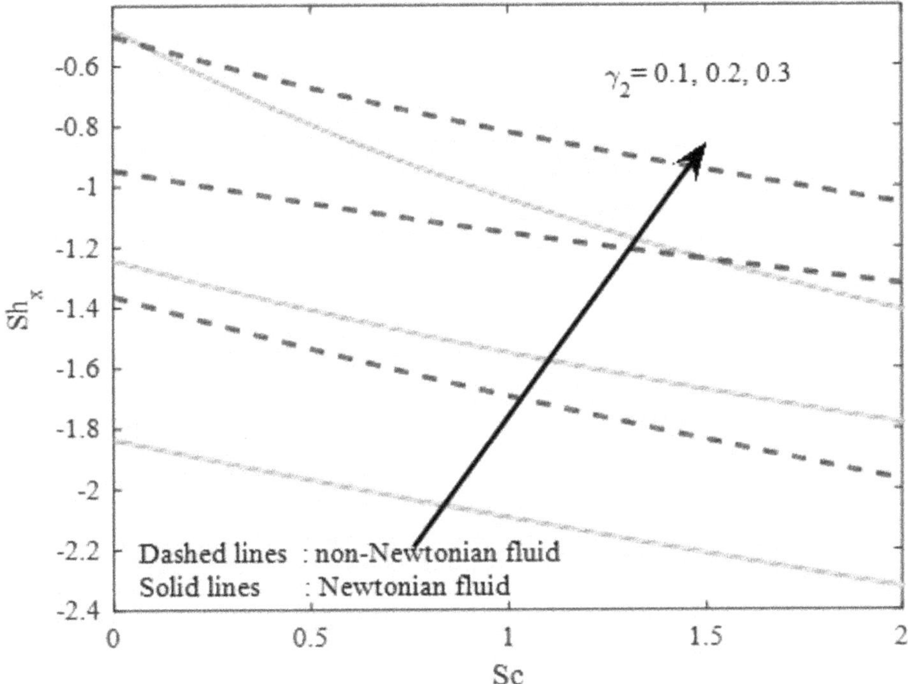

FIGURE 28.20 Impact of Sc versus γ_2 on Sh_x.

perceived from Figure 28.19 that the non-Newtonian nanofluid case has declined in $Nu_x \text{Re}_x^{-1/2}$ than that of the Newtonian nanofluid case. The behavior of local Sherwood number ($Sh_x \text{Re}_x^{-1/2}$) is portrayed in Figure 28.20 for $Sc(0:0.5:2)$ at $\gamma_2(0.1:0.1:0.3)$. For increasing Sc, the local Sherwood number near the wedge surface $Sh_x \text{Re}_x^{-1/2}$ is decreased, whereas for increasing γ_2, an opposite trend is observed. Further, it is also evident that the Newtonian nanofluid case has less in $Sh_x \text{Re}_x^{-1/2}$ than that of non-Newtonian nanofluid case.

28.5 CONCLUSIONS

In the current chapter, the convective buoyancy-driven incompressible magnetohydrodynamic nanofluid flow past a wedge surface has been explored numerically. The influence of the nonlinear thermal radiation, binary chemical reaction, heat generation/absorption, and buoyancy forces is accounted in the flow-governing boundary layer equations. The substantial observations made in this numerical scrutiny are presented as underneath.

1. The fluid momentum is enhanced for magnetic field, Weissenberg number, thermal buoyancy, and wedge angle parameters.
2. The thermal energy is accelerated for heat source, radiation, Biot number, and temperature ratio parameter, while an opposite trend is witnessed for thermophoresis and Brownian motion parameters.
3. The concentration of the fluid is augmented for Brownian motion, Arrhenius activation energy, and Biot number, whereas the reverse trend is noticed for thermophoresis and chemical reaction parameters.
4. The non-Newtonian nanofluid flow attains higher friction factor than the Newtonian nanofluid for magnetic field and wedge angle parameter.

5. The Newtonian fluid nanoflow attains greater local Nusselt number than the non-Newtonian nanofluid for Prandtl number and radiation parameter.
6. The non-Newtonian nanofluid flow attains a higher friction factor than the Newtonian nanofluid for Schmidt number and solutal Biot number.

REFERENCES

1. S.W. Skan, V.M. Falkner, Some approximate solutions of the boundary-layer equations, *Philos. Mag.* 12 (1931) 865–896.
2. D.R. Hartree, On an equation occurring in Falkner and Skan's approximate treatment of the equations of the boundary layer, *Math. Proc. Cambridge Philos. Soc.* 33 (1937) 223–239. https://doi.org/10.1017/S0305004100019575.
3. K.R. Rajagopal, A.S. Gupta, T.Y. Na, A note on the Falkner-Skan flows of a non-newtonian fluid, *Int. J. Non. Linear. Mech.* 18 (1983) 313–320. https://doi.org/10.1016/0020-7462(83)90028-8.
4. S.J. Liao, A uniformly valid analytic solution of two-dimensional viscous flow over a semi-infinite flat plate, *J. Fluid Mech.* 385 (1999) 101–128. https://doi.org/10.1017/S0022112099004292.
5. B. Yao, J. Chen, Series solution to the Falkner-Skan equation with streCOHing boundary, *Appl. Math. Comput.* 208 (2009) 156–164. https://doi.org/10.1016/j.amc.2008.11.028.
6. A.O. Elnady, M.F. Abd Rabbo, H.M. Negm, Solution of the Falkner–Skan equation using the chebyshev series in matrix form, *J. Eng.* 2020 (2020) 1–9. https://doi.org/10.1155/2020/3972573.
7. E. Alizadeh, M. Farhadi, K. Sedighi, H.R. Ebrahimi-Kebria, A. Ghafourian, Solution of the Falkner-Skan equation for wedge by Adomian decomposition method, *Commun. Nonlinear Sci. Numer. Simul.* 14 (2009) 724–733. https://doi.org/10.1016/j.cnsns.2007.11.002.
8. F.M. Allan, Q. Al Mdallal, Series solutions of the modified Falkner-Skan equation, *Int. J. Open Probl. Compt. Math.* 4 (2011).
9. H. Temimi, M. Ben-Romdhane, Numerical solution of Falkner-Skan equation by iterative transformation method, *Math. Model. Anal.* 23 (2018) 139–151. https://doi.org/10.3846/mma.2018.009.
10. W.A. Khan, I. Haq, M. Ali, M. Shahzad, M. Khan, M. Irfan, Significance of static–moving wedge for unsteady Falkner–Skan forced convective flow of MHD cross fluid, *J. Brazilian Soc. Mech. Sci. Eng.* 40 (2018). https://doi.org/10.1007/s40430-018-1390-3.
11. U. Khan, A. Shafiq, A. Zaib, A. Wakif, D. Baleanu, Numerical exploration of MHD Falkner-Skan-sutterby nanofluid flow by utilizing an advanced non-homogeneous two-phase nanofluid model and non-fourier heat-flux theory, *Alexandria Eng. J.* 51(6) (2020). https://doi.org/10.1016/j.aej.2020.08.048.
12. A.R. Bestman, Natural convection boundary layer with suction and mass transfer in a porous medium, *Int. J. Energy Res.* 14 (1990) 389–396. https://doi.org/10.1002/er.4440140403.
13. Z. Shafique, M. Mustafa, A. Mushtaq, Boundary layer flow of Maxwell fluid in rotating frame with binary chemical reaction and activation energy, *Results Phys.* 6 (2016) 627–633. https://doi.org/10.1016/j.rinp.2016.09.006.
14. U. Khan, A. Zaib, I. Khan, K.S. Nisar, Activation energy on MHD flow of titanium alloy (Ti6Al4V) nanoparticle along with a cross flow and streamwise direction with binary chemical reaction and non-linear radiation: Dual solutions, *J. Mater. Res. Technol.* 9 (2020) 188–199. https://doi.org/10.1016/j.jmrt.2019.10.044.
15. B.J. Gireesha, M. Archana, B. Mahanthesh, B.C. Prasannakumara, Exploration of activation energy and binary chemical reaction effects on nano Casson fluid flow with thermal and exponential space-based heat source, *Multidiscip. Model. Mater. Struct.* 15 (2019) 227–245. https://doi.org/10.1108/MMMS-03-2018-0051.
16. M. Ijaz, M. Ayub, Nonlinear convective stratified flow of Maxwell nanofluid with activation energy, *Heliyon.* 5 (2019). https://doi.org/10.1016/j.heliyon.2019.e01121.
17. M. Ramzan, N. Ullah, J.D. Chung, D. Lu, U. Farooq, Buoyancy effects on the radiative magneto Micropolar nanofluid flow with double stratification, activation energy and binary chemical reaction, *Sci. Rep.* 7 (2017) 1–15. https://doi.org/10.1038/s41598-017-13140-6.
18. A. Hamid, Hashim, M. Khan, Impacts of binary chemical reaction with activation energy on unsteady flow of magneto-Williamson nanofluid, *J. Mol. Liq.* 262 (2018) 435–442. https://doi.org/10.1016/j.molliq.2018.04.095.

19. M. Asma, W.A.M. Othman, T. Muhammad, F. Mallawi, B.R. Wong, Numerical study for magnetohydrodynamic flow of nanofluid due to a rotating disk with binary chemical reaction and arrhenius activation energy, *Symmetry (Basel)*. 11 (2019). https://doi.org/10.3390/sym11101282.
20. M. Asma, W.A.M. Othman, T. Muhammad, Numerical study for Darcy-Forchheimer flow of nanofluid due to a rotating disk with binary chemical reaction and Arrhenius activation energy, *Mathematics*. 7 (2019). https://doi.org/10.3390/math7100921.
21. A.S. Alshomrani, M.Z. Ullah, S.S. Capizzano, W.A. Khan, M. Khan, Interpretation of chemical reactions and activation energy for unsteady 3D flow of eyring–powell magneto-nanofluid, *Arab. J. Sci. Eng*. 44 (2019) 579–589. https://doi.org/10.1007/s13369-018-3485-7.
22. M. Ijaz, M. Ayub, H. Khan, Entropy generation and activation energy mechanism in nonlinear radiative flow of Sisko nanofluid: Rotating disk, *Heliyon*. 5 (2019) e01863. https://doi.org/10.1016/j.heliyon.2019.e01863.
23. T. Hayat, A.A. Khan, F. Bibi, S. Farooq, Activation energy and non-Darcy resistance in magneto peristalsis of Jeffrey material, *J. Phys. Chem. Solids*. 129 (2019) 155–161. https://doi.org/10.1016/j.jpcs.2018.12.044.
24. S. Ahmad, M. Farooq, A. Anjum, N.A. Mir, Squeezing flow of convectively heated fluid in porous medium with binary chemical reaction and activation energy, *Adv. Mech. Eng*. 11 (2019) 1–12. https://doi.org/10.1177/1687814019883774.
25. F. Haq, M.U. Rahman, M. Ijaz Khan, T. Hayat, Transportation of activation energy in the development of a binary chemical reaction to investigate the behavior of an Oldroyd-B fluid with a nanomaterial, *Phys. Scr*. 94 (2019) 105010. https://doi.org/10.1088/1402-4896/ab194a.
26. B.H. Babu, P.S. Rao, M.G. Reddy, S.V.K. Varma, Numerical modelling of activation energy and hydromagnetic non-Newtonian fluid particle deposition flow in a rotating disc. *Proc. Inst. Mech. Eng., Part E: J. Proc. Mech. Eng*. 237 (2021). http://doi.org/10.1177/09544089211045907.
27. M.E. Nasr, M. Gnaneswara Reddy, W. Abbas, A.M. Megahed, E. Awwad, K.M. Khalil, Analysis of non-linear radiation and activation energy analysis on hydromagnetic reiner–philippoff fluid flow with cattaneo–christov double diffusions. *Mathematics*. 10 (2022) 1534. https://doi.org/10.3390/math10091534.
28. N. Tarakaramu and P.V.S. Narayana, Nonlinear thermal radiation and joule heating effects on MHD stagnation point flow of nanofluid over a convectively heated streCOHing surface, *J. Nanofluids*. 8 (2019) 1066–1075.
29. S.R.R. Reddy, P. Bala Anki Reddy, K. Bhattacharyya, Effect of nonlinear thermal radiation on 3D magneto slip flow of Eyring-Powell nanofluid flow over a slendering sheet with binary chemical reaction and Arrhenius activation energy, *Adv. Powder Technol*. 30 (2019) 3203–3213. https://doi.org/10.1016/j.apt.2019.09.029.
30. M. Archana, M.G. Reddy, B.J. Gireesha, B.C. Prasannakumara, S.A. Shehzad, Triple diffusive flow of nanofluid with buoyancy forces and nonlinear thermal radiation over a horizontal plate, *Heat Transf.—Asian Res*. 47 (2018) 957–973. https://doi.org/10.1002/htj.21360.
31. O.K. Koriko, I.L. Animasaun, M.G. Reddy, N. Sandeep, Scrutinization of thermal stratification, nonlinear thermal radiation and quartic autocatalytic chemical reaction effects on the flow of three-dimensional Eyring-Powell alumina-water nanofluid, *Multidiscip. Model. Mater. Struct*. 14 (2018) 261–283. https://doi.org/10.1108/MMMS-08-2017-0077.
32. A. Kumar, R. Tripathi, R. Singh, V.K. Chaurasiya, Simultaneous effects of nonlinear thermal radiation and Joule heating on the flow of Williamson nanofluid with entropy generation, *Phys. A Stat. Mech. Its Appl*. 551 (2020) 123972. https://doi.org/10.1016/j.physa.2019.123972.
33. B. Mahanthesh, I.L. Animasaun, M. Rahimi-Gorji, I.M. Alarifi, Quadratic convective transport of dusty Casson and dusty Carreau fluids past a streCOHed surface with nonlinear thermal radiation, convective condition and non-uniform heat source/sink, *Phys. A Stat. Mech. Its Appl*. 535(C) (2019). https://doi.org/10.1016/j.physa.2019.122471.
34. C.S.K. Raju, N. Sandeep, Nonlinear radiative magnetohydrodynamic Falkner-Skan flow of Casson fluid over a wedge, *Alexandria Eng. J*. 55 (2016) 2045–2054. https://doi.org/10.1016/j.aej.2016.07.006.
35. M. Waqas, S. Jabeen, T. Hayat, S.A. Shehzad, A. Alsaedi, Numerical simulation for nonlinear radiated Eyring-Powell nanofluid considering magnetic dipole and activation energy, *Int. Commun. Heat Mass Transf*. 112 (2020) 104401. https://doi.org/10.1016/j.icheatmasstransfer.2019.104401.

36. A. Wakif, A novel numerical procedure for simulating steady MHD convective flows of radiative Casson fluids over a horizontal streCOHing sheet with irregular geometry under the combined influence of temperature-dependent viscosity and thermal conductivity, *Math. Probl. Eng.* 2020 (2020). https://doi.org/10.1155/2020/1675350.
37. A. Wakif, A. Chamkha, T. Thumma, I.L. Animasaun, R. Sehaqui, Thermal radiation and surface roughness effects on the thermo-magneto-hydrodynamic stability of alumina–copper oxide hybrid nanofluids utilizing the generalized Buongiorno's nanofluid model, *J. Therm. Anal. Calorim.* 141 (2020). https://doi.org/10.1007/s10973-020-09488-z.
38. B. Souayeh, M. Gnaneswara Reddy, P. Sreenivasulu, T. Poornima, Mohammad Rahimi-Gorji, Ibrahim M. Alarifi, Comparative analysis on non-linear radiative heat transfer on MHD Casson nanofluid past a thin needle, *J. Mol. Liq.* 284 (2019) 163–174.
39. M. Gnaneswara Reddy, P. Padma, M.V.V.N.L. Sudha Rani, Non-linear thermal radiative analysis on hydromagnetic nanofluid transport through a rotating cone. *Int. J. Appl. Comput. Math.* 5 (2019) 69. https://doi.org/10.1007/s40819-019-0654-7
40. A. Ishak, R. Nazar, I. Pop, Falkner-Skan equation for flow past a moving wedge with suction or injection, *J. Appl. Math. Comput.* 25(1) (2007) 67–83.
41. S. Nadeem, S. Ahmad, N. Muhammad, Computational study of Falkner-Skan problem for a static and moving wedge, *Sens. Actuators B.* 263 (2018) 69–76.
42. Imran Ullah, Ilyas Khan, Sharidan Shafie, Hydromagnetic Falkner-Skan flow of Casson fluid past a moving wedge with heat transfer, *Alex. Eng. J.* 55 (2016) 2139–2148.

29 Convection with Cu-MgO/Water Hybrid Nanofluid and Discrete Heating

Ines Chabani and Fateh Mebarek-Oudina

29.1 INTRODUCTION

Thermal energy enhancement is a topic of great interest, where several studies have developed techniques and approaches to intensify the heat exchange rate in order to reduce the consumption and the expenses. The review of Mebarek-Oudina and Chabani [1] presented the recently used methods, from exploiting the nanoparticles' great thermophysical properties in nanofluids, porosity of the materials, magnetic forces, to the geometrical configurations of the heat systems.

Aiming for a better insight of the employment of nanofluids in heat transfer systems, it is crucial to study the effect of the NP volume fraction on different parameters. Studies of researchers like Amidu et al. [2] and Abusorrah et al. [3] stated that this factor has a significant influence on thermal energy transmission. Also, the numerical results of Cieśliński et al. [4] revealed that augmenting the nanoparticles concentration improves the thermal conductivity of the nanofluid and provides up to 16% heat transfer enhancement compared to the base fluid.

Wu and Zhang [5] exploited the use of different nanofluids in micro-channels to conclude that the NP concentration strongly affects the system's temperature and heat transfer in general. Moreover, the work of Elias et al. [6] showed a decrease in the entropy generation rate of the heat exchanger studied after incrementing the nanoparticle's volume fraction.

Following this framework, Chaharborj and Moameni [7] executed an optimal study targeted at boosting heat transmission; their technique leaned towards nanoparticles combination, since incrementing their volume fraction offered major upgraded results separately. This approach has been shown to be advantageous from the perspective of thermal performance and expenses, thus promoting the idea of hybrid nanofluids that already attracted great interests. For instance, the results of Asadi et al. [8] showed that utilizing hybrid nanoparticles with increased volume fraction can yield a decent coolant fluid with higher thermal conductivity compared to standard coolants. Mebarek-Oudina et al. [9] explained the influence of Rayleigh on heat transfer that shows on the Nusselt values and therefore the flow's category; high values of Ra enable excellent heat transmission. Furthermore, it is also reported by the experimental results of [10] that the efficiency of the convective flow is strongly dependent on Ra. It has been stated by the results of Chabani et al. [11] that the hybrid nanofluid convective flow behavior can be highly influenced by the magnetic field; mainly, it is recommended to reduce the Hartmann values in order to accelerate the heat transfer. Moreover, the results of Hatami et al. [12] revealed that while increasing the concentration of nanoparticles improves heat transfer, this enhancement is diminished when a strong external magnetic field is applied. The high values of Ha decreased the Nu number, which led to 25% reduction in heat transfer. Also, scientists explored the importance of porosity on the performance of thermal systems [13–16], and they all showed that employing porous enclosures enhances Nusselt number and the flow rate.

Furthermore, many scholars studied the impact of the geometrical properties on thermal transmission. Das et al. [17] reviewed several research papers and revealed that square, triangular, and

DOI: 10.1201/9781003299608-29

trapezoidal shapes contribute to enhancing the heat transfer performance. Alguboori et al. [18] investigated the natural convection of a nanofluid and the influence of the inclination of an annular geometry with hot circles, while Mebarek-Oudina et al. [19] inspected an annular subjected to a magnetic field induction, and Bakhshi et al. [20] inspected the aspect ratio impact of the enclosure on the average Nusselt and reported an indirect relationship. Studies like [21–23] analyzed nanofluids flow in wavy enclosures and obstacles. Several scholars explored parallelogramic enclosures with different boundary conditions [24–26]; for instance, Majdi et al. [27] investigated the natural convection of a nanofluid in a parallelogram with a circular heated obstacle. Also, the MHD flow in a wavy parallelogram with a heated wall was studied by Yasin et al. [28]. Moreover, many research papers [29–31] investigated heat generation in cavities with nanofluids, and they all revealed and reported that placing heat sources in the upper section of the enclosure enhances heat transmission. It is crucial as well to investigate the irreversibility inside thermal configurations and to study the total entropy production [32–35]. Omri et al. [36] reported that lowering the nanoparticle's concentration reduced the generation of the flow's entropy due to frictions, while the results of Marzougui et al. [37] explain how increasing the magnetic field's intensity also decreases the entropy generation.

Alqaed et al. [38], Ferhi et al. [39], and Prasad et al. [40] showed that although augmenting Rayleigh number and the volume fraction of the hybrid nanofluid enhances the Nusselt number, it also augments the generated entropy in the system.

Based on the results acquired in the literature previously described, as well as many recent work papers, and aiming for great thermal efficiency, it is essential to consider MHD hybrid nanofluid flows in porous cavities and incorporate the contribution of the heat sources location in the enclosures and magnetic fields into the heat transfer process. Thus, as an attempt to provide extended significant details to the area of the natural convective heat transfer, which has numerous technical uses in nuclear reactors, solar collectors, and electronics, an investigation has been performed to examine the magneto-convective flow occurring in a porous 2D parallelogram filled with water-based Cu-MgO hybrid nanofluid and an elliptical cylinder, and their impact on the heat transfer characteristics.

29.2 PHYSICAL PROBLEM AND BOUNDARY CONDITIONS

The Darcy–Brinkmann–Forchheimer model is utilized to investigate the two-dimensional MHD free convection of Cu-MgO hybrid nanofluid in a parallelogram-shaped enclosure with L × H dimension and a porosity of $\varepsilon = 0.4$, an elliptical obstacle located at the center of the cavity, and also subjected to a uniform magnetic field. The motionless elliptical cylinder, the top, and the bottom walls of the cavity are supposed to be adiabatic, while the right and left walls are cold, with separately inserted partial heaters. The hybrid nanofluid is supposed to be Newtonian, enabling a laminar and incompressible flow.

The studied enclosure and the boundary conditions are depicted in Figure 29.1.

The thermophysical properties of Cu, TiO_2 nanoparticles, and water are presented in Table 29.1.

29.3 MATHEMATICAL MODELLING

29.3.1 Equations

The 2D dimensionless equations expressed by the Darcy–Brinkmann–Forchheimer model to describe the continuity, energy, and Navier–Stokes equations that study the stationary, laminar, and incompressible flow of a hybrid nanofluid inside a porous enclosure under the effect of a magnetic field are presented as follows [43–45]:

$$\frac{\partial U}{\partial X}+\frac{\partial V}{\partial Y}=0; \tag{29.1}$$

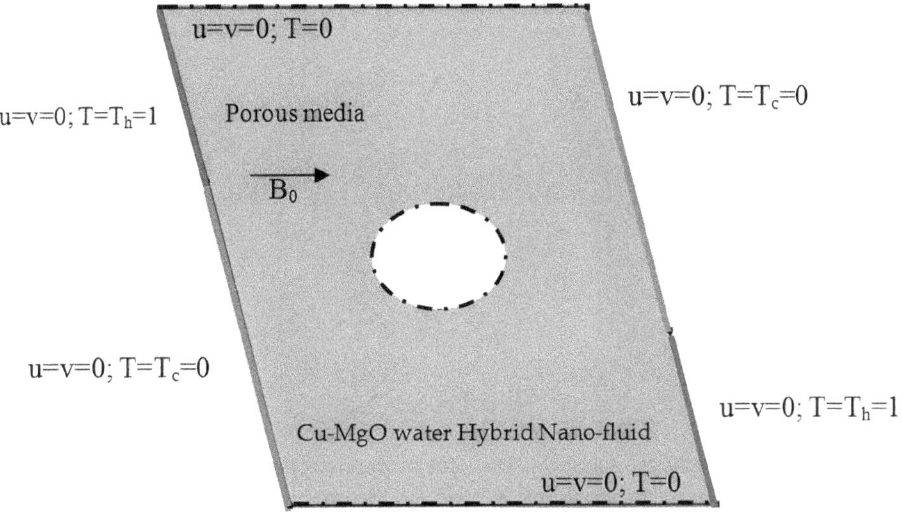

FIGURE 29.1 Configuration of the studied model with boundary conditions.

TABLE 29.1
Thermophysical Properties of Water and Nanoparticles [41, 42]

	Water	Cu	MgO
C_P (J. K^{-1}.Kg^{-1})	4179	385	879
ρ (Kg.m^{-3})	997.1	8,933	3,970
k (W. K^{-1}.m^{-1})	0.613	401	30
β (K^{-1})	21×10^{-5}	1.67×10^{-5}	3.36×10^{-5}
σ (Ohm.m)$^{-1}$	5.5×10^{-6}	5.96×10^{-7}	8×10^{-4}

$$U\frac{\partial \theta}{\partial X}+V\frac{\partial \theta}{\partial Y}=\frac{\alpha_{hnf}}{\alpha_{bf}}\left(\frac{\partial^2 \theta}{\partial X^2}+\frac{\partial^2 \theta}{\partial Y^2}\right); \tag{29.2}$$

$$\frac{1}{\varepsilon^2}\frac{\rho_{hnf}}{\rho_{bf}}\left(U\frac{\partial U}{\partial X}+V\frac{\partial U}{\partial Y}\right)=-\frac{\partial P}{\partial X}-\frac{\vartheta_{hnf}}{\vartheta_{bf}}\frac{P_r}{D_a\sqrt{R_a}}U-\frac{F_C}{\sqrt{D_a}}U|U|$$

$$+\frac{1}{\varepsilon}\frac{\vartheta_{hnf}}{\vartheta_{bf}}\frac{P_r}{\sqrt{R_a}}\left(\frac{\partial U}{\partial X}+\frac{\partial U}{\partial Y}\right); \tag{29.3}$$

$$\frac{1}{\varepsilon^2}\frac{\rho_{hnf}}{\rho_{bf}}\left(U\frac{\partial V}{\partial X}+V\frac{\partial V}{\partial Y}\right)=-\frac{\partial P}{\partial Y}-\frac{\vartheta_{hnf}}{\vartheta_{bf}}\frac{P_r}{D_a\sqrt{R_a}}V+P_r\frac{\beta_{hnf}}{\beta_{bf}}g\theta+$$

$$\frac{\sigma_{hnf}}{\rho_{hnf}}\frac{\rho_{bf}}{\rho_{hnf}}\frac{P_r}{\varepsilon}\frac{H_a}{\sqrt{R_a}}V-\frac{F_C}{\sqrt{D_a}}V|U|+\frac{1}{\varepsilon}\frac{\vartheta_{hnf}}{\vartheta_{bf}}\frac{P_r}{\sqrt{R_a}}\left(\frac{\partial V}{\partial X}+\frac{\partial V}{\partial Y}\right); \tag{29.4}$$

The entropy production is expressed as follows [43]:

$$S_t = \frac{k_{hnf}}{T^2}\left[\left(\frac{\partial \theta}{\partial X}\right)^2 + \left(\frac{\partial \theta}{\partial Y}\right)^2\right]; \tag{29.5}$$

$$S_f = \frac{\mu_{hnf}}{\mu_{bf}}\gamma \times \left\{ Da\left[2\left(\frac{\partial V}{\partial Y}\right)^2 + \left(\frac{\partial U}{\partial Y} + \frac{\partial V}{\partial X}\right)^2 + 2\left(\frac{\partial U}{\partial X}\right)^2\right] + \left(U^2 + V^2\right)\right\}; \tag{29.6}$$

$$S_{mf} = \frac{\sigma_{hnf}}{\sigma_{bf}}\gamma Ha^2\left(U - V\right); \tag{29.7}$$

$$S_{tot} = S_t + S_f + S_{mf}; \tag{29.8}$$

Where the total entropy production is the sum of the entropy due to thermal irreversibility S_t, the entropy due to the flow irreversibility S_f, and S_{mf} the entropy produced with the presence of a magnetic field.

Also, γ represents the irreversibility distribution ratio in the porous layer and is expressed as follows:

$$\gamma = \frac{\mu_{bf} T_{avg}}{k_{bf}}\left(\frac{\alpha_{bf}^{\ 2}}{K\left(\Delta T\right)^2}\right)$$

With the following dimensionless numbers and variables:

Numbers:

$$Ra = \frac{\beta_{bf} g\left(T_h - T_C\right)L^3}{\alpha_{bf}\vartheta_{bf}},\ Ha = LB_0\sqrt{\frac{\sigma_{bf}}{\mu_{bf}}},\ Da = \frac{K}{L},\ Pr = \frac{\vartheta_{bf}}{\alpha_{bf}}$$

Variables:

$$\theta = \frac{T - T_c}{T_h - T_c},\ Y = \frac{y}{L},\ X = \frac{x}{L},\ V = \frac{vL}{\alpha_{bf}},\ U = \frac{uL}{\alpha_{bf}},\ P = \frac{\left(p + \rho_{bf} g_y\right)L}{\alpha_{bf}\ \rho_{bf}}$$

The stream function, local, and the average Nusselt number characterizing the thermal transmission are expressed as follows [46–48]:

$$U = \frac{\partial \Psi}{\partial Y} \tag{29.9}$$

$$V = -\frac{\partial \Psi}{\partial X} \tag{29.10}$$

$$\frac{\partial \Psi}{\partial X} + \frac{\partial \Psi}{\partial Y} = \frac{\partial U}{\partial Y} - \frac{\partial V}{\partial X} \tag{29.11}$$

$$Nu_{local} = \frac{k_{hnf}}{k_{bf}}\frac{\partial T}{\partial y} \tag{29.12}$$

$$Nu_{average} = \frac{1}{L}\int_0^L Nu_{local} dL \tag{29.13}$$

The expressions of the thermophysical properties of the working nanoparticles and hybrid nanofluid are presented in Table 29.2, from density, heat capacity, thermal conductivity, diffusivity, and expansion, electrical conductivity, to the dynamic viscosity presented by Brinkmann. While the volume fraction of the hybrid nanofluid is expressed as following:

$$\phi = \phi_{Cu} + \phi_{MgO}$$

29.3.2 Grid Test

A validation test with previously published research is necessary to assess the relevance of the current working program implemented in COMSOL 5.5 *Multiphysics* software, which adapts the finite element method to solve the thermodynamical problem. Thus, a comparison of our results with those of [51] is shown in Figure 29.2, where the streamlines and isotherms of the natural convective flow formed in an elliptical cold cavity filled with a nanofluid and containing a hot square barrier are shown for 2% volume fraction.

The results of our program are proven to be considerably similar to the findings of the paper for both concentrations of the nanofluid.

Moreover, the accuracy of the findings is directly related to the type of mesh used to analyze the research; hence, a grid test indicating the divergence between the results is presented in Table 29.3, where the quality of the three types of meshes is explored. The average Nusselt number is computed and compared at Ra = 10^5, Ha = 0, and ϕ = 6%, and it is discovered that increasing the mesh quality and element number reduces the Nusselt deviation; in this context, the "extra-fine mesh" is chosen to pursue the investigation.

TABLE 29.2
Thermophysical Properties of Nanoparticle and Hybrid Nanofluid [27, 49–52]

Thermophysical Properties	Nanoparticles	Hybrid nNnofluid
Density	$\rho_{np} = \frac{\phi_{Cu}\rho_{Cu} + \phi_{MgO}\rho_{MgO}}{\phi}$	$\rho_{hnf} = (1-\phi)\rho_{bf} + \phi\rho_{np}$
Heat capacity	$Cp_{np} = \frac{\phi_{Cu}Cp_{Cu} + \phi_{MgO}Cp_{MgO}}{\phi}$	$(\rho Cp)_{hnf} = (1-\phi)(\rho Cp)_{bf} + \phi(\rho Cp)_{np}$
Thermal conductivity	$k_{np} = \frac{\phi_{Cu}k_{Cu} + \phi_{MgO}k_{MgO}}{\phi}$	$\frac{k_{hnf}}{k_{bf}} = \frac{k_{np} + (n-1)k_{bf} - (n-1)(k_{bf} - k_{np})\phi}{k_{np} + (n-1)k_{bf} - (k_{bf} - k_{np})\phi}$
Thermal expansion	$\beta_{np} = \frac{\phi_{Cu}\beta_{Cu} + \phi_{MgO}\beta_{MgO}}{\phi}$	$(\rho\beta)_{hnf} = (1-\phi)(\rho\beta)_{bf} + \phi(\rho\beta)_{np}$
Thermal diffusivity	$\alpha_{nf} = \frac{k_{nf}}{(\rho Cp)_{nf}}$	$\alpha_{hnf} = \frac{k_{hnf}}{(\rho Cp)_{hnf}}$
Electronic conductivity	$\sigma_{np} = \frac{\phi_{Cu}\sigma_{Cu} + \phi_{MgO}\sigma_{MgO}}{\phi}$	$\sigma_{hnf} = (1-\phi)\sigma_{bf} + \phi\sigma_{np}$
Dynamic viscosity	/	$\mu_{hnf} = \frac{\mu_{bf}}{(1-\phi)^{2.5}}$

FIGURE 29.2 Comparison of results with present program (top) and previous work (bottom) [51].

TABLE 29.3
Mesh Test for Ra = 10^5, Ha = 0, and ϕ = 6%

Mesh	Quality	Elements Number	Nusselt	Deviation
Fine	0.8014	3192	1.4001	3.53%
Finer	0.8221	7894	1.4476	0.54%
Extra-fine	0.8395	20904	1.4554	/

29.4 RESULTS AND DISCUSSION

At $\varepsilon = 0.4$, a numerical investigation of the efficiency of the heat transfer occurring in an annulus between a discretely heated parallelogram and an adiabatic elliptic cylinder was carried out, by altering the emplacement of the heating wall from top to bottom, in order to select the optimal thermal configuration, along with several parameters, such as Rayleigh ($10^4 \leq \text{Ra} \leq 10^6$), to explore the natural convection and buoyancy forces impact on the convective flow; Hartmann ($0 \leq \text{Ha} \leq 100$), to examine how the magnetic field induction influences the nanofluid's performance inside the enclosure; and the concentration of the hybrid nanofluid ($0.02 \leq \phi \leq 0.06$), to study the presence of the nanoparticles. The results present the hydrothermal flow and are depicted as streamlines, entropy profiles, and Nusselt numbers to demonstrate how the hybrid nanofluid influences heat transmission and how its behavior develops when subjected to a magnetic field.

29.4.1 Impact of the Volume Fraction and Heating Placement

Figure 29.3 and Figure 29.4 illustrate how the free convective transmission is influenced by adjusting the heating wall placement and the volumetric fraction of the Cu-MgO/H_2O hybrid nanofluid. The outcomes of the numerical analysis indicated that placing the heat source in the upper section

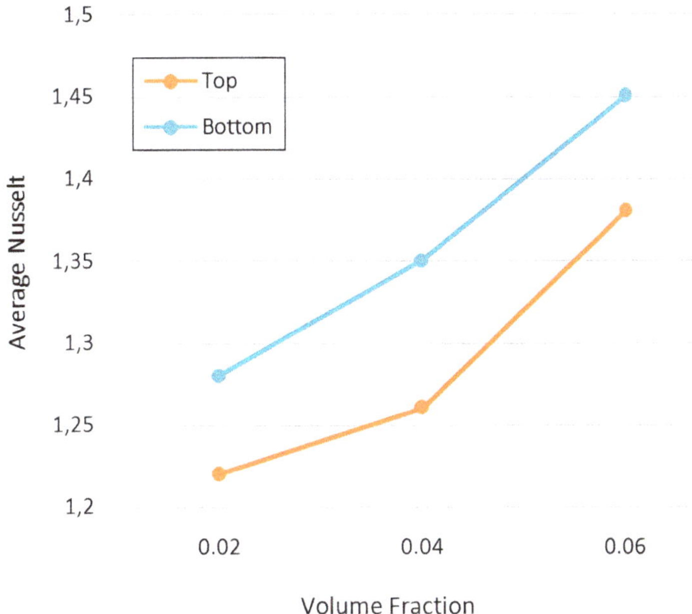

FIGURE 29.3 Average Nusselt at Ra = 10^5 and Ha = 0.

of the wall provides extended values of the average Nusselt number, especially when strengthening the presence of the hybrid nanofluid, by augmenting its concentration from 2% to 6%.

Figure 29.3 shows how the top heated section case presents augmented values of the average Nusselt; also, raising the volume fraction from 2% to 6% yielded a 14% improvement in Nu_{avg} compared to the other case. On the other hand, adapting the heating in the bottom section of the wall improved Nusselt by about 12% for the same enhancement of the concentration. It is concluded that placing the discretely heated wall in the upper section of the cavity provides better transmission; thus, the following study is adapted with this configuration.

Furthermore, in Figure 29.4, the streamlines and total entropy generation profiles at Ra = 10^5 and Ha = 0 are also presented, demonstrating how the concentration of the hybrid nanofluid and entropy is proportional, since augmenting the volume fraction of the hybrid nanofluid to 6% enables greater entropy production values compared to 2% and 4%, which explains how the stream values are minimally decremented with the volume fraction. Hence, although the great thermophysical properties of nanoparticles enhance the heat transfer and strengthen the average Nusselt, they also increment the total entropy production in the enclosure, which can lead to the deterioration of the thermal efficiency. In this context, an equilibrium must be imposed to manage the concentration of the hybrid nanofluid and its influence on the irreversibility that develops during the convective heat transfer.

29.4.2 Impact of Rayleigh

Figure 29.5 depicts the streamlines and entropy generation in the enclosure at Ha=0 and ϕ=2%. The stream values are considerably improved when Rayleigh is increased from 10^4 to 10^6, and the natural convective flow is accelerated as the hybrid nanofluid motion is dispersed along the cavity. The increase in Rayleigh number enhances the presence of buoyant forces, which reflects the production of large vortices, allowing the heated flow to circulate around the cylinder and so ensuring improved transmission as the flow is no longer attached to the discretely heated wall. This enhancement, on the other hand, increases the entropy creation in the enclosure by 32%, which might reduce the

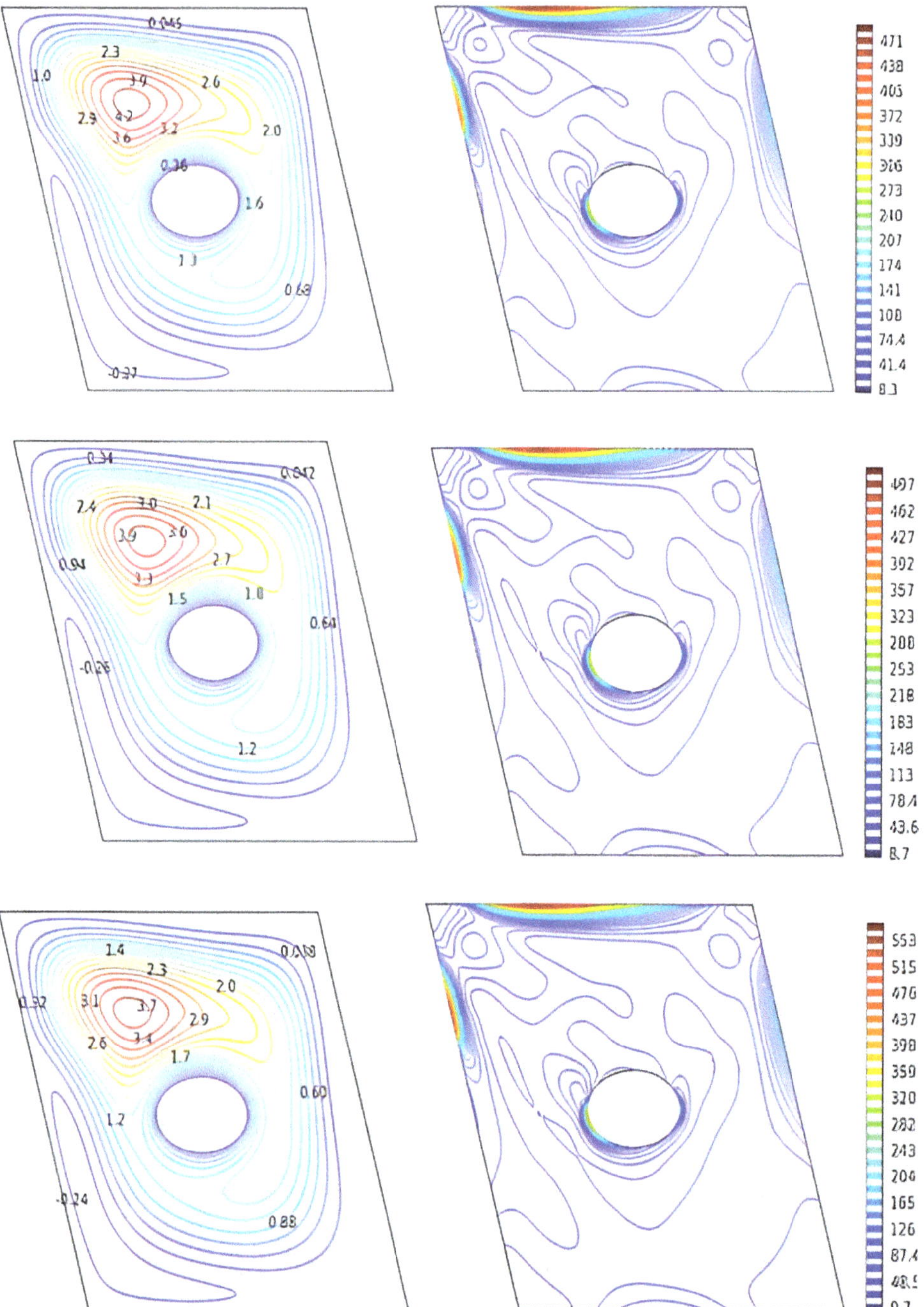

FIGURE 29.4 Streamlines (left) and entropy generation (right) at Ra = 10^5 and Ha = 0.

thermal performance of the convective flow as the irreversible heat transmission is accumulated all around the cavity.

Figure 29.6 presents how the average Nusselt number changes with the volume fraction and Rayleigh number. It is shown that in order to achieve an improvement of up to 6% in the Nusselt values, it is recommended to increase the concentration of the hybrid nanofluid along with great

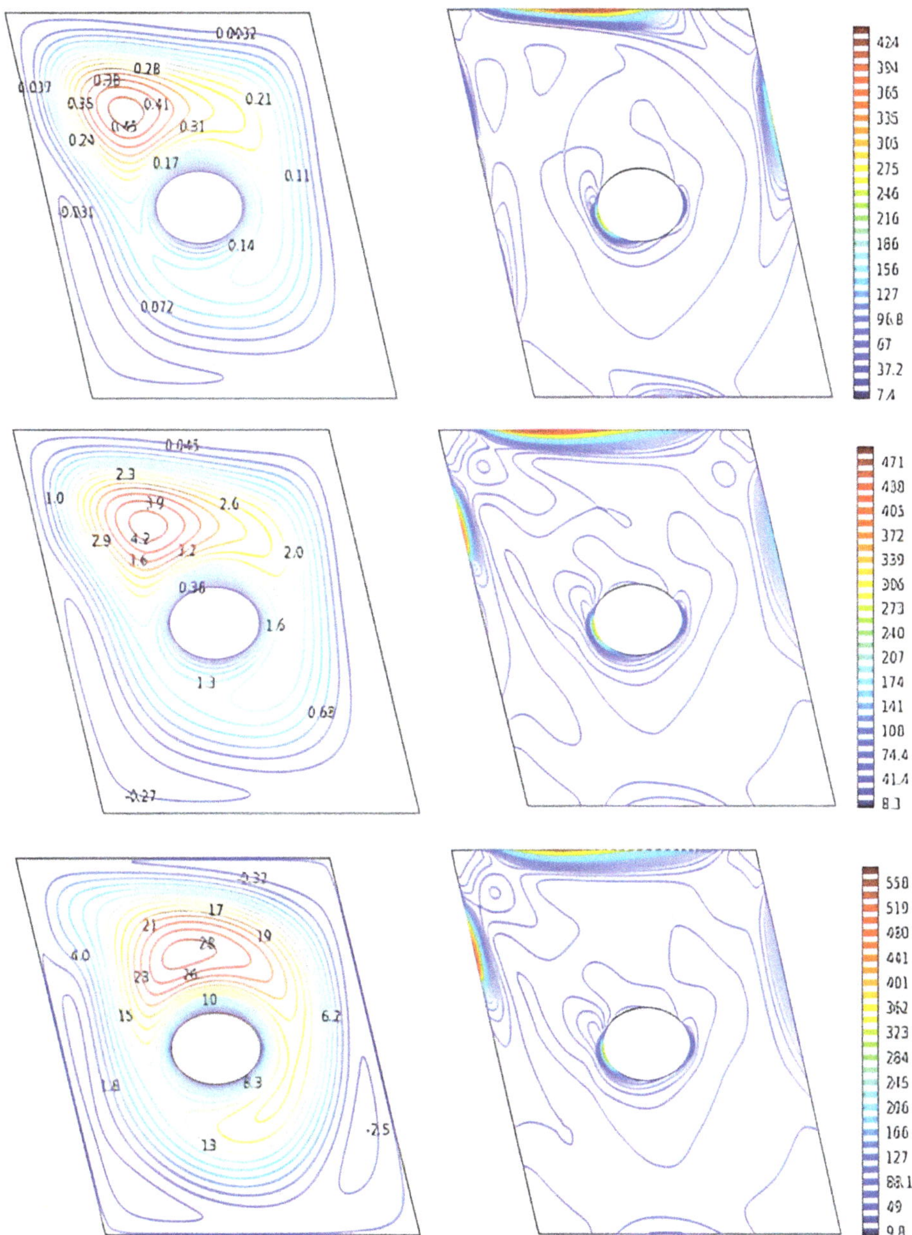

FIGURE 29.5 Streamlines (left) and entropy generation (right) at Ha=0 and φ=2%.

values of Ra. However, boosting the efficiency of this thermal system should be managed in order to avoid reaching high levels of entropy formation.

29.4.3 Impact of Hartmann

The variation of the average Nusselt number in the presence of a magnetic field is depicted in Figure 29.7. It is clear that the enhancement of Nusselt is indirectly related to the strength of the magnetic field for both volume fractions. For 2% and 6% of the hybrid nanofluid, altering the Hartmann

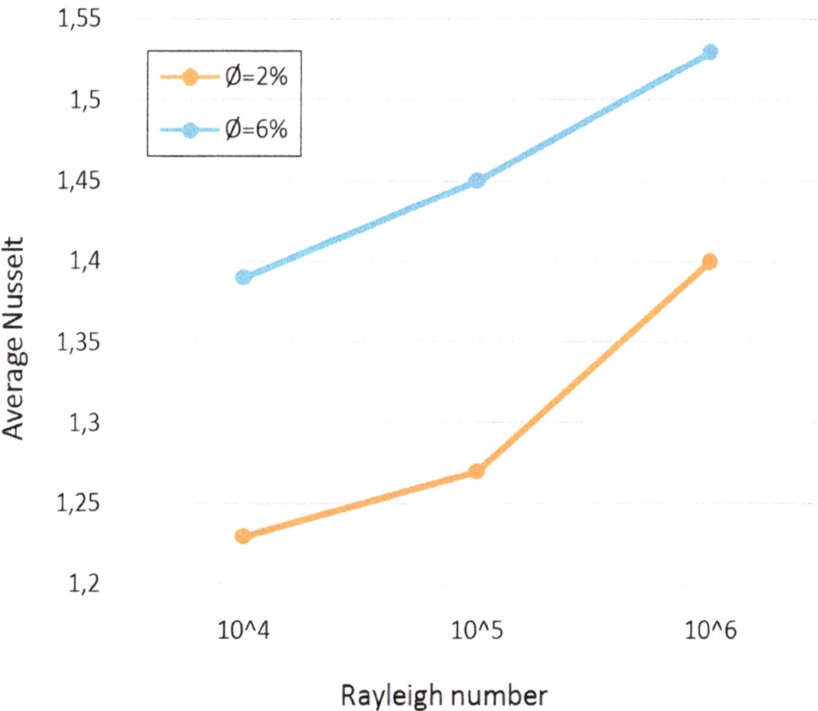

FIGURE 29.6 Average Nusselt at Ha = 0.

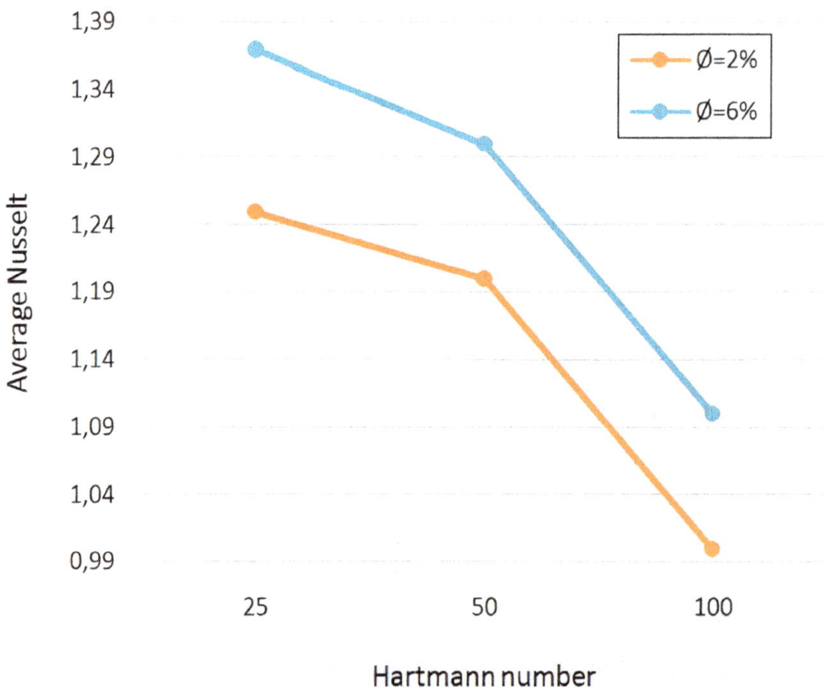

FIGURE 29.7 Average Nusselt at Ra = 10^5.

number from 25 to 100 lowered Nusselt by 25% and 20%, respectively, thus constraining the performance of the thermophysical properties of the nanoparticles.

Additionally, the magneto-convective flow of the hybrid nanofluid is explored by varying the Hartmann number from 25 to 100; the results are presented as streamlines and entropy production at Ra = 10^5 and $\phi = 2\%$ in Figure 29.8. The results show that augmenting Ha decreases the stream values, as the motion of the hybrid nanofluid is restricted, resulting in the flow ascending towards the upper heated section of the enclosure. Applying a strong intensity of a magnetic field which corresponds to Ha = 100 enhances the presence and the influence of the Lorentz forces, which limit

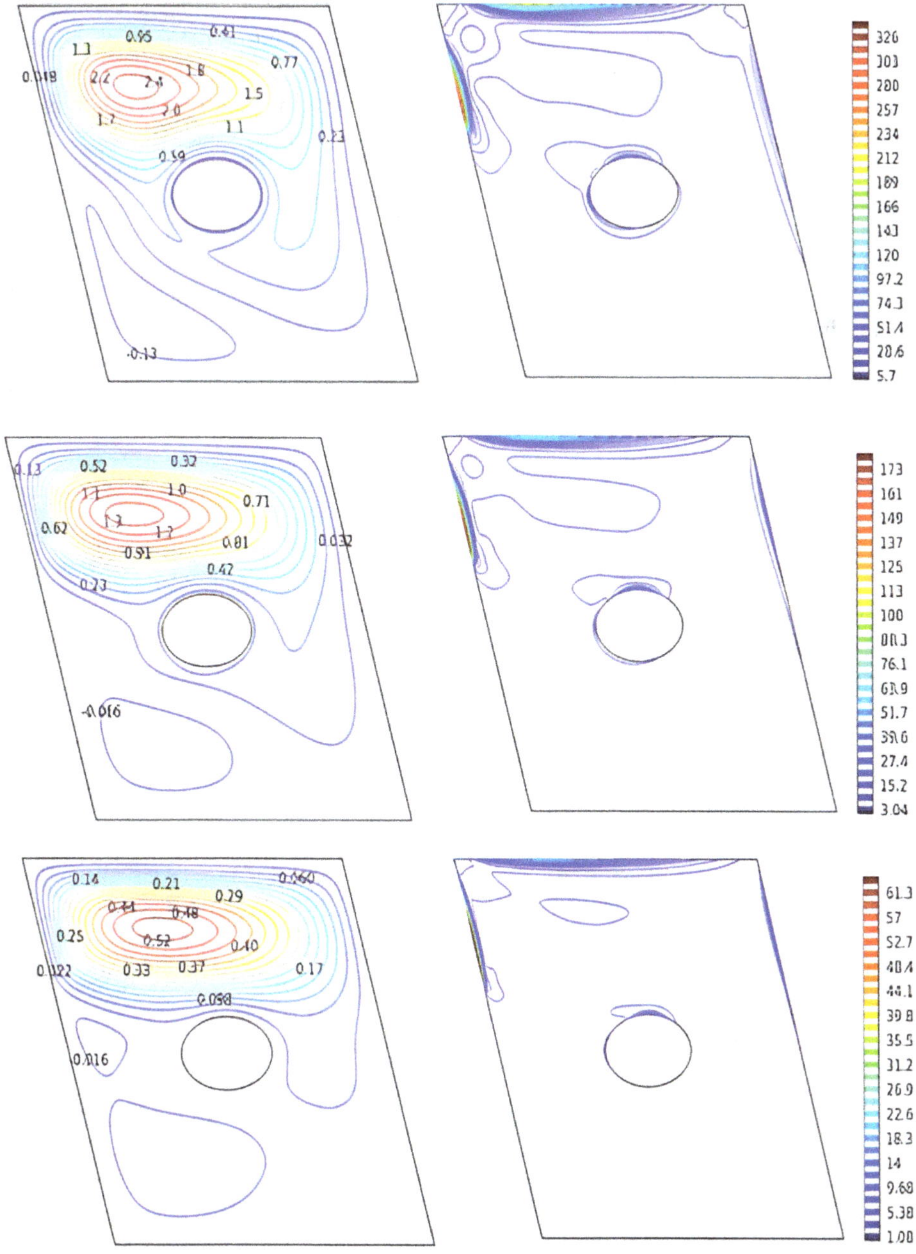

FIGURE 29.8 Streamlines (left) and entropy generation (right) at Ra = 10^5 and $\varphi = 2\%$.

the nanoparticles behavior and inhibit the convective movement. Thus, the entropy production is reduced with the slow motion of the hybrid nanofluid.

29.5 CONCLUSIONS

The magnetohydrodynamic flow of a Cu-MgO/H_2O hybrid nanofluid in a porous parallelogram was numerically evaluated in this chapter. The study conducted an investigation to assess the entropy production, the average Nusselt number, and the streamlines as function of the Hartmann number, Rayleigh number, and the volume fraction of the hybrid nanofluid. The presented findings demonstrate how augmenting the intensity of the magnetic field significantly lowers the performance of the hybrid nanofluid and decrements Nusselt number and the stream function values. On the other hand, to boost the thermal behavior, a discrete heating is applied in the top area of the enclosure on the inclined wall, featuring great values of Rayleigh number and the volume fraction to raise the buoyancy forces. Nonetheless, such augmentation necessitates a control evaluation in order to minimize entropy formation. Hence, we can conclude the following:

- Intensifying the natural convective flow along the cavity is persistent with enhancing Rayleigh.
- Augmenting the nanofluid's concentration significantly alters the average Nusselt number.
- The thermal efficiency improvement is proportional to no magnetic induction.
- Increasing the volume fraction and Rayleigh number significantly contributes in the growth of the entropy production.

Adapting heat sources in the top area of the enclosure amplifies the thermal performance.

Nomenclature

u, v	velocity components ($m \cdot s^{-1}$)
x, y	Cartesian coordinates (m)
p	pressure ($N \cdot m^{-2}$)
U, V:	dimensionless velocity components
X, Y	dimensionless Cartesian coordinates
P	dimensionless pressure
ρ	density ($Kg \cdot m^{-3}$)
g	gravitational acceleration ($m \cdot s^{-2}$)
T	temperature (K)
T_{avg}	average temperature (K)
θ	dimensionless temperature
α	thermal diffusivity ($m^2 \cdot s^{-1}$)
υ	kinematic viscosity ($m^2 \cdot s^{-1}$)
K	permeability ($H \cdot m^{-1}$)
ε	porosity
σ	electric conductivity $(Ohm \cdot m)^{-1}$
B_0	magnetic field density ($A \cdot m^{-1}$)
k	thermal conductivity ratio ($W \cdot K^{-1} \cdot m^{-1}$)
C_p	specific heat ($J \cdot K^{-1} \cdot Kg^{-1}$)
β	thermal expansion (K^{-1})
μ	dynamic viscosity ($Kg \cdot m^{-1} \cdot s^{-1}$)
ϕ	volume fraction of the nanoparticles
γ	irreversibility distribution ratio in the porous layer
S	entropy

L	length of the enclosure (m)
H	height of the enclosure (m)

Subscripts

h	hot
c	cold
tot	total
t	thermal
f	flow
mf	magnetic field
H_2O	water
Cu	copper
MgO	magnesium oxide
MHD	magnetohydrodynamic
Nf	nanofluids
Hnf	hybrid nanofluid
Bf	base fluid
Np	nanoparticle
max	maximum
Fc	Forchheimer coefficient
Ra	Rayleigh
Nu	Nusselt
Ha	Hartmann
Da	Darcy
Pr	Prandtl

REFERENCES

[1] Mebarek-Oudina, F., Chabani, I. Review on Nano-Fluids applications and heat transfer enhancement techniques in different enclosures. *Journal of Nanofluids*, 2022, 11(2): 155–168.

[2] Amidu, M.A., Addad, Y., Riahi, M.K., Abu-Nada, E. Numerical investigation of nanoparticles slip mechanisms impact on the natural convection heat transfer characteristics of nanofluids in an enclosure. *Scientific Reports*, 2021, 11(1): 15678.

[3] Abusorrah, A., Mebarek-Oudina, F., Ahmadian, A., Baleanu, D. Modeling of a MED-TVC desalination system by considering the effects of nanoparticles: Energetic and exergetic analysis. *Journal of Thermal Analysis and Calorimetry*, 2021, 144.

[4] Cieśliński, J.T., Smolen, S., Sawicka, D. Effect of temperature and nanoparticle concentration on free convective heat transfer of nanofluids. *Energies*, 2021, 14(12): 3566.

[5] Wu, H., Zhang, S. Numerical study on the fluid flow and heat transfer characteristics of Al_2O_3-Water nanofluids in microchannels of different aspect ratio. *Micromachines*, 2021, 12(8).

[6] Elias, M.M., Miqdad, M., Mahbubul, I.M., Saidur, R., Kamalisarvestani, M., Sohel, M.R., Hepbasli, A., Rahim, N.A., Amalina, M.A. Effect of nanoparticle shape on the heat transfer and thermodynamic performance of a shell and tube heat exchanger. *International Communications in Heat and Mass Transfer*, 2013, 44: 93–99.

[7] Seddighi Chaharborj, S., Moameni, A. Nano-particle volume fraction optimization for efficient heat transfer and heat flow problems in a nanofluid. *Heat Transfer Research*, 2018, 49.

[8] Asadi, M., Asadi, A., Aberoumand, S. An experimental and theoretical investigation on the effects of adding hybrid nanoparticles on heat transfer efficiency and pumping power of an oil-based nanofluid as a coolant fluid. *International Journal of Refrigeration*, 2018, 89.

[9] Mebarek-Oudina, F. Convective heat transfer of Titania nanofluids of different base fluids in cylindrical annulus with discrete heat source. *Heat Transfer*, 2019, 48(1): 135–147.

[10] Kumar, A., Hassan, M.A. An experimental investigation of Rayleigh Benard convective heat transport using MWCNT/water based nanofluids. *AIP Conference Proceedings*, 2021, 2341(1): 030031.

[11] Chabani, I., Mebarek-Oudina, F., Ismail, A.A.I. MHD flow of a hybrid nano-fluid in a triangular enclosure with zigzags and an elliptic obstacle. *Micromachines*, 2022, 13(2): 224.
[12] Hatami, N., Kazemnejad Banari, A., Malekzadeh, A., Pouranfard, A.R. The effect of magnetic field on nanofluids heat transfer through a uniformly heated horizontal tube. *Physics Letters A*, 2017, 381(5): 510–515.
[13] Saeed, A., Jawad, M., Alghamdi, W., Nasir, S., Gul, T., Kumam, P. Hybrid nanofluid flow through a spinning Darcy–Forchheimer porous space with thermal radiation. *Scientific Reports*, 2021, 11(1): 16708.
[14] Nield, D.A., Bejan, A. *Convection in Porous Media*. Springer, New York, Vol. 4, 2017.
[15] Alsabery, A.I., Chamkha, A.J., Saleh, H., Hashim, I. Natural convection flow of a nanofluid in an inclined square enclosure partially filled with a porous medium. *Scientific Reports*, 2017, 7(1): 2357.
[16] Hussein, A.K., Hamzah, H.K., Ali, F.H., Kolsi, L. Mixed convection in a trapezoidal enclosure filled with two layers of nanofluid and porous media with a rotating circular cylinder and a sinusoidal bottom wall. *Journal of Thermal Analysis and Calorimetry*, 2019, 141: 2061–2079.
[17] Das, D., Roy, M., Basak, T. Studies on natural convection within enclosures of various (non-square) shapes—A review. *International Journal of Heat and Mass Transfer*, 2017, 106: 356–406.
[18] Alguboori, A., Mousa, M., Kalash, R., Habeeb, L. Natural convection heat transfer in an inclind elliptic enclosure with circular heat source. *Journal of Mechanical Engineering Research and Developments*, 2020, 43: 207–222.
[19] Mebarek-Oudina, F., Bessaih, R., Mahanthesh, B., Chamkha, A. J., Raza, J. Magneto-thermal-convection stability in an inclined cylindrical annulus filled with a molten metal. *International Journal of Numerical Methods for Heat & Fluid Flow*, 2021, 31 (4): 1172–1189. https://doi.org/10.1108/HFF-05-2020-0321
[20] Bakhshi, H., Khodabandeh, E., Akbari, O., Toghraie, D., Joshaghani, M., Rahbari, A. Investigation of laminar fluid flow and heat transfer of nanofluid in trapezoidal microchannel with different aspect ratios. *International Journal of Numerical Methods for Heat & Fluid Flow*, 2019, 29(5): 1680–1698.
[21] Raizah, Z., Aly, A.M., Alsedais, N., Mansour, M.A. MHD mixed convection of hybrid nanofluid in a wavy porous cavity employing local thermal non-equilibrium condition. *Scientific Reports*, 2021, 11(1): 17151.
[22] Chabani, I., Mebarek-Oudina, F., Vaidya, H., and Ismail, A.I. Numerical analysis of magnetic hybrid Nano-fluid natural convective flow in an adjusted porous trapezoidal enclosure. *Journal of Magnetism and Magnetic Materials*, 2022, 564: 170142. https://doi.org/10.1016/j.jmmm.2022.170142.
[23] Mebarek-Oudina, F., Fares, R., Choudhari, R. Convection heat transfer of MgO-Ag/water magneto-hybrid nanoliquid flow into a special porous enclosure. *Algerian Journal of Renewable Energy and Sustainable Development*, 2020, 2(2): 84–95. https://doi.org/10.46657/ajresd.2020.2.2.1.
[24] Chamkha, A.J., Hussain, S.H., Ali, F. H., Shaker, A. Conduction-combined forced and natural convection in a lid-driven parallelogram-shaped enclosure divided by a solid partition. *Progress in Computational Fluid Dynamics*, 2012, 12: 309–321.
[25] Hussain, A.H., Al-Zamily, A.M.A., Ali, F.H., Hussain, S.H. Heatline visualization of natural convection heat transfer for nanofluids confined within parallelogrammic cavities in presence of discrete isoflux sources. *Journal of Applied Mathematics and Mechanics*, 2021, 101(10): e202000024.
[26] Dutta, S., Goswami, N., Biswas, A.K., Pati, S. Numerical investigation of magnetohydrodynamic natural convection heat transfer and entropy generation in a rhombic enclosure filled with Cu-water nanofluid. *International Journal of Heat and Mass Transfer*, 2019, 136: 777–798.
[27] Bandaru, M., Jangili, S., Krishna, G., Bég, O., Kadir, A. Spectral numerical study of entropy generation in magneto-convective viscoelastic biofluid flow through poro-elastic media with thermal radiation and buoyancy effects. *Journal of Thermal Science and Engineering Applications*, 2021, 14: 1–36.
[28] Yasin, A., Ullah, N., Nadeem, S., Ghazwani, H.A. Numerical simulation for mixed convection in a parallelogram enclosure: Magnetohydrodynamic (MHD) and moving wall-undulation effects. *International Communications in Heat and Mass Transfer*, 2022, 135: 106066. 17: 29 *Cu-MgO/Water hybrid Nanofluid and discrete heating*.
[29] Getachew Ushachew, E., Sharma, M.K., Rashidi, M.M. Heat transfer enhancement with nanofluid in an open enclosure due to discrete heaters mounted on sidewalls and a heated inner block. *International Journal of Numerical Methods for Heat & Fluid Flow*, 2021, 31(7): 2172–2196.
[30] Babar, H., Sajid, M.U., Ali, H. M. Chapter 4—Hybrid nanofluids as a heat transferring media. *Hybrid Nanofluids for Convection Heat Transfer*, 2020: 143–177.

[31] Mebarek-Oudina, F., Reddy, N., Sankar, M. Heat source location effects on buoyant convection of nanofluids in an annulus. *Lecture Notes in Mechanical Engineering, Advances in Fluid Dynamics*, 2021: 923–937.

[32] Sheikholeslami, M., Shah, Z., Shafee, A., Kumam, P., Babazadeh, H. Lorentz force impact on hybrid nanofluid within a porous tank including entropy generation. *International Communications in Heat and Mass Transfer*, 2020, 116: 104635.

[33] Warke, A.S., Ramesh, K., Mebarek-Oudina, F., Abidi, A. Numerical investigation of the stagnation point flow of radiative magnetomicropolar liquid past a heated porous streCOHing sheet. *Journal of Thermal Analysis and Calorimetry*, 2022, 147(12): 6901–6912.

[34] Marzougui, S., Bouabid, M., Mebarek-Oudina, F., Abu-Hamdeh, N., Magherbi, M., Ramesh, K. A computational analysis of heat transport irreversibility phenomenon in a magnetized porous channel. *International Journal of Numerical Methods for Heat & Fluid Flow*, 2021, 31(7): 2197–2222.

[35] Al-Chlaihawi, K.K., Alaydamee, H.H., Faisal, A.E., Al-Farhany, K., Alomari, M.A. Newtonian and non-Newtonian nanofluids with entropy generation in conjugate natural convection of hybrid nanofluid-porous enclosures: A review. *Heat Transfer*, 2022, 51(2): 1725–1745.

[36] Omri, M., Bouterra, M., Ouri, H., Kolsi, L. Entropy generation of nanofluid flow in hexagonal microchannel. *Journal of Taibah University for Science*, 2022, 16(1): 75–88.

[37] Marzougui, S., Mebarek-Oudina, F., Magherbi, M., McHirgui, A. Entropy generation and heat transport of Cu–water nanoliquid in porous lid-driven cavity through magnetic field. *International Journal of Numerical Methods for Heat & Fluid Flow*, 2022, 32(6): 2047–2069.

[38] Alqaed, S., Mustafa, J., Sharifpur, M. Numerical investigation and optimization of natural convection and entropy generation of alumina/H_2O nanofluid in a rectangular cavity in the presence of a magnetic field with artificial neural networks. *Engineering Analysis with Boundary Elements*, 2022, 140: 507–518.

[39] Ferhi, M., Djebali, R., Mebarek-Oudina, F., H. Abu-Hamdeh, N., Abboudi, S. Magnetohydrodynamic free convection through entropy generation scrutiny of eco-friendly nanoliquid in a divided L-shaped heat exchanger with lattice boltzmann method simulation. *Journal of Nanofluids*, 2022, 11(1): 99–112.

[40] Prasad, KV., Vaidya, H., Mebarek-Oudina, F., Ramadan, K.M., Khan, M.I., Choudhari, R., Gulab, R.K., Tlili, I., Guedri, K., Galal, A.M. Peristaltic activity in blood flow of Casson nanoliquid with irreversibility aspects in vertical non-uniform channel. *Journal of the Indian Chemical Society*, 2022, 28: 100617.

[41] Javed, T., Mehmood, Z., Siddiqui, M.A., Pop, I. Study of heat transfer in water-Cu nanofluid saturated porous medium through two entrapped trapezoidal cavities under the influence of magnetic field. *Journal of Molecular Liquids*, 2017, 240: 402–411.

[42] Azmi, W.H., Abdul Hamid, K., Ramadhan, A.I., Shaiful, A.I.M. Thermal hydraulic performance for hybrid composition ratio of TiO2–SiO2 nanofluids in a tube with wire coil inserts. *Case Studies in Thermal Engineering*, 2021, 25: 100899.

[43] Mebarek-Oudina, F., Fares, R., Aissa, A., Lewis, R.W., Abu-Hamdeh, N. Entropy and convection effect on magnetized hybrid nano-liquid flow inside a trapezoidal cavity with zigzagged wall. *International Communications in Heat and Mass Transfer*, 2021, 125: 105279.

[44] Chamkha, A., Miroshnichenko, I., Sheremet, M. Numerical analysis of unsteady conjugate natural convection of hybrid water-based nanofluid in a semicircular cavity. *Journal of Thermal Science and Engineering Applications*, 2017, 9.

[45] Mebarek-Oudina, F., Chabani, I. Review on nano enhanced PCMs: Insight on nePCM application in thermal management/storage systems. *Energies*, 2023, 16(3): 1066. https://doi.org/10.3390/en16031066

[46] Reddy, Y.D., Mebarek-Oudina, F., Goud, B.S., Ismail, A.I. Radiation, velocity and thermal slips effect toward MHD boundary layer flow through heat and mass transport of Williamson nanofluid with porous medium. *Arabian Journal for Science and Engineering*, 2022, 47(12): 16355–16369. https://doi.org/10.1007/s13369-022-06825-2

[47] Costa, V., Raimundo, A. Steady mixed convection in a differentially heated square enclosure with an active rotating circular cylinder. *International Journal of Heat and Mass Transfer*, 2010, 53: 1208–1219.

[48] Kanti, P., Sharma, K.V., Ramachandra, C.G., Gurumurthy, M., RaghundanaRaghava, B.M. Numerical study on fly ash–Cu hybrid nanofluid heat transfer characteristics. *IOP Conference Series: Materials Science and Engineering*, 2021, 1013(1): 012031.

[49] Dharmaiah, G., Mebarek-Oudina, F., Sreenivasa Kumar, M., Chandra Kala, K. Nuclear reactor application on Jeffrey fluid flow with Falkner-Skan factor, Brownian and thermophoresis, non linear thermal

radiation impacts past a wedge. *Journal of the Indian Chemical Society*, 2023, 100(2): 100907. https://doi.org/10.1016/j.jics.2023.100907.

[50] Hormozi Moghaddam, M., Karami, M. Heat transfer and pressure drop through mono and hybrid nanofluid-based photovoltaic-thermal systems. *Energy Science and Engineering*, 2022, 10(3): 918–931.

[51] Bouras, A., Taloub, D., Driss, Z., Debka, S. Heat transfer by natural convection from a heated square inner cylinder to its elliptical outer enclosure utilizing nanofluids. *International Journal of Applied Mechanics and Engineering*, 2022, 27: 22–34.

[52] Raza, J., Mebarek-Oudina, F., Ali Lund, L. The flow of magnetised convective Casson liquid via a porous channel with shrinking and stationary walls. *Pramana—Journal of Physics*, 2022, 96: 229. https://doi.org/10.1007/s12043-022-02465-1

30 Simulation of Magnetic Hybrid Nanofluid (Al_2O_3-Cu/H_2O) Effect on Natural Convection and Entropy Generation in a Square Enclosure with a Hot Obstacle

Bouchmel Mliki, Mohamed Ammar Abbassi, and Fateh Mebarek-Oudina

30.1 INTRODUCTION

Hybrid nanofluid is a new type of nanofluid obtained by dispersing two or more than two different nanoparticles in a base fluid [1–24]. Actually, the hybrid nanofluid has attracted more attention as a new method of enhancement heat transfer properties for various industrial applications.

Mliki et al. [25] studied the effect of nanoparticles' Brownian motion on natural convection heat transfer in an inclined C-shaped enclosure and L-shaped enclosure, respectively, under the effect of the external magnetic field. The obtained results showed that considering the role of Brownian motion caused a remarkable increase in thermal conductivity and viscosity of the nanofluid.

The applications of this new kind of nanofluid are in domestic refrigerator [25], solar water heating [26], grinding [27], nuclear plant [28], heat exchanger [29], etc. Numerous published literature reviews considered the construction, properties, and applications of hybrid nanofluid, such as Sidik et al. [30], Babu et Kumar [31], Hamzah and Sidik [32], Leong et al. [33], and Chabani et al. [34].

The main objective of synthesizing hybrid nanofluid is to improve the heat transfer of the nanofluid [35]. According to Suresh et al. [36], the enhancement is higher in terms of viscosity by hybridizing (Al_2O_3/Cu/H_2O). It was found that the maximum thermal conductivity is observed when introducing nanohybrid. Similar results were also reported by Takabi and Salehi [37]. Another innovative study was studied by Balla et al. [38] on thermal conductivity of hybrid nanofluid. They reported an enhancement in heat transfer by hybridizing (CuO/Cu/H_2O).

Moumni et al. [39] studied experimentally the mixed convection with (Al_2O_3/Cu/H_2O) hybrid nanofluid inside an inclined tube. Another experimental study on synthesized (Al_2O_3/Cu/H_2O) was studied experimentally by Suresh et al. [40]. The results revealed that the maximum value of the thermal conductivity is obtained at $\phi = 0.02$. The effect of (Al_2O_3/Cu/H_2O) on the thermophysical properties of the hybrid nanofluid was studied by Suresh and Venkitaraj [41]. They reported an enhancement in thermal conductivity by hybridizing the pure fluid.

Sheikholeslami et al. [42] studied the effect of Lorentz force on laminar natural convection heat transfer in nanofluid (CuO-H_2O). The results indicated that the interaction of electrically conducting fluids and magnetic fields caused a decrease in heat transfer and velocity.

DOI: 10.1201/9781003299608-30

This chapter presents the MHD effects on laminar natural convection of hybrid nanofluid (Al_2O_3-Cu/H_2O) in a square enclosure with a hot obstacle.

30.2 MATHEMATICAL FORMULATION

30.2.1 Geometry Description

The physical domain of the considered problem is shown in Figure 30.1. It consists of a square enclosure with a hot obstacle filled with hybrid nanofluid (Al_2O_3-Cu/H_2O) (see Table 30.1). The horizontal walls are assumed cold. In this cavity, a vertical hot block with a constant temperature, T_h, is situated at the center. The vertical walls are considered adiabatic.

30.2.2 Governing Equations

The dimensional governing equations for the problem are as follows [43]:

$$\frac{\partial u}{\partial x}+\frac{\partial v}{\partial y}=0 \tag{30.1}$$

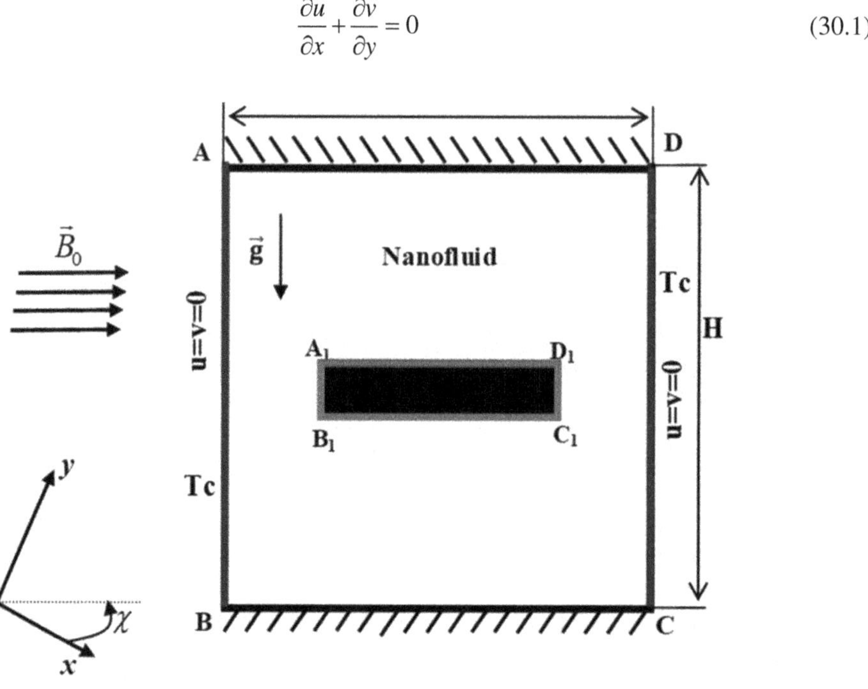

FIGURE 30.1 Geometry of the problem.

TABLE 30.1
Thermophysical Properties of Fluid and Nanoparticles [34]

Thermophysical Properties	H_2O	Al_2O_3	Cu
$C_p(J/kg.K)$	4179	765	385
$\rho(kg/m^3)$	997.1	3,970	8,933
$k(W/m.K)$	0.631	40	401
$\beta\times10^{-5}(1/K)$	21	0.85	1.67
$\sigma\,(\Omega/m)^{-1}$	0.05	1×10^{-10}	$5.69\ 10^{-7}$

$$\rho_{nf}(u\frac{\partial u}{\partial x}+v\frac{\partial u}{\partial y})=-\frac{\partial p}{\partial x}+\mu_{nf}(\frac{\partial^2 u}{\partial x^2}+\frac{\partial^2 u}{\partial y^2})+(\rho\beta_T)_{nf}\,g(T-T_c)\sin\gamma \tag{30.2}$$

$$\rho_{nf}(u\frac{\partial v}{\partial x}+v\frac{\partial v}{\partial y})=-\frac{\partial p}{\partial y}+\mu_{nf}(\frac{\partial^2 v}{\partial x^2}+\frac{\partial^2 v}{\partial y^2})+(\rho\beta_T)_{nf}\,g(T-T_c)\cos\gamma-B^2.\sigma_{nf}v \tag{30.3}$$

$$u\frac{\partial T}{\partial x}+v\frac{\partial T}{\partial y}=\alpha_{nf}(\frac{\partial^2 T}{\partial x^2}+\frac{\partial^2 T}{\partial y^2}) \tag{30.4}$$

Where $\Delta T=T_h-T_c$ is the temperature difference between the hot and the cold walls.

30.2.3 Thermophysical Properties of the Hybrid Nanofluid

The thermophysical properties of the hybrid nanofluid (Al_2O_3-Cu/H_2O) are defined as follows [44]:

The effective density is calculated by:

$$\rho_{hnf}=(1-\phi)\rho_f+\phi_{Al_2O_3}\rho_{Al_2O_3}+\phi_{Cu}\rho_{Cu} \tag{30.5}$$

The heat capacitance is defined by:

$$\left(\rho Cp_{hnf}\right)=(1-\phi)(\rho Cp)_f+\phi_{Al_2O_3}\left(\rho Cp\right)_{Al_2O_3}+\phi_{Cu}\left(\rho Cp\right)_{Cu} \tag{30.6}$$

The thermal expansion coefficient is:

$$(\rho\beta)_{hnf}=(1-\phi)(\rho\beta)_f+\phi_{Al_2O_3}\left(\rho\beta\right)_{Al_2O_3}+\phi_{Cu}\left(\rho\beta\right)_{Cu} \tag{30.7}$$

Where $\phi=\phi_{Al_2O_3}+\phi_{Cu}$.

The dynamic viscosity of the hybrid nanofluid is:

$$\mu_{hnf}=\frac{\mu_f}{\left(1-\left(\phi_{Al_2O_3}+\phi_{Cu}\right)\right)^{2.5}} \tag{30.8}$$

The thermal diffusivity is defined by:

$$\alpha_{hnf}=\frac{k_{hnf}}{(\rho C_p)_{hnf}} \tag{30.9}$$

The thermal conductivity and the electrical conductivity of the hybrid nanofluid are respectively defined by [40]:

$$\frac{k_{hnf}}{k_f}=\left[\frac{\left(\frac{\phi_{Al_2O_3}k_{Al_2O_3}+\phi_{Cu}k_{Cu}}{\phi}+2k_f+2\left(\phi_{Al_2O_3}k_{Al_2O_3}+\phi_{Cu}k_{Cu}\right)-2\phi k_f\right)}{\left(\frac{\phi_{Al_2O_3}k_{Al_2O_3}+\phi_{Cu}k_{Cu}}{\phi}+2k_f-\left(\phi_{Al_2O_3}k_{Al_2O_3}+\phi_{Cu}k_{Cu}\right)+\phi k_f\right)}\right] \tag{30.10}$$

$$\frac{\sigma_{hnf}}{\sigma_f}=\left[1+\frac{3\left(\frac{\phi_{Al_2O_3}\sigma_{Al_2O_3}+\phi_{Cu}\sigma_{Cu}}{\sigma_f}-\left(\phi_{Al_2O_3}+\phi_{Cu}\right)\right)}{\left(\frac{\phi_{Al_2O_3}\sigma_{Al_2O_3}+\phi_{Cu}\sigma_{Cu}}{\phi\sigma_f}+2\right)-\left(\frac{\phi_{Al_2O_3}\sigma_{Al_2O_3}+\phi_{Cu}\sigma_{Cu}}{\sigma_f}-\left(\phi_{Al_2O_3}+\phi_{Cu}\right)\right)}\right] \quad (30.11)$$

30.2.4 Nusselt Number Calculation

Local and average Nusselt numbers along the hot source ($A_1B_1C_1D_1$) are respectively defined by equations (30.12) and (30.13):

$$Nu_l=-\frac{k_{hnf}}{k_f}\left(\frac{\partial\theta}{\partial X}\right)\Bigg|_{Hot\ block} \quad (30.12)$$

$$\overline{Nu_l}=-\frac{1}{A_1B_1}\int_{A_1}^{B_1}\frac{k_{hnf}}{k_f}\left(\frac{\partial\theta}{\partial X}dY\right)\Bigg|_{X=L/4}-\frac{1}{B_1C_1}\int_{B_1}^{C_1}\frac{k_{hnf}}{k_f}\left(\frac{\partial\theta}{\partial Y}dX\right)\Bigg|_{Y=0.4H}$$
$$-\frac{1}{C_1D_1}\int_{C_1}^{D_1}\frac{k_{hnf}}{k_f}\left(\frac{\partial\theta}{\partial X}dY\right)\Bigg|_{X=3L/4}-\frac{1}{D_1A_1}\int_{B_1}^{C_1}\frac{k_{hnf}}{k_f}\left(\frac{\partial\theta}{\partial Y}dX\right)\Bigg|_{Y=0.6H} \quad (30.13)$$

30.2.5 Entropy Generation Equations

The local entropy generation S_{gen} is given by:

$$S_{gen}=\frac{k_{hnf}}{T_0^2}\left[\left(\frac{\partial T}{\partial x}\right)^2+\left(\frac{\partial T}{\partial y}\right)^2\right]+\frac{\mu_{hnf}}{T_0}\left[\begin{array}{c}2\left(\frac{\partial u}{\partial x}\right)^2+2\left(\frac{\partial v}{\partial y}\right)^2\\+\left(\frac{\partial u}{\partial y}+\frac{\partial v}{\partial x}\right)^2\end{array}\right]$$
$$+\frac{\sigma_{hnf}B_0^2}{T_0}v^2=S_{gen,h}+S_{gen,v}+S_{gen,M} \quad (30.14)$$

Where:

$S_{gen,h}$ is entropy generation due to the heat transport.
$S_{gen,v}$ is entropy generation due to the fluid friction.
$S_{gen,M}$ is entropy generation due to the application of the magnetic field.

The total entropy generation is calculated by:

$$S=\frac{1}{V}\int_V S_{gen}dV \quad (30.15)$$

Where V is the total volume of the physical domain.

30.2.5 Grid Refinement

Grid sensitivity tests have been examined for different uniform grids: 25 × 25, 75 × 75, and 100 × 100, 125 × 125, and 150 × 150 (Figure 30.2). It is clear from the results that a grid of 100 × 100 can be selected for low Rayleigh number ($Ra \leq 10^4$). At higher Rayleigh numbers, the chosen grid is equal to 120 × 120 for $Ra = 10^5$ and 200 × 200 for $Ra = 10^6$.

30.2.6 Code Validation

The present computer code has been tested with the numerical results presented by Lai and Yang [45]. The compared results are presented in Figure 30.3. In this case, excellent agreement is found.

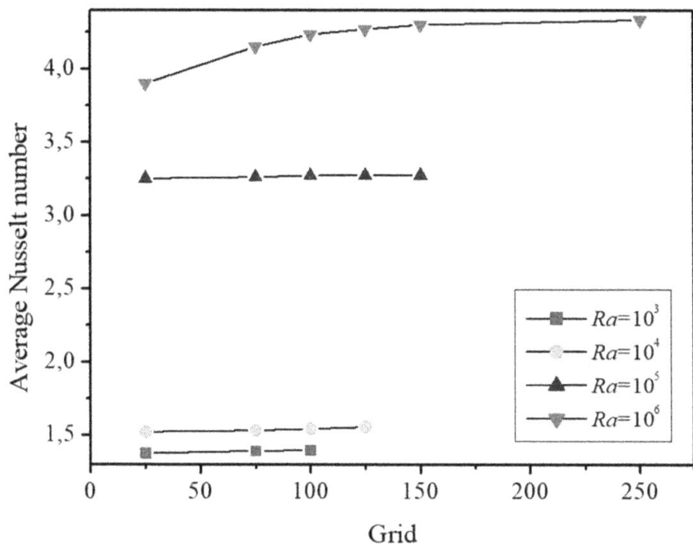

FIGURE 30.2 Grid independent test.

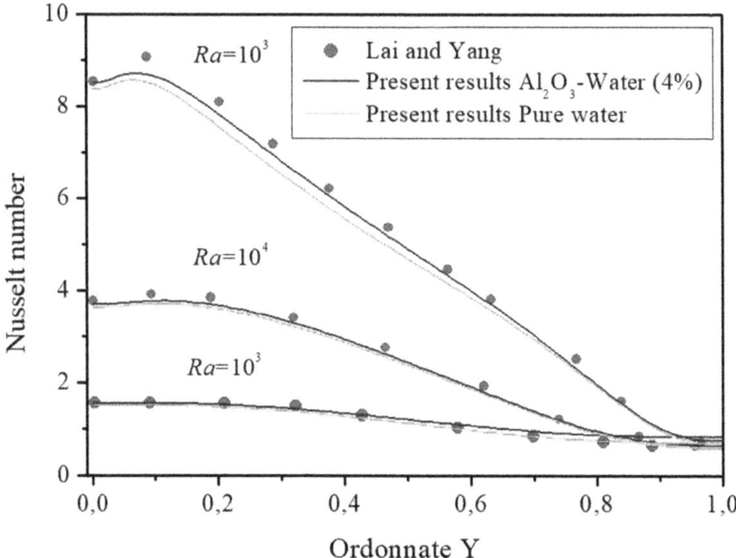

FIGURE 30.3 Local Nusselt number along the hot wall [46].

30.3 RESULTS AND DISCUSSION

Numerical simulations are performed for hybrid nanofluid (Al_2O_3-Cu/H_2O). The effect of Rayleigh number *Ra* is studied by considering four different values: 10^3, 10^4, 10^5, and 10^6. Volumetric fraction of nanoparticles and Hartmann number are varied in the ranges of ($0 \le \phi \le 0.04$) and ($0 \le Ha \le 100$).

30.3.1 Effect of Volumetric Fraction of Nanoparticles

For comparison, the streamlines, isotherms, and entropy generation contours for pure fluid and hybrid nanofluid (Al_2O_3-Cu/H_2O) are shown by a dashed line and a solid line, respectively, at $Ra = 10^4$. The volume fraction of nanoparticles taken is equal to $\phi = \phi_{Al2O3} + \phi_{Cu} = 4.10^{-2}$ when $\phi_{Al2O3} = 2.10^{-2}$ *and* $\phi_{Cu} = 2.10^{-2}$.

Regarding the structure of streamlines (Figure 30.4(a)), it is practically unaffected by the addition of (Al_2O_3-Cu) nanoparticles to a pure fluid (water). So the direction of the flow is not modified. At the same time, the maximum stream function $|\psi|_{max}$ decreases by increasing ϕ.

More precisely, for $\phi = 0.00$ and $\phi = 0.04$, the maximum stream function is equal to 8.05 and 5.59, respectively. This decrease can be explained by equation (30.18), which indicates that the increasing ϕ leads to an increase in the dynamic viscosity of the fluid.

Generally, increasing ϕ causes an increase of the thermal diffusivity coefficient and, consequently, enhanced convection heat transfer. At the same time, an increase of the viscosity dynamic of the hybrid nanofluid is prevented from suitable convection.

In the first region (Y < 0.5), the analysis of the structure of isotherms (Figure 30.4(b)) shows that the spacing of the isotherms 0.3 increases. Consequently, the total temperature inside this region is increased (Figure 30.4(b)). This can be explained by the the fact that the enhancement of heat transfer by the thermal conductivity coefficient is dominant on the decreasing effect of dynamic viscosity.

On the contrary, for (Y > 0.5), a reduction of the spacing of the isotherms 0.3 is observed. In this region, the effect of the effective viscosity dominates. Consequently, the total temperature inside this region is decreased.

Globally, the Nu_{moyen} is found to be increased by 20% when ϕ passes from 0 to 0.04. This is due to the fact that increasing thermal conductivity of Al_2O_3-Cu/H_2O leads to an increase of thermal diffusivity coefficient.

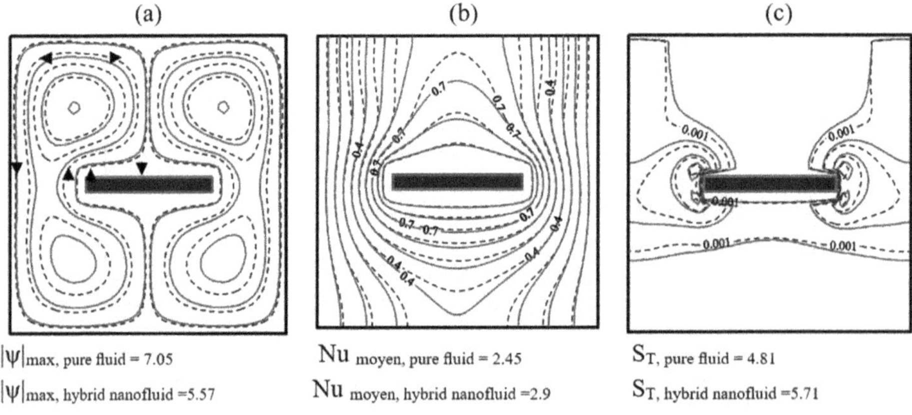

FIGURE 30.4 Comparison of the (a) streamlines, (b) isotherms, and (c) entropy generation between pure fluid, represented by dashed lines (---), and the hybrid nanofluid, represented by solid lines (—), $Ha = 0$, and $Ra = 10^4$ (Al_2O_3-Cu/H_2O).

Figure 30.4(c) presents the comparison of the local entropy generation between the pure fluid and the hybrid nanofluid. As can be seen in Figure 30.4(c), the distribution of $S_{gen,T}$ illustrates an intensification of S_{gen} along the hot source ($A_1B_1C_1D_1$) due to larger temperature gradients along this source.

The variation of $S_{gen,T}$ is similar to the variation of Nu_m with increasing ϕ. The minimum value of the total entropy is 4.81 and occurs for $\phi = 0.00$, whereas the maximum value is 5.71, which occurs at $\phi = 4.10^{-2}$.

30.3.2 Effect of *Ra* Number

In this part, the calculations are made for a different number of *Ra*. The volumetric fraction of nanoparticles taken is equal to $\phi = 4.10^{-2}$, which limits the value of the validity of the model de Maxwell.

The analysis of the structure of streamlines presented in Figure 30.5(a) shows that the circulations inside the cavity depend strongly on the *Ra*. For $Ra = 10^3$, the flow contour is characterized by the presence of two symmetric cells, turning in opposite rotating direction, having the same strength ($|\psi|_{max} = 0.25$). Regarding the structure of isotherms (Figure 30.5(b)), they are parallel to the isothermal walls and perpendicular to the adiabatic walls. Since for $Ra = 10^3$, the heat transfer regime inside the enclosure is dominated by the conduction.

At low *Ra*, the heat transfer conduction mode is practically unaffected; the direction of the flow is not modified.

With increasing Rayleigh numbers, when $Ra \geq 10^5$, the isotherms are compressed towards the hot source ($A_1B_1C_1D_1$). Therefore, the temperature gradient becomes higher near these heated walls. Consequently, the heat transfer effect is dominated by the convection for higher *Ra*; increasing the maximum of the stream function is another indication.

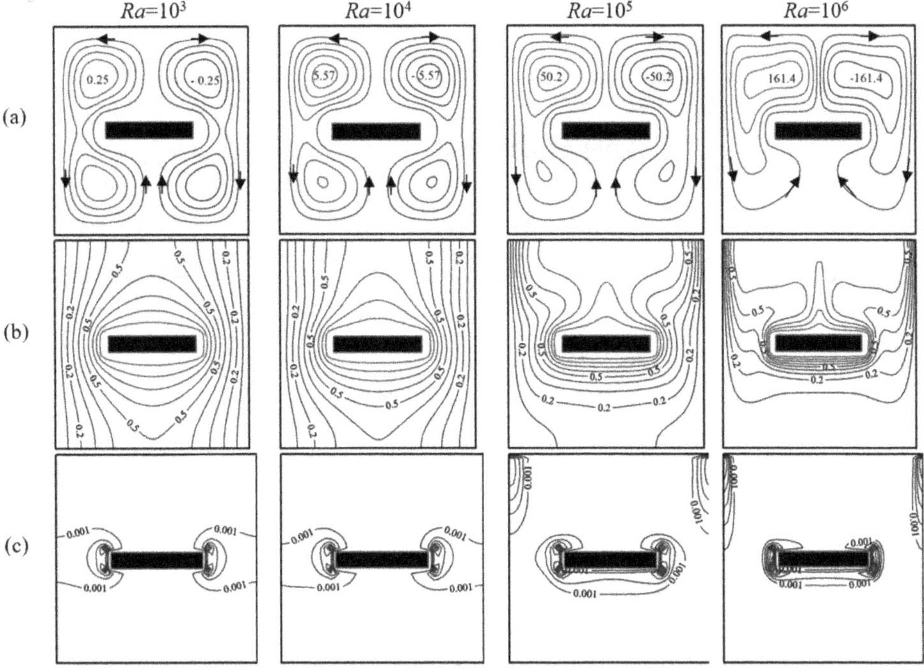

FIGURE 30.5 (a) Streamlines, (b) isotherms, and (c) entropy generation lines for different *Ra*, *Ha* = 0, and $\varphi = 0.04$ (Al_2O_3-Cu/H_2O).

Also, the local entropy generation is presented in this figure (Figure 30.5(c)) and is found to be maximum when $Ra = 10^6$. Generally, increasing the Rayleigh number causes a formation of the active regions for $S_{gen,T}$, especially in higher Ra. The cause of these changes is the development of boundary layers flow near the hot source ($A_1B_1C_1D_1$).

Figures 30.6(a, b) represent the variation in local and average Nusselt numbers at the hot part ($A_1B_1C_1D_1$) for different Rayleigh numbers. Globally, with increasing Ra, the heat transfer regime is enhanced. This is because the predominance of the convection effect by the conduction mechanism

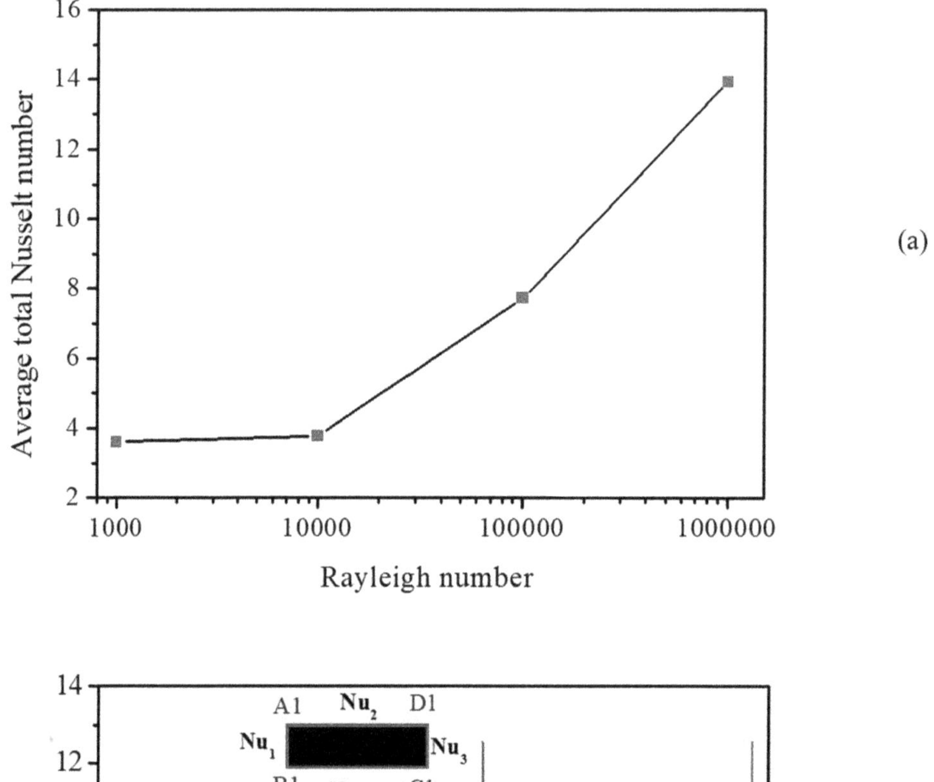

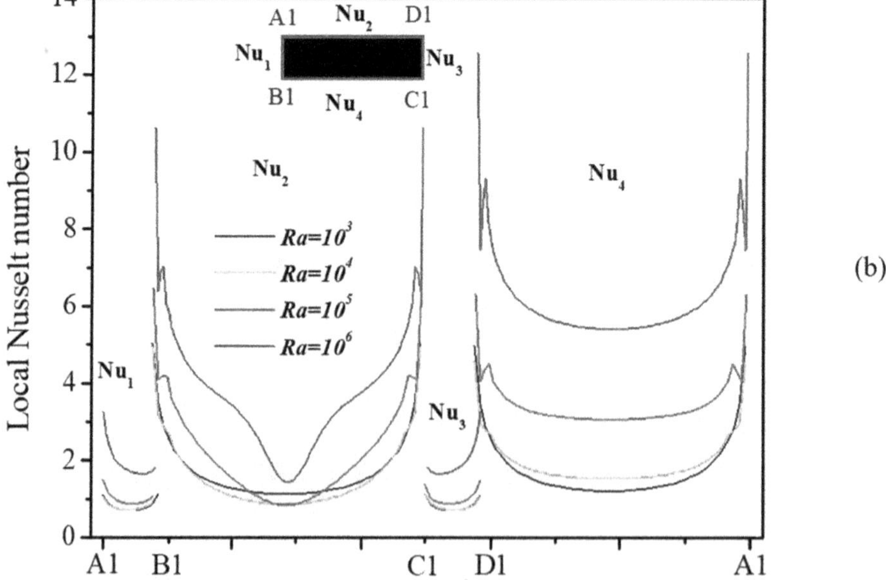

FIGURE 30.6 Effect of Rayleigh number on local and average Nusselt numbers on the hot source faces $Ha = 0$, $\phi = 4\%$.

in a hybrid nanofluid. Also, we note that the heat transfer is important near the two heated walls (A_1D_1 and B_1C_1). This can be explained by the fact that the temperature gradient becomes higher near these heated walls.

30.3.3 Effect of Hartmann Number

In this part, the calculations are made for $Ra = 10^5$ and $f = 4.10^{-2}$ at a different number of *Ha*.

The analysis of the structure of streamlines presented in Figure 8.7(a) shows that the heat transfer conduction mode is dominated when the number of Hartmann increases. This can be explained by the fact that the effect of Lorentz forces on the streamlines is obtained by the decrease in the maximum of the stream function. More precisely, for $Ha = 0$, $Ha = 30$, 60, and 100, the maximum of the stream function is equal to 50.2, 25.3, 9.17, and 3.39, respectively. The cause of these changes is the existence of a magnetic force in the term of volume forces.

For higher *Ha* values, the temperature contours are distributed parallel to the isothermal walls (Figure 30.7(b)). This is a clear point that the conduction mode heat transfer has the biggest role in the enclosure. Also, we note that the minimum entropy generation is obtained at $Ha = 100$ (Figure 30.7(c)).

Figures 30.8 and 30.9 represent the variation in average Nu_m and S_T for different values of *Ra* and *Ha*. The effect of the magnetic field is more noticeable when *Ra* is increased from 0 to 100.

30.3.4 Effects of Inclination Angle

For the selected case, $Ra = 10^5$ and $\phi = 4.10^{-2}$, the effects of inclination angle γ on the streamlines, isotherms, and local entropy generation are presented in Figure 30.10.

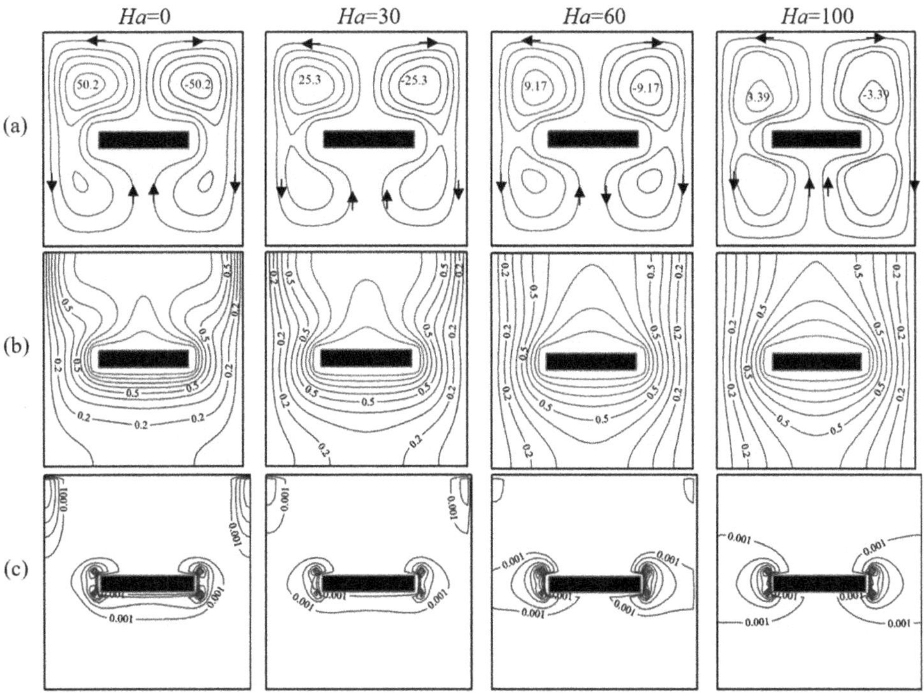

FIGURE 30.7 (a) Streamlines, (b) isotherms, and (c) entropy generation lines for different *Ha*, at $Ra = 10^5$, and $\phi = 0.04$ (Al_2O_3-Cu/H_2O).

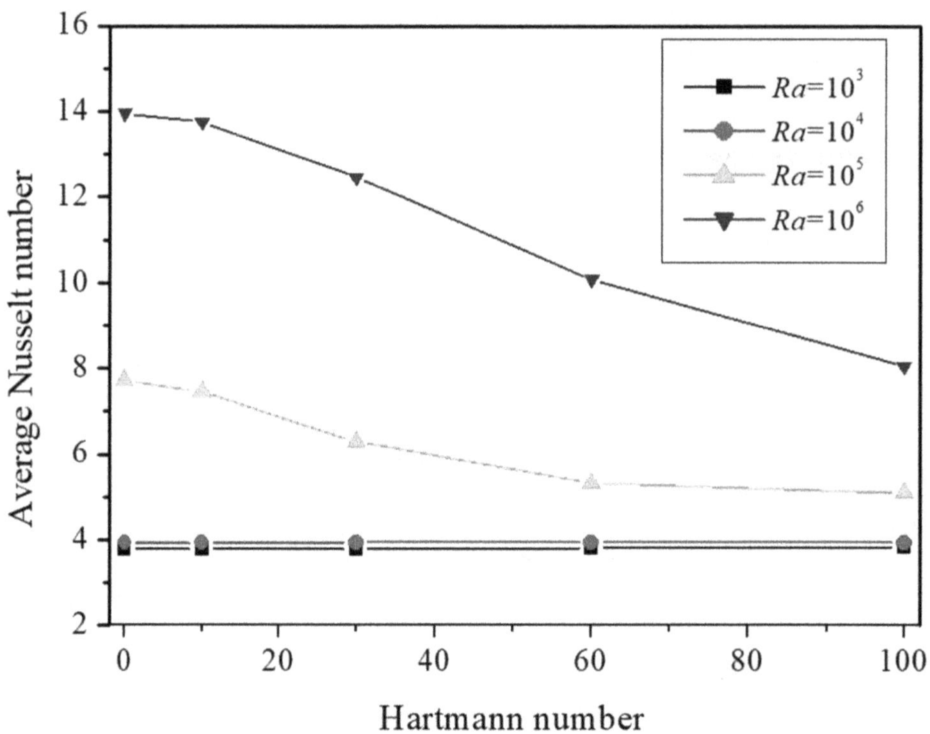

FIGURE 30.8 Effect of Hartmann number on average Nusselt number for different Rayleigh numbers ($Ra = 10^3$-10^6) at $\phi = 0.04$ (Al_2O_3-Cu/H_2O).

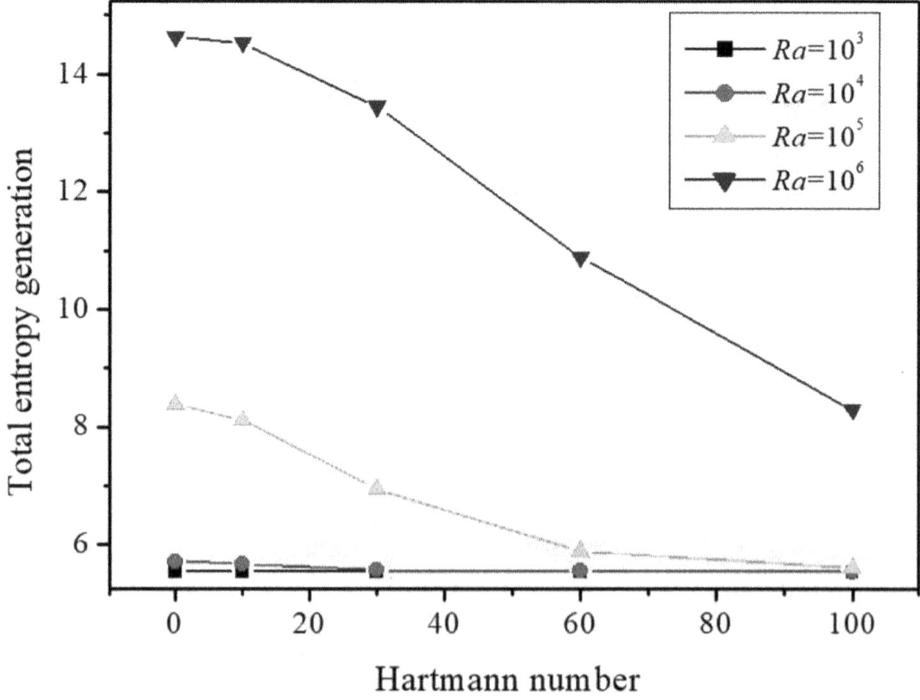

FIGURE 30.9 Total entropy generation for different *Ra* and *Ha*, at $\phi = 0.04$ (Al_2O_3-Cu/H_2O).

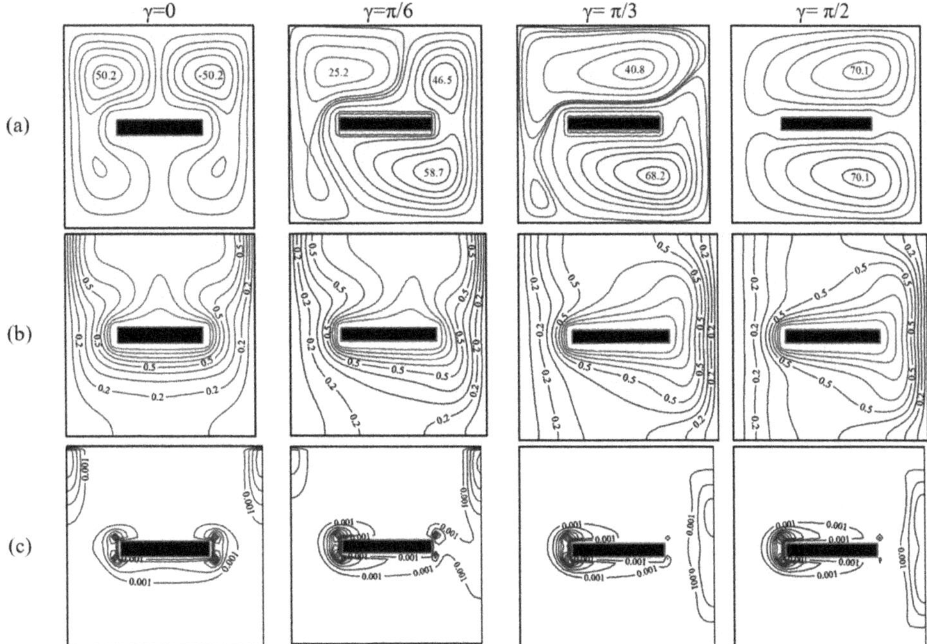

FIGURE 30.10 (a) Streamlines, (b) isotherms, and (c) entropy generation lines for different Ra at $\gamma = 0$, $Ha = 0$, and $\phi = 0.04$ (Al_2O_3-Cu/H_2O).

For $\gamma = 0$, we notice the existence of a two symmetrical cells with clockwise and anticlockwise rotations inside the cavity. Physically, this is true because of the symmetrical boundary conditions about the horizontal X-axis ($X = 0.5$). As γ increases to $\pi/6$, the maximum stream function $|\psi|_{max}$ cells that exist inside the right vertical portion of the enclosure becomes weaker. Conversely, an opposite effect occurs in the left vertical portion of the cavity.

At the same time, we notice that the temperature gradient along the walls (A_1D_1) of the heat source is reduced. So it leads to a reduction in the local entropy generation.

When the inclination angle of the cavity increases from $\gamma = 0$ to $\pi/2$, the streamlines and isotherm contours are symmetrical about the horizontal centerline (Y = 0.5) of the cavity. Consequently, a symmetrical local entropy generation is obtained.

The minimum value of the Nu_m is 7.491 and occurs for $\gamma = 0$, whereas the maximum value is 7.798, which occurs at $\gamma = \pi/2$. Moreover, as inclination angle γ increases from 0 to $\gamma = \pi/2$, the density of the total entropy generation is increased by 4.8% when the inclination angle passes from 0 to $\gamma = \pi/2$.

30.4 CONCLUSIONS

At present study, the magnetohydrodynamic (MHD) natural convection of Al_2O_3-Cu/H_2O nanohybrid in a square enclosure was investigated. Numerical results show that:

- The maximum value of $|\psi|_{max}$ is obtained for $Ra = 10^6$.
- The augmentation of the Rayleigh numbers leads to an increase in the Nu and S_T.
- Nu and S_T are in direct relation with Ra and ϕ.

- The minimum value of the stream function $|\psi|_{max}$ is observed when introducing nanohybrid.
- With applying the magnetic field and increasing the Hartmann number, the velocity of the nanohybrid and, subsequently, the maximum value of $|\psi|_{max}$ decrease because of the existence of Lorentz force in the term of volume forces.
- The heat transfer by the conduction effect is dominated for higher Hartmann number ($Ha = 100$).
- Nu and S_T are in indirect relation with Ha.
- The maximum value of the average Nusselt and the total entropy generation were obtained at $\gamma = \pi/2$.

30.5 ACKNOWLEDGMENTS

This work was supported by the Tunisian Ministry of Higher Education and Scientific Research under grant 20/PRD-22.

Nomenclature

B	magnetic field $\left(\text{unit: Tesla} = \text{N}/\left(\text{A} \bullet \text{m}^2\right)\right)$
c	lattice speed
c_s	speed of sound
c_i	discrete particle speed
c_p	specific heat at constant pressure (unit:$\text{J} \bullet \text{kg}^{-1} \bullet \text{K}^{-1}$)
F_i	external forces (unit: N)
Ha	hartmann number
k	thermal conductivity (unit: $(\text{W} \bullet \text{m}^{-}1)/\text{K}$)
Nu_m	average Nusselt number
Nu	local Nusselt number
P	pressure (unit:Pa)
pr	prandtl number
Ra	rayleigh number
S	entropy Generation
Ra	rayleigh number
T	temperature (unit:K)
u, v	velocities (unit:$\text{m} \bullet \text{s}^{-1}$)
x, y	lattice coordinates (unit:m)
H	height of cavity (unit:m)

Greek Letters

α	thermal diffusivity (unit:m^2s^{-1})
β	coefficient of thermal expansion (unit:K^{-1})
ϕ	solid volume fraction $\left(\text{unit: Pa}\right)$
μ	dynamic viscosity (unit:$\text{kgm}^{-1}\text{s}^{-1}$)
ρ	fluid density (unit:kgm^{-3})
θ	non-dimensional temperature
ν	kinematic viscosity (unit:m^2s^{-1})

σ electrical conductivity (unit: $(Wm)^{-1}$)

ψ stream function (unit:m^2s^{-1})

Subscripts

t_h cold temperature

t_c hot temperature

n_h hybrid-nanofluid

REFERENCES

1. Mejri, I., Mahmoudi, A., Abbassi, M.A., and Omri, A. Magnetic field effect on natural convection in a nanofluid filled enclosure with non-uniform heating on both side walls. *International Journal of Heat and Technology*, 2014, 32: 127–133.
2. Mahmoudi, A., Mejri, I., Abbassi, M.A., and Omri A. Analysis of MHD natural convection in a nanofluid-filled open cavity with non-uniform boundary condition in the presence of uniform heat generation/absorption. *Powder Technology*, 2015, 269: 275–289.
3. Chabani, I., Mebarek Oudina, F., and Ismail, A.I. MHD flow of a hybrid nano-fluid in a triangular enclosure with zigzags and an elliptic obstacle. *Micromachines*, 2022, 13(2): 224.
4. Asogwa, K., Mebarek-Oudina, F., and Animasaun, I. Comparative investigation of water-based Al_2O_3 Nanoparticles through water-based CuO nanoparticles over an exponentially accelerated radiative Riga plate surface via heat transport. *Arabian Journal for Science and Engineering*, 2022, 47(7): 8721–8738.
5. Marzougui, S., Mebarek-Oudina, F., Mchirgui, A., and Magherbi, M. Entropy generation and heat transport of Cu-water nanoliquid in porous lid-driven cavity through magnetic field. *International Journal of Numerical Methods for Heat & Fluid Flow*, 2022, 32(6): 2047–2069.
6. Djebali, R., Mebarek-Oudina, F., and Choudhari, R. Similarity solution analysis of dynamic and thermal boundary layers: Further formulation along a vertical flat plate. *Physica Scripta*, 2021, 96(8): 085206.
7. Pushpa, B.V., Sankar, M., and Mebarek-Oudina, F. Buoyant convective flow and heat dissipation of Cu-H_2O nanoliquids in an annulus through a thin baffle. *Journal of Nanofluids*, 2021, 10(2): 292–304.
8. Dhif, K., Mebarek-Oudina, F., Chouf, S., Vaidya, H., and Chamkha, A.J. Thermal analysis of the solar collector cum storage system using a hybrid-nanofluids. *Journal of Nanofluids*, 2021, 10(4): 634–644.
9. Warke, A.S., Ramesh, K., Mebarek-Oudina, F., and Abidi, A. Numerical investigation of nonlinear radiation with magnetomicropolar stagnation point flow past a heated streCOHing sheet. *Journal of Thermal Analysis and Calorimetry*, 2022, 147(12): 6901–6912.
10. Rajashekhar, C., Mebarek-Oudina, F., Vaidya, H., Prasad, K.V., Manjunatha, G., and Balachandra, H. Mass and heat transport impact on the peristaltic flow of Ree-Eyring liquid with variable properties for hemodynamic flow. *Heat Transfer*, 2021, 50(5): 5106–5122.
11. Reddy, Y.D., Mebarek-Oudina, F., Goud, B.S., and Ismail, A.I. Radiation, velocity and thermal slips effect toward MHD boundary layer flow through heat and mass transport of Williamson nanofluid with porous medium. *Arabian Journal for Science and Engineering*, 2022, 47: 16355–16369.
12. Fares, R., Mebarek-Oudina, F., Aissa, A., Bilal, S.M., and Öztop, H.F. Optimal entropy generation in darcy-forchheimer magnetized flow in a square enclosure filled with silver based water nanoliquid. *Journal of Thermal Analysis and Calorimetry*, 2022, 147: 1571–1582.
13. Swain, K., Mahanthesh, B., and Mebarek-Oudina, F. Heat transport and stagnation-point flow of magnetized nanoliquid with variable thermal conductivity with Brownian moment and thermophoresis aspects. *Heat Transfer*, 2021, 50(1): 754–764.
14. Bouchmel, M., and Abbassi, M.A. Entropy generation of MHD natural convectionheat transfer in a heated incinerator using hybrid-nanoliquid. *Propulsion and Power Research*, 2021, 10(2): 143–154.
15. Ahmed, S.E., Mansour, M.A., Hussein, A.K., and Sivasankaran, S. Mixed convection from a discrete heat source. *Engineering Science and Technology, an International Journal*, 2016, 19: 364–376.
16. Teamah, M.A., Sorour, M.M., El-Maghlany, W.M., and Afifi, A. Numerical simulation of double diffusive laminar mixed convection. *Alexandria Engineering Journal*, 2013, 52: 227–239.

17. Mliki, B., Abbassi, M.A., Omri, A., and Zeghmati, B. Lattice boltzmann analysis of MHD natural convection of CuO-water nanofluid in inclined C-shaped enclosures under the effect of nanoparticles Brownian motion. *Powder Technology*. 2017, 308: 70–83.
18. Mliki, B., Abbassi, M.A., and Omri, A. Lattice boltzmann simulation of magnethydrodynamics natural convection in an L-shaped enclosure. *International Journal of Heat and Technology*, 2012, 34(4).
19. Mliki, B., Abbassi, M.A., and Omri, A. Lattice boltzmann simulation of MHD double dispersion natural convection in a C-shaped enclosure in the presence of a nanofluid. *Fluid Dynamic and Material Processing*, 2015, 11(1): 87–114.
20. Mliki, B., Abbassi, M.A., Omri, A., Zeghmati, B. Effects of nanoparticles Brownian motion in a linearly/sinusoidally heated cavity with MHD natural convection in the presence of uniform heat generation/absorption. *Powder Technology*, 2016, 295: 69–83.
21. Abbassi, M.A., Safaei, M.R., Djebali, R., Guedri, K., Zeghmati, B., and Alrashed, A.A.A. LBM simulation of free convection in a nanofluid filled incinerator containing a hot block. *International Journal of Mechanical Sciences*, 2018, 148: 393–408.
22. Khan, U., Mebarek-Oudina, F., Zaib, A., Ishak, A., Abu Bakar, S., Sherif, E.M., and Baleanu, D. An exact solution of a Casson fluid flow induced by dust particles with hybrid nanofluid over a streCOHing sheet subject to Lorentz forces. *Waves in Random and Complex Media*, 2022, 32: 1–14.
23. Hassan, M., Mebarek-Oudina, F., Faisal, A., Ghafar, A., and Ismail, A.I. Thermal energy and mass transport of shear thinning fluid under effects of low to high shear rate viscosity. *International Journal of Thermofluids*, 2022, 15: 100176.
24. Mliki, B., Abbassi, M.A., Omri, A., and Zeghmati, B. Augmentation of natural convective heat transfer in linearly heated cavity by utilizing nanofluids in the presence of magnetic field and uniform heat generation/absorption. *Powder Technology*, 2015, 284: 312–325.
25. Mliki, B., Abbassi, M.A., and Omri, A. Lattice Boltzmann simulation of natural convection in an L-shaped enclosure in the presence of nanofluid. *Engineering Science and Technology, an International Journal*, 2015, 18: 503–511.
26. Bi, S.S., Shi, L., and Zhang, L. Application of nanoparticles in domestic refrigerators. *Applied Thermal Engineering*, 2008, 28: 1834–1843.
27. He, Q., Zeng, S., and Wang, S. Experimental investigation on the efficiency of flat-plate solar collectors with nanofluids. *Applied Thermal Engineering*, 2015, 88: 9–14.
28. Zhang, C.L., Zhang, Y., Jia, D., Li, B., Wang, Y., Yang, M., Hou, Y., and Zhang, X. Performances of Al_2O_3/SiC hybrid nanofluids in minimum-quantity lubrication grinding. *The International Journal of Advanced Manufacturing Technology*, 2016: 1–15.
29. Azwadi, C.S.N., and Adamu, I.M. Turbulent force convective heat transfer of hybrid nano-fluid in a circular channel with constant heat flux. *Journal of Advanced Research in Fluid Mechanics and Thermal Sciences*, 2016, 19.
30. Allahyar, H.R., Hormozi, F., and Nezhad, B.Z. Experimental investigation on the thermal performance of a coiled heat exchanger using a new hybrid nanofluid. *Experimental Thermal and Fluid Science*, 2016, 76: 324–329.
31. Sidik, N.A.C., Md. Adamu, I., Jamil, M.M., Kefayati, G.H.R., Mamat, R., and Najafi, G. Recent progress on hybrid nanofluids in heat transfer applications: A comprehensive review. *International Communications in Heat and Mass Transfer*, 2016, 78: 68–79.
32. Babu, J.A.R., Kumar, K.K., and Rao, S.S. State-of-art review on hybrid nanofluids. *Renewable and Sustainable Energy Reviews*, 2017, 77: 551–565.
33. Hamzah, M.H., Sidik, N.A.C., Ken, T.L., Mamat, R., and Najafi, G. Factors affecting the performance of hybrid nanofluids: A comprehensive review. *International Journal of Heat and Mass Transfer*, 2017, 115(A): 630–646.
34. Chabani, I., Mebarek-Oudina, F., Vaidya, H., and Ismail, A.I. Numerical analysis of magnetic hybrid Nano-fluid natural convective flow in an adjusted porous trapezoidal enclosure. *Journal of Magnetism and Magnetic Materials*, 2022, 564(2): 170142.
35. Leong, K.Y., Ku Ahmad, K.Z., Ong, H.C., Ghazali, M.J., and Baharum, A. Synthesis and thermal conductivity characteristic of hybrid nanofluids—A review. *Renewable and Sustainable Energy Reviews*, 2017, 75: 868–878.
36. Suresh, S., Venkitaraj, K.P., Selvakumar, P., and Chandrasekar, M. Synthesis of Al_2O_3/Cu/water hybrid nanofluids using two step method and its thermo physical properties. *Colloids and Surfaces A: Physicochemical and Engineering Aspects*, 2011, 388: 41–48.

37. Takabi, B., and Salehi, S. Augmentation of the heat transfer performance of a sinusoidal corrugated enclosure by employing hybrid Nanofluid. *Advances in Mechanical Engineering*, 2014, 6: 147059.
38. Balla, H.H., Abdullah, S., MohdFaizal, W., Zulkifli, R., and Sopian, K. Numerical study of the enhancement of heat transfer for hybrid CuO/Cu nanofluids flowing in a circular pipe. *Journal of Oleo Science*, 2013, 62: 533–539.
39. Momin, G.G. Experimental investigation of mixed convection with water–Al_2O_3& hybrid nanofluid in inclined tube for laminar flow. *IJSTR*, 2013, 2: 195–202.
40. Suresh, S., Venkitaraj, K., Selvakumar, P., and Chandrasekar, M. Synthesis of Al_2O_3–Cu/water hybrid nanofluids using two step method and its thermo physical properties. *Colloids and Surfaces A*, 2011, 388: 8–41.
41. Suresh, S., Venkitaraj, K., Selvakumar, P., and Chandrasekar, M. Effect of Al_2O_3–Cu/water hybrid nanofluid in heat transfer. *Experimental Thermal and Fluid Science*, 2012, 38: 54–60.
42. Sheikholeslami, M., Gorji-Bandpy, M., Ellahi, R., and Zeeshan, A. Simulation of MHD CuO-water nanofluid flow and convective heat transfer considering Lorentz forces. *Journal of Magnetism and Magnetic Materials*, 2014, 369: 69–80.
43. Hussain, S., Sameh, A., and Akbar, T. Entropy generation analysis in MHD mixed convection of hybrid nanofluid in an open cavity with a horizontal channel containing an adiabatic obstacle. *International Journal of Heat and Mass Transfer*, 2017, 114: 1054–1066.
44. Malik, S., and Nayak, A.K. MHD convection and entropy generation of nanofluid in a porousenclosure with sinusoidal heating. *International Journal of Thermal Sciences*, 2017, 111: 329–345.
45. Lai, F.H., and Yang, Y.T. Lattice Boltzmann simulation of natural convection heat transfer of Al2O3/ water nanofluids in a square enclosure. *International Journal of Thermal Sciences*, 2011, 50: 1930–1941.
46. Sheikholeslami, M., Bandpay, M.G., and Ganji, D.D. Numerical investigation of MHD effects on Al_2O_3 water nanofluid flow and heat transfer in a semi-annulus enclosure using LBM. *Energy*, 2013, 60: 501–510.

31 Heat Transfer Enhancement in a Shell and Tube Heat Exchanger Using Hybrid Nanofluid

Fatih Selimefendigil, Gürel Şenol, and Hakan F. Öztop

31.1 INTRODUCTION

The need for clean energy systems is growing. One of the applications of thermal systems includes the use of heat exchangers. Therefore, an energy-efficient design of heat exchangers becomes an important issue. Many methods have been offered to increase the effectiveness of heat exchangers. In one of the available methods, nanoparticles can be used in heat transfer fluid. The nanofluid technology has been successfully implemented in diverse energy systems and convective heat transfer applications [1–10]. Researchers have performed studies to examine the benefits of nanofluids in heat exchangers, especially in shell and tube heat exchangers. Bahrehmand and Abbassi [11] performed a study to examine the nanofluid effect in a shell and coil tube heat exchanger using the Al_2O_3 nanofluid. One of the remarkable points of their study was the concentration effect of the nanofluid on the heat transfer ratio. The concentration value 0.2 and 0.3% increased the heat transfer ratio by about 14 and 18%, respectively. Shahrul et al. [12] examined the performance evaluation of an STHX preparing different kinds of water-based nanofluids using ZnO, Fe3O4, Al2O3, CuO, TiO2, and SiO2 nanoparticles. They observed that except for the specific heat, other parameters, such as thermal conductivity, viscosity, and density, increased as the concentration increased, while the specific heat of the nanofluids decreased. Lotfi et al. [13] examined the heat enhancement in an STHX which was in a horizontal position by using multiwalled carbon nanotube (MWNT)/water nanofluid. They proved the effect of the nanofluid on heat enhancement compared to the base fluid, water. Albadr et al. [14] performed a study in a horizontal STHX with Al_2O_3 nanoparticles on different concentration values in water in turbulent flow. According to the results of their study, the increase of the nanofluid concentration had a positive effect on heat transfer coefficient, while this increase caused an increment in friction factor, increasing the viscosity of the nanofluid. Kumar et al. [15] used Al_2O_3/water nanofluid prepared on three different volume concentrations as 0.1%, 0.4%, and 0.8% in a shell and helically coiled tube heat exchanger. The Nusselt numbers were 28%, 36%, and 56% higher than water due to the concentrations, respectively. Due to the concentrations, the increase in pressure drop was 4%, 6%, and 9% of water, respectively. Ignoring the slight increase in pressure drop compared to water, they observed that the nanofluid had a remarkable effect on the heat enhancement. Anoop et al. [16] used SiO2–water nanofluid prepared on concentrations as 0.2%, 0.4%, and 0.6%. Due to their experimental study, some positive and negative results were obtained. The nanoparticles in the suspensions showed enhanced heat transfer behavior, while the pressure drop increased due to the increase in concentration, which made the researchers think that there might be some limits in the usage of nanofluids in such applications. Bahiraei and Monavari [17] examined the thermohydraulic performance in a mini STHX, using a boehmite nanofluid considering the shape of nanoparticles and the baffles in the system. In their study, the heating fluid was

 DOI: 10.1201/9781003299608-31

the nanofluid, while the coolant was water. According to their study, the heat transfer rate, performance index, and pressure drop for the MSTHX with baffles are larger than those for the MSTHX without baffles. The increment in Reynolds number from 500 to 2,000 in nanofluids consisting of flattened spherical nanoparticles caused an increase in effectiveness at a rate of 20% and a decrease in the performance index at a rate of 21.7%. Leong et al. [18] examined the entropy generation and heat transfer in STHXs using nanofluid with a low concentration value of copper nanoparticles. They considered the effect of 25° and 50° spiral baffles in their study. The highest heat transfer ratio was observed in the heat exchanger with 25° spiral baffles, and the lowest entropy generation was observed in the heat exchanger with 50° spiral baffles. This study aims to show the effect of Ag-MgO hybrid nanofluid in a shell and tube heat exchanger with baffles.

31.2 MATHEMATICAL MODEL

The k-ε turbulence model is used in the flow mechanics. The turbulence flow domains the system. The expressions related to the k-ε turbulence model are given next.

Continuity equation:

$$\nabla \cdot (\rho u) = 0 \tag{31.1}$$

Momentum equation:

$$(\nabla.u)\rho u = -\nabla p + \nabla.\mu\left(\nabla u + (\nabla u)^T\right) \tag{31.2}$$

Energy equation:

$$\rho c_p u(\nabla T) = \nabla.(k\nabla T) + Q \tag{31.3}$$

k-ε turbulence model and turbulent kinetic energy (k):

$$\rho(u.\nabla)\varepsilon = \nabla\left[\left(\mu + \frac{\mu T}{\sigma\varepsilon}\right)\nabla\varepsilon\right] + C_{e1}\frac{\varepsilon}{k}P_k - C_{e2}\rho\frac{\varepsilon^2}{k}, \varepsilon = eq \tag{31.4}$$

$$\rho(u.\nabla)k = \nabla\left[\left(\mu + \frac{\mu T}{\sigma k}\right)\nabla T\right] + P_k - \rho\varepsilon \tag{31.5}$$

The power term (P_k) seen in equation (31.4) and equation (31.5) is given here.

$$P_k = \mu_T\left[\nabla u :\right]\left(\nabla u + (\nabla u)^T\right) \tag{31.6}$$

Turbulent viscosity:

$$\mu_T = \rho C_\mu \frac{k^2}{\varepsilon} \tag{31.7}$$

The equation of the conductive heat transfer in the solid walls of the shell and tube heat exchanger:

$$\nabla^2 T = 0 \tag{31.8}$$

Pressure drop on the shell side:

$$\Delta P = f\frac{D_s}{D_e}(N_b + 1)\frac{1}{2}\rho V^2 \tag{31.9}$$

Effective diameter (D_e) is given next.

$$D_e = \frac{4\left(\frac{\sqrt{3}pt^2}{4} - \frac{\pi d_0^2}{8}\right)}{\frac{\pi \mathrm{d}_0}{2}} \tag{31.10}$$

Reynolds number is given as:

$$Re_s = \frac{\rho u D_e}{\mu} \tag{31.11}$$

The heat transfer coefficient is derived from the following equation:

$$Q = h.\ A.\ LMTD \tag{31.12}$$

Where $Q, h, A, LMTD$ are heat transfer, heat transfer coefficient, effective heat transfer area, and logarithmic temperature difference, respectively.

$$LMTD = \frac{(T_{h1} - T_{h2}) - (T_{h2} - T_{c1})}{ln\left(\frac{T_{h1} - T_{c2}}{T_{h2} - T_{c1}}\right)} \tag{31.13}$$

Where $T_{h1}, T_{h2}, T_{c1}, T_{c2}$ are inlet temperature of heating fluid, outlet (equilibrium) temperature of heating fluid, inlet temperature of the coolant, and outlet (equilibrium) temperature of the coolant, respectively.

The walls of the shell and tube heat exchanger are designed in shells in 3D. The heating fluid, water, and the coolant, nanofluid, are separated from each other by the interior wall boundary condition. The baffles are defined by the same boundary condition. With the k-ε turbulence model, this boundary condition is for simulating the walls. Thin layer boundary condition is applied to show the heat transfer in shell structures. The material of the heat exchanger's shell is steel, and it has a thickness of 5 mm. The system is thermally insulated from its surrounding. Therefore, all exterior boundaries are thermally insulated. The velocity of the nanofluid is determined by Reynolds number range, which has a variation of 2,500–12,000. Geometry parameters of the STHX are shown in Table 31.1.

31.3 SOLUTION METHODOLOGY AND MESH INDEPENDENCE

COMSOL Multiphysics 5.5 uses a finite element method in order to analyze the fluid flow and heat transfer of the system. Weak forms of the equations were obtained by using Galerkin weighted residual method. Lagrange FEMs of different orders are used for flow field variable approximations. Residuals are set to be zero in an average sense. The convergence of the solution is set to be 10^{-7}. The governing equations are solved by the PARDISO solver.

TABLE 31.1
Geometric Parameters of STHX

Parameter	Value
Shell diameter	200 mm
Inner and outer diameter of tubes	15 mm
Number of tubes	37
Number of baffles	4
Shell and tube length	500 mm

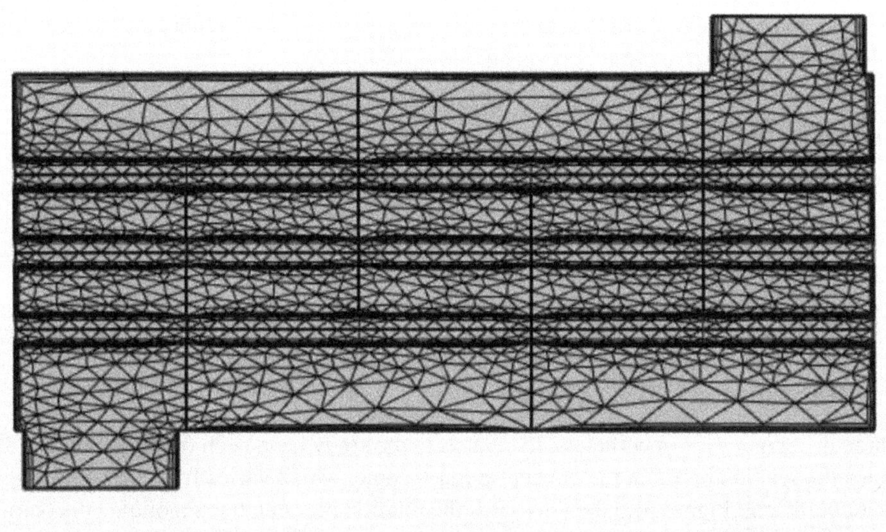

FIGURE 31.1 Mesh generation for the shell and tube heat exchanger.

The tetrahedral elements are used to create a mesh on the model. Normal mesh has been selected to avoid long simulation time and to obtain accurate results. Mesh independence study is carried out to obtain grid independence of the solution. Boundaries have been selected manually for the size of the mesh. Domain and boundary selections have been performed manually for corner refinement. Also, boundary layers have been selected for the entire geometry. Figure 31.1 shows the mesh of the geometry created on the simulation code.

31.4 NANOFLUID PROPERTIES

The main properties of the base fluid and nanoparticles are given in Table 31.2.

The density and specific heat capacity of nanofluid are derived from the equations given in the following [19]:

$$\rho_{nf} = \left[(1-\varphi_2)\left((1-\varphi_1)\rho_f + \varphi_1\rho_1\right)\right] + \varphi_2\rho_2 \tag{31.14}$$

TABLE 31.2
Main Properties of AgO-MgO and Water

	$\rho\left(\frac{kg}{m^3}\right)$	$C_P\left(\frac{J}{kg\,K}\right)$	$k\left(\frac{W}{m\,K}\right)$	$\mu\left(\frac{kg}{m\,s}\right)$
Water	997.1	4179	0.623	8.55×10^{-4}
Ag	10,500	235	429	—
MgO	3,560	955	45	—

$$\left(\rho c_p\right)_{nf} = \left[\left(1-\varphi_2\right)\left(\left(1-\varphi_1\right)\left(\rho c_p\right)_f + \varphi_1\left(\rho c_p\right)_1\right)\right] + \varphi_2\left(\rho c_p\right)_2 \tag{31.15}$$

where $\varphi_1, \varphi_2, \rho_1, \rho_2, \left(c_p\right)_1, \left(c_p\right)_2$ are defined as volume fractions, densities, and specific heat capacities of AgO and MgO nanoparticles, respectively. The thermal conductivity and the viscosity of the nanofluid are derived from the following equations [20]:

$$k_{nf} = \left(\left(0.1747x10^5 + \varphi\right) \Big/ \left(0.1747x10^5 - 0.1498x10^6\varphi + 0.1117x10^7 x\varphi^2 + 0.1997x10^8 x\varphi^3\right) \right) k_f \tag{31.16}$$

Where φ is defined as volume concentration of the nanofluid.

$$\mu_{nf} = \left(1 + 32.795\varphi - 7214\varphi^2 + 714600\varphi^3 - 0.1941x10^8\varphi^4\right)\mu_f \tag{31.17}$$

31.5 RESULTS AND DISCUSSION

For three different cases, specific results related to the study have been obtained. Figure 31.2 demonstrates the relation between the concentration volume, which varies from 0.01% to 2% and heat transfer coefficient. Figure 31.3 demonstrates the relation between the nanofluid concentration volume and pressure drop in the STHX. Due to the increasing concentration volume of the nanoparticles, the increment in both parameters is seen in the figures. At the same Reynolds number, the fluid velocity rises with the inclusion of nano-sized particles and increasing its loading amount in the base fluid. Even though thermal transport increases with higher nanoparticle solid volume fractions, the pressure drop increases as well. The heat transfer coefficient rises by about 17% with hybrid nanofluid at the highest nanoparticle volume fraction. There is also a significant rise of pressure drop with inclusion of nanoparticles. In this study, experimental data is used for the viscosity of the nanofluid. The increment in the thermal performance is fast from loading amount, from 0.01 to 0.015.

The relation of Reynolds number and heat transfer coefficient is seen in Figure 31.4, and the relation of Reynolds number and pressure drop is seen in Figure 31.5. In both figures, it's seen that the increment in Reynolds number causes an increase in both heat transfer coefficient and pressure drop. Using the maximum value of volume concentration ($\varphi = 2\%$) and constant inlet temperature of the nanofluid ($T_{c1} = 40°C$), the heat transfer coefficient is increased by about 136%, when the lowest and the highest Re cases are compared.

The streamlines and the temperature distributions of the nanofluid in the STHX are evaluated based on the variation of Reynolds numbers. Figure 31.6 and Figure 31.7 illustrate the streamline distributions under the presence of using baffles in STHX for the minimum value in and maximum range of Reynolds number. The presence of the baffles results in better mixing, while at higher

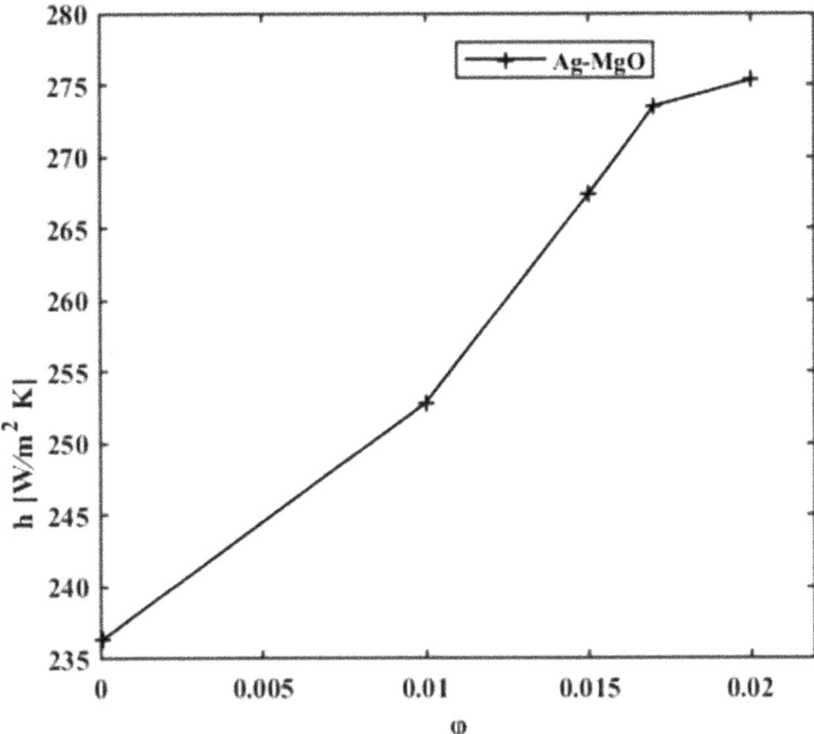

FIGURE 31.2 Heat transfer coefficient versus solid volume fraction at Re = 8,000, T_{c1} = 40°C.

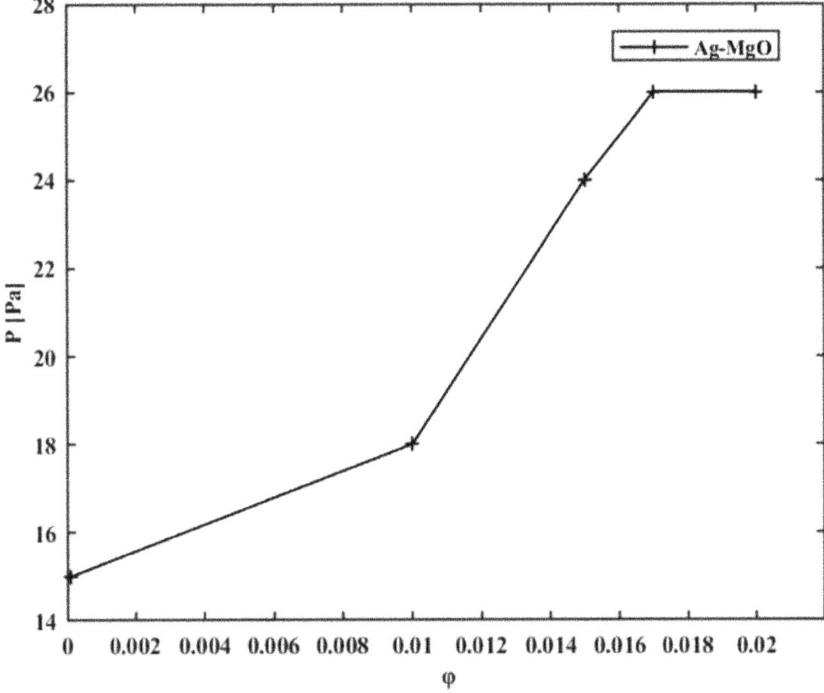

FIGURE 31.3 Pressure drop versus solid volume fraction at Re = 8,000, T_{c1} = 40°C.

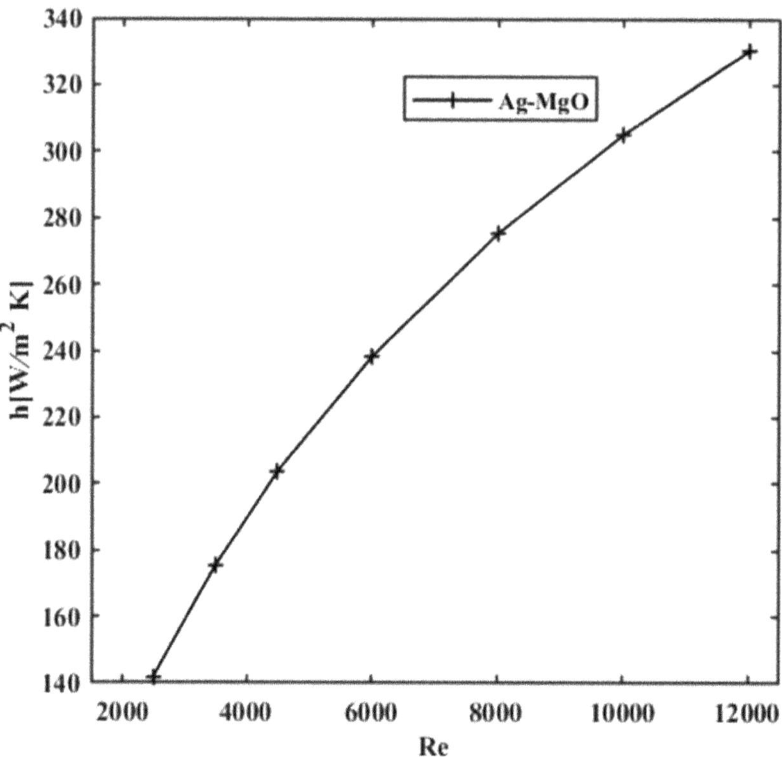

FIGURE 31.4 Heat transfer coefficient Reynolds number relation at $\varphi = 2\%$, $T_{c1} = 40°C$.

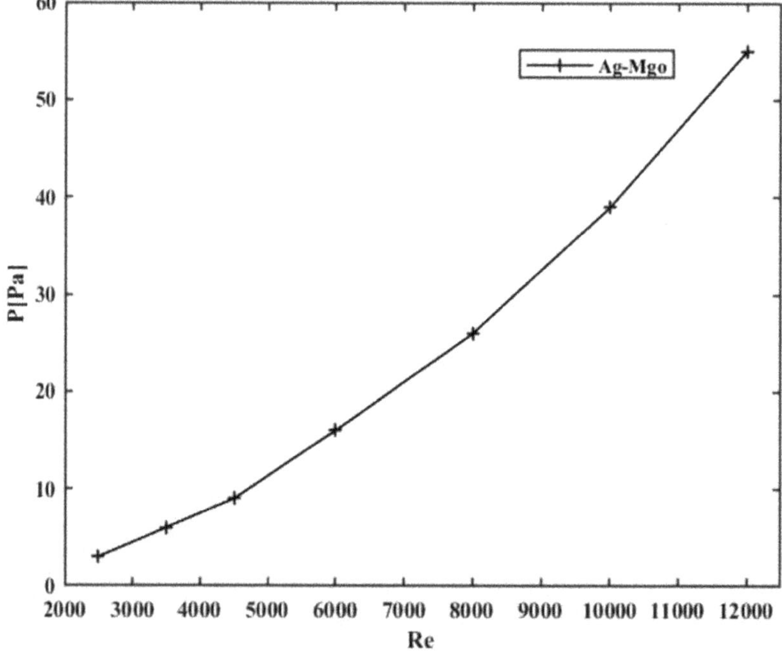

FIGURE 31.5 Pressure drop Reynolds number relation at $\varphi = 2\%$, $T_{c1} = 40°C$.

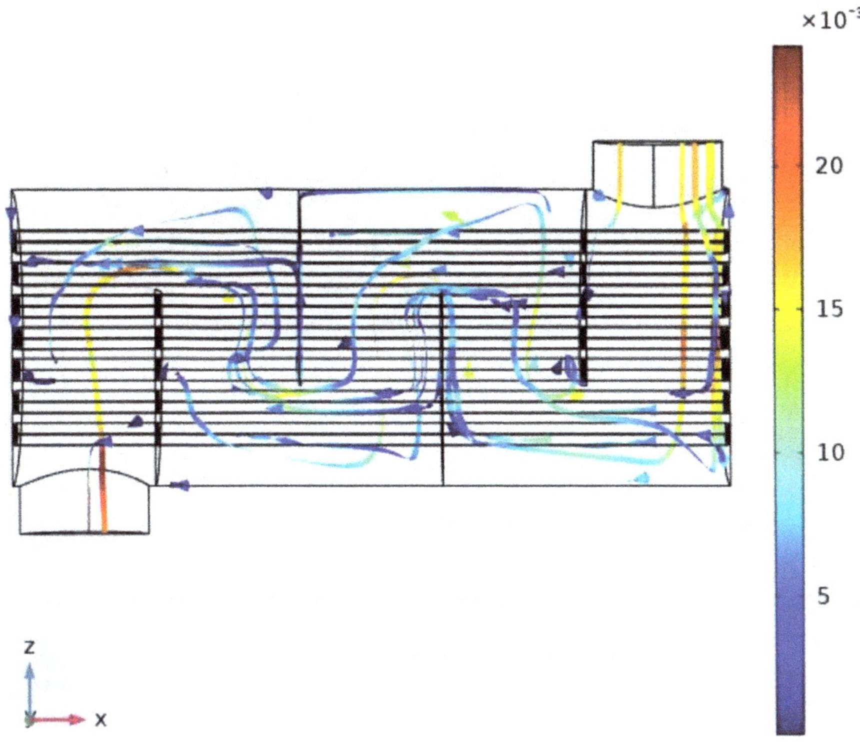

FIGURE 31.6 Velocity streamlines of the nanofluid for Re = 2,500, $\varphi = 2\%$, $T_{c1} = 40°C$.

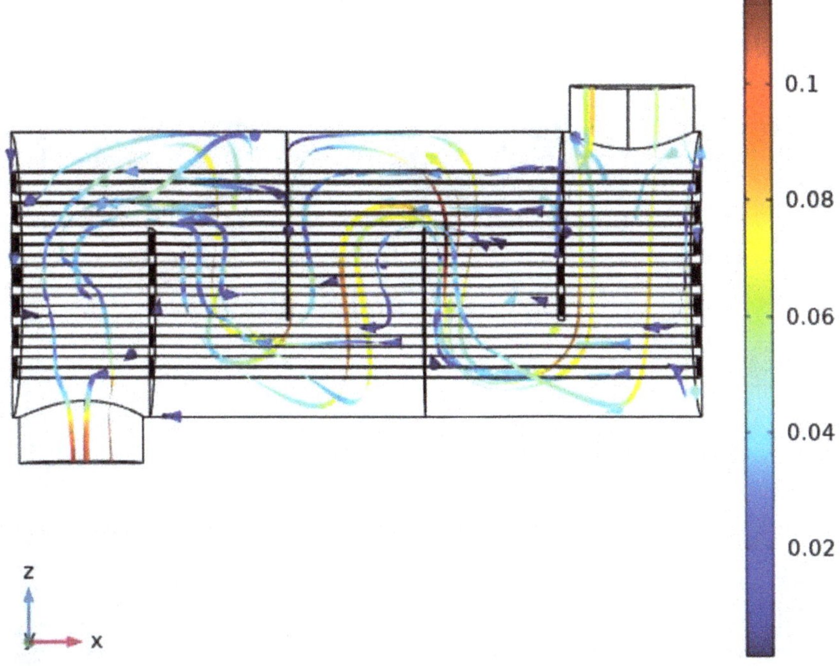

FIGURE 31.7 Velocity streamlines of the nanofluid for Re = 12,000, $\varphi = 2\%$, $T_{c1} = 40°C$.

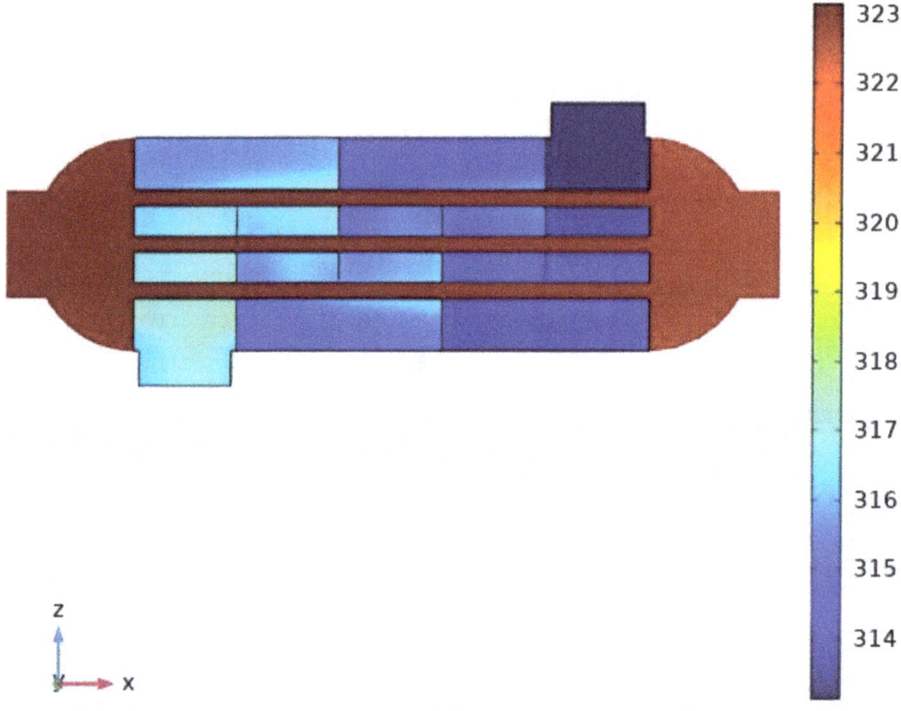

FIGURE 31.8 Temperature distributions of the nanofluid for Re = 2,500, $\varphi = 2\%$, $T_{c1} = 40°C$.

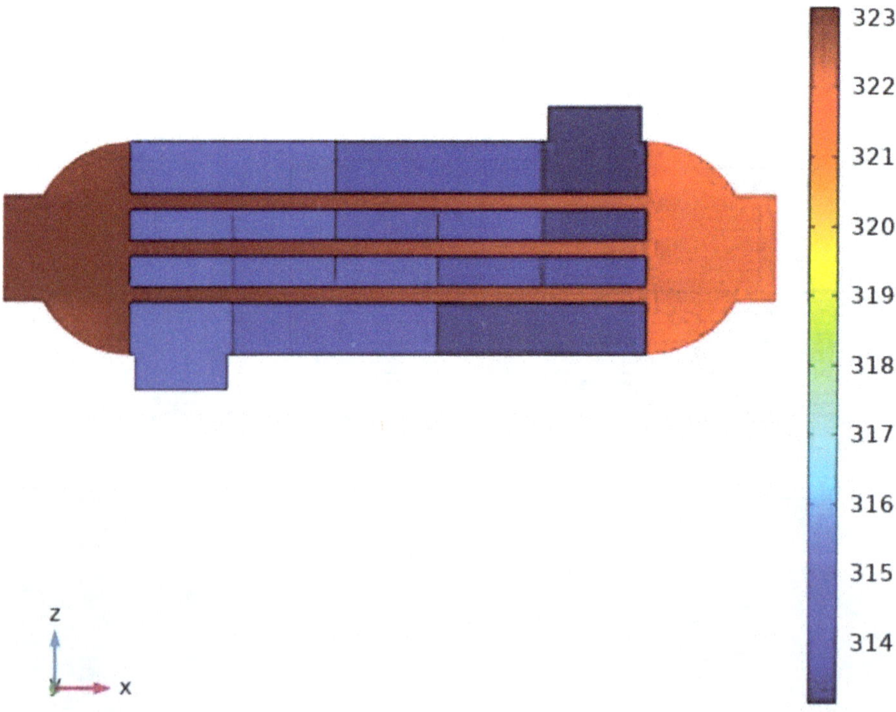

FIGURE 31.9 Temperature distributions of the nanofluid for Re = 12,000, $\varphi = 2\%$, $T_{c1} = 40°C$.

Reynolds number, thermal mixing is further enhanced. As the value of Re is increased, thermal transport in the shell side increases, which gives higher heat transfer coefficients.

Figure 31.8 and Figure 31.9 illustrate the temperature distributions in K for the minimum and maximum value in the range of Reynolds number. Better thermal enhancement with the increase of Re is seen where nanofluid velocity is higher.

31.6 CONCLUSIONS

In this study, the thermal performance of a shell and tube heat exchanger is improved by using Ag-MgO/water hybrid nanofluid. Using a hybrid nanofluid in the system has enhanced the heat transfer in the heat exchanger in a better way. Due to the increment in the volume concentration of the nanofluid, there has been a remarkable heat transfer enhancement in the system. At nanoparticle loading of 2%, there is a 16% rise of heat transfer coefficient. At the same time, pressure drop is also increased as compared to using water as heat transfer fluid. For the viscosity of the nanofluid, an experimental-based model is used instead of other widely used models, such as the Brinkman model. Therefore, higher pressure drops are obtained. The future models may be considered to account for the impacts of using different baffle geometrical parameters, shape effects of nanoparticles, and two-phase modelling approaches in the nanofluid.

Nomenclature

c_p specific heat capacity
h heat transfer coefficient
k thermal conductivity
p pressure
Re Reynolds number
T temperature
q heat flux
$\vec{u}$ velocity vector
V volume
x x-coordinate
y y-coordinate

Greek Letters

ρ density
φ solid volume fraction
μ dynamic viscosity

Subscripts

m mean
nf nanofluid
p nanoparticle
eff effective
h hot
c cold

REFERENCES

[1] Fares, R., Mebarek-Oudina, F., Aissa, A., Bilal, S.M. and Öztop, H.F. (2022). Optimal entropy generation in darcy-forchheimer magnetized flow in a square enclosure filled with silver based water nanoliquid. *Journal of Thermal Analysis and Calorimetry*, 147, 1571–1582.

[2] Mebarek-Oudina, F., Aissa, A., Mahanthesh, B. and Oztop, H.F. (2020). Heat transport of magnetized newtonian nanoliquids in an annular space between porous vertical cylinders with discrete heat source. *International Communications in Heat and Mass Transfer*, 117, 104737.

[3] Choudhari, R., Mebarek-Oudina, F., Öztop, H.F., et al. (2022). Electro-osmosis modulated peristaltic flow of non-newtonian liquid via a microchannel and variable liquid properties. *Indian Journal of Physics*, 96, 3853–3866.

[4] Can, A., Selimefendigil, F. and Oztop, H.F. (2022). A review on soft computing and nanofluid applications for battery thermal management. *Journal of Energy Storage*, 53, 105214.

[5] Okulu, D., Selimefendigil, F. and Oztop, H.F. (2022). Review on nanofluids and machine learning applications for thermoelectric energy conversion in renewable energy systems. *Engineering Analysis with Boundary Elements*, 144, 221–261.

[6] Selimefendigil, F., Oztop, H.F. and Abu-Hamdeh, N. (2022). Optimization of phase change process in a sinusoidal-wavy conductive walled cylinder with encapsulated-phase change material during magnetohydrodynamic nanofluid convection. *Journal of Energy Storage*, 55, 105512.

[7] Selimefendigil, F. and Oztop, H.F. (2022). Thermal management for conjugate heat transfer of curved solid conductive panel coupled with different cooling systems using non-Newtonian power law nanofluid applicable to photovoltaic panel systems. *International Journal of Thermal Sciences*, 173, 107390.

[8] Kakac, S. and Pramuanjaroenkij, A. (2009). Review of convective heat transfer enhancement with nanofluids. *International Journal of Heat and Mass Transfer*, 52, 3187–3196.

[9] Chakraborty, S. and Pradipta Kumar, P. (2020). Stability of nanofluid: A review. *Applied Thermal Engineering*, 174, 115259.

[10] Verma, S.K. and Tiwari, A.K. (2015). Progress of nanofluid application in solar collectors: a review. *Energy Conversion and Management*, 100, 324–346.

[11] Bahrehmand, S. and Abbassi, A. (2016). Heat transfer and performance analysis of nanofluid flow in helically coiled tube heat exchangers. *Chemical Engineering Research and Design*, 109, 628–637.

[12] Shahrul, I.M., Mahbubul, I.M., Saidur, R., Khaleduzzaman, S.S. and Sabri, M.F.M. (2016). Performance evaluation of a shell and tube heat exchanger operated with oxide based nanofluids. *Heat and Mass Transfer*, 52(8), 1425–1433.

[13] Lotfi, R., Rashidi, A.M. and Amrollahi, A. (2012). Experimental study on the heat transfer enhancement of MWNT-water nanofluid in a shell and tube heat exchanger. *International Communications in Heat and Mass Transfer*, 39(1), 108–111.

[14] Albadr, J., Tayal, S. and Alasadi, M. (2013). Heat transfer through heat exchanger using Al2O3 nanofluid at different concentrations. *Case Studies in Thermal Engineering*, 1(1), 38–44.

[15] Kumar, P.C., Kumar, J., Tamilarasan, R., Sendhil Nathan, S. and Suresh, S. (2014). Heat transfer enhancement and pressure drop analysis in a helically coiled tube using Al2O3/water nanofluid. *Journal of Mechanical Science and Technology*, 28(5), 1841–1847.

[16] Anoop, K., Cox, J. and Sadr, R. (2013). Thermal evaluation of nanofluids in heat exchangers. *International Communications in Heat and Mass Transfer*, 49, 5–9.

[17] Bahiraei, M. and Monavari, A. (2021). Thermohydraulic performance and effectiveness of a mini shell and tube heat exchanger working with a nanofluid regarding effects of fins and nanoparticle shape. *Advanced Powder Technology*, 32(12), 4468–4480.

[18] Leong, K.Y., Saidur, R., Khairulmaini, M., Michael, Z. and Kamyar, A. (2012). Heat transfer and entropy analysis of three different types of heat exchangers operated with nanofluids. *International Communications in Heat and Mass Transfer*, 39(6), 838–843.

[19] Ma, Y., Mohebbi, R., Rashidi, M.M. and Yang, Z. (2019). MHD convective heat transfer of Ag-MgO/water hybrid nanofluid in a channel with active heaters and coolers. *International Journal of Heat and Mass Transfer*, 137, 714–726.

[20] Esfe, M.H., Arani, A.A.A., Rezaie, M., Yan, W.M. and Karimipour, A. (2015). Experimental determination of thermal conductivity and dynamic viscosity of Ag-MgO/water hybrid nanofluid. *International Communications in Heat and Mass Transfer*, 66, 189–195.

Index

S

T

U

V

W

For Product Safety Concerns and Information please contact our EU
representative GPSR@taylorandfrancis.com
Taylor & Francis Verlag GmbH, Kaufingerstraße 24, 80331 München, Germany

www.ingramcontent.com/pod-product-compliance
Lightning Source LLC
Chambersburg PA
CBHW060949210326
41598CB00031B/4773
* 9 7 8 1 0 3 2 2 9 0 2 0 1 *